T0134993

Communications in Computer and Information Science 1568

More information about this series at https://link.springer.com/bookseries/7899

Balasubramanian Raman ·
Subrahmanyam Murala · Ananda Chowdhury ·
Abhinav Dhall · Puneet Goyal (Eds.)

Computer Vision and Image Processing

6th International Conference, CVIP 2021
Rupnagar, India, December 3–5, 2021
Revised Selected Papers, Part II

Editors
Balasubramanian Raman
Indian Institute of Technology Roorkee
Roorkee, India

Subrahmanyam Murala
Indian Institute of Technology Ropar
Ropar, India

Ananda Chowdhury
Jadavpur University
Kolkata, India

Abhinav Dhall
Indian Institute of Technology Ropar
Ropar, India

Puneet Goyal
Indian Institute of Technology Ropar
Ropar, India

ISSN 1865-0929 ISSN 1865-0937 (electronic)
Communications in Computer and Information Science
ISBN 978-3-031-11348-2 ISBN 978-3-031-11349-9 (eBook)
https://doi.org/10.1007/978-3-031-11349-9

This Springer imprint is published by the registered company Springer Nature Switzerland AG
The registered company address is: Gewerbestrasse 11, 6330 Cham, Switzerland

Preface

The sixth edition of the International Conference in Computer Vision and Image Processing (CVIP 2021) was organized by the Indian Institute of Technology (IIT) Ropar, Punjab, India. CVIP is a premier conference focused on image/video processing and computer vision. Previous editions of CVIP were held at IIIT Allahabad (CVIP 2020), MNIT Jaipur (CVIP 2019), IIIT Jabalpur (CVIP 2018), and IIT Roorkee (CVIP 2017 and CVIP 2016). The conference has witnessed extraordinary success with publications in multiple domains of computer vision and image processing.

In the face of COVID-19, the conference was held in virtual mode during December 3–5, 2021, connecting researchers from different countries around the world such as Sri Lanka, the USA, etc. The team—composed of Pritee Khanna (IIIT DMJ), Krishna Pratap Singh (IIIT Allahabad), Shiv Ram Dubey (IIIT Sri City), Aparajita Ojha (IIIT DMJ), and Anil B. Gonde (SGGSIET Nanded)—organized an online event with flawless communication through Webex. Moreover, the publicity for the submission of research articles by Shiv Ram Dubey (IIIT Sri City), Deep Gupta (VNIT Nagpur), Sachin Chaudhary (PEC Chandigarh), Akshay Dudhane (MBZUAI, Abu Dhabi, UAE), and Prashant Patil (Deakin University, Australia) made CVIP 2021 a great success, with the overwhelming participation of about 110 researchers. Also, the efficient teamwork by volunteers from IIT Ropar and PEC Chandigarh helped to overcome the challenges of virtual communication, thus resulting in the smooth running of the event.

CVIP 2021 received 260 regular paper submissions that went through a rigorous review process undertaken by approximately 500 reviewers from different renowned institutes and universities. The technical program chairs, Puneet Goyal (IIT Ropar), Abhinav Dhall (IIT Ropar), Narayanan C Krishnan (IIT Ropar), Mukesh Saini (IIT Ropar), Santosh K. Vipparthi (MNIT Jaipur), Deepak Mishra (IIST Trivandrum), and Ananda S. Chowdhury (Jadavpur University), coordinated the overall review process which resulted in the acceptance of 97 research articles.

The event was scheduled with one plenary talk and two keynote talk sessions each day. On the very first day, the event commenced with a plenary talk on "AI for Social Good" by Venu Govindaraju (State University of New York at Buffalo, USA) followed by the keynote talks by Shirui Pan (Monash University, Australia) and Victor Sanchez (University of Warwick, UK). On second day, Tom Gedeon (Curtin University, Australia) guided the audience with a plenary talk on "Responsive AI and Responsible AI". The keynote talks by Vitomir Štruc (University of Ljubljana, Slovenia) and Munawar Hayat (Monash University, Australia) enlightened the audience with informative discussion on computer vision. The last day of the conference, with the informative plenary talk on "Cognitive Model Motivated Document Image Understanding" by Santanu Chaudhury (IIT Jodhpur) and the keynote talk by Sunil Gupta (Deakin University), gave a deep insight to the audience on AI and its applications.

CVIP 2021 presented high-quality research work with innovative ideas. All the session chairs were invited to vote for four different categories of awards. For each award, three papers were nominated depending on the novelty of work, presentation skills, and the reviewer scores. Four different awards were announced: the IAPR Best Paper Award, the IAPR Best Student Paper Award, the CVIP Best Paper Award, and the CVIP Best Student Paper Award.

Also, CVIP 2021 awarded Prabir Kumar Biswas (IIT Kharagpur) with a CVIP Lifetime Achievement Award for his remarkable research in the field of image processing and computer vision. The awards were announced in the valedictory ceremony by General Co-chair Balasubramanian Raman (IIT Roorkee).

All the accepted and presented papers from CVIP 2021 are published in this volume in Springer's Communications in Computer and Information Science (CCIS) series. The proceedings of all previous editions of CVIP have also been successfully published in this series, and the papers are indexed by ISI Proceedings, EI-Compendex, DBLP, SCOPUS, Google Scholar, Springer link, etc. The organizers of the next event have given us a glimpse of their plan for CVIP 2022 at VNIT Nagpur: https://vnit.ac.in/cvip2022/.

December 2021

Balasubramanian Raman
Subrahmanyam Murala
Ananda Chowdhury
Abhinav Dhall
Puneet Goyal

Organization

Patron

Bidyut Baran Chaudhuri	ISI Kolkata, India

General Chairs

Venu Govindaraju	State University of New York at Buffalo, USA
Mohamed Abdel-Mottaleb	University of Miami, USA
Rajeev Ahuja	IIT Ropar, India

General Co-chairs

Balasubramanian Raman	IIT Roorkee, India
Javed N. Agrewala (Dean Research)	IIT Ropar, India

Conference Chairs

Subrahmanyam Murala	IIT Ropar, India
Satish Kumar Singh	IIIT Allahabad, India
Gaurav Bhatnagar	IIT Jodhpur, India
Sanjeev Kumar	IIT Roorkee, India
Partha Pratim Roy	IIT Roorkee, India

Technical Program Chairs

Santosh Kumar Vipparthi	MNIT Jaipur, India
Abhinav Dhall	IIT Ropar, India
Narayanan C. Krishnan	IIT Ropar, India
Deepak Mishra	IIST Trivandrum, India
Ananda S. Chowdhury	Jadavpur University, India
Puneet Goyal	IIT Ropar, India
Mukesh Saini	IIT Ropar, India

Conference Convenors

Sachin Chaudhary	PEC Chandigarh, India
Pritee Khanna	IIIT DMJ, India

Krishna Pratap Singh	IIIT Allahabad, India
Shiv Ram Dubey	IIIT Allahabad, India
Aparajita Ojha	IIIT DMJ, India
Anil B. Gonde	SGGSIET Nanded, India

Publicity Chairs

Shiv Ram Dubey	IIIT Allahabad, India
Deep Gupta	VNIT Nagpur, India
Akshay Dudhane	MBZUAI, Abu Dhabi, UAE
Prashant W. Patil	Deakin University, Australia

Website Chairs

Sachin Chaudhary	PEC Chandigarh, India
Prashant W. Patil	Deakin University, Australia

International Advisory Committee

Luc Van Gool	ETH Zurich, Switzerland
B. S. Manjunath	University of California, USA
Vishal M. Patel	Johns Hopkins University, USA
Richard Hartley	ANU, Australia
Mohammed Bennamoun	University of Western, Australia
Srinivasa Narasimhan	Carnegie Mellon University, USA
Daniel P. Lopresti	Lehigh University, USA
Victor Sanchez	University of Warwick, UK
Fahad Shahbaz Khan	MBZUAI, Abu Dhabi, UAE
Junsong Yuan	State University of New York at Buffalo, USA
Chenliang Xu	University of Rochester, USA
Xiaojun Chang	Monash University, Australia
Sunil Gupta	Deakin University, Australia
Naoufel Werghi	Khalifa University, Abu Dhabi, UAE
C.-C. Jay Kuo	University of Southern California, USA
Salman Khan	MBZUAI, Abu Dhabi, UAE
Hisham Cholakkal	MBZUAI, Abu Dhabi, UAE
Santu Rana	Deakin University, Australia
Zhou Wang	University of Waterloo, Canada
Paul Rosin	Cardiff University, UK
Bir Bhanu	University of California, Riverside, USA
Gaurav Sharma	University of Rochester, USA
Gian Luca Foresti	University of Udine, Italy

Mohan S. Kankanhalli	National University of Singapore, Singapore
Sudeep Sarkar	University of South Florida, USA
Josep Lladós	Autonomous University of Barcelona, Spain
Massimo Tistarelli	University of Sassari, Italy
Kiran Raja	NTNU, Norway
Alireza Alaei	Southern Cross University, Australia
Ankit Chaudhary	University of Missouri – St. Louis, USA
Ajita Rattani	Wichita State University, USA
Emanuela Marasco	George Mason University, USA
Thinagaran Perumal	Universiti Putra Malaysia, Malaysia
Xiaoyi Jiang	University of Münster, Germany
Paula Brito	University of Porto, Portugal
Jonathan Wu	University of Windsor, Canada

National Advisory Committee

P. K. Biwas	IIT Kharagpur, India
Sanjeev Kumar Sofat	PEC Chandigarh, India
Debashis Sen	IIT Kharagpur, India
Umapada Pal	ISI Kolkata, India
Chirag N. Paunwala	SCET Surat, India
Sanjay Kumar Singh	IIT (BHU) Varanasi, India
Surya Prakash	IIT Indore, India
A. S. Chowdhury	Jadavpur University, India
S. N. Singh	IIT Kanpur, India
K. R. Ramakrishnan	IISC Bangalore, India
Sushmita Mitra	ISI Kolkata, India
Puneet Gupta	IIT Indore, India
Somnath Dey	IIT Indore, India
M. Tanveer	IIT Indore, India
O. P. Vyas	IIIT Allahabad, India
G. C. Nandi	IIIT Allahabad, India
Aparajita Ojha	IIIT Jabalpur, India
U. S. Tiwari	IIIT Allahabad, India
Sushmita Gosh	Jadavpur University, India
D. S. Guru	University of Mysore, India
B. H. Shekhar	Mangalore University, India
Bunil Kumar Balabantaray	NIT Meghalaya, India
Munesh C. Trivedi	NIT Agartala, India
Sharad Sinha	IIT Goa, India

Reviewers

Aalok Gangopadhyay	IIT Gandhinagar, India
Abhimanyu Sahu	Jadavpur University, India
Abhinav Dhall	IIT Ropar, India
Abhirup Banerjee	University of Oxford, UK
Abhishek Sharma	IIIT Naya Raipur, India
Abhishek Sinha	Indian Space Research Organization, India
Adarsh Prasad Behera	IIIT Allahabad, India
Ahmed Elnakib	Mansoura University, Egypt
Ajita Rattani	Wichita State University, USA
Akshay Dudhane	IIT Ropar, India
Albert Mundu	IIIT Allahabad, India
Alireza Alaei	Southern Cross University, Australia
Alok Ranjan Sahoo	IIIT Allahabad, India
Amar Deshmukh	RMDSSOE, India
Amit Singhal	NSUT, India
Amitesh Rajput	BITS Pilani, India
Anamika Jain	IIIT Allahabad, India
Anand Singh Jalal	GLA University, India
Ananda Chowdhury	Jadavpur University, India
Angshuman Paul	IIT Jodhpur, India
Anil Kumar	MANIT Bhopal, India
Anirban Mukhopadhyay	IIT Gandhinagar, India
Anjali Gautam	IIIT Allahabad, India
Ankit Chaudhary	University of Missouri – St. Louis, USA
Ankur Gupta	IIT Roorkee, India
Anoop Jacob Thomas	IIIT Tiruchirappalli, India
Anshul Pundhir	IIT Roorkee, India
Ansuman Mahapatra	NIT Puducherry, India
Anuj Rai	IIT, Ropar, India
Anuj Sharma	PEC Chandigarh, India
Anukriti Bansal	LNMIIT Jaipur, India
Anup Nandy	NIT Rourkela, India
Anupam Agrawal	IIIT Allahabad, India
Anurag Singh	NIT Delhi, India
Aparajita Ojha	IIIT Jabalpur, India
Arindam Sikdar	Jadavpur University, India
Arnav Bhavsar	IIT Mandi, India
Aroof Aimen	IIT Ropar, India
Arun Chauhan	IIIT Dharwad, India
Arya Krishnan	IIITMK, India
Ashish Khare	University of Allahabad, India

Ashish Mishra	Jaypee Institute of Information Technology, India
Ashish Phophalia	IIIT Vadodara, India
Ashish Raman	NIT Jalandhar, India
Ashutosh Kulkarni	IIT Ropar, India
Asish Mishra	IIST, India
Avadh Kishor	Graphic Era University, India
Ayatullah Faruk Mollah	Aliah University, India
B. H. Shekar	Mangalore University, India
B. N. Chatterji	IIT Kharagpur, India
B. Surendiran	NIT Pondicherry, India
Babu Mehtre	IDRBT Hyderabad, India
Badri Subudhi	IIT Jammu, India
Balasubramanian Raman	IIT Roorkee, India
Balasubramanyam Appina	IIITDM Kancheepuram, India
Bibal Benifa	IIIT Kottayam, India
Bikash Sahana	NIT Patna, India
Bini Aa	IIIT Kottayam, India
Binsu C. Kovoor	Cochin University of Science and Technology, India
Bir Bhanu	University of California, Riverside, USA
Bishal Ghosh	IIT Ropar , India
Brijraj Singh	Sony Research, India
Buddhadeb Pradhan	NIT Jamshedpur, India
Bunil Balabantaray	NIT Meghalaya, India
Chandra Prakash	NIT Delhi, India
Chandra Sekhar	IIIT Sri City, India
Chandrashekhar Azad	NIT Jamshedpur, India
Chinmoy Ghosh	Jalpaiguri Government Engineering College, India
Chirag Paunwala	SCET Surat, India
Chiranjoy Chattopadhyay	IIT Jodhpur, India
D. Guru	Mysore University, India
Daniel Lopresti	Lehigh University, USA
Debanjan Sadhya	ABV-IIITM Gwalior, India
Debashis Sen	IIT Kharagpur, India
Debi Dogra	IIT Bhubaneswar, India
Debotosh Bhattacharjee	Jadavpur University, India
Deep Gupta	VNIT Nagpur, India
Deepak Ranjan Nayak	NIT Jaipur, India
Deepankar Adhikari	IIT Ropar, India
Deepika Shukla	National Brain Research Centre, India
Dileep A. D.	IIT Mandi, India
Dilip Singh Sisodia	NIT Raipur, India

Dinesh Vishwakarma	Delhi Technological University, India
Dipti Mishra	IIIT Allahabad, India
Diptiben Patel	IIT Gandhinagar, India
Durgesh Singh	IIITDM Jabalpur, India
Dushyant Kumar Singh	MNNIT Allahabad, India
Earnest Paul Ijjina	NIT Warangal, India
Ekjot Nanda	PEC Chandigarh, India
Emanuela Marasco	George Mason University, USA
G. C. Nandi	IIIT Allahabad, India
G. Devasena	IIIT Trichy, India
Gagan Kanojia	IIT Gandhinagar, India
Garima Sharma	Monash University, Australia
Gian Luca Foresti	University of Udine, Italy
Gopa Bhaumik	NIT Sikkim, India
Gopal Chandra Jana	IIIT Allahabad, India
Gourav Modanwal	IIT (BHU) Varanasi, India
Graceline Jasmine	VIT University, India
Gurinder Singh	IIT Ropar, India
H. Pallab Dutta	IIT Guwahati, India
Hadia Kawoosa	IIT Ropar, India
Hariharan Muthusamy	NIT Uttarakhand, India
Harsh Srivastava	IIIT Allahabad, India
Hemant Aggarwal	University of Iowa, USA
Hemant Kumar Meena	MNIT Jaipur, India
Hemant Sharma	NIT Rourkela, India
Himadri Bhunia	IIT Kharagpur, India
Himansgshu Sarma	IIIT Sri City, India
Himanshu Agarwal	Jaypee Institute of Information Technology, India
Himanshu Buckchash	IIT Roorkee, India
Hrishikesh Venkataraman	IIT Sri City, India
Indra Deep Mastan	LNMIIT Jaipur, India
Irshad Ahmad Ansari	IITDM Jabalpur, India
Ishan Rajendrakumar Dave	University of Central Florida, USA
J. V. Thomas	ISRO Bangalore, India
Jagat Challa	BITS Pilani, India
Jagdeep Kaur	NIT Jalandhar, India
Jasdeep Singh	IIT Ropar, India
Javed Imran	IIT Roorkee, India
Jayant Jagtap	SIT Pune, India
Jayanta Mukhopadhyay	IIT Kharagpur, India
Jaydeb Bhaumik	Jadavpur University, India
Jayendra Kumar	NIT Jamshedpur, India

Jeevaraj S	ABV-IIITM Gwalior, India
Jignesh Bhatt	IIIT Vadodara, India
Joohi Chauhan	IIT Ropar, India
Josep Llados	Computer Vision Center Barcelona, Spain
Juan Tapia	Hochschule Darmstadt, Germany
K. M. Bhurchandi	VNIT Nagpur, India
K. R. Ramakrishnan	IISc Bangalore, India
Kalidas Yeturu	IIT Tirupati, India
Kalin Stefanov	Monash University, Australia
Kapil Mishra	IIIT Allahabad, India
Kapil Rana	IIT Ropar, India
Karm Veer Arya	IIITM Gwalior, India
Karthick Seshadri	NIT Andhra Pradesh, India
Kaushik Roy	West Bengal State University, India
Kaustuv Nag	IIIT Guwahati, India
Kavitha Muthusubash	Hiroshima University, Japan
Kiran Raja	NTNU, Norway
Kirti Raj Bhatele	RJIT, BSF Academy, India
Kishor Upla	NIT Surat, India
Kishore Nampalle	IIT Roorkee, India
Komal Chugh	IIT Ropar, India
Koushlendra Singh	NIT Jamshedpur, India
Krishan Kumar	NIT Uttarakhand, India
Krishna Pratap Singh	IIIT Allahabad, India
Kuldeep Biradar	MNIT Jaipur, India
Kuldeep Singh	MNIT Jaipur, India
Lalatendu Behera	NIT Jalandhar, India
Lalit Kane	UPES Dehradun, India
Liangzhi Li	Osaka University, Japan
Lyla Das	NIT Calicut, India
M. Srinivas	NIT Warangal, India
M. Tanveer	IIT Indore, India
M. V. Raghunath	NIT Warangal, India
Mahua Bhattacharya	IIIT Gwalior, India
Malaya Dutta Borah	NIT Silchar, India
Malaya Nath	NIT Puducherry, India
Mandhatya Singh	IIT Ropar, India
Manish Khare	DA-IICT, India
Manish Okade	NIT Rourkela, India
Manisha Verma	Osaka University, Japan
Manoj Diwakar	Graphic Era University, India
Manoj Goyal	Samsung, India

Manoj K. Arora	BML Munjal University, India
Manoj Kumar	Babasaheb Bhimrao Ambedkar University, India
Manoj Kumar	GLA University, India
Manoj Rohit Vemparala	BMW Group, Germany
Manoj Singh	UPES, India
Mansi Sharma	IIT Madras, India
Massimo Tistarelli	University of Sassari, Italy
Michal Haindl	UTIA, Czech Republic
Mohammed Javed	IIIT Allahabad, India
Mohan Kankanhalli	National University of Singapore, Singapore
Mohit Singh	MNIT Jaipur, India
Monika Mathur	IGDTUW, India
Monu Verma	MNIT Jaipur, India
Mridul Gupta	Purdue University, USA
Mrinal Kanti Bhowmik	Tripura University, India
Muhammad Kanroo	IIT Ropar, India
Muneendra Ojha	IIIT Allahabad, India
Munesh C. Trivedi	NIT Agartala, India
Murari Mandal	National University of Singapore, Singapore
Muzammil Khan	MANIT Bhopal, India
N. V. Subba Reddy	Manipal Institute of Technology, India
Naga Srinivasarao Kota	NIT Warangal, India
Nagendra Singh	NIT Hamirpur, India
Nagesh Bhattu	NIT Andhra Pradesh, India
Namita Mittal	MNIT Jaipur, India
Namita Tiwari	NIT Bhopal, India
Nancy Mehta	IIT Ropar, India
Nand Kr Yadav	IIIT Allahabad, India
Nanda Dulal Jana	NIT Durgapur, India
Narasimhadhan A. V.	NITK Surathkal, India
Navjot Singh	NIT Allahabad, India
Navjot Singh	IIIT Allahabad, India
Nayaneesh Kumar	IIIT Allahabad, India
Neeru Rathee	MSIT, India
Neetu Sood	BRANIT, India
Neha Sahare	SIT Pune, India
Nehal Mamgain	Woven Planet, India
Nibaran Das	Jadavpur University, India
Nidhi Goel	IGDTUW, India
Nidhi Saxena	VIT University, India
Nilkanta Sahu	IIIT Guwahati, India

Nishant Jain	Jaypee University of Information Technology, India
Nitin Arora	IIT Roorkee, India
Nitin Kumar	NIT Uttarakhand, India
Nitish Andola	IIIT Allahabad, India
Oishila Bandyopadhyay	IIIT Kalyani, India
Om Prakaah	HNB Garhwal University, India
P. V. Sudeep	NIT Calicut, India
P. V. Venkitakrishnan	ISRO Bangalore, India
Pankaj Kumar	DA-IICT, India
Pankaj Kumar Sa	NIT Rourkela, India
Pankaj P. Singh	CIT, India
Parmeshwar Patil	SGGSIET Nanded, India
Parth Neshve	SVIT Satara, India
Partha Pakray	NIT Silchar, India
Parveen Kumar	NIT Uttarakhand, India
Paula Brito	University of Porto, Portugal
Piyush Kumar	NIT Patna, India
Poonam Sharma	VNIT Nagpur, India
Poornima Thakur	IITDM Jabalpur, India
Prabhu Natarajan	DigiPen Institute of Technology, Singapore
Prabhu Natarajan	University of Technology and Applied Sciences - Al Mussanah, Oman
Pradeep Kumar	Amphisoft, India
Pradeep Singh	NIT Raipur, India
Praful Hambarde	IIT Ropar, India
Prafulla Saxena	MNIT Jaipur, India
Pragati Agrawal	NIT Bhopal, India
Pragya Dwivedi	MNNIT Allahabad, India
Prashant Patil	Deakin University, Australia
Prashant Shukla	IIIT Allahabad, India
Prashant Srivastava	University of Allahabad, India
Prateek Keserwani	IIT Roorkee, India
Pratik Chattopadhyay	ITI (BHU) Varanasi, India
Pratik Narang	BITS Pilani, India
Pratik Shah	IIIT Vadodara, India
Pratik Somwanshi	IIT Jodhpur, India
Praveen Kumar Chandaliya	MNIT Jaipur, India
Praveen Sankaran	NIT Calicut, India
Praveen Tirupattur	University of Central Florida, USA
Pravin Kumar	IIIT Allahabad, India
Prerana Mukherjee	Jawaharlal Nehru University, India

Pritee Khanna	IITDM Jabalpur, India
Pritpal Singh	National Taipei University of Technology, Taiwan
Priya Kansal	Couger, Japan
Priyanka Singh	DA-IICT, India
Priyankar Choudary	IIT Ropar, India
Puneet Gupta	IIT Indore, India
Puneet Kumar	IIT Roorkee, India
Pushpendra Kumar	MANIT Bhopal, India
R. Malmathanraj	NIT Trichy, India
Rachit S. Munjal	Samsung, India
Ragendhu S. P.	IITM Kerala, India
Rahul Dixit	IIIT Pune, India
Rahul Kumar	IIT Roorkee, India
Rajeev Srivastava	IIT (BHU) Varanasi, India
Rajendra Nagar	IIT Jodhpur, India
Rajet Joshi	SIT Pune, India
Rajitha Bakthula	MNNIT Allahabad, India
Rajiv kumar Tripathi	NIT Delhi, India
Rajlaxmi Chouhan	IIT Jodhpur, India
Rameswar Panda	MIT-IBM Watson AI Lab, USA
Ramya Akula	University of Central Florida, USA
Ravindra Kumar Soni	MNIT Jaipur, India
Ridhi Arora	IIT Roorkee, India
Ripon Patgiri	NIT Silchar, India
Rishav Singh	NIT Delhi, India
Rohit Gupta	University of Central Florida, USA
Rohit Mishra	IIIT Allahabad, India
Rubin Bose S.	Madras Institute of Technology, India
Rukhmini Bandyopadhyay	University of Texas, USA
Rukhmini Roy	Jadavpur University, India
Rupam Bhattacharyya	IIIT Bhagalpur, India
Rusha Patra	IIIT Guwahati, India
S. H. Shabbeer Basha	IIIT Sri City, India
S. K. Singh	IIT (BHU) Varanasi, India
S. N. Tazi	RTU, India
S. Sumitra	IIST, India
Sachin Chaudhary	IIT Ropar, India
Sachin Dube	MNIT Jaipur, India
Sachit Rao	IIIT Bangalore, India
Sahana Gowda	BNMIT Bangalore, India
Sambhavi Tiwari	IIIT Allahabad, India
Sandeep Kumar	NIT Delhi, India

Sandesh Bhagat	SGGSIET Nanded, India
Sanjay Ghosh	University of California, San Francisco, USA
Sanjeev Kumar	IIT Roorkee, India
Sanjoy Pratihar	IIIT Kalyani, India
Sanjoy Saha	Jadavpur University, India
Sanoj Kumar	UPES, India
Santosh Kumar	IIIT Naya Raipur, India
Santosh Kumar Vipparthi	MNIT Jaipur, India
Santosh Randive	PCCOER Pune, India
Saravanan Chandran	NIT Durgapur, India
Saroj Kr. Biswas	NIT Silchar, India
Satendra Singh	IIT Jodhpur, India
Sathiesh Kumar V.	Madras Institute of Technology, India
Satish Singh	IIIT Alahabad, India
Satya Jaswanth Badri	IIT Ropar, India
Satya Prakash Sahu	NIT Raipur, India
Satyasai Jagannath Nanda	MNIT Jaipur, India
Satyendra Chouhan	MNIT Jaipur, India
Satyendra Yadav	NIT Meghalaya, India
Saugata Sinha	VNIT Nagpur, India
Saurabh Kumar	Osaka University, Japan
Sebastiano Battiato	Università di Catania, Italy
Shailza Sharma	Thapar Institute of Engineering and Technology, India
Shanmuganathan Raman	IIT Gandhinagar, India
Sharad Sinha	IIT Goa, India
Shashi Poddar	CSIR, India
Shashi Shekhar Jha	IIT Ropar, India
Shashikant Verma	IIT Gandhinagar, India
Shekhar Verma	IIIT Allahabad, India
Shirshu Verma	IIIT Allahabad, India
Shitala Prasad	NTU Singapore, Singapore
Shiv Ram Dubey	IIIT Allahabad, India
Shivangi Nigam	IIIT Allahabad, India
Shreya Ghosh	Monash University, Australia
Shreya Goyal	IIT Jodhpur, India
Shreya Gupta	MANIT Bhopal, India
Shrikant Malwiya	IIIT Allahabad, India
Shruti Phutke	IIT Ropar, India
Shubham Vatsal	Samsung R&D, India
Shubhangi Nema	IIT Bombay, India
Shyam Lal	NIT Karnataka, India

Shyam Singh Rajput	NIT Patna, India
Skand Skand	Oregon State University, USA
Slobodan Ribaric	University of Zagreb, Croatia
Smita Agrawal	Thapar Institute of Engineering and Technology, India
Snehasis Mukherjee	Shiv Nadar University, India
Somenath Das	IISc, India
Somnath Dey	IIT Indore, India
Sonali Agarwal	IIIT Allahabad, India
Soumen Bag	IIT Dhanbad, India
Soumendu Chakraborty	IIIT Lucknow, India
Sourav Pramanik	New Alipore College, India
Sri Aditya Deevi	IIST, India
Srimanta Mandal	DA-IICT, India
Subhas Barman	Jalpaiguri Government Engineering College, India
Subrahamanian K. S. Moosath	IIST, India
Subrahmanyam Murala	IIT Ropar, India
Sudhakar Kumawat	Osaka University, Japan
Sudhakar Mishra	IIIT Allahabad, India
Sudhish George	NIT Calicut, India
Sudipta Banerjee	Michigan State University, USA
Sukwinder Singh	NIT Jalandhar, India
Sule Yildirim-Yayilgan	NTNU, Norway
Suman Deb	NIT Surat, India
Suman Kumar Maji	IIT Patna, India
Suman Mitra	DA-IICT, India
Sumit Kumar	IIIT Allahabad, India
Suneeta Agarwal	MNNIT Allahabad, India
Suraj Sawant	COEP, India
Suranjan Goswami	IIIT Allahabad, India
Surendra Sharma	Indian Institute of Remote Sensing, India
Suresh Raikwar	Thapar Institute of Engineering and Technology, India
Suresh Raikwar	GLA University, India
Surya Prakash	IIT Indore, India
Sushil Ghildiyal	IIT Ropar, India
Sushmita Mitra	ISI Kolkata, India
Susmita Ghosh	Jadavpur University, India
Suvidha Tripathi	LNMIIT, India
Suvidha Tripathi	IIIT Allahabad, India
Swalpa Kumar Roy	Jalpaiguri Government Engineering College, India
Swarnima Singh Gautam	IIIT Allahabad, India

T. Veerakumar	NIT Goa, India
Tandra Pal	NIT Durgapur, India
Tannistha Pal	NIT Agartala, India
Tanushyam Chattopadhyay	TCS Pune, India
Tarun Chaudhary	NIT Jalandhar, India
Tasneem Ahmed	Integral University Lucknow, India
Thinagaran Perumal	Universiti Putra Malaysia, Malaysia
Tirupathiraju Kanumuri	NIT Delhi, India
Trilochan Panigrahi	NIT Goa, India
Tripti Goel	NIT Silchar, India
U. S. N. Raju	NIT Warangal, India
U. S. Tiwary	IIIT Allahabad, India
Umapada Pal	ISI Kolkata, India
Umarani Jayaraman	IIITDM Kancheepuram, India
Umesh Pati	NIT Rourkela, India
Upendra Pratap Singh	IIIT Allahabad, India
Varsha Singh	IIIT Allahabad, India
Varun P. Gopi	NIT Tiruchirppalli, India
Vibhav Prakash Singh	NIT Allahabad, India
Vibhav Prakash Singh	MNNIT Allahabad, India
Vibhor Kant	BHU Varanasi, India
Vidhya Kamakshi	IIT Ropar, India
Vijander Singh	NSIT Delhi, India
Vijay Kumar Yadav	IIIT Allahabad, India
Vijay N. Gangapure	Government Polytechnic Kolhapur, India
Vijay Semwal	MANIT Bhopal, India
Vinit Jakhetiya	IIT Jammu, India
Vinti Agarwal	BITS Pilani, India
Vishal Satpute	VNIT, India
Vishwambhar Pathak	BIT Jaipur, India
Vishwas Rathi	IIT Ropar, India
Viswanath P.	IIIT Sri City, India
Vivek Singh Verma	Ajay Kumar Garg Engineering College, India
Vivek Tiwari	IIIT Naya Raipur, India
Vivekraj K.	IIT Roorkee, India
Vrijendra Singh	IIIT Allahabad, India
W. Wilfred Godfrey	IIIT Gwalior, India
Watanabe Osamu	Takushoku University, Japan
Wei-Ta Chu	National Cheng Kung University, Taiwan
Xiaoyi Jiang	University of Münster, Germany
Zhixi Cai	Monash University, Australia
Ziwei Xu	National University of Singapore, Singapore

Contents – Part II

Handwritten Text Retrieval from Unlabeled Collections 1
Santhoshini Gongidi and C. V. Jawahar

Detecting Document Forgery Using Hyperspectral Imaging and Machine Learning 14
Vrinda Rastogi, Sahima Srivastava, Garima Jaiswal, and Arun Sharma

An Hour-Glass CNN for Language Identification of Indic Texts in Digital Images 26
Neelotpal Chakraborty, Ayatullah Faruk Mollah, Subhadip Basu, and Ram Sarkar

Single Frame-Based Video Dehazing with Adversarial Learning 36
Vijay M. Galshetwar, Prashant W. Patil, and Sachin Chaudhary

Spatio-Temporal Event Segmentation for Wildlife Extended Videos 48
Ramy Mounir, Roman Gula, Jörn Theuerkauf, and Sudeep Sarkar

Comparative Analysis of Machine Learning and Deep Learning Models for Ship Classification from Satellite Images 60
Abhinaba Hazarika, P. Jidesh, and A. Smitha

Channel Difference Based Regeneration Architecture for Fake Colorized Image Detection 73
Shruti S. Phutke and Subrahmanyam Murala

DenseASPP Enriched Residual Network Towards Visual Saliency Prediction 85
Shilpa Elsa Abraham and Binsu C. Kovoor

Brain MRI and CT Image Fusion Using Generative Adversarial Network 97
Bharati Narute and Prashant Bartakke

MFCA-Net: Multiscale Feature Fusion with Channel-Wise Attention Network for Automatic Liver Segmentation from CT Images 110
Devidas T. Kushnure and Sanjay N. Talbar

Automatic Double Contact Fault Detection in Outdoor Volleyball Videos 122
Pratibha Kumari, Anmol Kumar, Min-Chun Hu, and Mukesh Saini

Classroom Slide Narration System 135
K. V. Jobin, Ajoy Mondal, and C. V. Jawahar

Humanoid Robot - Spark 147
Kunal Gawhale, Aditya Chandanwar, Rujul Zalte, Suhit Metkar, Sahana Kulkarni, Abhijith Panikar, Labhesh Lalka, Arti Bole, and Santosh Randive

Attention-Based Deep Autoencoder for Hyperspectral Image Denoising 159
Shashi Kumar, Hazique Aetesam, Anirban Saha, and Suman Kumar Maji

Feature Modulating Two-Stream Deep Convolutional Neural Network for Glaucoma Detection in Fundus Images 171
Snehashis Majhi and Deepak Ranjan Nayak

Retinal Image Quality Assessment Using Sharpness and Connected Components 181
S. Kiruthika and V. Masilamani

$(MS)^2$EDNet: Multiscale Motion Saliency Deep Network for Moving Object Detection 192
Santosh Nagnath Randive, Kishor B. Bhangale, Rahul G. Mapari, Kiran M. Napte, and Kishor B. Wane

Video Enhancement with Single Frame 206
Vijay M. Galshetwar, Prashant W. Patil, and Sachin Chaudhary

Blind Video Quality Assessment Using Fusion of Novel Structural Features and Deep Features 219
Anish Kumar Vishwakarma and Kishor M. Bhurchandi

On-Device Spatial Attention Based Sequence Learning Approach for Scene Text Script Identification 230
Rutika Moharir, Arun D. Prabhu, Sukumar Moharana, Gopi Ramena, and Rachit S. Munjal

Post-harvest Handling of Mangoes: An Integrated Solution Using Machine Learning Approach 243
D. S. Guru, Anitha Raghavendra, and Mahesh K. Rao

Morphological Gradient Analysis and Contour Feature Learning for Locating Text in Natural Scene Images 254
B. H. Shekar and S. Raveeshwara

Introspecting Local Binary Feature Vectors for Classification of Radiographic Weld Images 262
Jayendra Kumar, Pratul Arvind, Prashant Singh, and Yamini Sarada

Performance Evaluation of Deep Learning Models for Ship Detection 273
Rahul Sharma, Harshit Sharma, Tamanna Meena, Padmavati Khandnor, Palak Bansal, and Paras Sharma

COVID-19 Social Distance Surveillance Using Deep Learning 288
Praveen Nair, Uttam Kumar, and Sowmith Nandan

Towards Semi-supervised Tree Canopy Detection and Extraction from UAV Images 299
Uttam Kumar, Anindita Dasgupta, Lingamallu S. N. Venkata Vamsi Krishna, and Pranav Kumar Chintakunta

Pose Guided Controllable Gesture to Gesture Translation 308
Mallika, Debashis Ghosh, and Pyari Mohan Pradhan

EDR: Enriched Deep Residual Framework with Image Reconstruction for Medical Image Retrieval 319
Rohini Pinapatruni and Shoba Bindu Chigarapalle

Depth Estimation Using Sparse Depth and Transformer 329
Roopak Malik, Praful Hambarde, and Subrahmanyam Murala

Analysis of Loss Functions for Image Reconstruction Using Convolutional Autoencoder 338
Nishant Khare, Poornima Singh Thakur, Pritee Khanna, and Aparajita Ojha

Fuzzy Entropy k-Plane Clustering Method and Its Application to Medical Image Segmentation 350
Puneet Kumar, Dhirendra Kumar, and Ramesh Kumar Agrawal

FMD-cGAN: Fast Motion Deblurring Using Conditional Generative Adversarial Networks 362
Jatin Kumar, Indra Deep Mastan, and Shanmuganathan Raman

Region Extraction Based Approach for Cigarette Usage Classification Using Deep Learning 378
Anshul Pundhir, Deepak Verma, Puneet Kumar, and Balasubramanian Raman

Fire Detection Model Using Deep Learning Techniques 391
Arun Singh Pundir and Balasubramanian Raman

Two Novel Methods for Multiple Kinect v2 Sensor Calibration 403
Sumit Hazra, Manasa Pisipati, Amrit Puhan, Anup Nandy, and Rafał Scherer

Residual Inception Cycle-Consistent Adversarial Networks 415
Ekjot Singh Nanda, Vijay M. Galshetwar, and Sachin Chaudhary

MAG-Net: A Memory Augmented Generative Framework for Video Anomaly Detection Using Extrapolation 426
Sachin Dube, Kuldeep Biradar, Santosh Kumar Vipparthi, and Dinesh Kumar Tyagi

Hand Gesture Recognition Using CBAM-RetinaNet 438
Kota Yamini Suguna, H Pallab Jyoti Dutta, M. K. Bhuyan, and R. H. Laskar

Elderly Patient Fall Detection Using Video Surveillance 450
Amartya Raghav and Sachin Chaudhary

OGGN: A Novel Generalized Oracle Guided Generative Architecture for Modelling Inverse Function of Artificial Neural Networks 460
V. Mohammad Aaftab and Mansi Sharma

Deep Learning Based DR Medical Image Classification 472
Preeti Deshmukh and Arun N. Gaikwad

Human Action Recognition in Still Images 483
Palak and Sachin Chaudhary

Enhancing Unsupervised Video Representation Learning by Temporal Contrastive Modelling Using 2D CNN 494
Vidit Kumar, Vikas Tripathi, and Bhaskar Pant

Deep Two-Stage LiDAR Depth Completion 504
Moushumi Medhi and Rajiv Ranjan Sahay

3D Multi-voxel Pattern Based Machine Learning for Multi-center fMRI Data Normalization 516
Anoop Jacob Thomas and Deepti R. Bathula

Efficient Approximation of Curve-Shaped Objects in $\mathbb{Z}^2$ Based on the Maximum Difference Between Discrete Curvature Values 529
Sutanay Bhattacharjee and Shyamosree Pal

Exploring the Role of Adversarial Attacks in Image Anti-forensics 542
Krishan Gopal Sharma, Gurinder Singh, and Puneet Goyal

A Novel Artificial Intelligence-Based Lung Nodule Segmentation and Classification System on CT Scans 552
Shubham Dodia, B. Annappa, and Mahesh A. Padukudru

Author Index 565

Contents – Part I

Classification of Brain Tumor MR Images Using Transfer Learning and Machine Learning Models ... 1
LillyMaheepa Pavuluri and Malaya Kumar Nath

Deep-TDRS: An Integrated System for Handwritten Text Detection-Recognition and Conversion to Speech Using Deep Learning ... 10
Bisakh Mondal, Shuvayan Ghosh Dastidar, and Nibaran Das

Computer Aided Diagnosis of Autism Spectrum Disorder Based on Thermal Imaging ... 21
Kavya Ganesh, Snekhalatha Umapathy, and Palani Thanaraj Krishnan

Efficient High-Resolution Image-to-Image Translation Using Multi-Scale Gradient U-Net ... 33
Kumarapu Laxman, Shiv Ram Dubey, Baddam Kalyan, and Satya Raj Vineel Kojjarapu

Generic Multispectral Image Demosaicking Algorithm and New Performance Evaluation Metric ... 45
Vishwas Rathi and Puneet Goyal

A Platform for Large Scale Auto Annotation of Scanned Documents Featuring Real-Time Model Building and Model Pooling ... 58
Komuravelli Prashanth, Boyalakuntla Kowndinya, Chilaka Vijay, Dande Teja, Vidya Rodge, Ramya Velaga, Reena Abasaheb Deshmukh, and Yeturu Kalidas

AC-CovidNet: Attention Guided Contrastive CNN for Recognition of Covid-19 in Chest X-Ray Images ... 71
Anirudh Ambati and Shiv Ram Dubey

Application of Deep Learning Techniques for Prostate Cancer Grading Using Histopathological Images ... 83
Mahesh Gour, Sweta Jain, and Uma Shankar

Dyadic Interaction Recognition Using Dynamic Representation and Convolutional Neural Network ... 95
R. Newlin Shebiah and S. Arivazhagan

Segmentation of Unstructured Scanned Devanagari Newspaper Documents 107
Rupinder Pal Kaur, Munish Kumar, and M. K. Jindal

Automatic Classification of Sedimentary Rocks Towards Oil Reservoirs Detection 118
Anu Singha, Priya Saha, and Mrinal Kanti Bhowmik

Signature2Vec - An Algorithm for Reference Frame Agnostic Vectorization of Handwritten Signatures 130
Manish Kumar Srivastava, Dileep Reddy, and Kalidas Yeturu

Leaf Segmentation and Counting for Phenotyping of Rosette Plants Using Xception-style U-Net and Watershed Algorithm 139
Shrikrishna Kolhar and Jayant Jagtap

Fast and Secure Video Encryption Using Divide-and-Conquer and Logistic Tent Infinite Collapse Chaotic Map 151
Jagannath Sethi, Jaydeb Bhaumik, and Ananda S. Chowdhury

Visual Localization Using Capsule Networks 164
Omkar Patil

Detection of Cataract from Fundus Images Using Deep Transfer Learning 175
Subin Sahayam, J. Silambarasan, and Umarani Jayaraman

Brain Tumour Segmentation Using Convolution Neural Network 187
Karuna Bhalerao, Shital Patil, and Surendra Bhosale

Signature Based Authentication: A Multi-label Classification Approach to Detect the Language and Forged Sample in Signature 198
Anamika Jain, Satish Kumar Singh, and Krishna Pratap Singh

A Data-Set and a Real-Time Method for Detection of Pointing Gesture from Depth Images 209
Shome S. Das

VISION HELPER: CNN Based Real Time Navigator for the Visually Impaired 221
Chetan Maheshwari, Pawan Kumar, Ajeet Gupta, and Oishila Bandyopadhyay

Structure-Texture Decomposition-Based Enhancement Framework for Weakly Illuminated Images 232
K. M. Haritha, K. G. Sreeni, Joseph Zacharias, and R. S. Jeena

Low Cost Embedded Vision System for Location and Tracking of a Color Object 244
Diego Ayala, Danilo Chavez, and Leopoldo Altamirano Robles

Towards Label-Free Few-Shot Learning: How Far Can We Go? 256
Aditya Bharti, N. B. Vineeth, and C. V. Jawahar

AB-net: Adult- Baby Net 269
Sahil Salim Makandar, Ashish Tiwari, Sahil Munaf Bandar, and Allen Joshey

Polarimetric SAR Classification: Fast Learning with k-Maximum Likelihood Estimator 281
Nilam Chaudhari, Suman K. Mitra, Srimanta Mandal, Sanid Chirakkal, Deepak Putrevu, and Arundhati Misra

Leveraging Discriminative Cues for Masked Face Recognition in Post COVID World 294
Hiren Pokar, Nilay Patel, Himansh Mulchandani, Ajitesh Singh, Rinkal Singh, and Chirag Paunwala

Pretreatment Identification of Oral Leukoplakia and Oral Erythroplakia Metastasis Using Deep Learning Neural Networks 306
Rinkal Shah and Jyoti Pareek

Soft Biometric Based Person Retrieval for Burglary Investigation 316
K. Iyshwarya Ratthi, B. Yogameena, and A. Jansi Rani

A Deep Learning Framework for the Classification of Lung Diseases Using Chest X-Ray Images 328
M. Vyshnavi, Bejoy Abraham, and Sivakumar Ramachandran

Scene Graph Generation with Geometric Context 340
Vishal Kumar, Albert Mundu, and Satish Kumar Singh

Deep Color Spaces for Fingerphoto Presentation Attack Detection in Mobile Devices 351
Emanuela Marasco, Anudeep Vurity, and Asem Otham

Cancelable Template Generation Using Convolutional Autoencoder and RandNet 363
Pankaj Bamoriya, Gourav Siddhad, Pritee Khanna, and Aparajita Ojha

Homogeneous and Non-homogeneous Image Dehazing Using Deep Neural Network 375
Manan Gajjar and Srimanta Mandal

Improved Periocular Recognition Through Blend of Handcrafted and Deep Features 387
Aryan Lala, Kalagara Chaitanya Kumar, Ritesh Vyas, and Manoj Sharma

Kernels for Incoherent Projection and Orthogonal Matching Pursuit 398
Himanshu Kandiyal and C. S. Sastry

AAUNet: An Attention Augmented Convolution Based UNet for Change Detection in High Resolution Satellite Images 407
P. S. Patil, R. S. Holambe, and L. M. Waghmare

A Large Volume Natural Tamil Character Dataset 425
M. Arun, S. Arivazhagan, and R Ahila Priyadharshini

Datasets of Wireless Capsule Endoscopy for AI-Enabled Techniques 439
Palak Handa, Nidhi Goel, and S. Indu

Moving Objects Detection in Intricate Scenes via Spatio-Temporal Co-occurrence Based Background Subtraction 447
Shweta Singh and Srikanth Vasamsetti

Script Identification in Natural Scene Text Images by Learning Local and Global Features on Inception Net 458
Kalpita Dutta, Shuvayan Ghosh Dastidar, Nibaran Das, Mahantapas Kundu, and Mita Nasipuri

Survey of Leukemia Cancer Cell Detection Using Image Processing 468
Tulasi Gayatri Devi, Nagamma Patil, Sharada Rai, and Cheryl Sarah Philipose

MS-Net: A CNN Architecture for Agriculture Pattern Segmentation in Aerial Images 489
Sandesh Bhagat, Manesh Kokare, Vineet Haswani, Praful Hambarde, and Ravi Kamble

Vision Transformer for Plant Disease Detection: *PlantViT* 501
Poornima Singh Thakur, Pritee Khanna, Tanuja Sheorey, and Aparajita Ojha

Evaluation of Detection and Segmentation Tasks on Driving Datasets 512
Deepak Singh, Ameet Rahane, Ajoy Mondal, Anbumani Subramanian, and C. V. Jawahar

Classification of Gender in Celebrity Cartoon Images 525
S. Prajna, N. Vinay Kumar, and D. S. Guru

Localization of Polyps in WCE Images Using Deep Learning Segmentation Methods: A Comparative Study 538
Samir Jain, Ayan Seal, and Aparajita Ojha

Evaluation of Deep Architectures for Facial Emotion Recognition 550
B. Vinoth Kumar, R. Jayavarshini, Naveena Sakthivel, A. Karthiga, R. Narmadha, and M. Saranya

Adaptive Rough-Fuzzy Kernelized Clustering Algorithm for Noisy Brain MRI Tissue Segmentation 561
Rudrajit Choudhuri and Amiya Halder

On the Prospects of Latent MasterPrints 574
Mahesh Joshi, Bodhisatwa Mazumdar, and Somnath Dey

Author Index 583

Handwritten Text Retrieval from Unlabeled Collections

Santhoshini Gongidi(✉) and C. V. Jawahar

Centre for Visual Information Technology, IIIT Hyderabad, Hyderabad, India
santhoshini.gongidi@research.iiit.ac.in, jawahar@iiit.ac.in

Abstract. Handwritten documents from communities like cultural heritage, judiciary, and modern journals remain largely unexplored even today. To a great extent, this is due to the lack of retrieval tools for such unlabeled document collections. This work considers such collections and presents a simple, robust retrieval framework for easy information access. We achieve retrieval on unlabeled novel collections through invariant features learned for handwritten text. These feature representations enable zero-shot retrieval for novel queries on unlabeled collections. We improve the framework further by supporting search via text and exemplar queries. Four new collections written in English, Malayalam, and Bengali are used to evaluate our text retrieval framework. These collections comprise *2957* handwritten pages and over *300K* words. We report promising results on these collections, despite the zero-shot constraint and huge collection size. Our framework allows the addition of new collections without any need for specific finetuning or labeling. Finally, we also present a demonstration of the retrieval framework. [Project Page].

Keywords: Document retrieval · Keyword spotting · Zero-shot retrieval

1 Introduction

Digitized handwritten documents are a colossal source of information. Many communities use handwritten records like cultural heritage collections, judicial records, and modern journals to gather information. Many of these documents still remain to be made search-friendly. This is where handwritten search tools come in handy to make the textual content of these documents easily accessible. However, there is a lack of these search tools in digital libraries even today. For Latin scripts, document retrieval demonstrations are primarily experimental. Such demonstrations rely on transcribed pages from the document collection to make it retrievable. However, due to the vast number of collections, there is a need for annotation-free approaches to develop retrieval solutions.

Keyword spotting (KWS), a recognition-free approach, is becoming increasingly popular for document retrieval. KWS is defined as the task of identifying

B. Raman et al. (Eds.): CVIP 2021, CCIS 1568, pp. 1–13, 2022.
https://doi.org/10.1007/978-3-031-11349-9_1

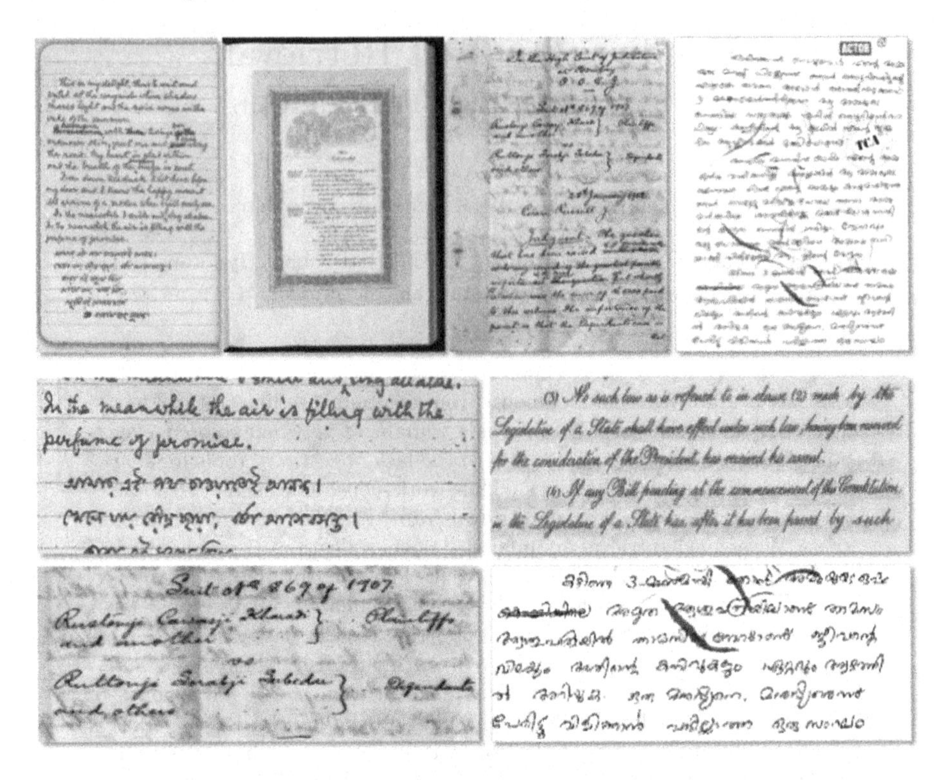

Fig. 1. Sample snippets from the historic and contemporary Indian handwritten collections: *Tagore's papers*, *Constitution of India manuscript*, *Bombay High Court judgements*, *Mohanlal writings*. Malayalam and Bengali scripts can be found in the bottom right and top left images, respectively. Challenges in these collections involve poor image quality, ink bleed, complex backgrounds, and layouts.

the occurrences of a given query in a set of documents. Existing state-of-the-art KWS solutions [2,10,14] are discussed majorly in the context of benchmark datasets like GW [12], IAM [11]. These datasets are neat and legible. It implies that the proposed approaches are evaluated on known handwriting styles and limited vocabulary sets. However, in practical settings, document retrieval tools must accommodate for new handwriting styles and open vocabulary. Additionally, the natural handwritten collections are associated with challenges like ink-bleed, illegible writing, various textures, poor image resolution, and paper degradation. This results in a huge domain gap between these collections and the benchmark datasets. This gap causes a drop in retrieval performance for unseen documents. Hence there arises a need to study the application of existing approaches for retrieval in unexplored documents. With this work, we propose an embedding-based framework that enables search over unexplored collections. Our framework performs retrieval on unseen collections without any need for finetuning or transcribed data. In other words, we perform zero-shot retrieval from unexplored collections. This is achieved by learning holistic embeddings for handwritten text and text strings. These embeddings are also invariant to writing styles, paper textures, degradation, and layouts.

Another limitation of existing works is the lack of studies on document retrieval from non-Latin collections. We study and evaluate the performance of our framework on untranscribed collections in English and 2 Indic scripts: Malayalam and Bengali. This setup can easily be extended to other Indic scripts. Handwritten document collections from public digital libraries are used to demonstrate the efficiency of our retrieval pipeline. These collections contain a total of 2,957 pages and 313K words. Figure 1 shows snippets from chosen collections discussed in this work. Search on these specific records is difficult as they come with significant issues like poor image resolution, unstructured layouts, different scripts on a single page, ink-bleed, ruled and watermark backgrounds. We address these challenges and discuss a retrieval approach developed for these unseen collections. We utilize complementary modules along with spotting approaches to tackle some of these challenges. Another crucial aspect of any retrieval framework is its scalability. Therefore, we study and discuss efficient approaches to improve retrieval speed and memory requirements. We perform experiments on these collections to study the efficiency of our retrieval pipeline for searching across large document collections.

To the best of our knowledge, this work is an initial effort to perform handwritten search in a zero-shot setting at a large scale in multiple scripts. Our framework also supports querying with both textual and exemplar queries. We present a real-time handwritten document retrieval demonstration[1] for handwritten collections from India. The major contributions of this work are as follows:

i. An embedding based framework for retrieval from unseen handwritten collections.
ii. Study and discussion on the approaches towards generic keyword spotting that can handle unseen words and unseen writing styles.
iii. Evaluation of efficient indexing, and query matching methods.
iv. A demonstration to validate our end-to-end retrieval pipeline on 4 collections written in English, Bengali and Malayalam.

2 Related Works

In this section, we discuss earlier works in KWS and existing end-to-end retrieval demonstrations for handwritten collections.

Keyword Spotting: Earlier works in KWS use various methods such as template matching, traditional feature extraction along with sequence matching, and feature learning methods. Works [1,5] survey numerous spotting techniques and discuss commonly adopted methods for document retrieval. Recently popular

[1] Demo links available at project page.

feature learning methods for KWS aim to learn a holistic fixed-length representation for word images. Current top trends that use this approach are: attribute learning in PHOCNet [14], verbatim and semantic feature learning in triplet CNN [17], and joint representation learning in HWNet [10]. All these approaches are highly data-driven and can be categorized as segmentation-based approaches. Segmented word images from collections are a prerequisite in the above methods to train and extract features. In segmentation-free techniques, the network identifies potential text regions and then spots a given keyword. Wilkinson et al. [18] propose one such method for neural word search. An end-to-end network comprising a region proposal network and a holistic feature learning network is presented in their work. In another segmentation-free approach proposed for Bengali script [4], a CNN trained to identify the class label of word images is used to spot keywords in a dataset of 50 pages. Due to limited class labels and the incapability to handle out-of-vocabulary words, this setup is not suitable for spotting in new and unseen collections. In this work, we adopt a segmentation-based feature learning approach for searching. This choice enables us to ensure easier indexing and reasonable accuracy on new collections.

Retrieval Demonstrations: Search engine for handwritten collections was first introduced in [12] for George Washington's letters containing 987 document images. Recently, another concept referred to as probabilistic indexing(PrIx) has been introduced. Multiple large-scale Latin collections are indexed using PrIx, such as Bentham papers [15], Carabela manuscripts [16], and Spanish TSO collection [15]. A case study presented in [18] discusses a segmentation-free EBR for Swedish court records consisting of 55K images. These existing demonstrations are developed explicitly for Latin scripts. No such retrieval demonstrations are available for non-Latin scripts. Existing methods rely on transcribed pages from the collection to make it retrievable. The number of pages to be transcribed varies for different methods. 558 and 1213 transcribed pages are used to build demonstrators for the Carabela manuscripts and Bentham papers. In contrast, the demonstrator for Swedish court records used only 11 transcribed pages. The transcription costs are unfeasible as there are millions of massive collections written in different scripts. Our proposed pipeline performs retrieval on unseen collections without any need for finetuning. At the same time, our collections are equally complex when compared to the above mentioned collections. As the pipeline is not fine-tuned for specific writing styles or vocabulary of a new collection, our work is closely related to zero-shot learning.

3 Proposed Framework

In this section, we discuss the setup employed for searching in a document collection. Figure 2 shows an overview of the retrieval pipeline. Given a historic collection, the document images are forwarded to a text detector. The output from the detector is processed to extract text regions from the images. Valid text regions are selected and forwarded to the embedding network to compute holistic embeddings. The computed embeddings and positions are used to create embedding index and position index for a collection. Pretrained networks are employed for embedding network and text detector. Further details about the modules are mentioned below.

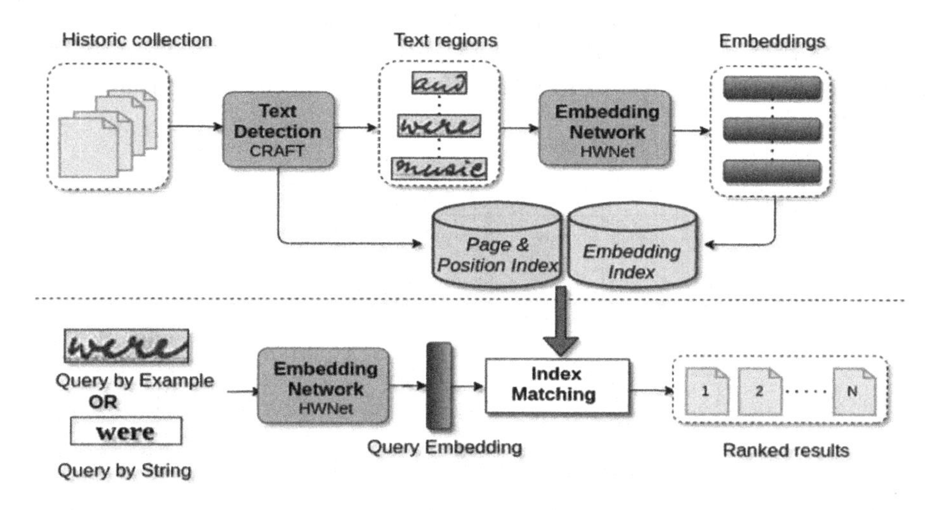

Fig. 2. Overview of the spotting and retrieval framework. (Top) Processing and indexing a handwritten collection to enable search operations. (Bottom) The flow of a query across the retrieval framework to retrieve relevant results.

The real-time flow of an input query is shown in the bottom image of Fig. 2. Text strings and exemplar images are supported as input query formats. These modes are referred to as QbS (Query by String) and QbE (Query by Example), respectively. Similarity search is performed using euclidean distance metric to retrieve relevant documents. Using brute force search in large indexes is costly due to a huge number of computations. We use approximate nearest neighbor approaches as these methods are non-exhaustive. This reduces the retrieval time for a given query and makes our framework efficient.

Characteristics: Our retrieval pipeline is designed so that the user can search within a specific collection with a single query. We believe that this is a reasonable assumption as the domains of the collection vary significantly and a user looking for specific information knows which collection to choose for querying. This choice enables to reduce the search index size and improve the retrieval time. Querying can be done through both text strings (QbS) and word image examples (QbE). The designed document retrieval pipeline is generic and can be extended to any collection. Our proposed setup is capable of dealing with unseen document images with reasonable confidence. Efficient representation learning enables this zero-shot retrieval from unexplored collections. In our setup, labeled handwritten data is only used during training the embedding network.

Embedding Network: Spotting a given keyword in an unseen document can be achieved by learning holistic feature representations for handwritten text. These representations must be invariant to writing styles, document degradation and poor image resolution. At the same time, the feature learning method should also generate discriminative representations for an open set of vocabulary. To achieve these goals, we employ the end-to-end deep embedding network, HWNetV2, discussed in [8]. This network learns a common subspace of embeddings for both word images and text strings. We utilize these embeddings as features for indexing and matching.

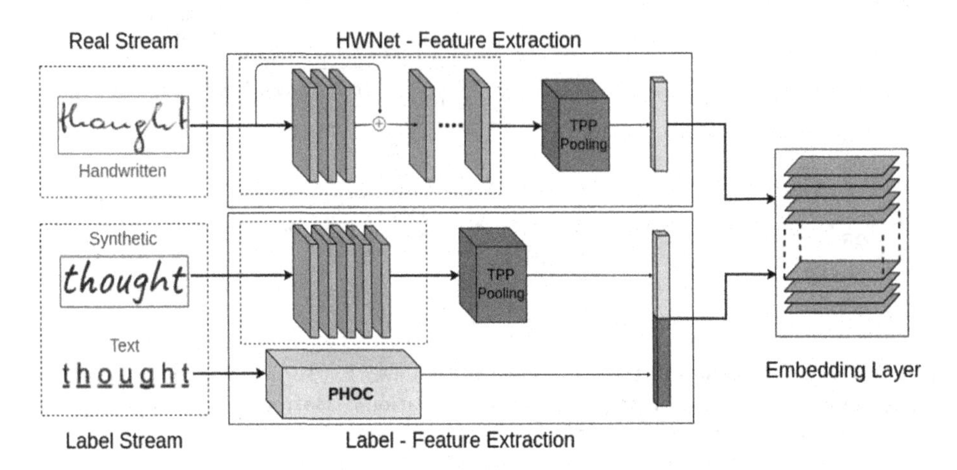

Fig. 3. Overview of the HWNetv2 embedding network for generating holistic handwritten text features. Embedding layer in the network is responsible for mapping the image features and text string features to a common subspace.

The overview of this architecture is shown in Fig. 3. The real stream, as shown in the figure, computes embeddings for word images from document collections using a deep network. Label stream is introduced to compute feature representations for text strings. It comprises of a PHOC [2] module and a shallow network. We request the readers to refer to [10] for technical details on architecture. The real stream and label stream embeddings are forwarded to an embedding layer. This layer learns to map these feature representations from two different modalities to a common embedding/representation space.

Feature Extraction: A collection is processed using Otsu thresholding to binarize and reduce the impact of different backgrounds and textures on the retrieval pipeline. We utilize a pretrained deep network CRAFT [3] to extract words from handwritten collections for text detection. This network is trained in a weakly supervised manner with pseudo ground truths on scene text datasets. Although CRAFT is trained on the scene text images, we observe reasonable detection rate for handwritten documents in Latin and Indic scripts. We believe that the underlying idea behind the CRAFT detector makes the network robust to detect handwritten text. The detector works by localizing individual character regions and linking closer character regions to form words. As the detection network is used in a zero-shot setting, the text bounding boxes have a certain degree of error due to over-segmentation or under-segmentation. To overcome this, extreme affine augmentation is applied while training the HWNetv2 network. This is done to imitate segmentation issues and enable embedding learning for wrongly segmented inputs. This strategy helps to overcome segmentation issues during retrieval. The extracted text image regions are forwarded to the pretrained HWNetv2 to extract features.

Retrieval: The computed embeddings for a collection are indexed using inverted file index (IVF). While indexing a collection, the embeddings are clustered to identify Voronoi cells and their centroids, which are the representatives of the cells. For an input query, the computed embedding is used to search for top N similar Voronoi cells by matching the query embedding to the centroids. The identified Voronoi cells are further matched to all the embeddings in these top N cells to obtain top k closest embeddings. Documents corresponding to these embeddings are the top k relevant matches for the query. Only valid embeddings are indexed to reduce the index size. Invalid embedding corresponding to over-segmented, under-segmented text regions are avoided. Extremely small text regions are also pruned off as they correspond to stop words mostly. We use FAISS [7] to make efficient indexes and perform search operations. In our experiments, we cluster the embeddings into varying cell sizes depending on the total embeddings for a collection.

Query Processing: Input query is forwarded to the pretrained embedding network. Image queries are resized, normalized and forwarded to the HWNetv2 real stream. For textual queries, synthetic image is rendered containing the query string. This rendered image and the text string are forwarded to the HWNetv2 label stream. The computed query embedding is matched with relevant indexes. The index keys obtained from similarity matching are used to retrieve document ids, keyword position details from the page and position index. Relevant lines associated with the spotted query word are cropped from the corresponding documents using these positions. For this, the bounding box of the spotted query is extended along the image width and this selected area is presented as relevant lines for the input query.

4 Experiments and Results

4.1 Training Phase

The retrieval framework is developed for documents written in English, Bengali and Malayalam. We train the HWNetv2 network with existing handwritten datasets. IAM [11], GW [13] datasets are used for training English embeddings. For Bengali and Malayalam scripts, we use IIIT-INDIC-HW-WORDS [6] dataset. Affine, elastic and color transformations are used to augment the training datasets. We also use the IIIT-HWS [9] synthetic dataset containing 1 million word images rendered using handwritten style fonts for pretraining the embedding network. The mean Average Precision (mAP) obtained on three training datasets are reported in Table 1. Note that the high evaluation scores are reported on rather clean, carefully collected datasets. This is unlike a real, practical setting where the handwriting, vocabulary and unseen layouts are prevalent.

Table 1. Evaluation metrics on the test split of IAM, IIIT-INDIC-HW-WORDS dataset. Evaluation done for both QbS and QbE settings. Full and OOV test refers to the complete test split and out-of-vocabulary test split respectively.

Dataset	QbS mAP		QbE mAP	
	Full test	OOV test	Full test	OOV test
IAM (English)	0.93	0.92	0.93	0.93
IIIT-INDIC-HW-WORDS (Malayalam)	0.99	0.98	0.99	0.99
IIIT-INDIC-HW-WORDS (Bengali)	0.95	0.95	0.95	0.94

4.2 Evaluation Datasets

Four collections from public websites and digital libraries are chosen to evaluate our pipeline: a collection consisting of handwritten poems and plays written by Rabindranath Tagore, a historic collection of Bombay High Court law judgements from the year 1864, a modern collection of handwritten blogs by an actor and a handwritten manuscript of Constitution of India (CoI). Source links for these documents are linked in Table 2. The table also lists the scripts used, total pages and estimated words in these collections. Figure 1 shows sample blocks from these collections. The pages in these collections have printed text, handwritten text, watermarks and backgrounds along with illegible text due to degradation, difficult writing styles. Our retrieval framework is capable of retrieving meaningful results for most of the queries despite these challenges. The low image resolution in CoI manuscript and Tagore's papers is also handled implicitly without the need for additional processing.

Table 2. Digital handwritten collections demonstrated and evaluated in this work. Digital sources are linked in the collection title column.

Collection title	Script	Pages	Words (est.)
Bombay High Court (BHC) records	English	1,964	207,028
Constitution of India (CoI) manuscript	English	479	42,040
Tagore's Papers (TP)	English, Bengali	125	10,990
Mohanlal Writings (LAL)	Malayalam	389	53,111

4.3 Evaluation and Discussion

In this section, we discuss the results obtained on the four collections mentioned above. Unlike the training datasets mentioned in Table 1, these collections are unexplored and unlabeled. Therefore, reporting exhaustive evaluation metrics is not feasible. Labeled data is a prerequisite to compute mAP. In place of mAP, we report the precision (P) of top-k ranked results at $k = 1, 10, 25$. This evaluation method is also followed in similar case study discussed in [18]. This web-scale metric does not require the knowledge of all relevant instances for a given query. We pick random queries from each of the collections and report the top-k precision (P_k) in Tables 3, 4, 5 and 6. The results are reported for QbS retrieval setting in these tables. We observe similar results for QbE query retrieval as well. The retrieval framework achieves similar performance for both seen and unseen out-of-vocabulary (OOV) queries. For queries like *vacancy* and *sudden* low values of P_k at $k = 10, 25$ are reported as these are rare words that are used in the collection. Evaluating on CoI manuscript for extremely rare OOV words like *Madras, Travancore,* and *surcharge*, the P_k for $k = 1$ is 1.00. The framework is also capable of retrieving related terms for a given query. For example, for the query *vacancy* relevant results include the term *vacant* as well. More such pairs are *discharge-surcharge, clause-subclause, constitution-constitute,* and *Vice President-President.*

Table 3. Precision for queries in *Constitution of India* collection. Unseen vocabulary during training time are marked in blue color.

	Constitution	India	Right	Vacancy	Rajpramukh	Deputy	Discharge
$P_{k=1}$	1.00	1.00	1.00	1.00	1.00	1.00	1.00
$P_{k=10}$	1.00	1.00	1.00	0.60	0.80	1.00	1.00
$P_{k=25}$	1.00	0.96	0.92	0.24	0.76	1.00	0.88

Table 4. Precision for queries in *Bombay High Court records* collection. Unseen queries during training time are marked in blue color.

	Bombay	Court	August	Judge	Defendant	Registrar	Plaintiff
$P_{k=1}$	1.00	1.00	1.00	1.00	1.00	1.00	1.00
$P_{k=10}$	0.90	1.00	1.00	1.00	1.00	0.90	1.00
$P_{k=25}$	0.88	1.00	0.88	1.00	1.00	0.92	1.00

Table 5. Precision for queries in *Tagore's papers.* Unseen queries during training time are marked in blue color.

	heart	light	নেই	sudden	thy	ছিল	অনুবাদ
$P_{k=1}$	1.00	1.00	1.00	1.00	1.00	1.00	1.00
$P_{k=10}$	1.00	1.00	0.70	0.60	0.90	0.40	1.00
$P_{k=25}$	1.00	0.92	0.52	0.24	0.88	0.16	0.64

Table 6. Precision for queries in *Mohanlal writings.* Unseen queries during training time are marked in blue color.

	സിനിമ	ജീവിതം	കുടുംബം	മോഹൻലാൽ	നടൻ	ഇന്ത്യ
$P_{k=1}$	1.00	1.00	1.00	1.00	1.00	1.00
$P_{k=10}$	1.00	1.00	0.50	1.00	0.70	1.00
$P_{k=25}$	1.00	1.00	0.20	0.44	0.28	0.72

We also show the qualitative results obtained for the four collections in different settings. Figures 4, 5 and 6 shows the top retrieval results for both seen and unseen queries. Our framework retrieves accurate results despite the drastic style variations. For example, positive results shown for 3 English handwriting styles in both Fig. 4 and Fig. 5. For the query *registrar* shown in Fig. 4(a), the retrieval results contain both printed text images and handwritten images. It shows the robustness of the learnt embeddings. Without any prior training for these specific handwriting styles, the retrieval results are promising.

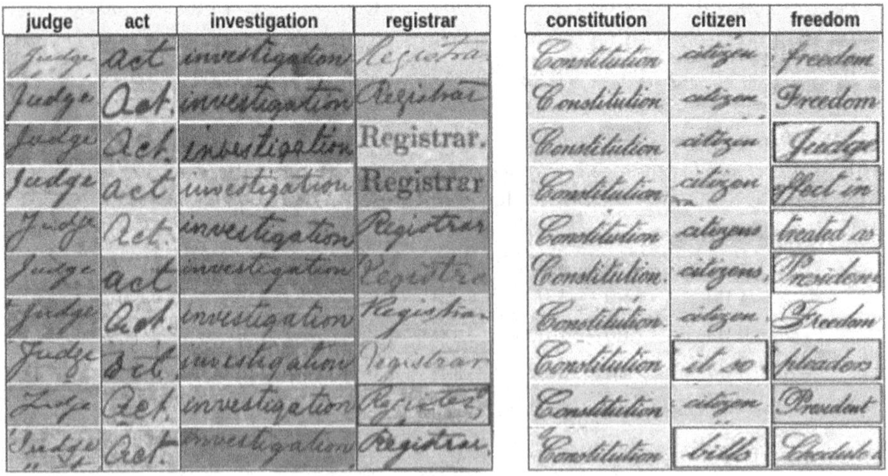

(a) Bombay High Court records

(b) Constitution of India manuscripts

Fig. 4. Top-10 qualitative search results from collections: BHC records and CoI manuscript. Queries are shown at the top and incorrect retrievals are highlighted in red. Querying is done in QbS setting. Poor image quality of the collections is also shown here. (Color figure online)

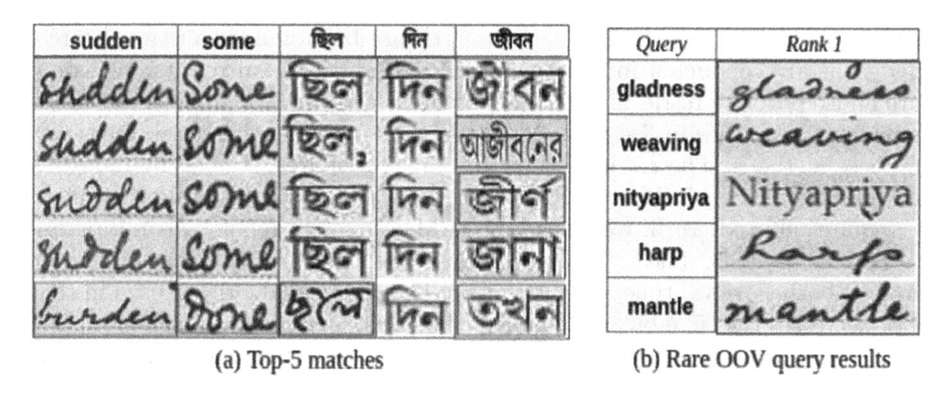

Fig. 5. Qualitative search results from Tagore's papers collection for QbS setting. (a) Top-5 results for English and Bengali queries. OOV queries are highlighted in blue and incorrect retrievals are highlighted in red. (b) Showing top-1 result for a few rare OOV words from the collection. (Color figure online)

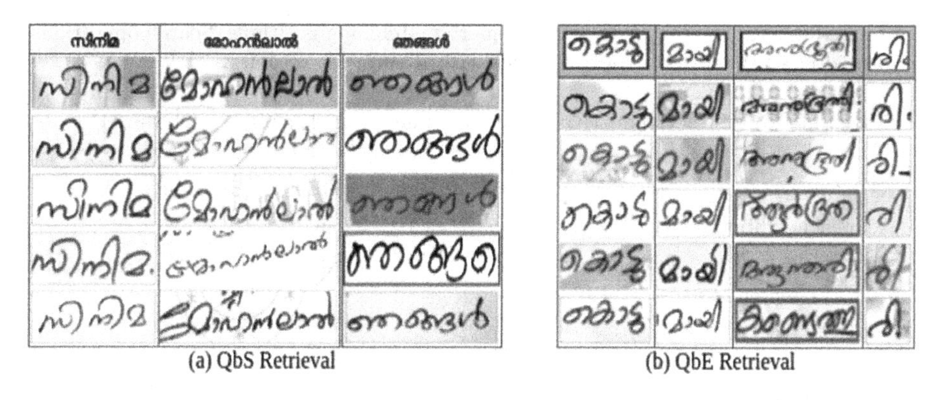

Fig. 6. Qualitative search results from Mohanlal's writings collection for QbS and QbE setting. Top row shows the queries. OOV queries are highlighted in blue and incorrect retrievals are highlighted in red. (Color figure online)

In Figs. 5 and 6, the incorrect retrievals match closely to the query keyword. We also show promising results in QbE retrieval mode for the Malayalam collection in Fig. 6(b). Despite the watermarks present in the images, the obtained results are accurate and relevant. QbE mode of querying is especially helpful while searching for signatures and unknown symbols encountered in a collection.

4.4 Comparative Results for Retrieval

This section discusses the time complexity and memory requirements of the indexing and retrieval module. We compare our index matching algorithm to two other approaches using K-D Tree and nearest neighbour approach. Nearest neighbour approach involves an exhaustive search over all the samples in a given index. K-D Tree and IVF based search are both non-exhaustive. Search operations are conducted in selective lists to obtain retrieval results. We compare and report the retrieval time and the mAP obtained for these three index matching approaches in Table 7. We also report the index sizes(disk memory) for these methods. The retrieval time and index size are obtained by indexing and querying the fairly large BHC records collection. As mAP evaluation metric requires ground truth, we utilize the IAM test set to report this metric for the search methods. The search method used in this work is both efficient in terms of both retrieval time and QbS mAP. The retrieval algorithm used in this work decreases the time by 86% compared to the simple nearest neighbour approach. The choice of retrieval algorithm is really important when the indexes are huge. Therefore, we also study the effect of index sizes on the search time. We observe that time required to perform brute-force search increases rapidly with an increase in index size compared to the inverted index based approach. Therefore, we use the inverted index based approach for searching.

Table 7. Retrieval time vs. Accuracy for query matching algorithms. Search operations performed in an index with 200K samples with feature dimension as 2048.

Search method	Exhaustive	Retrieval time (sec/query)	Index Size (GB)	QbS mAP (IAM test)
K-D tree	No	0.59	3.5	0.95
Nearest neighbour*	Yes	0.15	1.6	0.96
This work*	No	**0.02**	1.6	0.95

*Note: * Computed using FAISS library.*

5 Conclusion

Document retrieval from complex handwritten records containing an unseen vocabulary set is challenging. In this work, we discuss a document retrieval framework for unseen and unexplored collections. This framework can perform search operations on these collections without any fine-tuning for a specific collection. We discuss and evaluate our method both quantitatively and qualitatively on four different collections written in three scripts, of which two are Indic. Our framework performs reasonably well on these sizeable collections. Even for rare words, the top results are accurate. Utilising this simplified framework we plan to introduce more Indic collections written in other Indic scripts. Finally, we also present a demonstration to showcase the usefulness of the retrieval framework.

Acknowledgments. The authors would like to acknowledge the funding support received through IMPRINT project, Govt. of India to accomplish this project.

References

1. Ahmed, R., Al-Khatib, W.G., Mahmoud, S.: A Survey on handwritten documents word spotting. Int. J. Multimed. Inf. Retr. **6**(1), 31–47 (2016). https://doi.org/10.1007/s13735-016-0110-y
2. Almazán, J., Gordo, A., Fornés, A., Valveny, E.: Word spotting and recognition with embedded attributes. TPAMI **36**, 2552–2566 (2014)
3. Baek, Y., Lee, B., Han, D., Yun, S., Lee, H.: Character region awareness for text detection. In: CVPR (2019)
4. Das, S., Mandal, S.: Keyword spotting in historical Bangla handwritten document image using CNN. In: ICACCP (2019)
5. Giotis, A.P., Sfikas, G., Gatos, B., Nikou, C.: A survey of document image word spotting techniques. PR **68**, 310–332 (2017)
6. Gongidi, S., Jawahar, C.V.: IIIT-INDIC-HW-WORDS: a dataset for Indic handwritten text recognition. In: Lladós, J., Lopresti, D., Uchida, S. (eds.) ICDAR 2021. LNCS, vol. 12824, pp. 444–459. Springer, Cham (2021). https://doi.org/10.1007/978-3-030-86337-1_30
7. Johnson, J., Douze, M., Jégou, H.: Billion-scale similarity search with GPUs. IEEE Trans. Big Data **7**(3), 535–547 (2019)
8. Krishnan, P., Dutta, K., Jawahar, C.: Word spotting and recognition using deep embedding. In: DAS (2018)
9. Krishnan, P., Jawahar, C.: Generating synthetic data for text recognition. arXiv preprint arXiv:1608.04224 (2016)
10. Krishnan, P., Jawahar, C.V.: HWNet v2: an efficient word image representation for handwritten documents. IJDAR **22**(4), 387–405 (2019). https://doi.org/10.1007/s10032-019-00336-x
11. Marti, U.V., Bunke, H.: The IAM-database: an English sentence database for offline handwriting recognition. IJDAR **5**, 39–46 (2002). https://doi.org/10.1007/s100320200071
12. Rath, T.M., Manmatha, R., Lavrenko, V.: A search engine for historical manuscript images. In: SIGIR (2004)
13. Rath, T.M., Manmatha, R.: Word spotting for historical documents. IJDAR **9**, 139–152 (2007). https://doi.org/10.1007/s10032-006-0027-8
14. Sudholt, S., Fink, G.A.: PHOCNet: a deep convolutional neural network for word spotting in handwritten documents. In: ICFHR (2016)
15. Toselli, A.H., Romero, V., Sánchez, J.A., Vidal, E.: Making two vast historical manuscript collections searchable and extracting meaningful textual features through large-scale probabilistic indexing. In: ICDAR (2019)
16. Vidal, E., et al.: The CARABELA project and manuscript collection: large-scale probabilistic indexing and content-based classification. In: ICFHR (2020)
17. Wilkinson, T., Brun, A.: Semantic and verbatim word spotting using deep neural networks. In: ICFHR (2016)
18. Wilkinson, T., Lindström, J., Brun, A.: Neural word search in historical manuscript collections. ArXiv (2018)

Detecting Document Forgery Using Hyperspectral Imaging and Machine Learning

Vrinda Rastogi(✉), Sahima Srivastava, Garima Jaiswal, and Arun Sharma

Indira Gandhi Delhi Technical University for Women, Kashmere Gate, Delhi, India
rastogi.vrinda@gmail.com

Abstract. Forgery is the process of fabricating, transforming or imitating writings, objects, or documents. It is a white-collar crime. Investigating forged cheques, wills or modified documents frequently involves analysing the inks used in these write-ups. Hyperspectral imaging can be used to identify various types of materials. This technology paired with powerful classifiers can be implemented to identify the various types of inks used in a document. This study leveraged the UWA Writing Ink Hyperspectral Images database (WIHSI) to carry forth ink detection by applying three different dimension reduction algorithms namely: Principal Component Analysis, Factor Analysis, and Independent Component Analysis. After which, a comparative study was carried forth between different processes applied in this study and existing methods. In essence, this work aims to integrate the use of hyperspectral imagery with machine learning and dimension reduction to detect document forgery.

Keywords: Document forgery · Hyperspectral imaging · Ink analysis · Machine learning · Dimension reduction

1 Introduction

Document forgery is a ubiquitous and noteworthy problem in today's age. It is the process of fabricating, transforming or imitating writings, objects, or documents [16]. False making entails fraudulently writing on an official document or making subtle changes to the original document. False making brings the authenticity of a document into question. More often than not it involves falsifying or forging signatures [28]. This type of forgery directly correlates to the theft of the identity of an individual. Forgery can also include backdating, overwriting, and creating addendums to the document. Thus, a different ink is used to modify an already existing document resulting in a forgery of ink.

Broadly speaking, there are two methods to pin-point or identify forgery namely: destructive methods and non-destructive methods. Destructive methods carry out analysis of the ink, usually in laboratories [1] or similar settings.

V. Rastogi and S. Srivastava—Equal contribution.

B. Raman et al. (Eds.): CVIP 2021, CCIS 1568, pp. 14–25, 2022.
https://doi.org/10.1007/978-3-031-11349-9_2

However, this method doesn't always abide by the ethical code of conduct [11]. This is because they are by default intrusive and involve taking a sample of the document, like a punch in the paper, and analysing it separately or disturbing the ink on the entire document [10]. Analysis of such kind can damage or alter the entire document permanently. The most common method of destructive ink testing is thin layer chromatography (TLC) [1].

Albeit, non-destructive methods are non-intrusive and do not change the document fundamentally [35]. The document is left intact and can be re-evaluated in the future. This method is faster and cheaper [32]. The most prevalent technique for non-destructive analysis is spectrophotometry [6]. Spectroscopy, which is the study of the relation between matter and radiation is an umbrella term for spectrophotometry [5].[1]

Hyperspectral imaging (HSI) is a type of spectral imaging capable of leveraging both spatial and spectral data from materials or objects [18]. With the ability to analyse a wide range of the electromagnetic spectrum [7,14], HSI was first utilised in remote sensing technology [29]. Unlike normal imaging, it can extend the number of bands scanned to up to hundreds and thousands [4,34]. As the bands analysed are narrow and contiguous in nature, HSI effectively records subtle variations in reflected energy.

2 Literature Survey

This section explains some of the previous work done in detecting forgery in documents using hyperspectral technology.

Khan et al. [25] created the first publicly available dataset for document forgery detection using hyperspectral imaging which is known as "UWA Writing Ink Hyperspectral Image Database". In order to reduce dimensionality and extract crucial features from the dataset, they used Joint Sparse Principal Component Analysis (JSPCA). Similarly, to identify different inks, Joint Sparse Band Selection (JSBS) technique was implemented. This technique achieved an accuracy of around 85%.

Devassy et al. [12] created their own document dataset using hyperspectral technology. The data was then normalised using the standard reference target technique. After which, this data was fed into a one dimensional convolutional neural network (CNN) and achieved 91% accuracy.

In [22], six different CNN's were implemented using the UWA Writing Ink Hyperspectral Image Database. It achieved 99.2% accuracy for blue inks and 99.6% accuracy for black inks. Yet this approach is limited as it requires prior knowledge for training the neural networks.

Luo et al. [30] also utilised the UWA's Image Database where ink distinction in a local area was carried out using anomaly detection. The study concluded that Influenced Outlierness (INFLO) anomaly detection method integrated with

[1] https://socratic.org/questions/what-is-difference-between-spectrophotometry-and-spectroscopy.

point-to-point (P2P) criterion for feature selection gave the highest results. This technique also tackles the apriori problem via clustering.

UWA's Image Database is also implemented in [33], yet in this technique orthogonal and graph regularised Non-negative Matrix Factorisation Model is applied. It achieves an accuracy of around 85%, but lacks in finding the optimal hyperparameter selection.

Hyperspectral data was fetched using the neighboring pixel's spectral responses in [20]. They also took advantage of the WIHSI database. They fed the responses to convolutional neural networks (CNN) to garner the writer's identity with the highest accuracy of 71%. However, an extreme limit of this work is that the dataset picked was not huge enough to affirm the model's certainty.

The utilisation of Least Square Support Vector Machines (LS-SVM) for ink detection using hyperspectral data was purposed by Morales et al. [31]. To carry out their work, they formed a dataset from scratch where they created hyperspectral curves by removing the background. Smoothening procedures were applied to the curves to extract 21 crucial pixels. Feature pairs were created by finding the area and features of the slope. Finally, the SVM algorithm was applied on this data. The highest accuracy achieved was of 80%.

Another work whom created their own dataset is of Wang et al. [36]. They focused on noise removal and reducing dimensions by applying Principal Component Analysis (PSA). Then, the dataset underwent psedudo colour synthesis. It created a technique to detect the writer's identity. The accuracy rate altered due to the changing ink patterns.

3 Dataset Description

For this work, the UWA Writing Inks Hyperspectral Images (WIHSI) database [26] was selected. This database consists of 33 visible band spectrums via 70 hyperspectral images. It contains ten different types of inks, five blue and five black, along with seven different subjects. One one page, the sentence 'The quick brown fox jumps over the lazy dog' was written by each subject using five different inks, from varying manufacturers. This was done to ensure they appeared visually similar as they were the same colour (blue or black) yet remained unique.

4 Methodology

The methodology followed to do the work is depicted in Fig. 1.

4.1 Preprocessing

On the dataset, background was removed and Sauvola thresholding was applied. It was chosen as it considers unequal illumination and can take crucial data from hyperspectral information efficiently. After which, the five sentences in each document were decomposed into singular sentences for easy analysis in regards

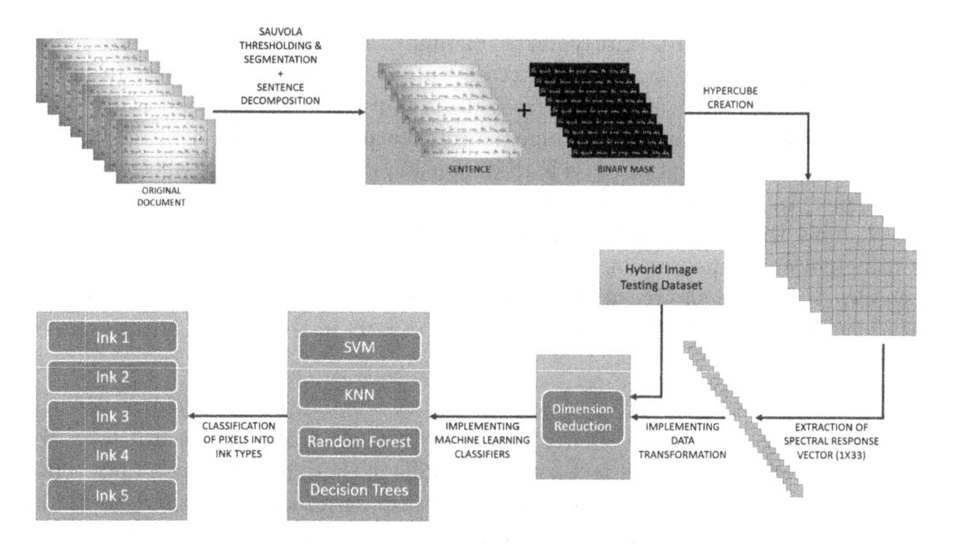

Fig. 1. Methodology followed in this work

to classification. These sentences have been transformed into hypercubes which are 3-dimensional matrices that reflects the 33 spectral bands. Finally, spectral response vector is created of dimensions 1×33.

In order to carry forth testing of ink detection, hypercubes were created with different inks in various ratios as showed in Table 1.

Table 1. Ratios in which inks were mixed

Combinations name	Number of inks mixed	Ratio of mixing
1by1	2	1:1
1by4	2	1:4
1by8	2	1:8
1by16	2	1:16
1by32	2	1:32
3_hybrid	3	1:1:1
4_hybrid	4	1:1:1:1
5_hybrid	5	1:1:1:1:1

4.2 Dimension Reduction

Dimensionality Reduction is a process that creates subsets of important features which then in turn act like one new attribute. The significance of this technique is that less data is lost in comparison to feature selection but it takes up more space as well. The following elaborates on the various dimension reduction algorithms applied:

Principal Component Analysis. Principal Component Analysis (PCA) is a technique for acquiring significant factors (in the form of components) from an enormous arrangement of factors accessible in an informational collection [37]. It separates low dimensional arrangement of features by taking a projection of unessential measurements from a high dimensional informational collection with a thought process to catch however much data as could be expected. In the end, it aims to take the crucial data from the overall dataset and portrays it as principal components which are essentially a collection of novice orthogonal variables [3]. These showcase the similarity pattern and creates a map of point variables. It is a widely implemented tool, especially in the field of unsupervised applications [13].

Factor Analysis. In this Factor Analysis strategy, correlations are used to gather the variables, i.e., all factors in a specific gathering will have a high relationship among themselves, yet a low relationship with factors of other group(s). In factor analysis, the scale of the variables do not come into play, unlike the orthogonal rotations of each of the factors [17]. According to [21], there are two steps in this process: initially the solution is found, then it is rotated. It is not primarily limited to dimension reduction, but also to understand the various dimensions and to test hypothesis [27].

Independent Component Analysis. Independent Component Analysis (ICA) depends on data hypothesis and is additionally quite possibly the most generally utilized dimensionality decreasing procedures [19]. The significant contrast among PCA and ICA is that PCA searches for uncorrelated components while ICA searches for free factors. On the off chance that two factors are uncorrelated, it implies there is no straight connection between them. In the event that they are free, it implies they are not subject to different factors [9]. For instance, the age of an individual is free of what that individual eats, or how much TV one watches. This calculation expects that the given factors are straight combinations of some obscure idle factors. It likewise accepts that these dormant factors are commonly free, i.e., they are not subject to different factors and thus they are known as the independent segments of the noticed information [8].

4.3 Machine Learning

In this method, there are 10 different inks utilised which are known beforehand. Therefore, supervised machine learning classifiers are applied. Decision Trees and Random Forest were chosen to check how crucial different bands are in detecting inks. K-Nearest Neighbors was chosen to encapsulate document forgery inspection by leveraging neighbors in dimensionality. Similarly, Support Vector Machines (SVM) was selected and their more powerful as they use soft margins and complicated hyper-planes.

5 Results

Table 2 and 3 summarises the quantitative results obtained by applying various machine learning classifiers and dimension reduction algorithms on HSI data. The accuracy used was segmentation accuracy as displayed by Eq. 1 and is defined as the crossing point or association metric, which ascertains the quantity of accurately marked pixels of an ink isolated by the quantity of pixels marked with that ink in either reality or predicted [15]. Overall, better results were obtained for blue inks as compared to black inks, keeping the processes same.

$$Accuracy = \frac{TruePositives}{TruePositives + FalsePositives + FalseNegatives} \tag{1}$$

Table 2. Black ink accuracy when applied with different processes.

Ratios	Dimension reduction	KNN (n = 5)	KNN (n = 10)	KNN (n = 15)	Decision Tree	SVM (linear)	SVM (poly)	SVM (RBF)	Random Forest
1:1	None	0.755	0.762	0.778	0.706	0.766	0.607	0.806	0.798
	PCA	0.19	0.19	0.19	0.17	0.17	0.17	0.17	0.16
	Factor analysis	0.8	0.8	0.83	0.73	0.76	0.79	0.81	0.8
	ICA	0.78	0.8	0.8	0.72	0.55	0.79	0.82	0.8
1: 4	None	0.697	0.706	0.718	0.673	0.642	0.568	0.787	0.751
	PCA	0.24	0.24	0.24	0.22	0.24	0.25	0.23	0.22
	Factor analysis	0.75	0.76	0.76	0.7	0.76	0.73	0.79	0.78
	ICA	0.74	0.76	0.77	0.69	0.49	0.74	0.79	0.77
1:8	None	0.675	0.686	0.696	0.656	0.752	0.553	0.777	0.733
	PCA	0.2	0.19	0.2	0.19	0.2	0.21	0.19	0.19
	Factor analysis	0.73	0.74	0.74	0.68	0.75	0.7	0.77	0.78
	ICA	0.72	0.74	0.75	0.67	0.45	0.71	0.78	0.76
1:16	None	0.664	0.674	0.685	0.642	0.743	0.547	0.77	0.72
	PCA	0.22	0.22	0.22	0.22	0.22	0.21	0.22	0.2
	Factor analysis	0.73	0.73	0.74	0.68	0.74	0.69	0.77	0.75
	ICA	0.7	0.72	0.73	0.65	0.43	0.7	0.77	0.74
1:32	None	0.627	0.631	0.64	0.591	0.698	0.527	0.729	0.665
	PCA	0.18	0.18	0.18	0.18	0.19	0.16	0.18	0.17
	Factor analysis	0.73	0.72	0.72	0.68	0.74	0.68	0.76	0.75
	ICA	0.68	0.7	0.71	0.63	0.41	0.69	0.75	0.72
1:1:1	None	0.753	0.756	0.775	0.692	0.74	0.592	0.781	0.798
	PCA	0.17	0.17	0.17	0.15	0.15	0.18	0.14	0.15
	Factor analysis	0.77	0.78	0.85	0.7	0.72	0.78	0.79	0.78
	ICA	0.76	0.77	0.78	0.69	0.51	0.77	0.79	0.77
1:1:1:1	None	0.728	0.729	0.748	0.664	0.705	0.57	0.749	0.769
	PCA	0.14	0.15	0.14	0.14	0.12	0.14	0.12	0.14
	Factor analysis	0.74	0.75	0.75	0.67	0.7	0.74	0.75	0.75
	ICA	0.73	0.74	0.75	0.65	0.48	0.74	0.75	0.73
1:1:1:1:1	None	0.75	0.765	0.785	0.675	0.688	0.585	0.755	0.799
	PCA	0.09	0.09	0.09	0.1	0.09	0.11	0.08	0.12
	Factor analysis	0.74	0.75	0.76	0.67	0.68	0.77	0.78	0.75
	ICA	0.73	0.75	0.76	0.65	0.45	0.75	0.77	0.74

5.1 Machine Learning Without Dimension Reduction

After investigating the classification accuracy of classifiers for both inks without dimension reduction the following insights were gained.

1. Out of the three SVM's kernels implemented, the one with the best performance is the 'RBF' kernel for both blue and black inks.

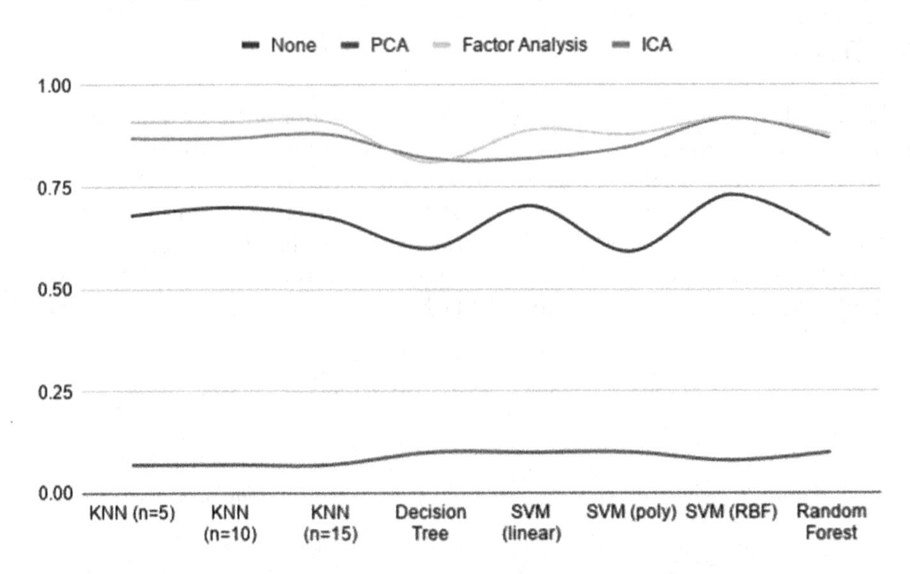

Fig. 2. 1:32 ratio for blue inks which displays that dimension reduction helped to increase the accuracy in this study. (Color figure online)

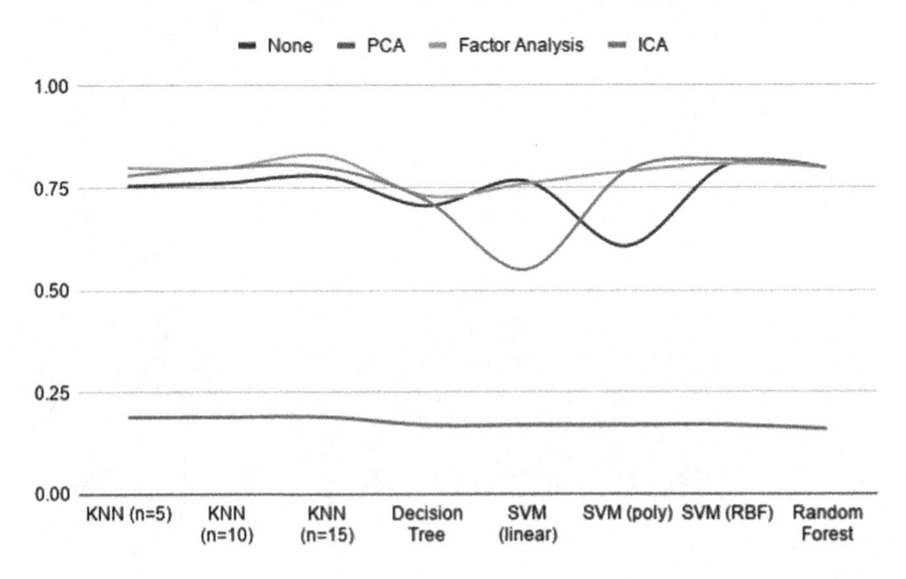

Fig. 3. 1:1 ratio for black inks which showcases that dimension reduction helped to increase the accuracy in this study.

2. After the 'RBF' kernel, the next best kernel was 'linear' kernel for both blue and black inks in regards to accuracy. The kernel which had the least performance was 'poly' kernel.
3. Random Forest outperformed decision trees for both inks in terms of accuracy. This can be due to the combination of many decision trees in random forest.
4. In K-nearest neighbours, the best performance for black inks was when 'k' was set to 15.
5. For blue inks, in KNN, the performance varied. For some ratios, 'k' was best to be at 5 and for other ratios, it was for when 'k' was set to 10.
6. From all of the ML classifiers applied, the global trend was that the blue ink had a pronounced accuracy in comparison to black inks.
7. In all of the accuracies compared, another global trend was that 1:32 had an overall lower accuracy in comparison to the other ratios. This can be due to the fact that in this ratio, the first ink has a very small impact, making it difficult to classify the ink.

5.2 Machine Learning with Dimension Reduction

After applying dimension reduction to the data, the classifiers accuracy performance showcased the following:

1. Out of the three dimension algorithms applied, the best performance was by Factor Analysis.
2. After factor analysis, the next best algorithm was independent component analysis. In fact, both their performances were neck-to-neck. This could be as both algorithms are very similar. The only exception is that while both focus on finding basis vectors, ICA takes to that vector and finds its kurtosis.
3. PCA performed the worst out of all the methods applied. It couldn't even match the performance of the process with only machine learning. This could be due as PCA doesn't work with specific vectors nor is it generative like factor analysis.
4. Factor analysis and ICA helped to increase the performance, especially in the 1:32 ratio and overall in black inks. This is depicted in Figs. 2 and 3 respectfully.

5.3 Comparison with Existing Methods

Table 4 showcases this method's accuracy in contrast with existed processes which also implements the same dataset for ink detection. The proposed technique outflanks the past strategies by accomplishing the most elevated exactness on the blended ink mixes which contains unequal proportions.

Table 3. Blue ink accuracy when applied with different processes.

Ratios	Dimension reduction	KNN (n = 5)	KNN (n = 10)	KNN (n = 15)	Decision tree	SVM (linear)	SVM (poly)	SVM (RBF)	Random Forest
1:1	None	0.979	0.98	0.98	0.941	0.987	0.852	0.991	0.976
	PCA	0.22	0.22	0.22	0.21	0.23	0.2	0.24	0.22
	Factor analysis	0.99	0.99	0.99	0.96	0.99	0.97	0.99	0.98
	ICA	0.98	0.98	0.98	0.96	0.96	0.96	0.99	0.98
1: 4	None	0.966	0.967	0.965	0.927	0.98	0.825	0.984	0.963
	PCA	0.15	0.14	0.14	0.16	0.17	0.13	0.16	0.16
	Factor analysis	0.98	0.98	0.98	0.94	0.98	0.96	0.98	0.97
	ICA	0.97	0.97	0.97	0.95	0.95	0.94	0.98	0.97
1:8	None	0.954	0.956	0.955	0.911	0.97	0.8	0.976	0.952
	PCA	0.12	0.12	0.12	0.13	0.14	0.12	0.13	0.14
	Factor analysis	0.97	0.97	0.97	0.92	0.97	0.95	0.98	0.96
	ICA	0.95	0.95	0.96	0.93	0.93	0.92	0.98	0.96
1:16	None	0.926	0.925	0.925	0.868	0.942	0.726	0.949	0.927
	PCA	0.09	0.09	0.09	0.11	0.11	0.12	0.1	0.11
	Factor analysis	0.94	0.95	0.94	0.87	0.94	0.91	0.95	0.93
	ICA	0.92	0.92	0.92	0.87	0.87	0.89	0.95	0.92
1:32	None	0.68	0.7	0.675	0.6	0.704	0.592	0.731	0.63
	PCA	0.07	0.07	0.07	0.1	0.1	0.1	0.08	0.1
	Factor analysis	0.91	0.91	0.91	0.81	0.89	0.88	0.92	0.88
	ICA	0.87	0.87	0.88	0.82	0.82	0.85	0.92	0.87
1:1:1	None	0.973	0.973	0.971	0.931	0.985	0.807	0.988	0.969
	PCA	0.24	0.24	0.24	0.22	0.24	0.19	0.25	0.23
	Factor analysis	0.99	0.99	0.99	0.96	0.98	0.97	0.99	0.98
	ICA	0.98	0.98	0.98	0.96	0.96	0.95	0.99	0.98
1:1:1:1	None	0.966	0.965	0.961	0.922	0.983	0.786	0.985	0.962
	PCA	0.35	0.35	0.35	0.33	0.35	0.21	0.37	0.37
	Factor analysis	0.99	0.99	0.99	0.96	0.98	0.96	0.99	0.98
	ICA	0.98	0.98	0.98	0.96	0.96	0.95	0.99	0.98
1:1:1:1:1	None	0.962	0.964	0.958	0.917	0.972	0.814	0.983	0.957
	PCA	0.06	0.06	0.06	0.1	0.1	0.23	0.07	0.08
	Factor analysis	0.99	0.99	0.99	0.94	0.98	0.96	0.98	0.97
	ICA	0.98	0.98	0.98	0.95	0.95	0.94	0.98	0.97

Table 4. Comparative study of proposed method

Method	Accuracy (%)		Greatest number of inks	Unbalanced ink amounts
	Blue	Black		
Proposed	99	85	5	Yes
Abbas et al. [2]	86.2	83.4	4	Yes
Luo et al. [30]	89	82.3	2	Yes
Khan et al. [23]	86.7	81.9	2	No
Khan et al. [24]	85.6	81.4	2	No

6 Conclusions and Future Work

From the investigations carried out, one can safely assume that spectral data is a good enough discriminator of inks and plays an eminent role in document forgery. It can also be concluded that though machine learning classifiers got a great accuracy while identifying ink mismatch.

In the future, more complex models and pre-analysis techniques should be explored for classification such as deep learning. This approach uses only spatial data, hence, the performance may improve using both spatial and spectral data. Another drawback that can be worked upon is the disparity in performance of black and blue inks. To further improve document forgery detection, writer detection can also be studied in the future.

References

1. World of forensic science. encyclopedia.com, 16 October 2020, December 2020. https://www.encyclopedia.com/science/encyclopedias-almanacs-transcripts-and-maps/ink-analysis
2. Abbas, A., Khurshid, K., Shafait, F.: Towards automated ink mismatch detection in hyperspectral document images. In: 2017 14th IAPR International Conference on Document Analysis and Recognition (ICDAR), vol. 1, pp. 1229–1236. IEEE (2017)
3. Abdi, H., Williams, L.J.: Principal component analysis. Wiley Interdiscip. Rev. Comput. Stat. **2**(4), 433–459 (2010)
4. Barnes, M., Pan, Z., Zhang, S.: Systems and methods for hyperspectral imaging, US Patent 9,117,133, 25 Aug 2015
5. Braz, A., López-López, M., García-Ruiz, C.: Raman spectroscopy for forensic analysis of inks in questioned documents. Forensic Sci. Int. **232**(1–3), 206–212 (2013)
6. Causin, V., Casamassima, R., Marruncheddu, G., Lenzoni, G., Peluso, G., Ripani, L.: The discrimination potential of diffuse-reflectance ultraviolet-visible-near infrared spectrophotometry for the forensic analysis of paper. Forensic Sci. Int. **216**(1–3), 163–167 (2012)
7. Chang, C.I.: Hyperspectral Imaging: Techniques for Spectral Detection and Classification, vol. 1. Springer, Cham (2003). https://doi.org/10.1007/978-1-4419-9170-6
8. Comon, P.: Independent Component Analysis (1992)
9. Comon, P.: Independent component analysis, a new concept? Signal Process. **36**(3), 287–314 (1994)
10. Craddock, P.: Scientific Investigation of Copies, Fakes and Forgeries. Routledge, Milton Park (2009)
11. Deep Kaur, C., Kanwal, N.: An analysis of image forgery detection techniques. Stat. Optim. Inf. Comput. **7**(2), 486–500 (2019)
12. Devassy, B.M., George, S.: Ink classification using convolutional neural network. NISK J. **12** (2019)
13. Ding, C., He, X.: k-means clustering via principal component analysis. In: Proceedings of the Twenty-First International Conference on Machine Learning, ICML 2004, p. 29. Association for Computing Machinery, New York (2004). https://doi.org/10.1145/1015330.1015408
14. ElMasry, G., Sun, D.W.: Principles of hyperspectral imaging technology. In: Hyperspectral Imaging for Food Quality Analysis and Control, pp. 3–43. Elsevier (2010)
15. Everingham, M., Eslami, S.A., Van Gool, L., Williams, C.K., Winn, J., Zisserman, A.: The pascal visual object classes challenge: a retrospective. Int. J. Comput. Vis. **111**(1), 98–136 (2015)
16. Farid, H.: Image forgery detection. IEEE Signal Process. Mag. **26**(2), 16–25 (2009)

17. Fodor, I.K.: A survey of dimension reduction techniques. Technical report, Citeseer (2002)
18. Gowen, A.A., O'Donnell, C.P., Cullen, P.J., Downey, G., Frias, J.M.: Hyperspectral imaging-an emerging process analytical tool for food quality and safety control. Trends Food Sci. Technol. **18**(12), 590–598 (2007)
19. Hyvärinen, A.: Survey on independent component analysis (1999)
20. Islam, A.U., Khan, M.J., Khurshid, K., Shafait, F.: Hyperspectral image analysis for writer identification using deep learning. In: 2019 Digital Image Computing: Techniques and Applications (DICTA), pp. 1–7. IEEE (2019)
21. Joliffe, I.T., Morgan, B.: Principal component analysis and exploratory factor analysis. Stat. Meth. Med. Res. **1**(1), 69–95 (1992)
22. Khan, M.J., Khurshid, K., Shafait, F.: A spatio-spectral hybrid convolutional architecture for hyperspectral document authentication. In: 2019 International Conference on Document Analysis and Recognition (ICDAR), pp. 1097–1102. IEEE (2019)
23. Khan, M.J., Yousaf, A., Khurshid, K., Abbas, A., Shafait, F.: Automated forgery detection in multispectral document images using fuzzy clustering. In: 2018 13th IAPR International Workshop on Document Analysis Systems (DAS), pp. 393–398. IEEE (2018)
24. Khan, Z., Shafait, F., Mian, A.: Hyperspectral imaging for ink mismatch detection. In: 2013 12th International Conference on Document Analysis and Recognition, pp. 877–881. IEEE (2013)
25. Khan, Z., Shafait, F., Mian, A.: Automatic ink mismatch detection for forensic document analysis. Pattern Recogn. **48**(11), 3615–3626 (2015)
26. Khan, Z., Shafait, F., Mian, A.S.: Towards automated hyperspectral document image analysis. In: AFHA, pp. 41–45 (2013)
27. Kim, J.O., Ahtola, O., Spector, P.E., Mueller, C.W., et al.: Introduction to Factor Analysis: What It Is and How to Do It, no. 13. Sage (1978)
28. Koppenhaver, K.M.: Forgery. Forensic Document Examination: Principles and Practice, pp. 55–60 (2007)
29. Lu, G., Fei, B.: Medical hyperspectral imaging: a review. J. Biomed. Opt. **19**(1), 010901 (2014)
30. Luo, Z., Shafait, F., Mian, A.: Localized forgery detection in hyperspectral document images. In: 2015 13th International Conference on Document Analysis and Recognition (ICDAR), pp. 496–500. IEEE (2015)
31. Morales, A., Ferrer, M.A., Diaz-Cabrera, M., Carmona, C., Thomas, G.L.: The use of hyperspectral analysis for ink identification in handwritten documents. In: 2014 International Carnahan Conference on Security Technology (ICCST), pp. 1–5. IEEE (2014)
32. Polikreti, K.: Detection of ancient marble forgery: techniques and limitations. Archaeometry **49**(4), 603–619 (2007)
33. Rahiche, A., Cheriet, M.: Forgery detection in hyperspectral document images using graph orthogonal nonnegative matrix factorization. In: Proceedings of the IEEE/CVF Conference on Computer Vision and Pattern Recognition Workshops, pp. 662–663 (2020)
34. Schultz, R.A., Nielsen, T., Zavaleta, J.R., Ruch, R., Wyatt, R., Garner, H.R.: Hyperspectral imaging: a novel approach for microscopic analysis. Cytometry **43**(4), 239–247 (2001)
35. Springer, E., Bergman, P.: Applications of non-destructive testing (NDT) in vehicle forgery examinations. J. Forensic Sci. **39**(3), 751–757 (1994)

36. Wang, W., Zhang, L., Wei, D., Zhao, Y., Wang, J.: The principle and application of hyperspectral imaging technology in detection of handwriting. In: 2017 9th International Conference on Advanced Infocomm Technology (ICAIT), pp. 345–349. IEEE (2017)
37. Wells, M.R., Vaidya, U.: RNA transcription in axotomized dorsal root ganglion neurons. Mol. Brain Res. **27**(1), 163–166 (1994)

An Hour-Glass CNN for Language Identification of Indic Texts in Digital Images

Neelotpal Chakraborty[1](✉), Ayatullah Faruk Mollah[2], Subhadip Basu[1], and Ram Sarkar[1]

[1] Department of Computer Science and Engineering, Jadavpur University, Kolkata 700032, India
chakraborty.neelotpal@gmail.com

[2] Department of Computer Science and Engineering, Aliah University, Kolkata 700160, India

Abstract. Understanding multi-lingual texts in any digital image calls for identifying the corresponding languages of the localized texts. India houses a multilingual ambience which necessitates the pursuit of an efficient model that is robust against various complexities and successfully identifies the language of Indic texts. This paper presents a deep learning based Convolutional Neural Network (CNN) model having an hour-glass like structure, for classifying texts in popular Indic languages like Bangla, English and Hindi. A new dataset, called Indic Texts in Digital Images (ITDI), is also presented which is a collection of text images, both scene and born-digital, written in Bangla, English and Hindi. The performance of the hour-glass CNN is evaluated upon standard Indic dataset like AUTNT giving an accuracy of 90.93% which is higher than most state-of-the-art models. The proposed model is also used to benchmark the performance on ITDI dataset with a reasonable accuracy of 85.18%. Sample instances of the proposed ITDI dataset can be found at: https://github.com/NCJUCSE/ITDI

Keywords: Multi-lingual · Indic text · Natural scene image (NSI) · Born-digital image (BDI) · Language identification · Hour-glass CNN (HgCNN)

1 Introduction

India being a country with a multi-cultural ambience, is home to people communicating in more than two languages, unlike other nations, where communication is put forth using an officially accepted language, at most two languages. Thus, Indic texts appearing in all sorts of digital images [1], be it Natural Scene Images (NSI) [1] or Born-digital Images (BDI) [1], may be expected to carry a multi-lingual [1] aspect. NSI denote images captured in the wild which can be considered as digital images since nowadays, the camera-based devices are digital in nature. BDI denote images that are synthetically produced by some computer systems or digital devices. Both NSI and BDI may have multi-lingual texts in the Indian scenario as shown in Fig. 1.

B. Raman et al. (Eds.): CVIP 2021, CCIS 1568, pp. 26–35, 2022.
https://doi.org/10.1007/978-3-031-11349-9_3

Fig. 1. Multi-lingual Indic texts in NSI and BDI. In all the cases, texts are written in Bangla, English and Hindi.

The ultimate goal of text understanding in digital images is to perform Optical Character Recognition (OCR) [2] of the text retrieved from an image and represent it from image to textual form. Most standard OCR engines are dedicated to recognizing characters of a specific language only, thereby necessitating the development of a Language Identification Module (LIM) [3] which works as an intermediate sub-system in any multi-lingual text retrieval cum recognition system.

Language identification [4] of texts in an image is essentially a classification task where the text image is assigned a class label denoted by the corresponding language. Recent researches in this domain have contributed a certain number of LIMs which are based on various machine learning [4] and deep learning [5] approaches. These paradigms of approaches rely on feature extraction from a set of images belonging to different classes and "learning" from these feature maps to identify their relevant classes. Although several machine learning models have been employed in standard LIMs, mostly do not achieve much language identification accuracy. With the advent of deep learning models, this accuracy of the state-of-the-art has considerably improved, thereby making them preferable among the research community. Deep learning models like Convolutional Neural Network (CNN) [3] usually comprise of multiple sequential layers of convolutional feature generators where the final layer produces a feature map for a neural network classifier to assign a class label to the image instance. A number of CNN based LIMs have been contributed to the domain which have significantly elevated the state-of-the-art performance. However, it is observed that very few works have been done in the quest for an efficient LIM for language identification of Indic texts in digital images.

Indic texts offer more challenging aspects to put the robustness of any standard LIMs to test. As depicted in Fig. 2, for a given language, Indic text components illustrate high variance in font style, size, stroke pattern, color usage and foreground (text)-background heterogeneity [2]. A tendency is observed among Indian communities who in general, prefer to enhance the texture vibrance in the above properties to gain more attention of the viewers. Thus, most of the current standard methods owing to being trained on non-Indic texts, occasionally tend to falter in robustness against the complexities associated with Indic texts.

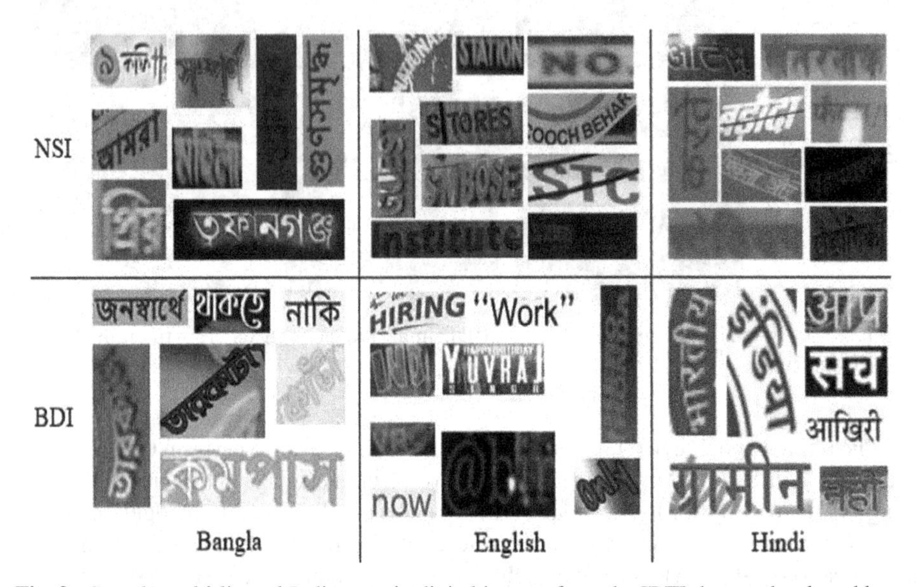

Fig. 2. Sample multi-lingual Indic texts in digital images from the IDTI dataset developed here.

To address the above challenges put forth by multi-lingual Indic texts in identifying their corresponding languages, in this paper, an Hour-glass CNN (HgCNN) model is implemented. This model is designed in resemblance to the shape of an hour-glass [6] using a lightweight VGG model [7] having two parts: encoder and decoder [3], each comprising of a set of convolutional layers with decreasing and increasing size respectively. Furthermore, to address the issue of Indic dataset scarcity, a new dataset, called Indic Texts in Digital Images (ITDI), is introduced which comprises of digital images of Bangla, English and Hindi words. The proposed HgCNN model is evaluated on standard Indic dataset like AUTNT [8] and is also used to benchmark the language identification performance on the proposed ITDI dataset.

2 Related Works

In this research domain, many LIMs have been contributed to ameliorate the challenges involved in text analytics for NSI images. However, the problems associated with BDI has been inadequately addressed. Moreover, LIM development for multi-lingual texts in BDIs in still in a nascent stage and hardly any datasets are found in the literature. Some of the recent developments in this domain are discussed below.

In the work [9], a typical CNN model is employed to perform language identification of scene texts at component level. A patch-based framework is implemented in

[10] to classify scene texts into their respective language classes. A CNN ensemble is constructed in [11] where multiple color channels serve as individual inputs to each base models and their confidence scores are combined to determine the final language class of the scene text. Some popular machine learning models are experimented with in [12] for language identification of Indic texts where Multi-layer Perceptron (MLP) model performed better than other models. In the work [13], another ensemble is employed where a global CNN classifies among the text proposals along with a local CNN that classifies among patches of those text proposal to determine the final language class. An arrangement of CNN and Recurrent Neural Network (RNN) is designed in [14] where the RNN module receives feature maps of random length from the CNN module to predict the language of a scene text. Language identification is performed in [15] using the Inception model by encoding an image of a text line into a feature sequence. The sequences generated are aggregated for language classification. Similarly, other latest deep learning models like Xception [16] are seen to be used for predicting the corresponding language of any input text image.

Most of the standard LIMs discussed above have elevated the language identification performance in the recent years. However, relatively a few models are dedicated to addressing multi-lingual Indic texts. Additionally, these approaches mostly explored multi-lingual texts in NSI and relatively a few LIMs have been developed for multi-lingual texts in BDI owing to the lack of dedicated datasets. Hence, this paper contributes an HgCNN model for effective language identification of Indic texts in both NSI and BDI. Also, the first version of a new dataset called ITDI is introduced where images of cropped words in Bangla, English and Hindi languages from both NSI and BDI are present.

3 Proposed Model and Dataset

In this paper, HgCNN model is deployed to predict the language class of any Indic text image as either Bangla, English or Hindi, all of which are popular Indic languages. In contrast to the hour-glass model in [6] that uses a ResNet model as the base model, the architecture of the proposed work resembles the shape of a typical hour-glass as shown in Fig. 3, with two lightweight VGG bases for each encoder and decoder.

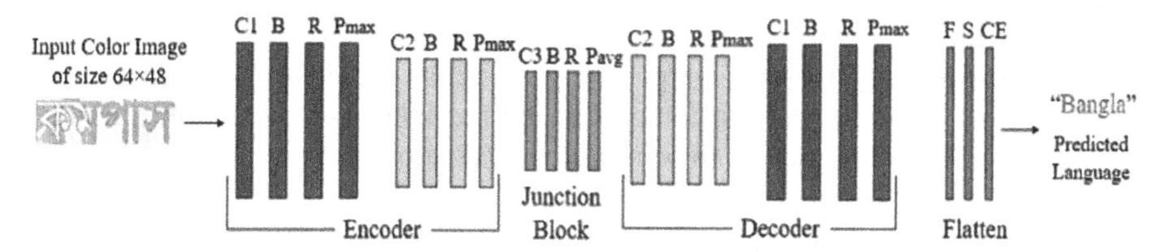

Fig. 3. Architecture of the proposed HgCNN model for language identification of Indic texts in digital images. Here, C1, C2 and C3 respectively denote 128, 64 and 32 3 × 3 convolutions with stride [1, 1] and zero padding, B for batch normalization layer, R for ReLU, P_{max} and P_{min} denote 1 × 1 max and average pooling layers respectively with stride [2, 2] and zero padding, F denotes 3 fully convolutional layers followed by a Softmax layer S and cross entropy denoted as CE.

The encoder accepts an input word image of size 64 × 48 from which a feature map is generated by 128 convolutional layers of size 3 × 3, a batch normalization layer, a Rectified Linear Unit (ReLU) and a max-pooling layer of size 1 × 1 with stride 2 [3]. This is followed by a converging sub-block of 64 convolutional layers of size 3 × 3 and other successive layers of the same dimensions. The decoder block on the other hand diverges from 64 to 128 convolutional layers of size 3 × 3 each followed by a batch normalization layer, a ReLU and a max-pooling layer of size 1 × 1 with stride 2. A junction block joins both the encoder as well as the decoder blocks using a 32 × 32 convolutional layer succeeded by a batch normalization layer, a ReLU layer and a layer of 1 × 1 average pooling with stride as 2. This layer operates as a part of both the encoder and decoder by generating a normalized feature map form the output of the encoder. Finally, for language class prediction, the feature map generated is flattened a classification block successively comprising of 3 fully connected layers, a softmax layer and cross entropy loss function [3] is appended to the decoder.

Apart from the proposed HgCNN module, this paper introduces a new dataset comprising of Bangla, English and Hindi words cropped from digital images of both NSI and BDI types. Sample images of this dataset are illustrated in Fig. 2. The categorical composition of the proposed ITDI dataset is given in Table 1.

Table 1. Summary of the ITDI dataset.

Image type	Bangla	English	Hindi
NSI	1595	2571	972
BDI	3172	2762	1758
NSI + BDI	4767	5333	2730

As mentioned earlier, Indic texts in most cases are subject to various complexities like stroke pattern, font style, size and color. Moreover, there is a stark difference between the textures of any NSI as well as BDI. NSI usually depicts texts written in the real world with vibrant color usage and often any kind of wall or hoarding damage along with the surrounding environment collaterally contributes to the complexity of a scene text. Contrarily, BDI is a digital device generated image where the text is typed on a digitally generated background. Unlike NSI, although the text components are well-defined and void of any possible occlusions or warping, BDI in various social media as well as other random internet sources may have complex background depicted by multiple arbitrary colors. Thus, ITDI dataset serves to be useful in evaluating the robustness of any LIM to efficiently determine the language of any Indic text.

4 Experiments and Results

The proposed LIM framework based on HgCNN architecture is trained and evaluated on Indic texts in digital images. Also, this system is utilized for benchmarking the language identification performance on the newly developed ITDI dataset. To highlight its lightweight nature, the proposed HgCNN model is implemented using MATLAB (R2018a) upon a hardware having Intel i5-8400 CPU @ 2.80 GHz, 2808 MHz, 6 Core(s), 6 Logical Processor(s) and 8 GB RAM.

4.1 Training and Testing

The training of HgCNN is performed for 300 epochs, with an initial learning rate of 0.001 and Adam optimization [17]. From each language class of the training set, 10% of it is reserved as for validation. The training cum validation progress for each dataset is graphically illustrated in Fig. 4.

The data pool called "Scene" of a standard recent Indic dataset AUTNT is employed to understand the effectiveness of HgCNN in accurately predicting the language class of the text image. The HgCNN model is trained on the "Train" set comprising of Bangla scene texts, 1759 English scene texts and 280 Hindi scene texts. The trained model is then tested on the "Test" set comprising of 251 Bangla scene texts, 439 English scene texts and 71 Hindi scene texts.

In case of the ITDI datasets, the HgCNN is trained and tested on each data pool "NSI", "BDI" and their combination. In each of these pools, 80% of the image instances from each of the language classes "Bangla", "English" "and Hindi" are used for training the HgCNN model. The trained model is then tested on the remaining 20% image instances of each language class.

Additionally, a combined pool of both AUTNT (scene) and ITDI is utilized to obtain a trained model that is again applied on the test set of this new data pool. It is to be noted that the test set for AUTNT is used as available along with the 20% instances in the ITDI test set.

4.2 Analysis of Results

The proposed HgCNN model is evaluated on standard Indic dataset AUTNT and is also used for benchmarking the language identification performance on the newly developed ITDI dataset. As previously discussed, the HgCNN model is trained and validated individually on AUTNT and ITDI datasets along with their combined pool. The respective trained models are used for language classification on the test sets of individual pools and the resulting confusion matrices are obtained as shown in Fig. 5. Accuracies obtained by the HgCNN model for the datasets individually and combined, are compared with some recent standard LIMs as depicted in Table 2.

As can be seen in Table 2, HgCNN model achieves the highest accuracies for AUTNT, ITDI and the combination of both datasets in comparison of the methods considered here for comparison. This itself establishes the efficiency of the proposed LIM compared to that of other standard models. A drop of $\approx$10%, 15% and 14% in the accuracy for AUTNT, ITDI and their combination, can be attributed to the stroke similarities between the Bangla and Hindi texts. Additionally, the texture complexities of both the NSI as well as BDI affect the feature learning process of HgCNN. Despite some limitations, the proposed lightweight model can be trained easily and be deployed for efficient language identification of Indic texts extracted from any digital image.

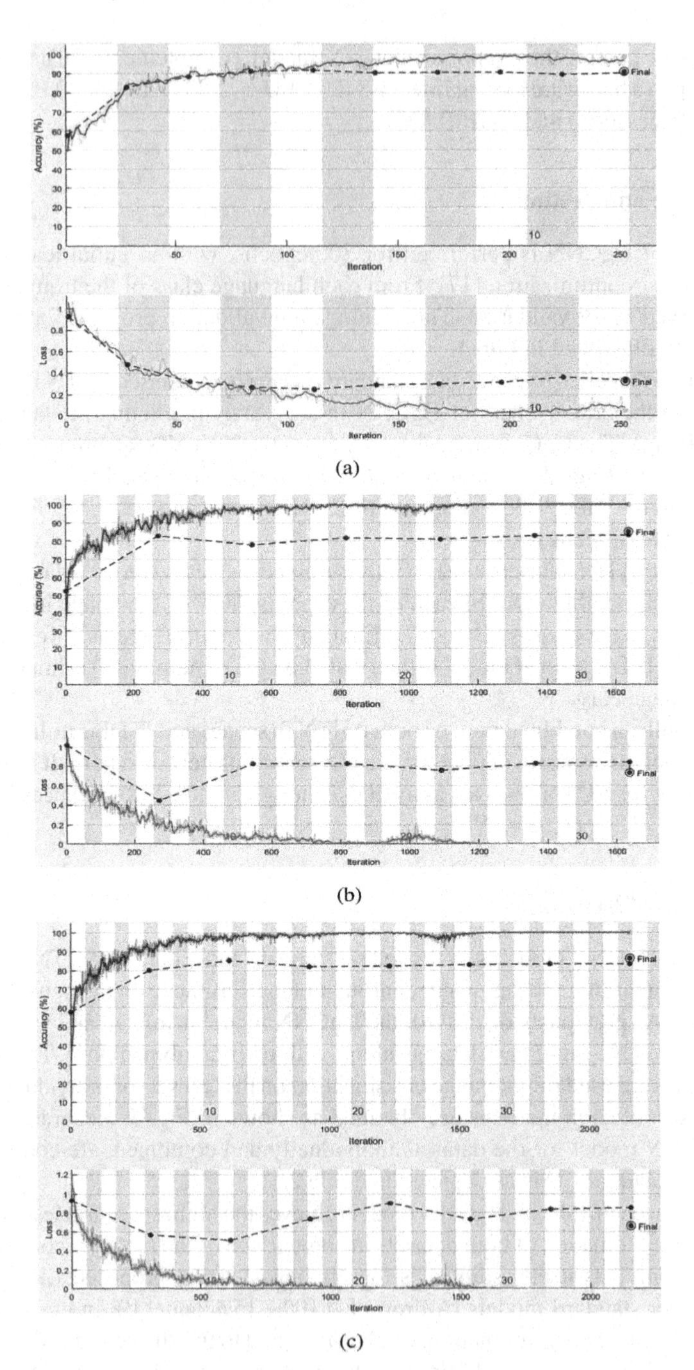

Fig. 4. Training progress of HgCNN model for (a) AUTNT (b) ITDI and (c) AUTNT and ITDI. Here, the blue and orange curves denote training convergence in terms of accuracy and loss respectively. The black curve denotes the same for validation.

	Bangla	English	Hindi
Bangla	**220**	15	16
English	19	**420**	0
Hindi	13	6	**52**

(a)

	Bangla	English	Hindi
Bangla	**2060**	191	331
English	204	**2862**	110
Hindi	76	22	**448**

(b)

	Bangla	English	Hindi
Bangla	**2852**	248	464
English	271	**4607**	188
Hindi	71	28	**517**

(c)

Fig. 5. Confusion matrices obtained for the test sets of (a) AUTNT (b) ITDI and (c) AUTNT and ITDI.

Table 2. Comparison of accuracies of some standard LIMs with that of the proposed HgCNN model for AUTNT and ITDI datasets.

Dataset	Model	Accuracy (%)
AUTNT	Jajoo et al., 2019 [12]	82.63
AUTNT	Saha et al., 2020 [3]	90.56
AUTNT	Khan & Mollah, 2021 [9]	89.49
AUTNT	**HgCNN (Proposed)**	**90.93**
ITDI	Xception [16]	83.18
ITDI	Jajoo et al., 2019 [12]	80.86
ITDI	Saha et al., 2020 [3]	82.61
ITDI	Chakraborty et al., 2021 [4]	84.06
ITDI	**HgCNN (Proposed)**	**85.18**
AUTNT + ITDI	Fujii et al., 2017 [15]	86.00
AUTNT + ITDI	Xception [16]	80.38
AUTNT + ITDI	Jajoo et al., 2019 [12]	81.68
AUTNT + ITDI	Saha et al., 2020 [3]	86.11
AUTNT + ITDI	**HgCNN (Proposed)**	**86.26**

5 Conclusion

This work presents an LIM based on a lightweight hour-glass shaped HgCNN model for effectively identifying the corresponding languages of Indic texts cropped from various digital images of NSI and BDI types. Also, a new dataset called ITDI is introduced which comprises Indic texts written in popular Indic languages namely, Bangla, English and Hindi, cropped from various NSI and BDI. The proposed model is evaluated on standard Indic dataset AUTNT and achieves an impressive accuracy of 90.93% which is higher than many recently proposed models. The HgCNN model is also used to benchmark the performance on ITDI dataset with an impressive accuracy of 85.18%. Additionally, an accuracy of 86.26% is achieved by the HgCNN model for the dataset combining AUTNT and ITDI, which is higher than some recent models. The deficiency in the accuracy may be attributed to the similarity that exists between Bangla and Hindi texts. Also, certain image instances have text foreground that becomes hard to distinguish, thereby resulting in erroneous classification. Overall, the HgCNN can be viewed as an effective model for implementing an LIM for Indic texts. In future, attempts will be made to achieve better classification accuracy and the proposed ITDI dataset will be updated with digital images of texts in other Indic languages.

Acknowledgements. This work is partially supported by the CMATER research laboratory of the Computer Science and Engineering Department, Jadavpur University, India.

References

1. Joan, S.F., Valli, S.: A survey on text information extraction from born-digital and scene text images. Proc. Nat. Acad. Sci. India Sec. A Phys. Sci. **89**(1), 77–101 (2019)
2. Kanagarathinam, K., Sekar, K.: Text detection and recognition in raw image dataset of seven segment digital energy meter display. Energy Rep. **5**, 842–852 (2019)
3. Saha, S., Chakraborty, N., Kundu, S., Paul, S., Mollah, A.F., Basu, S., Sarkar, R.: Multi-lingual scene text detection and language identification. Pattern Recogn. Lett. **138**, 16–22 (2020)
4. Chakraborty, N., Chatterjee, A., Singh, P.K., Mollah, A.F., Sarkar, R.: Application of daisy descriptor for language identification in the wild. Multimedia Tools Appl. **80**(1), 323–344 (2021)
5. Long, S., He, X., Yao, C.: Scene text detection and recognition: the deep learning era. Int. J. Comput. Vis. **129**(1), 161–184 (2021)
6. Melekhov, I., Ylioinas, J., Kannala, J., Rahtu, E.: Image-based localization using hourglass networks. In: Proceedings of the IEEE International Conference on Computer Vision Workshops, pp. 879–886 (2017)
7. Liu, S., Shang, Y., Han, J., Wang, X., Gao, H., Liu, D.: Multi-lingual scene text detection based on fully convolutional networks. In: Pacific Rim Conference on Multimedia, pp. 423–432. Springer, Cham (2017). https://doi.org/10.1007/978-3-319-77380-3_40
8. Khan, T., Mollah, A.F.: AUTNT-A component level dataset for text non-text classification and benchmarking with novel script invariant feature descriptors and D-CNN. Multimedia Tools Appl. **78**(22), 32159–32186 (2019)
9. Khan, T., Mollah, A.F.: Component-level script classification benchmark with CNN on AUTNT Dataset. In: Bhattacharjee, D., Kole, D.K., Dey, N., Basu, S., Plewczynski, D. (eds.) Proceedings of International Conference on Frontiers in Computing and Systems and Computing, vol. 1255, pp. 225–234. Springer, Singapore (2021). https://doi.org/10.1007/978-981-15-7834-2_21
10. Cheng, C., Huang, Q., Bai, X., Feng, B., Liu, W.: Patch aggregator for scene text script identification. In: 2019 International Conference on Document Analysis and Recognition (ICDAR), pp. 1077–1083. IEEE (2019)
11. Chakraborty, N., Kundu, S., Paul, S., Mollah, A.F., Basu, S., Sarkar, R.: Language identification from multi-lingual scene text images: a CNN based classifier ensemble approach. J. Ambient Intell. Hum. Comput. **12**, 7997–8008 (2020)
12. Jajoo, M., Chakraborty, N., Mollah, A.F., Basu, S., Sarkar, R.: Script identification from camera-captured multi-script scene text components. In: Kalita, J., Balas, V., Borah, S., Pradhan, R. (eds.) Recent Developments in Machine Learning and Data Analytics. Advances in Intelligent Systems and Computing, vol. 740, pp. 159–166. Springer, Singapore (2019). https://doi.org/10.1007/978-981-13-1280-9_16

13. Lu, L., Yi, Y., Huang, F., Wang, K., Wang, Q.: Integrating local CNN and global CNN for script identification in natural scene images. IEEE Access **7**, 52669–52679 (2019)
14. Mei, J., Dai, L., Shi, B., Bai, X.: Scene text script identification with convolutional recurrent neural networks. In: 2016 23rd International Conference on Pattern Recognition (ICPR), pp. 4053–4058. IEEE (2016)
15. Fujii, Y., Driesen, K., Baccash, J., Hurst, A., Popat, A. C.: Sequence-to-label script identification for multilingual ocr. In: 2017 14th IAPR International Conference on Document Analysis and Recognition (ICDAR), vol. 1, pp. 161–168. IEEE (2017)
16. Chollet, F.: Xception: Deep learning with depthwise separable convolutions. In: Proceedings of the IEEE conference on computer vision and pattern recognition, pp. 1251–1258 (2017)
17. Kingma, D.P., Ba, J.: Adam: A method for stochastic optimization. arXiv:1412.6980 (2014)

Single Frame-Based Video Dehazing with Adversarial Learning

Vijay M. Galshetwar[1(✉)], Prashant W. Patil[2], and Sachin Chaudhary[1]

[1] Punjab Engineering College, Chandigarh, India
{vijaymadhavraogalshetwar.phdcse20,sachin.chaudhary}@pec.edu.in
[2] Applied Artificial Intelligence Institute, Deakin University, Geelong, Australia
prashant.patil@deakin.edu.in

Abstract. Hazy environmental conditions degrade the quality of captured videos which leads to poor visibility and color distortion in videos. Such deterioration of captured video quality is mainly because of the attenuation caused by the scattering of light due to the haze particles present in the environment. In this paper, we propose an adversarial learning based single frame video dehazing encoder-decoder network. The proposed method comprises of Dilated Residual Block (DRB) used as encoder and Skip connection. DRB module is used to gain more contextual information by achieving large receptive fields. Skip connections are established between each encoder and decoder, which helps to detect and give more attention to haze specific features by adjusting the weights of learned feature maps automatically. This helps to extract the haze-relevant feature maps and recovers the haze-free video frame by using Channel Attention Block (CAB) and Residual Block (ResB) respectively. An extensive quantitative and qualitative analysis of the proposed method is done on benchmark synthetic hazy video database namely DAVIS-16 and NYU depth. Experimental result shows that the proposed method outperforms the other existing state-of-the-art (SOTA) approaches for video dehazing.

Keywords: Adversarial learning · Dilated residual block · Video enhancement

1 Introduction

Haze is an usual atmospheric phenomenon caused due to the presence of very small particles in the surroundings such as smoke, dust, fog, humid (water droplets), and sand particulates. Video captured by camera in such surroundings have poor quality in terms of low contrast, poor visibility, dull colors, and also color imbalance. Scattering effects are responsible for the deterioration of video when light intensity reflected by scene object falls on hazy particulates which results in scattering and leads to degradation of its original view quality.

Computer vision applications like object detection, autonomous vehicle tracking, depth estimation [12,16], human action recognition [3,10,22,25], moving

B. Raman et al. (Eds.): CVIP 2021, CCIS 1568, pp. 36–47, 2022.
https://doi.org/10.1007/978-3-031-11349-9_4

object segmentation [14,15], Image classification, and marine surveillance [33] require good quality videos to achieve good performance. The presence of haze in the video degrades the performance capability of the computer vision algorithms [3] and hence the output result. The requirement of haze removal in such hazy videos is essential. The main challenge is to remove haze from the hazy video frames and enhance the video quality. Many computer vision-based dehazing algorithms are proposed to solve the problem of video dehazing. Researchers [4,13,20,23,31] used optical model for the hazy-free scene reconstruction. Because the optical model has the physical properties of transmission of light through air medium and is commonly used to describe video frame formation. Equation 1 shows the optical model for hazy video frame formation,

$$I(x) = J(x)t(x) + A(1 - t(x)) \tag{1}$$

where $I(x)$ is the radiance of hazy video frame at pixel x, $J(x)$ is the radiance of clean frame at pixel x, A is the atmospheric light, and $t(x)$ is the outdoor or indoor scene transmission map. With the help of Eq. 1, it is possible to recover haze free scene, if the estimation of air light and transmission map scene is done properly. Transmission map is defined as,

$$t(x) = e^{(-\beta d(x))} \tag{2}$$

where β denotes the attenuation coefficient and $d(x)$ is the distance of the scene at pixel x. Degradation of Video frame for every pixel is different and depends on the distance between the scene point from the camera. Transmission coefficients control the scene attenuation and amount of haze in every pixel. The major contributions of this proposed architecture are summarized as follows:

1. We develop an adversarial learning based encoder-decoder network for single frame video dehazing.
2. We designed a Dilated Residual Block (DRB) module as encoder, which learns more contextual feature information by increasing the receptive field from single frame.
3. Further, channel attention block with residual learning is implemented in the decoder while reconstructing the dehazed frame.
4. The outdoor video dehazing database is generated synthetically with the help of DAVIS-2016 [35] video database. Experimental analysis is carried out on proposed synthetic video dehazing and NYU depth [36] database.

The rest of the paper is organized as follows: Literature review on video haze removal is illustrated in Sect. 2. In Sect. 3, we have discussed our proposed method for video frame haze removal. Section 4 describes quantitative and qualitative results and their comparison with the state of the art result of other methods. The paper is concluded in Sect. 5.

2 Literature Review

The dehazing techniques are essentially focused on the restoration of the haze-free images or videos from the hazy images or videos. Dehazing approaches are mainly categorized into three main groups: prior based [2], multi-image fusion based [5], and learning-based [6]. Based on better priors or assumptions noteworthy progress has taken place in the field of single image haze removal. The factors responsible for dehazing, using prior based technique [2] are outdoor scene transmission maps, haze-specific priors (e.g. thickness of haze), and surrounding light intensity. The multi-fusion-based methods are using fusion techniques to combine different domain-specific information of the image. Learning-based techniques are used by many researchers for finding estimates of outdoor scene transmission map based on the optical model and also use depth or semantic information.

2.1 Prior Based

Initially, He et al. [13] implemented dark channel prior (DChP) to acquire a haze-free scene. It consists of dark pixels whose intensity is very low as compared to other color channels of the haze-free video frame. The haze density and atmospheric light are estimated by using this approach and generates a haze-free scene. However, this approach fails when atmospheric light and the object have similar colors. Berman et al. [2] introduced the non-local image dehazing hypothesis and stated that an image can be assumed as a collection of multiple dissimilar colors. Zhu et al. [30] implemented color attenuation prior (CAP) for linear mapping of local priors, other prior based hazy models [13,31] etc. are used for hazy scene removal task. Tan et al. [23] implemented contrast improvement of the hazy scene. They made a haze-free image from the hazy image by increasing the local contrast of the hazy scene. But, this technique fails due to the presence of a discontinuity in the hazy image and is responsible to generate blocking artifacts.

2.2 Multi-image Fusion

Multi-image fusion techniques are used to retain useful features by combining multiple images in spatial domains or the domains having chromatic, luminance, and saliency features. Most of the methods of image restoration uses feature concatenation [26], element-wise summation [29] or dense connection [27] to fuse features. Existing work focuses on merging images with low quality into final result [1,17]. Ancuti et al. [1] proposed a multi-scale fusion-based contrast-enhancing and white balancing procedural strategy for the haze free scene by considering two hazy images as input. Kim et al. [17] proposed a real-time algorithm for video dehazing, which helps to recover high-quality video at the output by reducing flickering artifacts. Dudhane et al. [9] proposed RYFNet, whose first focus was to generate two transmission maps for RGB and YCbCr color spaces and then integrated these two transmission maps (i.e. RGB and YCbCr). This results in one robust transmission map to tackle the hazy-free scene task. First, haze-free videos were introduced by Zhang et al. [28], he considered frame-by-frame videos to improve the temporal coherence using optical flow and Markov Random Field.

2.3 Learning-Based

The main focus of the learning-based dehazing approach is to map hazy images directly to haze-free images based on learning models with the assistance of convolutional neural network (CNN) and Generative adversarial network (GAN) architectures. To resolve this dehazing task, many researchers developed architectures based on CNN [24] and GAN [11] architectures. MSCNN [20] and GFN [21] were implemented by Ren et al. to tackle the haze-free image task using holistic transmission map estimate by incorporating fine-scale local refinement and uses a gated mechanism to optimize multiple inputs. Cai et al. [4] implemented DehazeNet, which helps for intermediate map estimation and their relationship with hazy-image patches. He also proposed an extended version of the rectified linear unit and named it a bilateral rectified linear unit (BReLU). BReLU helps to improve convergence, and is responsible for accurate image restoration. However, Li et al. [18] introduced AOD-Net, where he focused on scattering model reformulation instead of using intermediate map calculation. It is the first model to optimize hazy images to clear images based on an end-to-end pipeline. Zhang et al. proposed DCPDN [26], which uses a dense encoder-decoder network-based edge-preserving pyramid for estimating accurate transmission maps and atmospheric light to optimize hazy-images to clear images. GAN-based end-to-end cycle-consistency approaches RI-GAN [7] and CD-Net [8] implemented by Dudhane et al. where RI-GAN incorporates novel generator and discriminator by using residual and inception modules. Whereas CD-Net follows estimation of transmission map based on unpaired image data and optical model so that haze-free scene recovered effectively.

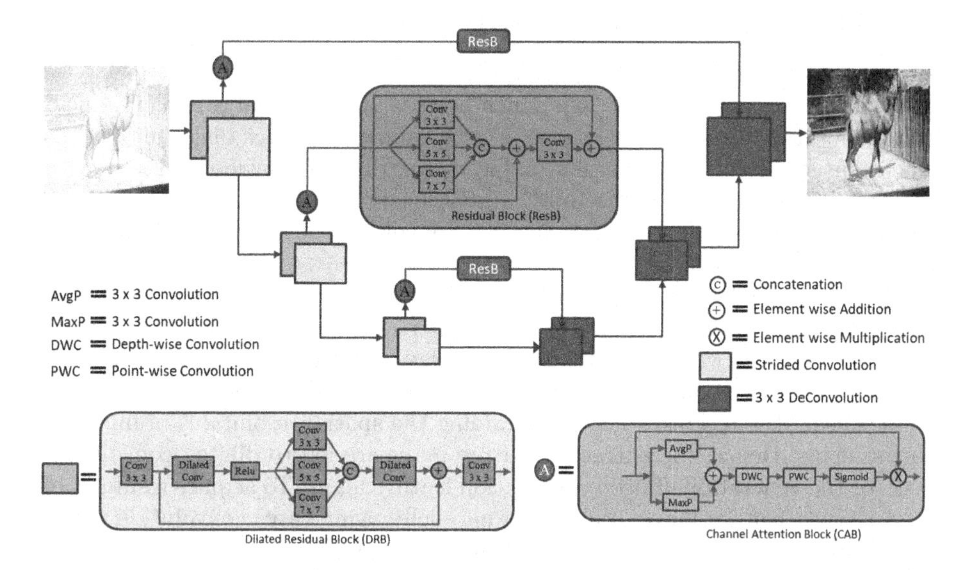

Fig. 1. Proposed framework of encoder-decoder network for single video dehazing.

3 Proposed Method

The proposed framework is a single video frame adversarial learning based encoder-decoder network architecture, where the encoder is responsible for spatial dimension reductions at every layer and increases the channels. The decoder is responsible for the increase in spatial dimensions and reduces the channels. The aim is to predict pixels in the input video frame from restored spatial dimensions. Our focus is to use this framework to predict each haze-free pixel in the hazy input video. Figure 1 shows the proposed architecture of single frame video dehazing. To build the proposed model we used pix2pix [42] as base model.

The proposed framework starts with the encoder. The encoder consists of three DRBs followed by a strided convolution. DRB exploits the hazy input video frame to produce local features. Dilated convolutions of DRB are responsible for increasing the receptive fields [19] by retaining the spatial feature resolutions. Then, application of strided convolution with a stride rate of 2 helps to compress and downsample the extracted feature maps from the hazy input video frame. The decoder consists of a $3 * 3$ convolution layer and the upsampling of the incoming learned feature map by using a $3 * 3$ deconvolution with $2 * 2$ stride size, which reduces the number of feature channels to half. Then the concatenation layer integrates the learned feature map of each encoder coming via skip connection (which comprises of CAB and ResB) and inputted feature map of last encoder. Then output of concatenation layer is provided to a $3 * 3$ convolution layer followed by a ReLU activation function of decoder block. The last convolution layer is used for mapping the channels with the required number of classes.

3.1 Dilated Residual Block (DRB)

A large receptive field helps in gaining more contextual information. Hence, we developed DRB motivated by [32], which helps to achieve a large receptive field by incorporating dilated convolutions. Figure 1 shows the block diagram of DRB, which consists of two $3 * 3$ convolution layers, two dilated convolution layers with the same dilated rate of 2, one ReLU activation function, and a residual block, which is a concatenation of three parallel convolution layers with spatial sizes $3 * 3$, $5 * 5$, and $7 * 7$ of input feature map.

The roles of DRB are to extract feature maps of the hazy video frame and obtain contextual information by widening the receptive field of incoming feature maps. DRB starts with a $3 * 3$ convolution layer which extracts features from the hazy video frame are called feature maps. As dilated convolutions are useful for increasing the receptive field by retaining the spatial resolutions of inputted feature maps. Hence, the extracted feature maps are fed to dilated convolution layer for the generation of deep convolution feature maps. To acquire residuals of these learned feature maps, scale them parallelly using three convolution layers with spatial sizes $3 * 3$, $5 * 5$, and $7 * 7$, and integrate them using concatenation operation. Generated feature maps are then applied to dilated convolution for the generation of deep convolution feature maps for acquiring more contextual information. We added these learned feature maps output with the initially extracted feature maps by a $3 * 3$ convolution layer to retain low-level information of video frame at the output and inputted to a $3 * 3$ convolution layer.

3.2 Skip Connection

A skip connection is developed by using CAB and ResB to provide low-level learned features generated at initial convolution layers of encoder to convolution layers of decoder for generating noticeable edge information.

Channel Attention Block (CAB): Figure 1 shows the flow chart of CAB [34]. The CAB is used in the skip connection to detect and give more attention to learn haze feature maps by adjusting the weights of learned feature maps automatically. The CAB starts with global average pooling to capture common features and maximal pooling for capturing distinctive features, these features are then aggregated as spatial information. Depth-wise convolution is then used to learn and acquire the weights by applying a convolution filter to each channel independently. The operation of channel attention is based on squeeze and excitation for weight re-calibration of the feature map. The squeeze operates on spatial information with the help of both global average pooling and max pooling and transforms into channel-wise information. On the other hand, excitation operates using depth-wise convolution layer, which enables channel separable mechanism. Point-wise convolution layer used to project output channels of depth-wise convolution layer onto new channel space, and then sigmoid activation.

Residual Block (ResB): Residual block helps in widening the learning capabilities of deep networks by using a parallel combination of convolution layers. ResB module increases the capability of the proposed network to extract the haze-relevant feature maps and helps to recover the haze-free frame. ResB is developed using parallel connection of three convolution layers with spatial sizes $3 * 3$, $5 * 5$, and $7 * 7$, and integrated their individual learned feature outputs using concatenation layer. Output is then added with the initial input of ResB via short skip connection. A $3 * 3$ convolution layer is used to process this output and then output is added with initial input of ResB via long skip connection.

3.3 Loss Function

The adversarial loss is used for optimization of the network parameters and then edge loss and Vgg loss are added in the overall loss function.

Adversarial Loss: Equation 3 is the equation of adversarial loss [42] used for mapping G: H to C.

$$\begin{aligned} l_{GAN}(G, D, H, C) = {} & E_{c \sim p_{data}}(c)\left[\log\left(D_C\left(C\left(c\right)\right)\right)\right] \\ & + E_{h \sim p_{data}}(h)\left[\log\left(1 - D_Y\left(G\left(h\right)\right)\right)\right] \end{aligned} \tag{3}$$

where, G is the generator and D_C is the discriminator. G is developed to convert hazy frame to hazy free-frame where it tries to generate frame G_h same as that of hazy-free frame. D_C responds generated frame G_h is real or fake by comparing it with ground truth frame.

Edge Loss: We use sobel edge detector for edge map computation. Equation 4 depicts the formula for edge loss calculation [7].

$$l_{Edge}(G) = \| E_{g(x)} - E_{y(x)} \| \tag{4}$$

where, $E_{g(x)}$ denotes edge map of generated scene and $E_{y(x)}$ denotes edge map of ground truth scene.

Vgg Loss: The perceptual loss is helpful to guide the network by acquiring information related to texture and structure, and it is calculated by considering generated and ground truth frames. Then it is passed through the pre-trained model of VGG19 [43].

$$l_p = \sum_{i=1}^{X}\sum_{j=1}^{Y}\sum_{k=1}^{Z} \frac{1}{XYZ} \| \Theta_l(G(h))_{i,j,k} - \Theta_l(C(c))_{i,j,k} \| \tag{5}$$

where, C_c is ground truth frame, G_h is the generated frame by proposed generator network. X, Y, Z are the feature map dimensions.

For training of the network, following is the overall loss equation

$$l_{Total} = \lambda_{Edge} l_{Edge} + \lambda_p l_p + \lambda_{GAN} l_{GAN} \tag{6}$$

where, λ_{Edge}, λ_p and λ_{GAN} are the assigned loss function weights for l_{Edge}, l_P and l_{GAN} respectively.

4 Experimental Results

The training is done on synthetic and Real-world hazy video frames and haze-free video frames. The outdoor video dehazing database is generated synthetically using Eq. 1 with the help DAVIS 2016 [35] video database. Experimental analysis is carried out on proposed synthetic video dehazing and NYU depth [36] database. The video haze removal model is trained on NVIDIA DGX station having the configuration NVIDIA Tesla V100 4 × 32 GB GPU.

This section elaborates experimental analysis of the proposed method based on qualitative and quantitative results. We carried out the evaluation of the proposed method by comparing them with the existing state-of-the-art methods. Performance is evaluated based on structural similarity index (SSIM), peak signal to noise ratio (PSNR) parameters. We divide the comparison of experimental results of the proposed method with the existing state-of-the-art methods into two parts (1) Quantitative Analysis (2) Qualitative Analysis.

4.1 Quantitative Analysis

We adopted quantitative analysis by using two existing benchmarks video datasets: DAVIS-2016 [35] and NYU depth [36]. The DAVIS-2016 dataset consists of 50 videos which are divided into 30 videos (2079 frames) for training and 20 videos (1376 frames) for testing. The NYU-Depth dataset consists of 45 videos which are divided into 25 videos (28,222 frames) for training and 20 videos (7528 frames) for testing.

The SOTA methods used for comparative evaluation with proposed method are TCN [37], FFANet [38], MSBDN [39], GCANet [40], RRO [41], CANCB [44] and FME [45]. We trained and tested the proposed model separately on the DAVIS 2016 [35] and NYU-Depth [36] datasets. We tested all the trained SOTA models using DAVIS 2016 and NYU depth. The results obtained from testing of the proposed method have compared with the testing results of SOTA methods (Tables 1 and 2).

Table 1. Quantitative analysis of proposed method and SOTA for single video frame haze removal on DAVIS-16 [35] database

Method	TCN [37]	FFANet [38]	MSBDN [39]	GCANet [40]	RRO [41]	FMENet [45]	Proposed
PSNR	16.608	12.19	14.64	17.66	15.09	16.16	21.463
SSIM	0.619	0.6501	0.7319	0.7185	0.7604	0.8297	0.8034

Table 2. Quantitative analysis of proposed method and SOTA for single video frame haze removal on NYU depth [36] database

Method	TCN [37]	FMENet [45]	GCANet [40]	RRO [41]	CANCB [44]	Proposed
PSNR	18.837	19.81	22.55	19.47	20.87	23.58
SSIM	0.6142	0.8429	0.9013	0.8422	0.8903	0.9083

4.2 Qualitative Analysis

Figures 2 and 3 shows qualitative analysis of proposed method against SOTA methods using DAVIS 2016 [35] and NYU depth [36] video datasets respectively. By observing these analysis we can state that the proposed method effectively reduces the effect of haze in the single hazy video frame.

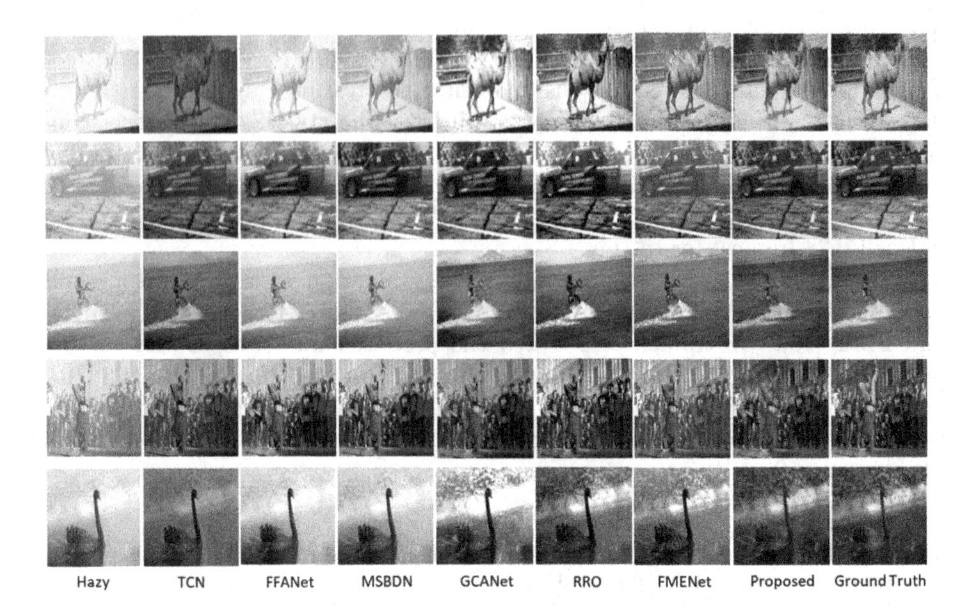

Fig. 2. Analysis of proposed method and SOTA for video dehazing on DAVIS-16 [35] Database. (TCN [37], FFANet [38], MSBDN [39], GCANet [40], RRO [41], and FME [45]

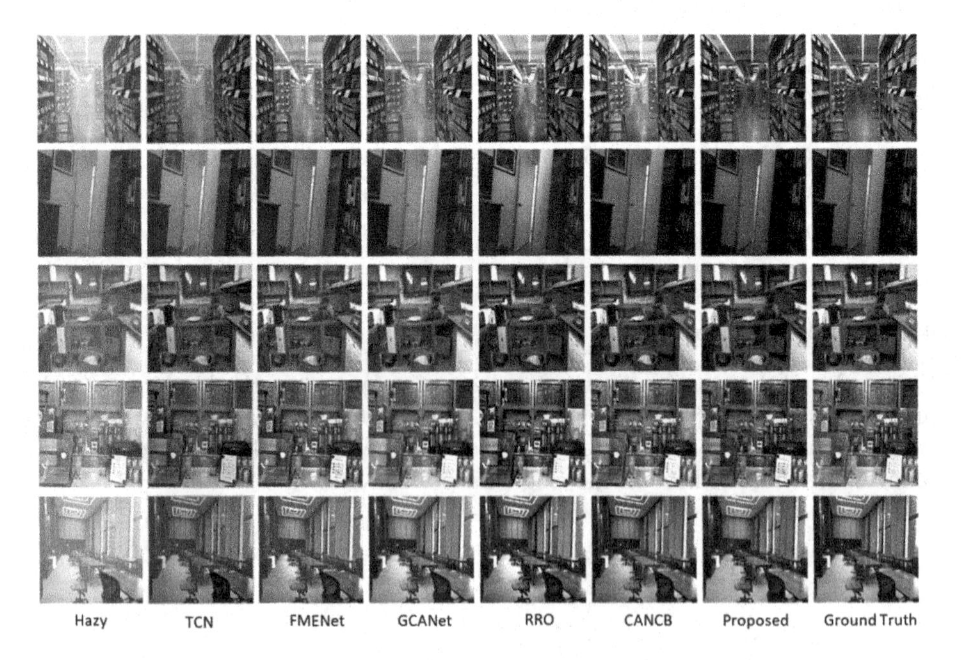

Fig. 3. Analysis of proposed method and SOTA for video dehazing on NYU Depth [36] Database. (TCN [37], FME [45], GCANet [40], RRO [41], and CANCB [44])

5 Conclusion

The proposed adversarial learning based encoder-decoder framework is an end-to-end learning approach for single frame video dehazing. The DRB is used as an encoder to extract feature maps of the hazy video frame and obtain contextual information by widening the receptive field of incoming feature maps. A skip connection is developed by using CAB and ResB to carry information from encoder to decoder. The skip connections were used to provide low-level learned features generated at initial convolution layers of encoder to convolution and deconvolution layers of decoder, which helps to generate the noticeable edge information. Performance evaluation of the proposed method is done on two benchmark datasets: DAVIS 2016 and NYU Depth. The qualitative analysis of the proposed method is analyzed and compared with the state-of-the-art video dehazing methods. Experimentally we analyzed that the results of the proposed method out performs other dehazing methods.

References

1. Ancuti, C.O., Ancuti, C.: Single image dehazing by multi-scale fusion. IEEE Trans. Image Process. **22**(8), 3271–3282 (2013)
2. Berman, D., treibitz, T., Avidan, S.: Non-local image dehazing. In: Proceedings of the IEEE Conference on Computer Vision and Pattern Recognition (CVPR), June 2016
3. Chaudhary, S., Murala, S.: TSNet: deep network for human action recognition in hazy videos. In: 2018 IEEE International Conference on Systems, Man, and Cybernetics (SMC), pp. 3981–3986 (2018). https://doi.org/10.1109/SMC.2018.00675
4. Cai, B., Xu, X., Jia, K., Qing, C., Tao, D.: DehazeNet: an end-to-end system for single image haze removal. IEEE Trans. Image Process. **25**(11), 5187–5198 (2016)
5. Choi, L.K., You, J., Bovik, A.C.: Referenceless prediction of perceptual fog density and perceptual image defogging. IEEE Trans. Image Process. **24**(11), 3888–3901 (2015)
6. Zhang, J., et al.: Hierarchical density-aware dehazing network. IEEE Trans. Cybern., 1–13 (2021). https://doi.org/10.1109/TCYB.2021.3070310
7. Dudhane, A., Aulakh, H.S., Murala, S.: RI-GAN: an end-to-end network for single image haze removal. In: 2019 IEEE/CVF Conference on Computer Vision and Pattern Recognition Workshops (CVPRW), pp. 2014–2023 (2019)
8. Dudhane, A., Murala, S.: CDNet: single image dehazing using unpaired adversarial training, pp. 1147–1155 (2019)
9. Dudhane, A., Murala, S.: RYF-Net: deep fusion network for single image haze removal. IEEE Trans. Image Process. **29**, 628–640 (2020)
10. Chaudhary, S., Murala, S.: Deep network for human action recognition using Weber motion. Neurocomputing **367**, 207–216 (2019)
11. Engin, D., Genc, A., Ekenel, H.K.: Cycle Dehaze: enhanced cycleGAN for single image dehazing. In: Proceedings of the IEEE Conference on Computer Vision and Pattern Recognition (CVPR) Workshops, June 2018

12. Hambarde, P., Dudhane, A., Patil, P.W., Murala, S., Dhall, A.: Depth estimation from single image and semantic prior. In: 2020 IEEE International Conference on Image Processing (ICIP), pp. 1441–1445. IEEE (2020)
13. He, K., Sun, J., Tang, X.: Single image haze removal using dark channel prior. IEEE Trans. Pattern Anal. Mach. Intell. **33**(12), 2341–2353 (2011)
14. Patil, P.W., Dudhane, A., Kulkarni, A., Murala, S., Gonde, A.B., Gupta, S.: An unified recurrent video object segmentation framework for various surveillance environments. IEEE Trans. Image Process. **30**, 7889–7902 (2021)
15. Patil, P.W., Biradar, K.M., Dudhane, A., Murala, S.: An end-to-end edge aggregation network for moving object segmentation. In: Proceedings of the IEEE/CVF Conference on Computer Vision and Pattern Recognition, pp. 8149–8158 (2020)
16. Chaudhary, S., Murala, S.: Depth-based end-to-end deep network for human action recognition. IET Comput. Vis. **13**(1), 15–22 (2019)
17. Kim, J.H., Jang, W.D., Park, Y., HahkLee, D., Sim, J.Y., Kim, C.S.: Temporally x real-time video dehazing. In: 2012 19th IEEE International Conference on Image Processing, pp. 969–972. IEEE (2012)
18. Li, B., Peng, X., Wang, Z., Xu, J., Feng, D.: AOD-Net: all-in-one dehazing network. In: Proceedings of the IEEE International Conference on Computer Vision (ICCV), October 2017
19. Phutke, S.S., Murala, S.: Diverse receptive field based adversarial concurrent encoder network for image inpainting. IEEE Signal Process. Lett. **28**, 1873–1877 (2021)
20. Ren, W., Liu, S., Zhang, H., Pan, J., Cao, X., Yang, M.-H.: Single image dehazing via multi-scale convolutional neural networks. In: Leibe, B., Matas, J., Sebe, N., Welling, M. (eds.) ECCV 2016. LNCS, vol. 9906, pp. 154–169. Springer, Cham (2016). https://doi.org/10.1007/978-3-319-46475-6_10
21. Ren, W., et al.: Gated fusion network for single image dehazing. In: Proceedings of the IEEE Conference on Computer Vision and Pattern Recognition, pp. 3253–3261 (2018)
22. Chaudhary, S., Dudhane, A., Patil, P., Murala, S.: Pose guided dynamic image network for human action recognition in person centric videos. In: 2019 16th IEEE International Conference on Advanced Video and Signal Based Surveillance (AVSS), pp. 1–8 (2019). https://doi.org/10.1109/AVSS.2019.8909835
23. Tan, R.: Visibility in bad weather from a single image (2008)
24. Yang, D., Sun, J.: Proximal DehazeNet: a prior learning-based deep network for single image dehazing. In: Proceedings of the European Conference on Computer Vision (ECCV), September 2018
25. Chaudhary, S.: Deep learning approaches to tackle the challenges of human action recognition in videos. Dissertation (2019)
26. Zhang, H., Patel, V.M.: Densely connected pyramid dehazing network. In: Proceedings of the IEEE Conference on Computer Vision and Pattern Recognition (CVPR), June 2018
27. Zhang, H., Sindagi, V., Patel, V.M.: Multi-scale single image dehazing using perceptual pyramid deep network. In: 2018 IEEE/CVF Conference on Computer Vision and Pattern Recognition Workshops (CVPRW), pp. 1015–101509 (2018)
28. Zhang, J., Li, L., Zhang, Y., Yang, G., Cao, X., Sun, J.: Video dehazing with spatial and temporal coherence. Vis. Comput. **27**(6), 749–757 (2011). https://doi.org/10.1007/s00371-011-0569-8
29. Zhang, X., Dong, H., Hu, Z., Lai, W., Wang, F., Yang, M.: Gated fusion network for joint image deblurring and super-resolution. arXiv preprint arXiv:1807.10806 (2018)

30. Zhu, H., Peng, X., Chandrasekhar, V., Li, L., Lim, J.H.: DehazeGAN: when image dehazing meets differential programming. In: Proceedings of the Twenty-Seventh International Joint Conference on Artificial Intelligence, IJCAI-18, pp. 1234–1240. International Joint Conferences on Artificial Intelligence Organization (2018)
31. Zhu, Q., Mai, J., Shao, L.: Single image dehazing using color attenuation prior. In: BMVC. Citeseer (2014)
32. Zhu, L., et al.: Learning gated non-local residual for single-image rain streak removal. IEEE Trans. Circ. Syst. Video Technol. **31**(6), 2147–2159 (2021). https://doi.org/10.1109/TCSVT.2020.3022707
33. Kulkarni, A., Patil, P.W., Murala, S.: Progressive subtractive recurrent lightweight network for video deraining. IEEE Signal Process. Lett. **29**, 229–233 (2022). https://doi.org/10.1109/LSP.2021.3134171
34. Li, P., Tian, J., Tang, Y., Wang, G., Wu, C.: Model-based deep network for single image deraining. IEEE Access **8**, 14036–14047 (2020). https://doi.org/10.1109/ACCESS.2020.2965545
35. Perazzi, F., et al.: A benchmark dataset and evaluation methodology for video object segmentation. In: Proceedings of the IEEE Conference on Computer Vision and Pattern Recognition (2016)
36. Silberman, N. Fergus, R.: Indoor scene segmentation using a structured light sensor. In: 2011 IEEE International Conference on Computer Vision Workshops (ICCV Workshops), pp. 601–608. IEEE (2011)
37. Shin, J., Park, H., Paik, J.: Region-based dehazing via dual-supervised triple-convolutional network. IEEE Trans. Multimedia **24**, 245–260 (2021). https://doi.org/10.1109/TMM.2021.3050053
38. Qin, X., Wang, Z., Bai, Y., Xie, X., Jia, H.: FFA-Net: feature fusion attention network for single image dehazing. In: Proceedings of the AAAI Conference on Artificial Intelligence, vol. 34, pp. 11908–11915 (2020). https://doi.org/10.1609/aaai.v34i07.6865
39. Dong, H., et al.: Multi-scale boosted dehazing network with dense feature fusion, pp. 2154–2164 (2020). https://doi.org/10.1109/CVPR42600.2020.00223
40. Chen, D., et al.: Gated context aggregation network for image dehazing and deraining, pp. 1375–1383 (2019). https://doi.org/10.1109/WACV.2019.00151
41. Shin, J., Kim, M., Paik, J., Lee, S.: Radiance-reflectance combined optimization and structure-guided ℓ_0-norm for single image dehazing. IEEE Trans. Multimedia **22**(1), 30–44 (2020). https://doi.org/10.1109/TMM.2019.2922127
42. Isola, P., et al.: Image-to-image translation with conditional adversarial networks. In: 2017 IEEE Conference on Computer Vision and Pattern Recognition (CVPR), pp. 5967–5976 (2017)
43. Simonyan, K., Zisserman, A.: Very deep convolutional networks for large-scale image recognition. CoRR arXiv:1409.1556 (2015)
44. Dhara, S.K., Roy, M., Sen, D., Biswas, P.K.: Color cast dependent image dehazing via adaptive airlight refinement and non-linear color balancing. IEEE Trans. Circuits Syst. Video Technol. **31**(5), 2076–2081 (2021). https://doi.org/10.1109/TCSVT.2020.3007850
45. Zhu, Z., Wei, H., Hu, G., Li, Y., Qi, G., Mazur, N.: A novel fast single image dehazing algorithm based on artificial multiexposure image fusion. IEEE Trans. Instrum. Meas. **70**, 1–23 (2021). https://doi.org/10.1109/TIM.2020.3024335. Art no. 5001523

Spatio-Temporal Event Segmentation for Wildlife Extended Videos

Ramy Mounir[1(✉)], Roman Gula[2], Jörn Theuerkauf[2], and Sudeep Sarkar[1]

[1] University of South Florida, Tampa, Florida 33647, USA
{ramy,sarkar}@usf.edu

[2] Museum and Institute of Zoology, PAS, Warsaw, Poland
{rgula,jtheuer}@miiz.eu

Abstract. Using offline training schemes, researchers have tackled the event segmentation problem by providing full or weak-supervision through manually annotated labels or self-supervised epoch-based training. Most works consider videos that are at most 10's of minutes long. We present a self-supervised perceptual prediction framework capable of temporal event segmentation by building stable representations of objects over time and demonstrate it on long videos, spanning several days at 25 FPS. The approach is deceptively simple but quite effective. We rely on predictions of high-level features computed by a standard deep learning backbone. For prediction, we use an LSTM, augmented with an attention mechanism, trained in a self-supervised manner using the prediction error. The self-learned attention maps effectively localize and track the event-related objects in each frame. The proposed approach does not require labels. It requires only a single pass through the video, with no separate training set. Given the lack of datasets of very long videos, we demonstrate our method on video from 10 d (254 h) of continuous wildlife monitoring data that we had collected with required permissions. We find that the approach is robust to various environmental conditions such as day/night conditions, rain, sharp shadows, and windy conditions. For the task of temporally locating events at the activity level, we had an 80% activity recall rate for one false activity detection every 50 min. We will make the dataset, which is the first of its kind, and the code available to the research community. Project page is available at https://ramymounir.com/publications/EventSegmentation/.

Keywords: Self-supervised event segmentation · Spatial object localization · Streaming input

1 Introduction

One of the tasks involved in wild-life monitoring, or even in the video monitoring of other contexts, is detecting significant events in long videos, spanning several days. The goal is to flag temporal segments and highlight possible events in the video snippets flagged, i.e., spatial and temporal localization of possible events, such as bird {leaving,

---

Supplementary Information The online version contains supplementary material available at https://doi.org/10.1007/978-3-031-11349-9_5.

B. Raman et al. (Eds.): CVIP 2021, CCIS 1568, pp. 48–59, 2022.
https://doi.org/10.1007/978-3-031-11349-9_5

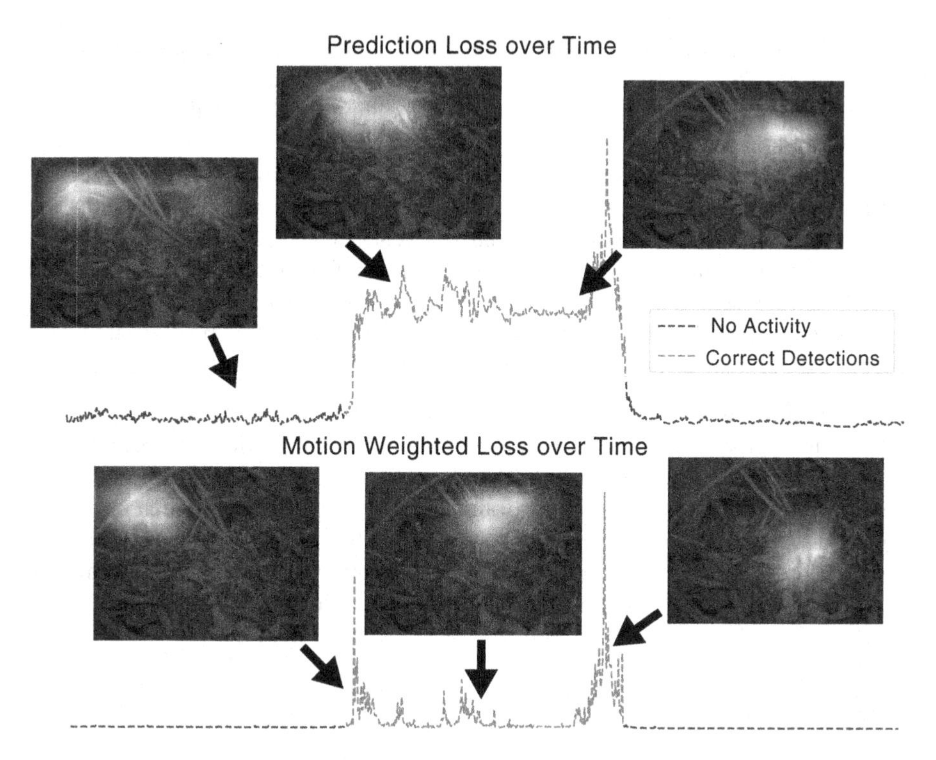

Fig. 1. Plots of the two kinds of errors before, during, and after an activity: (top) feature prediction loss over the frames (bottom) motion weighted feature prediction loss over the frames. Errors for some selected frames are shown for both plots, overlaid with the corresponding attention map.

entering, building} a nest. One cannot rely on low-level features for this task as they may change due to environmental conditions. We need high-level features that are sufficient to capture object-level representations and a model to capture these features' temporal evolution over time. There are very few works in the literature that show performance on video spanning several days.

Event segmentation research has largely focused on offline epoch-based training methods which requires training the model on the entire training dataset prior to testing its performance. This poses a challenge for many real world applications, where the entire dataset is simply non-existent and has to be collected sequentially in a stream over time [24]. Our training scheme completely disregards datapoints after being processed by the network. Training and inference are done simultaneously, alleviating the need for epoch-based training in order to appeal to more practical applications and reduce training time.

Our framework follows key ideas from the perceptual prediction line of work in cognitive psychology [15,34–36]. Research has shown that "event segmentation is an ongoing process in human perception, which helps form the basis of memory and learning". Humans can identify event boundaries, in a purely bottom up fashion, using a biological perceptual predictive model which predicts future perceptual states based on the current perceived sensory information. Experiments have shown that human perceptual system identifies event boundaries based on the appearance and motion cues in the video [22,26,33]. Our model implements this perceptual predictive framework and introduces a motion weighted loss function to allow for the localization and processing of motion cues.

Our approach uses a feature encoding network to transform low-level perceptual information to higher level feature representation. The model is trained to predict the future perceptual encoded input and signal an event if the prediction is significantly different from the future perceived features. The prediction signal also incorporates a higher level representation of the movement cues within frames. The intuition of our approach is presented in Fig. 1 for a "walk in and out" event. The error signal is used for temporal event segmentation, while an attention map segments a frame spatially.

Novel contributions: To the best of our knowledge, we are among the first to (1) introduce the attention-based mechanism to temporal event segmentation models, allowing the model to localize the event in each processed frame, in a purely self-supervised manner, without the need for labels or training data; (2) introduce the idea of motion weighted loss function to stabilize the attention maps that works even when the object of interest does not move; and (3) evaluate and report the performance of temporal segmentation on a remarkably long dataset (over ten days of continuous wildlife monitoring).

2 Relevant Work

Supervised Temporal Event Segmentation uses direct labelling (of frames) to segment videos into smaller constituent events. Fully supervised models are heavily dependent on vast amount of training data to achieve good segmentation results. Different model variations and approaches have been tested, such as using an encoder-decoder temporal convolutional network (ED-TCN) [13], or a spatiotemporal CNN model [14]. To alleviate the need for expensive direct labelling, weakly supervised approaches [7,9,12,23] have emerged with an attempt to use metadata (such as captions or narrations) to guide the training process without the need for explicit training labels [4,18]. However, such metadata are not always available as part of the dataset, which makes weakly supervised approaches inapplicable to most practical applications.

Self-supervised Temporal Event Segmentation attempts to completely eliminate the need for annotations [20,25]. Many approaches rely heavily on higher level features clustering of frames to sub-activities [6,28]. The performance of the clustering algorithms in unsupervised event segmentation is proportional to the performance of the embedding/encoding model that transforms frames to higher level feature representations. Clustering algorithms can be highly computationally expensive depending on the number of frames to be clustered. Recent work [2] uses a self-supervised perceptual predictive model to detect event boundaries; we improve upon this model to include attention unit, which helps the model focus on event-causing objects. Other work [19] uses a self-supervised perceptual prediction model that is refined over significant amount of reinforcement learning iterations.

Frame Predictive Models have attempted to provide accurate predictions of the next frame in a sequence [10,16,21,29,30]; however, these models are focusing on predicting future frames in raw pixel format. Such models may generate a prediction loss that only captures frame motion difference with limited understanding of higher level features that constitutes event boundaries.

Attention Units have been applied to image captioning [31], and natural language processing [5,8,17,27,32] fully supervised applications. Attention is used to expose different temporal - or spatial - segments of the input to the decoding LSTM at every time step using fully supervised model architectures. We use attention in a slightly different form, where the LSTM is decoded only once (per input frame) to predict future features and uses attention weighted input to do so. Unlike [5,8,17,27,31,32], our attention weights and biases are trained using an unsupervised loss functions.

Recent work [3] has used the prediction loss, with the assistance of region proposal networks (RPNs) and multi-layer LSTM units, to localize actions. We eliminate the need for RPNs and multi-layer LSTM units by extracting Bahdanau [5] attention weights prior to the LSTM prediction layer, which allows our model to localize objects of interest, even when stationary. From our experiments, we found out that prediction loss attention tends to fade away as moving objects become stationary, which makes its attention map more similar to results extracted from background subtraction or optical flow. In contrast, our model proves to be successful in attending to moving and stationary objects despite variations in environmental conditions, such as moving shadows and lighting changes, as presented in the supplementary videos.

3 Methodology

The proposed framework is inspired by the works of Zacks *et al.* on perceptual prediction for events segmentation [34]. The proposed architecture, summarised in Fig. 2, can be divided into several individual components. In this section, we explain the role of each component starting by the encoder network and attention unit in Sect. 3.1 and 3.2, followed by a discussion on the recurrent predictive layer in Sect. 3.3. We conclude by introducing the loss functions (Sect. 3.4) used for self-supervised learning as well as the adaptive thresholding function (Sect. 3.5). Full pseudocode is provided in the appendix.

3.1 Input Encoding

The raw input images are transformed from pixel space into a higher level feature space by utilizing an encoder (CNN) model. This encoded feature representation allows the network to extract features of higher importance to the task being learned. We denote the output of the CNN layers by $I'_t = f(I_t, \theta_e)$ where θ_e is the learnable weights and biases parameters and I_t is the input image.

3.2 Attention Unit

In this framework, we utilize Bahdanau attention [5] to spatially localize the event in each processed frame. The attention unit receives as an input the encoded features and

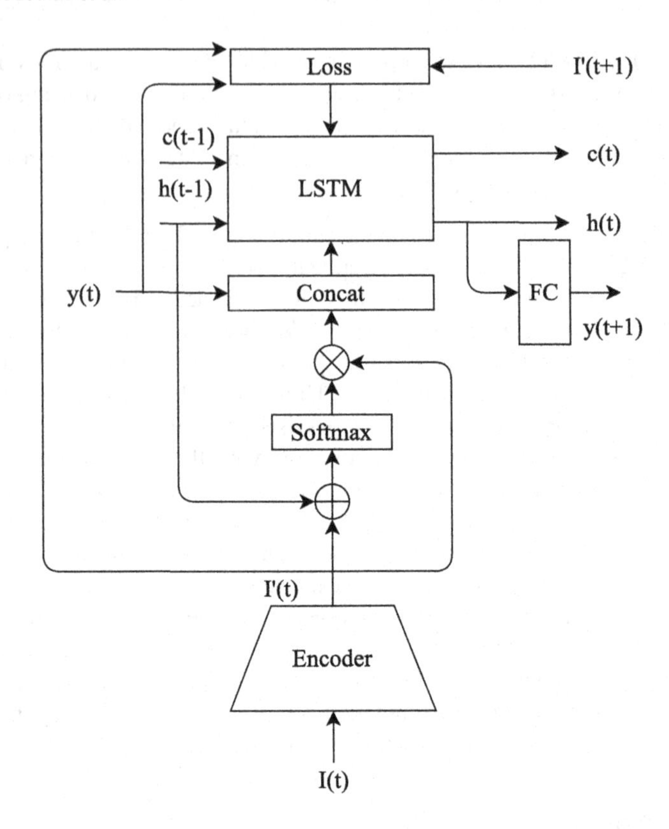

Fig. 2. The architecture of the self-learning, perceptual prediction algorithm. Input frames from each time instant are encoded into high-level features using a deep-learning stack, followed by an attention overlay that is based on inputs from previous time instant, which is input to an LSTM. The training loss is composed based on the predicted and computed features from current and next frames.

outputs a set of attention weights (A_t) with dimensions $N \times N \times 1$. The hidden feature vectors (h_{t-1}) from the prediction layer of the previous time step is used to calculate the output set of weights using Eq. 1, expressed visually in Fig. 2.

$$A_t = \gamma(\,FC(\,\varphi(FC(h_{t-1}) + FC(I'_t))\,)\,) \tag{1}$$

where φ represents hyperbolic tangent ($\tanh$) function, and γ represents a softmax function. The weights (A_t) are then multiplied by the encoded input feature vectors (I'_t) to generate the masked feature vectors (I''_t).

3.3 Future Prediction Layer

The process of future prediction requires a layer capable of storing a flexible internal state (event model) of the previous frames. For this purpose, we use a recurrent layer, specifically long-short term memory cell (LSTM) [11], which is designed to

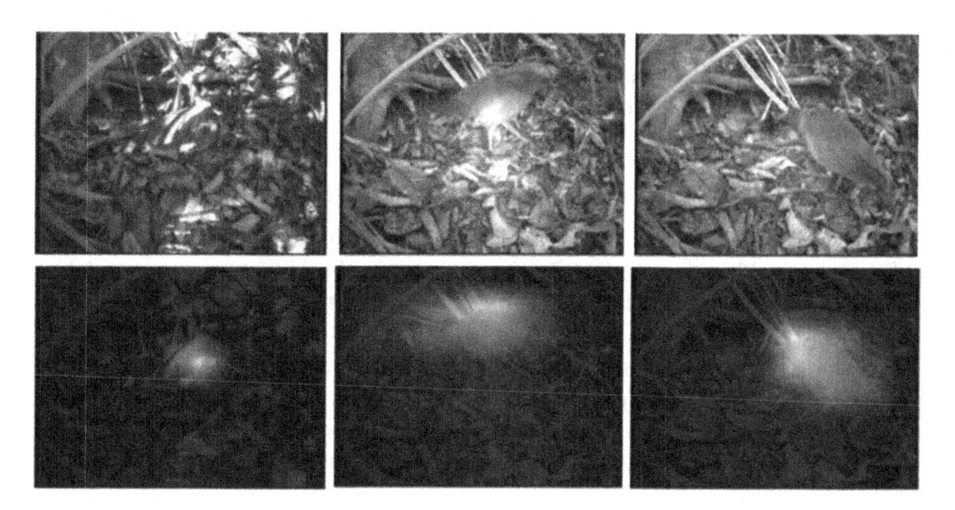

Fig. 3. Qualitative Results. **(Top):** Samples of images from the Kagu bird wildlife monitoring dataset. **(Bottom):** Corresponding attention maps overlayed on the input image

output a future prediction based on the current input and a feature representation of the internal state. More formally, the LSTM cell can be described using the function $h_t = g(I''_t, W_{lstm}, h_{t-1})$, where h_t and h_{t-1} are the output hidden state and previous hidden state respectively, I''_t the attention-masked input features at time step t and W_{lstm} is a set of weights and biases vectors controlling the internal state of the LSTM. The input to the LSTM can be formulated as:

$$FC(y_{t-1} \oplus I''_t) \tag{2}$$

where I''_t is the masked encoded input feature vector and h_{t-1} is the hidden state from the previous time step. The symbol $\oplus$ represents vectors concatenation.

3.4 Loss Function

The perceptual prediction model aims to train a model capable of predicting the feature vectors of the next time step. We define two different loss functions, prediction loss and motion weighted loss.

Prediction Loss. This function is defined as the L2 Euclidean distance loss between the output prediction y_t and the next frame encoded feature vectors I'_{t+1}.

$$e_t = ||(I'_{t+1} - y_t)||^2 \tag{3}$$

Motion Weighted Loss. This function aims to extract the motion related feature vectors from two consecutive frames to generate a motion dependent mask, which is applied to the prediction loss. The motion weighted loss function allows the network to benefit from motion information in higher level feature space rather than pixel space. This function is formally defined as:

$$e_t = ||(I'_{t+1} - y_t)^{\odot 2} \odot (I'_{t+1} - I'_t)^{\odot 2}||^2 \tag{4}$$

where $\odot$ denotes Hadamard (element-wise) operation.

3.5 Error Gate

The error gating function receives, as an input, the error signal defined in Sect. 3.4, and applies a thresholding function to classify each frame. In this framework, we define two types of error gating functions. A simple threshold function $f(e_t, \psi)$ and an adaptive threshold function $f(\{e_{t-m} \dots e_t\}, \psi)$. Equation 5 formally defines the smoothing function for the adaptive error gating implementation. Both error gating functions use Eq. 6 to threshold the error signal. Equations 5 and 6 apply the smoothing function to the full loss signal for analyses purposes; however, the convolution operation can be reduced to element-wise multiplication to calculate a single smoothed value at time step t.

$$\left.\begin{aligned} e &= \{e_{t-m} \dots e_t\} \in \mathbb{R}^m \\ e_s &= e - [e \circledast [\{\tfrac{1}{n} \dots \tfrac{1}{n}\} \in \mathbb{R}^n]] \end{aligned}\right\} n < m \tag{5}$$

$$f(e_s(t)) = \begin{cases} 1, & \text{if } e_s(t) \geq \psi \\ 0, & \text{otherwise} \end{cases} \tag{6}$$

where $\circledast$ represents a 1D convolution operation.

4 Experimental Evaluation

In this section, we present the results of our experiments for our approach defined in Sect. 3. We begin by defining the wildlife extended video dataset used for testing, followed by explaining the evaluation metrics used to quantify performance. We discuss the model variations evaluated and conclude by presenting quantitative and qualitative results in Sects. 4.4 and 4.5.

4.1 Dataset

We analyze the performance of our model on a wildlife monitoring dataset. The dataset consists of 10 d (254 h) continuous monitoring of a nest of the Kagu, a flightless bird of New Caledonia. The labels include four unique bird activities, {feeding the chick, incubation/brooding, nest building while sitting on the nest, nest building around the nest}. Start and end times for each instance of these activities are provided with the annotations. We modified the annotations to include walk in and walk out events representing the transitioning events from an empty nest to incubation and vice versa. Our approach can flag the nest building (on and around the nest), feeding the chick, walk in and out events. Other events based on climate, time of day, lighting conditions are ignored by our segmentation network. Figure 3 (Top) shows a sample of images from the dataset.

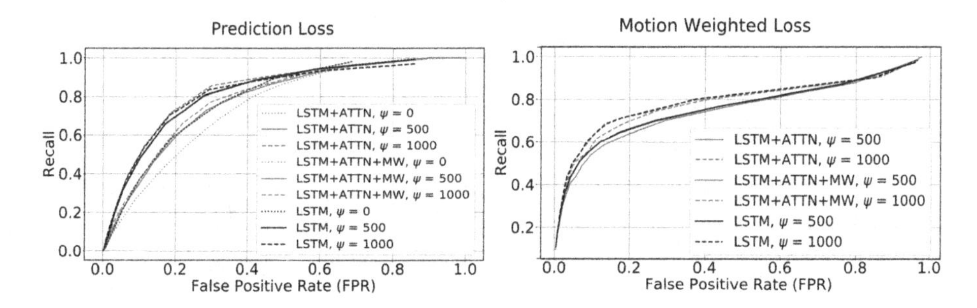

Fig. 4. Frame-level event segmentation ROCs when activities are detected based on simple thresholding of the prediction and motion weighted loss signals. Plots are shown for different ablation studies.

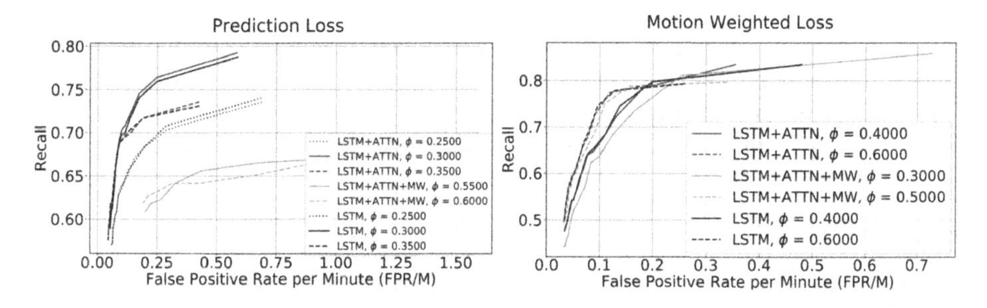

Fig. 5. Activity-level event segmentation ROCs when activities are detected based on simple thresholding of the prediction and motion weighted loss signals. Plots are shown for different ablation studies.

4.2 Evaluation Metrics

We provide quantitative ROC results for both frame level (Fig. 4) and activity level (Figs. 5 and 6) event segmentation. Frame window size (ψ) is defined as the maximum joining window size between events; a high ψ value can causes separate detected events to merge, which decreases the overall performance.

Frame Level. The recall value in frame level ROC is calculated as the ratio of true positive frames (event present) to the number of positive frames in the annotations dataset, while the false positive rate is expressed as ratio of the false positive frames to the total number of negative frames (event not present) in the annotation dataset. Threshold value (ϕ) is varied to obtain a single ROC line, while varying the frame window size (ψ) results in a different ROC line.

Activity Level. The Hungarian matching (Munkres assignment) algorithm is utilized to achieve one to one mapping between the ground truth labeled events and the detected events. Recall is defined as ratio of the number of correctly detected events (overlapping frames) to the total number of groundtruth events. For the activity level ROC chart, the recall values are plotted against the false positive rate per minute, defined as the ratio of

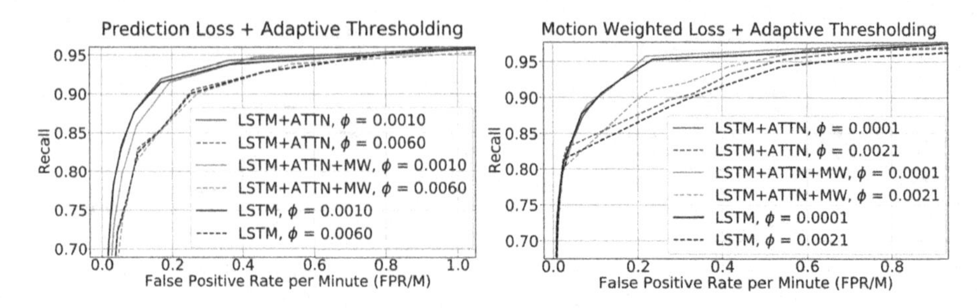

Fig. 6. Activity-level event segmentation ROCs when activities are detected based on adaptive thresholding of the prediction and motion weighted loss signals. Plots are shown for different ablation studies.

the total number of false positive detected events to the total duration of the dataset in minutes. The false positive rate per minute evaluation metric is also used in the ActEV TRECVID challenge [1]. Frame window size value (ψ) is varied to obtain a single ROC line, while varying the threshold value (ϕ) results in a different ROC line.

4.3 Ablative Studies

Different variations of our framework (Sect. 3) have been evaluated to quantify the effect of individual components on the overall performance. In our experiments, we tested the base model, which trains the perceptual prediction framework - including attention unit - using the prediction loss function for backpropagation of the error signal. We refer to the base model as *LSTM+ATTN*. We also experimented with the effect of removing the attention unit, from the model architecture, on the overall segmentation performance; results of this variation are reported under the model name *LSTM*. Further testing includes using the motion weighted loss for backpropagation of the error signal. We refer to the motion weighted model as *LSTM+ATTN+MW*. Each of the models has been tested extensively; results are reported in Sects. 4.4 and 4.5, as well as visually expressed in Figs. 3, 4, 5 and 6.

Comparing the results shown in Figs. 5 and 6 indicate a significant increase of overall performance when using an adaptive threshold for loss signal gating. The efficacy of adaptive thresholding is evident when applied to activity level event segmentation. Comparing the results (LSTM & LSTM+ATTN) show that the model can effectively generate attention maps (Sect. 4.5) without impacting the segmentation performance.

4.4 Quantitative Evaluation

We tested three different models, *LSTM*, *LSTM+ATTN*, and *LSTM+ATTN+MW*, for frame level and activity level event segmentation. Simple and adaptive gating functions (Sect. 3.5), were applied to prediction and motion weighted loss signals (Sect. 3.4) for frame level and activity level experiments. For each model we vary parameters such as the threshold value ψ and the frame window size ϕ to achieve the ROC charts presented in Figs. 4, 5 and 6.

It is to be noted that thresholding a loss signal does not necessarily imply that the model was trained to minimize this particular signal. In other words, loss functions used for backpropagating the error to the models' learnable parameters are identified only in the model name (Sect. 4.3); however, thresholding experiments have been conducted on different types of loss signals, regardless of the backpropagating loss function used for training.

The best performing model, for frame level segmentation (*LSTM+ATTN*, $\psi = 1000$) is capable of achieving $\{40\%, 60\%, 70\%\}$ frame recall value at $\{5\%, 10\%, 20\%\}$ frame false positive rate respectively. Activity level segmentation can recall $\{80\%, 90\%, 95\%\}$ of the activities at $\{0.02, 0.1, 0.2\}$ activity false positive rate per minute, respectively, for the model (*LSTM+ATTN+MW*, $\phi = 0.0001$) as presented in Fig. 6. A 0.02 false positive activity rate per minute can also be interpreted as one false activity detection every 50 min of training (for detecting 80% of the groundtruth activities).

4.5 Qualitative Evaluation

A sample of the qualitative attention results is presented in Fig. 3. The attention mask, extracted from the model, has been trained to track the event in all processed frames. Our results show that the events are tracked and localized in various lighting (shadows, day/night) and occlusion conditions. Attention has also learned to indefinitely focus on the bird regardless of its motion state (stationary/Non-stationary), which indicates that the model has acquired a high-level temporal understanding of the events in the scene and learned the underlying structure of the bird. Supplementary results[1] display a timelapse of attention weighted frames during illumination changes and moving shadows. We also provide a supplementary video showing the prediction loss signal, motion weighted loss signal and attention mask during a walk in and out event (summarized in Fig. 1). Additional qualitative results are provided in the appendix.

5 Conclusion

We demonstrate a self-supervised approach to temporal event segmentation. Our framework can effectively segment a long sequence of activities (video) into a set of individual events. We introduce a novel approach to extract attention results from unsupervised temporal event segmentation network. Gating the loss signal with different threshold values can result in segmentation at different granularities. Quantitative and qualitative results are presented in the form of ROC charts and attention weighted frames. Our results demonstrate the effectiveness of our approach in understanding the higher level spatio-temporal features required for practical temporal event segmentation.

Acknowledgment. This research was supported in part by the US National Science Foundation grants CNS 1513126 and IIS 1956050. The bird video dataset used in this paper was made possible through funding from the Polish National Science Centre (grant NCN 2011/01/M/NZ8/03344 and 2018/29/B/NZ8/02312). Province Sud (New Caledonia) issued all permits - from 2002 to 2020 - required for data collection.

[1] Available at https://ramymounir.com/publications/EventSegmentation/.

References

1. Actev: Activities in extended video. https://actev.nist.gov/
2. Aakur, S.N., Sarkar, S.: A perceptual prediction framework for self supervised event segmentation. In: Proceedings of the IEEE Conference on Computer Vision and Pattern Recognition, pp. 1197–1206 (2019)
3. Aakur, S.N., Sarkar, S.: Action localization through continual predictive learning (2020). arXiv preprint arXiv:2003.12185
4. Alayrac, J.B., Bojanowski, P., Agrawal, N., Sivic, J., Laptev, I., Lacoste-Julien, S.: Unsupervised learning from narrated instruction videos. In: Proceedings of the IEEE Conference on Computer Vision and Pattern Recognition, pp. 4575–4583 (2016)
5. Bahdanau, D., Cho, K., Bengio, Y.: Neural machine translation by jointly learning to align and translate (2014). arXiv preprint arXiv:1409.0473
6. Bhatnagar, B.L., Singh, S., Arora, C., Jawahar, C.: CVIT K: Unsupervised learning of deep feature representation for clustering egocentric actions. In: IJCAI, pp. 1447–1453 (2017)
7. Bojanowski, P., et al.: Weakly supervised action labeling in videos under ordering constraints. In: Fleet, D., Pajdla, T., Schiele, B., Tuytelaars, T. (eds.) ECCV 2014. LNCS, vol. 8693, pp. 628–643. Springer, Cham (2014). https://doi.org/10.1007/978-3-319-10602-1_41
8. Devlin, J., Chang, M.W., Lee, K., Toutanova, K.: Bert: pre-training of deep bidirectional transformers for language understanding (2018). arXiv preprint arXiv:1810.04805
9. Ding, L., Xu, C.: Weakly-supervised action segmentation with iterative soft boundary assignment. In: Proceedings of the IEEE Conference on Computer Vision and Pattern Recognition, pp. 6508–6516 (2018)
10. Finn, C., Goodfellow, I., Levine, S.: Unsupervised learning for physical interaction through video prediction. In: Advances in Neural Information Processing Systems, pp. 64–72 (2016)
11. Hochreiter, S., Schmidhuber, J.: Long short-term memory. Neural Comput. **9**(8), 1735–1780 (1997)
12. Huang, D.A., Fei-Fei, L., Niebles, J.C.: Connectionist temporal modeling for weakly supervised action labeling. In: European Conference on Computer Vision, pp. 137–153. Springer (2016). https://doi.org/10.48550/arXiv.1607.08584
13. Lea, C., Flynn, M.D., Vidal, R., Reiter, A., Hager, G.D.: Temporal convolutional networks for action segmentation and detection. In: Proceedings of the IEEE Conference on Computer Vision and Pattern Recognition, pp. 156–165 (2017)
14. Lea, C., Reiter, A., Vidal, R., Hager, G.D.: Segmental spatiotemporal CNNs for fine-grained action segmentation. In: Leibe, B., Matas, J., Sebe, N., Welling, M. (eds.) ECCV 2016. LNCS, vol. 9907, pp. 36–52. Springer, Cham (2016). https://doi.org/10.1007/978-3-319-46487-9_3
15. Loschky, L.C., Larson, A.M., Smith, T.J., Magliano, J.P.: The scene perception & event comprehension theory (spect) applied to visual narratives. Top. Cogn. Sci. **12**(1), 311–351 (2020)
16. Lotter, W., Kreiman, G., Cox, D.: Deep predictive coding networks for video prediction and unsupervised learning (2016). arXiv preprint arXiv:1605.08104
17. Luong, M.T., Pham, H., Manning, C.D.: Effective approaches to attention-based neural machine translation (2015). arXiv preprint arXiv:1508.04025
18. Malmaud, J., Huang, J., Rathod, V., Johnston, N., Rabinovich, A., Murphy, K.: What's cookin'? interpreting cooking videos using text, speech and vision (2015). arXiv preprint arXiv:1503.01558
19. Metcalf, K., Leake, D.: Modelling unsupervised event segmentation: learning event boundaries from prediction errors. In: CogSci. (2017)
20. Garcia del Molino, A., Lim, J.H., Tan, A.H.: Predicting visual context for unsupervised event segmentation in continuous photo-streams. In: Proceedings of the 26th ACM International Conference on Multimedia, pp. 10–17 (2018)

21. Qiu, J., Huang, G., Lee, T.S.: A neurally-inspired hierarchical prediction network for spatiotemporal sequence learning and prediction (2019). arXiv preprint arXiv:1901.09002
22. Radvansky, G.A., Krawietz, S.A., Tamplin, A.K.: Walking through doorways causes forgetting: further explorations. Quart. J. Exper. Psychol. **64**(8), 1632–1645 (2011)
23. Richard, A., Kuehne, H., Gall, J.: Weakly supervised action learning with RNN based fine-to-coarse modeling. In: Proceedings of the IEEE Conference on Computer Vision and Pattern Recognition, pp. 754–763 (2017)
24. Sahoo, D., Pham, Q., Lu, J., Hoi, S.C.: Online deep learning: Learning deep neural networks on the fly (2017). arXiv preprint arXiv:1711.03705
25. Sener, F., Yao, A.: Unsupervised learning and segmentation of complex activities from video. In: Proceedings of the IEEE Conference on Computer Vision and Pattern Recognition, pp. 8368–8376 (2018)
26. Speer, N.K., Swallow, K.M., Zacks, J.M.: Activation of human motion processing areas during event perception. Cogn. Aff. Behav. Neurosci. **3**(4), 335–345 (2003)
27. Vaswani, A., et al.: Attention is all you need. In: Advances in Neural Information Processing Systems, pp. 5998–6008 (2017)
28. VidalMata, R.G., Scheirer, W.J., Kuehne, H.: Joint visual-temporal embedding for unsupervised learning of actions in untrimmed sequences (2020). arXiv preprint arXiv:2001.11122
29. Wang, Y., Gao, Z., Long, M., Wang, J., Yu, P.S.: Predrnn++: towards a resolution of the deep-in-time dilemma in spatiotemporal predictive learning (2018). arXiv preprint arXiv:1804.06300
30. Wichers, N., Villegas, R., Erhan, D., Lee, H.: Hierarchical long-term video prediction without supervision (2018). arXiv preprint arXiv:1806.04768
31. Xu, K., et al.: Show, attend and tell: neural image caption generation with visual attention. In: International Conference on Machine Learning, pp. 2048–2057 (2015)
32. Yang, Z., Dai, Z., Yang, Y., Carbonell, J., Salakhutdinov, R.R., Le, Q.V.: Xlnet: generalized autoregressive pretraining for language understanding. In: Advances in Neural Information Processing Systems, pp. 5754–5764 (2019)
33. Zacks, J.M.: Using movement and intentions to understand simple events. Cogn. Sci. **28**(6), 979–1008 (2004)
34. Zacks, J.M., Swallow, K.M.: Event segmentation. Curr. Dir. Psychol. Sci. **16**(2), 80–84 (2007)
35. Zacks, J.M., Tversky, B.: Event structure in perception and conception. Psychol. Bull. **127**(1), 3 (2001)
36. Zacks, J.M., Tversky, B., Iyer, G.: Perceiving, remembering, and communicating structure in events. J. Exp. Psychol. Gen. **130**(1), 29 (2001)

Comparative Analysis of Machine Learning and Deep Learning Models for Ship Classification from Satellite Images

Abhinaba Hazarika, P. Jidesh(✉), and A. Smitha

Department of Mathematical and Computational Sciences,
National Institute of Technology Karnataka, Mangalore, India
{abhinabahazarika.203cd001,jidesh}@nitk.edu.in

Abstract. The automatic detection of the ship from satellite image analysis is the limelight of research in recent years due to its widespread applications. In this paper, a handful of traditional machine learning and deep learning models are compared based on their performance to classify the satellite images available in the public repository as a ship or other categories. The Support Vector Machine(SVM), Decision Trees, Random Forest, K-Nearest Neighbor (KNN), Gaussian Naive Bayes (GaussianNB), and Logistic Regression are machine learning models used in the present work. Histogram of Gradient (HoG) features are used as feature descriptors considering the diverse size and shape of ships in the satellite image dataset. Transfer learning is applied using the deep learning models namely, Inception and ResNet, that are fine-tuned for various learning rates and optimizers. The meticulous experimentation carried out reveals that traditional machine learning performs well when trained and tested on a single dataset. However, there is a drastic change in the performance of machine learning models when tested on a different ship dataset. The results show that the deep learning models have better feature detection and thus have better performance when transfer learning is used on various datasets.

Keywords: Ship classification · Machine learning · Deep learning

1 Introduction

Ship detection and categorization play a significant role for various range of applications such as military naval warfare, fishery management and other civilian applications. This is usually done using Automated Identification System (AIS), which uses Very High Frequency (VHF) radio signals to broadcast to the ship location and the destination identifies nearby receiver devices which might be on other ships and land-based systems [1,2]. AIS is very effective, however, it fails to identify ships which are not installed with the VHF transponder or ships where the

B. Raman et al. (Eds.): CVIP 2021, CCIS 1568, pp. 60–72, 2022.
https://doi.org/10.1007/978-3-031-11349-9_6

transponders are disconnected illegally [3,4]. It is a daunting task to detect such uncooperative ships. Satellite imagery can be highly beneficial in this regard. Satellites are collecting oodles of images which can be analyzed using machine learning algorithms for detecting the ships and also categorization. The difference in weather conditions, diverse land environment scenario, assorted sizes, angle of orientation and shapes of the ship hinders the automatic detection of ships from satellite images. Furthermore, the speculation due to cluttered scenes, clouds, waves and islands between the ocean poses an additional burden to precisely classify the type of ship. Although plethora of literature exists on automatic ship detection, a comparative analysis on machine learning models and deep learning models can provide better insights. Deep learning models require enormous training data, and involve hyperparameters [5]. The training of these models are computationally expensive as well. On the other hand, the machine learning models can work great with limited amount of training data, but employing appropriate feature descriptors and classifiers is an open problem [6]. An effective, fast and precise method is required in order to work with a huge amount of satellite imagery dataset. The need to identify the best approach is the motivation for this work. The major contributions of our work are listed below.

- To verify the performance of machine learning models such as Support Vector Machine (SVM), logistic regression, K-Nearest Neighbors (KNN), decision tree, random forest and Gaussian Naive Bayes (GaussianNB) for ship detection.
- To assess the state-of-the-art deep learning models in ship detection from satellite images using transfer learning approach.

The rest of the document is organized as follows. Section 2 discusses the recent related works. The proposed approach is elaborately explained in Sect. 3. Section 4 discusses the results obtained and the concluding remarks are presented in Sect. 5.

2 Literature Review

Researchers have explored various approaches to detect the ship using satellite images from assorted datasets. Zhangxia Zou and Zhenwei Shi [7] proposed an algorithm using singular value decomposition (SVD) network and convolutional neural netoworks. Their proposed algorithm overcomes the difficulty of interferences due to clouds and waves, etc. Future work is concerned with combining global and local environments in order to increase the accuracy. Chua et al. [8] compared 3 classical machine learning algorithms, namely Histogram of Oriented Gradients (HOG), exemplar-Support Vector Machine (SVM) and latent SVM [9], and found that exemplar SVM is good for specificity measure. Some other deep learning methods explored for ship detection is summarized in Table 1.

Table 1. Existing literature on ship detection from satellite images using deep learning approaches.

Author	Model used	Dataset used	Remarks
Cordova A.W. et al. (2020) [10]	You Only Look Twice (YOLT) and YOLOv4	PERUSAT-1 and Google Earth satellite images	Mean average precision of 95.94%
Li Z. et al. (2020) [11]	YoloV3 and DenseNet	Web-crawled ship images	Lightweight ship detection model
Liu C et al. (2020) [12]	RetinaNet	open SAR image ship detection	Efficient weighted bidirectional feature fusion network—BiFPN works well on complex background
Stofa M.M. et al.(2020) [13]	Densenet	Kaggle ship dataset	More than 90% accuracy
Wang Y. et al. (2019) [14]	You Only Look Once(YOLO)V3	MS coco2017, Pascal VOC and 1500 data sets collocated	74.8% and a detection rate of 29.8 frames
Gallego et al. (2018) [15]	Combined Convolutional neural network with k-Nearest Neighbor	MWPU VHR-10 & MASATI dataset	86% accuracy on cross validated datasets
Liu Y. et al. (2017)[16]	Convolutional Neural network	Google Earth dataset with 0.5 m resolution	Achieved upto 96% accuracy
Zou Z. et al. (2016) [17]	Convolutional Neural Network with Singular Value Decomposition	GaoFen-1	Suppresses undesired background and works well with offshore & inshore ships
Tang J. et al. (2015) [18]	Deep Neural Networks and Extreme Learning	Spaceborne optical image dataset	Faster and better generalization with 97.58% accuracy

From the detailed literature review, it is evident that majority of the works are based on advance deep learning with hybrid or complex architecture. However, the stability of these models on multiple datasets is unpredictable. Since limited work exists on the machine learning models, the proposed work aims to carryout a comparative analysis of state-of-the art models on multiple datasets.

3 Methodology

The overall process layout of the proposed work is shown in Fig. 1, which is explained in the following subsections.

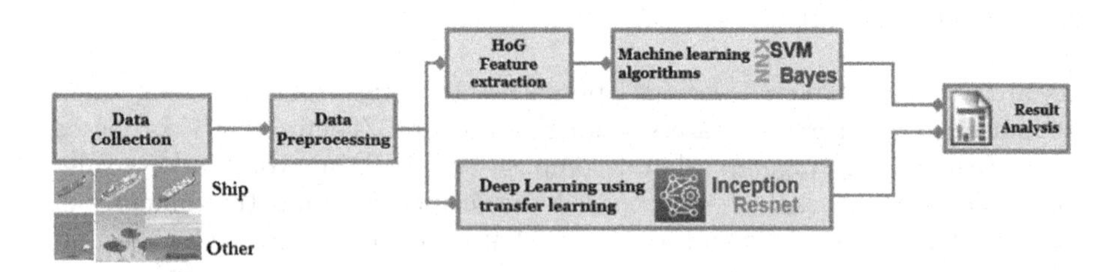

Fig. 1. Process layout.

3.1 The Dataset

Multiple datasets are collected from kaggle. The first dataset [19] is available through Planet's Open California dataset, which is openly licensed. It consists of two classes, 'ship' and 'non-ship' class. They are labelled as 1 and 0 respectively. The input image size is 80 × 80 pixel matrix. This dataset has 3000 'non-ship' images with plain sea, clouds or land etc., while there are 1000 'ship' images with different shape and orientation of the ships. The second repository is named as Airbus ship dataset [20]. It has images of ships which differs in size and is located on the open sea, at docks and marinas etc. The ships include tankers, commercial and fishing ships of various shapes and sizes. The original images are 768 × 768 pixels. However, while testing on the trained models these images are converted into 80 × 80 pixel matrix to ensure homogeneity in the input data. The sample of images from both datasets are shown in Fig. 2 and Fig. 3 respectively.

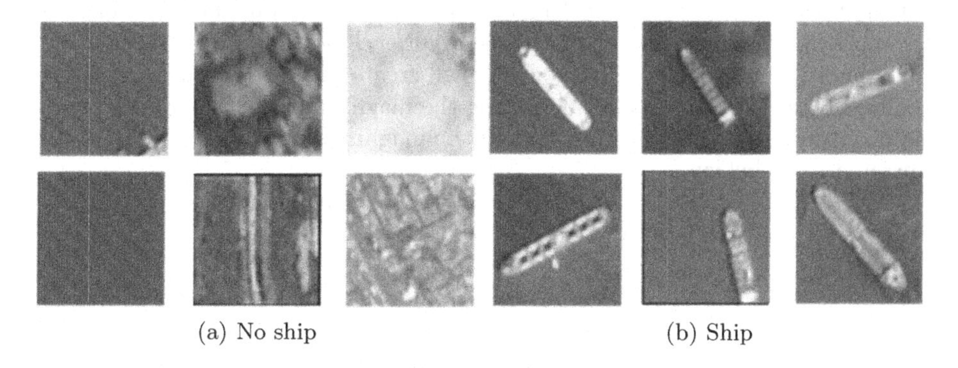

(a) No ship (b) Ship

Fig. 2. Sample images from Planet's Open California dataset [19].

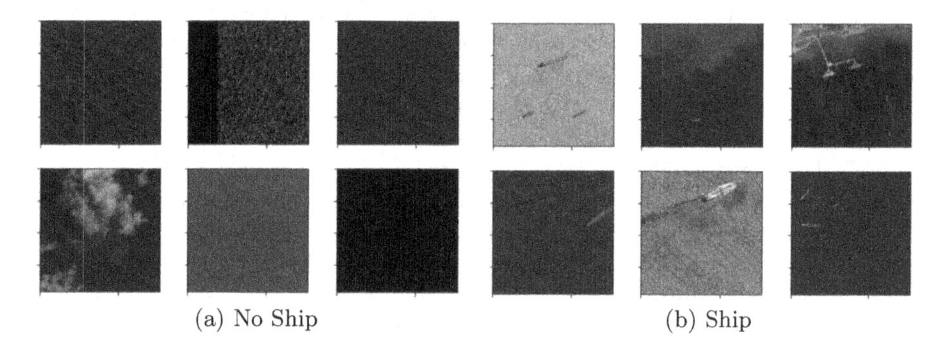

(a) No Ship (b) Ship

Fig. 3. Sample images from the Airbus dataset [20].

3.2 Pre-processing

In the data pre-processing phase, the images of the training dataset consists of 3 channels (RGB). It is converted into grayscale. Data augmentation is not incorporated as there are thousands of training images. For deep learning models the input image size is 224 × 224 images since transfer learning is used. The training and validation split for both machine learning and deep learning models is done in a 70:30 ratio.

3.3 Machine Learning Approach

For machine learning approach, the grayscaled images are converted into Histogram of Oriented Gradients (HOG) features [21]. HOG features are used as feature extraction technique as HOG feature descriptor helps to extract the edge features such as magnitude as well as the orientation of the edges. The magnitude and the orientations are calculated in localized portions. i.e. the image is decomposed into small squared portions and for each portion the magnitude and the orientation is obtained. Finally HOG will create a histogram for each of the localized regions which is calculated using the magnitude and the orientation of the pixel values. The formula for finding the magnitude and the orientation for a pixel is given below:

$$\begin{aligned} Magnitude &= \sqrt{G_x^2 + G_y^2}, \\ Orientation &= tan(\frac{G_y}{G_x}). \end{aligned} \tag{1}$$

where G_x and G_y are the change in gradient in x and y direction respectively. The HOG images for a 4 × 4 block with 8 pixels per block/cell and 4 pixels per block for a randomly selected input images are depicted in Fig. 4. Accordingly, the first row depicts the input images from both datasets. The corresponding gray-scaled images are shown in the second row. The third and the fourth row represents the HOG images with pixels per cell (PPC) equal to 8 and 4 respectively. It is observed that the features with 4 × 4 blocks and 4 pixels per block gives better performance. For classification, the machine learning models are trained on the features extracted from HOG. The six classifiers used are Support Vector Machines (SVM) [22], Logistic Regression [23], K-Nearest Neighbors (KNN), Decision Tree, Random Forest and Gaussian NB [24]. Radial Basis Function (RBF) kernel is chosen for SVM as it is non-linear. The K-value for KNN is set to 2, as the error rate is lowest with this K value, as inferred from the empirical analysis depicted in Fig. 5. Furthermore, this supports the two-class classification where the input images are labeled as 'ship' and 'noship'. All the machine learning models are trained and resampled using k-fold cross validation procedure, where the value of k = 10.

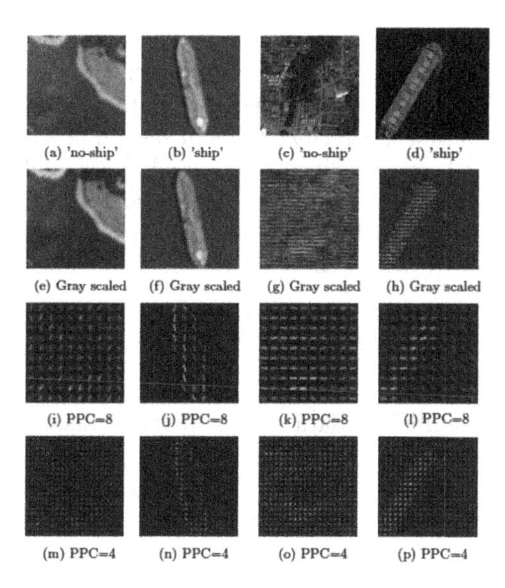

Fig. 4. HOG features on randomly selected images from the two datasets.

3.4 Deep Learning Approach

In deep learning approach, ResNet-101 [25] and Inceptionv3 [26] models are used to fine-tune the considered datasets. DenseNet [27], which was used in the paper [13] for the dataset [19] is also used using transfer learning for a comparative analysis. Transfer learning approach refers to training the input images on a base dataset, and further utilizing the pretrained model on various other image datasets. The performance of such models can be further fine-tuned during training process. The benefit of using such transfer learning process is reduced training time. Specifically, designing a novel deep learning architecture and training the

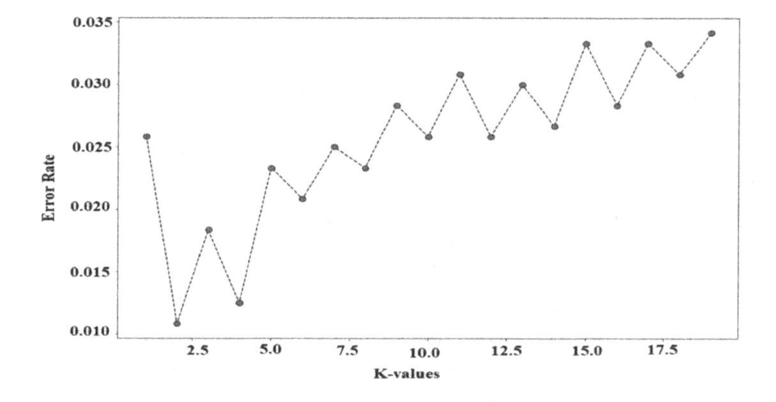

Fig. 5. Error Rate vs k-value for KNN

model requires massive input images for training and validation. Since about thousand images are available, in the proposed work, transfer learning approach is considered as highly relevant. ResNet-101 and Inceptionv3 models are pre-trained on Imagenet database which consists of several objects found in nature. The ResNet-101 has 101 layers defined by residual blocks and convolutional layers. Inceptionv3 is the third edition of inception convolutional neural network which has a total of 48 layers. Binary Cross Entropy Loss is employed to train the deep learning models, defined by:

$$J = -\sum_{i=1}^{n}(y_i \log(h_0(x_i)) + (1 - y_i)\log(1 - (h_0(x_i))) \quad (2)$$

where n is the total number of data points in the dataset y_i is the true label and $h_0(x_i)$ is the predicted value. The hyperparameters are set empirically as follows: batch size = 16, learning rate = 0.001, epochs = 10, and Adam and RMSProp are engaged as optimizer.

4 Results and Discussions

The implementation is carried out on python platform. The deep learning models are implemented using tensorflow. The experiment was conducted on an Intel Core i5 7th Gen 7300HQ processor of 2.5 GHz with 8 GB of RAM and a 4 GB of Nvidia GeForce GTX 1050 GPU. The standard performance metrics of classification such as precision, recall, F-1 score and accuracy are measured to compare the machine learning models. The metrics are defined as follows:

$$Accuracy = \frac{TP + TN}{TP + TN + FP + FN}. \quad (3)$$

$$Precision = \frac{TP}{TP + FP}. \quad (4)$$

$$Recall = \frac{TP}{TP + FN}. \quad (5)$$

$$F1 = \frac{2 * Precision * Recall}{Precision + Recall}. \quad (6)$$

where, TP, TN, FP and FN are the True Positive, True Negative, False Positive and False Negative values obtained from the confusion matrix of a two category classification (ship or other).

The accuracy values using machine learning models for training dataset [19] is shown in Table 2. Accordingly, the SVM model resulted the highest accuracy of 99.41%. Decision Tree algorithm gave an accuracy of 90.46%, which is the least of all the other algorithms. Figure 6a shows the accuracy comparison between these models. The F1-score, precision and recall score plots are shown in Fig. 6b, Fig. 6c and Fig. 6d respectively. From this figure, it is evident that the SVM outperformed other classifiers in terms of F1-score, precision, and recall scores as well.

Table 2. Accuracy scores on training dataset [19]

Model	ML/DL	Accuracy
SVM	**ML**	**99.41**
Logistic Regression	ML	98.12
KNN	ML	97.96
Decision Tree	ML	90.46
Random Forest	ML	95.78
GaussianNB	ML	94.75
ResNet 101 - Adam	DL	94.31
Inception V3 - Adam	DL	91.76
ResNet 101 - RMSProp	DL	94.55
Inception V3 - RMSProp	DL	92.73
DenseNet - Adam	DL	76.25
DenseNet - RMSProp	**DL**	**99.75**

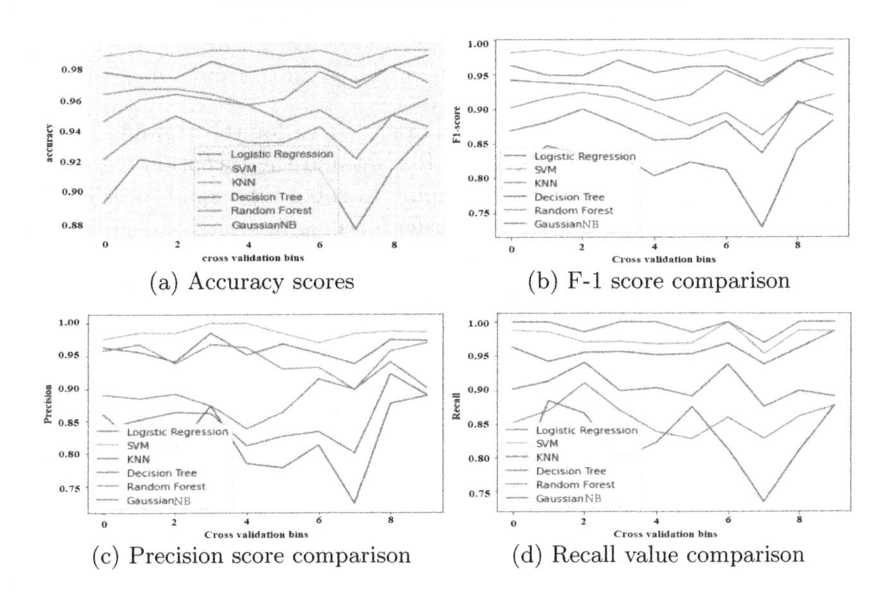

(a) Accuracy scores (b) F-1 score comparison

(c) Precision score comparison (d) Recall value comparison

Fig. 6. Performance metrics of different machine learning classifiers.

Furthermore, Cohen's kappa score measures reliability of the two classes. Cohen's kappa score is also to be found the highest for the SVM classifier and least for decision tree classifier, valued 0.985 and 0.817 respectively. The rest of the models are in between these two models. The classifiers trained using first dataset [19] are then tested on second dataset [20]. The accuracy on this test dataset for the traditional machine learning models is shown in the Table 3. The highest accuracy is achieved by the SVM valued 78.25%, proving its stability against other classifiers.

Table 3. Accuracy scores on the test dataset [20] using the models trained on dataset [19]

Model	ML/DL	Accuracy
SVM	**ML**	**78.25**
Logistic Regression	ML	78.12
KNN	ML	78.19
Decision Tree	ML	76.75
Random Forest	ML	77.32
GaussianNB	ML	73.67
ResNet 101 - Adam	DL	78.14
Inception V3 - Adam	DL	88.94
ResNet 101 - RMSProp	**DL**	**89.45**
Inception V3 - RMSProp	DL	87.69
DenseNet - Adam	DL	78.64
DenseNet - RMSProp	DL	78.14

For the deep learning models, the accuracy curve for the training and the validation is shown in Fig. 7 for both the ResNet and the Inception model with 'Adam' and 'RMSProp' as optimizer, keeping learning rate equal to 0.001. The accuracy values for dataset [19] using transfer learning methods of deep learning models are also showed in Table 2. The accuracy of the ResNet model with 'RMSProp' as optimizer is 94.55%, which is the highest among all the deep learning models. The Inception model with 'Adam' as an optimizer achieved an accuracy of 91.76% which is the least for the deep learning models. Overall, the machine learning models performed well based on the HOG features for the first dataset.

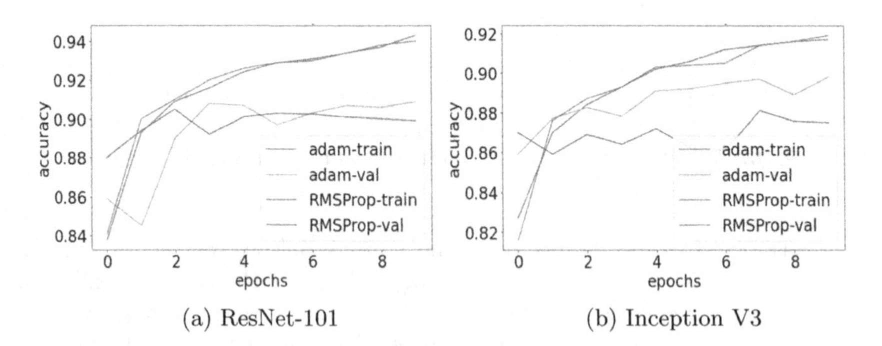

(a) ResNet-101 (b) Inception V3

Fig. 7. Accuracy plot during training and validation of deep learning models using different optimizers against the number of epochs for dataset [19].

For a comparative analysis, DenseNet model is also used and it is found to be giving better accuracy of 99.75% when tested on the same dataset, more than all of the trained models. Similar to machine learning models, the deep learning models trained using first dataset [19] are also then tested on second dataset [20]. The accuracy of the deep learning models achieved on this test dataset [20] is represented in Table 3. Compared to the machine learning models, deep learning models performed well to classify the ships images. This could be possibly due to the fact that, in the second dataset multiple tiny ships are present in single image and the HoG features are solely not able to capture the details.

Further, the explainability of deep learning models is visualized using Grad-CAM (Gradient-weighted Class Activation Mapping) [28] as shown in Fig. 8. Accordingly, the first column represents the input images, second column is the Grad-CAM heat map and the third column of each subfigure is Guided Grad-CAM. Grad-CAM uses class-specific gradient information that is flowing into the last layer of the convolutional layer of the model. It helps to visualize the region of interest of the model for a particular class. Guided Grad-Cam combines guided backpropagation with Grad-Cam which obtain more fine-grained activation maps. From the heatmaps generated, we can infer that the activation region to classify the given image as 'ship' or 'other' category is precise on both datasets. Thus, the deep learning models are capable to achieve stable results across multiple datasets.

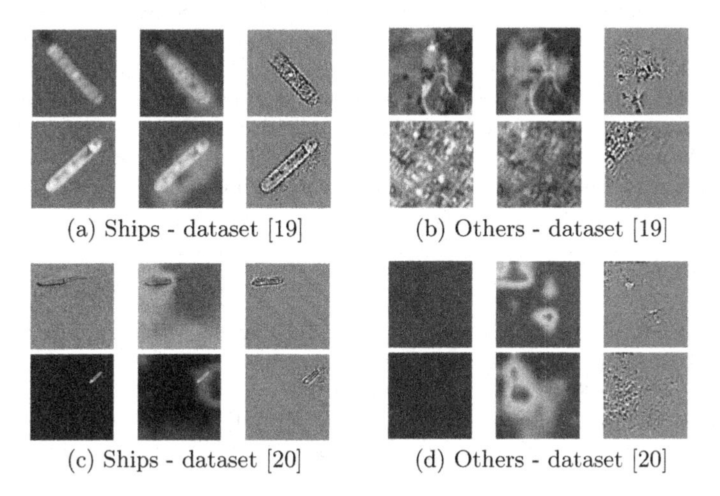

(a) Ships - dataset [19] (b) Others - dataset [19]

(c) Ships - dataset [20] (d) Others - dataset [20]

Fig. 8. Grad-Cam and Guided Grad-Cam visualization.

5 Conclusion

This paper describes a comparative analysis of machine learning and deep learning algorithms for classification of ships from satellite image dataset. SVM model with RBF kernel achieved an accuracy of 99.41% when tested on the trained ship dataset. However, when model is cross validated with another ship dataset it has drastically reduced to 78.25%. Similar trend is observed in other machine learning classifiers. This could be possibly due to the fact that the ship located on the second dataset images was different in size and shape (sometimes significantly) and there were many interference due to waves, clouds, etc. It is well known fact that the HoG features are scale and shape invariant. However, in images such as those from Aibus dataset, the distinction between ship and other closely related objects is challenging to the presence of multiple ships and other floating objects in the scene. Hence, the performance of the machine learning classifiers might be boosted when more feature descriptors are used such as texture based filters. This will be taken up as future work along with grid search and AutoML for the selection of best hyperparameters in these classifiers. On the other hand, ResNet-101 with RMSProp as optimizer showed an accuracy of 89.45% on the airship dataset while it's accuracy on the trained dataset was 93.03%. Therefore, the present comparative analysis reveals the capabilities of deep learning models against the trade-off of massive training image requirement and huge computational resources. Moreover, transfer learning approaches might not be suitable for categorization of ship type (cargo/non-cargo) since the intuitive understanding for the feature extraction greatly depends on the ground truth data. This leaves room to further explore a better deep learning model and opens up wider research in this domain.

Acknowledgements. Dr. P. Jidesh wish to thank the Science and Engineering Research Board, India for providing financial support under the research grant No. CRG/2020/000476. The other authors (Mr. Abhinaba and Ms. Smitha) wish to thank Ministry of Education (MoE) for providing financial support under the scholarship scheme to carryout the research at National Institute of Technology Karnataka, Surathkal. Authors would like to acknowledge the creators of the datasets [19,20] mentioned in this paper and making their dataset publicly available.

References

1. Lázaro, F., Raulefs, R., Wang, W., Clazzer, F., Plass, S.: VHF data exchange system (VDES): an enabling technology for maritime communications. CEAS Space J. **11**(1), 55–63 (2019). https://doi.org/10.1007/s12567-018-0214-8
2. Tetreault, B.: Use of the automatic identification system (AIS) for maritime domain awareness (MDA). In: Proceedings of OCEANS 2005 MTS/IEEE, vol. 2, pp. 1590–1594 (2005)
3. Ford, J.H., Peel, D., Kroodsma, D., Hardesty, B.D., Rosebrock, U., Wilcox, C.: Detecting suspicious activities at sea based on anomalies in automatic identification systems transmissions. PLOS ONE **13**(8), e0201640 (2018). https://doi.org/10.1371/journal.pone.0201640
4. Kamirul, K., Hasbi, W., Hakim, P., Syafrudin, A.H.: Automatic ship recognition chain on satellite multispectral imagery. IEEE Access **8**, 221918–221931 (2020). https://doi.org/10.1109/ACCESS.2020.3042702
5. Probst, P., Boulesteix, A.L., Bischl, B.: Tunability: importance of hyperparameters of machine learning algorithms. J. Mach. Learn. Res. **20**(53), 1–32 (2019)

6. Sarker, I.: Machine learning: algorithms, real-world applications and research directions. SN Comput. Sci. **2**, 160 (2021)
7. Yang, G., Li, B., Ji, S., Gao, F., Xu, Q.: Ship detection from optical satellite images based on sea surface analysis. IEEE Geosci. Remote Sens. Lett. **11**(3), 641–645 (2014). https://doi.org/10.1109/LGRS.2013.2273552
8. Chua, M., Aha, D.W., Auslander, B., Gupta, K., Morris, B.: Comparison of object detection algorithms on maritime vessels. Tech. rep, Naval Research Lab, Washington, D.C. (2014)
9. Mohamed, N.A., Zulkifley, M.A., Zaki, W.M.D.W., Hussain, A.: An automated glaucoma screening system using cup-to-disc ratio via simple linear iterative clustering superpixel approach. Biomed. Signal Process. Control **53**, 101454 (2019). https://doi.org/10.1016/j.bspc.2019.01.003
10. Cordova, A.W.A., Quispe, W.C., Inca, R.J.C., Choquehuayta, W.N., Gutierrez, E.C.: New approaches and tools for ship detection in optical satellite imagery. J. Phys. Conf. Ser. **1642**, 012003 (2020). https://doi.org/10.1088/1742-6596/1642/1/012003
11. Li, Z., Zhao, L., Han, X., Pan, M.: Lightweight ship detection methods based on yolov3 and densenet. Math. Prob. Eng. **2020**, 4813183 (2020). https://doi.org/10.1155/2020/4813183
12. Liu, C., Zhu, W.: An improved algorithm for ship detection in SAR images based on CNN. In: Pan, Z., Hei, X. (eds.) Twelfth International Conference on Graphics and Image Processing (ICGIP 2020), vol. 11720, pp. 63–71. International Society for Optics and Photonics, SPIE (2021). https://doi.org/10.1117/12.2589421.
13. Stofa, M.M., Zulkifley, M.A., Zaki, S.Z.M.: A deep learning approach to ship detection using satellite imagery. IOP Conf. Ser. Earth Environ. Sci. **540**, 012049 (2020). https://doi.org/10.1088/1755-1315/540/1/012049
14. Wang, Y., Ning, X., Leng, B., Fu, H.: Ship detection based on deep learning. In: 2019 IEEE International Conference on Mechatronics and Automation (ICMA), pp. 275–279 (2019). https://doi.org/10.1109/ICMA.2019.8816265
15. Gallego, A.J., Pertusa, A., Gil, P.: Automatic ship classification from optical aerial images with convolutional neural networks. Remote Sens. **10**(4) (2018). https://doi.org/10.3390/rs10040511
16. Liu, Y., Cui, H.Y., Kuang, Z., Li, G.: Ship detection and classification on optical remote sensing images using deep learning. ITM Web Conf. **12**, 05012 (2017). https://doi.org/10.1051/itmconf/20171205012
17. Zou, Z., Shi, Z.: Ship detection in spaceborne optical image with SVD networks. IEEE Trans. Geosci. Remote Sens. **54**(10), 5832–5845 (2016). https://doi.org/10.1109/TGRS.2016.2572736
18. Tang, J., Deng, C., Huang, G.B., Zhao, B.: Compressed-domain ship detection on spaceborne optical image using deep neural network and extreme learning machine. IEEE Trans. Geosci. Remote Sens. **53**(3), 1174–1185 (2015). https://doi.org/10.1109/TGRS.2014.2335751
19. Rhammell. Ships in satellite imagery-v1 (2018). https://www.kaggle.com/rhammell/ships-in-satellite-imagery/metadata. Accessed 06 Apr 2021
20. Airbus. Airbus ship detection challenge (2018). https://www.kaggle.com/c/airbus-ship-detection/overview. Accessed 25 Apr 2021
21. Eum, H., Bae, J., Yoon, C., Kim, E.: Ship detection using edge-based segmentation and histogram of oriented gradient with ship size ratio. IJFIS **15**, 251–259 (2015). https://doi.org/10.5391/IJFIS.2015.15.4.251

22. Evgeniou, T., Pontil, M.: Support vector machines: theory and applications. In: Paliouras, G., Karkaletsis, V., Spyropoulos, C.D. (eds.) ACAI 1999. LNCS (LNAI), vol. 2049, pp. 249–257. Springer, Heidelberg (2001). https://doi.org/10.1007/3-540-44673-7_12
23. Peng, C.Y.J., Lee, K.L., Ingersoll, G.M.: An introduction to logistic regression analysis and reporting. J. Educ. Res. **96**(1), 3–14 (2002). https://doi.org/10.1080/00220670209598786
24. Kaviani, P., Dhotre, S.: Short survey on Naive Bayes algorithm. Int. J. Adv. Res. Comput. Sci. Manag. **4** (2017)
25. He, K., Zhang, X., Ren, S., Sun, J.: Deep residual learning for image recognition. CVPR **2016**, 770–778 (2016). https://doi.org/10.1109/CVPR.2016.90
26. Szegedy, C., Vanhoucke, V., Ioffe, S., Shlens, J., Wojna, Z.: Rethinking the inception architecture for computer vision. CVPR2016 (2016). https://doi.org/10.1109/CVPR.2016.308
27. Huang, G., Liu, Z., Weinberger, K.Q.: Densely connected convolutional networks. arXiv preprint arXiv:1608.06993 (2016)
28. Selvaraju, R.R., Cogswell, M., Das, A., Vedantam, R., Parikh, D., Batra, D.: Grad-cam: visual explanations from deep networks via gradient-based localization. In: 2017 IEEE International Conference on Computer Vision (ICCV), pp. 618–626 (2017). https://doi.org/10.1109/ICCV.2017.74

Channel Difference Based Regeneration Architecture for Fake Colorized Image Detection

Shruti S. Phutke(✉) and Subrahmanyam Murala

Indian Institute of Technology Ropar, Punjab, India
2018eez0019@iitrpr.ac.in

Abstract. Selling of counterfeit/fraud goods like publishing the fake images of the products is increasing now a days due to advancement of multimedia technology for image editing applications. This work aims to detect the fake colorized images with a novel observation of channel difference maps. Here, we propose a novel deep channel difference network for fake colorized image detection/classification. Initially, an effective channel difference map (CDM) based auto-encoder is proposed for image regeneration. Here, to get distinguishing edge and color information between real and fake colorized images, CDM is proposed. After effective regeneration, the abstract version of learned features can be used for classification. Thus, we have proposed the transfer learning based network for fake colorized image detection (FCID) with learned encoder features of regeneration network. To the best of our knowledge, this is the first work with CDM based auto-encoder for FCID. The performance of the proposed network is tested on benchmark datasets and compared with the existing state-of-the-art methods for FCID in terms of half total error rate (HTER). The experimental results demonstrate that the proposed network is superior than the state-of-the-art approaches for FCID.

Keywords: Colorization · Channel difference maps · Image regeneration · Classification

1 Introduction

Advancement in image editing methods like image retouching [1], image inpainting [2,3], image watermarking [4], copy move or region duplication, splicing, and image colorization [6], etc. allows the user to modify or enhance the images easily. The image editing methods are providing the enhanced version of an image for many applications. But these may be used to falsify the identity of the scene in an image called as tampering an image for illegal purpose i.e., image forgery. Now-a-days, the researchers are working towards image forgery/manipulation detection [9,12,31]. One of the image editing technique i.e., image colorization can be used for advertising the products with different colors than original one (to make it visibly good) so that the customer will be encouraged to buy the fraud products. It becomes difficult to identify the actual and fake colorized image

B. Raman et al. (Eds.): CVIP 2021, CCIS 1568, pp. 73–84, 2022.
https://doi.org/10.1007/978-3-031-11349-9_7

easily because image colorization methods tamper the image in an efficient way. Hence, there is a dire need of detection of fake colorized images because they may be misused for illegal purpose such as selling of counterfeit/fraud products.

2 Literature Survey

Image colorization [33] is now-a-days being used for generating the fake images to throw someone of the scent. Addition of colors in the gray-scale image is termed as the colorization. This colorization can be done in two ways i.e., by human imagination (learning representations) or computerized (algorithm based) colorization. With human imagination, we can colorize an image by recalling the predefined colors, for example: the trees are green, the sky is blue, etc. The another way is digital colorization where we have to colorize a gray image using some algorithms. Wilson Markle invented the computerized colorization in 1970s [9]. The algorithm based colorization methods can be categorized into two different types i.e., supervised and fully automatic colorization [9]. The supervised digital colorization can be of scribble-based or dictionary based. In scribble based [17] colorization, the user scribbles are provided whereas in dictionary based colorization dictionary of color image is provided. In dictionary based method, the dictionary is prepared first and according to that colorizaiton of an image or video is done. In fully automatic method, the colorization can be done by modifying different color channels of input image [13,16,25,37] by using different modelling approaches (histogram regression) [23], by using regression approaches [19]. In [25], the authors modified the YCbCr color space to generate colorized images. The Markov random fields is used to model a color image in [23]. As the color in an image is considered an important content for various applications like color based classification [15], region segmentation [18], etc., the fake colorization degrades performance of these applications.

In literature [9,13,16,25,37] it is observed that the fake colorization methods adjust some colors of either RGB, HSV or YCbCr color channels. During this process, these methods enhances or disturbs some of the colors which are not visually distinguishable by the observer. Hence, the algorithms for fake colorization detection should identify these minuet changes present in the fake images. In [9], two hand-crafted methods (histogram and encoding based methods) have been proposed to detect (classify) the fake colorized images. Deep learning algorithms are gaining attention of the researchers due to its success in different applications such as action recognition [22,28,34], object segmentation [24,26,32,36], image super-resolution [40], image dehazing [8,11], depth estimation [14,20,30], image inpainting [39], *etc.* In [38], the authors utilized the WiSERNet [35] for detection of fake colorized images. They used pre-trained kernels of WiserNet to train the model for detection of fake colorized images. Here, we have utilized the concept of channel-wise parallel convolution [38] in the proposed framework. Even though fake image does not reflect any changes in its given form (RGB image), there will be some changes reflected in different maps (channel difference) of image which can be utilized to differentiate it from real one. For this, we have observed these changes in its channel difference map (CDM) which are then used to detect the real and fake colorized image.

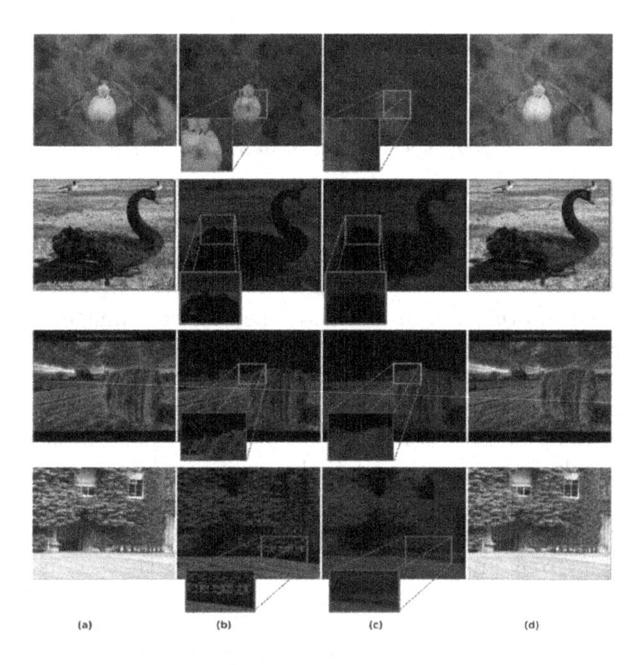

Fig. 1. (a) Real (RGB) images, (b) channel difference map of real images, (c) channel difference map of fake images, (d) fake (RGB) images (Color figure online)

2.1 Contributions of the Work

- A novel framework with channel difference map (CDM) is proposed for fake colorized image detection.
- An effective channel difference map based auto-encoder is proposed for image regeneration.
- The transfer learning based network is proposed for fake colorized image detection with learned encoder features of regeneration network.

Performance of the proposed network is tested on benchmark datasets and compared with the state-of-the-art methods for fake colorized image

3 Proposed Framework

3.1 Proposed Channel Difference Map

The advancement of technology in the field of computer vision for image generation has enabled to generate fake colorized images which are not visually distinguishable from real images. If the real image and corresponding fake image are displayed at the same time, person may be able to identify the fake and real one. But, if only a single image is provided, it is difficult to identify whether it is fake or real. So, it is necessary to make this distinguish ability independent of any reference. Here, to make this, it is observed that the channel difference map provides more information to distinguish between real and fake image as compared to normal color (RGB) image. The channel difference map (CDM) is formed as:

$$CDM = \Upsilon\{CDM_{RG}, CDM_{GB}, CDM_{RB}\} \tag{1}$$

where, Υ is concatenation operation, $CDM_{RG} = R - G$, $CDM_{GB} = G - B$, $CDM_{RB} = R - B$ and R, G, B are Red, Green and Blue channels of input image respectively. The sample CDM for real and fake colorized images are shown in Fig. 1. From Fig. 1, it is easily observed that the channel difference map provides more distinguished information in real and fake colorized image. Thus, multiple observations are obtained from generated CDM:

- **Observation I:** The edges of CDM for the fake images are blurred as compared to real images. This is may be because the colorization methods may undergo some filtering operations to color an image (refer 1st and 2nd row of Fig. 1).
- **Observation II:** The color shade distinguishing ability of fake image is less as compared to real image. This may be because there is no color purity present in the fake images (refer 3rd and 4th row of Fig. 1).

These observations motivated us to develop an architecture, based on difference image for fake colorized image detection. We have integrated these observations with auto-encoder based image regeneration followed by fake colorized image detection framework.

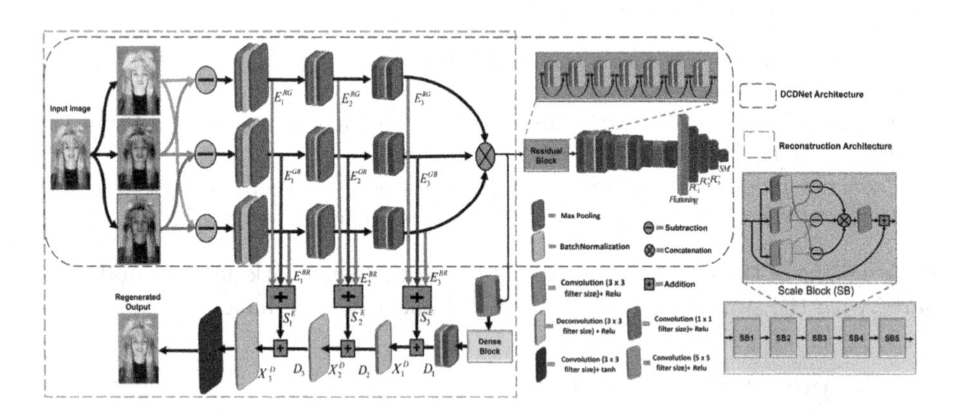

Fig. 2. Proposed architecture for image regeneration and fake colorized image detection. (Color figure online)

3.2 Image Regeneration Network

Auto-encoders are being used in various applications such as image de-noising [10], image enhancement [27], image retrieval [5], image classification [21], etc. In the proposed approach, we have designed a CDM based auto-encoder for image regeneration (real and fake colorized) task. Figure 2 depicts the proposed auto-encoder network for image regeneration (see red dotted coloured box). Each channel of CDM from input image gives distinguished information for real and fake colorized image (observation II). So, to avail this advantage, we have proposed the channel-wise parallel encoders to learn the efficient color features. Each encoder layer comprises of `Convolution` → `BatchNormalization` → `Maxpool`. The lth level encoder in P^{th} path (P ∈ (RG, GB, BR)) with channels ($f = 16$), max-pooling with $l \times f$ stride factor is represented as $E^{P}_{l\_l \times f}$.

These concatenated feature maps are forwarded to the dense module which consists of five cascaded scale blocks (SB) (please refer scale block in Fig. 2). The purpose behind designing the SB is to correlate the color and edge information from each channel for effective image regeneration. The decoder module is used after the dense block to regenerate the input image back. As CDM based convolution learns color based features, the skip connections are provided from respective parallel encoder layers to subsequent decoder. The sample visual results of the proposed CDM based auto encoder are shown in Fig. 3.

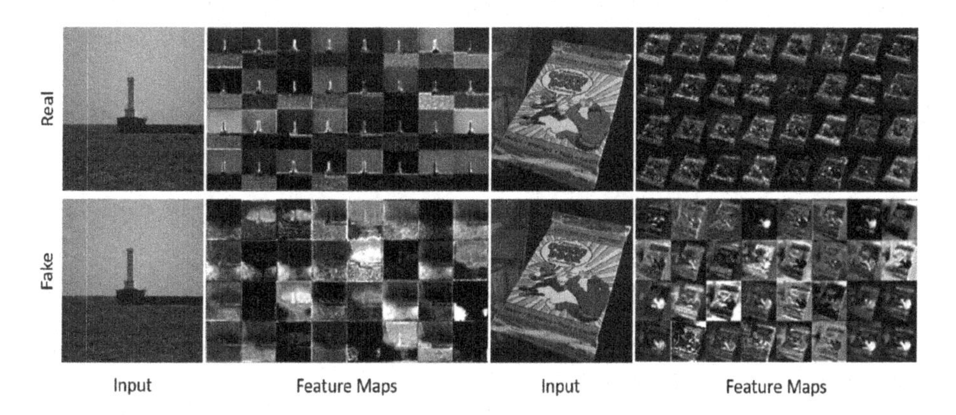

Fig. 3. Sample Feature maps (at the output of CDM before dense block) for real and fake colorized images.

3.3 Fake Colorized Image Detection Network (DCDNet)

Training of the detection network with random weights may not give effective detection results (*see Sect.* 6.2 *for more details*). To overcome this limitation, we have used the concept of transfer learning. Also, the proposed regeneration network is able to reconstruct image back very accurately. Thus, we have used the trained encoder weights of regeneration network (explained in Sect. 3.2) in the network for fake colorized image detection (*see blue dotted box from Fig.* 2). The learned independent encoder feature maps need to correlate with each other to form efficient feature vector for classification/detection. As observed in Fig. 1, there are blurred edges in the fake colorized images. This will provide the efficient features for detecting the real and fake colorized images. So, concatenated output of the three learned parallel paths is then passed through the residual block (RB) to extract correlated features (*as shown in blue box of Fig.* 2). The output of RB is then processed through the three encoder layers (as explained in Sect. 3.2) followed by fully connected layers for detection of real and fake colorized images.

4 Datasets and Evaluation Metric

Datasets: For the training of the regeneration and detection network, we have collected the datasets from [9]. In [9], authors have labelled the datasets as D1 to D7. The D1 consists of 10k natural images from ImageNet dataset [7] and their respective fake colorized images from ctest10k dataset (colorized using [16]). This category consists furniture, animals, outdoor scenes and humans, etc. images. The D2, D3 and D4 consists of randomly selected images from ImageNet and their respective fake colorized images which are colorized using [16,37] and [13] respectively. There is no overlap between images in D2 to D4. Similarly, the fake colorized images in dataset D5, D6 and D7 are generated using [16,37] and [13] respectively. The natural images for these datasets (D5–D7) are selected from Oxford building dataset [29]. Each of the dataset from D2–D7 consist of 2k images for training and 1k for testing.

Evaluation Parameter: The sofmax layer with two neurons is used at last stage of detection network to classify the input image into two classes (real and fake colorized). In all the experiments, the evaluation parameter, Half Total Error Rate (HTER) is considered same as that of [9]. The HTER can be calculated using Eq. (2).

$$HTER = \frac{\frac{FP}{TN+FP} + \frac{FN}{TP+FN}}{2} \tag{2}$$

where, TP, TN, FP and FN are true positive, true negative, false positive and false negative values respectively.

5 Training of the Proposed Networks

Training of the proposed approach is divided into two steps:

- End-to-end training of image regeneration network (real to real and fake to fake).
- Training of detection network by using learned encoder weighs of regeneration network, as initial weights for fake colorized image detection.

Training of Image Regeneration Network: The regeneration network is trained on training set of D2–D7 datasets for real-to-real and fake-to-fake image regeneration. Here, the maximum number of epochs are 100 with the batch size of 16, stochastic gradient descent (SGD) optimizer, MSE loss are used. The encoder weights of the trained regeneration network on every dataset D2–D7 are stored and used transfer learning of the detection network.

Training of Detection Network: For respective dataset training of the detection network, the respective dataset trained encoder weights of regeneration network are used as initial weights. For training of the detection network, SGD optimizer with 0:01 learning rate and mean squared error (MSE) loss is used. As the transfer learning is used, the maximum number of epochs are kept 50 with batch size = 16 and early stopping criteria with validation accuracy as monitoring parameter. Weight parameters are updated on NVIDIA DGX station with processor 2.2 GHz, Intel Xeon $E5-2698$, NVIDIA Tesla V100 1×16 GB GPU, tensorflow library.

Table 1. Comparison of state-of-the-art methods for fake colorized image detection. Lower value of % HTER is better. *Note: PM is Proposed Method*

Original images from →		ImageNet dataset			Oxford building dataset		
Testing ↓	Training →	D2	D3	D4	D5	D6	D7
	Methods ↓	Intra			Inter		
D2	[9]	22.3	25.1	38.15	43.5	–	–
	[38]	0.65	6.2	5.2	9.45	22.25	16.6
	PM	0.79	7.5	**2.4**	**3.75**	**10.5**	**3.86**
D3	[9]	23.65	22.85	36.15	–	30.25	–
	[38]	12.75	1.6	3.7	12.75	11.8	11.35
	PM	16.07	1.76	6.0	13.38	**5.47**	**8.3**
D4	[9]	31.7	34.25	17.3	–	–	23.15
	[38]	5.65	1.9	1.0	9.4	12.15	9.35
	PM	**4.5**	2.74	1.18	**4.23**	**6.74**	**3.66**
	Methods ↓	Inter			Intra		
D5	[9]	22.85	–	–	–	–	–
	[38]	1.85	8.7	5.9	1.0	5.4	1.9
	PM	**1.57**	**6.71**	**0.84**	**0.1**	**3.16**	**0.8**
D6	[9]	–	21.5	–	–	–	–
	[38]	14.3	6.15	7.6	4.15	0.95	1.55
	PM	**11.29**	**0.2**	2.13	**3.64**	**0.3**	2.6
D7	[9]	–	–	20.2	–	–	–
	[38]	9.7	6	5.5	2.5	1.6	1.1
	PM	**2.36**	**5.18**	**1.53**	**2.28**	**5.1**	1.8

6 Results and Discussion

6.1 Result Analysis

In this section, the evaluation of the proposed network for fake colorized image detection is conducted on D2–D7 datasets. The performance of the proposed detection network is evaluated in terms of HTER (%) and compared with the existing state-of-the- art methods [9,38] for fake colorized image detection. The model trained on training splits of images from ImageNet D2–D4 are tested on testing splits of D2–D4. Similarity, model trained on training splits of images from Oxford building dataset D5–D7 are tested on testing spits D5–D7 respectively. This experiment is conducted to verify the effectiveness of proposed detection network on **intra-dataset** i.e., provided train-test splits of D2–D4 (original images from ImageNet) and train-test of D5–D7 (original images from Oxford Building Dataset).

Along with the performance evaluation on intra-dataset, the **inter-dataset** (trained on images from Imagenet and tested on images from Oxford Building dataset, vice-versa) performance evaluation is conducted. For this, each of the trained detection model on D2–D4 is tested on each of the testing splits D5-D7 and vice versa. All these intra and inter-datasets quantitative results are given in Table 1. Also, Fig. 4 shows some of the detection results on D2–D7 datasets. From Table 1 it is clear that the proposed detection network performs better than existing state-of-the-art methods.

Fig. 4. The detection results for fake colorized image detection on D2–D7 ((a)–(f)) respectively. The actual class of the image is written on the image on right-up corner. The boundary color indicates the correct detection (green) and incorrect detection (red). (Color figure online)

6.2 Ablation Study

Ablation study is performed on effect of weight initialization and number of scale blocks in dense module. For both analysis, the training splits of D2–D7 and testing split of D5 is used. **Does the learned encoder weights of regeneration network benefits the detection network?** To do this, the performance of detection network is examined with and without learned weights of regeneration network. The effectiveness comparison of detection network without (Random) and with (PM) learned encoder weights of regeneration network is given in Table 2. From Table 2, it is evident that the learned encoder weights help the detection network to effectively distinguish between real and fake colorized images.

Table 2. Comparison of detection network tested on dataset D5

Train	D2	D3	D4	D5	D6	D7
Random	1.58	10.50	6.20	0.45	17.20	1.48
PM	**1.57**	**6.71**	**0.84**	**0.1**	**3.16**	**0.8**
SB3	**1.47**	14.26	2.2	0.2	**2.99**	2.43
SB5	1.57	**6.71**	**0.84**	**0.1**	3.16	**0.8**
SB7	1.69	8.65	1.98	0.4	4.04	0.92

Now, it is very important to have superior performance of image regeneration network. The regeneration network will give the effective performance if we are able to efficiently correlate the parallel processed encoder features. To do this, we have proposed a novel dense module consisting cascaded scale blocks. **How many scale blocks work effectively?** This is verified by conducting the experimental analysis with three (SB3), five (SB5) and seven (SB7) scale blocks in the dense module of regeneration network. So, the regeneration network is trained with SB3, SB5 and SB7. These learned encoder weights of regeneration network (consisting of SB3/SB5/SB7) are then used to train the detection network. The evaluation of the detection network with learned encoder weights of regeneration network with SB3, SB5 and SB7 is given in Table 2. From Table 2, it is evident that, the detection network with encoder weights from regeneration network consisting of SB5 gives superior performance.

7 Conclusion

In this paper, we have designed a network for fake colorized image detection. The concept of the channel difference map with auto-encoder is proposed to distinguish the real and fake colorized image. To get the superior performance of the detection network, the learned encoder weights of regeneration network are used as initial weights for proposed detection network. The effectiveness of the proposed network is examined on benchmark datasets for fake colorized image detection. The experimental analysis and ablation study demonstrates that the proposed network is superior as compared to the existing state-of-the-art methods for fake colorized image detection.

References

1. Cao, G., Zhao, Y., Ni, R., Li, X.: Contrast enhancement-based forensics in digital images. IEEE Trans. Inf. Forensics Secur. **9**, 515–525 (2014)
2. Ghorai, M., Mandal, S., Chanda, B.: A group-based image inpainting using patch refinement in MRF framework. IEEE Trans. Image Process. **27**, 556–567 (2017)
3. Liu, J., Yang, S., Fang, Y., Guo, Z.: Structure-guided image inpainting using homography transformation. IEEE Trans. Multimedia **20**, 3252–3265 (2018)
4. Liu, X.L., Lin, C.C., Yuan, S.M.: Blind dual watermarking for color images authentication and copyright protection. IEEE Trans. Circ. Syst. Video Technol. **28**, 1047–1055 (2016)
5. Bhunia, A.K., Perla, S.R.K., Mukherjee, P., Das, A., Roy, P.P.: Texture synthesis guided deep hashing for texture image retrieval. In: 2019 IEEE Winter Conference on Applications of Computer Vision (WACV), pp. 609–618. IEEE (2019)
6. Cheng, Z., Yang, Q., Sheng, B.: Colorization using neural network ensemble. IEEE Trans. Image Process. **26**, 5491–5505 (2017)
7. Deng, J., Dong, W., Socher, R., Li, L.J., Li, K., Fei-Fei, L.: ImageNet: a large-scale hierarchical image database. In: 2009 IEEE Conference on Computer Vision and Pattern Recognition, pp. 248–255. IEEE (2009)
8. Akshay, D., Hambarde, P., Patil, P., Murala, S.: Deep underwater image restoration and beyond. IEEE Signal Process. Lett. **27**, 675–679 (2020)
9. Guo, Y., Cao, X., Zhang, W., Wang, R.: Fake colorized image detection. IEEE Trans. Inf. Forensics Secur. **13**, 1932–1944 (2018)
10. Hashisho, Y., Albadawi, M., Krause, T., von Lukas, U.F.: Underwater color restoration using U-Net denoising autoencoder. arXiv preprint arXiv:1905.09000 (2019)
11. Akshay, D., Biradar, K.M., Patil, P.W., Hambarde, P., Murala, S.: Varicolored image de-hazing. In: Proceedings of the IEEE/CVF Conference on Computer Vision and Pattern Recognition, pp. 4564–4573 (2020)
12. Hosseini, M.D.M., Kirchner, M.: Unsupervised image manipulation localization with non-binary label attribution. IEEE Signal Process. Lett. **26**, 976–980 (2019)
13. Iizuka, S., Simo-Serra, E., Ishikawa, H.: Let there be color!: joint end-to-end learning of global and local image priors for automatic image colorization with simultaneous classification. ACM Trans. Graph. (TOG) **35**, 110 (2016)
14. Praful, H., Dudhane, A., Murala, S.: Single image depth estimation using deep adversarial training. In: 2019 IEEE International Conference on Image Processing (ICIP), pp. 989–993. IEEE (2019)
15. Khan, R., Hanbury, A., Stottinger, J., Bais, A.: Color based skin classification. Pattern Recogn. Lett. **33**, 157–163 (2012)
16. Larsson, G., Maire, M., Shakhnarovich, G.: Learning representations for automatic colorization. In: Leibe, B., Matas, J., Sebe, N., Welling, M. (eds.) ECCV 2016. LNCS, vol. 9908, pp. 577–593. Springer, Cham (2016). https://doi.org/10.1007/978-3-319-46493-0_35
17. Levin, A., Lischinski, D., Weiss, Y.: Colorization using optimization. ACM Trans. Graph. (ToG) **23**, 689–694 (2004)

18. Lie, M.M., Borba, G.B., Neto, H.V., Gamba, H.R.: Joint upsampling of random color distance maps for fast salient region detection. Pattern Recogn. Lett. **114**, 22–30 (2018)
19. Liu, S., Zhang, X.: Automatic grayscale image colorization using histogram regression. Pattern Recogn. Lett. **33**, 1673–1681 (2012)
20. Hambarde, P., Dudhane, A., Patil, P.W., Murala, S., Dhall, A.: Depth estimation from single image and semantic prior. In: 2020 IEEE International Conference on Image Processing (ICIP), pp. 1441–1445. IEEE (2020)
21. Luo, W., Li, J., Yang, J., Xu, W., Zhang, J.: Convolutional sparse autoencoders for image classification. IEEE Trans. Neural Netw. Learn. Syst. **29**, 3289–3294 (2017)
22. Chaudhary, S., Murala, S.: Deep network for human action recognition using Weber motion. Neurocomputing **367**, 207–216 (2019)
23. Noda, H., Korekuni, J., Niimi, M.: A colorization algorithm based on local map estimation. Pattern Recogn. **39**, 2212–2217 (2006)
24. Patil, P.W., Dudhane, A., Chaudhary, S., Murala, S.: Multi-frame based adversarial learning approach for video surveillance. Pattern Recogn. **122**, 108350 (2022)
25. Noda, H., Niimi, M.: Colorization in YCbCr color space and its application to jpeg images. Pattern Recogn. **40**, 3714–3720 (2007)
26. Patil, P.W., Biradar, K.M., Dudhane, A., Murala, S.: An end-to-end edge aggregation network for moving object segmentation. In: Proceedings of the IEEE/CVF Conference on Computer Vision and Pattern Recognition, pp. 8149–8158 (2020)
27. Park, S., Yu, S., Kim, M., Park, K., Paik, J.: Dual autoencoder network for retinex-based low-light image enhancement. IEEE Access **6**, 22084–22093 (2018)
28. Chaudhary, S., Murala, S.: Depth-based end-to-end deep network for human action recognition. IET Comput. Vis. **13**(1), 15–22 (2019)
29. Philbin, J., Chum, O., Isard, M., Sivic, J., Zisserman, A.: Object retrieval with large vocabularies and fast spatial matching. In: Proceedings of the IEEE Conference on Computer Vision and Pattern Recognition (2007)
30. Hambarde, P., Murala, S.: S2DNet: depth estimation from single image and sparse samples. IEEE Trans. Comput. Imaging **6**, 806–817 (2020)
31. Shi, Z., Shen, X., Chen, H., Lyu, Y.: Global semantic consistency network for image manipulation detection. IEEE Sig. Process. Lett. **27**, 1755–1759 (2020)
32. Patil, P.W., et al.: An unified recurrent video object segmentation framework for various surveillance environments. IEEE Trans. Image Process. **30**, 7889–7902 (2021)
33. Sugawara, M., Uruma, K., Hangai, S., Hamamoto, T.: Local and global graph approaches to image colorization. IEEE Sig. Process. Lett. **27**, 765–769 (2020)
34. Chaudhary, S., Murala, S.: TSNet: deep network for human action recognition in hazy videos. In: 2018 IEEE International Conference on Systems, Man, and Cybernetics (SMC), pp. 3981–3986 (2018). https://doi.org/10.1109/SMC.2018.00675
35. Zeng, J., Tan, S., Liu, G., Li, B., Huang, J.: WISERNet: wider separate then-reunion network for steganalysis of color images. IEEE Trans. Inf. Forensics Secur. **14**, 2735–2748 (2019)
36. Patil, P.W., Murala, S.: MSFgNet: a novel compact end-to-end deep network for moving object detection. IEEE Trans. Intell. Transp. Syst. **20**(11), 4066–4077 (2018)
37. Zhang, R., Isola, P., Efros, A.A.: Colorful image colorization. In: Leibe, B., Matas, J., Sebe, N., Welling, M. (eds.) ECCV 2016. LNCS, vol. 9907, pp. 649–666. Springer, Cham (2016). https://doi.org/10.1007/978-3-319-46487-9_40

38. Zhuo, L., Tan, S., Zeng, J., Lit, B.: Fake colorized image detection with channel-wise convolution based deep-learning framework. In: 2018 Asia-Pacific Signal and Information Processing Association Annual Summit and Conference (APSIPA ASC), pp. 733–736. IEEE (2018)
39. Phutke, S.S., Murala, S.: Diverse receptive field based adversarial concurrent encoder network for image inpainting. IEEE Sig. Process. Lett. **28**, 1873–1877 (2021)
40. Mehta, N., Murala, S.: MSAR-Net: multi-scale attention based light-weight image super-resolution. Pattern Recogn. Lett. **151**, 215–221 (2021)

DenseASPP Enriched Residual Network Towards Visual Saliency Prediction

Shilpa Elsa Abraham(✉) and Binsu C. Kovoor

Division of Information Technology, Cochin University of Science and Technology, Kerala, India
shilpamat75@gmail.com

Abstract. Predicting regions of interest, otherwise called salient regions tend to top out with the rise of deep learning techniques. Although convolutional neural networks have evaded the domain to let it reach newer heights, there still exists room for improvement on how to integrate the hierarchical features efficiently. In fact, the rich features at multiple spatial scales are found to be powerful towards accurate prediction. This paper proposes a novel end-to-end visual saliency prediction technique, based on DenseASPP (Dense Atrous Spatial Pyramid Pooling) and residual connections. It enriches the multi-scale contextual features via DenseASPP module, that gathers information via dense connections across multiple scales. Further, incorporation of residual connections between encoder and decoder blocks allow learning of more robust features that can result in better prediction. The model is trained on the largest dataset for saliency prediction, SALICON and experimental results on two public datasets, OSIE and PASCAL-S verify the effectiveness of the proposed framework compared with state-of-the-art results.

Keywords: Saliency prediction · Eye fixation prediction · Convolutional neural networks · Atrous spatial pyramid pooling · Residual connection

1 Introduction

Human visual System (HVS) is so impressive that it can quickly prioritize scenes and objects in a single glance. Focus is selectively inclined towards most distinctive regions in the scene, while ignoring trivial ones. This is termed as visual attention and is an active research area in areas of psychology, neuroscience and computer vision. In computer vision, it is often referred to as visual saliency prediction.

Early studies on saliency [1,9,17,47] targeted low level visual features such as colour, contrast, intensity, orientation to predict the regions of interest. However, advent of deep learning reformed the research in the domain of computer vision. Vision tasks like image classification [22] and semantic segmentation [32] are the spheres where deep neural networks were initially explored. Inspired by

B. Raman et al. (Eds.): CVIP 2021, CCIS 1568, pp. 85–96, 2022.
https://doi.org/10.1007/978-3-031-11349-9_8

Fig. 1. A visualization of the saliency prediction task. Top row illustrates input images and bottom row represents corresponding saliency maps as probability distributions.

their success, initial attempts of applying deep neural networks for eye fixation prediction have been made by saliency researchers [26,39]. Convolutional neural networks (CNN) are efficient towards extracting rich features from raw images. The availability of large scale annotated datasets like SALICON [19] boosted the performance further. Figure 1 illustrates some examples of saliency prediction.

Contextual information is crucial towards efficient saliency prediction. Better learning of contextual features allow the network to perceive the entirety of visual scene. This can best be captured via hierarchical representations at multiple spatial scales. This paper proposes an effective encoder-decoder network coupled with DenseASPP module. The DenseASPP module connects multiple atrous convolutional layers in a dense manner, so as to consider larger scale range densely. This enables extracting of complex semantic features at multiple spatial scales in a dense range to result in better saliency inference. Further, to fully exploit the powerful hierarchical features, residual connections are employed to capture multi-level saliency response. The prediction accuracy on unseen test images reports that the model is capable enough to compete with state-of-the-art approaches.

The remainder of this paper is organised as follows: Sect. 2 presents a review of related saliency works. Section 3 details the proposed architecture and Sect. 4 outlines the experimental results along with ablation studies. Finally, conclusion is made in Sect. 5.

2 Related Work

Multi-level Feature Integration: The hierarchical features are best extracted with the aid of CNNs. The convolutional layers at multiple levels capture image representations at multiple information level, which when fused coherently can thrive saliency prediction to greater heights. Cornia *et al.* [6] utilizes feature maps from three levels, constituted as medium and high level features to learn saliency features efficiently. Another model, FUCOS [2] is implemented as a fully convolutional network, along with skip connections across certain convolution blocks to better transfer relevant features.

Encoder-decoder (ED) structure is a prominent kind of saliency inference architectural style [6,16,34,40,45] that utilizes multi-level features. This kind of network is associated with feature extraction in the encoder phase and reconstruction in the decoder segment. As seen in literature, different variants of encoder-decoder structures exist for saliency prediction including symmetric ED with skip connections [34], asymmetric ED [45], fused features from encoder blocks to decoder part [40], residual decoder [16] etc.

Multi-scale Feature Integration: The task of saliency prediction can be addressed better if semantic context at multiple scales is efficiently handled. Inspired by the success of GoogLeNet [38], saliency researchers [12,24,25,41, 44] attempt on applying inception module to capture multi-scale features. It works by applying filters of different receptive fields (3×3, 5×5, 7×7) on the input feature tensor, hence generating a multi-scale representation of the hierarchical features. Another prominent module that is capable of capturing multi-scale features is Atrous Spatial Pyramid Pooling (ASPP) [4]. Kroner *et al.* [23] implements ASPP for saliency prediction and is able to achieve promising results, while diminishing the overhead of heavy computation cost associated with large sized kernel. The proposed model is formulated utilizing multi-level feature integration and multi-scale feature integration.

3 Proposed Model

The model follows an encoder- decoder based structure, augmented with a module designed for semantic segmentation [43] customized towards the task of saliency prediction to predict fixation density maps. The overview of the model is illustrated in Fig. 2. In general, the model can be viewed as composed of three basic modules: an encoder network to capture semantically rich features from the input images, a DenseASPP to aggregate relevant contextual information and a decoder part to efficiently generate saliency map as probability distributions.

Encoder Network. The encoder part transforms the three dimensional input image into multiple stacks of feature maps through levels of convolution operations. This corresponds to the features of the input image at different hierarchical levels, ranging from edges and blobs in the initial level to semantic information in the final levels.

The proposed model begins with a VGG-16 network [37] pretrained on ImageNet dataset, to act as the primary feature extractor. An input image fed through *five blocks* of VGG-16 backbone network produces feature representations at five levels, denoted as f_i where i = 1, 2,...,5. As image to image understanding requires spatial dimension to be preserved, the model discourages excessive down sampling operations in feature extraction network and therefore

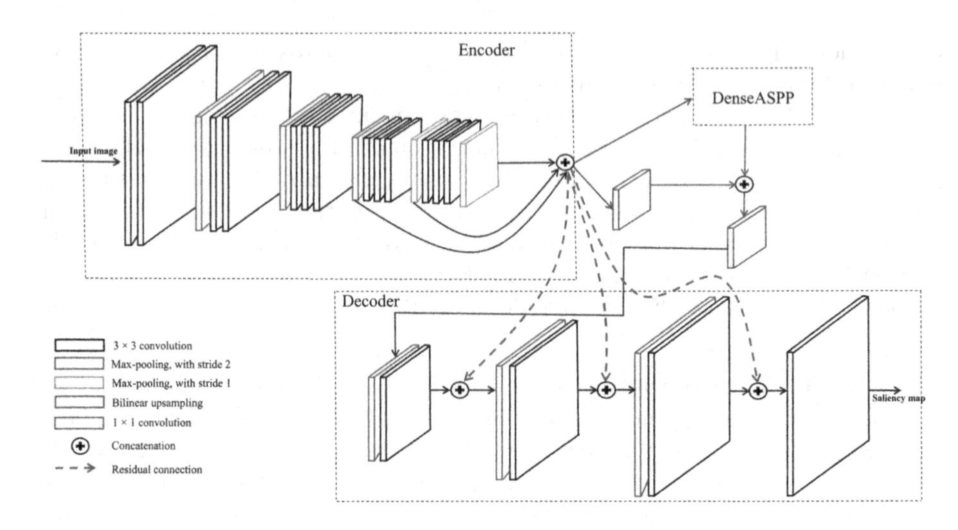

Fig. 2. Architecture of the proposed model

uses maxpooling with stride 1 in the last two convolutional blocks. Also, fully connected layers at the end of VGG-16 network, that is inherent to image classification task is dropped to cater the need for reconstruction.

For saliency prediction task, it is high-level semantic information that is relevant and hence the encoder output, E is framed as concatenating feature maps from third, fourth and fifth convolutional blocks. i.e.,

$$E = f_3 \oplus f_4 \oplus f_5 \tag{1}$$

where f_i represents feature representation at i^{th} level and $\oplus$ represents concatenation operation. Each of the f_i represents tensor (feature map) obtained after maxpooling in the corresponding block. This encoder representation forms the input to the next module, DenseASPP.

Dense Atrous Spatial Pyramid Pooling (DenseASPP)

Atrous Spatial Pyramid Pooling (ASPP): In real scenes, the salient objects usually vary in their sizes. Therefore, it is sound to consider multiple scales of receptive fields to get a global contextual view of the input image. Atrous spatial pyramid pooling proposed by Chen *et al.* [4] is effective at resampling such semantic information distributed across multiple scales. This is facilitated by applying atrous convolutions with different dilation rates in parallel upon the input feature maps. This enables the network to learn from wider receptive fields while maintaining resolution of feature maps and not increasing the number of parameters.

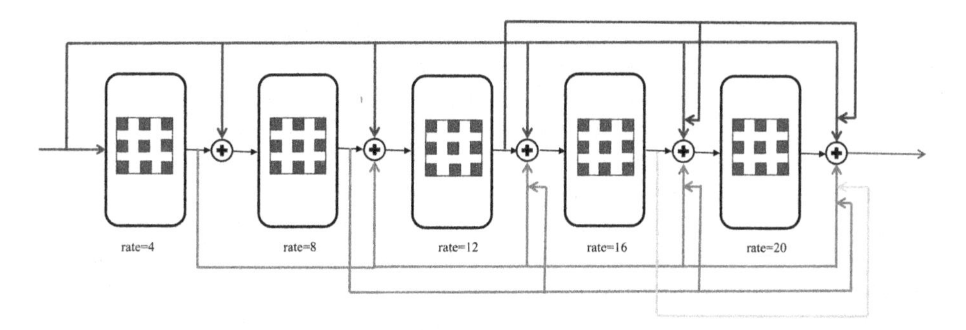

Fig. 3. Structure of DenseASPP used in the model. The concatenation operation after each atrous convolution layer takes into account encoder output as well as all previous outputs

DenseASPP: In fact, the notion of ASPP arose from semantic segmentation task, from where the module has undergone several updations subsequently to result in more robust structures. Yang *et al.* [43] put forward DenseASPP that cascades layers of atrous convolutions in a linear dense manner. Figure 3 illustrates the structure of DenseASPP employed in the proposed architectural framework. The network is formed with convolutions of increasing dilation rates arranged in a linear fashion and output of each atrous convolutional layer is concatenated with the input feature map and outputs from all preceding layers. If $A_{W,d}()$ represents atrous convolution with W denoting filter size and d representing dilation rate, each unit block (atrous convolutional layer) in DenseASPP may be formulated as follows:

$$y_l = A_{W,d_l}\left([y_{l-1} \oplus y_{l-2} \oplus ... \oplus y_0]\right) \tag{2}$$

where d_l denotes the dilation rate of layer l and $\oplus$ represents concatenation of feature maps at different layers.

It may be noted that DenseASPP is not just associated with capturing multi-scale features at a larger scale range, but also covers those scales densely. Hence, it caters the issues caused due to variable object sizes and positions efficiently, without increasing model size drastically.

In the proposed model, the encoder representation is passed through DenseASPP module with dilation rates 4, 8, 12, 16 and 20 arranged in successive manner. Furthermore, a global average pooling is done on encoder output to facilitate incorporating scene context [5]. This context information is integrated to the multi-scale feature representation of DenseASPP block.

Decoder Network. The decoder network operates on low resolution feature maps from DenseASPP module and performs upsampling in order to restore the original input image dimension. A couple of upsampling techniques have been adopted in previous saliency prediction works. Wang *et al.* [40], He *et al.* [16] and Zaibihi *et al.* incorporated transposed convolutions to upsample the coarse feature tensor. Bilinear interpolation is yet another powerful upsampling technique [6,30] that is found to prevent checkerboard artifacts in the upsampled feature tensor that arises with transposed convolution.

The proposed framework employs three upsampling blocks, each consisting of bilinear upsampling and 3 × 3 convolution. The upsampling technique doubles the size of the feature map in each block and hence restores the dimension 8 times to recover the dimension lost in encoding stage.

The residual connections/skip connections constitute a coherent architectural style towards better learning of features. As stated by He *et al.* [15], it enables convolutional layers to learn residual functions that accelerates the training process, yet reducing computation costs. In the proposed model, there have been skip connections from encoder output towards each decoder block. This allows feature representation at semantic levels (third, fourth and fifth convolution blocks of VGG backbone) to propagate to saliency map reconstruction phase. Finally, a 3 × 3 convolution is applied with a single filter to yield continuous saliency distribution.

4 Experimental Analysis

4.1 Saliency Prediction Datasets

For training and testing the saliency model, three public datasets have been considered, which vary in image content and diversity.

SALICON [19]: It contains 10000 training, 5000 validation and 5000 test images of rich contextual information taken from Microsoft COCO dataset [29]. Currently, this is the largest dataset available for the task and therefore best suits training of deep CNN models.

OSIE [42]: It is composed of 700 images, collected by allowing 15 human subjects to gaze at each image. It is characterized by the presence of objects from multiple domains, which makes the dataset more demanding.

PASCAL-S [28]: It contains 850 natural images obtained from the validation set of PASCAL VOC 2010 [10]. All images are subject to viewing for 8 observers and are represented by cluttered background and complex foreground aspects.

4.2 Implementation Details

A pretrained VGG-16 network is used as the backbone of the network. For training, the SALICON images are resized to a reduced dimension of 240 × 320 pixels. A batch size of one is assumed, along with initial learning rate of 10^{-5}. The network attempts to minimize the loss function utilizing Adam optimizer [21]. No data augmentation has been used for training purpose. The network is implemented in Python on Tensorflow backend.

Loss Function: The selection of loss function is crucial towards efficient training. The model uses Kullback-Leibler Divergence (KL-Div) as the loss function. It computes the difference between the probability distribution corresponding to predicted saliency map (S) and the ground truth (GT). Assuming that the distribution of G is denoted as P with a probability distribution function p_k and that of S is represented by Q with q_k density function, the KL-Divergence is defined as follows:

$$KL(P, Q) = \sum_{k} q_k \, log \, \frac{p_k}{q_k} \tag{3}$$

The network is trained with the objective to bring down the metric value closer to 0, inferring that the two distributions are as close as possible.

4.3 Comparison Results

The model is evaluated based on five classic saliency prediction metric: Area Under Curve (AUC), Correlation Coefficient (CC), Normalized Scanpath Saliency (NSS), shuffled AUC (sAUC) and Similarity measure (Sim). The readers can refer [36] for deeper understanding on these metric. The quantitative comparison of results are made on two independent datasets, OSIE and PASCAL-S to see how well the model behaves on unseen images. Figure 4 summarizes the results of the proposed model on OSIE dataset, compared with traditional saliency models, including Itti [17], GBVS [14], Judd [20], AWS [11],BMS [46] as well as deep learning models including eDN [39], DeepGaze I [26], MrCNN [31], DeepFix [24], ConEd [23], ACNet [27] and Cross [35]. It is evident that the proposed approach outperforms other saliency prediction methods with respect to most metric except NSS.

The model is further evaluated on PASCAL-S and the results are shown in Fig. 5. This also involves classical saliency models (Itti [17], GBVS [14], AIM [3], CAS [13], BMS [46]) and deep learning models (eDN [39], JN [33], SU [25], DeepFix [24], DVA [40], SAM [7], MxSalNet [8], Sal-DCNN [18], DINet [44], ConED [23] and SalED [41]. The results indicate that the proposed model fails only a marginal value to attain state-of-the-art performance on any evaluation metric. In fact, similarity measure stands top. The performance value of other models are taken from the corresponding paper or have been cited from other trusted literature.

However, it can be noticed that all the metric do not behave equally towards the model. This is the inherent pitfall associated with saliency prediction evaluation metric. This is because different metric are borrowed from different domains and may expect inputs in different formats. For example, Pearson's Correlation Coefficient expects random variables, Similarity metric looks for unnormalized densities and so on. Keeping this issue apart, the model achieved competitive results. Figure 6 presents a few visualization results of the model on PASCAL-S and OSIE datasets.

Ablation Analysis: To analyze the effect of DenseASPP and residual connections in the saliency prediction model, four variants of the model are constructed:

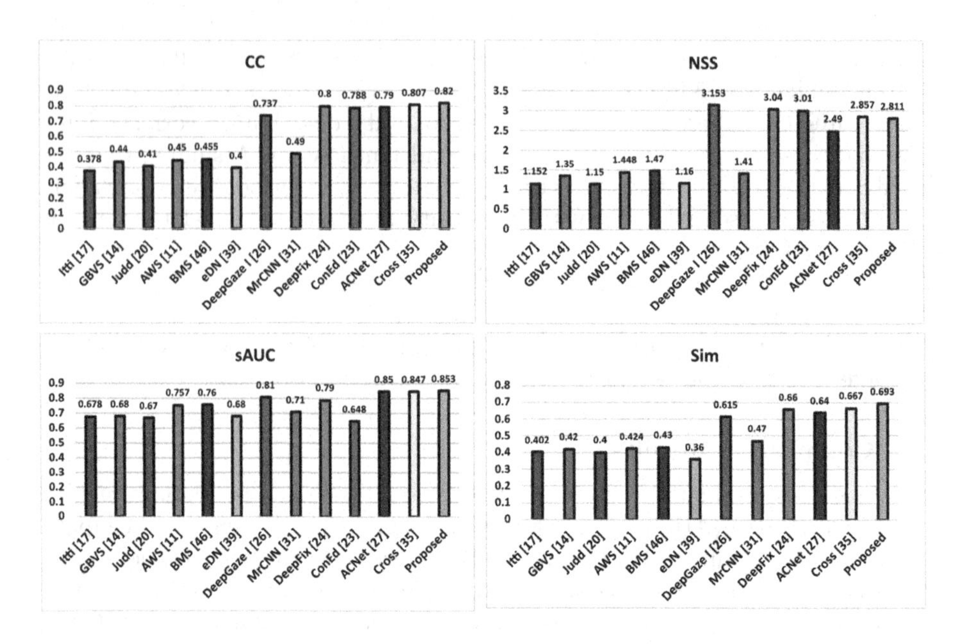

Fig. 4. Quantitative results of the proposed model on OSIE dataset in the context of previous works.

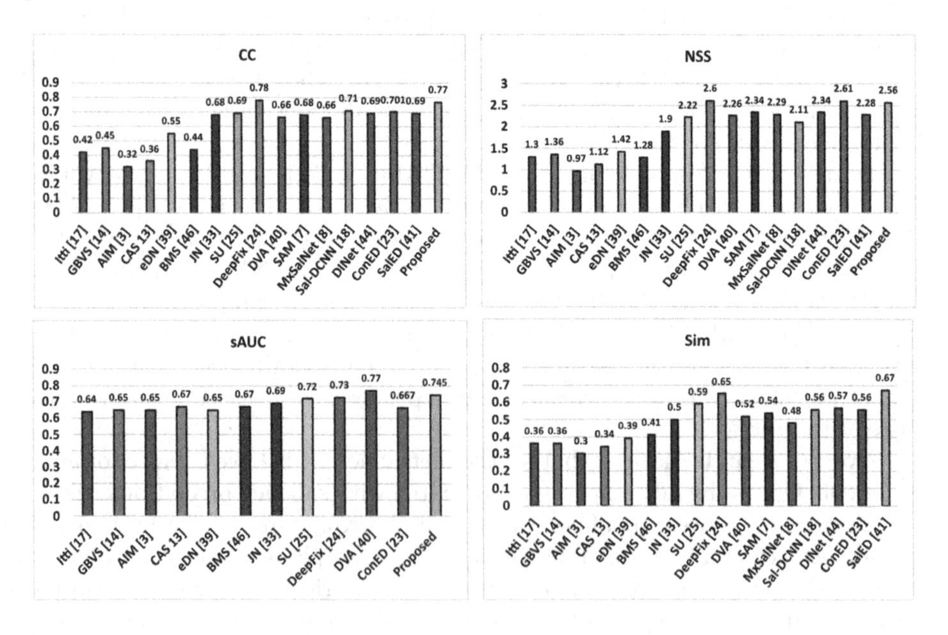

Fig. 5. Quantitative results on PASCAL-S compared with previous works.

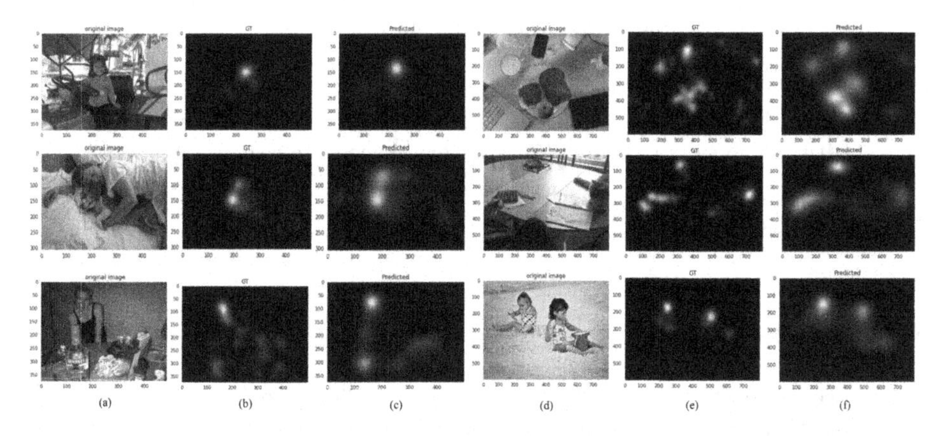

Fig. 6. Visual saliency results: (a), (b) and (c) illustrates original image, ground truth and predicted saliency map from PASCAL-S dataset. (d), (e) and (f) shows original image, ground truth and predicted saliency map that of OSIE

1. Model without ASPP/DenseASPP module: This is considered the baseline variant and it behaves as a general encoder-decoder structure with skip connections across the blocks.
2. Model with ASPP module in lieu of DenseASPP: ASPP is still able to capture contextual information to a good extent.
3. Model with no residual connections: Essentially, this variant is *Proposed model minus skip connections*. Hence it simply works with DenseASPP between encoder-decoder blocks.
4. Model with DenseASPP and residual connections: This constitutes the proposed architectural framework.

The above mentioned variants are all trained under the same settings as that of the proposed model and the experimental results of the ablation studies are presented in Tables 1 and 2. The results highlight that learning multi-scale contextual information in a dense manner, along with skip connections result in better fixation prediction. The performance gain can be credited to the usage of DenseASPP module, that considers dense relationship across information at multiple scales to generate saliency inference.

Table 1. Ablation results on OSIE

Model name	AUC (↑)	CC (↑)	NSS (↑)	sAUC (↑)	Sim (↑)
No ASPP/DenseASPP	0.8912	0.7952	2.6490	0.8344	0.6785
ASPP	0.9024	0.8036	2.6547	0.8393	0.6855
No residual	0.8999	0.8124	2.7454	0.8452	0.6902
Proposed	0.9120	0.8235	2.811	0.8530	0.6934

Table 2. Ablation results on PASCAL-S

Model name	AUC (↑)	CC (↑)	NSS (↑)	sAUC (↑)	Sim (↑)
No ASPP/DenseASPP	0.8824	0.7555	2.2323	0.7206	0.6614
ASPP	0.8878	0.7585	2.2333	0.7285	0.6630
No residual	0.8902	0.7674	2.4566	0.7396	0.6676
Proposed	0.8912	0.7748	2.5612	0.7450	0.6733

5 Conclusion

This work proposes a contextual aware deep saliency model for predicting human eye fixations on the images. The model incorporates DenseASPP between encoder-decoder blocks to efficiently capture contextual information utilizing atrous convolutions connected in a dense manner. The module is capable of generating multi-scale features at a larger scale range densely, allowing network to cater objects that vary in scale. Moreover, the residual connections encourage encoder feature representation to help decoder part generate efficient saliency inference. Compared with models that utilize hand crafted features and deep learning methods, the proposed framework accomplishes competitive objective results.

References

1. Bruce, N., Tsotsos, J.: Saliency based on information maximization. In: Advances in Neural Information Processing Systems, pp. 155–162 (2005)
2. Bruce, N.D., Catton, C., Janjic, S.: A deeper look at saliency: feature contrast, semantics, and beyond. In: Proceedings of the IEEE Conference on Computer Vision and Pattern Recognition, pp. 516–524. IEEE, Las Vegas (2016)
3. Bruce, N.D., Tsotsos, J.K.: Saliency, attention, and visual search: an information theoretic approach. J. Vis. **9**(3), 5 (2009)
4. Chen, L.C., Papandreou, G., Kokkinos, I., Murphy, K., Yuille, A.L.: DeepLab: semantic image segmentation with deep convolutional Nets, Atrous Convolution, and fully connected CRFs. IEEE Trans. Pattern Anal. Mach. Intell. **40**(4), 834–848 (2017)
5. Chen, L.C., Papandreou, G., Schroff, F., Adam, H.: Rethinking atrous convolution for semantic image segmentation. arXiv preprint arXiv:1706.05587 (2017)
6. Cornia, M., Baraldi, L., Serra, G., Cucchiara, R.: A deep multi-level network for saliency prediction. In: 2016 23rd International Conference on Pattern Recognition (ICPR), pp. 3488–3493. IEEE (2016)
7. Cornia, M., Baraldi, L., Serra, G., Cucchiara, R.: Predicting human eye fixations via an LSTM-based saliency attentive model. IEEE Trans. Image Process. **27**(10), 5142–5154 (2018)
8. Dodge, S.F., Karam, L.J.: Visual saliency prediction using a mixture of deep neural networks. IEEE Trans. Image Process. **27**(8), 4080–4090 (2018)
9. Erdem, E., Erdem, A.: Visual saliency estimation by nonlinearly integrating features using region covariances. J. Vis. **13**(4), 11–11 (2013)

10. Everingham, M., Van Gool, L., Williams, C.K., Winn, J., Zisserman, A.: The Pascal visual object classes (VOC) challenge. Int. J. Comput. Vis. **88**(2), 303–338 (2010)
11. Garcia-Diaz, A., Leboran, V., Fdez-Vidal, X.R., Pardo, X.M.: On the relationship between optical variability, visual saliency, and eye fixations: a computational approach. J. Vis. **12**(6), 17 (2012)
12. Ghariba, B.M., Shehata, M.S., McGuire, P.: A novel fully convolutional network for visual saliency prediction. PeerJ. Comput. Sci. **6**, e280 (2020)
13. Goferman, S., Zelnik-Manor, L., Tal, A.: Context-aware saliency detection. IEEE Trans. Pattern Anal. Mach. Intell. **34**(10), 1915–1926 (2011)
14. Harel, J., Koch, C., Perona, P.: Graph-based visual saliency (2007)
15. He, K., Zhang, X., Ren, S., Sun, J.: Deep residual learning for image recognition. In: Proceedings of the IEEE Conference on Computer Vision and Pattern Recognition, pp. 770–778 (2016)
16. He, S., Borji, A., Mi, Y., Pugeault, N.: What catches the eye? Visualizing and understanding deep saliency models. arXiv preprint arXiv:1803.05753 (2018)
17. Itti, L., Koch, C., Niebur, E.: A model of saliency-based visual attention for rapid scene analysis. IEEE Trans. Pattern Anal. Mach. Intell. **20**(11), 1254–1259 (1998)
18. Jiang, L., Wang, Z., Xu, M., Wang, Z.: Image saliency prediction in transformed domain: a deep complex neural network method. In: Proceedings of the AAAI Conference on Artificial Intelligence, vol. 33, pp. 8521–8528 (2019)
19. Jiang, M., Huang, S., Duan, J., Zhao, Q.: SALICON: saliency in context. In: Proceedings of the IEEE Conference on Computer Vision and Pattern Recognition, pp. 1072–1080 (2015)
20. Judd, T., Ehinger, K., Durand, F., Torralba, A.: Learning to predict where humans look. In: 2009 IEEE 12th International Conference on Computer Vision, pp. 2106–2113. IEEE (2009)
21. Kingma, D.P., Ba, J.: Adam: a method for stochastic optimization. arXiv preprint arXiv:1412.6980 (2014)
22. Krizhevsky, A., Sutskever, I., Hinton, G.E.: ImageNet classification with deep convolutional neural networks. Adv. Neural. Inf. Process. Syst. **25**, 1097–1105 (2012)
23. Kroner, A., Senden, M., Driessens, K., Goebel, R.: Contextual encoder-decoder network for visual saliency prediction. Neural Netw. **129**, 261–270 (2020)
24. Kruthiventi, S.S., Ayush, K., Babu, R.V.: DeepFix: a fully convolutional neural network for predicting human eye fixations. IEEE Trans. Image Process. **26**(9), 4446–4456 (2017)
25. Kruthiventi, S.S., Gudisa, V., Dholakiya, J.H., Babu, R.V.: Saliency unified: a deep architecture for simultaneous eye fixation prediction and salient object segmentation. In: Proceedings of the IEEE Conference on Computer Vision and Pattern Recognition, pp. 5781–5790 (2016)
26. Kümmerer, M., Theis, L., Bethge, M.: Deep Gaze I: boosting saliency prediction with feature maps trained on ImageNet. arXiv preprint arXiv:1411.1045 (2014)
27. Li, P., Xing, X., Xu, X., Cai, B., Cheng, J.: Attention-aware concentrated network for saliency prediction. Neurocomputing **429**, 199–214 (2021)
28. Li, Y., Hou, X., Koch, C., Rehg, J.M., Yuille, A.L.: The secrets of salient object segmentation. In: Proceedings of the IEEE Conference on Computer Vision and Pattern Recognition, pp. 280–287 (2014)
29. Lin, T.Y., et al.: Microsoft COCO: common objects in context. In: Fleet, D., Pajdla, T., Schiele, B., Tuytelaars, T. (eds.) ECCV 2014. LNCS, vol. 8693, pp. 740–755. Springer, Cham (2014). https://doi.org/10.1007/978-3-319-10602-1_48
30. Liu, N., Han, J.: A deep spatial contextual long-term recurrent convolutional network for saliency detection. IEEE Trans. Image Process. **27**(7), 3264–3274 (2018)

31. Liu, N., Han, J., Zhang, D., Wen, S., Liu, T.: Predicting eye fixations using convolutional neural networks. In: Proceedings of the IEEE Conference on Computer Vision and Pattern Recognition, pp. 362–370 (2015)
32. Long, J., Shelhamer, E., Darrell, T.: Fully convolutional networks for semantic segmentation. In: Proceedings of the IEEE Conference on Computer Vision and Pattern Recognition, pp. 3431–3440 (2015)
33. Pan, J., Sayrol, E., Giro-i Nieto, X., McGuinness, K., O'Connor, N.E.: Shallow and deep convolutional networks for saliency prediction. In: Proceedings of the IEEE Conference on Computer Vision and Pattern Recognition, pp. 598–606 (2016)
34. Reddy, N., Jain, S., Yarlagadda, P., Gandhi, V.: Tidying deep saliency prediction architectures. In: 2020 IEEE/RSJ International Conference on Intelligent Robots and Systems (IROS), pp. 10241–10247. IEEE (2020)
35. Ren, D., Wen, X., Jia, T., Chen, J., Li, Z.: Saliency detection via cross-scale deep inference. J. Vis. Commun. Image Represent. **75**, 103031 (2021)
36. Riche, N., Duvinage, M., Mancas, M., Gosselin, B., Dutoit, T.: Saliency and human fixations: state-of-the-art and study of comparison metrics. In: Proceedings of the IEEE International Conference on Computer Vision, pp. 1153–1160 (2013)
37. Simonyan, K., Zisserman, A.: Very deep convolutional networks for large-scale image recognition. arXiv preprint arXiv:1409.1556 (2014)
38. Szegedy, C.: Going deeper with convolutions. In: Proceedings of the IEEE Conference on Computer Vision and Pattern Recognition, pp. 1–9 (2015)
39. Vig, E., Dorr, M., Cox, D.: Large-scale optimization of hierarchical features for saliency prediction in natural images. In: Proceedings of the IEEE Conference on Computer Vision and Pattern Recognition, pp. 2798–2805 (2014)
40. Wang, W., Shen, J.: Deep visual attention prediction. IEEE Trans. Image Process. **27**(5), 2368–2378 (2017)
41. Wang, Z., Liu, Z., Wei, W., Duan, H.: SALED: saliency prediction with a pithy encoder-decoder architecture sensing local and global information. Image Vis. Comput. **109**, 104149 (2021)
42. Xu, J., Jiang, M., Wang, S., Kankanhalli, M.S., Zhao, Q.: Predicting human gaze beyond pixels. J. Vis. **14**(1), 28–28 (2014)
43. Yang, M., Yu, K., Zhang, C., Li, Z., Yang, K.: DenseASPP for semantic segmentation in street scenes. In: Proceedings of the IEEE Conference on Computer Vision and Pattern Recognition, pp. 3684–3692 (2018)
44. Yang, S., Lin, G., Jiang, Q., Lin, W.: A dilated inception network for visual saliency prediction. IEEE Trans. Multimedia **22**(8), 2163–2176 (2019)
45. Zabihi, S., Tavakoli, H.R., Borji, A.: A compact deep architecture for real-time saliency prediction. arXiv preprint arXiv:2008.13227 (2020)
46. Zhang, J., Sclaroff, S.: Saliency detection: a Boolean map approach. In: Proceedings of the IEEE International Conference on Computer Vision, pp. 153–160 (2013)
47. Zhang, L., Tong, M.H., Marks, T.K., Shan, H., Cottrell, G.W.: SUN: a Bayesian framework for saliency using natural statistics. J. Vis. **8**(7), 32–32 (2008)

Brain MRI and CT Image Fusion Using Generative Adversarial Network

Bharati Narute(✉) and Prashant Bartakke

College of Engineering, Pune 411001, MH, India
bharati.narute@mescoepune.org, ppb.extc@coep.ac.in

Abstract. The generative adversarial networks (GAN), complete model, is used to fuse computed tomography (CT) and magnetic resonance imaging (MRI) brain images in this research paper. To create a resultant fused image with bone structures from CT images and soft tissues from MRI images, our method develops an adversarial game between a generator and a discriminator. To make a stable training process, we use GAN instead of conventional fusion methods, and our architecture can handle different resolutions of multi-source medical images. The efficacy of the proposed procedure is demonstrated using several evaluation metrics. The proposed algorithms provide the best fused images without distortion and false artefacts. Comparison of proposed methods is done with the conventional techniques. The images obtained by fusing both sources' content with the help of the above algorithm gives the best with respect to visualization and diagnosis of the condition.

Keywords: CT · MRI · Deep learning · GAN · Image fusion

1 Introduction

Image fusion is a technique fusing different modality images into a single image with enhanced features. Both military and civilian fields have seen rapid advancements in infrared and visible light fusion, remotely sensed imaging, and multiple image fusion. Imaging methods are also important in medical diagnosis. It cannot meet the clinical diagnosis requirements due to the minimal details present in unimodal medical images. It necessitates additional information. Medical images are divided into two types: single modal and multimodal. Since multimodal fusion offers more details than single modal fusion, it is often used in medical image diagnosis. The image fusion technique aims to combine the image acquired by different sensors to create a more transparent and informative image that can help in decision-making [1, 2]. For instance, the doctor can manually combine the patient's different modes of images to make an accurate analysis, but this task is more tedious and time-consuming. The decision of the manual system may vary according to the experience of the doctor. Hence, there is a requirement to develop a robust and accurate medical image fusion system to enhance diagnosis consistency and decrease the doctor's workload. Image fusion has wide applications in clinical diagnosis.

B. Raman et al. (Eds.): CVIP 2021, CCIS 1568, pp. 97–109, 2022.
https://doi.org/10.1007/978-3-031-11349-9_9

MRI images capture soft tissue structure data of organs (eg. texture description information), computed tomography images provide bone structures, PET images provide functional and metabolic information, and X-Ray, DSA, ECT, and SPECT are some of the clinical research modalities used in medical imaging. PET and SPECT are low-resolution images that provide the organ metabolism [3, 4]. Thus, the different imaging techniques offer further details about the organ, but they interconnect. As a result, image fusion requires extensive use of additional images obtained through different imaging methods to acquire substantial complementary details, which is critical for clinical diagnosis and more precise and perfect treatment. To fuse source images from different sensors. Multi-scale transform schemes [5, 6], sparse [7, 8], Artificial Neural Network (ANN) [9, 10], subspace [11, 12], and saliency methods [13, 14], hybrid models [15, 16], and other methods [17, 18] have all been used to solve the image fusion many issues. The key fusion method, on the other hand, consists of three main components: image transformation, activity level estimation, and fusion rule design [19].

Current approaches usually employ the same transformation or representation during the fusion process for different source images. However, MRI and CT images may not be sufficient to resolve the MRI, and CT images demonstrate two separate phenomena. In most current methods, estimating the activity level and designing the fusion rule is manually drawn up and has become increasingly complex, restricting the complexity of implementation and computation cost [17]. To conquer the problem, we present a brain MRI and CT image fusion approach using a GAN that formulates the adversarial game between keeping the MRI's sharp details and preserving the detailed structure of the CT images. It is presented as the nominal relation between generator and discriminator. The GAN is used to fuse brain MRI and CT images. The generator's aim is to use a different visible gradient to fuse images with high MRI intensities. The discriminator, on the other hand, attempts to push the fused image to provide more detail in CT images. It helps the MRI and structure to keep the sharp details from the CT images in the fused image. In contrast to traditional fusion approaches, the proposed GAN algorithm complete model that avoids the manual/physical process of precise measurements of activity level fusion laws [27, 28]. With the CT texture, the architecture generator tries to create a fused image with high MRI intensities. The architecture's discriminator tries to get the merge the image to provide more texture detail. Furthermore, the GAN escapes the manual fusion law.

The rest of this paper framed as follows. Section 2 covers conventional multimodal fusion approaches as well as a simple introduction to GAN. Section 3 coordinated in themethod of proposed GAN algorithm for fusing brain MRI and CT images. The output of our system on various types of MRI and CT images is examined in Sect. 4 using various subjective and objective evaluation metrics. Followed by the concluding remark drawn in Sect. 5.

2 Related Works

We offer a short-lived overview based on conventional methods in this section. Vani M et al. [18] use the DTDWT algorithm to combine multi-focus and multimodal images. By choosing the average approximation coefficients and informative coefficients limit,

DTDWT coefficients are fused from two source images. By taking inverse DTDWT, fused images are obtained. Bhavana, et al. [19] use spatial filtering techniques such as Gaussian filter, MRI, and PET images to preprocess and enhance the input image quality that is noisy and unreadable. The improved image is fused for brain regions with different activity levels using DWT. A new DWT and SR fusion algorithm had proposed by Aishwarya N et al. [20]. The images are first decayed using DWT to remove the LF and HF components. Variance-based HF and LF components are combined as activity level calculation using the SR-based fusion technique. Nirmala, et al. [21] presents standard deviation and wavelet-based image fusion methods. Initially, the DWT is applied over the source image to decompose LF and HF components. A fusion framework is also developed to select the approximation band and fusion rules for a detailed band.

Chatterjee et al. [22] presented a novel, enhanced DWT-based image fusion process with Artificial Bee Colony Optimization. This approach focuses primarily on visible and infrared image fusion due to light variation, manifestation, etc. It's difficult to tell the difference between fine-point information in a visual image and fine-point information in an infrared image. Fusion is used to track or detect an item. A multimodal image fusion-based deep learning technique was proposed by Rajalingam et al. [23]. They used the CNN algorithm to create an image weight map. They used MRI, PET, and CT images that were 256x256 pixels in size. To build the weight diagram, they used a Siamese convolution network. It solves the problems caused by artificial architecture. It improves image quality and out performs the competition. Talbar S.N.et al. [30] presents NSCWT image fusion. Talbi et al. [31] presents image fusion algorithm based on the Predator–Prey Optimizer and DTCWT. They used this technique with 256x256 MRI-PET and MR-T1-PET images. This approach is best for extracting more details from images. This paper uses CNN features to fuse images. CNN is a deep learning algorithm for extracting features.

3 Proposed Method

The image fusion algorithm for MRI and CT images using GAN is described in this section, as well as the architecture of the generator and discriminator. The Training

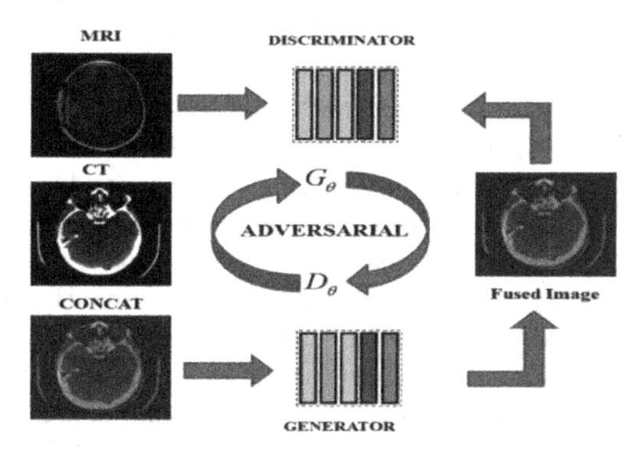

Fig. 1. Trainingframework of MRI-CT image fusion

framework of the MRI and CT fusion technique using the GAN algorithm is shown in Fig. 1 and Testing framework is shown in Fig. 2. Fusion of MRI and CT scans is formulated as an adversarial issue. Concatenate the brain MR and CT images in the channel dimension and feed the resulting image to the generator Gg, which produces the fused image I_f. The fused picture appears to retain the sharper features of the MRI while retaining the complex structure of the CT images due to the loss function of the generator without discriminator (D). Second, the discriminator was fed the fused image and MRI image, which discriminated against the MRI from the fused image.

In the training framework Fig. 1. once the generator generates the fused sample called I_f that is not distinguishable by the discriminator, so we can say that the expected fused image is obtained. This trained generator saved for the testing purpose. In the testing process, MRI and CT images are provided to the system by concatenation into a trained generator, and it produces the output fused image. A detailed explanation of this approach is discussed in this section.

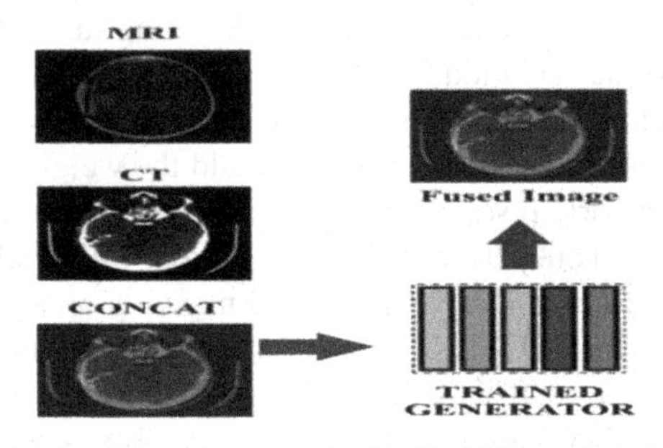

Fig. 2. Testingframework of MRI-CT image fusion

Network Architecture

This method's network architecture is split into two parts: generator and discriminator. Different layers of CNN are used to build this architecture.

Generator's Architecture: Fig. 3 depicts the generator architecture for this framework.

The generator architecture of this system consists of five-layered CNN. It consists of convolutional layers with $5x5$ filters in the initial two layers, $3x3$ filters in the middle layers, and $1x1$ in the final layer. Stride in each layer is set to 1 without padding pixels. The generator input of this system is a concatenated and noise-free image. The feature map extracted by convolutional operation also reconstructed by transposing the convolution layer to the same input image to improve the generated images diversity [29].

For MRI and CT images, each down-sampling process will reveal some extensive data in source images necessary for fusion [27]. We just add the convolution layer. It also holds input and output size the same, making the transformed layer redundant in our network. We also obey GAN [28] rules for batch normalization in the initial four layers to resolve sensitivity to data initialization and activation features to prevent declining gradients. The batch normalization will make the model stable and effectively propagate gradients to each layer. The 'Leaky ReLu' is used as an activation function in the initial four-layer and Tanh in the last layer.

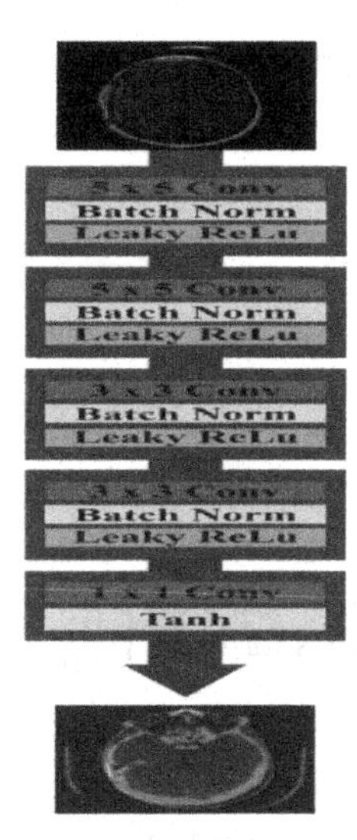

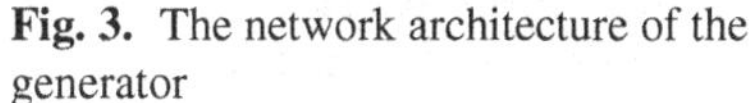

Fig. 3. The network architecture of the generator

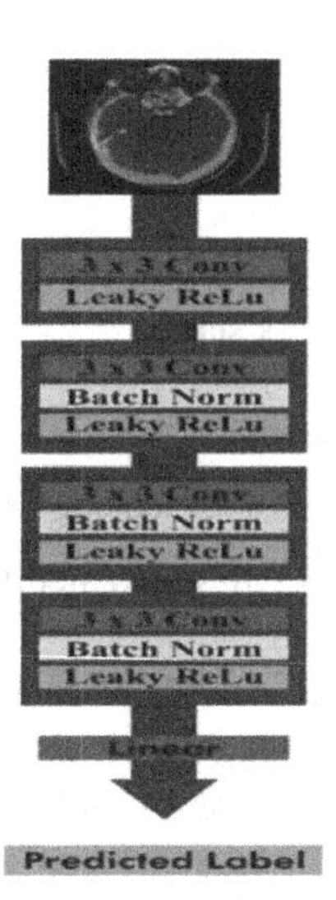

Fig. 4. The network architecture of discriminator

The architecture discriminator: Our discriminator network architecture is a simple five-layer CNN in Fig. 4. The architecture discriminator system consists of five-layered CNN. It consists of convolutional layers with $3x3$ filters in the initial four layers, setting the stride of 2 without padding. The discriminator is the classifier that first extracts input image feature map. The pooling layer with the stride of 2 is applied over the feature map. The noise's effect is minimized by adding padding operation on the very first convolution layer while no padding is performed on other convolution layers. Each convolutional layer is followed by the leaky ReLu layer to increase the non-linearity in the features. The batch normalization is applied over the second to fourth layers of the discriminator. The final layer is the linear layer used for the classification and the linear layer is the last layer.

Training and Testing

In this approach, MRI and CT image of 31 patients are considered for the training process. The training images resize to the same size $120x120$ and set stride to 14. So, can get 64,381 patches of MRI and CT images, centralize them to $[-1, 1]$. The pair in m number of MRI and CT patches picked from the training dataset and padded them to $132x132$ scale, used as generator input. The output of fused image patch is of size $120x120$. Next, we use MRI pairs and patches as discriminatory input. First, the discriminator is trained k times by using Adam as an optimizer [26]. Then, the generator is trained with maximum iterations. The algorithm below explains the system's training process.

Algorithm Training process of MRI and CT image fusion using GAN

Input:
1: Do the following for the number of training iterations:
2: Do this for k stages.

3: Select m fusion patches $\left\{I_f^{(1)}, \ldots, I_f^{(m)}\right\}$ from $G_{\theta g}$;
4: Select m MRI patches $\left\{I_{mri}^{(1)}, \ldots, I_{mri}^{(m)}\right\}$;
5: Adam Optimizer's modified discriminator:
6: end
7: Choose from m CT patches. $\{I_{ct}^{1}, \ldots, I_{ct}^{m}\}$ and m Patches for MRI $\left\{I_{mri}^{(1)}, \ldots, I_{mri}^{(m)}\right\}$for data used in training;
8: Adam Optimizer's Update Generator
9: end
The new MRI and CT images are the first crops in the testing phase by setting 14 strides and $120x120$ patch size and fed to the generator as a batch. Then the system generates the output fused image according to the sequence of the cropping.

4 Results and Discussions

The presented brain MRI-CT image fusion algorithm results are presented using qualitative and quantitative analysis. Visual inspection is an important variable while evaluating the fused multimodality medical images. However, fusion quality evaluation can't completely depend on visual inspection alone.

Dataset
Images are the largest source of data in healthcare and one of the most challenging sources to analyze This proposed GAN is verified on publicly available database on the radiopaedia.org and Harvard school. This proposed GAN is verified on publicly available database [32, 33]. We worked on dataset of brain MRI, CT images of 38 individuals. MRI and CT register images were available. In the proposed method, out of 38 persons, we use 31 person data (MRI, CT) for training while seven persons for testing. The input size of all the images is 256*x*256. A comparison based analysis of the presented fusion algorithm is carried out subjectively and objectively. A complete quality assessment is done and compared with several different traditional fusion techniques. Subjective and objective evaluation of this multi-modality medical fusion imaging system is done with the help of several metrics as listed below.

(i) Fusion Factor
The comparison of the quantity of data in the fused image and the input images is calculated by fusion factor. This is done by calculating mutual information between original medical images. It is calculated by the Eq. (1) [24, 25].

$$FusFact = MIinput1fus + MIinput2fus \tag{1}$$

$$\mathrm{MI}_{\mathrm{X,Y}} = \sum\nolimits_{\mathrm{k,l}} \mathrm{P}_{\mathrm{X,Y}}(\mathrm{k,l}) \log \frac{\mathrm{P}_{\mathrm{X}}(\mathrm{k,l})}{\mathrm{P}_{\mathrm{X}}(\mathrm{k})\mathrm{P}_{\mathrm{Y}}(\mathrm{l})} \tag{2}$$

Here, MI_{XY} indicates common information among images X and Y, $P_x^{(k)}$,$P_y^{(l)}$ are marginalprobabilitiesand $P_{X,Y}^{(k,l)}$ is joint probability distribution function. Increased FusFact values indicate enhanced fused image standard.

(ii) Cross Entropy

The cross entropy is the parameter for estimating the output quality of medical image compared to input medical image. This is calculated using Eq. (3).

$$\mathrm{CE(X, Y)} = \sum\nolimits_{\mathrm{k=0}}^{\mathrm{L-1}} \mathrm{P}_{\mathrm{X}}(\mathrm{k}) \log_2 \frac{\mathrm{P}_{\mathrm{X}}(\mathrm{k})}{\mathrm{P}_{\mathrm{Y}}(\mathrm{l})} \tag{3}$$

The two input images are X and Y, N is highest gray value of image. CE value and quality of image are inversely proportional. Hence lower CE value indicates better quality fused image.

(iii) Standard Deviation (STD)

This determines the difference of data from the average or mean value. STD value will be bigger if input data is clearer. STD can be found by using this Eq. (4): [26]

$$\mathrm{STD} = \frac{\sqrt{\sum_{\mathrm{i=1}}^{\mathrm{R}} \sum_{\mathrm{j=1}}^{\mathrm{S}} |\mathrm{f(i, j)} - \mu|^2}}{\mathrm{RS}} \tag{4}$$

where, R and S indicate the dimensions of the image $f(i, j)$, and the mean is denoted by μ.

(iv) Mutual Information (MI)

The extent at which two images (R, S) depend on each other is calculated by MI index. It provides the joint distribution detachment between them with the help of the below given equation:

$$\mathrm{I(r, s)} = \sum\nolimits_{\mathrm{y \epsilon R}} \sum\nolimits_{\mathrm{r \epsilon R}} \mathrm{p(r, s)} \log\left(\frac{\mathrm{p(r, s)}}{\mathrm{p(r)p(s)}}\right) \tag{5}$$

where, $p(r)$ and $p(s)$ indicate the marginal PDF of the both images, and $p(r, s)$ indicate the joint PDF.

$$\mathrm{MI(r, s, f)} = \frac{\mathrm{I(r, s)} + \mathrm{I(r, f)}}{\mathrm{H(r)} + \mathrm{H(s)}} \tag{6}$$

where, H(r), H(s) are the entropies of images r and s.

(v) PSNR

This is a quantitative value. It is calculated by Root Mean Square Error(RSME) with:

$$PSNR = 10 \text{ x} \log \left(\frac{(fmax)^2}{RMSE^2}\right) \tag{7}$$

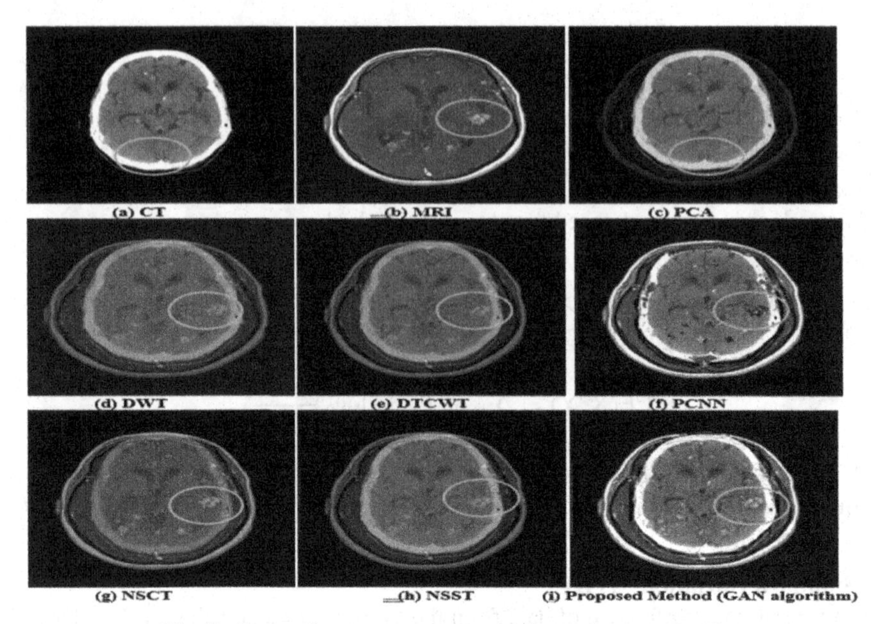

Fig. 5. Output for neurocysticercosis disease images (Set 1)

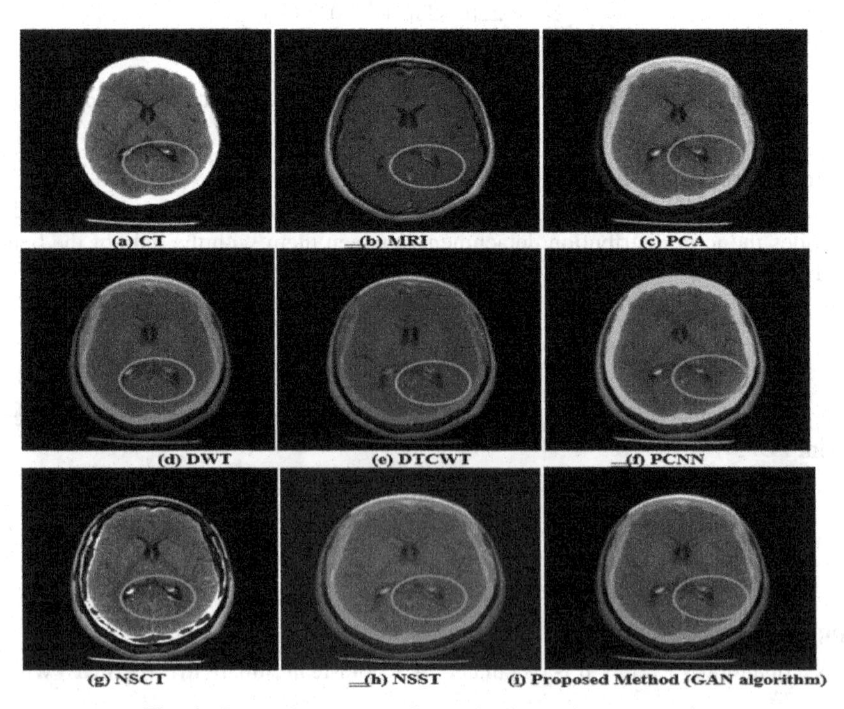

Fig. 6. Output for neurocysticercosis disease images (Set 2)

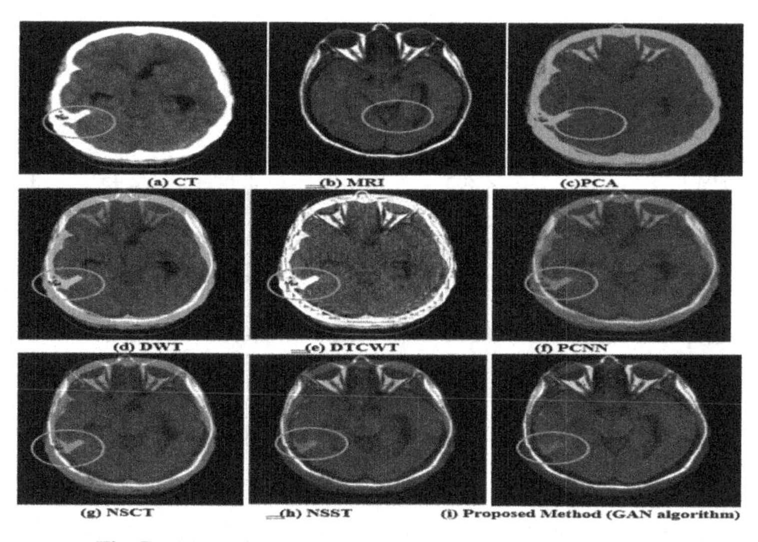

Fig. 7. Output for neurocysticercosis disease images (Set 3)

where f_{max} denotes the maximum pixel gray level value in the reconstructed image.

The three sets of source images are CT and MRI scans of patients with neurocysticercosis. In Figs. 5, 6, and 7, a and b are the source images, the fused output images using known methods are c, d, e, f, g, and h, and the fused picture using the proposed method is fig i.

Table 1 shows a comparison of the traditional and proposed approaches' performance measurement results for fusion factor, cross entropy, MI, PSNR, and standard deviation.

Table 1. Comparative study analysis for evaluation metrics with various fusion methods

Set	Metrics/Method	FusFact	CE	MI	PSNR	STD
Set 1	PCA	1.582	2.502	1.820	24.03	20.34
	DWT	1.716	2.072	1.899	25.90	22.23
	DTCWT	1.862	1.928	1.903	26.30	24.39
	NSCT	2.012	1.898	2.030	28.60	26.50
	NSST	2.162	1.807	2.230	29.88	28.34
	Proposed Method	**3.201**	**0.887**	**2.630**	**34.20**	**32.39**
Set 2	PCA	1.576	1.263	2.304	35.38	34.58
	DWT	2.389	1.378	2.505	37.08	36.20
	DTCWT	2.479	1.359	2.805	39.64	40.63
	NSCT	2.171	1.295	2.603	38.30	38.40
	NSST	3.145	1.582	2.905	44.69	43.55
	Proposed Method	**4.191**	**0.773**	**3.340**	**55.06**	**49.50**

(*continued*)

Table 1. (*continued*)

Set	Metrics/Method	FusFact	CE	MI	PSNR	STD
Set 3	PCA	2.187	2.261	2.903	29.60	15.30
	DWT	2.265	2.174	2.998	32.43	17.20
	DTCWT	2.471	2.017	2.801	32.53	16.40
	NSCT	2.693	1.997	2.703	33.70	23.54
	NSST	2.880	1.378	3.205	37.70	25.78
	Proposed Method	**4.950**	**0.756**	**3.730**	**50.43**	**31.20**

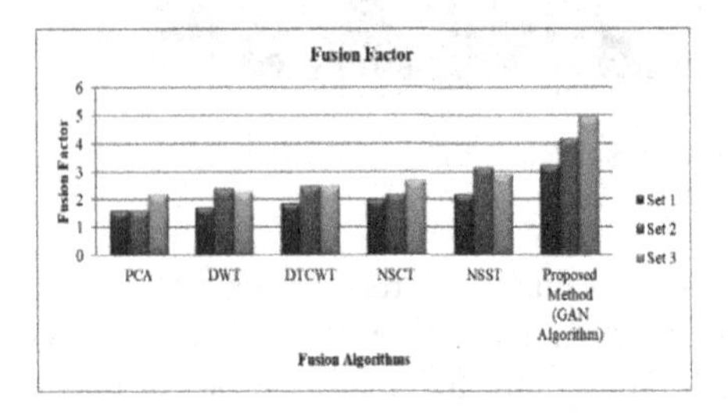

Fig. 8. Comparative Study: FusFact

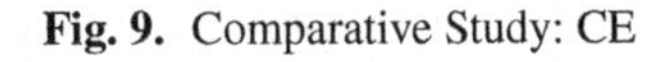

Fig. 9. Comparative Study: CE

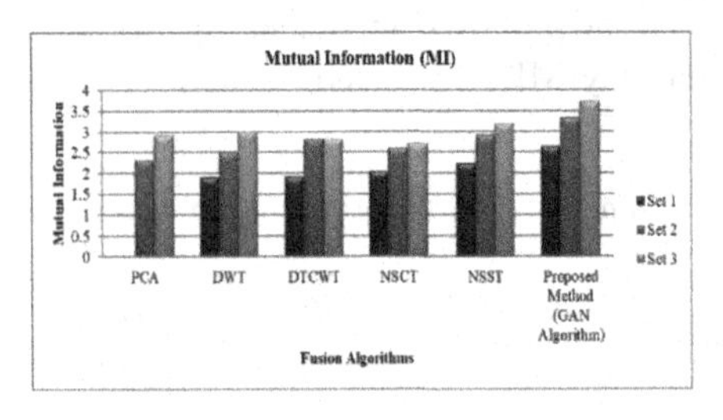

Fig. 10. Comparative Study: MI

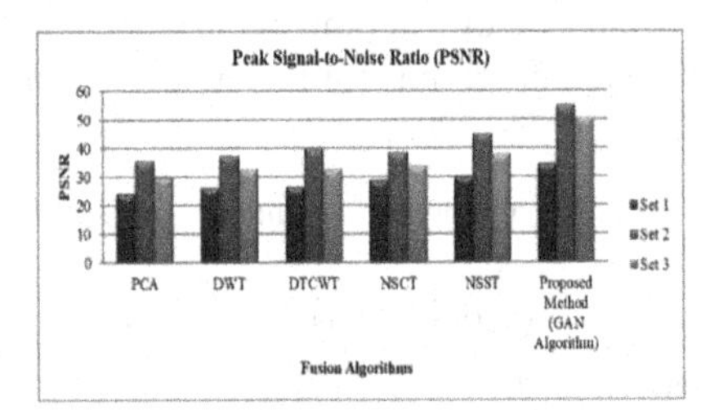

Fig. 11. Comparative Study: PSNR

The Fusion factor, Cross Entropy, standard deviation, MI, and PSNR for three sets of neurocysticercosis illness affected (CT-MRI) graphs are shown in Figs. 8, 9, 10, 11, and 12. When compared to other existing conventional techniques, the suggested GAN method has greater values for Fusion factor, standard deviation, MI, and PSNR, but lower values for cross entropy.

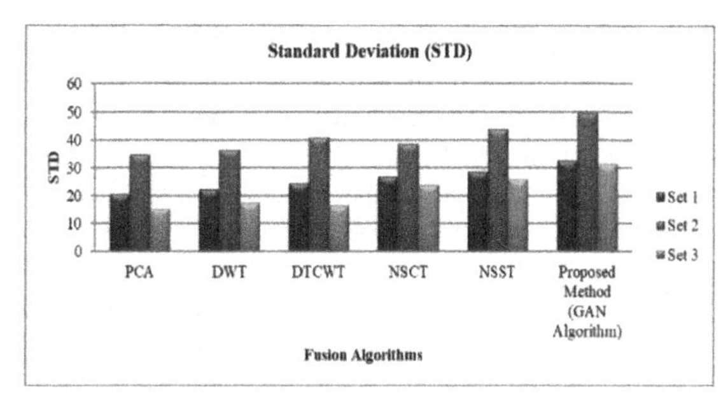

Fig. 12. Comparative Study: STD

5 Conclusion

In this paper, a GAN-based method for MRI and CT image fusion is proposed. It can simultaneously retain sharp MRI information and structure CT images. Our proposed GAN algorithm is a complete model like traditional fusion strategies that can manually design comprehensive level calculation and fusion law. Our fusion results can preserve soft tissue structural details in organs from high resolution MRI images while also retaining functional information from low resolution CT images with specific targets and details. The quantitative analysis of the proposed system using five measurement metrics, i.e., FusFact, CE, PSNR, SD & MI confirm that our GAN can produce better visual effects and hold the most enormous or nearest amount of information in the images.

References

1. Dogra, A., Goyal, B., Agrawal, S.: From multi-scale decomposition to non-multi-scale decomposition methods: a comprehensive survey of image fusion techniques and its applications. IEEE Access **5**, 16040–16067 (2017)
2. Ma, Y., Chen, J., Chen, C., Fan, F., Ma, J.: Infrared and visible image fusion using total variation model. Neurocomputing **202**, 12–19 (2016)
3. Sruthy, S, Parameswaran, L., Sasi, A.P.: Image Fusion Technique using DT-CWT. IEEE 978–1–4673–5090–7/13/2013
4. Li, S., Yang, B., Hu, J.: Performance comparison of different multi-resolution transform for image fusion. Information Fusion **12**(2), 74–84 (2011)
5. Zhang, Z., Blum, R.S.: A categorization of multiscale-decomposition-based image fusion schemes with a performance study for a digital camera application. Proceedings of the IEEE **87**(8), 1315–1326 (1999)
6. Wang, J., Peng, J., Feng, X., He, G., Fan, J.: Fusion method for infrared and visible images by using non-negative sparse representation. Infrared Physics & Technology **67**, 477–489 (2014)
7. Li, S., Yin, H., Fang, L.: Group-sparse representation with dictionary learning for medical image denoising and fusion. IEEE Trans. Biomedi. Eng. **59**(12), 3450–3459 (2012)
8. Xiang, T., Yan, L., Gao, R.: A fusion algorithm for infrared and visible images based on adaptive dual-channel unit-linking pcnn in nsct domain. Infrared Physics & Technology **69**, 53–61 (2015)

9. Kong, W., Zhang, L., Lei, Y.: Novel fusion method for visible light and infrared images based on nsst-sf-pcnn. Infrared Physics & Technology **65**, 103–112 (2014)
10. Bavirisetti, D.P., Xiao, G., Liu, G.: Multi-sensor image fusion based on fourth order partial differential equations. In: International Conference on Information Fusion, pp. 1–9 (2017)
11. Kong, W., Lei, Y., Zhao, H.: Adaptive fusion method of visible light and infrared images based on non-subsampled shearlet transform and fast non-negative matrix factorization. Infrared Physics & Technology **67**, 161–172 (2014)
12. Zhang, X., Ma, Y., Fan, F., Zhang, Y., Huang, J.: Infrared and visible image fusion via saliency analysis and local edge-preserving multi-scale decomposition. JOSA A **34**(8), 1400–1410 (2017)
13. Zhao, J., Chen, Y., Feng, H., Xu, Z., Li, Q.: Infrared image enhancement through saliency feature analysis based on multi-scale decomposition. Infrared Physics & Technology **62**, 86–93 (2014)
14. Liu, Y., Liu, S., Wang, Z.: A general framework for image fusion based on multi-scale transform and sparse representation. Information Fusion **24**, 147–164 (2015)
15. Ma, J., Zhou, Z., Wang, B., Zong, H.: Infrared and visible image fusion based on visual saliency map and weighted least square optimization. Infrared Physics & Technology **82**, 8–17 (2017)
16. Ma, J., Chen, C., Li, C., Huang, J.: Infrared and visible image fusion via gradient transfer and total variation minimization. Information Fusion **31**, 100–109 (2016)
17. Liu, Y., Chen, X., Wang, Z., Wang, Z.J., Ward, R.K., Wang, X.: Deep learning for pixel-level image fusion: Recent advances and future prospects. Information Fusion **42**, 158–173 (2018)
18. Vani, M., Saravanakumar, S.: Multi focus and multi modal image fusion using wavelet transform. In: 3rd International Conference on Signal Processing, Communication and Networking (ICSCN). In proceeding of IEEE (2015)
19. Bhavana, V., Krishnappa, H.K.: Multi-modality medical image fusion using discrete wavelet transform. In: International Conference on Eco-friendly Computing and Communication Systems. Procedia Computer Science **70**, pp. 625–631. Elsevier (2015)
20. Aishwarya, N., Abirami, S., Amutha, R.: Multifocus image fusion using discrete wavelet transform and sparse representation. In: IEEE WiSPNET conference (2016)
21. Paramanandham, N., Rajendiran, K.: A simple and efficient image fusion algorithm based on standard deviation in wavelet domain. In: IEEE WiSPNET conference (2016)
22. Chatterjee, A., Biswas, M., Maji, D.: Discrete wavelet transform based V-I image fusion with artificial bee colony optimization. In: proceeding of IEEE, 7th annual Computing and Communication Workshop and Conference (CCWC) (2017)
23. Rajalingam, B., Priya, R.: Multimodal medical image fusion based on deep learning neural network for clinical treatment analysis. Int. J. ChemTech Res. **11**(06), 160–176 (2018)
24. Rajalingam, B., Priya, R., Bhavani, R.: Hybrid multimodal medical image fusion algorithms for astrocytoma disease analysis. Emerging Technologies in Computer Engineering: Microservices in Big Data Analytics **985**, pp. 336–348. Springer (2019)
25. Rajalingam, B., Priya, R., Bhavani, R.: Hybrid multimodal medical image fusion using combination of transform techniques for disease analysis. Procedia Computer Science **152**, pp. 150–157. Elsevier (2019)
26. Rajalingam, B., Priya, R., Bhavani, R.: Multimodal medical image fusion using hybrid fusion techniques for neoplastic and alzheimer's disease analysis. J. Computati. Theoreti. Nanoscience **16**, 1320–1331 (2019)
27. Santhoshkumar, R., Kalaiselvi Geetha, M.: Deep learning approach: emotion recognition from human body movements. J. Mechani. Continua Mathemati. Sci. (JMCMS) **14**(3), pp. 182–195 (June 2019). ISSN: 2454-7190

28. Santhoshkumar, R., Kalaiselvi Geetha, M.: Vision based human emotion recognition using HOG-KLT feature' advances in intelligent system and computing. Lecture Notes in Networks and Systems **121**, 261–272. Springer. ISSN: 2194–5357. https://doi.org/10.1007/978-981-15-3369-3_20
29. Santhoshkumar, R., Kalaiselvi Geetha, M.: Deep learning approach: emotion recognition from human body movements. J. Mechani. Continua and Mathemati. **14**(3), 182–195 (2019)
30. Talbi, H., KhireddineKholladi, M.: Predator Prey Optimizer and DTCWT for Multimodal Medical Image Fusion. 978–1–5386–4690–8/18/$31.00 IEEE (2018)
31. Talbar, S.N., Satishkumar, S.C., Pawar, A.: Non-subsampled complex wavelet transform based medical image fusion. In: Proceedings of the Future Technologies Conference, pp. 548–556. Springer, Cham (2018)
32. https://radiopaedia.org
33. http://www.med.harvard.edu

MFCA-Net: Multiscale Feature Fusion with Channel-Wise Attention Network for Automatic Liver Segmentation from CT Images

Devidas T. Kushnure[1,2](✉) and Sanjay N. Talbar[1]

[1] Department of Electronics and Telecommunication Engineering, Shri Guru Gobind Singhji Institute of Engineering and Technology, Nanded, MS, India
devidas.kushnure@gmail.com, sntalbar@sggs.ac.in

[2] Department of Electronics and Telecommunication Engineering, Vidya Pratishthan's Kamalnayan Bajaj Institute of Engineering and Technology, Baramati, MS, India

Abstract. Radio imaging has become instrumental in examining internal anomalies and body organs. The medical expert manually delineates the images in routine practice and reaches the decision. However, manual intervention leads to errors due to human limitations, and the decision becomes operator-dependent. Therefore, automatic liver segmentation from computed tomography (CT) images plays a decisive role in detecting hepatic anomalies, treatment planning, liver transplantation, liver cancer treatment, and post-treatment assessment. Recently deep learning (DL) techniques have proven competence in medical image segmentation. The proposed method is DL-based multiscale feature fusion with channel-wise attention network (MFCA-Net) designed using the computationally efficient Res2Net (R2N) layer, which has the ability to enhance the receptive field of convolutional neural network (CNN) and extract the multiscale information at a more granular level.

Further, we reconstructed the multiscale low-level features and fused them with high-level features that enhance the semantic details in the features. In addition, we employed the channel-wise attention mechanism that renovates the features by modelling the interdependencies between the channels and focusing on the prominent features. Also, we altered the low-level features by fusing the renovated features that augment the contextual information, which simplifies the network learning potential. The efficacy of the proposed network was demonstrated on the CHAOS challenge test dataset, where the network attained the DICE of 96.67 $\pm$ 0.72%. Thus, the proposed MFCA-Net is computationally efficient, and liver segmentation performance is comparable with state-of-the-art methods.

Keywords: Multiscale features · Feature fusion · Channel attention · Liver segmentation · CNN · CT images

1 Introduction

Radio imaging is a non-invasive procedure to perceive the internal anomalies and structures of the organs. It plays a contributory role in disease diagnosis, treatment planning,

B. Raman et al. (Eds.): CVIP 2021, CCIS 1568, pp. 110–121, 2022.
https://doi.org/10.1007/978-3-031-11349-9_10

and post-treatment assessment in the medical realm. The medical practitioner prefers the different imaging modalities for the disease diagnosis and clinical relevance, comprising ultrasound, X-ray, computed tomography (CT), and magnetic resonance imaging (MRI). The CT has become the choice of medical experts for diagnosing hepatic diseases due to its ease and speedy acquisition protocol, robustness, high spatial resolution, and vast accessibility [2]. The delineation of the liver from CT images is essential for numerous hepatic complications and investigations such as liver resection planning, liver transplant, deciding radiation dose in radioembolization for treating the liver tumor, continuous assessment for deciding the progress of the disease, post-treatment assessment, liver volume measurement, and designing of computer-assisted diagnosis (CAD) systems. Currently, medical experts delineate the images manually. However, manual delineation is a laborious and tedious process that leads to operator-dependent segmentation and is susceptible to errors due to human limitations [10]. Because of these grounds, automatic segmentation of the liver has been of interest to many researchers.

In the last decade, numerous frameworks have been proposed to automatically segment the liver, which could be the second opinion for the medical practitioner to reach a precise decision quickly. Still, automatic liver segmentation from the CT images remains challenging due to irregular shape, fewer intensity variations between the liver and neighbouring organs, overlap with nearby organs, and unclear liver boundaries. Further, the CT images are acquired by injecting a contrast agent for enhancement; the quantity of contrast decides the noise in CT images, which is already noisy without contrast. Therefore, these complications lead to automatic liver segmentation from CT images is remains challenging [10].

2 Related Literature

Liver segmentation is the technique of extracting the liver parenchyma from radio images. For delineating the liver from CT images, many conventional algorithms were proposed based on the distribution of HU values in CT images utilized to extract the liver features, shape information, and texture of liver region such as grey level, structure-based, and texture-based methods [9]. However, the performance of these methods depends on the model parameters decided by the expert with prior knowledge that significantly affects the performance.

Recently, in the medical domain, deep learning-based methods gained popularity and have shown significant improvement in medical image analysis than the traditional methods. Specifically, convolutional neural networks (CNN) have become a technique for examining radio images for numerous applications, as summarized by Litjens et al. [7]. Furthermore, the use of CNN has become widespread in computer vision applications due to its nonlinear feature extraction property by using several filters at different network layers and its capability to process massive data volumes [16]. Furthermore, the performance of CNN architectures outdid in numerous visual recognition applications such as image classification, object recognition, and action recognition. Specifically, Alexnet, VGGNet, and GoogleNet proved capability in ImageNet Large Scale Visual Recognition Challenge (ILSVRC) for visual recognition applications [14]. Later, researchers extended the classification backbone network for semantic segmentation by modifying the fully dense layers with convolutional layers. The modified architectures were exploited for the semantic segmentation, and those networks were referred to as fully convolutional neural networks (FCN) [8]. In FCN, the decoder network was designed to obtain the output segmentation maps same as that of the input image size.

The most competent encoder-decoder FCN-based UNet architecture was designed [11] for biomedical image segmentation. Later the UNet gained much attention, and it became a choice for medical image segmentation. In UNet, the encoding path extracts the high-level details of the object at each stage of the network using successive convolution and downsampling operations. However, feature size becomes reduced at each stage, and the network learns contextual details of the object, but spatial details are lost at each downsampling stage. Finally, the feature maps rebuild to the identical size of the input image by upsampling the low-resolution encoded feature using a decoding path. The skip connections were utilized to bring high-level feature maps in the encoder to the decoder at the respective stages to construct the segmentation maps and recover the spatial loss. However, skip connections can not recover the losses fully. Therefore UNet performance is restricted up to a certain extent [13].

Recently, the performance of the UNet was uplifted by deriving the many architectures using the baseline UNet. The UNet derived multiscale information extraction, and fusion architectures represent the spatial and contextual information of the input features more precisely, which boosts the segmentation performance. Many architectures based on multiscale information extractions and fusion for medical image segmentation have proven their competence [17, 18]. These architectures exploited the information from different stages of the network and fused that information to obtain rich semantic features [5]. Moreover, the attention mechanisms were employed to improve the segmentation performance further by concentrating more on the essential features and reducing redundant details, which benefit in refining the model's learning and generalization capability. The spatial and channel-wise attention mechanisms were employed in the architectures [6, 12, 15] for medical image segmentation.

This paper proposed the UNet derived architecture based on the multiscale information extraction and fusion with a channel-wise attention mechanism.

3 Methodology

The proposed methodology is derived from the baseline UNet architecture [11] with a novel encoder design and improved skip-connection features shown in Fig. 1. The encoder architecture is modified to extract the multiscale features and channel-wise attention mechanism utilized to renovate the features in the skip-connection. The multiscale features extracted using the Res2Net (R2N) backbone layer depicted in Fig. 2 can represent the multiscale information of the object at a more granular level by increasing the cumulative receptive field of CNN [1]. We employed the multiscale R2N backbone layer instead of convolution operations for representing the multiscale features. Next, the low-level multiscale features of the next stage are fused by reconstructing them using the deconvolution operation, followed by nonlinear Relu activation with high-level features from the previous stage [5]. These results in the rich semantic information abstraction at each stage. The fused features are further transformed through the channel-wise attention block (CAB). The CAB is the Squeeze-and-Excitation network (SENet) [3] that can renovate the input features by modelling the channel-wise interdependencies to focus more on the prominent details in the input features indicated in Fig. 3. Finally, the renovated features are concatenated with the respective decoder stages through skip connections that enhance the decoder's capability to recover the segmentation map.

Moreover, the renovated features of CAB have prominent high-level information, which is again fused with the low-level multiscale features by passing it through the (1×1) convolution filter with Relu activation to improve the contextual details in the low-resolution features. Thus, the low-resolution feature fusion enhances the learning ability of the successive network layers. These novel modifications extracts and fuse multiscale information and improve the encoder's high-level and low-level features representation ability, resulting in rich contextual and semantic details about the liver region that enhance the learning and generalization potential of the network.

The process of multiscale feature fusion and channel-wise attention features of the proposed methodology is explained in the subsequent section.

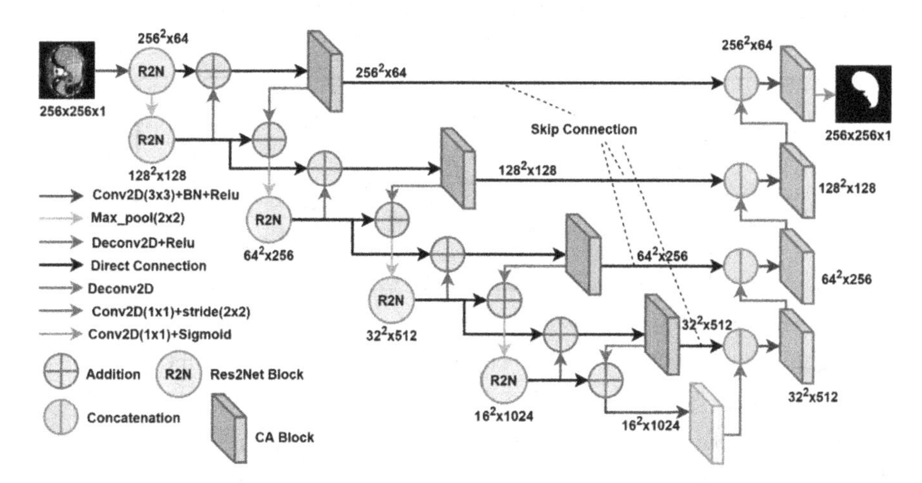

Fig. 1. Proposed multiscale feature fusion with channel-wise attention network (MFCA-Net).

3.1 Multiscale Feature Representation

The R2N block shown in Fig. 2 has the potential to strengthen the CNN by extracting granular details by improving the receptive field that leads to the multiscale feature representation [1]. The multiscale features are extracted by splitting the N-input features into equal s-subgroups that have n -identical size features such that $n = N/s$ without loss of generality. Where s- is a scaling factor, here we employed $s = 4$. Each subgroup has the $(1/s)$ times features, and it is designated by X_o where $o \epsilon \{1, 2, 3, \ldots, s\}$. Then, it is passed through the small convolution filters with size 3×3 except for the first group feature X_1 representing the set of input features, and the output features are denoted $L_o()$. The set of filters connected in a residual hierarchical fashion ensures different scale features, which leads to receptive field improvement of each subgroup due to multiple identical features aggregation. The accumulated features represent the multiscale features (M_i) indicated as (Eq. 1).

$$M = \begin{cases} X_o, & o = 1 \\ L_o(X_o), & o = 2 \\ L_o(X_o + M_{o-1}), & 2 < o \leq s \end{cases} \tag{1}$$

Finally, all the features from the subgroups representing distinct features of different resolutions and varying combinations of the receptive field are concatenated. Then, aggregated features are transformed through the 1×1 filter to represent the complete multiscale information. Thus, the R2N reduces the computational cost of the network due to the use of small-size convolution filters and enhances feature learning competence. Also, in the R2N, the features are split and concatenated. The split aggregated multiscale features, which advanced both global and local features characterization of input features and concatenation, better fuse the information with varying scales. The split and concatenation enable the CNN to process the features more efficiently.

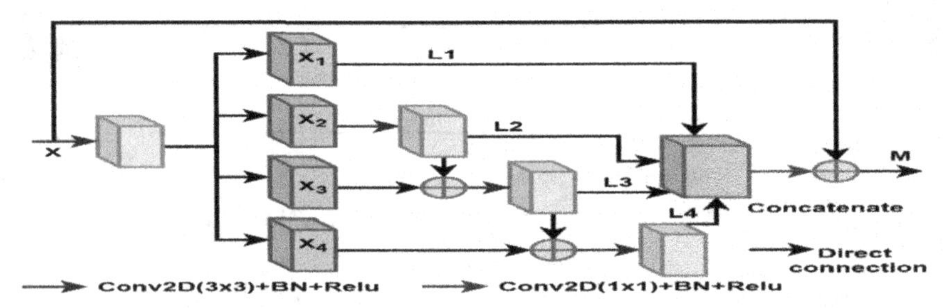

Fig. 2. Multiscale Res2Net (R2N) bottleneck layer employed in the MFCA-Net.

Moreover, the low-level multiscale features ($M^{(i+1)}$) of $(i+1)^{th}$-stage are indicated in (Eq. 2), which is reconstructed using deconvolution and nonlinear Relu activation to abstract the more semantic details and fuse them with the high-level multiscale features of ith-stage. The channel-wise fused high-level features (F^i) is depicted as (Eq. 3),

$$M^{(i+1)} = \begin{cases} \Upsilon^i(M^i), i = 1 \\ \Upsilon^i(F'^{\,i}_l), i \neq 1 \end{cases} \tag{2}$$

$$F^i = \left\{M^i \oplus \left\{\mathcal{R}(\mho^i(M^{(i+1)}))\right\}\right\} \tag{3}$$

where $i = 1, 2, 3, \ldots\ldots, k$ k-is number of encoding stages, Υ^i-downsampled features of ith, $\mho^i$- Upsampled features, $\mathcal{R}$- nonlinear Relu activation function, symbol $\oplus$-represent the channel-wise feature fusion at i^{th} stage, $F'^{\,i}_l$- is the low-level fused features at the ith-stage (Refer to Eq. 7).

The fused features are further transformed using the CAB, where we utilized the SENet [3] shown in Fig. 3. It models the channel-wise interdependencies by performing a simple two-step operation named squeeze and excitation on the input features.

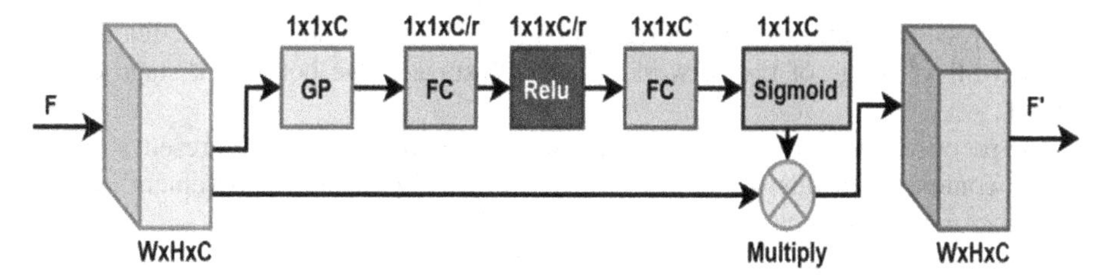

Fig. 3. Channel-wise Attention Block (CAB) utilized in MFCA-Net.

In squeeze operation, the input features are passed through the average global pooling (GP) that converts the input features of size $(H \times W \times C)$ with the one-dimensional vector of size $1 \times 1 \times C$, where C- represents the number of input channels, H- height and W-width of input features. The GP features are representative of each spatial channel. The channel-wise GP features (s_q) represented by (Eq. 4),

$$\mathrm{s_q} = \frac{1}{\mathrm{H} \times W} \sum_{i=1}^{\mathrm{H}} \sum_{j=1}^{W} \mathrm{F^i}(i,j) \tag{4}$$

where the squeeze features, $S_Q = [s_1, s_2, \ldots., s_C]$. Afterwards, the GP output is passed through the two fully connected (FC) layers using the Relu $(\mathcal{R})$ and Sigmoid (σ) activation with dimensionality reduction factor (r) that decides the computational cost. Here we set $r = 8$, which shows the significant result for the medical image segmentation [12]. The features (e_x) per channel is denoted by (Eq. 5),

$$\mathrm{e_x} = \sigma\left(\mathcal{R}\left(s_\mathrm{q}, \mathrm{W}\right)\right) = \sigma\left(\mathrm{W_2}\mathcal{R}\left(\mathrm{W_1 s_q}\right)\right) \tag{5}$$

The exciting features of the input channels are $E_X = [e_1, e_2, \ldots, e_C]$ represent the channel-wise descriptors of the spatial features lies between the [0, 1]. Where $W_1 \in \mathbb{R}^{\frac{C}{r} \times C}$ and $W_2 \in \mathbb{R}^{C \times \frac{C}{r}}$. The excitation operation adaptively models the input features by learning interdependencies between the channels. The final output of the CAB is the renovated features $(F^{'})$ which focus more on the prominent details of the fused features are represented by (Eq. 6),

$$F^{'i} = F \times E_X \tag{6}$$

The renovated features with enhanced high-level details are concatenated with the corresponding decoding stages to rebuild the segmentation maps. Further, it is fused with multiscale features of the $(i + 1)$ stage that better flow the high-level details to the succeeding stages and enhance the learning potential of the network layers. The CAB output features are passed through the convolution operation with kernel size one, Batch normalization, Relu, and stride set to two for maintaining the identical dimension denoted as $\mathbb{C}$ and low-level fused features $(F_i^{'})$ are indicated by (Eq. 7)

$$F_i^{'i} = \left\{M^i \oplus (\mathbb{C}(\mathrm{F}^{'(\mathrm{i}-1)}))\right\} \tag{7}$$

The process of multiscale feature fusion of high-level and low-level features is followed till k^{th}-stage of the network. where k^{th}-stage is the bottleneck stage of the network.

The proposed network can process the multiscale information to represent the spatial and contextual details of the object. As a result, significant improvement in liver segmentation performance.

4 Experimental Settings and Result Analysis

4.1 Dataset and Preprocessing

The CHAOS - Combined Healthy Abdominal Organ Segmentation challenge dataset [4] was utilized for training and testing the model. It includes 40 3D CT volumes (20 Training and 20 Test) acquired at the portal venous phase after injecting contrast agents of different patients with healthy livers and annotated by the expert radiologist team. The images in the volume have 512×512 spatial resolution, and the number of axial slices varies from 78 to 294 in each CT volume.

We employed the Hounsfield Units (HU) value windowing with window size [-250,200] to suppress the irrelevant intensities from the CT volume. Secondly, scale down the 512×512 size images into 256×256 to reduce the computational burden and dataset normalized, which simplifies the network learning process and improves the network's convergence speed. Finally, we accomplished image enhancement to get an enhanced liver region for segmentation. These preprocessing steps provide clean images for training.

4.2 Training Settings and Performance Metrics

The proposed network was trained end-to-end by an Adam optimizer with decreasing learning rate on the plateau from 1×10^{-3} to 1×10^{-6}. The weight regularization with a decay rate of 1×10^{-5} employed to evade overfitting, and a batch size of 8 is employed. The training process is optimized with the dice loss function [6]. Also, we augmented the dataset by employing the geometric transformation and elastic deformation to diminish the risk of overfitting and enhance the model's generalization capability. The Tensorflow and Keras high-level neural network API frameworks were utilized for implementation. The terminal used for training and testing has Intel(R)Xeon(R) CPU and 12 GB NVIDIA GEFORCE TITAN-X GPU.

The CHAOS challenge metrics [4] were used to evaluate the segmentation outcome of the model, which includes dice coefficients (DICE), Relative absolute volume difference (RAVD), Average symmetric surface distance (ASSD), and Maximum symmetric surface distance (MSSD).

4.3 Result Analysis and Discussion

The results presented in Table 1 demonstrated the effect of multiscale feature fusion with and without CAB on the segmentation performance. The multiscale information fusion represents the rich semantic details that boost the segmentation performance. Also, the channel-wise attention mechanism employed in the network architecture further uplifts the segmentation performance by modelling channel-wise interdependencies and highlighting the essential details in the input features. As a result, the DICE increased by 0.63%, RAVD reduced by 1.45%, ASSD reduced by 3.61 mm, and MSSD by 26.14 mm than the network without CAB. However, the CAB increases network parameters by 87,040 compared to the network without CAB configuration. The CHAOS challenge performance for 20 CT test volumes is shown in Fig. 4, which indicates that the CAB is significantly refining the segmentation outcome of the proposed model.

Table 1. Results on the CHAOS challenge test dataset and network parameters.

Method	DICE (%)	RAVD (%)	ASSD (mm)	MSSD (mm)	# parameters
Proposed model without CAB	96.04 ± 1.06	4.05 ± 2.56	6.69 ± 6.79	127.82 ± 61.97	30,301,701
Proposed model with CAB	96.67 ± 0.72	2.60 ± 2.04	3.08 ± 2.99	101.68 ± 45.71	30,388,741

Figure 5 shows sample results on the validation dataset that indicates the FMCA-Net can classify the liver voxels accurately for the complex and discontinue liver boundaries, small size liver region, and overlapped liver with adjacent organs. In addition, the difference in the GT and segmentation maps indicated in Fig. 6 indicates that the proposed model has the competence to segment complex liver parenchyma from the CT images. However, the network under-segments the liver region where discontinue liver regions and fuzzy liver boundaries.

The performance of the proposed method compared with the state-of-the-art methods proposed in the CHAOS challenge for liver segmentation from the CT test volumes provided by the challenge organizer [4] is shown in Table 2. The method proposed by the MedianCHAOS6 outperforms, which exploited the ensembled method derived from five different computationally expensive models to obtain the segmentation results. However, our model exploited the multiscale feature extraction potential of the computationally efficient R2N layer and channel-wise attention mechanism, leading to significant performance gain with reduced computational expenses.

The automatic liver segmentation performance obtained by MFCA-Net is comparable with the state-of-the-art methods. However, the R2N module reduces the number of parameters due to small size convolutional filters connected in a hierarchical manner that increases the cumulative receptive field, resulting in multiscale features with fine-grained object details. The low-level reconstructed features are fused with high-level features that enrich the semantic details. Moreover, the channel-wise attention mechanism is used to renovate the features by focusing on the essential features that advance the decoder's capability to recover the accurate segmentation maps. Further, the renovated features are fused with low-level multiscale features that enhance contextual details and uplift network learning potential. Therefore, the proposed MFCA-Net leverages multiscale information processed by the R2N layer is fused with the low-level and high-level

Table 2. Comparative analysis of different methods with the proposed method on the CHAOS test dataset.

Methods	DICE (%)	RAVD (%)	ASSD (mm)	MSSD (mm)
MedianCHAOS6	97.55 ± 0.42	1.54 ± 1.22	0.90 ± 0.24	23.71 ± 13.66
OvGUMEMoRIAL	90.18 ± 21.25	$9 \times 10^3 \pm 4 \times 10^3$	4.89 ± 12.05	55.99 ± 38.47
IITKGP-KLIV	91.51 ± 21.54	8.36 ± 21.62	27.55 ± 114.04	102.37 ± 110.9
ISDUE	87.08 ± 20.60	13.27 ± 7.61	3.25 ± 1.64	27.99 ± 9.99
Lachinov	68.00 ± 40.45	13.91 ± 20.4	11.47 ± 22.34	93.70 ± 9.40
Proposed Method	96.67 ± 0.72	2.60 ± 2.04	3.08 ± 2.99	101.68 ± 45.71

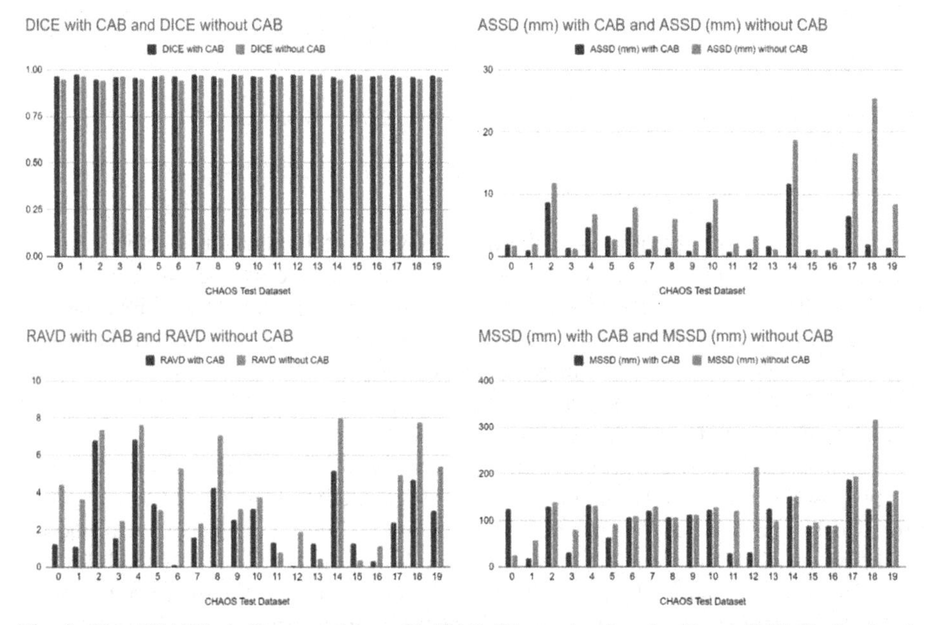

Fig. 4. The CHAOS challenge metrics with CAB (Blue colour) and without CAB (Red colour) on all the CHAOS test volumes. (Color figure online)

features for extracting rich semantic and contextual details of the object. In addition, the CAB block provides renovated features by modelling interdependencies between the channels that lead to attention on the prominent features. The proposed modifications offered the state-of-the-art performance for automatic liver segmentation from CT images with reduced computational complexity.

Nevertheless, the proposed model's performance and computational burden can be varied by modifying the scaling factor (s) of the R2N layer and the value of the dimensionality reduction factor (r) of CAB. Moreover, the performance of the supervised algorithms is sensitive to hyperparameters, data volume, and data quality.

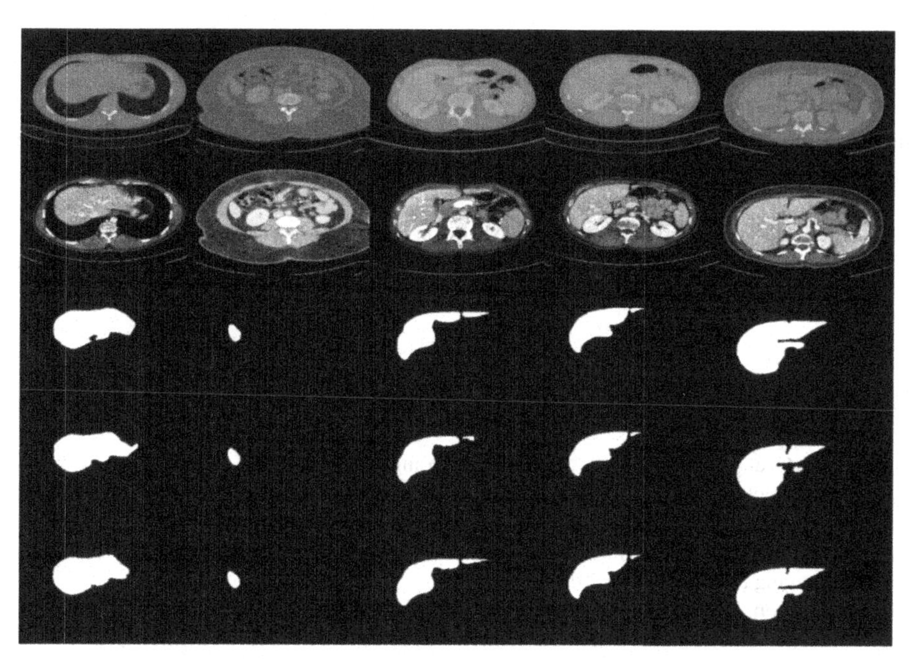

Fig. 5. Segmentation results. Row1: Input samples, Row2: Preprocessed samples, Row3: GT, Row4: Proposed method results without CAB, Row5: Proposed method results with CAB.

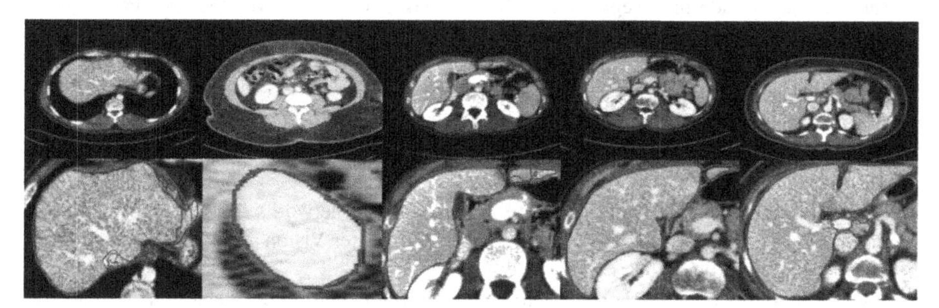

Fig. 6. Segmentation difference between the GT (Blue colour), Proposed method without CAB (Green colour), and Proposed method with CAB (Red colour). (Color figure online)

5 Conclusion

The proposed MFCA-Net exploited the multiscale feature representation property of the R2N layer, which extracts the fine-grained information and enhances the receptive field of CNN. The multiscale feature fusion boosts the sematic and contextual information in the features that improve the learning potential of the model. Furthermore, the channel-wise attention mechanism is utilized to remap the features by modelling the interdependencies between the channels to focus on the essential details in the features, which advances the decoder performance. The liver segmentation performance of the proposed model is evaluated on the CHAOS challenge test dataset, and it attained comparable performance with the state-of-the-art methods. Further, the proposed method is computationally more efficient than the outperforming methods proposed in the CHAOS challenge.

Further performance gain can be achieved by fine-tuning the hyperparameters and training strategies. In addition, the proposed MFCA-Net can be extended for the segmentation of different organs and anomalies using other medical imaging modalities.

Acknowledgements. The authors would like to thank the Faculty and Management of VPKBIET, Baramati, and VIIT, Baramati, for enabling the essential resources to accomplish the proposed research work.

References

1. Gao, S., Cheng, M.M., Zhao, K., Zhang, X.Y., Yang, M.H., Torr, P.H.: Res2net: a new multi-scale backbone architecture. IEEE transactions on pattern analysis and machine intelligence (2019)
2. Gotra, A., et al.: Liver segmentation: Indications, techniques and future directions. Insights into imaging **8**(4), 377–392 (2017)
3. Hu, J., Shen, L., Sun, G.: Squeeze-and-excitation networks. In: Proceedings of the IEEE Conference on Computer Vision and Pattern Recognition, pp. 7132–7141 (2018)
4. Kavur, A.E., et al.: Chaos challenge-combined (ct-mr) healthy abdominal organ segmentation. Med. Image Anal. **69**, 101950 (2021)
5. Kushnure, D.T., Talbar, S.N.: HFRU-Net: high-level feature fusion and recalibration unet for automatic liver and tumor segmentation in CT images. Computer Methods and Programs in Biomedicine, 106501 (2021)
6. Kushnure, D.T., Talbar, S.N.: MS-UNet: a multi-scale unet with feature recalibration approach for automatic liver and tumor segmentation in ct images. Comput. Med. Imaging Graph. **89**, 101885 (2021)
7. Litjens, G., et al.: A survey on deep learning in medical image analysis. Med. Image Anal. **42**, 60–88 (2017)
8. Long, J., Shelhamer, E., Darrell, T.: Fully convolutional networks for semantic segmentation. In: Proceedings of the IEEE Conference on Computer Vision and Pattern Recognition, pp. 3431–3440 (2015)
9. Luo, S.: Review on the methods of automatic liver segmentation from abdominal images. J. Comp. Commu. **2**(02), 1 (2014)
10. Moghbel, M., Mashohor, S., Mahmud, R., Saripan, M.I.B.: Review of liver segmentation and computer assisted detection/diagnosis methods in computed tomography. Artif. Intell. Rev. **50**(4), 497–537 (2018)
11. Ronneberger, O., Fischer, P., Brox, T.: U-net: convolutional networks for biomedical image segmentation. In: International Conference on Medical Image Computing and Computer-Assisted Intervention, pp. 234–241. Springer (2015)
12. Rundo, L., et al.: Use-net: Incorporating squeeze-and-excitation blocks into u-net for prostate zonal segmentation of multiinstitutional mri datasets. Neurocomputing **365**, 31–43 (2019)
13. Seo, H., Huang, C., Bassenne, M., Xiao, R., Xing, L.: Modified u-net (mu-net) with incorporation of object-dependent high level features for improved liver and liver tumor segmentation in ct images. IEEE Trans. Med. Imaging **39**(5), 1316–1325 (2019)
14. Ueda, D., Shimazaki, A., Miki, Y.: Technical and clinical overview of deep learning in radiology. Jpn. J. Radiol. **37**(1), 15–33 (2019)

15. Xia, H., Ma, M., Li, H., Song, S.: Mc-net: multi-scale context-attention network for medical ct image segmentation. Applied Intelligence, 1–12 (2021)
16. Yamashita, R., Nishio, M., Do, R.K.G., Togashi, K.: Convolutional neural networks: an overview and application in radiology. Insights Imaging **9**(4), 611–629 (2018)
17. Zhang, J., Jin, Y., Xu, J., Xu, X., Zhang, Y.: Mdu-net: Multi-scale densely connected u-net for biomedical image segmentation. arXiv preprint arXiv:1812.00352 (2018)
18. Zhou, Z., Siddiquee, M.M.R., Tajbakhsh, N., Liang, J.: Unet++: redesigning skip connections to exploit multiscale features in image segmentation. IEEE Trans. Med. Imaging **39**(6), 1856–1867 (2019)

Automatic Double Contact Fault Detection in Outdoor Volleyball Videos

Pratibha Kumari[1](✉), Anmol Kumar[1], Min-Chun Hu[2], and Mukesh Saini[1]

[1] Indian Institute of Technology Ropar, Rupnagar, India
{2017csz0006,2017csb1070,mukesh}@iitrpr.ac.in
[2] National Tsing Hua University, Hsinchu, Taiwan
anitahu@cs.nthu.edu.tw

Abstract. One of the common faults in volleyball is double contact while setting the ball for a spike. It is hard to detect this fault by the players. Even the referees sometimes find it difficult to observe. In this work, we propose an automatic double contact fault detection approach using a single camera in outdoor volleyball video. The video is first analyzed to detect and track the ball; the bounding boxes are then processed to extract a deep Spatio-temporal representation using a state-of-the-art 3D-convolution-based neural network, which is finally fed to a multilayer perceptron for classification. To the best of our knowledge, this is the first work on volleyball double-contact detection. The proposed framework achieves an average accuracy of 77.16% on 5-fold cross-validation. The framework is useful for players during training and for referees as a decision-support tool.

Keywords: Volleyball tracking · Fault detection · Double contact

1 Introduction

In volleyball, double contact fault is a common event, yet it can be difficult to recognize it manually. Double contact occurs when the same person touches the ball twice in succession [20]. Particularly in the case of a setting event, the player touches the ball with soft hands; hence there is a good chance for the fault to occur. However, the two touches are so close in time that even the referees find it difficult to detect them at times [18]. Having a highly skilled professional referee is expensive as well as inconvenient during training sessions. Therefore, there is a need to automatically detect this fault. In this work, we propose a framework to automatically detect double-contact fault in volleyball videos using a neural network framework.

1.1 What is a Double Contact Fault?

In volleyball, players are allowed a maximum of three contacts before passing the ball to the other side of the court. The Federation of International Volleyball

B. Raman et al. (Eds.): CVIP 2021, CCIS 1568, pp. 122–134, 2022.
https://doi.org/10.1007/978-3-031-11349-9_11

defines double contact fault as: "when a player hits the ball twice in succession or the ball contacts various parts of his/her body in succession" [7]. While this fault can occur during any touch, it is relatively less common during the first and third touches. In fact, recent rules allow double contact to some extent during the first touch [20]. The double contact fault is most commonly found during the second touch, as the setter releases the ball for the spike with soft hands. As can be observed from Fig. 1, the ball has released the left hand while the right hand is still in contact which result in a fault. Generally, because of the uneven forces on the opposite surfaces, the volleyball starts spinning after a *set* with a double touch fault (effect of spin due to a double touch fault is observed in second row of Fig. 2).

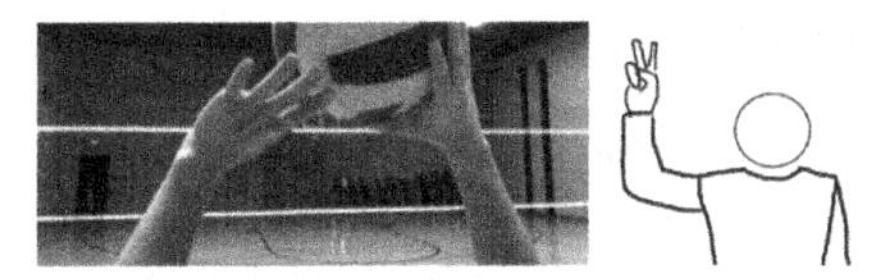

Fig. 1. A double contact fault while setting the ball. The ball releases the left hand before the right hand.

Fig. 2. The motion view of a correct *set* (top row) and a faulty *set* (bottom row). We observe different motion pattern in both the cases.

1.2 Our Solution

It is very hard to automatically detect the exact point of contact and its precise timing in the video unless it is recorded at an extremely high frame rate. Even in a video with a high frame rate, the hand/finger detection may not be accurate enough to detect the point of contact. Therefore we rely on motion of the volleyball post *set*. Figure 2 shows the zoomed view of the ball post *set* in case of a faulty *set* and a correct *set*. Different motion patterns in both cases can be observed. The volleyball is first detected in each frame. A generic object detector may fail to detect volleyball in all the frames; hence, we propose a method to estimate volleyball location in the missing frames. Specifically, we propose a YOLOv4 [1] based hybrid tracking for correctly detecting volleyballs in each frame. After this, we consider a certain number of frames on the upward trajectory and take their volleyball regions (bounding boxes). We further map these bounding boxes to a deep Spatio-temporal descriptor using a state-of-the-art 3D convolution network. Further, for classification into faulty and non-faulty *set*, a multi-layer perceptron (MLP) based classifier is trained using these descriptors. An overview of the framework is given in Fig. 3.

1.3 Contributions

Our contributions are three-fold:

- We have proposed a novel framework that is able to detect double contact fault with good average accuracy of 77.16% on 5-fold cross-validation. To the best of our knowledge, this is the first work on automatic double contact fault detection in volleyball videos.
- We have designed a YOLOv4 based hybrid tracking that outperforms baseline YOLOv4 for volleyball detection in an outdoor environment.
- We have built a challenging dataset of outdoor volleyball video *set* events along with the ground truth (see sample frames in Fig. 6). The dataset can be used for the performance evaluation of double contact fault detection frameworks. The dataset can be accessed from here[1].

The rest of the paper is organized as follows. In the next section, we review the literature to position our work. The details of the proposed framework for double contact fault detection are given in Sect. 3. Section 4 contains a detailed description of the dataset and the performance evaluation. Finally, we conclude our paper in Sect. 5.

2 Literature Review

In the field of sports video analysis, most of the works have focused on ball and player tracking [21], particularly in soccer. There have been only a few attempts on tracking a volleyball [15]. Chen et al. [6] propose a method to track volleyball by employing physics rules. Based on the ball trajectory, the authors detect different *set* types. However, it is not mentioned how the *set* trajectory is actually obtained. Chakraborty and Mehar [4] detect volleyball trajectories and analyze them for shot type detection *(pass, set, spike)*. The following rule is used to classify: *"a ball that originates near the net and follows a short parabolic motion"*. Kurowski et al. [15] present a volleyball tracking method using Adaptive Gaussian Mixture Model based background subtraction. Since there will be multiple players as well as audience moving in a real volleyball video, the method has limited application.

There are a few works on detecting other types of volleyball game activities in the video as well. Cuspinera et al. [8] detect the type of serve in a beach

[1] DatasetLink.

volleyball video. Similarly, Hu et al. [13] detect spiking events, and Gageler et al. [10] detect jump action. Kautz et al. [14] detect player activities at a finer level using wearable sensors (worn by players). Recently, Hidden Markov Model (HMM) has been used for player motion analysis in volleyball videos [17,26]. Szelag et al. [23] propose to track volleyball using 12 cameras to build a referee support system for professional games. The authors evaluate the method on a synthetic dataset. Our goal is to use a single normal camera (not high frame rate) for double contact detection. We evaluate our framework on real video data.

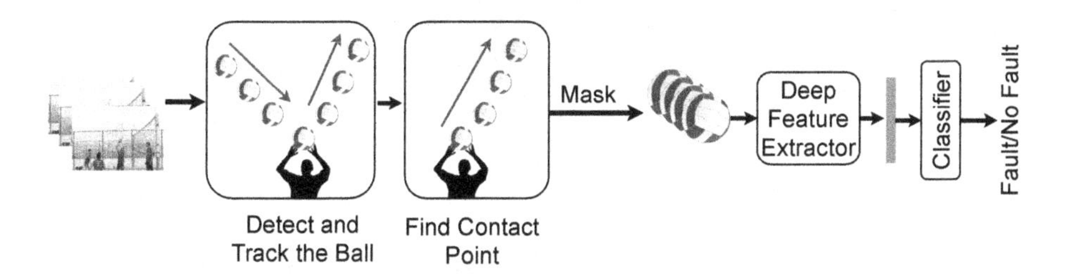

Fig. 3. Overview of the proposed framework

3 Proposed Work

The overview of the proposed framework is given in Fig. 3. Our dataset consists of *set* events. The volleyball supply enters from the left upper half of the frame, makes contact with the player, and goes up again in the right upper direction. So the first task is to detect and track the ball. Further, if the player makes a double contact, this will be reflected in the frames after the player touches the ball. Hence, the second step is to identify the contact point and separate the frames after the contact point. In the frames immediately after the player contact, we extract the ball regions. A video is created using the extracted ball regions. We map this video into a motion and appearance-based deep feature using a state-of-the-art 3D-convolution-based model, I3D [3]. Finally, we train an MLP network on these feature vectors. Below, we provide details of each step.

3.1 Detect and Track the Ball

Since, there can be other moving objects in the camera view along with the volleyball, therefore, we do not use background-foreground subtraction techniques for finding and tracking the ball. Instead, we use the pre-trained YOLOv4 neural network, which consists of CSPDarknet53 [25] as backbone, SPP [11] as additional module, PANet [16] as path-aggregation neck, and YOLOv3 [19] (anchor-based) as head in its architecture. By running YOLOv4 on our dataset, we made two observations:

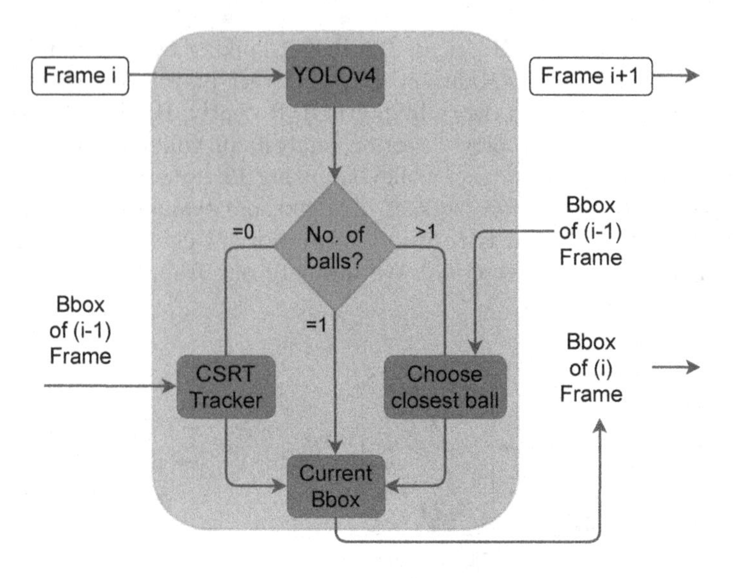

Fig. 4. Overview of hybrid-tracking algorithm.

1. YOLOv4 detects the ball with extremely high precision. In our dataset, we could not find even one false positive.
2. Depending on the complexity of the background and distance of the ball from the camera, YOLOv4 may fail to detect the ball in some frames.

While the proposed approach would work in the case of missed detections as well, it is desirable to detect balls in as many frames as possible. Particularly after the contact, we need to detect the ball in sufficient consecutive frames in order to capture the appearance and motion of a spin event. Therefore, we fine-tune YOLOv4 in order to reduce miss detections. The details of fine-tuning are provided in Sect. 4.2. After fine-tuning, we observed the improvement in detection, but still, there were several miss detections. For further improvement, we consider the first observation to detect the ball in missing frames. Let us assume that fine-tuned YOLOv4 is able to detect ball in $\{k^{th}, m^{th}, \ldots n^{th}, \ldots\}$ frames of a *set* event. There may be one or more missed detections among these frames. In order to find the missing bounding boxes, we perform a hybrid tracking. We initialize a CSRT tracker [2] using the bounding box at k^{th} frame and predict the location of the ball for $(k+1)^{th}$ frame. We continue to predict the bounding boxes in subsequent frames using CSRT until m^{th} frame is encountered. Since the precision of YOLOv4 is very high, we re-initialize the CSRT tracker with detection at m^{th} frame. This way, we continue to fill up missing detections and re-initializing CSRT until the last frame (l^{th}) of the *set* event is reached. There may be the case when there are two or more detections (due to the presence of extra balls in the scene) by YOLOv4 in one frame. In such a case, we choose the bounding box that is closest (lowest Euclidean distance) to the previous frame's detection. The overview of the hybrid-tracking is given in Fig. 4. With the hybrid-tracking strategy, we are able to predict all frames missed in-between. Figure 5 shows an example output of our proposed hybrid-tracking. The red boxes are the YOLOv4 detections. Blue boxes indicate frames where YOLOv4 fails to detect but fine-tuned YOLOv4 succeeds. Green boxes depict the detection, using the hybrid approach, that were otherwise missed by fine-tuned YOLOv4. Again, with this approach, we achieved detection with high precision. We avoided tracking the ball as there were a large number of false detections, mainly due to the challenging outdoor environment. The existing tracking methods mostly deal with indoor videos.

Fig. 5. Red boxes shows the bounding boxes given by YOLOv4 detector. The boxes which were missed from YOLOv4 but detected in fine-tuned YOLOv4 are displayed by blue boxes. Green boxes are bounding boxes which was missed by fine-tuned VOLOv4 but predicted by our hybrid tracking approach. (Color figure online)

3.2 Find the Contact Point

In a volleyball game, the ball is either moving upwards or downwards. During a rally (continuous play of the game), the ball changes the trajectory from downward to upward at each contact point, except for the spike. Particularly during the *set* event, which is where the double contact fault occurs, the ball moves almost vertically upward after the contact, see Fig. 5. The effect of a double contact fault is reflected in the frames after the contact, see Fig. 2. Frames before the contact can be simply ignored. However, they are still needed to detect the contact point. The point where the ball changes trajectory from downward to upward would give us a contact point. If we consider the origin (0,0) at the top left corner of the frame, the contact point will be in the frame where the y-coordinate of the bounding box attains the largest value. We refer to the frame with contact point as c^{th} frame.

3.3 Deep Feature-Based Classifier

Figure. 2 shows zoomed view of a typical correct *set* in the top row and a typical faulty *set* in the bottom row. We observe that when the ball is *set* properly, it moves smoothly. However, when there is a double contact fault due to differential forces on the opposite surfaces of the ball, the ball movement follows a complex pattern. Thus, the ball's motion and appearance are different in the two cases. Hence, we aim to get an appearance and motion-based feature descriptor. We consider the bounding boxes of the volleyball region (extracted in Sect. 3.1) from the frame having contact point (c^{th}) to the last frame (l^{th}) of the *set* event. Since, the bounding boxes can be of different shapes; hence we first resize them to equal shapes and then create a video. After this, the video is mapped to a deep feature using the I3D network, a widely adopted video classification network. I3D was proposed to improve the C3D [24] model by inflating from 2D models such as Inception [22], ResNet [12], etc. It uses 3D convolution to learn Spatio-temporal information directly from the RGB videos. It produces a descriptor of size 1×2048.

Once we have the descriptor for each created video, we can train any classifier for fault detection. We use a multi-layer perceptron network having two hidden layers with 16 and 8 neurons. The input layer size is equal to the dimensionality of the descriptor. We use binary cross-entropy as loss function and sigmoid function in the last output layer. The network is trained using an SGD optimizer with a Nesterov momentum of 0.9. The starting learning rate is kept 1e-03. The network is trained for 200 epochs with a batch size of 32, and the learning rate decays using the time-step decay function.

Fig. 6. Samples frames from the dataset showing different lighting conditions and ball sizes.

4 Experiments and Results

In this section, we conduct experiments and evaluate the overall framework. We begin with dataset description and then evaluate our model.

Table 1. Dataset description.

Number of videos	959
Resolution	2028 ×2704
Recording device	Panasonic Lumix FZ200
Length of each clip	≈0.5–1.5 s each
Environment	Outdoor
Lighting	Evenings, multiple
Players	20
Faulty sets	472
Correct sets	487
Ground truth	Double contact fault

4.1 Dataset

We employed a 'Panasonic Lumix FZ200' camera to record non-staged volleyball games being played in an outdoor playground of an educational institute. The camera has a frame rate of 60 frames and a resolution of 2028 × 2704. The game videos were recorded from multiple viewpoints on different days of data collection. However, they are mostly focused on the net area to record *set* events. The camera was placed on a static tripod at the height of 5 ft. The dataset was collected over a period of 6 months and involved 20 players. Different level of illumination and size of balls makes our dataset more challenging and realistic. Sample frames are given in Fig. 6. We manually process the game video to extract only the *set* events. After this process, we achieve a total of 959 *set* events, each spanning 0.5–1.5 s. The ground truth of these *set* events is generated by a professional volleyball referee. The dataset contains 472 faulty and 487 correct sets. The statistics of the dataset are summarized in Table 1.

4.2 Results of Ball Tracking and Detection

Original YOLOv4 is trained on 1000 classes of objects from the ImageNet dataset [9]. We are, however, only interested in the ball object. Therefore, we fine-tuned the YOLOv4 detection network for a single class, volleyball. We took 2124 images from raw full game volleyball videos (not the *set* events) and annotated these frames for volleyball location with center coordinates, width, and height. The training configurations are as follows. The batch size is taken as 64, the learning rate is 1e−3, the network height and width are *set* to 640 each with three channels. The initial pre-trained network weights used are yolov4.conv.137. The network is retrained with the above mentioned configurations and pre-trained weights for 6000 iterations. The training results are given in Fig. 7. We observe that after 1800 iterations, the training loss becomes near zero. On the test dataset, the detection accuracy increased from 89.9% to 99.1%. The main reason for this dramatic increase in the accuracy is a constrained capturing scenario. For double contact detection, the camera has to focus on the net area to capture the *set* event. Therefore, the camera orientation is mostly similar in all frames. In this way, we are able to obtain a highly accurate object detector for volleyball detection.

Despite a fine-tuned YOLOv4 volleyball detector, we still missed the ball in a few frames. In order to further improve ball detection, we consider a hybrid tracking algorithm in which we combine detections of fine-tuned YOLOv4 as well as CSRT based tracking. If YOLOv4 does not detect a ball, we initialize a CSRT tracker with the help of the bounding box in the previous frame. The tracker tracks the ball in the current frame almost accurately. If YOLOv4 gives a detection, we accept it as the ball bounding box. Hence, by incorporating both the YOLOv4 and object tracker, we are able to increase the accuracy from 99.1% to 99.8%.

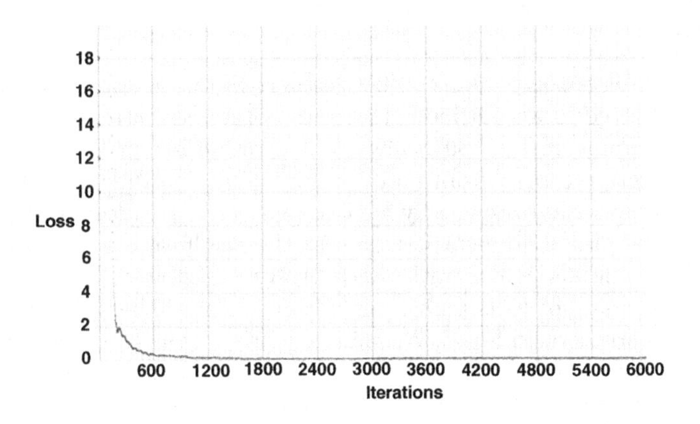

Fig. 7. Training loss during fine-tuning YOLOv4 for volleyball detection.

4.3 Results for the Double Contact Fault Classifier

In order to examine what region from the video should be focused on for double touch fault detection, we experiment on three different sets of videos, namely Type-1, Type-2, and Type-3. Type-1 videos are processed *set* events, i.e., videos from the dataset are considered in their original form. Type-3 represents the videos created by taking bounding boxes of the ball, while for Type-2, the bounding boxes are expanded in the vertical direction up to the image top. Also, note that Type-1 spans a whole *set* event while Type-2 and Type-3 span c^{th} to l^{th} frames only. The goal of feeding these distinct segments of video to the classifier is to determine whether the classifier can perform effectively on the bounding boxes containing only the ball or on the ball with any background region. Figure 8 demonstrates the creation of these three types of videos. Sample frames from Type-1, Type-2, and Type-3 are given in Fig. 9.

We train our MLP network using 5-fold cross-validation and report maximum and average accuracy with deviation in Table 2. We also compared the performance using Support Vector Machine (SVM) [5] classifiers with different kernels. These SVM classifiers are trained on the same deep features as MLP. We observe that all the classifiers perform better when the video of only the ball region, i.e., Type-3, is considered followed by Type-2, and then with Type-1. The best average accuracy on 5-fold cross-validation with Type-1, Type-2, and Type-3 videos are 63.82%, 64.44%, and 77.16%, respectively. Thus, the highest average accuracy is observed when Type-3 videos are used, followed by Type-2 videos and then Type-1 videos using the MLP classifier. The presence of other objects and their movement (in the case of Type-1 and Type-2) confuses the model and hence result in poor performance. Overall, the best performance is attained with the MLP network on Type-3 videos. However, the performance of SVM with a linear kernel (average accuracy is 76.33%) is comparable to the MLP network (average accuracy is 77.16%). We conclude that by sacrificing some accuracy, SVM, a lightweight model, can also be employed instead of a neural network-based classifier.

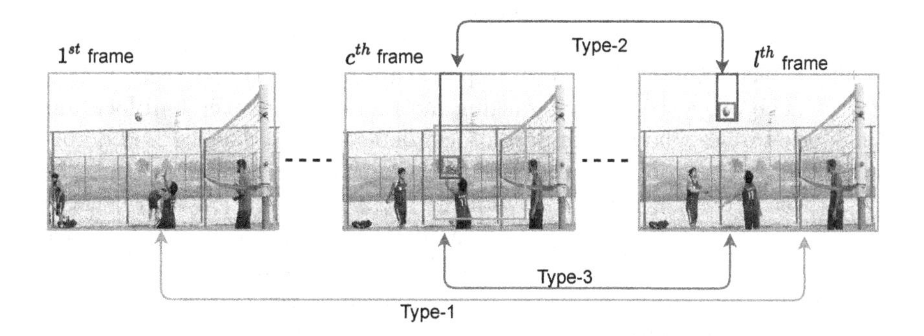

Fig. 8. Creation of Type-1, Type-2, and Type-3 videos

Table 2. Performance comparison of SVM and MLP based classifiers on Type-1, Type-2, and Type-3 videos

Classifier	Type-1	Type-2	Type-3
	(max, avg) accuracy (deviation)	(max, avg) accuracy, (deviation)	(max, avg) accuracy, (deviation)
SVM-linear	(64.58, 62.77), (±1.77)	(65.10, 64.13), (±1.17)	(**80.21, 76.33**), (±2.88)
SVM-rbf	(51.04, 50.89), (±0.21)	(73.44, 62.99), (±7.32)	(78.65, 76.74), (±1.74)
SVM-poly	(64.06, 61.84), (±1.85)	(68.75, 63.60), (±4.11)	(79.17, 75.91), (±2.34)
SVM-sigmoid	(58.85, 56.83), (±1.77)	(65.10, 56.82), (±5.26)	(70.83, 68.51), (±1.39)
MLP	(67.71, 63.82), (±2.45)	(68.23, 64.44), (±2.90)	(**80.21, 77.16**), (±2.39)

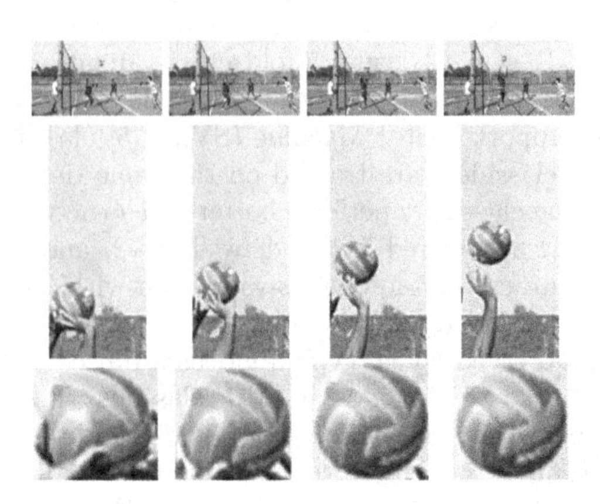

Fig. 9. The top row shows Type-1, middle row Type-2, and the bottom row Type-3 videos.

5 Conclusions

Detecting double contact fault automatically in a volleyball video is useful, but challenging. We have proposed a neural network based framework that detects the ball in each frame, obtains a deep feature vector, and detects if it is a faulty set. The proposed hybrid tracking method estimates the ball location in intermediate frames where a standard object detector fails. The method is able to estimate the ball location in most cases. Further, motion and appearance-based deep features extracted from a state-of-the-art 3D convolution neural network provide effective discrimination for the classifier. The proposed framework is able to detect double contact fault with 77.16% average accuracy on 5-fold cross-validation. In the future, we want to automatically detect the rotation fault based on the camera placed at the back of a volleyball court. It coordinates with another camera monitoring the referee and also uses the whistle signal to detect the service instance. We also want to separate the second touch from the first and the third touch in a given volleyball video.

Acknowledgements. This work is supported by the grant received from DST, Govt. of India for the Technology Innovation Hub at the IIT Ropar in the framework of National Mission on Interdisciplinary Cyber-Physical Systems. We also thank Devendra Raj for his help in data collection.

References

1. Bochkovskiy, A., Wang, C.Y., Liao, H.Y.M.: Yolov4: optimal speed and accuracy of object detection. arXiv preprint arXiv:2004.10934 (2020)
2. Bradski, G.: The OpenCV library. Dr. Dobb's J. Software Tools **25**, 120–125 (2000)
3. Carreira, J., Zisserman, A.: Quo Vadis, action recognition? A new model and the kinetics dataset. In: Proceedings of the IEEE Conference on Computer Vision and Pattern Recognition, pp. 6299–6308. IEEE, Honolulu, HI, USA (2017)

4. Chakraborty, B., Meher, S.: A trajectory-based ball detection and tracking system with applications to shot-type identification in volleyball videos. In: International Conference on Signal Processing and Communications, pp. 1–5. IEEE (2012)
5. Chang, C.C., Lin, C.J.: LibSVM: a library for support vector machines. ACM Trans. Intell. Syst. Technol. **2**(3), 1–27 (2011)
6. Chen, H.T., Tsai, W.J., Lee, S.Y., Yu, J.Y.: Ball tracking and 3D trajectory approximation with applications to tactics analysis from single-camera volleyball sequences. Multimedia Tools Appl. **60**(3), 641–667 (2012)
7. Chesaux, S.: Official volleyball rules 2017–2020 (2016). https://www.fivb.org/
8. Cuspinera, L.P., Uetsuji, S., Morales, F.O., Roggen, D.: Beach volleyball serve type recognition. In: Proceedings of the 2016 ACM International Symposium on Wearable Computers, pp. 44–45 (2016)
9. Deng, J., Dong, W., Socher, R., Li, L.J., Li, K., Fei-Fei, L.: ImageNet: a large-scale hierarchical image database. In: IEEE Conference on Computer Vision and Pattern Recognition, pp. 248–255. IEEE (2009)
10. Gageler, H.W., Wearing, S., James, A.D.: Automatic jump detection method for athlete monitoring and performance in volleyball. Int. J. Perform. Anal. Sport **15**(1), 284–296 (2015)
11. He, K., Zhang, X., Ren, S., Sun, J.: Spatial pyramid pooling in deep convolutional networks for visual recognition. IEEE Trans. Pattern Anal. Mach. Intell. **37**(9), 1904–1916 (2015)
12. He, K., Zhang, X., Ren, S., Sun, J.: Deep residual learning for image recognition. In: Proceedings of the IEEE Conference on Computer Vision and Pattern Recognition, pp. 770–778 (2016)
13. Hsu, C.C., Chen, H.T., Chou, C.L., Lee, S.Y.: Spiking and blocking events detection and analysis in volleyball videos. In: IEEE International Conference on Multimedia and Expo, pp. 19–24. IEEE (2012)
14. Kautz, T., Groh, B.H., Hannink, J., Jensen, U., Strubberg, H., Eskofier, B.M.: Activity recognition in beach volleyball using a deep convolutional neural network. Data Mining Knowl. Discov. **31**(6), 1678–1705 (2017)
15. Kurowski, P., Szelag, K., Zaluski, W., Sitnik, R.: Accurate ball tracking in volleyball actions to support referees. Opto-Electron. Rev. **26**(4), 296–306 (2018)
16. Liu, S., Qi, L., Qin, H., Shi, J., Jia, J.: Path aggregation network for instance segmentation. In: Proceedings of the IEEE Conference on Computer Vision and Pattern Recognition, pp. 8759–8768 (2018)
17. Lu, Y., An, S.: Research on sports video detection technology motion 3D reconstruction based on Hidden Markov Model. In: Cluster Computing, pp. 1–11 (2020)
18. NCAA: Division I women's volleyball annual committee (2013). http://www.ncaa.org/sites/default/files/Materials+to+post+on+website.pdf
19. Redmon, J., Farhadi, A.: Yolov3: an incremental improvement. arXiv preprint arXiv:1804.02767 (2018)
20. Schmidt, R.: Volleyball: Steps to Success. Human Kinetics (2015)
21. Shih, H.C.: A survey of content-aware video analysis for sports. IEEE Trans. Circ. Syst. Video Technol. **28**(5), 1212–1231 (2017)
22. Szegedy, C., et al.: Going deeper with convolutions. In: Proceedings of the IEEE Conference on Computer Vision and Pattern Recognition, pp. 1–9 (2015)
23. Szelag, K., Kurowski, P., Bolewicki, P., Sitnik, R.: Real-time camera pose estimation based on volleyball court view. Opto-Electron. Rev. **27**(2), 202–212 (2019)
24. Tran, D., Bourdev, L., Fergus, R., Torresani, L., Paluri, M.: Learning spatiotemporal features with 3D convolutional networks. In: Proceedings of the IEEE International Conference on Computer Vision, pp. 4489–4497 (2015)

25. Wang, C.Y., Liao, H.Y.M., Wu, Y.H., Chen, P.Y., Hsieh, J.W., Yeh, I.H.: CSPNet: a new backbone that can enhance learning capability of CNN. In: Proceedings of the IEEE/CVF Conference on Computer Vision and Pattern Recognition Workshops, pp. 390–391 (2020)
26. Zhou, L.: Sports video motion target detection and tracking based on Hidden Markov Model. In: 2019 11th International Conference on Measuring Technology and Mechatronics Automation, pp. 825–829. IEEE (2019)

Classroom Slide Narration System

K. V. Jobin(✉), Ajoy Mondal, and C. V. Jawahar

IIIT-Hyderabad, Hyderabad, India
jobin.kv@research.iiit.ac.in, {ajoy.mondal,jawahar}@iiit.ac.in
http://cvit.iiit.ac.in/research/projects/cvit-projects/csns

Abstract. Slide presentations are an effective and efficient tool used by the teaching community for classroom communication. However, this teaching model can be challenging for the blind and visually impaired (VI) students. As such student required a personal human assistance for understand the presented slide. This shortcoming motivates us to design a Classroom Slide Narration System (CSNS) that generates audio descriptions corresponding to the slide content. This problem poses as an image-to-markup language generation task. The initial step is to extract logical regions such as title, text, equation, figure, and table from the slide image. In the classroom slide images, the logical regions are distributed based on the location of the image. To utilize the location of the logical regions for slide image segmentation, we propose the architecture, Classroom Slide Segmentation Network (CSSN). The unique attributes of this architecture differs from most other semantic segmentation networks. Publicly available benchmark datasets such as WiSe and SPaSe are used to validate the performance of our segmentation architecture. We obtained 9.54% segmentation accuracy improvement in WiSe dataset. We extract content (information) from the slide using four well-established modules such as optical character recognition (OCR), figure classification, equation description, and table structure recognizer. With this information, we build a Classroom Slide Narration System (CSNS) to help VI students understand the slide content. The users have given better feedback on the quality output of the proposed CSNS in comparison to existing systems like Facebook's Automatic Alt-Text (AAT) and Tesseract.

Keywords: Slide image segmentation · Logical regions · Location encoding · Classroom slide narration

1 Introduction

Slide presentations play an essential role in classroom teaching, technical lectures, massive open online courses (MOOC), and many formal meetings. Presentations are an efficient and effective tool for the students to understand the subject better. However, the visually impaired (VI) students do not benefit from this type of delivery due to their limitations in reading the slides. Human assistants could help VI students to access and interpret the slide's content. However,

B. Raman et al. (Eds.): CVIP 2021, CCIS 1568, pp. 135–146, 2022.
https://doi.org/10.1007/978-3-031-11349-9_12

it is impossible and impractical for the VI students to have a human reader every time; because of the cost and the limited availability of trained personnel. From this perspective, this would be a primary issue in modern classrooms.

The automatic extraction of logical regions from these slide images is the initial step for understanding the slide's content. This can be defined as, segmentation of semantically similar regions, closely related to natural image semantic segmentation in computer vision [5,16,32]. The goal is to assign logical labels like title, paragraph, list, equation, table, and figure to each pixel in the slide images. Due to the large variability in theme (i.e., the layout), style, and slide content, extracting meaningful regions becomes a challenging task. The text extraction from these regions is beneficial for the VI student only if the text is tagged with its logical function. The left image of Fig. 1, the plain text "*Discovering Attention Patterns*" is more meaningful to the VI student only if he knows the text is the heading of the slide. Similarly, detecting figures, equations, and table regions is also essential for understanding the slides. Our goal is to develop a system that meaningfully describes the classroom slides with a markup text, as shown in Fig. 1.

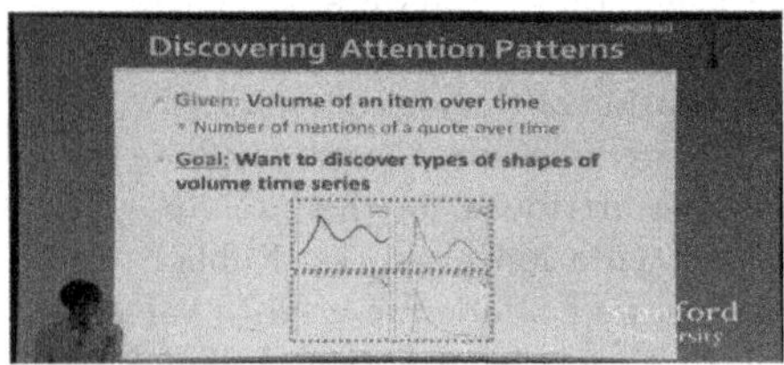

Alt text: A screen shot of a person

Tesseract OCR: Volume of an item over time tions of a quote over time

Ours: <heading> Discovering Attention Patterns </heading> <list> = Given: Volume of an item over time Number of mentions of a quote over time </list> <list> Goal: Want to discover types of shapes of volume time series </list> <fig- ure>Line graph</figure>

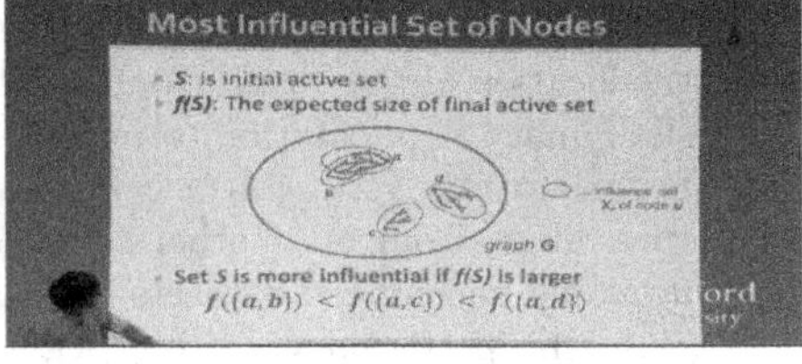

Alt text: A screenshot of a cell phone

Tesseract OCR:

Ours: <heading> Most Influential Set of Nodes </heading> <list> S is initial active set f of s: The expected size of final active set </list> <list> Set S is more influential if f(S) is larger </list> <Figure> Venn Diagram </Figure> <Figuretext> O ... influence set X, of node u </Figure- text><equation> f(a,b") < f(a,c") < f(a,d)</equation>

Fig. 1. Illustrate qualitative comparison between outputs obtained by the proposed CSNS and the existing assistive systems—Automatic Alt-Text (AAT) [30] and Tesseract OCR [26]. The proposed system generates a markup text, where each logical content is tagged with its appropriate logical function. The audio file generated using this markup text gives a clear picture of the classroom slide's content to the blind or VI students.

The classroom slides have a complex design. The exact text could appear in various labels such as "heading", "paragraph", and "list". Hence, only the visual features learned by the existing semantic segmentation networks (e.g., FCN [16], Deeplab [5], and PSPNet [32]) are not sufficient to distinguish various logical regions. Limited works [11,20] on classroom slide segmentation utilize (x, y) coordinate values to enhance the quality of segmentation results. However, all these works [11,20] do not utilize location information efficiently for segmentation. In contrast, Choi *et al.* [6] utilize the position encoding [27] to impose the height information in HANet for semantic segmentation of urban-scene images. In the case of slide images, the height and width information of each logical region are equally crucial for segmenting regions accurately. In this work, we propose a Classroom Slide Segmentation Network, called CSSNet to segment slide images accurately. The proposed network consists of (i) Attention Module, which utilizes the location encoding of the logical region (height and width of a region), and (ii) Atrous Spatial Pyramid Pooling (ASPP) module, which extracts multi-scale contextual features. The experiments on publicly available benchmark datasets WiSe [11] and SpaSe [20] establish the effectiveness of the proposed LEANet over state-of-the-art techniques—Haurilet *et al.* [11], Haurilet *et al.* [20], HANet [6], DANet [9], and DRANet [8].

The proposed Classroom Slide Narration System (CSNS) creates audio content for helping VI students to read and understand the slide's content. The audio created corresponds to the segmented region's content in reading order. We use four existing recognizers —OCR [26], equation descriptor [19], table recognizer [23], and figure classifier [12] to recognize content from the segmented regions. We evaluate the effectiveness of the developed system on the WiSe dataset [11] and compare the user experiences with the existing assistive systems AAT [30] and Tesseract OCR [26]. The proposed system attains a better user experience than the existing assistive systems.

The contributions of this work are as follows:

- Proposes a CSSNet to efficiently segment the classroom slide images by utilizing the location encoding.
- Presents a CSNS as an use case of slide segmentation task to narrate slide's content in reading order. It helps the VI students to read and understand the slide's content.
- Perform experiments on publicly available benchmark datasets WiSe and SPaSe establish the effectiveness of the proposed CSSNet and CSNS over existing slide segmentation and assistive techniques.

2 Related Work

Visual Assistance System: According to the World Health Organization (WHO), at least 2.2 billion people have a vision impairment or blindness worldwide. The computer vision community has undertaken significant efforts to deploy automated assistive systems for such people. Some of the unique assistive systems based on computer vision are—(i) vOICe[1] [2] vision technology is a sensory substitution system for blind people which gives visual experience through live camera views by image-to-sound renderings, (ii) Electro-Neural Vision System (ENVS) [18] provides a virtual perception of the three-dimensional profile and color of the surroundings through pulses, (iii) a wearable travel aid for environment perception and navigation [3], (iv) a smartphone-based image captioning for the visually and hearing impaired people [17]. (v) Currency Recognition using

[1] https://www.seeingwithsound.com.

Mobile Phones [25] (vi) Facebook's Automatic Alt-Text (AAT) [30] system is one of the most widely used assistive features, which automatically recognizes a set of 97 objects present in Facebook photos, (vii) FingerReader [24] is a wearable text reading device to assist blind people and dyslexic readers. All these existing assistive systems fail to properly narrate the classroom slides for a VI student due to various logical regions present in the slides.

Document Image Segmentation: Textual regions and graphics regions dominate document images. Graphics include natural images, various types of graphs, plots, and tables. Historically, the document image segmentation task is considered as dividing a document into various regions. Existing techniques [1,10] are either bottom-up or top-down. In the case of bottom-up [1], individual words are first detected based on hand-crafted features. Therefore, detected words are grouped to form lines and paragraphs. Segmenting textual regions (like paragraphs) into lines and words helps to find semantic regions present in the document in the case of top-down [10]. These methods mostly use feature engineering and do not necessitate any training data while applying clustering or thresholding.

With the recent advancement in deep convolutional neural networks (CNNs), several neural-based models [4,22,28,29,31] have been proposed to segment document images. Chen *et al.* [4] considered a convolutional auto-encoder to learn features from cropped document image patches. Thereafter, those features are used to train an SVM classifier [7] to segment document patches. Vo *et al.* [28] and Renton *et al.* [22] proposed lines detection algorithm in the handwritten document using FCN. In [31], Yang *et al.* proposed an end-to-end, multi-modal, fully convolutional network for extracting semantic structures from document images. Inspired by document image segmentation and natural image segmentation using deep CNNs, Haurilet *et al.* [11,20] explore DeepLab [5] architecture on classroom slide segmentation task. More recently, DANet [9], HANet [6], and DRANet [8] capture long-range dependency to improve performance by extending the self-attention mechanism. The proposed LEANet is strongly influenced by DANet [9], HANet [6], and DRANet [8].

3 Classroom Slide Narration System

We develop a CSNS as a use case of slide segmentation to help VI students to read and understand slide's content without human assistance. The proposed CSNS consists of three modules—(i) **Slide Segmentation Module:** to identify logical regions present in the slide, (ii) **Information Extraction Module:** to extract meaningful information from each segmented region, and (iii) **Audio Creation Module:** to create an audio narration of extracted information in proper order. We discuss each of these modules in detail.

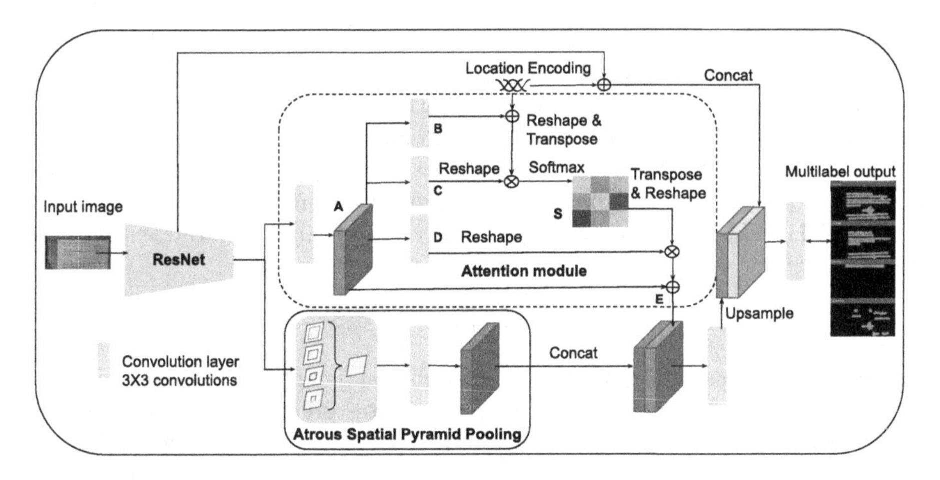

Fig. 2. Illustrate the architecture of the proposed CSSNet for classroom slide segmentation. The network consists of three modules—(i) attention module (upper dotted region), (ii) multi-scale feature extraction module (lower region), (iii) feature concatenation module. Here, $\oplus$ and $\otimes$ represent the element-wise summation and multiplication of features, respectively.

3.1 Slide Segmentation Module

CSSNet is the segmentation network utilized in this work to segment classroom slide images accurately. The CSSNet aims to learn more discriminating features for semantic associations among pixels through the location encoding attention mechanism. We propose a location encoded attention module that utilizes the location encoding of logical regions of slide image. To enhance the segmentation accuracy, we concatenate the attention module output feature with the multi-scale features through an Atrous Spatial Pyramid Pooling (ASPP) layer [5] as shown in the Fig. 2. Finally, a simple decoder network predicts the high-resolution, multi-label, dense segmentation output.

Attention Module: The previous attempts [5,32] suggest that the local features generated by traditional Fully Convolutional Networks (FCNs) could not capture rich context, leading to erroneous object detection. While Choi *et al.* [6] discussed that urban-scene images have common structural priors depending on a spatial position. The authors also discussed height-wise contextual information to estimate channels weights during pixel-level classification for urban-scene segmentation. Fu *et al.* [9] also discussed the self-attention mechanism to capture the spatial dependencies between any two positions of the feature maps. Inspired by the works [6,9], we introduce an attention module that utilizes location encoding to enhance the local feature's contextual information and relative position. The attention module selectively encodes a broader range of contextual and location information into each pixel according to semantic-correlated relations, thus enhancing the discrimination power for dense predictions.

As illustrated in Fig. 2 (inside the dotted box), we use a convolution layer to obtain features of reduced dimension $\mathbf{A} \in \mathbb{R}^{C\times H\times W}$. We feed $\mathbf{A}$ into a convolution layer to generate two new feature maps $\mathbf{B}$ and $\mathbf{C}$, respectively. $\{\mathbf{B},\mathbf{C}\} \in \mathbb{R}^{C\times H\times W}$. We add the location encoding $\mathbf{L}$ (refer Sect. 3.1) to the $\mathbf{B}$ and obtain $\mathbf{L}+\mathbf{B}$. Then we reshape it to $\mathbb{R}^{C\times N}$, where $N = H \times W$ is the number of pixels. We perform a matrix multiplication between the transpose of $\mathbf{C}$ and $\mathbf{B}+\mathbf{L}$, and apply a soft-max layer to obtain spatial attention map $\mathbf{S} \in \mathbb{R}^{N\times N}$:

$$s_{ji} = \frac{\exp\left((B_i + L_i)\cdot C_j\right)}{\sum_{i=1}^{N}\exp\left((B_i + L_i)\cdot C_j\right)}, \tag{1}$$

where s_{ji} measures the impact of ith position on j^{th} position. A more similar feature representation of two positions contributes greater correlation between them. Meanwhile, we feed feature $\mathbf{A}$ into a convolution layer to generate a new feature map $\mathbf{D} \in \mathbb{R}^{C\times H\times W}$ and reshape it to $\mathbb{R}^{C\times N}$. Then we perform a matrix multiplication between $\mathbf{D}$ and transpose of $\mathbf{S}$ and reshape the result to $\mathbb{R}^{C\times H\times W}$. Finally, we multiply it by a scale parameter α and perform an element-wise sum with the features $\mathbf{A}$ to obtain the final output $\mathbf{E} \in \mathbb{R}^{C\times H\times W}$ using:

$$E_j = \alpha \sum_{i=1}^{N} (s_{ji} D_i) + (1-\alpha) \times A_j, \tag{2}$$

where α is initialized as 0 and learned gradually.

Location Encoding: We follow strategy proposed by [6,27] for location encoding and use sinusoidal position encoding [27] as $\mathbf{L}$ matrix. In our approach, we give importance to both height and width location encoding. The dimension of the positional encoding is the same as the channel dimension C of the intermediate feature map that combines with $\mathbf{L}$. The location encoding is defined as

$$\begin{aligned} PE_v(v,2i) &= \sin\left(v/100^{2i/C}\right); \quad PE_v(v,2i+1) &&= \cos\left(v/100^{2i/C}\right), \\ PE_h(h,2i) &= \sin\left(h/100^{2i/C}\right); \quad PE_h(h,2i+1) &&= \cos\left(h/100^{2i/C}\right), \end{aligned}$$

where v and h denote the vertical and horizontal location index in the image ranging from 0 to $H/r-1$ and 0 to $W/r-1$, and i is the channel location index ranging from 0 to $C-1$. The r is called the reduction ratio, which determines the frequency of sin or cos wave. Here, we apply sin and cos location encoding for intermediate layers of the features. The new representation $\mathbf{L}$ incorporating location encoding is formulated as

$$\mathbf{L} = \beta PE_h \oplus (1-\beta) PE_v,$$

where $\oplus$ is an element-wise sum, β is initialized with 0 and learned gradually. Adding both the horizontal and vertical position encoding enables the attention module to learn the location information in both directions. The β value is learned based on the importance of width and height locations. Figure 3 shows the location encoding value of a layer in **L** matrix. Finally, we add random jitters to the location encoding PE_v and PE_h for the better generalization of location encoding. The height and width of the location encoding matrix **L** are adjusted to the dimension of the feature vector.

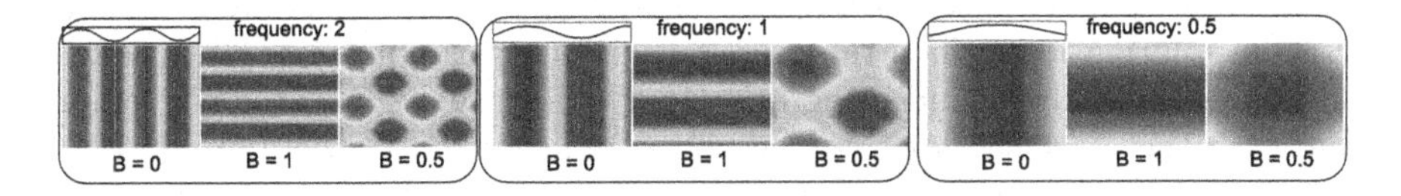

Fig. 3. Presents visualization of various location encoding with frequency 2, 1, and 0.5 with β values 0, 0.5, and 1.

3.2 Information Extraction Module

We coarsely group the classroom slide image into four logical regions—text, figure, equation, and table and use existing recognizers to recognize the content of the regions.

OCR: We use well-known Optical Character Recognition (OCR), Tesseract [26] to recognize the content of the text regions present in slide images.

Figure Classification: Figures mainly different types of plots, sketches, block diagrams, maps, and others appear in slides. We use DocFigure [12] to find the category of each figure in slide images.

Equation Description: The equation frequently occurs in the classroom slides. We follow MED model [19] for describing equation in natural language to interpret the logical meaning of equation for VI students.

Table Structure Recognition: Recognition of the table present in the slide includes detecting cells, finding an association between cells, and recognizing cell's content. We use TabStruct-Net [23] to recognize the physical structure of the table and then Tesseract [26] to recognize content.

The information from the four modules is combined and saved as a JSON file. These processes are done on the server and then sent to the JSON file to the mobile app, and the VI student understands the slide's content by interacting with the app.

3.3 Audio Creation Module

Based on a study that empirically investigates the difficulties experienced by VI internet users [21], a mobile application is developed for VI students to read and understand the classroom slide's content. For this, we place a camera in the classroom connected to the server for capturing slide images. The VI students use this mobile app to connect to the server through a WiFi connection. Using the mobile application (**CAPTURE** button), the student can request the server to get information on the current slide. The server captures the current slide image through the camera, extracts the information as JSON file, and sends it back to the mobile application. Figure 4(a) shows a screenshot of the developed mobile app with one example. The app has two modes—(i) interactive and (ii) non-interactive. In the interactive mode, the user can navigate various logical regions of slides by themselves. The app shows the captured slide image with all extracted information. When the user touches one particular area, it automatically plays the audio sound using android TTS library of that region's content along with the category label. In the non-interactive mode, the app automatically plays the audio corresponding to the slide's content. In this case, the screen contains the **READ ALL** button; when the user touches this particular area, the app automatically plays the audio sound of the full slide's content. These features are demonstrated in the demo video[2]

4 Experiments

4.1 Experimental Setup

We use the residual network (e.g., ResNet-101) pre-trained on ImageNet [13] as the backbone network for all the experiments. We replace a single 7×7 convolution with three of 3×3 convolutions in the first layer of ResNet-101. We use We also adopt an auxiliary cross-entropy loss in the intermediate feature map. Since for the multi-label segmentation task we need to predict multiple classes per pixel, we replace the softmax output layer in the previous deep models with a sigmoid activation function and train these models using binary cross entropy loss. We train the network using SGD optimizer with an initial learning rate of $2e-2$ and momentum of 0.9. We set the weight decays of $5e-4$ and $1e-4$ for backbone networks and segmentation networks, respectively. The learning rate schedule follows the polynomial learning rate policy [15]. The initial learning rate is multiplied by $(1 - \frac{iteration}{max\ iteration})^{0.9}$. We use typical data augmentations such as random scaling in the range of $[0.5, 2]$, Gaussian blur, color jittering, and random cropping to avoid over-fitting. We set crop size and batch size of 540×720 and 2, respectively. We use mean Intersection over Union (mIoU) [5] and Pixel Accuracy (PA) [11,20] for evaluation purpose.

4.2 Dataset

WiSe: The WiSe [11] dataset consists of 1300 slide images. It is divided into a training set, validation set, and test set consisting of 900, 100, and 300 slide images annotated with 25 overlapping region categories.

SPaSe: The SpaSe [20] dataset contains dense, pixel-wise annotations of 25 classes for 2000 slide images [20]. It is split into training, validation, and test sets containing 1400, 100, and 500 images, respectively.

[2] https://youtu.be/PnPYrA8ykF0.

4.3 Ablation Study

In the ablation study, we use the ResNet-101 backbone with an output stride of 16 and evaluate the WiSe dataset. First, we examine the best feature vector for adding location encoding in the LEANet architecture. Based on Fig. 2 of CSSNet architecture, the location encoding can be added to any of the feature vectors including, **A**, **B**, **C**, and **D**. We empirically find the best feature for adding location encoding. In addition to it, we also conduct experiments by changing the frequency of the sin wave encoding in the **L** matrix. Finally, we also compare with CoordConv [14] approach.

Table 1. Shows ablation studies and the impact of hyper-parameters on a validation set of the WiSe dataset. Table (a) shows the result of segmentation by adding location encoding to the features **A**, **B**, **C**, and **D**. Table (b) shows the effect of manual β values and as a learnable parameter.

Features				Frequency	mIoU
A	**B**	**C**	**D**		
				1	51.07
✓				1	51.44
	✓			1	52.26
		✓		1	52.12
			✓	1	50.04
	✓			2	51.14
	✓			0.5	50.32
A + CoordConv [14](width+ height)					51.12

(a)

β	mIoU
0	51.73
1	50.25
0.5	51.83
learning	52.26

(b)

Table 1 shows the ablation studies conducted in the proposed network architecture by changing the various hyper-parameters. The Table 1(a) shows that the location encoding on the feature vector outperformed the traditional CoordConv [14] approach. The experiment also shows that adding location encoding to the intermediate feature **B** and **C** is comparatively better than adding it to **A** and **D**. The best frequency of sin wave in the **L** is found as 1. While Table 1 (b) shows the learnable β value outperforms the manually assigned β value.

4.4 Comparison with State-of-the-Art Techniques

Comparison results of the proposed network with state-of-the-art techniques are presented in Table 2. From the table, we observe that none of the attention mechanisms (e.g., HANet [6], DANet [9], and DRANet [8]) is better than Haurilet *et al.* [20] on SPaSe dataset. While in the case of WiSe dataset, all attention architectures HANet [6], DANet [9], and DRANet [8] are better than Haurilet *et al.* [11]. DRANet [8] obtains the best performance (35.77% mean IoU and 78.64% PA) among all existing techniques on SPaSe dataset. While DANet [9] obtains the best performance (44.85% mean IoU and 88.80% PA) among all existing techniques on WiSe dataset. The proposed network obtains the best results (46.74% mean IoU and 89.80% PA) on Wise and (36.17% mean IoU and 79.11% PA) on SPaSe dataset compared to the existing techniques.

Table 2. Illustrates the comparison of the proposed method with state-of-the-art techniques on benchmark datasets WiSe and SPaSe datasets.

Methods	Datasets			
	SPaSe		WiSe	
	Mean IoU	PA	Mean IoU	PA
Haurilet *et al.*	35.80	77.40	–	–
Haurilet *et al.*	–	–	37.20	88.50
HANet	32.37	76.54	39.35	88.72
DANet	33.12	77.39	44.85	88.80
RANet	35.77	78.64	43.31	88.77
Ours	36.17	79.11	46.74	89.80

4.5 Narration System Evaluation

We establish the effectiveness of the developed slide narration system by comparing the performance with Facebook's Automatic Alt-Text (AAT) [30] and Tesseract OCR [26]. We use 300 test images of the WiSe dataset for evaluating the system. 30 volunteers who can read and write English are selected. Each volunteer takes 300 slide images and evaluates corresponding audio outputs generated by various techniques. The volunteer gives a grade (value in the range [0, 10]) to the generated audio files by various techniques by comparing them with the original slide content. We collect all these grades and plot them as a bar chart (shown in Fig. 4(b)). The user gives the best grade for the audio file corresponding to the ground truth segmentation of the slide. However, the proposed CSNS also performs comparatively better than the existing techniques.

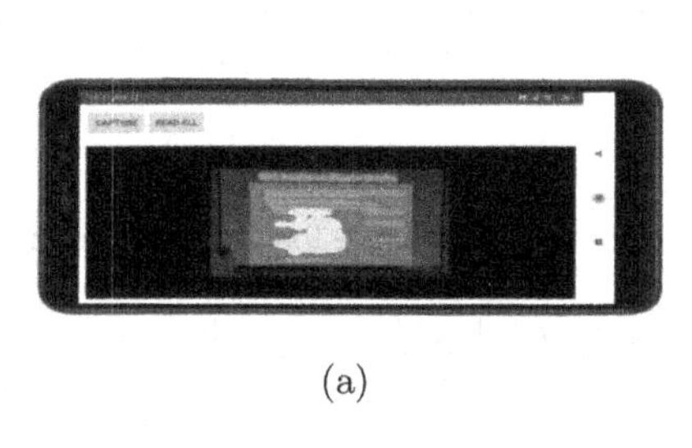

(a)

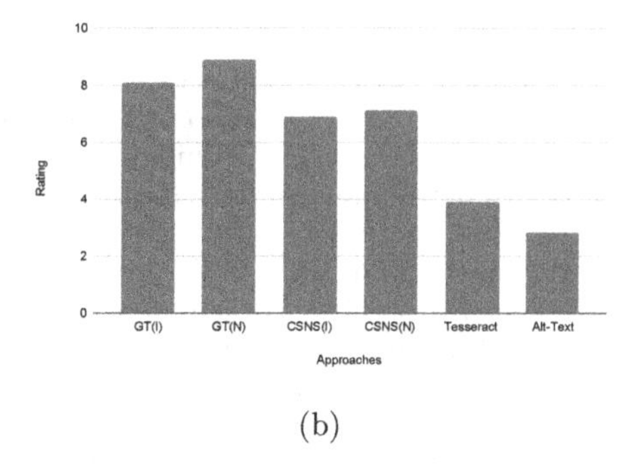

(b)

Fig. 4. **(a):** presents user interface of the developed mobile application. The mobile screen displays the camera-captured slide image. Various colors highlight the slide's logical regions for better visualization. The READ ALL button plays generated audio sound corresponding to the slide's content. **(b):** presents user rating on performance of various approaches. **I** and **N** indicate **interactive** and **non-interactive** modes, respectively.

5 Summary

This paper presents a classroom slide segmentation network to segment classroom slide images more accurately. The attention module in the proposed network enriches the learned contextual features by utilizing the location encoding of the logical region. We develop a Classroom Slide Narration System as an application of classroom slide segmentation for helping VI students to read and understand the slide's content. Experiments on benchmark datasets WiSe and SpaSe conclude that the proposed LEANet achieves the best slide segmentation results as compared to the state-of-the-art methods. We also demonstrate the effectiveness of our developed Classroom Slide Narration System over existing assistive tools in terms of user experiences.

References

1. Amin, A., Shiu, R.: Page segmentation and classification utilizing bottom-up approach. In: IJIG (2001)
2. Auvray, M., Hanneton, S., O'Regan, J.K.: Learning to perceive with a visuo-auditory substitution system: localisation and object recognition with 'the voice'. In: Perception (2007)
3. Bai, J., Liu, Z., Lin, Y., Li, Y., Lian, S., Liu, D.: Wearable travel aid for environment perception and navigation of visually impaired people. In: Electronics (2019)
4. Chen, K., Seuret, M., Liwicki, M., Hennebert, J., Ingold, R.: Page segmentation of historical document images with convolutional autoencoders. In: ICDAR (2015)
5. Chen, L.C., Papandreou, G., Kokkinos, I., Murphy, K., Yuille, A.L.: Deeplab: Semantic image segmentation with deep convolutional nets, atrous convolution, and fully connected CRFs. In: PAMI (2017)
6. Choi, S., Kim, J.T., Choo, J.: Cars can't fly up in the sky: improving urban-scene segmentation via height-driven attention networks. In: CVPR (2020)
7. Cortes, C., Vapnik, V.: Support-vector networks. In: ML (1995)

8. Fu, J., Liu, J., Jiang, J., Li, Y., Bao, Y., Lu, H.: Scene segmentation with dual relation-aware attention network. IEEE Trans. Neural Netw. Learn. Syst. (2020)
9. Fu, J., et al.: Dual attention network for scene segmentation. In: CVPR (2019)
10. Ha, J., Haralick, R.M., Phillips, I.T.: Recursive xy cut using bounding boxes of connected components. In: ICDAR (1995)
11. Haurilet, M., Roitberg, A., Martinez, M., Stiefelhagen, R.: WiSe - slide segmentation in the wild. In: ICDAR (2019)
12. Jobin, K., Mondal, A., Jawahar, C.: Docfigure: a dataset for scientific document figure classification. In: ICDARW (2019)
13. Krizhevsky, A., Sutskever, I., Hinton, G.E.: Imagenet classification with deep convolutional neural networks. In: NIPS (2012)
14. Liu, R., et al.: An intriguing failing of convolutional neural networks and the coordconv solution. arXiv preprint arXiv:1807.03247 (2018)
15. Liu, W., Rabinovich, A., Berg, A.C.: Parsenet: looking wider to see better. arXiv preprint arXiv:1506.04579 (2015)
16. Long, J., Shelhamer, E., Darrell, T.: Fully convolutional networks for semantic segmentation. In: CVPR (2015)
17. Makav, B., Kılıç, V.: Smartphone-based image captioning for visually and hearing impaired. In: ELECO (2019)
18. Meers, S., Ward, K.: A vision system for providing 3d perception of the environment via transcutaneous electro-neural stimulation. In: ICIV (2004)
19. Mondal, A., Jawahar, C.V.: Textual description for mathematical equations. In: ICDAR (2019)
20. Monica Haurilet, Z.A.H., Stiefelhagen, R.: Spase - multi-label page segmentation for presentation slides. In: WACV (2019)
21. Murphy, E., Kuber, R., McAllister, G., Strain, P., Yu, W.: An empirical investigation into the difficulties experienced by visually impaired internet users. In: UAIS (2008)
22. Renton, G., Chatelain, C., Adam, S., Kermorvant, C., Paquet, T.: Handwritten text line segmentation using fully convolutional network. In: ICDAR (2017)
23. Raja, S., Mondal, A., Jawahar, C.V.: Table structure recognition using top-down and bottom-up cues. In: ECCV (2020)
24. Shilkrot, R., Huber, J., Meng Ee, W., Maes, P., Nanayakkara, S.C.: Fingerreader: a wearable device to explore printed text on the go. In: ACM Conference on HFCS (2015)
25. Singh, S., Choudhury, S., Vishal, K., Jawahar, C.: Currency recognition on mobile phones. In: ICPR. IEEE (2014)
26. Smith, R.: An overview of the Tesseract OCR engine. In: ICDAR (2007)
27. Vaswani, A., et al.: Attention is all you need. arXiv preprint arXiv:1706.03762 (2017)
28. Vo, Q.N., Lee, G.: Dense prediction for text line segmentation in handwritten document images. In: ICIP (2016)
29. Wick, C., Puppe, F.: Fully convolutional neural networks for page segmentation of historical document images. In: IWDAS (2018)
30. Wu, S., Wieland, J., Farivar, O., Schiller, J.: Automatic alt-text: computer-generated image descriptions for blind users on a social network service. In: ACM Conference on CSCWSC (2017)
31. Yang, X., Yumer, E., Asente, P., Kraley, M., Kifer, D., Giles, C.L.: Learning to extract semantic structure from documents using multimodal fully convolutional neural networks. In: CVPR (2017)
32. Zhao, H., Shi, J., Qi, X., Wang, X., Jia, J.: Pyramid scene parsing network. In: CVPR (2017)

Humanoid Robot - Spark

Kunal Gawhale, Aditya Chandanwar, Rujul Zalte, Suhit Metkar, Sahana Kulkarni, Abhijith Panikar, Labhesh Lalka, Arti Bole, and Santosh Randive(✉)

Department of Electronics and Telecommunication Engineering, Pimpri Chinchwad College of Engineering and Research, Pune, Maharashtra, India
santosh.randive@pccoer.in

Abstract. This paper describes the system overview and biped walking mechanism of the humanoid robot, Spark. Spark is able to mimic various human expressions, hand gestures, walking actions, recognize various human faces and objects and perform vocal communication. Spark has a total of 66 Degrees of Freedom (DOF). Each arm has 20 DOF and each leg has 6 DOF. The head has 12 DOF and Trunk has 2 DOF. The Broadcom BCM2711 processor is used for all the major operations. It is also acting as a Master control unit for other microcontrollers present in different parts of Spark's body. This system is based on the Raspbian Operating System. Mechanical assembly is based on a hybrid system which includes both serial and parallel operations.

Keywords: Humanoid robots · Gesture · Posture · Social spaces and facial expressions · Recognition · Autonomous · Human machine interaction · Biped walking

1 Introduction

Human life is of utmost importance. More and more inventions are made to avoid direct association of human life with tasks that are life risking. Also, the world today is driven by cost and time effective solutions to problems ranging from the simplest tasks such as counting money to one of the most difficult tasks such as deep marine research. Automation is required in almost every field possible. Humanoid is possibly the best solution mankind can offer itself at this hour of need. A humanoid is basically a robot which can perform all the tasks a human can do. The most important of a humanoid's capabilities is that it can think, learn and process just like a human i.e., it has intelligence which matches human beings. We have created a machine that is able to walk, see, listen and talk. Spark can recognize and respond to it accordingly. Most of the humanoids present today are either good at facial expressions or biped walking or hand gestures, etc. but with Spark, our aim was to develop a fully functional humanoid which can perform all the actions together with proper synchronization. In the past decade, the development of humanoid robots has boosted significantly. There are many humanoid robots such as ASIMO [1], ATLAS [2], KHR3 [3], etc. who are currently leading the industry but these are very costly robots compared to

B. Raman et al. (Eds.): CVIP 2021, CCIS 1568, pp. 147–158, 2022.
https://doi.org/10.1007/978-3-031-11349-9_13

Spark. Spark is a low cost optimally designed robot in the field of humanoid robots. The face recognition algorithm is based on OpenCV [4] and Python. Spark's listening and talking algorithm is based on speech recognition [5] and python's text to speech library features [6]. Spark walks by collecting data from accelerometer, gyroscope, pressure sensors in unison with the gimbal mechanism and the corresponding leg actuators. A Raspberry Pi 4 [7] acts as the master controller, controlling all the other ATmega CH2560 [8] microcontrollers which act as slave controllers, controlling actuators at various joints of the robot's body. Spark sees through two image sensors [9] connected to the Raspberry Pi 4. Spark listens and talks using a microphone and a speaker respectively. Spark's walking algorithm is based on a series parallel hybrid mechanism [10].

2 Block Diagram

The robot is divided into four sub-systems and their functioning is interdependent on each other.

2.1 Spark Face

Figure 1 describes the block diagram and circuit configuration of Spark's face. The face consists of the main control board - a Raspberry Pi 4. This controller acts as the 'brain' of the robot controlling the entire body. The microphone, speaker and camera act as the ears, mouth and eyes of Spark respectively. An Arduino Mega is connected to this master controller which acts as a slave controller. The master and slave controller communicate with each other using serial communication [11]. The motors are connected through a motor driver [12] which drives the motor through Pulse Width Modulated (PWM) signal. The Arduino Mega communicates with the motor driver using I2C communication protocol [13].

2.2 Spark Torso and Arm

Figure 2 describes the block diagram and circuit configuration of Spark's torso and arms. The arms and torso are also controlled by another Arduino Mega board which is serially connected to the Raspberry Pi 4. The motors in the arms are driven using the motor driver by PWM signals. The power supply is common and it is taken through SMPS only and the voltage is regulated as per the device requirement.

2.3 Spark Gimbal

Figure 3 describes the block diagram and circuit configuration of the gimbal mechanism in Spark. The gimbal is an independent system which is used to balance the upper body. It is also used as an interconnection between the upper and lower body. It consists of an Arduino Nano board [14] which controls the

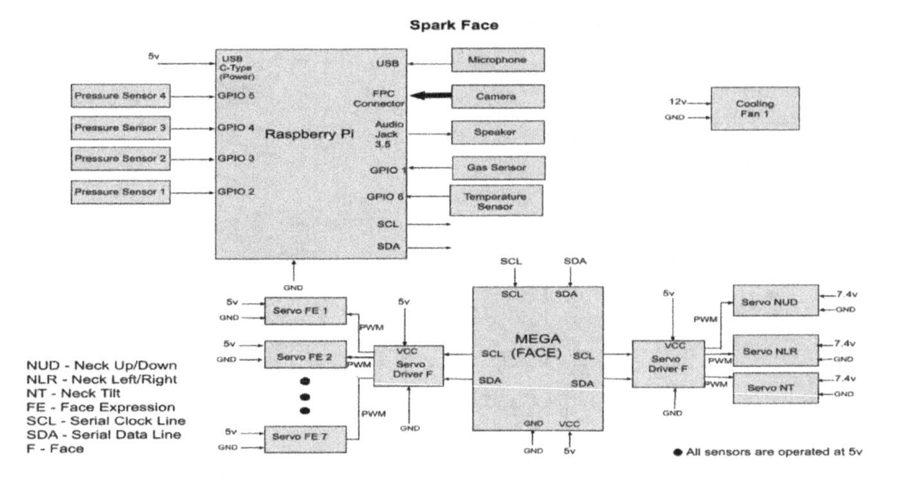

Fig. 1. Block diagram of face.

servo motor attached to the platform. Gyroscope and accelerometer sensor which senses the acceleration and direction is also connected to it. The system is a closed loop system. Continuous Feedback is taken from the IMU sensor [15] and the servomotors.

2.4 Spark Legs

Figure 4 describes the block diagram and circuit configuration of Spark's legs. Legs consists of an Arduino Mega board which takes input from Raspberry Pi 4 through serial communication. When the voice command to walk is received, the motors in the leg move in a specific pattern as defined in the algorithm loaded into the master controller. Whenever there is an additional load at one side, the signals to motors are adjusted so that the weight is shifted to the other side of the body and pressure is equalised on both sides [16].

3 Methodology

3.1 Hardware Architecture

Controller Network. The brain of Spark is a Raspberry Pi 4 motherboard acting as a master unit which is further connected with different Arduino boards acting as slave controllers. Each slave controller is connected with a motor driver which further gives commands to different servo motors.

Master Controller. The master controller of Spark is a Raspberry Pi 4 microcomputer which is based on Broadcom BCM2711 processor [7]. The main reason for using this as a master controller is that it has a 64 bit Quadcore processor

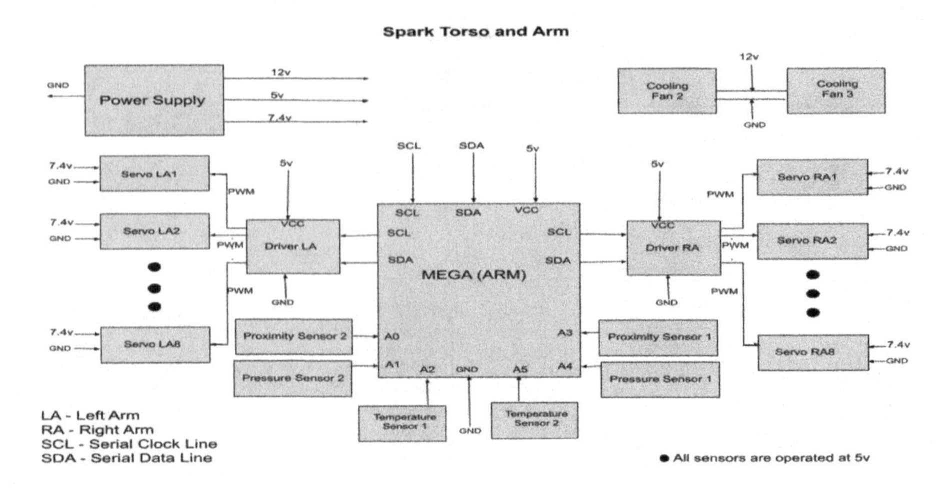

Fig. 2. Block diagram of Torso and Arms.

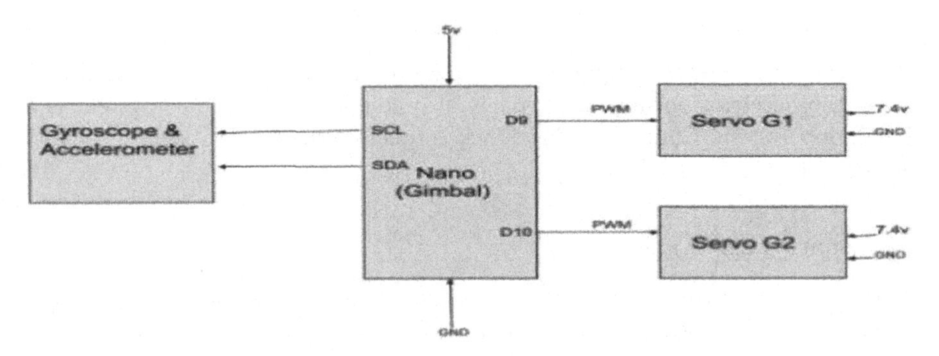

Fig. 3. Block diagram of gimbal mechanism.

consisting of 4 GB RAM which can help in faster execution of the programs. It can also sustain heavy processes executed on it. Raspberry Pi provides a wide range of peripheral interfacing. As a result, multiple hardware could be interfaced with it.

Slave Controller. An Arduino controller board in different parts of Spark's body is acting as a slave controller. Slave controller is nothing but a controller which accepts commands from the master control board (i.e. Raspberry Pi in this case) and acts according to the instructions which are received. There are conditions defined in the slave board and if those conditions satisfy then only the boards react.

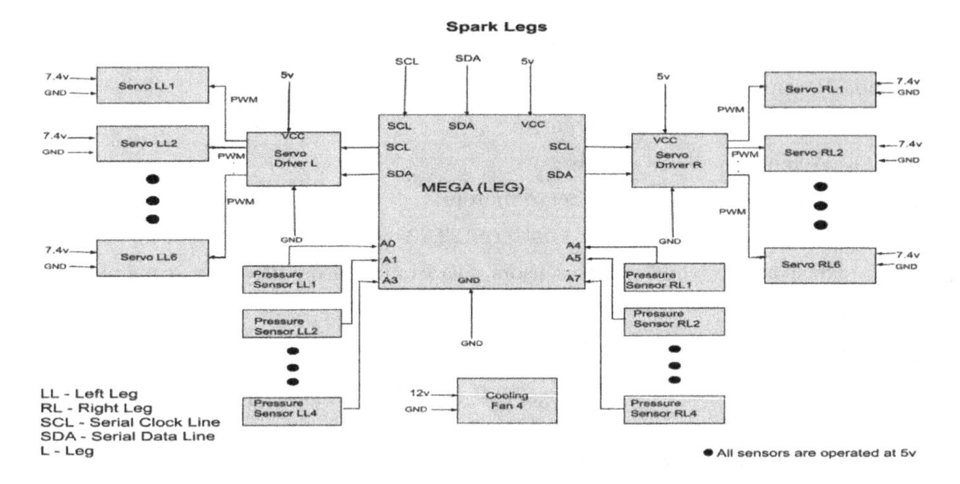

Fig. 4. Block diagram of Spark legs.

Serial Communication/I2C Communication. The communication setup between Arduino and different sensors is based on I2C communication. For communication between master controller and slave controller, serial communication (I2C protocol) [13] is used.

Power Consumption. Spark's power source is a 24 V/15 A Switch Mode Power Supply. When portability is required, this power supply is replaced by two 22.2 V, 4500 mAh Lithium Polymer batteries. Most of the circuitry is based on 5 V and 7.4 V power ratings. To achieve this, a mini adjustable voltage regulator at low power consumption regions and 300 W 20 A DC to DC Buck converter at high power consumption regions are used.

3.2 Features

Table 1 describes the system specifications of Spark and Table 2 shows the degrees of freedom for different joints of Spark's body.

3.3 Software Architecture

Chatbot System

Design of Chatbot. A chatbot is a computer program that simulates a conversation with one or more human users. Chatbots can be of many types such as simple chatbot, smart chatbot and hybrid chatbot [17]. Simple chatbot is a rule based chatbot in which the communication is through preset rules [18]. While smart chatbots are supposed to "think like humans". These chatbots work on the principle of Natural Language Processing (NLP). Hybrid chatbot is a combination of both simple and smart chatbot. These chatbots have some rule based

Table 1. System specifications.

Sr. no.	Parameters	Values
1	Height	180 cm
2	Weight	48 Kg
3	Actuator	40 Servo motors
4	Power supply	LiPo battery (22.2 V, 4500 mAh), SMPS (24 V, 15 A)
5	Sensors	Gas sensors, ultrasonic sensor, IR sensor, camera, temperature sensor, pressure sensor, gyroscope and accelerometer (IMU)
6	Control unit	Broadcom BCM2711, Atmel ATmega CH2560, ATmega328P CH340, PCA9685 16 channel servo motor driver
7	Total load capacity	5 kg

Table 2. Degrees of Freedom of Spark.

Sr. no.	Body part	Parameter	Degree of Freedom (DOF)
1	Head	Eyebrows	1 DOF × 2 = 2 DOF
		Eyes	2 DOF × 2 = 4 DOF
		Cheeks	1 DOF × 2 = 2 DOF
		Chin	1 DOF
		Neck	3 DOF
2	Hand	Wrist	1 DOF × 2 = 2 DOF
		Finger	15 DOF × 2 = 30 DOF
3	Arm	Elbow	2 DOF × 2 = 4 DOF
		Shoulder	2 DOF × 2 = 4 DOF
4	Trunk		2 DOF
5	Leg	Hip	3 DOF × 2 = 6 DOF
		Knee	1 DOF × 2 = 2 DOF
		Ankle	2 DOF × 2 = 4 DOF
6	Total		66 DOF

algorithms and are also able to understand intent and context based on NLP [17]. Spark consists of a chatbot which is based on a hybrid chatbot system.

Design Flow

1. Selection of OS
   Spark is based on a Raspberry Pi OS (Raspbian) that is built on top of the Linux Kernel.
2. Creating a chatbot

(a) Speech Recognition
Speech Recognition or Speech-to-Text is a capability that enables a computer program to process human speech into a text format. Spark Uses the 'speech recognition' library of Python that is based on Google Speech's API [5].
(b) Text-to-Speech conversion
Text-to-Speech conversion is done to convert speech in text format to human-like voices. Spark uses the 'pyttsx3' library of Python to convert text to speech. The advantage of this library is that it works in offline mode [6].

3. Creating a chat
A chat is created for the user to understand the flow of the conversation. Spark first greets the user after booting up and asks the user what help he/she needs. Spark speaks different dialogues at various points in time to create and maintain the flow of conversation. Spark showcases a chatbot that is able to converse with multiple users at the same time.
4. Pattern matching
Different algorithms, that is, rule based algorithms and AI-based algorithms are used in the chatbot that helps Spark respond to a query and maintain the flow of conversation accordingly. Rule based algorithms consist of a database that is used to match the input query and decide on an appropriate response. The AI-based algorithms use an online database to accurately recognize and match the input query and to find an appropriate response.

Image Processing System. One of the most important characteristics of a humanoid robot is its ability to analyse its environment. This mainly includes the vision system of the robot. Spark uses machine learning algorithms based on Support Vector Machine (SVM) for face detection [19], Single-Shot Multibox Detection (SSD) algorithm for object detection [20].

Person Recognition and Object Detection. The person recognition algorithm identifies humans based on their facial features by comparing it with previously en-coded data and identifies the person. This algorithm is based on the 'face recognition' library [21]. The face recognition library uses Open-cv and deep learning to find facial embeddings using the 'Deep metric learning' algorithm [22]. The face recognition library with the help of dlib's facial recognition network identifies the person [23]. The network architecture for face recognition is based on ResNet 34 from the 'Deep residual learning for image recognition' with few layers and reduced filters [24]. The object detection algorithm is based on the 'Single Shot multibox Detection' (SSD) network which is designed to perform the detection of objects. MobileNetSSD is a lightweight deep neural network architecture designed for mobile and embedded vision applications which helps in fast and efficient processing of images [20]. For object detection in Spark, a pre-trained neural network based on the open source dataset named 'COCO' [25] is used. The coco dataset consists of a total of 80 different object classes of segmented and labelled images. Hence, it is possible to identify 80 different objects within a single image [26]. The real time image is obtained and processed using open-CV python library [4].

Integrated Chatbot and Vision System. Spark showcases an integrated system of chatbot and image processing. Both the systems run parallel, interacting with each other on specific occasions depending on certain vocal inputs fed to the system. On receiving certain voice inputs, the chatbot integrates with the vision system module and an appropriate response is returned based on the algorithms.

3.4 Balancing Mechanism

A two axis gimbal mechanism is used to balance the upper body of Spark in synchronization with the lower body movements. Two servo motors and an IMU along with an Arduino Nano controller are used to precisely measure the position of centre of mass and stabilize the upper body to maintain balance and stability. For the inertial measurement unit, an MPU 6050 sensor chip [15] is used.

3.5 Walking Mechanism

Series-Parallel Hybrid Mechanism. Spark's walking mechanism is based on a Hybrid Leg mechanism. It is a bipedal mechanism that is a combined structure of series and parallel mechanisms [10]. Figure 5(a) shows the side view of Spark's leg. Spark adopts a pelvis structure with a yaw angle offset to enlarge the feet workspace, inspired by the toe-out angle of the human feet. It is designed to have a large workspace for agile bipedal locomotion. Figure 5(b) shows the joint structure of Spark. It has a total of 12 DOF, i.e., the hip has 6 DOF and the pair of legs has 6 DOF. Bearings and suspension are used to improve its structural rigidity while supporting its weight.

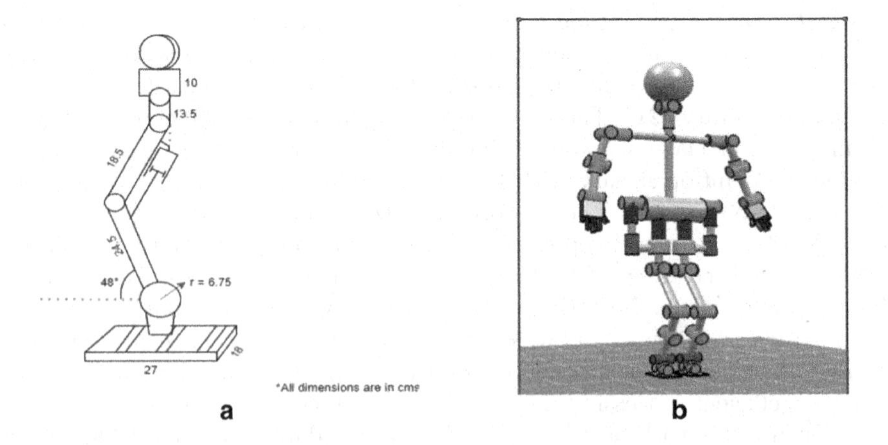

Fig. 5. (a) Side view of the leg. (b) Joint structure of Spark.

4 Results

4.1 Image Processing Results

Figure 6(a) shows the results of an object detection algorithm in Spark, where it is detecting a person, a bottle and a cell phone followed by the accuracy with which the objects are being detected. Figure 6(b) shows the outcomes of the person identification algorithm in Spark, where it is identifying a person based on their facial structure by comparing it with the encoded training data.

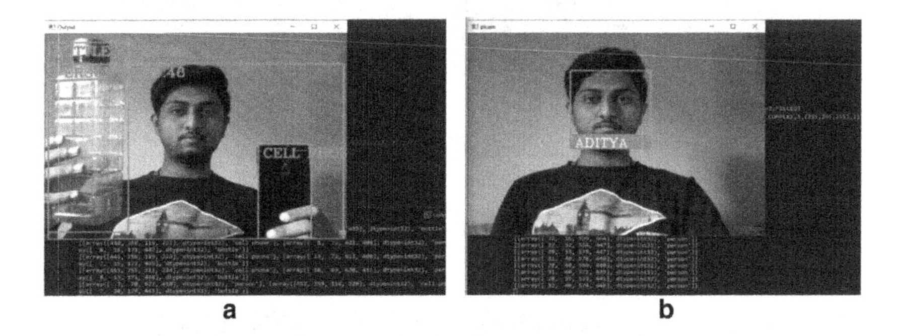

Fig. 6. a) Result of vision system b) result of face recognition system

4.2 Balancing Results

In Fig. 7(a) and (b) respectively show the front view and side view of Spark's movement while walking.

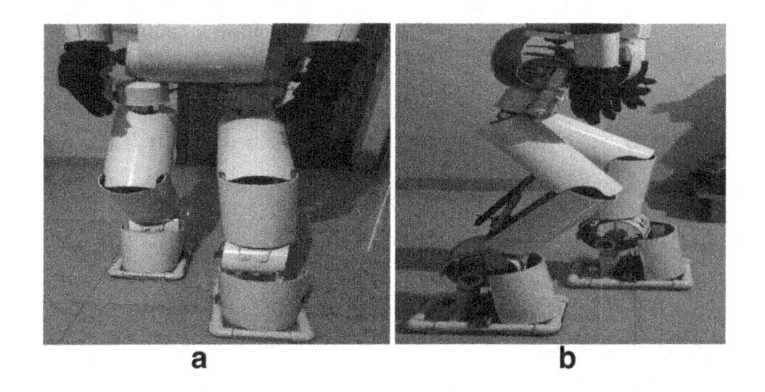

Fig. 7. a) Front view while walking. b) Side view while walking.

4.3 Gesture Results

Figure 8 and Fig. 9 show Spark doing various gestures.

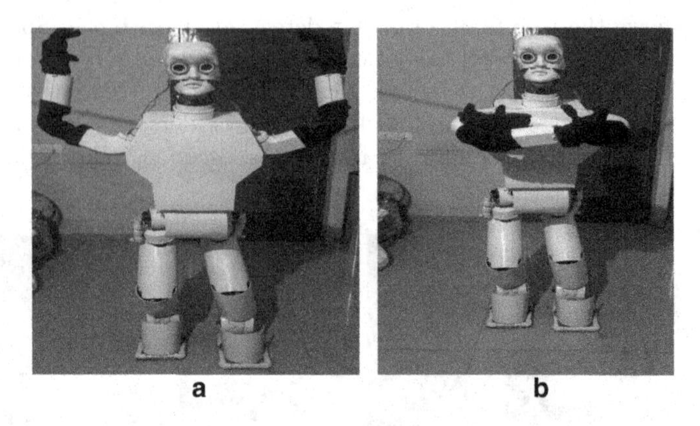

Fig. 8. a) Spark doing a 'hands-up' gesture. b) Spark folding hands.

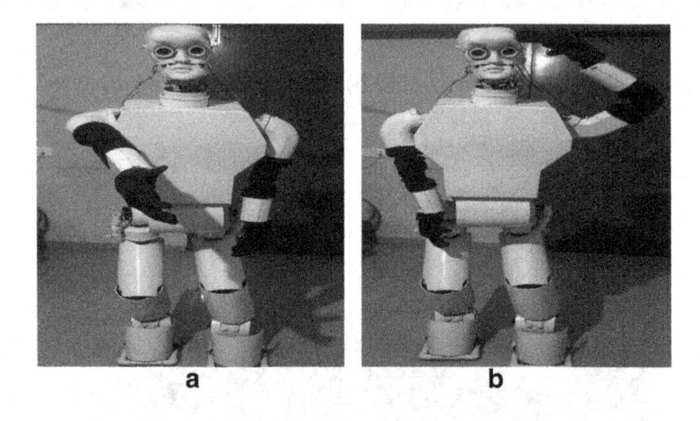

Fig. 9. a) Spark shaking hand. b) Spark saluting.

5 Applications

Spark can act as a companion to humans in day-to-day life such as a domestic helper. Spark can also serve as a service robot just like a service dog for physically or emotionally challenged people. It can be also used in the field of education to make learning more interesting for students and to encourage them towards the field of robotics and artificial intelligence. Spark also has a scope to be used in hospitals and old age homes as an aid to the patients and elderly people respectively. Future development of an advanced version of Spark would pave a way for it to be used in the field of defense thereby reducing the chances of human casualties.

6 Conclusion

In this article, we have demonstrated the system overview of a humanoid robot, Spark. Spark is basically developed for using it as a companion in daily life. Spark uses a vocal communication system for a user-friendly experience and to make the user feel as if he/she is talking to an actual human. Equipped with the vision system, Spark can recognize various human faces and a range of different objects around it. Spark's auditory system is useful for understanding vocal commands and interacting with people. Spark showcases an integrated chatbot-vision system for continuous analysis of its surroundings.

Acknowledgement. This research is mainly supported by PCCOER (Pimpri Chinchwad College of Engineering & Research). We thank Mr. Walmik R. Gawhale for the constant guidance, support and initial funding in this research. We thank Dr. H.U. Tiwari (Principal, PCCOER) and Dr. Rahul Mapari (H.O.D., E&TC, PCCOER) for their guidance and support throughout this research.

References

1. Sakagami, Y., Watanabe, R., Aoyama, C., Matsunaga, S., Higaki, N., Fujimura, K.: The intelligent ASIMO: system overview and integration, vol. 3, pp. 2478–2483 (2002). https://doi.org/10.1109/IRDS.2002.1041641
2. Feng, S., Whitman, E., Xinjilefu, X., Atkeson, C.G.: Optimization based full body control for the atlas robot, pp. 120–127 (2014). https://doi.org/10.1109/HUMANOIDS.2014.7041347
3. Park, I.W., Kim, J.Y., Lee, J., Oh, J.H.: Mechanical design of humanoid robot platform KHR-3 (KAIST humanoid robot 3: HUBO), pp. 321–326 (2005). https://doi.org/10.1109/ICHR.2005.1573587
4. OpenCV Library. https://docs.opencv.org/master/d6/d00/tutorial_py_root.html
5. Speech recognition library. https://pypi.org/project/SpeechRecognition/
6. pyttsx3 library. https://pypi.org/project/pyttsx3/
7. Raspberry Pi 4 (BCM2711). https://datasheets.raspberrypi.org/bcm2711/bcm2711-peripherals.pdf
8. ATMEGA CH2560. https://docs.arduino.cc/hardware/mega-2560
9. Megapixel Sony IMX219PQ. http://www.camera-module.com/product/8mpcameramodule/
10. Gim, K.G., Kim, J., Yamane, K.: Design and fabrication of a bipedal robot using serial-parallel hybrid leg mechanism, pp. 5095–5100 (2018). https://doi.org/10.1109/IROS.2018.8594182
11. Serial communication. https://roboticsbackend.com/raspberry-pi-arduino-serial-communication/
12. Adafruit PCA9685 16-channel servo driver. https://learn.adafruit.com/16-channel-pwm-servo-driver?view=all
13. Pandey, V., Kumar, S., Kumar, V., Goel, P.: A review paper on I2C communication protocol. Int. J. Adv. Res. Ideas Innov. Technol. **4**, 340–343 (2018). https://www.ijariit.com/manuscript/a-review-paper-on-i2c-communication-protocol/
14. Arduino Nano. https://docs.arduino.cc/hardware/nano
15. MPU 6050 gyroscope and accelerometer sensor. https://invensense.tdk.com/products/motion-tracking/6-axis/mpu-6050/

16. Kajita, S., Nagasaki, T., Kaneko, K., Hirukawa, H.: ZMP-based biped running control. IEEE Robot. Autom. Mag. **14**(2), 63–72 (2007). https://doi.org/10.1109/MRA.2007.380655
17. Types of chatbots. https://www.freshworks.com/live-chat-software/chatbots/three-types-of-chatbots/
18. Dahiya, M.: A tool of conversation: chatbot. Int. J. Comput. Sci. Eng. **5**, 158–161 (2017). https://www.researchgate.net/publication/321864990_A_Tool_of_Conversation_Chatbot
19. Real-time face detection and recognition with SVM and HOG features. https://www.eeweb.com/real-time-face-detection-and-recognition-with-svm-and-hog-features/
20. GitHub - chuanqi305/MobileNet-SSD: caffe implementation of Google MobileNet SSD detection network, with pretrained weights on VOC0712 and mAP = 0.727. https://github.com/chuanqi305/MobileNet-SSD
21. Machine learning is fun! part 4: modern face recognition with deeplearning. https://medium.com/@ageitgey/machine-learning-is-fun-part-4-modern-face-recognition-with-deep-learning-c3cffc121d78
22. dlib C++ library: high quality face recognition with deep metric learning. http://blog.dlib.net/2017/02/high-quality-face-recognition-with-deep.html
23. Github - davisking/dlib: a toolkit for making real world machine learning and data analysis applications in C++. https://github.com/davisking/dlib
24. He, K., Zhang, X., Ren, S., Sun, J.: Deep residual learning for image recognition, pp. 770–778 (2016). https://doi.org/10.1109/CVPR.2016.90
25. COCO dataset - common objects in contexts. https://cocodataset.org/#home
26. SSD MobileNet OpenVINO toolkit. https://docs.openvinotoolkit.org/2021.2/omz_models_public_ssd_mobilenet_v1_coco_ssd_mobilenet_v1_coco.html

Attention-Based Deep Autoencoder for Hyperspectral Image Denoising

Shashi Kumar, Hazique Aetesam, Anirban Saha, and Suman Kumar Maji(✉)

Department of Computer Science and Engineering, Indian Institute of Technology Patna, Patna 801103, India
{shashi_1911cs13,hazique.pcs16,anirban_2021cs13,smaji}@iitp.ac.in

Abstract. Hyperspectral Image (HSI) denoising is a crucial pre-processing task due to its widespread applications in areas that include geology, medicine, agriculture, surveillance, and the food industry. Denoising HSI data improves high-level vision tasks like classification, object tracking, and video surveillance. In this work, we propose a novel attention-based deep auto-encoder for HSI denoising which learns a mapping from noisy observation to clean the data. To exploit the spatial-spectral information present in HSI data, 3D symmetric convolution and deconvolution are used. The 3D features extracted from 3D convolutional layers (CL) helps the restoration across the depth dimension. However, feature maps obtained from the convolutional layers are localized and are not able to capture long-range dependencies in the data. Furthermore, CL does not differentiate between low and high degradation levels; producing smoothed results in low-textured areas of the image and artifacts in the high-textured regions. To avoid these adverse effects, attention blocks are plugged into the network to focus on more relevant features during the restoration process. Features obtained from attention blocks are fed as symmetric skip connections into the corresponding deconvolution layers. The proposed model is fully convolutional; it can process 3D images of arbitrary dimensions during training and inference. The experimental results show that our proposed model: *Attention-based Deep Auto-encoder (AbDAE)* outperforms the state-of-the-art methods in terms of visual as well as quantitative results.

Keywords: Hyperspectral image denoising · Spatial-spectral · Auto-encoder · Attention network · 3D convolution and deconvolution

1 Introduction

Images obtained from conventional RGB cameras work under the visible part of the electromagnetic spectrum (400 Å–800 Å). However more details, about the image, can be obtained beyond the visible range of human eyes. With the advancements in imaging technology, hyperspectral cameras can acquire images in the non-visible parts within a broader range of 400 nm to 2500 nm. This data gets recorded as fine spectra in the form of contiguously separated wavelength bands in contrast to multi-spectral imaging where measured spectral bands may not be contiguous and finely separated. The

B. Raman et al. (Eds.): CVIP 2021, CCIS 1568, pp. 159–170, 2022.
https://doi.org/10.1007/978-3-031-11349-9_14

resultant image is 3D with spatial dimension recording pixel intensities over a particular wavelength level and the spectral signature of a pixel recording intensity levels over the entire hyperspectral camera acquisition range. The degradation in HSI occurs due to several factors, like relative object-camera motion, blur due to defocused camera, narrow wavelength binning, etc. All these factors are responsible for the incorporation of mixed Gaussian-impulse noise, contaminating the HSI data [10,20].

There is a multitude of hyperspectral image denoising methodologies, that can be broadly classified into two categories: conventional model-driven approaches and recent learning-based techniques. In the work by Zhang et al. [20], authors proposed an efficient hyperspectral image restoration technique using low-rank matrix recovery (LRMR). By exploiting the spatial-spectral information uniformly among all the bands, He. et al. [10] proposed a novel total-variation regularized low-rank matrix factorization for hyperspectral image restoration (LRTV). Low-rank-based methods are used to model data with the assumption that clean HSI data has low-rank. However, the noise intensity may vary from band to band. Zhang et al. [9] proposed a noise-adjusted iterative low-rank matrix approximation (NAILRMA), which is based on a low-rank property with an iterative regularization framework. It utilizes spectral variance in noise levels using an adaptive iteration factor selection. This selection is done based on the noise variance of each HSI band. He et al. [8] claimed that a global spatial-spectral total variation regularized restoration strategy is utilized to ensure the global spatial-spectral smoothness of the reconstructed image from the low-rank patches. Fan et al. [6] proposed a novel spatial-spectral TV regularized Low-rank tensor factorization (SSTV-LRTF) method to remove mixed noise in HSIs. A mixed-norm fidelity model for HSI denoising is explored in [2]; while a variational approach is explored in [1] by splitting the image formation model into two parts and moulding the resultant optimization framework. A proximal approach to denoising HSI data under mixed noise is proposed in [3]. The major drawback of these conventional techniques is that they are time-consuming due to the complex prior-based iterative optimization process.

Recently, deep learning has opened new vistas into HSI restoration. Zhang et al. [21] proposed an efficient Gaussian Denoiser, entitled DnCNN. The architecture of DnCNN uses the power of deep learning through convolution network, batch normalization, and residual mapping. Ghose et al. [7] proposed the basic idea of deep learning using Image Denoising using Deep Learning. C. Tian et al. [16] proposed a deep learning approach on image denoising in which they tried to explore the feature extraction techniques. Apart from this, Dey et al. [5] proposed unsupervised learning-based image restoration techniques. Sidorov et al. [15] proposed a novel approach named Deep Hyperspectral Prior: Single-Image Denoising, Inpainting, Super-Resolution where they are using both 2D and 3D convolution and claimed that the 2D network outperforms the 3D network. Recently Zhao et al. [23] proposed a novel approach named Attention-based deep residual network for HSI denoising (ADRN) which uses the channel attention to focus on the more meaningful information and trained their model on the Washington DC Mall dataset. Another recent novel deep learning approach is proposed by Wei et al. [17] which is based on 3D quasi-recurrent neural network for HSI Denoising. The work by Z. Kan et al. [11] demonstrates an Attention-Based Octave Dense Network (AODN), in which the attention module fine-tunes the spatial-spectral features while the octave network learns the high-frequency noise. Beside this, Q. Shi et al. [14] noticed that the existing HSI denoising models neither consider the global dependence nor the spatial-spectral correlation. They developed a dual-attention denoising network that addresses the limitation of processing spatial and spectral information in parallel. Moreover, H. Ma et al. [12] discussed an enhanced non-local cascading network-based HSI denoising algorithm with attention mechanism, extracting the joint spatial-spectral feature more effectively. In addition to this, several efforts have been made in recent times to effectively denoise hyperspectral images [19,22,24].

This paper proposes a novel attention-based deep auto-encoder for HSI denoising. The proposed model learns a mapping function that maps the noisy observation to clean data. With the autoencoder, comprising of 3D symmetric convolution and deconvolution layers, the model can exploit the spatial-spectral information present in HSI data. The 3D convolutional layers are competent to extract depth features, while the attention blocks capture the relevant long-range feature dependencies. These features, analyzed by the convolution and the attention blocks, are fed to the corresponding symmetric deconvolution layers. With a complete convolutional architecture, the proposed model can adapt to process images of any arbitrary dimension. Experimentations over several simulated and real data indicates the potentiality of the proposed model.

The organisation of this paper is as given. Image formation model is presented in Sect. 2 followed by Sect. 3 that discusses the proposed methodology. Further in Sect. 4, quantitative and qualitative results of the experiments are shown. Then the paper concludes in Sect. 5.

2 Image Formation Model

Degradation of HSI can be expressed by the equation:

$$Y = X + \mathcal{G} + \mathfrak{s}; \tag{1}$$

The parameters $\mathcal{G}$ and $\mathfrak{s}$ represent the additive gaussian noise and impulse noise components, respectively. Y is the noisy image formed after adding the noisy components with the clean image X, such that $\{X, Y, \mathcal{G}, \mathfrak{s}\} \in \mathbb{R}^{H \times W \times B}$. Here, the variables H, W, and B indicate the spatial height, spatial width, and spectral bands, respectively. In the context of denoising, we consider Gaussian and Impulse noise or a mixture of these two noises which affects HSI the most. Our proposed technique takes a noisy HSI cube and outputs a denoised HSI cube.

3 Proposed Methodology

In this section, the proposed *Attention-based Deep Autoencoder (AbDAE)* model with skip connection is discussed in details. The overall architecture of the model is shown in Fig. 1(a). There are 8 pairs of symmetric convolution and deconvolution layers including attention networks. Each convolution and deconvolution operation is followed by a ReLU activation function. The set of convolution operation is expected to be the feature extractor preserving the useful image information. And the deconvolution blocks are

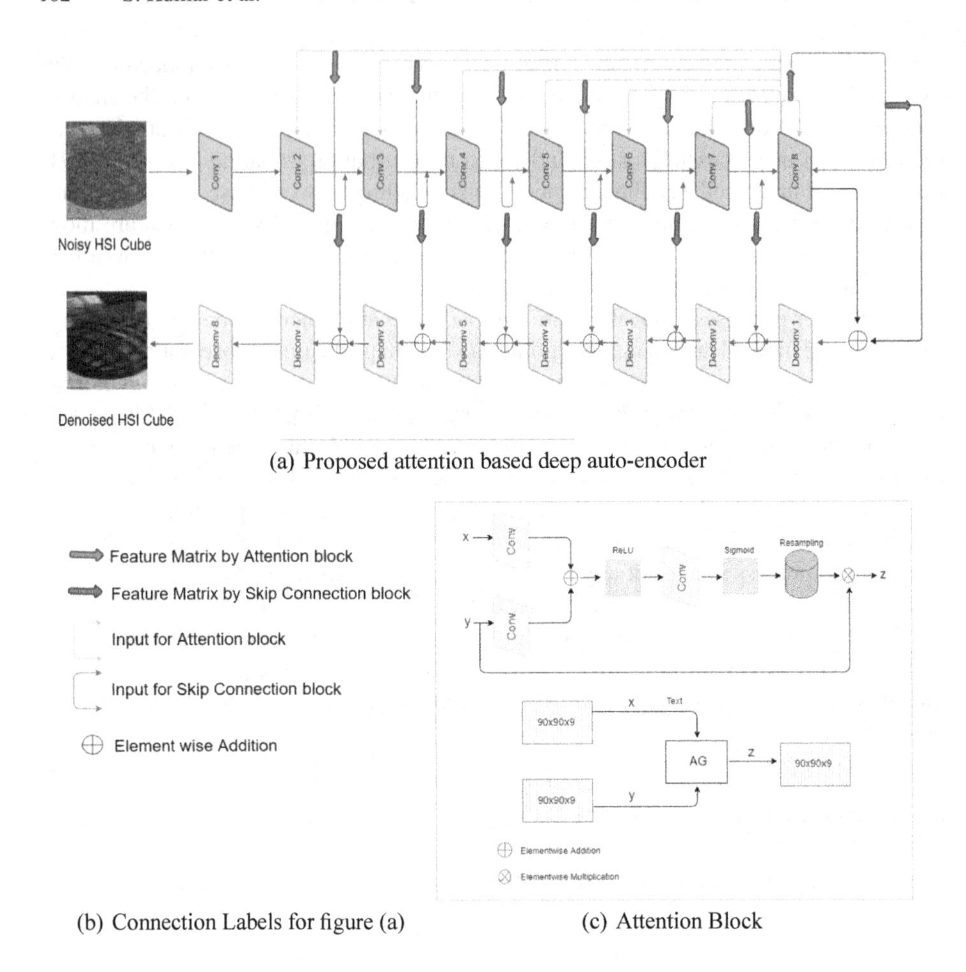

(a) Proposed attention based deep auto-encoder

(b) Connection Labels for figure (a)

(c) Attention Block

Fig. 1. Detailed representation of the proposed model.

responsible to reconstruct the denoised image. Apart from this, an attention network has been incorporated just before each skip connection as an input to it.

In the proposed network architecture, convolutional layers are analogous to down-sampling layers where the input image content is compressed into abstracted feature vectors. Symmetrically, the deconvolution layers upsample the abstraction blocks back into its original dimension. Each layer in the encoder and decoder segment has the same kernel size of $3 \times 3 \times 3$. And the stride is set to 1 in both convolution and deconvolution operation to extract large number of features with minimal computational burden. The detailed configuration of the convolution and deconvolution blocks are represented in Table 1.

Table 1. Conv-Deconv configuration of the proposed model

Config criteria	Encoder block	Decoder block	Attention block
No. of convolution layers	8	0	2+1
No. of deconvolution layers	0	8	0
Kernel size (each layer)	$3 \times 3 \times 3$	$3 \times 3 \times 3$	$3 \times 3 \times 3$
Stride size (each layer)	$1 \times 1 \times 1$	$1 \times 1 \times 1$	$1 \times 1 \times 1$
No. of filters (each layer)	64	64	64, 64, 1

3.1 Layer-Wise Attention Block

The purpose of using this block is to obtain a significant feature matrix for denoising process. It takes two identical shape feature matrices as input and provides a feature matrix of shape that corresponds to the input shape. Use of layer-wise attention block indicates the presence of this block at different layers. The attention block takes input from the last two convolution layers and feeds its output to the next layer. Initially, a replicated version of the output from the last convolution layer acts as the input to the attention block. Further, the inputs for the remaining attention blocks are extracted from the outputs of the last and the second last convolutional instance.

As per Fig. 1(b), the bidirectional yellow arrows show the input and the thick pink unidirectional arrows depict the output of attention blocks. Figure 1(c) illustrates that the attention block takes two input feature matrices (x and y) of identical shape and yields an output (z) of same size. Internal processing of the attention block can be segmented into several sub-processes. Firstly, both the inputs x and y are convoluted separately. Then these outputs are combined together, using an element-wise addition convolution component, and passed through a ReLU activation function. Now the outcome is processed with a convolutional block comprising of two sequential operations: convolution, and sigmoid activation. After this, the output of the convolutional block is resampled as per the input shape using trilinear interpolation in the resampling layer. Lastly, an element-wise multiplication is performed with the resampled outcome and the input feature matrix y to compute the output z.

3.2 Layer-Wise Skip Connection Block

It is seen that a deeper network suffers from performance degradation in image denoising tasks due to two major reasons: loss of information with increase in number of layers, and slow update or even loss of gradient affecting the training process in deeper layers. To get rid of these problems, skip connections are introduced which are inspired by highway networks and deep residual networks. The benefit of using such a connection is that the feature matrix, passed by skip connections, carry more informative details from the initial layers that helps to perform deconvolution and recover the noisy image in a better way. These networks are easy to train and show an early convergence providing superior quality results during inference under image restoration tasks. Further benefits of using skip connection is presented in [13].

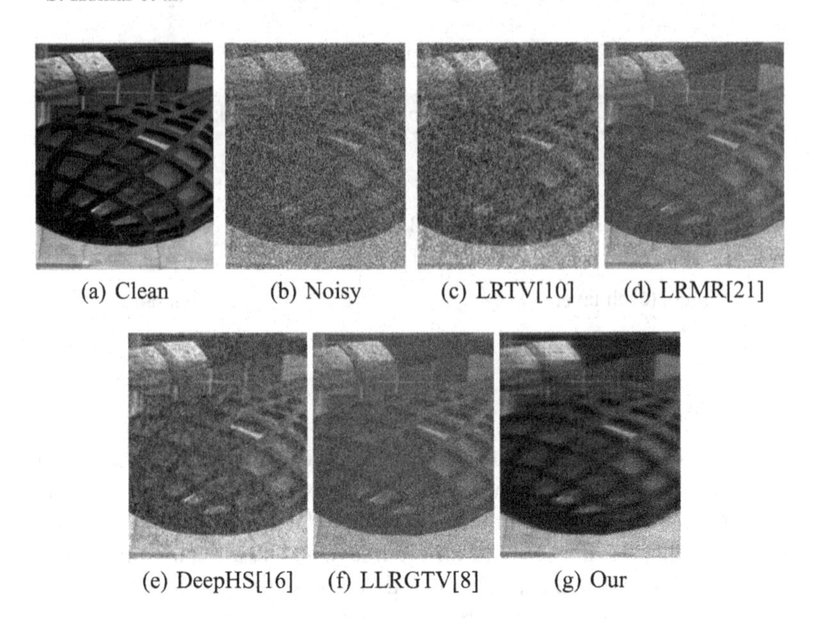

(a) Clean (b) Noisy (c) LRTV[10] (d) LRMR[21]

(e) DeepHS[16] (f) LLRGTV[8] (g) Our

Fig. 2. Visual results over simulated ICVL data with noise level $\{(snr, p)\} = \{(2\,\text{dB}, 30\%)\}$

The proposed model incorporates the skip connection block just after the attention block. In the model architecture, it is shown that this block takes inputs from both the attention and the convolution blocks. Initially, the outputs of the attention block and the last convolutional layer form the input to the first skip connection block, which is then followed by the subsequent blocks in a symmetric pattern. As per Fig. 1(b), the blue bidirectional arrow acts as the input to the skip connection, and the thick green unidirectional arrow indicates the skip connection output.

3.3 Training Details

The proposed attention-based deep auto-encoder for HSI denoising is trained over ICVL dataset [4]. Original size of an ICVL image sample is $1300 \times 1392 \times 31$. However, patch-based training [25] is carried out, with patch size of $90 \times 90 \times 9$. This led to a total of 12300 image patches of size $90 \times 90 \times 9$ on which the model is trained upon. Out of these, 2460 image patches are utilized for validation. The complete training took place on Nvidia GPU with 100 epochs and batch-size of 16. Adaptive Moment Estimation (ADAM) is used as the optimizer during training and the learning rate is set to be 0.0001. Mean square error (MSE), expressed by Eq. (2), is considered as the loss function. In the equation, the variables Θ and N represents the trainable parameters and the number of training instances, respectively.

$$L(\Theta) = \frac{1}{2N}\sum_{i=1}^{N}\|\hat{X}_i - X_i\|_2^2 \tag{2}$$

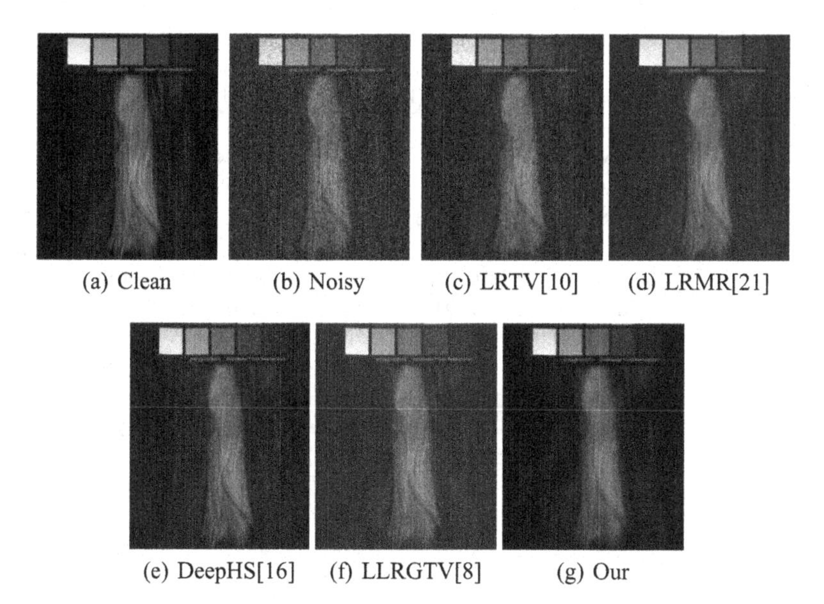

(a) Clean (b) Noisy (c) LRTV[10] (d) LRMR[21]

(e) DeepHS[16] (f) LLRGTV[8] (g) Our

Fig. 3. Visual results over simulated Cave data with noise level $\{(snr, p)\} = \{(5\,\text{dB}, 20\%)\}$

4 Experimental Results

With the objective to visualize and quantify the performance of the proposed technique, a comparative study has been carried out applying the proposed model along with some of the successful HSI denoising models over three different datasets. These datasets includes two synthetic datasets: Interdisciplinary Computer Vision Laboratory (ICVL) dataset [4] and Cave dataset [18], and one real datasets: Salinas dataset[1] acquired by Airborne Visible/Infrared Imaging Spectrometer (AVIRIS) sensors. For ICVL and CAVE datasets, the spatial dimension of 256×256 and spectral dimension of 31 is considered for experimentation. In contrast to this, the real dataset, Salinas is of different dimension with the spatial dimension of 217×217 and the spectral dimension of 224. The efficiency of the proposed technique is studied and compared with the performance of five other state-of-the-art approaches present in the literature. These comparison models include three model-driven approaches: Low-rank with Total-variation Regularization (LRTV) [10], Low-Rank Matrix Recovery (LRMR) [20], and Local Low-Rank Matrix Recovery and Global Spatial-Spectral Total Variation (LLRGTV) [8]; and one learning-based approach: Deep HSI Prior [15].

To quantify the quality of the denoised results, four quality evaluation metrics are considered. One of these metrics include peak-signal-to-noise ratio (PSNR), which is the most widely used evaluation criteria to compare image quality. Equation (3) gives the mathematical formulation to compute PSNR of denoised $\hat{X}$, considering the reference image X.

[1] Images obtained from: http://www.ehu.eus.

Table 2. Evaluation metric comparison for simulated ICVL dataset [4]

Noise Level	Index	Noisy	LRTV [10]	LRMR [20]	DeepHS [15]	LLRGTV [8]	Our
$\{snr = 2\,\text{dB}, p = 30\%\}$	PSNR	11.69	17.71	13.69	14.45	14.56	**23.65**
	SSIM	0.0729	0.467	0.2345	0.3194	0.3018	**0.5811**
	SAM		19.03	22.43	39.87	15.45	**12.18**
	EARGAS		2910.86	1202.21	3360.51	1048.16	**213.57**
$\{snr = 5\,\text{dB}, p = 20\%\}$	PSNR	13.47	18.86	15.86	18.04	16.63	**24.45**
	SSIM	0.1162	0.497	0.3391	0.4745	0.4017	**0.6326**
	SAM		17.23	19.22	41.61	12.50	**10.49**
	EARGAS		1968.67	741.25	3226.19	795.21	**284.07**
$\{snr = 10\,\text{dB}, p = 15\%\}$	PSNR	14.85	19.75	17.75	20.67	18.32	**25.07**
	SSIM	0.1592	0.5553	0.4224	0.5802	0.4794	**0.6931**
	SAM		14.51	16.66	41.95	11.26	**8.65**
	EARGAS		1932.89	555.24	3100.14	557.65	**204.12**
$\{snr = 15\,\text{dB}, p = 10\%\}$	PSNR	17.10	21.91	20.33	24.36	21.41	**27.92**
	SSIM	0.2447	0.6337	0.5463	0.7080	0.5966	**0.7681**
	SAM		10.96	13.16	42.51	8.98	**8.16**
	EARGAS		1615.79	430.34	2960.73	419.45	**132.55**
$\{snr = 20\,\text{dB}, p = 6\%\}$	PSNR	20.59	24.51	24.59	28.71	25.22	**28.96**
	SSIM	0.3934	0.7333	0.6997	0.8120	0.7219	**0.8212**
	SAM		8.26	9.37	43.93	**6.57**	6.77
	EARGAS		1415.70	223.82	2769.85	220.75	**158.50**
$\{snr = 25\,\text{dB}, p = 3\%\}$	PSNR	25.65	29.64	30.65	33.27	31.29	**34.84**
	SSIM	0.6197	0.8474	0.8441	0.9018	0.8501	**0.8989**
	SAM		5.46	5.54	43.45	**4.22**	5.35
	EARGAS		642.12	119.91	2540.72	**114.84**	137.44

$$\text{PSNR} = 20\log_{10}\frac{X_{max}}{\sqrt{E[(X_i - \hat{X}_i)^2]}} \tag{3}$$

The structural similarity index measure (SSIM) is reviewed as the second evaluation parameter. It captures the variation of the local patterns in pixel intensities as compared to the reference image. Equation (4) is utilized to record the SSIM values.

$$\text{SSIM} = \frac{(2E[X_i]E[\hat{X}_i] + \alpha_1)(2Cov[X_i, \hat{X}_i] + \alpha_2)}{(E[X_i^2] + E[\hat{X}_i^{\,2}] + \alpha_1)(Var[X_i] + Var[\hat{X}_i] + \alpha_2)} \tag{4}$$

The third evaluation metric, termed as spectral angle mapper (SAM), apprehends the spectral similarity between two spectra. And provides a numerical result that determines the quantity of spectral distortion. Equation (5) defines the mathematical computation of SAM.

$$\text{SAM} = \arccos\left(\frac{Cov[X_i, \hat{X}_i]}{\|X_i\|_2\|\hat{X}_i\|_2}\right) \tag{5}$$

Lastly, the metric relative dimensionless global error (ERGAS) provides a brief idea about the change in image quality, as compared to the reference image, in terms of

Table 3. Evaluation metric comparison for simulated cave dataset [18]

Noise level	Index	Noisy	LRTV[10]	LRMR[20]	DeepHS[15]	LLRGTV[8]	**Our**
$\{snr = 2dB, p = 30\%\}$	PSNR	14.84	24.91	17.13	20.43	17.46	**27.90**
	SSIM	0.0918	0.5627	0.2689	0.4154	0.3138	**0.6585**
	SAM		38.04	38.61	32.12	31.09	**26.79**
	ERGAS		733.55	803.58	711.98	863.39	**249.20**
$\{snr = 5\,\text{dB}, p = 20\%\}$	PSNR	17.36	27.08	19.50	26.43	20.35	**30.65**
	SSIM	0.1464	0.6352	0.371	0.6406	0.4237	**0.7406**
	SAM		34.29	35.22	27.79	26.51	**25.24**
	ERGAS		613.2	691.45	505.03	607.15	**200.92**
$\{snr = 10\,\text{dB}, p = 15\%\}$	PSNR	19.31	29.82	21.65	25.22	23.06	**32.2**
	SSIM	0.2025	0.7147	0.4599	0.6473	0.5202	**0.7969**
	SAM		30.45	32.24	25.55	23.61	**19.67**
	ERGAS		467.34	646.14	449.17	467.62	**141.41**
$\{snr = 15\,\text{dB}, p = 10\%\}$	PSNR	22.22	31.2	25.16	33.77	27.28	**34.99**
	SSIM	0.3112	0.7664	0.5914	0.8442	0.6595	**0.8615**
	SAM		26.6	27.76	24.36	19.45	**18.98**
	ERGAS		330.3	387.21	262.66	310.58	**101.49**
$\{snr = 20\,\text{dB}, p = 6\%\}$	PSNR	26.26	**36.82**	29.93	36.99	32.33	36.23
	SSIM	0.4951	**0.8908**	0.746	0.9114	0.8052	0.883
	SAM		20.706	21.75	23.58	**14.38**	15.98
	ERGAS		204.53	258.11	190.11	157.36	**93.51**
$\{snr = 25\,\text{dB}, p = 3\%\}$	PSNR	31.94	40.73	36.97	39.28	40.64	**43.65**
	SSIM	0.7444	0.9479	0.8994	0.939	0.9433	**0.9671**
	SAM		12.55	14.04	22.49	**9.09**	11.83
	ERGAS		94.98	90.66	178.40	78.49	**41.20**

normalized average error for each band. Evaluation of ERGAS values can be expressed by Eq. (6).

$$\text{ERGAS} = 100\sqrt{E\left[\frac{E[(X_i - \hat{X}_i)^2]}{E[X_i^2]}\right]} \tag{6}$$

Moreover, the statistical functions $E[\bullet]$, $Var[\bullet]$, and $Cov[\bullet,\bullet]$, that are used in the Eqs. (3)–(6), represents the spatial average, variance and co-variance functions, respectively.

Tables 2 and 3 records the metric values that compare the output of the proposed model with that of other aforementioned approaches when applied over simulated ICVL and Cave data. Analyzing these values, it is evident that the proposed model performs better in most cases. It provides the best outcome or the outcome which is so close to the best one that the difference is almost negligible. Apart from this quantitative analysis, traces of visual outcomes are also presented in Figs. 2, 3, and 4. Figure 2 presents the visual results of different methodologies when applied over an ICVL data, simulated with noise label $\{(snr, p)\} = \{(2\,\text{dB}, 30\%)\}$. Figure 3 shows the similar visual comparison on a CAVE data, contaminated with noise label $\{(snr, p)\} = \{(5\,\text{dB}, 20\%)\}$.

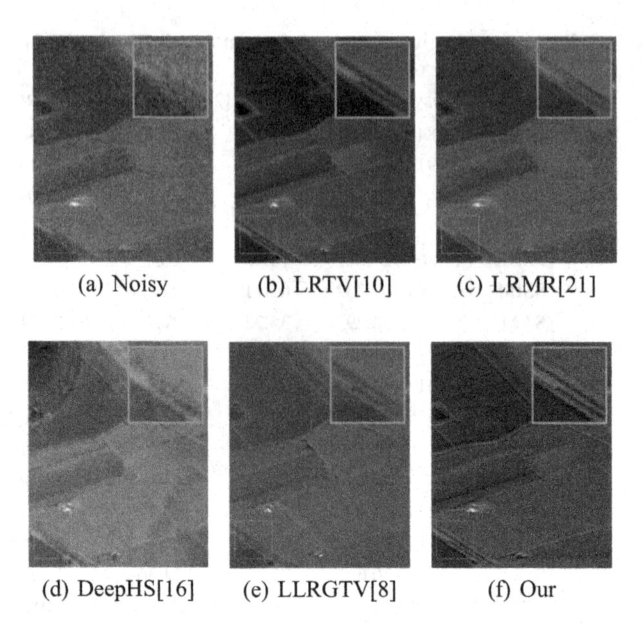

Fig. 4. Real-world unknown noise removal results at 1st band of AVIRIS Salinas dataset.

The noise parameters snr and p denotes the signal-to-noise ratio of gaussian noise and the impulse noise probability, respectively. Whereas, the visual analysis, given by Fig. 4, shows the performance of the comparing models over the 1st band of the real AVIRIS acquired Salinas data. Upon close analysis of all the visual samples, it becomes obvious that though the proposed model has resulted in slight improvement over simulated data, it outperforms when applied over the real data.

5 Conclusion and Future Work

In this work, we have proposed an attention-based deep auto-encoder for hyperspectral image denoising. Our main contribution is the usage of a layer-wise attention network to extract the more relevant features. The model is efficient and capable of modelling spatial-spectral dependency, where the 3D architecture explores spectral information. The auto-encoder in the proposed model consists of 3D convolution and deconvolution layers. We have applied our model on different levels of Gaussian and impulse noise. Our model achieves faster speed and better performance. The model is trained on the ICVL dataset and tested on artificially corrupted Cave and ICVL datasets, and also on a real-world Salinas dataset, which is infeasible for many existing deep learning and conventional techniques. The experimental results show that our proposed attention-based deep auto-encoder model outperforms several existing mainstream methods in both quantitative and visual outcomes. Our proposed model consists of an auto-encoder, which uses convolution and deconvolution for the restoration process. It will be interesting to explore the effects of architectural modifications in improving the restoration quality. U-Net architecture is an option that can be explored.

References

1. Aetesam, H., Maji, S.K., Boulanger, J.: A two-phase splitting approach for the removal of gaussian-impulse noise from hyperspectral images. In: Singh, S.K., Roy, P., Raman, B., Nagabhushan, P. (eds) Computer Vision and Image Processing, CVIP 2020. Communications in Computer and Information Science, vol. 1376, pp. 179–190. Springer, Singapore (2020). https://doi.org/10.1007/978-981-16-1086-8_16
2. Aetesam, H., Poonam, K., Maji, S.K.: A mixed-norm fidelity model for hyperspectral image denoising under Gaussian-impulse noise. In: 2019 International Conference on Information Technology (ICIT), pp. 137–142. IEEE (2019)
3. Aetesam, H., Poonam, K., Maji, S.K.: Proximal approach to denoising hyperspectral images under mixed-noise model. IET Image Process. **14**(14), 3366–3372 (2020)
4. Arad, B., Ben-Shahar, O.: Sparse recovery of hyperspectral signal from natural RGB images. In: Leibe, B., Matas, J., Sebe, N., Welling, M. (eds.) ECCV 2016. LNCS, vol. 9911, pp. 19–34. Springer, Cham (2016). https://doi.org/10.1007/978-3-319-46478-7_2
5. Dey, B., et al.: SEM image denoising with unsupervised machine learning for better defect inspection and metrology. In: Adan, O., Robinson, J.C. (eds.) Metrology, Inspection, and Process Control for Semiconductor Manufacturing XXXV. vol. 11611, pp. 245–254. International Society for Optics and Photonics, SPIE (2021). https://doi.org/10.1117/12.2584803
6. Fan, H., Li, C., Guo, Y., Kuang, G., Ma, J.: Spatial-spectral total variation regularized low-rank tensor decomposition for hyperspectral image denoising. IEEE Trans. Geosci. Remote Sens. **56**(10), 6196–6213 (2018). https://doi.org/10.1109/TGRS.2018.2833473
7. Ghose, S., Singh, N., Singh, P.: Image denoising using deep learning: convolutional neural network. In: 2020 10th International Conference on Cloud Computing, Data Science Engineering (Confluence), pp. 511–517 (2020). https://doi.org/10.1109/Confluence47617.2020.9057895
8. He, W., Zhang, H., Shen, H., Zhang, L.: Hyperspectral image denoising using local low-rank matrix recovery and global spatial-spectral total variation. IEEE J. Sel. Topics Appl. Earth Observ. Remote Sens. **11**(3), 713–729 (2018). https://doi.org/10.1109/JSTARS.2018.2800701
9. He, W., Zhang, H., Zhang, L., Shen, H.: Hyperspectral image denoising via noise-adjusted iterative low-rank matrix approximation. IEEE J. Sel. Topics Appl. Earth Observ. Remote Sens. **8**(6), 3050–3061 (2015). https://doi.org/10.1109/JSTARS.2015.2398433
10. He, W., Zhang, H., Zhang, L., Shen, H.: Total-variation-regularized low-rank matrix factorization for hyperspectral image restoration. IEEE Trans. Geosci. Remote Sens. **54**, 176–188 (2016). https://doi.org/10.1109/TGRS.2015.2452812
11. Kan, Z., Li, S., Zhang, Y.: Attention-based octave dense network for hyperspectral image denoising. In: 2021 IEEE 4th International Conference on Big Data and Artificial Intelligence (BDAI), pp. 230–235 (2021). https://doi.org/10.1109/BDAI52447.2021.9515262
12. Ma, H., Liu, G., Yuan, Y.: Enhanced non-local cascading network with attention mechanism for hyperspectral image denoising. In: ICASSP 2020–2020 IEEE International Conference on Acoustics, Speech and Signal Processing (ICASSP), pp. 2448–2452 (2020). https://doi.org/10.1109/ICASSP40776.2020.9054630
13. Mao, X.J., Shen, C., Yang, Y.B.: Image restoration using convolutional auto-encoders with symmetric skip connections. ArXiv, June 2016
14. Shi, Q., Tang, X., Yang, T., Liu, R., Zhang, L.: Hyperspectral image denoising using a 3-D attention denoising network. IEEE Trans. Geosci. Remote Sens. **59**(12), 10348–10363 (2021). https://doi.org/10.1109/TGRS.2020.3045273
15. Sidorov, O., Hardeberg, J.Y.: Deep hyperspectral prior: single-image denoising, inpainting, super-resolution. In: 2019 IEEE/CVF International Conference on Computer Vision Workshop (ICCVW), pp. 3844–3851 (2019). https://doi.org/10.1109/ICCVW.2019.00477

16. Tian, C., Fei, L., Zheng, W., Xu, Y., Zuo, W., Lin, C.W.: Deep learning on image denoising: an overview. Neural Netw. **131**, 251–275 (2020). https://doi.org/10.1016/j.neunet.2020.07.025, https://www.sciencedirect.com/science/article/pii/S0893608020302665
17. Wei, K., Fu, Y., Huang, H.: 3-D quasi-recurrent neural network for hyperspectral image denoising. IEEE Trans. Neural Netw. Learn. Syst. **32**(1), 363–375 (2021). https://doi.org/10.1109/TNNLS.2020.2978756
18. Yasuma, F., Mitsunaga, T., Iso, D., Nayar, S.K.: Generalized assorted pixel camera: postcapture control of resolution, dynamic range, and spectrum. IEEE Trans. Image Process. **19**(9), 2241–2253 (2010). https://doi.org/10.1109/TIP.2010.2046811
19. Zhang, H., Chen, H., Yang, G., Zhang, L.: LR-Net: low-rank spatial-spectral network for hyperspectral image denoising. IEEE Trans. Image Process. **30**, 8743–8758 (2021). https://doi.org/10.1109/TIP.2021.3120037
20. Zhang, H., He, W., Zhang, L., Shen, H., Yuan, Q.: Hyperspectral image restoration using low-rank matrix recovery. IEEE Trans. Geosci. Remote Sens. **52**(8), 4729–4743 (2014). https://doi.org/10.1109/TGRS.2013.2284280
21. Zhang, K., Zuo, W., Chen, Y., Meng, D., Zhang, L.: Beyond a gaussian denoiser: residual learning of deep CNN for image denoising. IEEE Trans. Image Process. **26**(7), 3142–3155 (2017). https://doi.org/10.1109/TIP.2017.2662206
22. Zhang, T., Fu, Y., Li, C.: Hyperspectral image denoising with realistic data. In: Proceedings of the IEEE/CVF International Conference on Computer Vision (ICCV), pp. 2248–2257, October 2021
23. Zhao, Y., Zhai, D., Jiang, J., Liu, X.: ADRN: attention-based deep residual network for hyperspectral image denoising. In: ICASSP 2020–2020 IEEE International Conference on Acoustics, Speech and Signal Processing (ICASSP), pp. 2668–2672 (2020). https://doi.org/10.1109/ICASSP40776.2020.9054658
24. Zhuang, L., Fu, X., Ng, M.K., Bioucas-Dias, J.M.: Hyperspectral image denoising based on global and nonlocal low-rank factorizations. IEEE Trans. Geosci. Remote Sens. **59**(12), 10438–10454 (2021). https://doi.org/10.1109/TGRS.2020.3046038
25. Zoran, D., Weiss, Y.: From learning models of natural image patches to whole image restoration. In: 2011 International Conference on Computer Vision, pp. 479–486. IEEE (2011)

Feature Modulating Two-Stream Deep Convolutional Neural Network for Glaucoma Detection in Fundus Images

Snehashis Majhi and Deepak Ranjan Nayak(✉)

Department of Computer Science and Engineering,
Malaviya National Institute of Technology, Jaipur 302017, Rajasthan, India
drnayak.cse@mnit.ac.in

Abstract. Detection of ocular disorders like glaucoma needs immediate actions in clinical practice to prevent irreversible vision loss. Although several detection methods exist, they usually perform segmentation of optic disc followed by classification to detect glaucoma in clinical fundus images. In such cases, the important visual features are ignored and the classification accuracy is majorly affected by inaccurate segmentation. Further, the deep learning-based existing methods demand the down-sampling of fundus images from high-resolution to low-resolution which results in loss of image details, thereby degrading the glaucoma classification performance. To handle these issues, we propose a feature modulating two-stream deep convolutional neural network (CNN). The network accepts the full fundus image and the region of interest (ROI) outlined by clinicians as input to each stream to capture more detailed visual features. A feature modulation technique is also proposed as an intrinsic module in the proposed network to further enrich the feature representation by computing the feature-level correlation between two streams. The proposed model is evaluated on a recently introduced large-scale glaucoma dataset, namely *G1020* and has achieved state-of-the-art glaucoma detection performance.

Keywords: Glaucoma · Two-stream CNN · Feature modulation

1 Introduction

Glaucoma is a neuro-degenerative chronic eye disorder that can cause subtle, gradual, and eventually complete blindness unless timely detected and treated [17]. The primary causes of glaucoma are usually the build-up of intraocular pressure (IOP) in the eye which blocks the drainage of intraocular fluid. Moreover,

This work is supported by the Science and Engineering Research Board (SERB), Department of Science and Technology, Govt. of India under project No. SRG/2020/001460.

B. Raman et al. (Eds.): CVIP 2021, CCIS 1568, pp. 171–180, 2022.
https://doi.org/10.1007/978-3-031-11349-9_15

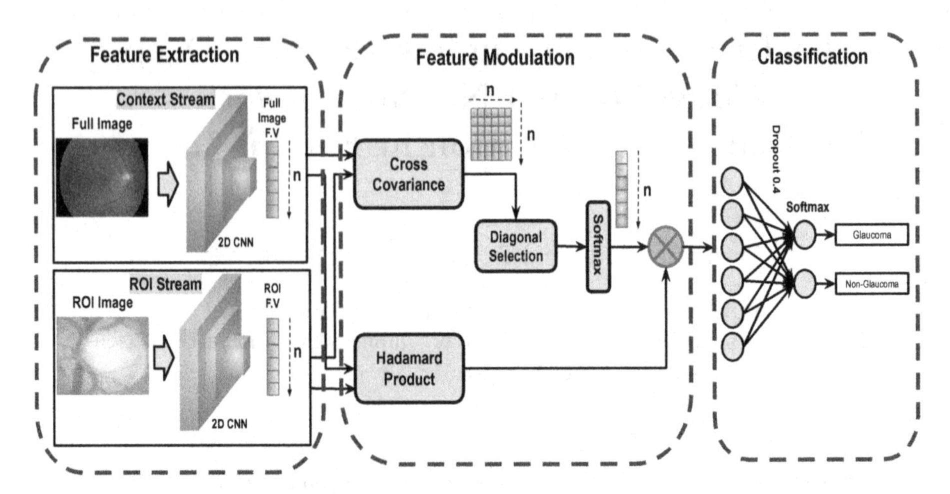

Fig. 1. Proposed network for automated glaucoma detection in fundus images. It comprises of two streams, namely, *context and ROI stream* along with a feature modulation block.

this disease can be linked to old age, ethnicity, steroid medication, and other diseases like diabetes. Unlike other eye disorders such as myopia and cataracts, glaucoma can lead to irreversible vision loss [6]. Hence, early diagnosis of glaucoma is a crucial task to safeguard the vision. In clinical practice, optic nerve head (ONH) assessment is a major inspection for glaucoma diagnosis which is often time-consuming and costly although a trained ophthalmologist is involved. Thus, automated diagnosis of glaucoma in real-world clinical practice is the utmost need of the hour.

Recently, several methods [1,8,11,14] have been reported to diagnose glaucoma using optical fundus images. Most of these methods operate in a two-step fashion *i.e., optic disc and cup segmentation followed by classification.* This procedure is majorly adopted since the abnormality is predominantly seen in optic disc and cup region of the fundus images. A commonly used classification method by clinicians is cup to disc ratio (CDR) to classify fundus images as glaucoma or non-glaucoma. For instance, authors in [3,4,7,9,18] proposed different methods to segment the optic disc and cup, and calculated the CDR to detect glaucoma. Recently, one approach [1] used Mask-RCNN first to segment the optic disc and cup, and the Inception model was then employed to classify the segmented region as glaucoma and non-glaucoma. However, such methods heavily rely on segmentation accuracy which can easily be affected by the inaccurate pathological regions and low-contrast images. The development of glaucoma classification methods that are independent of segmentation accuracy is thus a pressing need.

Motivated by this, a segmentation-independent deep convolutional neural network (CNN) approach is proposed in this paper to detect glaucoma in fundus images. Figure 1 shows the block diagram of the proposed deep network. It is inspired by the two-stream CNN [16], which takes the global fundus image and a region of interest (ROI) outlined by the clinician into account for glaucoma classification. The major motivations behind choosing two-stream CNN are: (1) clinical fundus images have significantly higher resolutions than the input resolutions required by most CNN models. Therefore, to meet the input dimensions, the original (global) fundus images need to be downsampled, resulting in loss of image fine details that are crucial for identifying abnormalities, and (2) a glaucoma patient may persist in multiple ocular disorders. Hence, considering only global images may affect the detection performance of glaucoma. On the other hand, the classification network with only ROI images may miss the global representation of the image, thereby leading to an incorrect diagnosis of glaucoma. In addition to the two-stream deep network, we propose a feature modulation technique to enrich the extracted visual features further by considering the feature-level correlation between two streams. The proposed feature modulation technique encourages those features which persist a strong correlation between the ROI and global image feature representation. In summary, the contributions of the paper include:

- A feature modulating two-stream deep CNN model is proposed in this paper for glaucoma screening in fundus images.
- The proposed model considers both the full (global) and ROI image to capture detailed visual representations from the fundus images. A feature modulation technique is also introduced to further enrich the feature representation which uses the feature-level correlation between two streams.
- The proposed deep network along with the feature modulation technique is validated on a large-scale recently introduced glaucoma dataset, namely *G1020* [1] and has achieved the state-of-the-art classification results.

The remaining of the paper is structured as follows: An overview of the existing studies is summarized in Sect. 2. Section 3 discusses the proposed two-stream dep CNN model for glaucoma detection. The experimental results and ablation studies are provided in Sect. 4. Finally, the conclusions are drawn in Sect. 5.

2 Related Work

Numerous automated methods have been developed for glaucoma screening using fundus images over the past decade. Most of the earlier approaches performed segmentation of optic cup and disc to compute CDR followed by glaucoma classification [3,9]. However, the performance of these approaches strongly relies on their segmentation results, thereby suffering from low sensitivity issues. An ample amount of methods have been designed later using a variety of handcrafted features like clinical and multiresolution features [10], empirical wavelet transform (EWT) and correntropy based descriptors [13], local binary pattern based features [12], etc., and classifiers like support vector machine, multi-layer perceptron, k-nearest neighbor, etc. However, the selection of an appropriate feature extractor, classifier, and limited detection performance has remained a concern. Furthermore, most of these methods are tested over limited fundus images.

On the other hand, deep learning techniques, mainly CNNs have achieved dramatic success in many medical image classification tasks, and therefore, have recently become an essential alternative for glaucoma detection in fundus images. Li *et al.* [11] derived the local as well as holistic features from ROI images using pre-trained CNN models for glaucoma detection. Raghavendra *et al.* [15] designed a customized CNN for glaucoma screening using full fundus images. Recently, Bajwa *et al.* [1] utilized Mask-RCNN with ResNet-50 for segmentation of optic disc and cup followed by Inception for classification of the segmented regions as glaucoma or healthy. However, this method ignored the global information of the full fundus images.

3 Proposed Method

An overview of the proposed two-stream deep network is shown in Fig. 1. It has three major stages such as *feature extraction, feature modulation, and classification* which are executed sequentially in the network to achieve the glaucoma classification task. A detailed description of each stage is given below.

3.1 Feature Extraction

Feature extraction stage performs a prime functionality in the two-stream deep network. It extracts visual features from the context and ROI stream by taking the global and ROI images as input, respectively. It can be visualized from Fig. 1 that two individual CNNs are used in the context and ROI stream to learn independent global and local representations from the fundus images. The feature vectors obtained from context and ROI streams are represented by F_C and F_R, respectively. The features are extracted from the last convolutional block of the CNN used in each stream and are then fed to a `Global Average Pooling` layer to obtain a single feature value per channel, resulting in n-dimensional feature vectors $F_C \in \mathbb{R}^{1 \times n}$ and $F_R \in \mathbb{R}^{1 \times n}$ in each stream. The feature modulation stage subsequently uses these two feature vectors to derive the salient features.

3.2 Feature Modulation

The objective of the feature modulation stage is to enhance the quality of visual features obtained from the context and ROI stream of the feature extraction stage. It is achieved by learning the correlation between the feature vectors F_C and F_R. As shown in Fig. 1, at first, a `Hadamard product` is performed between F_C and F_R to obtain a single feature vector F per fundus image. The `Hadamard product` between F_C and F_R is mathematically defined as,

$$F = F_C^i \odot F_R^j, \forall i, j \in [1, n] \tag{1}$$

where, $\odot$ denotes the point-wise multiplication operator. Now, in order to highlight the salient features in F, the feature-level cross-correlation (F_{Cross_Cor})

between F_C and F_R is taken into consideration by computing the cross-covariance between them. In F_{Cross_Cor}, only *diagonal* elements are chosen, since they define the one-to-one correlation between F_C and F_R. The F_{Cross_Cor} is expressed as,

$$F_{Cross\_Cor} = diag([(F_C - \overline{F_C}) \otimes (F_R - \overline{F_R})^T]/n]) \tag{2}$$

where, $\otimes$ indicates the Kronecker product operator, and $\overline{F_C}$ and $\overline{F_R}$ are the mean value of F_C and F_R vectors, respectively. In F_{Cross_Cor}, positive higher values correspond to the stronger correlation between F_C and F_R, and hence can facilitate in generating the salience map to highlight the prominent features in F for the classification of fundus images. Then, the F_{Cross_Cor} is passed through the `softmax` activation to assign a higher weight to those features possessing a stronger correlation between F_C and F_R. Subsequently, these feature-level weights are point-wise multiplied with F to obtain a modulated feature vector F_{Mod} which is defined as,

$$F_{Mod} = F \odot softmax(F_{Cross\_Cor}) \tag{3}$$

where, $\odot$ denotes the point-wise multiplication operator. The obtained modulated feature vector F_{mod} is then used by the classification stage to detect glaucoma.

3.3 Classification

The classification stage aims at classifying the input modulated feature vector F_{Mod} corresponding to a given fundus image to healthy or glaucoma category. The F_{Mod} is passed through a two-layer perceptron where the first and second layer contains 32 and 2 neurons, respectively. In addition, 20% `dropout` is also provided between the two-layer perceptron to avoid overfitting. The final layer is *softmax* activated to generate a high prediction score for a specific class and hence, decides the predicted label as healthy or glaucoma.

The proposed two-stream deep network is end-to-end trainable given a batch of fundus images. The network is optimized using Adam optimizer with the *categorical cross entropy* loss function. In addition, to address the class imbalance problem, a weighting factor equivalent to inverse class frequency is employed in the loss which is given as:

$$Loss_C = \sum_{i=1}^{2}(1 - f_i)(y_i \log \hat{y_i}) \quad \text{where} \quad f_i = \frac{\#\text{samples in class}_i}{\text{Total\#of samples}}. \tag{4}$$

4 Experimental Results and Analysis

4.1 Dataset Used

For the experimental evaluation of the proposed two-stream deep network, a recently introduced glaucoma classification dataset, namely **G1020** [1] has been

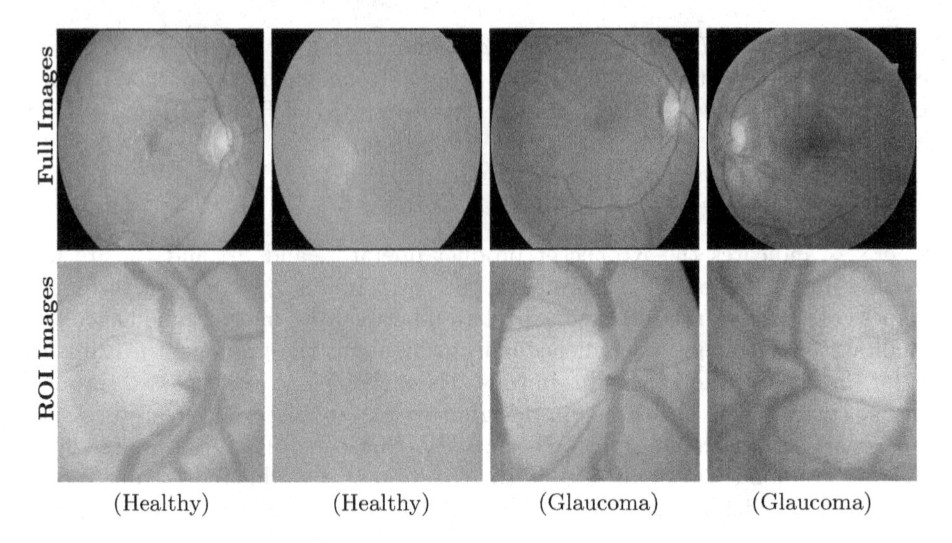

Fig. 2. Visualization of sample fundus images and their corresponding ROI for healthy and galucoma class of G1020 dataset.

used. It consists of 1020 images from 432 patients where each patient has number of images, ranging from 1 to 12. Among 1020 images, 296 images belong to glaucoma class and 724 images belong to healthy class which were collected from 110 and 322 patients, respectively. All images were gathered at a private clinical practice in Kaiserslautern, Germany between year 2005 and 2017 with 45° field of view and were kept in .JPG format. The images only with fundus region were preserved which vary in size from 1944 × 2108 to 2426 × 3007 pixels.

The major reason behind choosing **G1020** dataset is due to the availability of large number of full fundus images and their corresponding ROIs outlined by the clinicians. The sample fundus images and the ROIs from both the categories (healthy and glaucoma) are shown in Fig. 2.

4.2 Implementation Details and Performance Measures

The implementation details of the proposed method can be divided into *training* and *testing*. At first, the whole G1020 dataset containing 1020 fundus images from 432 patients is divided into training and testing by stratified random sampling with a train-test ratio of 70:30. It is worth noting that 70 % and 30% of the patients are considered from each of the classes (*i.e., healthy and glaucoma*) for training and testing, respectively. In order to train and test the proposed network, Inception-V3 model pre-trained with ImageNet [5] dataset was used in both context and ROI streams. The proposed network is trained with 0.0001 learning rate for 100 epochs with a batch size of 10 for error back propagation. Since in G1020 dataset there are patients with 12 images, so to ensure effective optimization equal number of healthy and glaucoma images are gathered in each batch of images for training the network. Once the proposed network is trained, it is tested with the images from the test patient list for performance evaluation.

The performance measures used for evaluation of the proposed network include precision, recall, F1-score and balanced accuracy (BA). The BA [2] has been used in this study in place of conventional accuracy to deal with the imbalanced G1020 dataset which is computed as the average of recall obtained on each class.

4.3 Ablation Study

In order to quantify the robustness of the proposed two-stream deep network, a detailed experimental ablation study is outlined in Table 1.

- **Experiment-1:** In this experiment, we verify the impact of the single stream for glaucoma classification. At first, only the full (global) fundus images are taken into account in the context stream as a baseline experiment which obtained an F1-score of 36.42% and BA of 50.13%. Since the optic disc and cup region in fundus images contain significant information for glaucoma diagnosis, we take only the ROI portion of the global fundus images outlined by the clinicians in the ROI stream which results in a significant improvement (36.42⟶41.34) in F1-score and (50.13⟶59.77) in balanced accuracy.
- **Experiment-2:** In this experiment, we verify the effect of the feature modulation technique in the proposed network. As the objective of this work is to utilize both global and local features of the fundus image, we first concatenate the features of both context and ROI streams which results in a degradation of F1-score by 2.53% (41.34⟶38.81) and BA by 3.65% (59.77⟶56.12). It is mainly due to the presence of contrasting features in full (global) images for glaucoma classification. Therefore, we choose to incorporate the feature modulation technique in the two-stream network, and a Hadamard product is performed between context and ROI stream features, instead of simple concatenation to highlight the salient features in the modulated feature vector. This network not only achieves superior performance (56.12⟶61.98) than the concatenation-based two-stream network, but also outperforms the single ROI stream-based network by 2.21% (59.77⟶61.98). It can also be observed that a higher precision of 32.16% and recall of 67.64% is achieved by the proposed two-stream deep network among all ablation experiments. The confusion matrices for each of the ablation experiments are shown in Fig. 3.

4.4 State-of-the-Art Performance Comparison

The glaucoma classification performance of the proposed two-stream deep network on G1020 [1] dataset is compared with a few state-of-the-art methods. It is due to the unavailability of methods validated using G1020. We considered the approach proposed by Bajwa *et al.* [1] for the state-of-the-art comparison since it is the only deep learning-based method validated on G0120. However, sufficient

Table 1. Results of ablation study on G1020 dataset

Method	Precision (%)	Recall (%)	F1-score (%)	BA (%)
Context stream only	23.50	80.88	36.42	50.13
ROI stream only	30.71	63.23	41.34	59.77
Concatenate context + ROI stream	27.21	67.64	38.81	56.12
Proposed method	**32.16**	**67.64**	**43.60**	**61.98**

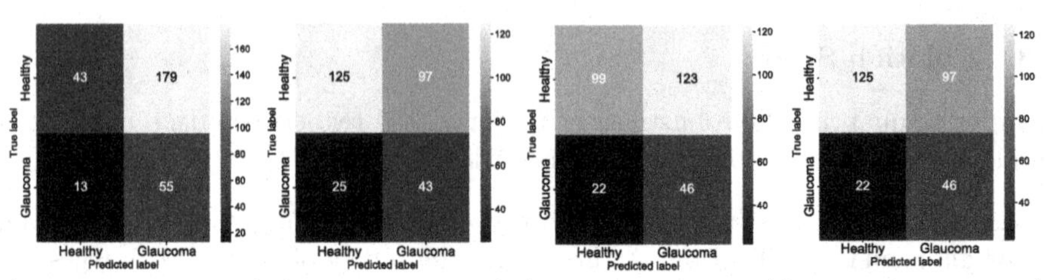

(Only context stream) (Only ROI stream) (Concatenate stream)(Proposed two-stream)

Fig. 3. Confusion matrices obtained for each case of the ablation study

information regarding the training of Inception V3 model was not provided in their paper. Therefore, we validated their method with our experimental setup using two training strategies such as *off-the-shelf and fine-tuned*, and the results are shown in Table 2. In the off-the-shelf strategy, only ImageNet pre-trained weights of Inception V3 are used for feature extraction from fundus images followed by the classification; whereas, in the fine-tuned strategy, the ImageNet pre-trained weights are retrained using transfer learning on G1020 dataset. It can be seen from Table 2 that the proposed two-stream deep network utilizing Inception V3 architecture achieves superior performance than the state-of-the-art approaches. It is due to the incorporation of context and ROI stream along with the feature modulation technique in it.

Table 2. State-of-the-art performance comparison on G1020 dataset

Method	Deep network	Precision (%)	Recall(%)	F1-Score (%)	BA (%)
Bajwa *et al.* [1]	Off-the-shelf Inception V3	30.43	30.88	30.65	54.63
Bajwa *et al.* [1]	Fine-tuned Inception V3	30.71	63.23	41.34	59.77
Proposed method	**Two-stream Inception V3 + Feature modulation**	**32.16**	**67.64**	**43.60**	**61.98**

5 Conclusion

In this paper, a feature modulating two-stream deep CNN model is proposed for glaucoma detection in fundus images. The proposed network takes the global as well as ROI fundus images as the input to each stream for extraction of fine detailed visual features. Subsequently, a feature modulation technique is also proposed as an essential block of the two-stream network to further enhance the visual feature representation. The proposed network achieves state-of-the-art glaucoma detection performance on a large-scale glaucoma dataset. Detailed ablation studies are also performed to confirm the superiority of the network. In future, the glaucoma classification performance could be improved further by designing more effective feature enhancement modules.

References

1. Bajwa, M.N., Singh, G.A.P., Neumeier, W., Malik, M.I., Dengel, A., Ahmed, S.: G1020: a benchmark retinal fundus image dataset for computer-aided glaucoma detection. In: 2020 International Joint Conference on Neural Networks (IJCNN), pp. 1–7. IEEE (2020)
2. Brodersen, K.H., Ong, C.S., Stephan, K.E., Buhmann, J.M.: The balanced accuracy and its posterior distribution. In: 20th International Conference on Pattern Recognition, pp. 3121–3124. IEEE (2010)
3. Cheng, J., Liu, J., Xu, Y., Yin, F., Wong, D.W.K., et al.: Superpixel classification based optic disc and optic cup segmentation for glaucoma screening. IEEE Trans. Med. Imaging **32**(6), 1019–1032 (2013)
4. Cheng, J., et al.: Superpixel classification based optic disc and optic cup segmentation for glaucoma screening. IEEE Trans. Med. Imaging **32**(6), 1019–1032 (2013)
5. Deng, J., Dong, W., Socher, R., Li, L.J., Li, K., Fei-Fei, L.: Imagenet: a large-scale hierarchical image database. In: IEEE Conference on Computer Vision and Pattern Recognition, pp. 248–255 (2009)
6. Dervisevic, E., Pavljasevic, S., Dervisevic, A., Kasumovic, S.S.: Challenges in early glaucoma detection. Med. Archiv. **70**(3), 203 (2016)
7. Fu, H., Cheng, J., Xu, Y., Wong, D.W.K., Liu, J., Cao, X.: Joint optic disc and cup segmentation based on multi-label deep network and polar transformation. IEEE Trans. Med. Imaging **37**(7), 1597–1605 (2018)
8. Fu, H., et al.: Disc-aware ensemble network for glaucoma screening from fundus image. IEEE Trans. Med. Imaging **37**(11), 2493–2501 (2018)
9. Joshi, G.D., Sivaswamy, J., Krishnadas, S.: Optic disk and cup segmentation from monocular color retinal images for glaucoma assessment. IEEE Trans. Med. Imaging **30**(6), 1192–1205 (2011)
10. Kausu, T., Gopi, V.P., Wahid, K.A., Doma, W., Niwas, S.I.: Combination of clinical and multiresolution features for glaucoma detection and its classification using fundus images. Biocybern. Biomed. Eng. **38**(2), 329–341 (2018)
11. Li, A., Cheng, J., Wong, D.W.K., Liu, J.: Integrating holistic and local deep features for glaucoma classification. In: 2016 38th Annual International Conference of the IEEE Engineering in Medicine and Biology Society (EMBC), pp. 1328–1331. IEEE (2016)
12. Maheshwari, S., Kanhangad, V., Pachori, R.B., Bhandary, S.V., Acharya, U.R.: Automated glaucoma diagnosis using bit-plane slicing and local binary pattern techniques. Comput. Biol. Med. **105**, 72–80 (2019)
13. Maheshwari, S., Pachori, R.B., Acharya, U.R.: Automated diagnosis of glaucoma using empirical wavelet transform and correntropy features extracted from fundus images. IEEE J. Biomed. Health Inf. **21**(3), 803–813 (2017)

14. Nayak, D.R., Das, D., Majhi, B., Bhandary, S.V., Acharya, U.R.: ECNet: an evolutionary convolutional network for automated glaucoma detection using fundus images. Biomed. Signal Process. Control **67**, 102559 (2021)
15. Raghavendra, U., Fujita, H., Bhandary, S.V., Gudigar, A., Tan, J.H., Acharya, U.R.: Deep convolution neural network for accurate diagnosis of glaucoma using digital fundus images. Inf. Sci. **441**, 41–49 (2018)
16. Simonyan, K., Zisserman, A.: Two-stream convolutional networks for action recognition in videos. arXiv preprint arXiv:1406.2199 (2014)
17. Tham, Y.C., Li, X., Wong, T.Y., Quigley, H.A., Aung, T., Cheng, C.Y.: Global prevalence of glaucoma and projections of glaucoma burden through 2040: a systematic review and meta-analysis. Ophthalmology **121**(11), 2081–2090 (2014)
18. Yin, F., et al.: Model-based optic nerve head segmentation on retinal fundus images. In: 2011 Annual International Conference of the IEEE Engineering in Medicine and Biology Society, pp. 2626–2629. IEEE (2011)

Retinal Image Quality Assessment Using Sharpness and Connected Components

S. Kiruthika(✉) and V. Masilamani

Indian Institute of Information Technology, Design and Manufacturing, Kancheepuram, Chennai, India
{coe18d003,masila}@iiitdm.ac.in

Abstract. Mobile application based diagnosis has become an aid nowadays. For better diagnosis, the quality of image needs to be good. Automatic assessment of images will help the ophthalmologists to focus more on the diagnosis. To assist the experts, an automated retinal image quality assessment method has been proposed. The proposed method make use of the features extracted from the sharpness and connected components of the fundus image. In particular, the image is divided into patches and the features are extracted. Those extracted features are used to train a machine learning model. The proposed model has achieved comparable results on the private dataset and outperformed the existing methods on public datasets.

Keywords: Retinal fundus image · Sharpness · Connected components · Machine learning · XGBoost · Classification

1 Introduction

Image is the most powerful element in the digital world. Particularly, medical image plays a vital role for better diagnosis in the clinical environment. Medical image gives the information that helps to do an in-depth and intra-operative inspection of the internal organs. This leads to an apparent growth in medical imaging as well as in the field of medical image processing. Starting from the visualization till the reconstruction, image processing plays an integral role with the clinical people.

Human eye is a very sensible part where the early symptoms for many diseases will not be explicitly visible even when there is an ail in the vision. That eventually reflects in the failure of human visual system. It is mandatory to identify the degradation of visual system in the early stage itself. Retinal fundus images play a significant part in identifying and diagnosing the retinal diseases. The retinal fundus image is the representation of bilateral cylinderical structures with semitransparent retinal tissues projected on the imaging plane [2].

Retinal fundus images aid in detecting diabetic retinopathy, degeneration of macula due to age factor, glaucoma, retinal vascular lesions, etc. In addition, the retinal fundus image screening test helps clinical people to diagnose high blood

B. Raman et al. (Eds.): CVIP 2021, CCIS 1568, pp. 181–191, 2022.
https://doi.org/10.1007/978-3-031-11349-9_16

pressure and diseases that affect the kidney, cerebral, etc. [32]. Such retinal fundus image has to be of good quality for better diagnosis when diagonasis is done by both expert and machine learning algorithm. In the initial stage, the separation of good and bad images were classified by the well trained humans involved in the clinical environment. Now, this has to be automated because a large number of people are approaching the ophthalmologist irrespective of the age factor. So, it will be a time taken process if only the humans are involved. Even if the imaging system is good there are many factors that will affect the quality of a retinal fundus image like an event of haze, blinking the eye during image capturing, etc. Due to this degradation, the prominent structure - optic disc, macula, and blood vessels will not be visible even for the eye which is physically fit [23]. Anatomy on such prominent structure in the poor quality retinal fundus image will lead to a wrong diagnosis. It is mandatory to check whether the retinal fundus image is of good visual quality or not before diagnosis.

It impulses us to develop an efficient retinal fundus image quality assessment model with high efficacy in real time. The objective of this paper is to build a model which assists clinical persons in choosing the good quality retinal fundus image for diagnosing purposes. Our objective is that given retinal image check if it normal or abnormal. Overview for achieving the objective are listed below

1. Data preprocessing - Imaging devices will vary for each clinical structure, this leads to segment foreground fundus image from the imaging system's output (discussed in the Sect. 3.1)
2. Extracting the features - extracting the image quality related features from the segmented image (discussed in the Sect. 3.2)
3. Develop a model - An efficient machine learning model has to be trained with the features extracted from the fundus image along with the corresponding label (discussed in the Sect. 3).
4. Validation - The model is validated using both public and private dataset (discussed in the Sect. 4)

The next section of the paper describes the related work and the Sect. 3 elaborates the proposed work with the detailed flow of feature extraction. Followed by, the experimental results that are analysed with public and private fundus image quality dataset in the Sect. 4. Finally, the work has been concluded in the Sect. 5.

2 Related Work

In recent days, particularly the pandemic situation leads us to prefer the telemedicine [6], diagnosis through the mobile application [6]. In such cases, it will be difficult for the ophthalmologist to predict and/or observe the severity of disease in the retinal fundus image. Only good and high quality retinal fundus images are forwarded by the ophthalmic photographer to the patients as well as to the ophthalmologist. Thus an efficient automated retinal fundus image quality assessment has to be done before forwarding those images for the diagnosis in the clinical structure. However, for a reliable diagnosis, the quality of a fundus image must be ensured. Assessing retinal fundus image is not a straightforward process, it is assessed based on the disease diagnosis. The presence of a dark region will be suitable for detecting glaucoma but not for detecting diabetic retinopathy [12].

In general, the quality of an image can be assessed by any one of the following ways: 1) by comparing with a reference image 2) by comparing with the attributes of the reference image (instead of using the reference image itself) 3) by analysing the image itself (without any knowledge about the reference images and its attributes). For retinal fundus image quality assessment, the last two methods are studied well in the literature. Lee and Wang have proposed the first automatic retinal image quality assessment based on template matching. The template considered here is the intensity histogram generated from twenty good quality retinal fundus images. Based on the similarity the quality score is defined [16].

Bartling et al. have proposed the quality evaluation model based on sharpness and illuminance. As a preprocessing, the fundus image is normalized and subdivided into the smallest square regions. Thus the sharpness score and illuminance label are generated for each square in the image [4]. Fleming et al. have proposed the model based on image clarity and field definition. Detection of macular vessels and their total length is the measure of image clarity. Clarity of the test image is good only if the macular length is greater than the threshold. The field definition of the image is good if the image has satisfied the four different metrics based on the distance between the optic disc, fovea and nearest edge, the length of superior arcade, angle of fovea from the center of optic disc [9]. Later, the same authors assessed the image clarity assessment using supervised classification. The vessel visibility and statistical measures are evaluated at the high saliency locations of an image [10].

Honggang et al. have proposed the model based on the features extracted from the histogram, texture, and from the density of the vessel. In addition, the cumulative probability of blur detection at every edge is used as the local sharpness feature score for the model [36]. Wang et al. have proposed the model based on blur, contrast, and sensation information in the multiple channels of the image [29]. Abdel-Hamid et al. have proposed the model based on image sharpness. Wavelet decomposed images are used for evaluating the sharpness measure [1]. Welikala et al. have proposed a model that predicts the image which is suitable for the studies related to epidemiology. The model works with the vessel segmented image for the prediction [30]. Feng et al. have proposed the model based on the illumination, structure, and naturalness of the retinal fundus image [26].

Not only machine learning models but deep learning models have also been used to detect the quality of the retinal fundus images. The first deep learning model for the fundus image quality has been proposed by Mahapatra et al. The model classifies the gradable and ungradable images. The model is trained with 101 images [19]. Yu et al. have proposed the model by combining both machine learning and deep learning. The features are extracted using convolution neural networks later classified using support vector machine [35]. Aditya et al. have proposed the multivariate regression based convolution neural network model for assessing the image quality. The above stated model uses the label along with six subjective scores of the image to train the model. The model gives the quality score of an image [22].

The conclusion made from the previous work is that most of the models have used private datasets to assess the quality in addition to the public datasets. Thus comparisons on the performance of the system are less in nature. Even though the deep learning models perform well with the less amount of data, the performance in the cross data set validation is not good enough [23]. Thus to build an efficient deep learning based system needs a large amount of fundus image dataset along with the quality label is mandatory. For machine learning based model, the feature extraction has to be done in an efficient manner, in order to work well in the real world dataset.

3 Proposed Method

The detailed description of preprocessing, feature extraction, and machine learning model used for the proposed work is given in this section.

3.1 Overview

The objective of our retinal fundus image quality is assessed particularly for the diabetic retinopathy image dataset. Where optic disc, fovea, and blood vessels are considered as the important elements for the diagnosis or detecting the disease. The retinal fundus image affected by diabetes has exudates and blood hemorrhages [27]. Both normal and abnormal images are considered in this work. The normal images are the images obtained from normal or healthy people, the abnormal images are obtained from the people affected by diabetic retinopathy. The preprocessing done before the assessment is that auto cropping the image in such a way that only the retinal fundus image portion will be visible and then resizing it to the uniform size of 256 * 256. By doing this the images obtained from the different imaging sources will be of uniform size and contain only the retinal fundus image portion alone instead of varying background black portion in the image.

The features based on the sharpness and connected components are extracted from the cropped, resized retinal fundus image. Then XGBoost machine learning model [7] is used for binary classification by training the features along with its label. Then the trained model is used for testing purposes. The reason behind the selection of the XGBoost classifier is due to its faster execution and outstanding performance by the regularised loss functions to avoid over-fitting. Similar to the random forest, column sampling is done in addition to that scaled down weights which are assigned to the new trees, for reducing the impact of that tree on the final score [18].

3.2 Feature Extraction

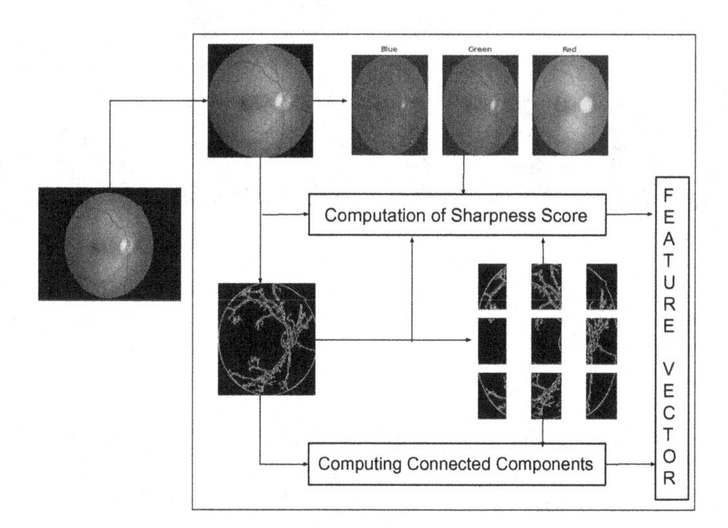

Fig. 1. Feature extraction

As shown in the Fig. 1 the feature extraction has two different sets of features, extracted from the retinal fundus image. They are based on the connected components and sharpness. These features are not directly computed from the image instead, the preprocessed image of size 256 * 256 is considered for the feature extraction process to maintain the scale invariant property. Then the image is divided into 9 patches (3 rows and 3 columns). Each patch is of the size $int((width/3)) * int((height/3))$ where the $height$ and $width$ of an image considered here is of 256 each. The reason behind dividing the image into patches is that the feature values we are considering here have to be high in the middle row particularly in the left or right column than the center column in the middle row the optic disc is located there. In the retinal fundus images, the presence of optic disc will be either in the vertical left or right portion of the image with the locations around the horizontal middle in the image. Thus the number of connected components and the sharpness will be very high in those regions. Figure 2 shows the same observation. These patches are formed from the edge map of a fundus image. The canny edge detection algorithm is used to extract the edges.

Extraction of structural information particularly, sharpness information from the retinal fundus image plays a major role in fundus image quality assessment [1,4,20,33,36]. In the proposed method, the sharpness features are computed by the gradient average of an image. As mentioned earlier, the sharpness feature is extracted for 1) Color image 2) Edge map and 3) Nine patches. For color image, the sharpness is computed for each channel ($three\ features$), then the cumulative sum is calculated. The same is represented as $featureset1$ with four features.

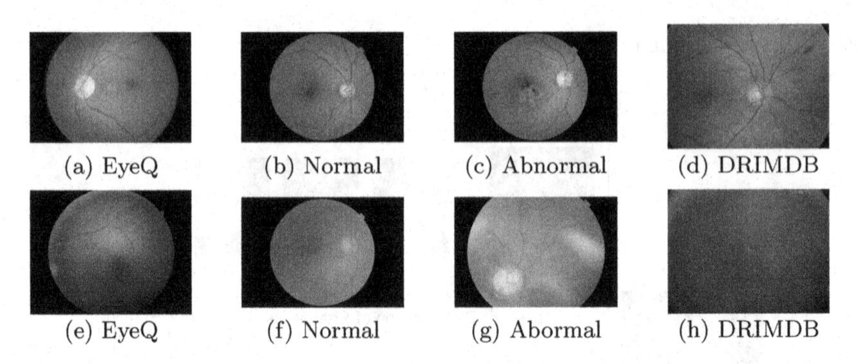

Fig. 2. Sample images from EyeQ, Private Dataset (normal and abnormal) and DRIMDB datasets (From left to right). Images from Fig. 2a–d are belongs to accept class and the images from Fig. 2e–h are the samples from reject class

For $featureset2$, sharpness computed for the nine patches ($nine\ features$) and one for the overall edge map, totally 10 features.

The next set of features is extracted from the connected components [8,24]. Each object in the image is assigned with a unique label, this process is known as labelling the connected component. Objects present in the image are referred as the connected components. This label plays a major role in many applications like document skew detection [3], cursive script writing [15], not only in these domains even in medical domains for segmenting cerebral cortical in brain images [28], and for breast cancer diagnosis by analysing mammography images [34]. As far as we know, in the literature, connected component labelling is not used for assessing the retinal fundus image quality. The speeded up connected components labelling [5] is used for extracting the count of connected components in the 9 patches as well as in the edge map of the retinal fundus image. Federico et al. have speeded up the improved classical two scan labelling algorithms [13,31] by creating the chunks or stripes from the image. Later the first scan for images is done using parallel scanning of the chunks of an image. Then for merging the border labels sequential and logarithmic merging is used. Again the parallel scan is performed on the chunks for labelling the connected components. Figure 3 shows how well the feature is discriminating the accept and reject class images.

4 Results and Discussion

The public as well as private datasets are used for the experiments. The performance of the model during different experiments is discussed in this section.

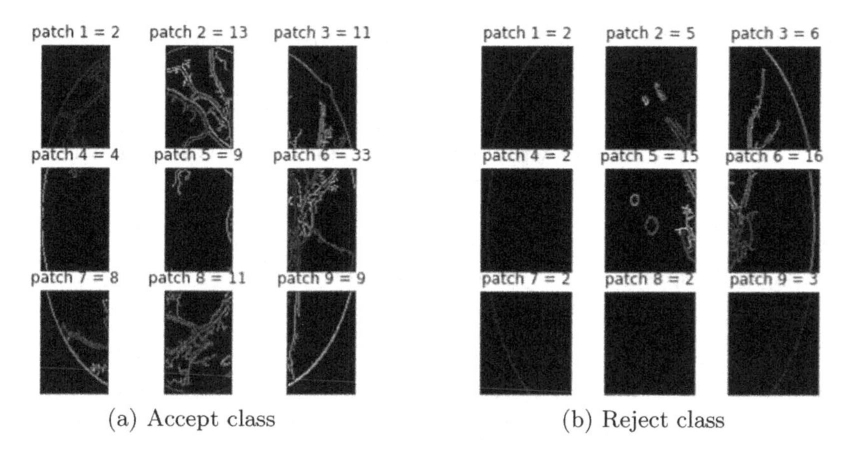

(a) Accept class (b) Reject class

Fig. 3. Number of connected components for each patch of an image. Total number of connected components in Fig. 3a is 100 and in Fig. 3b is 53

4.1 Dataset Description

Two publicly available fundus image quality datasets are used. They are EyeQ [11] and DRIMDB [25]. Both datasets are using the normal and diabetic retinopathy affected images for the abnormal category. The accept and reject classes will be available for both normal and abnormal images. Total of 216 images are available in the DRIMDB dataset with 125 for accept class and 69 for reject class and 22 images are in the outlier class which is not considered here. EyeQ dataset contains 12,543 training images and 16249 testing images including the usable class which is not considered here. In addition to these datasets, we have collected 500 normal and 1000 abnormal images from the Southern Railways Headquarters hospital, Chennai. The abnormal images are contain many diseases not only diabetic retinopathy. Three image processing subject experts with the help of doctors, have labelled the images manually either to accept or reject class. The subjective labelling is 0 for accept and 1 for reject class. Not only overall score but also 5 subcategory metrics are considered while labelling. They are 1) Illumination 2) Structure (presence of any contour like structure) 3) Visibility of nerve system 4) Visibility of optic disc and 5) Visibility of macula. Individual subjective labelling for an image is assigned based on the majority of subcategory labelling. The final labelling for the image is done if two or more experts agreed on the same label then that label is finalized for the image.

4.2 Performance Analysis on Public Dataset

The performance of the model on DRIMDB dataset is shown in the Table 1. The metric value shown in the Table 1 are of in percentage (%). An average of 100 - 10 fold validation value is reported for each metric for our model. Randomly chosen 80% of data is used for training and the remaining 20% of data is used for testing the model. Our model outperforms the existing models in three metrics on DRIMDB dataset, the same can be observed from the table.

Table 1. Performance analysis with existing system for DRIMDB dataset. The highest value in each metric is highlighted in bold text.

Model	Sensitivity	Specificity	Accuracy	AUC
Decision tree [26]	97.10	81.60	87.11	86.31
Support vector machine [26]	94.14	83.25	89.58	88.69
Dictionary learning [26]	75.82	63.30	71.05	71.06
Fine tuned random forest [14]	**99.3**	95.8	98.1	**99.9**
Gcforest [17]	96.7	98.0	97.4	Not available
Proposed method	97.77	**99.677**	**98.997**	99.77

Performance of the model on EyeQ dataset is reported in the Table 2. The 10 fold validation is not done in the experiment because the separate test dataset has been given along with the dataset. The model has been trained only with the given train dataset and tested using the given test dataset. The proposed model outperforms the existing method in all the metrics except sensitivity. Followed by these experiments we have assessed the performance of the proposed model with the cross data set validation. Where EyeQ train dataset is used for training the model and DRIMDB dataset is used for testing the model. The model achieves sensitivity as 100%, specificity as 80.8%, accuracy as 87.63%, and AUC score as 99.96%. The above results show that the model is not biased to the dataset.

Table 2. Performance analysis with existing system for EyeQ dataset. The highest value in each metric is highlighted in bold text.

Model	Sensitivity	Specificity	Accuracy	AUC
MFQ net (Binary) [21]	**94.21**	88.53	91.17	95.95
Proposed Method	88.29	**98.35**	**95.58**	**98.74**

4.3 Performance Analysis on Private Dataset

The proposed model is trained and tested with the dataset obtained from the Southern Railways Headquarters hospital, Chennai. The detailed dataset split-up is shown in the Table 3. The process used for assigning the label is discussed in the Sect. 4.1. For the experimental results, an average of 100 - 10 fold validation value is reported for each metric for our model. Randomly chosen 80% of data is used for training and the remaining unused 20% of data is used for testing the model. The model achieves the sensitivity as 84.65%, precision as 83.24%, accuracy as 81.47%, and the AUC score as 88.61%. The reason behind the slightly less performance when compared to the public dataset is that the abnormal category in the public dataset contains only diabetic retinopathy affected images but for the own dataset we have considered the images affected by all kinds of vision related diseases.

Table 3. Own dataset split-up

Label/Category	Normal	Abnormal	Total
Accept	354	295	1000
Reject	146	705	500
Total	649	851	1500

5 Conclusion

In this paper, a novel method to assess the quality of the retinal fundus images has been proposed. It has been observed that sharpness and connected components of the image well discriminate the good quality and bad quality fundus images. Using these features, a machine learning model, in particular, the XGBoost model has been trained. The proposed model has been evaluated on the private dataset and also on the public datasets. Although the proposed method outperforms the existing methods on public datasets, the method struggle with the private dataset. The reason for this is that the private dataset contains images that are affected by different eye diseases whereas the public dataset contains images that are only affected by diabetic retinopathy images. In future, more images with different abnormalities can be collected, and different image features can be explored to build an effective model.

References

1. Abdel-Hamid, L., El-Rafei, A., Michelson, G.: No-reference quality index for color retinal images. Comput. Biol. Med. **90**, 68–75 (2017)
2. Abràmoff, M., Kay, C.N.: Image processing. In: Ryan, S.J., et al.: (eds.) Retina (5th Edn.), Chap. 6, pp. 151–176. W.B. Saunders, London (2013)
3. Amin, A., Fischer, S.: A document skew detection method using the Hough transform. Pattern Anal. Appl. **3**(3), 243–253 (2000)
4. Bartling, H., Wanger, P., Martin, L.: Automated quality evaluation of digital fundus photographs. Acta Ophthalmol. **87**(6), 643–647 (2009)
5. Bolelli, F., Cancilla, M., Grana, C.: Two more strategies to speed up connected components labeling algorithms. In: Battiato, S., Gallo, G., Schettini, R., Stanco, F. (eds.) ICIAP 2017. LNCS, vol. 10485, pp. 48–58. Springer, Cham (2017). https://doi.org/10.1007/978-3-319-68548-9_5
6. Bourouis, A., Feham, M., Hossain, M.A., Zhang, L.: An intelligent mobile based decision support system for retinal disease diagnosis. Decis. Supp. Syst. **59**, 341–350 (2014)

7. Chen, T., Guestrin, C.: Xgboost: a scalable tree boosting system. In: Proceedings of the 22nd ACM SIGKDD International Conference on Knowledge Discovery and Data Mining, pp. 785–794 (2016)
8. Dillencourt, M.B., Samet, H., Tamminen, M.: A general approach to connected-component labeling for arbitrary image representations. J. ACM **39**(2), 253–280 (1992)
9. Fleming, A.D., Philip, S., Goatman, K.A., Olson, J.A., Sharp, P.F.: Automated assessment of diabetic retinal image quality based on clarity and field definition. Invest. Ophthalmol. Vis. Sci. **47**(3), 1120–1125 (2006)
10. Fleming, A.D., Philip, S., Goatman, K.A., Sharp, P.F., Olson, J.A.: Automated clarity assessment of retinal images using regionally based structural and statistical measures. Med. Eng. Phys. **34**(7), 849–859 (2012)
11. Fu, H., et al.: Evaluation of retinal image quality assessment networks in different color-spaces. In: Shen, D., Liu, et al. (eds.) MICCAI 2019. LNCS, vol. 11764, pp. 48–56. Springer, Cham (2019). https://doi.org/10.1007/978-3-030-32239-7_6
12. Giancardo, L., Meriaudeau, F., Karnowski, T.P., Chaum, E., Tobin, K.: Quality assessment of retinal fundus images using elliptical local vessel density. New Develop. Biomed. Eng. **11** (2010)
13. Grana, C., Borghesani, D., Cucchiara, R.: Optimized block-based connected components labeling with decision trees. IEEE Trans. Image Processing **19**(6), 1596–1609 (2010)
14. Karlsson, R.A., Jonsson, B.A., Hardarson, S.H., Olafsdottir, O.B., Halldorsson, G.H., Stefansson, E.: Automatic fundus image quality assessment on a continuous scale. Comput. Biol. Med. **129**, 104114 (2021)
15. Kim, J.H., Kim, K.K., Suen, C.Y.: An HMM-MLP hybrid model for cursive script recognition. Pattern Anal. Appl. **3**(4), 314–324 (2000)
16. Lee, S.C., Wang, Y.: Automatic retinal image quality assessment and enhancement. In: Medical Imaging 1999: Image Processing, vol. 3661, pp. 1581–1590. International Society for Optics and Photonics (1999)
17. Liu, H., Zhang, N., Jin, S., Xu, D., Gao, W.: Small sample color fundus image quality assessment based on GCforest. Multim. Tools Appl. **80**(11), 17441–17459 (2020). https://doi.org/10.1007/s11042-020-09362-y
18. Luckner, M., Topolski, B., Mazurek, M.: Application of XGBoost algorithm in fingerprinting localisation task. In: Saeed, K., Homenda, W.ł., Chaki, R. (eds.) CISIM 2017. LNCS, vol. 10244, pp. 661–671. Springer, Cham (2017). https://doi.org/10.1007/978-3-319-59105-6_57
19. Mahapatra, D., Roy, P.K., Sedai, S., Garnavi, R.: A CNN based neurobiology inspired approach for retinal image quality assessment. In: 2016 38th Annual International Conference of the IEEE Engineering in Medicine and Biology Society (EMBC), pp. 1304–1307. IEEE (2016)
20. Paulus, J., Meier, J., Bock, R., Hornegger, J., Michelson, G.: Automated quality assessment of retinal fundus photos. Int. J. Comput. Assist. Radiol. Surg. **5**(6), 557–564 (2010)
21. Pérez, A.D., Perdomo, O., González, F.A.: A lightweight deep learning model for mobile eye fundus image quality assessment. In: 15th International Symposium on Medical Information Processing and Analysis. vol. 11330, p. 113300K. International Society for Optics and Photonics (2020)
22. Raj, A., Shah, N.A., Tiwari, A.K., Martini, M.G.: Multivariate regression-based convolutional neural network model for fundus image quality assessment. IEEE Access **8**, 57810–57821 (2020)

23. Raj, A., Tiwari, A.K., Martini, M.G.: Fundus image quality assessment: survey, challenges, and future scope. IET Image Processing **13**(8), 1211–1224 (2019)
24. Samet, H., Tamminen, M.: Efficient component labeling of images of arbitrary dimension represented by linear bintrees. IEEE Trans. Pattern Anal. Mach. Intell. **10**(4), 579–586 (1988)
25. Sevik, U., Kose, C., Berber, T., Erdol, H.: Identification of suitable fundus images using automated quality assessment methods. J. Biomed. Opt. **19**(4), 046006 (2014)
26. Shao, F., Yang, Y., Jiang, Q., Jiang, G., Ho, Y.S.: Automated quality assessment of fundus images via analysis of illumination, naturalness and structure. IEEE Access **6**, 806–817 (2017)
27. Sisodia, D.S., Nair, S., Khobragade, P.: Diabetic retinal fundus images: preprocessing and feature extraction for early detection of diabetic retinopathy. Biomed. Pharmacol. J. **10**(2), 615–626 (2017)
28. Suri, J.S., Singh, S., Reden, L.: Computer vision and pattern recognition techniques for 2-d and 3-d mr cerebral cortical segmentation (part i): a state-of-the-art review. Pattern Anal. Appl. **5**(1), 46–76 (2002)
29. Wang, S., Jin, K., Lu, H., Cheng, C., Ye, J., Qian, D.: Human visual system-based fundus image quality assessment of portable fundus camera photographs. IEEE Trans. Med. Imaging **35**(4), 1046–1055 (2015)
30. Welikala, R., et al.: Automated retinal image quality assessment on the UK biobank dataset for epidemiological studies. Comput. Biol. Med. **71**, 67–76 (2016)
31. Wu, K., Otoo, E., Suzuki, K.: Optimizing two-pass connected-component labeling algorithms. Pattern Anal. Appl. **12**(2), 117–135 (2009)
32. Yao, Z., Zhang, Z., Xu, L.Q., Fan, Q., Xu, L.: Generic features for fundus image quality evaluation. In: 2016 IEEE 18th International Conference on e-Health Networking, Applications and Services (Healthcom), pp. 1–6 (2016)
33. Yao, Z., Zhang, Z., Xu, L.Q., Fan, Q., Xu, L.: Generic features for fundus image quality evaluation. In: 2016 IEEE 18th International Conference on e-Health Networking, Applications and Services (Healthcom), pp. 1–6. IEEE (2016)
34. Yapa, R.D., Koichi, H.: A connected component labeling algorithm for grayscale images and application of the algorithm on mammograms. In: Proceedings of the 2007 ACM symposium on Applied computing, pp. 146–152 (2007)
35. Yu, F., Sun, J., Li, A., Cheng, J., Wan, C., Liu, J.: Image quality classification for DR screening using deep learning. In: 2017 39th Annual International Conference of the IEEE Engineering in Medicine and Biology Society (EMBC), pp. 664–667. IEEE (2017)
36. Yu, H., Agurto, C., Barriga, S., Nemeth, S.C., Soliz, P., Zamora, G.: Automated image quality evaluation of retinal fundus photographs in diabetic retinopathy screening. In: 2012 IEEE Southwest Symposium on Image Analysis and Interpretation, pp. 125–128. IEEE (2012)

(MS)^{2}EDNet: Multiscale Motion Saliency Deep Network for Moving Object Detection

Santosh Nagnath Randive(✉), Kishor B. Bhangale, Rahul G. Mapari, Kiran M. Napte, and Kishor B. Wane

Department of E&TC, PimpriChinchwad College of Engineering & Research, Ravet, Pune, India
santosh.randive@pccoer.in

Abstract. Foreground segmentation in videos is a perplexing task. Infrequent motion of objects, illumination, shadow, camouflage, etc. are major factors which degrades the quality of segmentation. Usage of visual features like color, texture or shape, deficiencies the acquaintance of semantic evidence for foreground segmentation. In this paper, a novel compact multiscale motion saliency encoder-decoder learning network, ((MS)^{2}EDNet) is presented for moving object detection (MOD). Initially, the lengthy streaming video is split into several small video streams (SVS). The background for each SVS is estimated using proposed network. Further, the saliency map is estimated via the input frames and estimated background for each SVS. Further, a compact multiscale encoder–decoder network (MSEDNet) is presented to extract the multiscale foregrounds from saliency maps. The extracted multiscale foregrounds are integrated to estimate the final foreground of the video frame. The effectiveness of the proposed (MS)^{2}EDNet is estimated on three standard datasets (CDnet-2014 [1], and Wallflower [3]) for MOD. The compactness of the (MS)^{2}EDNet is analyzed based on computational complexity and compared with the present approaches. Experimental study shows that proposed network outpaces the present state-of-the-art approaches on three standard datasets for MOD in terms of both detection accuracy and computational complexity.

Keywords: Background estimation · Convolution neural network · Deep learning · Foreground estimation · Multi-scale motion saliency

1 Introduction

The moving object (background or *foreground*) is vital stage in any smart video processing applications, such as vehiclenavigation, video surveillance, traffic monitoring, action recognition, re-identification, categorization, and human tracking etc. The conventional moving object detection (MOD) approaches divide each pixel of video frame into foreground (moving object) and background (stationary object). Estimation of the foreground and background is very challenging due to illumination changes, dynamic background, night video, camera jitter, camouflage, infrequent motion of object, turbulence and thermal effects [1–3].

B. Raman et al. (Eds.): CVIP 2021, CCIS 1568, pp. 192–205, 2022.
https://doi.org/10.1007/978-3-031-11349-9_17

The existing state-of-the-art methods for MOD is broadly categorized into six groups as follows: frame difference based [4], region based [5–7], optical flow based [8], background subtraction based [9], saliency based [10], deep learning based [11] techniques. Deep learning based approaches are proposed in recent work for moving object detection [11]. Learning based algorithms are more powerful in extracting low-level, mid-level and high-level features from images or video sequences. Recently, deep learning based methodology achieved good accuracy in several image and video segmentation applications.

Based on the above discussion, in this paper, the concepts of deep learning based background estimation and multiscale motion saliency estimation are proposed for MOD in videos. The comprehensive literature on these approachesis given in next Section.

2 Related Work

Very basic and old method for MOD is differencing of two consecutive frames and thresholding. Hung*et al.* [12] presented a system for MOD in video surveillance system using hysteresis thresholding and motion compensation. It is difficult due to illumination changes, infrequent object motion, and camouflage, etc. To avoid the problem of thresholding, some of the researchers use optical flow based method. Chung*et al.* [8] used the static appearance based features of histograms of oriented gradients (Hog) and adaptive block-based gradient intensities for MOD.

Currently, background subtraction is most popular technique for foreground estimation Liao *et al.* [13] collaborated the global and local information for background estimation. They used Scale invariant local ternary pattern (SILTP) operatorto tackle the problem of illumination variations. They presented kernel density estimation (KDE) to estimate the probability distribution of extracted local patterns.Wang *et al.* [14] employed split Gaussian model and flux tensor for background subtraction and motion detectionrespectively. But, it suffered from infrequent object motion and the dynamic background. The temporal information along with spatial information of the video frames plays crucial role in case of scenarios likedebauched weather, irregular motion of objects, camouflage, shelters, shadow, and dynamic background.The combination of color information and spatio-temporal information is successfully used by Pierre *et al.* [15] for foreground estimationthroughvariation detection.Various machine learning based techniques has been presented which are less efficient, subjected to shodow, illumination changes and infrequent motion of object [16–20].

In recent work, the investigators are utilizing convolutional neural networks (CNN) for MOD. Braham *et al.* [21] proposed the unsupervised method to estimate the background and supervised approach to learn the spatial features between estimated background and current video frame for foreground segmentation. Various deep learning based architectures are utilized for the saliency detection in videos such as Fully Convolutional network [22], Atrous Convolution [23], ConvLSTM [24], and VGG16 net [25]. From literature, we observed few problems in the existing MOD algorithms due to which these methods are not able to get good detection accuracy and computationally very expensive. The identified problems are discussed as follows:

Problem I: To estimate the temporal saliency with the help of two consecutives frames is not enough to get sufficient temporal information.

Problem II: The classification (segmentation) of smaller *uniformregions* of bigger moving object is very challenging task. Because the semantic feature, extraction from those smaller regions is very difficult to classify them as foreground.

Problem III: The network given in [24] is proposed for pixel wise semantic segmentation. This network used EDNet (with VGG-16) which is computationally very heavy.

The main contributions of the proposed work to address the identified problems are given below.

Contribution I: The CNN based temporal/motion saliency map is estimated using *N*frames of small video streaming (SVS). The proposed network reduces the partial effect of illumination change, camouflage, small motion in background and weather degraded scenarios (*Problem I*).

Contribution II: The multiscale feature analysis can provide better visualization of moving object instead of single scale feature analysis. To overcome problem II, we proposed multiscale encoder-decoder network (MSEDNet) for MOD.

Contribution III: To address the problem of computational complexity, a compact MSEDNet(*with two convolution filters*) is proposed for foreground extraction from multiscale motion saliency maps (*Problem III*).

The overall network which is proposed for MOD is named as multiscale motion saliency encoder-decoder network ($(MS)^2$EDNet). The effectiveness of the proposed network is estimated on two benchmark datasets of CDnet-2014 and Wallflower for MOD in videos.

The remaining article is structured as follows: Sects. 1 and 2 describe the introduction and related work on MOD. Section 3 provides implementation details of the proposed system framework for motion saliency network (MSNet) and MS^2EDNet. Section 4 depicts training of proposed network. Section 5 focuses on result analysis. Finally, Section 6 concludes the article and presents future direction for enhancement of the work.

3 Proposed Network Framework

Encouraged from literature [11, 19, 20, 22], in this article, a novel compact multiscale deep network, $(MS)^2$EDNet is proposed for MOD in videos. The $(MS)^2$EDNet is further divided in to MSNet and MSEDNet. The proposed MSNet estimates the CNN based background for a given SVS. The motion saliency [20] gives more useful information regarding foreground and disregards the background information. The output of MSNetis given to the input of MSEDNet for multiscale semantic feature extraction for MOD. The use of multi-scale information is incorporated, to classify the small uniform regions of bigger objects. Figure 1 demonstratesthe flow diagram of $(MS)^2$EDNet. The framework of the proposed $(MS)^2$EDNet is shown in Fig. 2 and stepwise architecture is discussed in the following subsections.

3.1 Motion-Saliency Network(MSNet)

In the recent years, pre-trained CNN architectures [11] are widely used to extract pixel-wise semantic features. The saliency estimation (SE) is one of the important pre-processing step in recent algorithms for MOD. Some learning based techniques are used to estimate the saliency from current and previous frame [22]. Another most widely used approach for SE is through background estimation followed by pixel-wisesubtraction from the video frames. Further, the SEis employed for foreground estimation in videos.

Recently, Roy *et al.* [26] proposed the hand crafted features based model for background subtraction. They computed channel wise pixel level histogram of video frames. As motion is important property for SE, the grayscale histogram is enough for background estimation instead of channel wise histogram. Inspired from this, we introduced temporal histogram based technique thatis illustrated in subsequent sections.

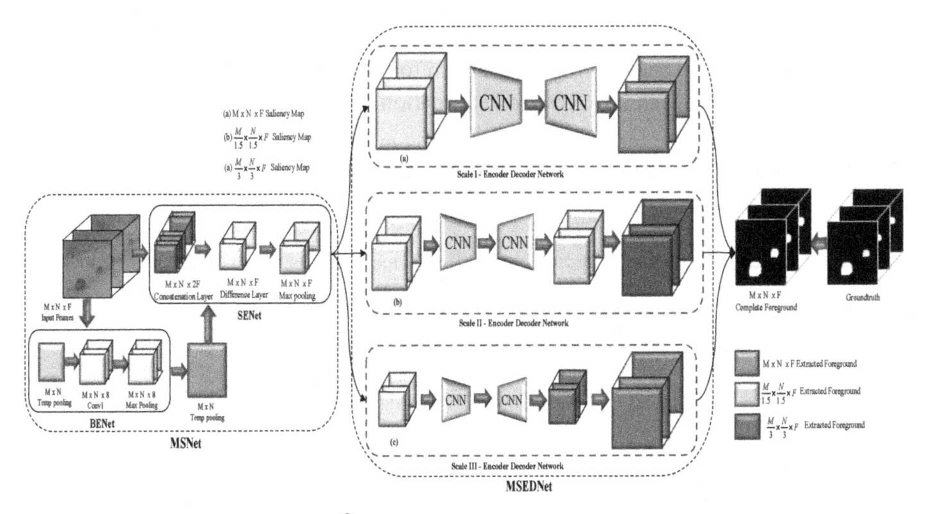

Fig. 1. Proposed network ((MS)^{2}EDNet) architecture for moving object detection in videos.

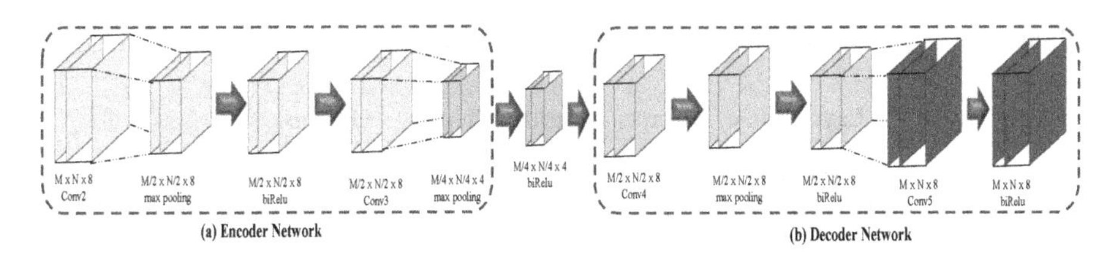

Fig. 2. Proposed encoder-decoder network architecture.

3.2 Background Estimation (BE)

The long video is split in sevral SVS V_n; $n \in [1, P]$ for background estimation. where, P represents the total number of SVS. Let Vn is the nth SVS having N video frames. The temporal histogram for the pixel at location (x, y) is calculated using Eq. 1 and 2.

$$H_n^{(x,y)}(l) = \sum_{t=1}^{N} f_1(V_n(x, y, t), l); \; l \in [0, 255] \tag{1}$$

$$f_1(a, b) = \begin{cases} 1 \; a = b \\ 0 \; else \end{cases} \tag{2}$$

The background pixel intensity for (x, y) is computed using Eq. 3.

$$B_n(x, y) = \arg\max_{l} \left(H_n^{(x,y)}(l) \right); \; l \in [0, 255] \tag{3}$$

where, argmax(.) consider bin index of histogram having higher value. However, temporal histogram based background estimation gives poor performance for the video frames having infrequent object movement. Hence, there is necessity of algorithm that treats the foreground as static object. The glitcheselevated in temporal histogram method (THM) encouraged us to improve a learning mechanism for background estimation using CNN. In this paper, we propose the background estimation network (BENet) based on CNN.

3.3 Background Estimation Using CNN (BENet)

Table 1. The architecture of proposed benet (Inputs and outputs are in number of frames)

BeNet				
Layer Name	Inputs	Kernel	Stride	Outputs
Pool1	50	---	50	1
Conv1	1	3 × 3	1 × 1	8
Pool2	8	3 × 3	1 × 1	8
Pool3	8	---	8	1

Table 2. The architecture of proposed SENet (Inputs and outputs are in number of frames)

Layer Name	Kernel	Stride	Inputs	Outputs
Replica	---	---	1	50
Concat	---	---	100	100
Difference	---	2	100	50
Pool4	3 × 3	1 × 1	50	50

Inspired from [26] and temporal histogram method, we proposeBENet that tackle the issue of infrequent object movements. The BENetarchitecture includes temporal pooling (pool1), convolutional layer (conv1), spatial pooling (pool2) and temporal pooling (pool3). The detailed discussion about each layer is given below,

***Temporal pooling (pool1)*:** In this temporal average pooling is used to approximate the temporal histogram using Eq. 4.

$$O_1^n(x, y) = \underset{t \in [1,N]}{avg} \{V_n(x, y, t)\} \tag{4}$$

***Convolutional (conv1)*:** Conv1 layer is added after pool1 to tackle the issue of infrequent object movements that can be represented using Eq. 5.

$$O_2^n = W_1^k \otimes O_1^n + b_1^k \tag{5}$$

where, W_1^k and b_1^k stands for the filters kernels and biases respectively. $W_1^k \in \mathbb{R}^{m \times m \times c}$ is one of the total number of filters ($F_1 = 8$) and $k \in [1, F_1]$, where "c" is number of channels and $m \times m$ is the convolutional filter size.

***SpatialPooling (pool2)*:** The spatial pooling (pool3) is used for the background estimation and to maintain local invariance that is computed using Eq. 6.

$$O_3^n(x, y) = \max_{\Sigma(x,y)} \left\{O_2^n(x, y)\right\} \tag{6}$$

where, $\Sigma(x, y)$ is an $m_1 \times n_1$ pattern size centered at (x, y).

The grouping of spatial pooling (*pool*2) and temporal pooling (*pool*3) overturn the consequence of infrequent object movements using convolution filters (*conv*1). Table 1 describes the parameters of BENet.

3.4 SaliencyEstimation Network (SENet)

The output of BENet obtained using Eq. (6) isprovided as input to the SENetas shown in Fig. 1. The SENetincludes four layers such as replica (R-layer), concatenation (C-layer), difference (D-layer) and pooling layer.The proposed SENetis based on difference computation and repetition of datausing generalized deep learning layers like, convolutional, pooling etc. The description of each layer is given as follows.

***Replica Layer (R-layer)*:** In this, the background estimated using BENET is repeated for 'N' times; and further utilized for temporal concatenation layer along with SVS frames.

***Concatenation Layer (C-layer)*:** It generates the temporal map by concatenating R-layer output with SVS frames.

***Difference Layer (D-layer)*:** It finds the difference between R-layer output and SVS frames temporally using convolution layer. Finally, spatial pooling is used to obtain the local invariance. SENet produces the motion saliency in SVS. Table 2 gives the details of parameters of SENet.

3.5 Multiscale Encoder-Decoder Network (MSEDNet)

The encoder network is set of convolution layers that is used to maintain the salient edges that may belong to foreground object. The decoder network is set of deconvolution layers that retain the original size from low-resolution feature map.

The issue of larger computational complexity of huge networks and single scaled input-output has motivated us to improve the existing encoder-decoder network [11] with fewer number of layers and enhanced to multiscale analysis. In this paper, a multiscale encoder-decoder network which called as *MSEDNet*is sued pixel-wise semantic features of estimated motion saliency. The block diagram of proposed *MSEDNet*is shown in Fig. 1 and detailed explanation is given below.

Table 3. The architecture of proposed three scale encoder network

Input	Layer	Kernel	Stride	Output
$\frac{M}{S} \times \frac{N}{S} \times 1$	*Conv2*	3x3	1x1	$\frac{M}{S} \times \frac{N}{S} \times 8$
$\frac{M}{S} \times \frac{N}{S} \times 8$	*Pool5*	3x3	2x2	$\frac{M}{2S} \times \frac{N}{2S} \times 8$
$\frac{M}{2S} \times \frac{N}{2S} \times 8$	*BiReLU1*	1x1	1x1	$\frac{M}{2S} \times \frac{N}{2S} \times 8$
$\frac{M}{2S} \times \frac{N}{2S} \times 8$	*Conv3*	3x3	1x1	$\frac{M}{2S} \times \frac{N}{2S} \times 4$
$\frac{M}{2S} \times \frac{N}{2S} \times 4$	*Pool6*	3x3	2x2	$\frac{M}{4S} \times \frac{N}{4S} \times 4$
$\frac{M}{4S} \times \frac{N}{4S} \times 4$	*BiReLU2*	1x1	1x1	$\frac{M}{4S} \times \frac{N}{4S} \times 4$

Scale $S \in \{1, 1.5, 3\}$; frame size is $M \times N$

Table 4. The architecture of proposed three scale decoder network

Input	Layer	Kernel	Stride	Output
$\frac{M}{4S} \times \frac{N}{4S} \times 4$	*Convt4*	3x3	1x1	$\frac{M}{2S} \times \frac{N}{2S} \times 4$
$\frac{M}{2S} \times \frac{N}{2S} \times 4$	*Pool7*	3x3	1x1	$\frac{M}{2S} \times \frac{N}{2S} \times 4$
$\frac{M}{2S} \times \frac{N}{2S} \times 4$	*BiReLU3*	1x1	1x1	$\frac{M}{2S} \times \frac{N}{2S} \times 4$
$\frac{M}{2S} \times \frac{N}{2S} \times 4$	*Convt5*	3x3	1x1	$\frac{M}{S} \times \frac{N}{S} \times 8$
$\frac{M}{S} \times \frac{N}{S} \times 8$	*Pool8*	3x3	1x1	$\frac{M}{S} \times \frac{N}{S} \times 8$
$\frac{M}{S} \times \frac{N}{S} \times 8$	*BiReLU4*	1x1	1x1	$\frac{M}{S} \times \frac{N}{S} \times 1$

Scale $S \in \{1, 1.5, 3\}$; frame size is $M \times N$

a. **Multi-scale Encoder Network**

The proposed multi-scale encoder network implemented with 2 convolutional layers followed by max pooling with a stride of 2 to extract multi-scale low resolution feature as shown in Fig. 2 (a). The encoder network consists of {*Convolutional (conv2) →max pooling (pool5) →Convolutional (conv3) →max pooling (pool6)*}. The main advantage of encoder network is that it retains only prominent edges from saliency map, andprobably these edges belong to the foreground. In this work, we used *three*scaleEDNet for multiscale semantic feature extraction. The detailed architecture of proposed multi-scale encoder network is given in Table 3. The non-linear activation function of BiReLU [27, 28] is used instead of ReLU.

The output of multiscale encoder network is given to the multiscale decoder network. The detailed explanation for multiscale decoder network is given in next sub-section.

b. **Multi-scale Decoder Network**
The proposed multiscale decoder network having 2 de-convolutional layers followed by max pooling with a stride of 1 as shown in Fig. 2(b). The decoder network consists of {*De-Convolutional (convt4)* → *max pooling (pool7)* → *De-Convolutional (convt5)* → *max pooling (pool8)*}. Here, decoder is used to map the extracted feature from encoder to original scale space. The output of three decoder networks are mapped to the original size of input video frame using bilinear interpolation. The detailed architecture of proposed decoder network is given in Table 4.

3.6 Computational Complexity of MSEDNet

In this section, the computational complexity of proposed MSEDNet of (MS)²EDNet is analyzed and compared with the existing EDNet [11] using *VGG-16* [24], *GoogLeNet* [29], *ResNet50* [30]. Table 5 shows the comparison between proposed MSEDNet and existing [11] in terms of computational complexity. From Table 5, it is clearly observed that our MSEDNethandles very less number of parameters as compared to [11] with *VGG, GoogLeNet*and*ResNet.* On the other hand, the proposed (MS)²EDNet shows better performance (0.9064) as compared to SFEN (*VGG*) + PSL + CRF (0.8292), SFEN (*GoogLeNet*) + PSL + CRF (0.7360) and SFEN (*ResNet50* [30]) + PSL + CRF (0.8772) in terms of average F-measure on CDNet-2014 database (*See in* Sect. 4).

We have also compared the computational complexity of entire proposed (MS)²EDNet network with the [11] (*VGG-16* [24], *GoogLeNet* [29], *ResNet50* [30]) in terms of execution time. The execution time for complete foreground extraction process of [11] is 203 *ms./frame*, whereas proposed (MS)²EDNet is only 49.8 *ms./frame*. From this above two observations (*execution time and number of filter parameters*), it is evident that our proposed (MS)²EDNet is very compact and faster than the existing approaches of MOD in videos.

4 Experimental Results and Discussions

The proposed network is implemented on personal Computer having core i7 processor and NVIDIA GTX 1080 11GB GPU. For training of proposed network, stochastic gradient descent (SDG) back propagation algorithm with random initial weight and mean square error as loss function is used. The qualitative as well asquantitative result analysis of proposed (MS)²EDNet for foreground extraction in videos is evaluated on three benchmark datasets, *CDnet*-2014 [1], and *Wallflower* [2] based on precision, recall and F-measure.

4.1 Results Analysis on CDnet-2014 Dataset

The segmentation performance of the proposed (MS)²EDNetis evaluated on *CDnet*-2014 dataset based on F-measure. The CDnet-2014 dataset encompasses 53 videos with 11 categories such as Bad Waether (BW), baseline (BL), Shado (SD), Camera Jitter (CJ), Pan Tilt Zoom (PLZ), Dynamic Background (DB), Low Frame Rate (LFR), Thermal (TH), Night Video (NV), and Intermittent Object Motion (IOM) along with its ground truths. Figure 3 and Table 5 gives the qualitative and quantitative comparison of various traditional techniques and proposed $(MS)^2$EDNetfor foreground estimation in CDnet-2014 dataset.

Table 5. Comparison of proposed MSEDNet with Ednet [11] In terms of computational complexity.

Parameter	MSEDNet	VGG-16 [24]	GoogLeNet [29]	ResNet [30]
Number of *conv* layers	6	13	22	17
Number of *deconv* layers	6	13	22	17
Number of filters	$12 \times 3 \times 2$	2688×2	256×2	3776×2
Filter size	3×3	3×3	7×7	$7 \times 7, 3 \times 3$
Number of filter parameters	$3 \times 3 \times 12 \times 2 \times 3$	$3 \times 3 \times 2688 \times 2$	$7 \times 7 \times 256 \times 2$	$(7 \times 7 \times 192 + 3 \times 3 \times 3584) \times 2$
Complexity of EDNet [11] over MSEDNet	1.0	74.6	38.7	128.6

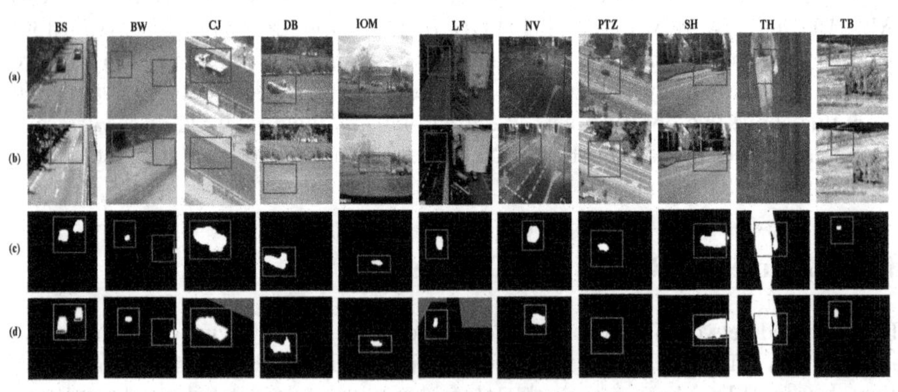

Fig. 3. The foreground extraction results of proposed $(MS)^2$EDNet on 11 categories of CDnet-2014. (a) Input video frame, (b) Estimated background using proposed BENet, (c) Extracted foreground using $(MS)^2$EDNet and (d) ground truth

Table 6 shows the comparison of proposed $(MS)^2$EDNetwith other existing methods in terms of category-wise average F-measure on CDnet-2014 Hence, we compare our $(MS)^2$EDNetwith these networks. Thequantitative analysis states, that proposed approach has lower computational complexity, better F-measure and outperform traditional state of arts on CDNet-2014 dataset.

Table 6. Comparison of proposed (MS)^{2}EDNet (category-wise) with the conventional methods for mod in terms of average F-measure on CDNet-2014 dataset.

Methods	BS	BW	CJ	DB	IOM	LF	NV	PTZ	SD	TH	TB	Avg.
IUTIS [16]	0.95	0.83	0.83	0.89	0.73	0.79	0.51	0.47	0.91	0.83	0.85	0.78
PAWCS [31]	0.94	0.81	0.81	0.89	0.78	0.64	0.42	0.45	0.89	0.83	0.77	0.75
SuBSENCE[15]	0.95	0.86	0.82	0.82	0.66	0.66	0.49	0.39	0.90	0.82	0.84	0.75
SFEN(VGG) [11]	0.92	0.85	0.91	0.60	0.57	0.59	0.51	0.74	0.89	0.72	0.73	0.73
DeepBS[32]	0.94	0.86	0.90	0.88	0.61	0.59	0.64	0.33	0.93	0.76	0.90	0.76
SFEN(VGG)+PSL	0.94	0.88	0.93	0.62	0.61	0.61	0.52	0.77	0.90	0.73	0.74	0.75
SFEN(VGG)+PSL+CRF [11]	0.94	0.89	0.94	0.74	0.75	0.62	0.75	0.78	0.91	0.85	0.92	0.83
SFEN(GoogLe)+PSL+CRF [11]	0.86	0.80	0.89	0.66	0.65	0.59	0.61	0.71	0.80	0.72	0.76	0.73
SFEN(ResNet)+PSL+CRF [11]	0.93	0.95	0.95	0.83	0.85	0.81	0.86	0.78	0.96	0.94	0.80	0.88
(MS)^{2}EDNet	0.95	0.95	0.88	0.92	0.88	0.86	0.91	0.83	0.93	0.94	0.92	0.91

4.2 Results Analysis on Wallflower Database

From last experiment, it is noticed that our method can handle conditions like illumination changes, dynamic background, night video, camera jitter, infrequent motion of object, turbulence and thermal effects. Further, the effectiveness of proposed (MS)^{2}EDNet is tested on Wallflower dataset [3] with other scenarios like foreground aperture, camouflage.The wallflower dataset includes six distinct category videos such as Bootstarp (BS), Camouflage (CF), Light Switch (LS), Foreground Aperture (FA), Time of Day (ToD), and Waving Trees (WT).

Table 7. Category-wise comparison of proposed (MS)^{2}EDNeT with state-of-the-art methods in terms of F-measure on wallflower dataset for MOD.

Method	RED [35]	GRA [36]	ISC [37]	MOD [38]	LRGB [17]	(MS)^{2}EDNet
CF	0.92	0.25	0.70	0.96	0.95	**0.97**
FA	0.58	0.31	0.41	0.61	0.83	**0.90**
LS	0.39	0.23	0.27	0.78	0.84	**0.88**
BS	0.68	0.61	0.68	0.73	0.78	**0.88**
TOD	0.16	0.12	0.13	0.40	0.76	**0.78**
WT	0.51	0.47	0.47	**0.95**	0.91	**0.91**
Avg.	0.54	0.33	0.44	0.74	0.84	**0.89**

The comparison of various methods in terms of average F-measure for MOD is illustrated in Table 7. From Table 7, it is observed that current state-of-the-art methods are able to achieve 0.8435 accuracy, whereas proposed (MS)^{2}EDNetachieved 0.8864 accuracy in terms of F-measure on Wallflower dataset for MOD. The qualitative stepwise result analysis of proposed (MS)^{2}EDNet with ground truth is depicted in Fig. 4. In literature, the visual analysis of MOD is available on four categories (*BS, FA, CF* and *WT*) of Wallflower dataset. Figure 5 illustrates the visual results comparison between proposed $(MS)^2$EDNet and other existing state-of-art methods [17, 33, 34, 39–43] for MOD. From the qualitative and quantitative analysis, it is clearly observed that the proposed $(MS)^2$EDNetgives significant improvement for foreground detection on Wallflower dataset.

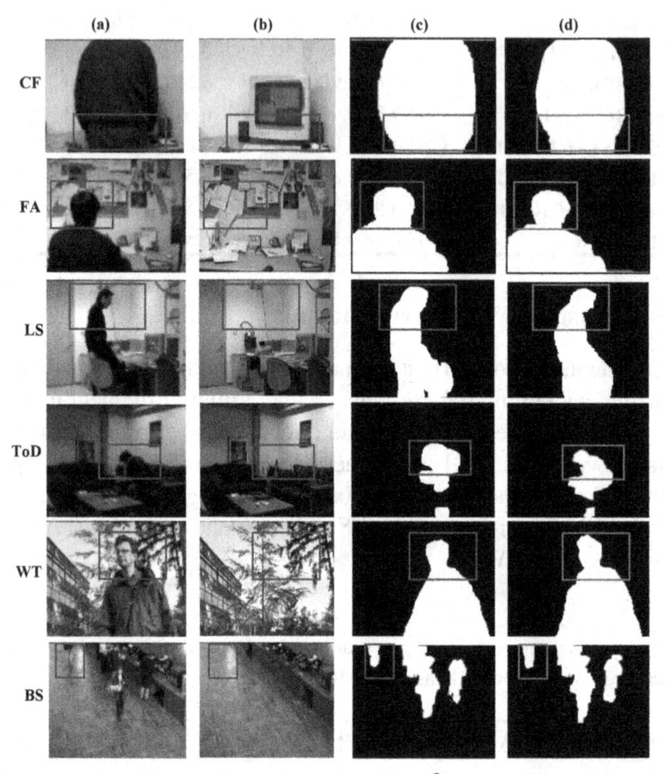

Fig. 4. The foreground extraction results of proposed $(MS)^2$EDNet on 6 categories of Wallflower database (a) Input video frame, (b) Estimated background using proposed BENet, (c) Extracted foreground and (d) Ground truth.

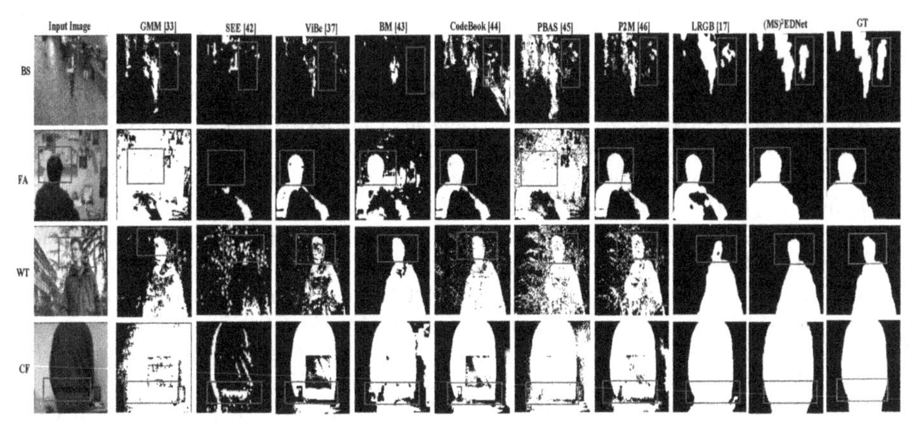

Fig. 5. Visual comparison of various methods on Wallflower dataset for foreground detection. (One result for each category) (*GT: ground truth*)

5 Conclusions

In this paper, the deep learning based multi-scale motion saliency is incorporated with compact multi-scale encoder-decoder network for foreground extraction. The proposed (MS)^{2}EDNetuses motion saliency to overcome the problems of weather degradation (*snow fall*), illumination change, dynamic background (*waving tree*), etc. in videos. The multi-scale information is used for the classification of small uniform regions of bigger object and encoder-decoder network is used to learn the low resolution features. The performance of the proposed (MS)^{2}EDNetis tested on CDNet-2014and WallFlowerbenchmark datasets for moving object detection in videos. The effectiveness of the proposed method is compared with the state-of-the-art methods for MOD in terms of precision, recall and F-measure. Experimental analysis shows that the proposed (MS)^{2}EDNet is faster than SFEN + PSL + CRF [11] with *VGG-16, GoogLeNet, ResNet*method based on execution time and outpaces the current state-of-the-art approaches on CDnet-2014 datasets in terms detection accuracy for MOD.

References

1. Y. Wang, P. M. Jodoin, F. Porikli, J. Konrad, Y. Benezeth, and P. Ishwar.: An expanded change detection benchmark dataset. In: IEEE Conference on Computer Vision and Pattern Recognition, pp. 393–400 (2014)
2. Toyama, K., Krumm, J., Brumitt, B., Meyers, B.: Wallflower: principles and practice of background maintenance. In: Proceedings of the IEEE International Conference on Computer Vision, vol. 1(1), pp. 255–261 (1999)
3. Gupta, A.K., Seal, A., Prasad, M., Khanna, P.: Salient object detection techniques in computer vision—a survey. Entropy **22**(10), 1174 (2020)
4. Wu, Z., Dahua Lin, X.T.: Adjustable bounded rectifiers: towards deep binary representations. arXiv Prepr. arXiv:1511.06201, pp. 1–11 (2015)
5. Lin, H., Member, S., Liu, T., Chuang, J., Member, S.: Learning a scene background model via classification. IEEE Trans. Signal Process **57**(5), 1641–1654 (2009)

6. Agarwala, A., et al.: Interactive digital photomontage. ACM SIGGRAPH 2004 Pap. - SIGGRAPH '04 **1**(212), 294 (2004)
7. Xu, X., Huang, T.S.: A loopy belief propagation approach for robust background estimation. In: 26th IEEE Conference on Computer Vision and Pattern Recognition. CVPR (2008)
8. Liang, C.W., Juang, C.F.: Moving object classification using a combination of static appearance features and spatial and temporal entropy values of optical flows. IEEE Trans. Intell. Transpor. Syst. **16**(6), 3453–3464 (2015)
9. Jiang, S., Lu, X.: WeSamBE: a weight-sample-based method for background subtraction. IEEE Trans. Circu. Syst. Video Technol. **8215**, 1–10 (2017)
10. Xi, T., Zhao, W., Wang, H., Lin, W.: Salient object detection with spatiotemporal background priors for video. IEEE Trans. Image Process. **26**(7), 3425–3436 (2017)
11. Chen, Y., Wang, J., Zhu, B., Tang, M., Lu, H.: Pixel-wise deep sequence learning for moving object detection. IEEE Trans. Circuits Syst. Video Technol. **8215**, 1–13 (2017)
12. Yeh, C., Member, S., Lin, C., Muchtar, K., Lai, H., Motivation, A.: Three-pronged compensation and hysteresis thresholding for moving object detection in real-time video surveillance. IEEE Trans. Ind. Electron. **64**(6), 4945–4955 (2017)
13. Liao, S., Zhao, G., Kellokumpu, V., Pietikäinen, M., Li, S.Z.: Modeling pixel process with scale invariant local patterns for background subtraction in complex scenes. In: Proceedings of the IEEE Computer Society Conference on Computer Vision and Pattern Recognition, pp. 1301–1306 (2010)
14. Wang, R., Bunyak, F., Seetharaman, G., Palaniappan, K.: Static and moving object detection using flux tensor with split gaussian models. In: IEEE Conference on Computer Vision and Pattern Recognition Workshops, pp. 420–424 (2014)
15. St-Charles, P.-L., Bilodeau, G.-A., Bergevin, R.: SuBSENSE: a universal change detection method with local adaptive sensitivity. IEEE Trans. Image Process. **24**(1), 359–373 (2015)
16. Bianco, S., Ciocca, G., Schettini, R.: Combination of video change detection algorithms by genetic programming. IEEE Trans. Evol. Comput. **21**(6), 914–928 (2017)
17. Romero, J.D., Lado, M.J., Mendez, A.J.: A background modeling and foreground detection algorithm using scaling coefficients defined with a color model called lightness-red-green-blue. IEEE Trans. Image Process. **27**(3), 1243–1258 (2017)
18. Lin, Y., Tong, Y., Cao, Y., Zhou, Y., Wang, S.: Visual-attention-based background modeling for detecting infrequently moving objects. IEEE Trans. Circuits Syst. Video Technol. **27**(6), 1208–1221 (2017)
19. Aytekin, C., Possegger, H., Mauthner, T., Kiranyaz, S., Bischof, H., Gabbouj, M.: Spatiotemporal saliency estimation by spectral foreground detection. IEEE Trans. Multim. **20**(1), 82–95 (2018)
20. Pang, Y., Member, S., Ye, L., Li, X., Pan, J.: Incremental learning with saliency map for moving object detection. IEEE Trans. Circu. Sys. Video Technol. (TCSVT) **1**, 1–12 (2016)
21. Braham, M., Van Droogenbroeck, M.: Deep background subtraction with scene-specific convolutional neural networks. In: International Conference on Systems, Signals, and Image Processing, pp. 1–4 (2016)
22. Wang, W., Shen, J., Shao, L.: Video salient object detection via fully convolutional networks. IEEE Trans. Image Process. **27**(1), 38–49 (2018)
23. Yang, L., Li, J., Member, S., Luo, Y., Member, S.: Deep background modeling using fully convolutional network. IEEE Trans. Intell. Transp. Syst. **19**(1), 254–262 (2018)
24. Simonyan, K., Zisserman, A.: Very Deep Convolutional Networks for Large-Scale Image Recognition. arXiv Prepr. arXiv:1409.1556, pp. 1–14 (2014)
25. Badrinarayanan, V., Kendall, A., Cipolla, R.: SegNet: a deep convolutional encoder-decoder architecture for image segmentation. IEEE Trans. Pattern Anal. Mach. Intell. **39**(12), 2481–2495 (2017)

26. Roy, S.M., Ghosh, A.: Real-time adaptive histogram min-max bucket (HMMB) model for background subtraction. IEEE Trans. Circuits Syst. Video Technol. **8215**(c), 1–1 (2017)
27. Cai, B., Xu, X., Jia, K., Qing, C.: DehazeNet : an end-to-end system for single image haze removal. IEEE Trans. Image Proce.ss **25**(11), 1–13 (2016)
28. Wu, Z., Lin, D., Tang, X.: Adjustable bounded rectifiers: towards deep binary representations. arXiv Prepr. arXiv1511.06201, pp. 1–11 (2015)
29. Szegedy, C., et al.: Going deeper with convolutions. In: IEEE Conference on Computer Vision and Pattern Recognition, vol. 07–12–June, pp. 1–9 (2015)
30. He, K., Zhang, X., Ren, S., Sun, J.: Deep residual learning for image recognition. In: IEEE Conference on Computer Vision and Pattern Recognition, pp. 770–778 (2016)
31. St-Charles, P.L., Bilodeau, G.A., Bergevin, R.: A self-adjusting approach to change detection based on background word consensus. In: IEEE Winter Conference on Applications of Computer Vision, WACV 2015, pp. 990–997 (2015)
32. Babaee, M., Dinh, D.T., Rigoll, G.: A Deep Convolutional Neural Network for Background Subtraction. arXiv preprint arXiv:1702.01731 (2017)
33. Zivkovic, Z., Van Der Heijden, F.: Efficient adaptive density estimation per image pixel for the task of background subtraction. Pattern Recognit. Lett. **27**(7), 773–780 (2006)
34. Barnich, O., Van Droogenbroeck, M.: ViBe : a universal background subtraction algorithm for video sequences. IEEE Trans. Image Process. **20**(6), 1709–1724 (2011)
35. Schindler, K., Wang, H.: Smooth foreground-background segmentation for video processing. In: Asian Conference on Computer Visio, pp. 581–590 (2006)
36. Pan, J., Li, X., Li, X., Pang, Y.: Incrementally detecting moving objects in video with sparsity and connectivity. Cognit. Comput. **8**(3), 420–428 (2016)
37. He, J., Balzano, L., Szlam, A.: Incremental gradient on the Grassmannian for online foreground and background separation in subsampled video. In: IEEE Conference on Computer Vision and Pattern Recognition, pp. 1568–1575 (2012)
38. Guo, X., Wang, X., Yang, L., Cao, X., Ma, Y.: Robust foreground detection using smoothness and arbitrariness constraints. In: European Conference on Computer Vision, pp. 535–550 (2014)
39. Dikmen, M., Huang, T.S.: Robust estimation of foreground in surveillance videos by sparse error estimation. In: 19th International Conference on Pattern Recognition, pp. 1–4 (2008)
40. Sheikh, Y., Shah, M.: Bayesian modeling of dynamic scenes for object detection. IEEE Trans. Pattern Anal. Mach. Intell. **27**(11), 1778–1792 (2005)
41. Kim, K., Chalidabhongse, T.H., Harwood, D., Davis, L.: Real-time foreground-background segmentation using codebook model. Real-Time Imaging **11**(3), 172–185 (2005)
42. Hofmann, M., Tiefenbacher, P., Rigoll, G.: Background segmentation with feedback: the pixel-based adaptive segmenter. In: IEEE Computer Society Conference on Computer Vision and Pattern Recognition Work, pp. 38–43 (2012)
43. Yang, L., Cheng, H., Su, J., Li, X.: Pixel-to-model distance for robust background reconstruction. IEEE Trans. Circuits Syst. Video Technol. **26**(5), 903–916 (2016)

Video Enhancement with Single Frame

Vijay M. Galshetwar[1(✉)], Prashant W. Patil[2], and Sachin Chaudhary[1]

[1] Punjab Engineering College, Chandigarh, India
{vijaymadhavraogalshetwar.phdcse20,sachin.chaudhary}@pec.edu.in
[2] Applied Artificial Intelligence Institute, Deakin University, Geelong, Australia
prashant.patil@deakin.edu.in

Abstract. Hazy scenes captured by the camera in a hazy environment are the main cause of the deterioration of captured video quality. Hazy environmental conditions degrade the quality of captured videos which leads to poor visibility and color distortion in videos. In this paper, we propose a video enhancement technique with a single frame dehazing encoder-decoder network. The proposed method focuses to restore the haze-free video from the hazy video by developing an encoder-decoder-based dehazing network. The framework comprises of encoder block (EB) and decoder block (DB). Both EB and DB are used to extract the haze-relevant feature maps and to give more attention to haze-specific features by re-calibrating the weights of learned feature maps automatically and helps to recover the haze-free frame. The use of dilated residual module (DRM) helps to improve output results and gain more contextual information by widening the receptive field. The extensive analysis of the proposed method is carried out by considering benchmark synthetic hazy video databases and analyzed quantitative results. Experimental results show that the proposed method out performs the other state-of-the-art (SOTA) existing approaches for video frame dehazing.

Keywords: Encoder block · Decoder block · Video enhancement

1 Introduction

Video captured by a camera in hazy surrounding may deteriorate the video quality in terms of low contrast, poor visibility, dull colors, and color imbalance. As a result, captured video contributes hazy scene view of the original view. Such hazy surrounding is the presence of haze particles in the air medium. Haze particle is a common phenomenon initiated by very small surrounding particles like dust, fog, humid (water droplet), smoke, and sand particles.

Deterioration of image or video takes place when reflected light intensity from scene objects fall on hazy particulate resulting in scattering of light and leads to degrading its original view quality. This effect is called as scattering effect. Researchers [10,27] used scattering model for the reconstruction of clear images from its hazy counterpart. The scattering model is commonly used to describe

B. Raman et al. (Eds.): CVIP 2021, CCIS 1568, pp. 206–218, 2022.
https://doi.org/10.1007/978-3-031-11349-9_18

an image or video frame formation and is based on the physical properties of transmission of light through the air medium. Equation 1 shows the scattering model for hazy image or video frame formation.

$$I(x) = J(x)t(x) + A(1 - t(x)) \quad (1)$$

where $I(x)$ represents the hazy video frame radiance at pixel x, $J(x)$ represents the clean frame radiance at pixel x, A represents the atmospheric light, and $t(x)$ represents the indoor or outdoor scene transmission map. With the help of Eq. 1, we can recover haze free scenes by estimating both the air light and scene transmission map. Transmission map is defined as,

$$t(x) = e^{(-\beta d(x))} \quad (2)$$

where β represents the attenuation coefficient and $d(x)$ represents the distance of the scene at pixel x. Video frame degradation at every pixel is different and depends on the distance between the scene point and camera. The transmission coefficient controls the amount of haze in every pixel and scene attenuation.

The presence of hazy scene in the video degrades the performance capability and results in the poor quality of output. The applications of computer vision such as depth estimation [6,16], moving object segmentation [17,18], human action recognition [9,14,19,22], object detection, autonomous vehicle tracking, and marine surveillance [24]requires good quality of videos to achieve good performance. The requirement of haze removal in such hazy videos is an essential task. However, the main challenge is to remove haze from the hazy video and enhance the video quality. Many computer vision-based dehazing algorithms are designed to solve hazy-free scene problem. The key contributions of the proposed method are explained below:

1. We develop an encoder-decoder-based video enhancement network for single frame video dehazing.
2. We developed a EB module as encoder, which extracts feature maps of the hazy video frame. To improve the representational capability of this model CAB and scaling are used to re-calibrate learned weights of channel feature responses.
3. We develop a DB module as decoder, which is a video frame restoration subnetwork for video dehazing. DB is used to restore the hazy free video frames from incoming extracted feature maps of EB. The use of CAB in DB to re-calibrate learned weights of channel feature responses for gaining a hazy free scene.

The rest of the paper is organized as follows: Image or video frame haze removal based literature review is illustrated in Sect. 2. The detailed discussion on the proposed method for video frame haze removal is done in Sect. 3. Section 4 describes quantitative and qualitative analysis and their comparison with the SOTA results of other methods. Finally, the paper is concluded in Sect. 5.

2 Literature Review

Restoration of the haze-free images or videos from the hazy images or videos is the main aim of all the dehazing techniques. Approaches used for proposing dehazing models are categorized based on three main groups: prior based [1], multi-image fusion [2], and learning-based [3]. Priors or assumptions based single image hazy scene removal achieve remarkable progress. For dehazing using prior based technique [1] some factors are responsible, which are surrounding light intensity, haze-specific priors (e.g. thickness of haze), and outdoor scene transmission maps. The fusion techniques are used to combine different domain-specific information of the image or video using multi-fusion-based methods. Many researchers find learning-based dehazing techniques are used for estimating the outdoor scene transmission maps based on the scattering model and depth or semantics information.

2.1 Prior Based

Dark channel prior (DChP) was initially implemented by He et al. [10] to acquire a clear scene. DchP uses dark pixels whose pixel intensity is very low compared to any other color channel of the hazy-free scene. The haze density and atmospheric light are estimated by using this approach, which procures the haze-free scene. However, this approach fails when both atmospheric light and the object have similar colors. The non-local image dehazing hypothesis introduced by Berman et al. [1], which states that an image can be assumed as a collection of multiple dissimilar colors. Other prior based hazy models [8,10,27] etc. are used for hazy scene removal.

2.2 Multi-image Fusion

Multi-image fusion techniques are used for retaining mainly useful features by combining multiple image features in spatial domains or the domains having chromatic, luminance, and saliency features. Most of the image restoration methods used feature concatenation [21], element-wise summation [26] or dense connection [23] to fuse features. Dudhane et al. [7] proposed RYFNet, where the first focus was to generate two transmission maps for RGB and YCbCr color spaces and then integrate these two transmission maps (i.e. RGB and YCbCr) resulting in one robust transmission map to tackle the haze-free scene task. First, haze-free videos were introduced by Zhang et al. [25], he considered frame-by-frame videos to improve the temporal coherence using optical flow and Markov Random Field.

2.3 Learning-Based

Learning-based dehazing approach uses CNNs and GANs for network architecture building, where the main focus is to map hazy images directly to hazy-free images. To resolve this dehazing task, many researchers developed architectures based on CNN [20] and GAN [12]. Zhang et al. proposed DCPDN [21], where he uses dense encoder-decoder network-based edge-preserving pyramid for estimating accurate transmission map and atmospheric light to optimize hazy-images to clear images. GAN-based end-to-end cycle-consistency approaches RI-GAN [4] and CD-Net [5] implemented by Dudhane et al. RI-GAN incorporates novel generator and discriminator by using residual and inception modules. Whereas CD-Net follows estimation of transmission map based on unpaired image data and optical model so that haze-free scene recovered effectively.

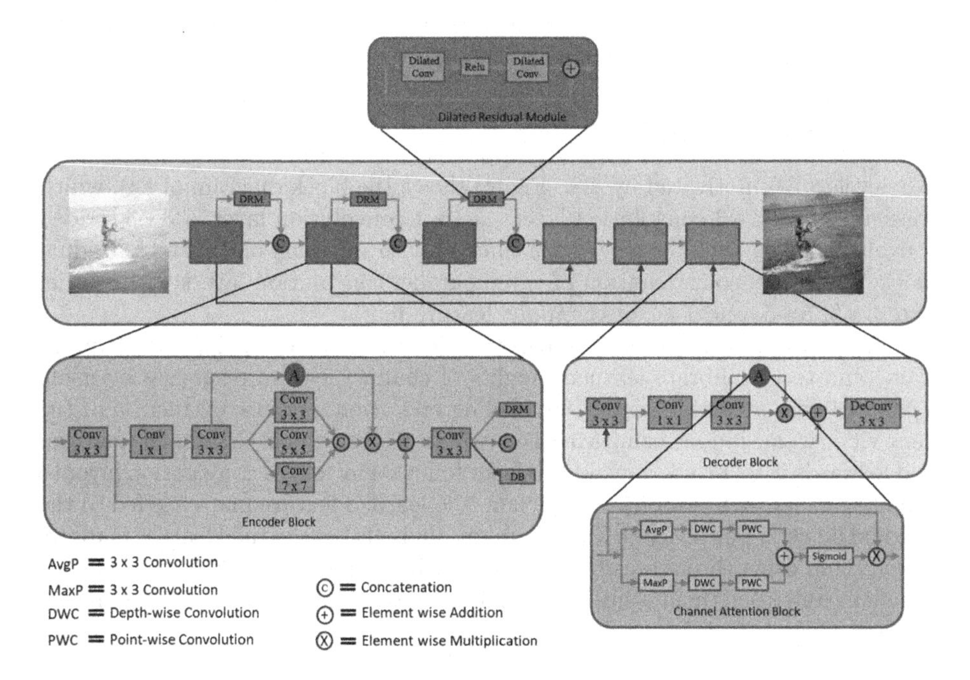

Fig. 1. Proposed framework of encoder-decoder network for single video frame haze removal.

3 Proposed Method

The proposed method is a video enhancement framework using single-frame-based encoder-decoder network architecture. The encoder reduces spatial dimension at every layer and increases the channels. The decoder increases the spatial dimension at every layer and reduces the channels. We use an encoder-decoder framework to predict each haze-free pixel in the hazy input video frame by using pix2pix [37] as a base model. Figure 1 shows the proposed architecture of single frame video dehazing.

Input hazy video frame is inputted to encoder where feature extraction take place and decoder reconstructs the haze free video frame using the proposed model. The encoder consists of three encoder blocks (EBs); EB is responsible for the extraction of feature maps of the hazy input video frame, re-calibration of learned weights of channel feature response, and downsampling of learned feature maps of the hazy input video frame. The decoder consists of three decoder blocks (DBs). Learned feature maps of the encoder are integrated with output features of dilated residual module (DRM) using concatenation layer. The increased receptive field [11] of learned feature maps by retaining the spatial feature resolutions is fed to the first decoder block followed by the second and third. Then the concatenation layer integrates the learned features of each encoder coming via skip connection and inputted feature maps. Each DB reduces the number of feature channels to half. The last convolution layer is used for mapping the channels with the required number of classes.

3.1 Encoder Block (EB)

We develop EB motivated by [28]. Figure shows the block diagram of EB, which consists of two $3*3$ convolution layers, a $1*1$ convolution layer, a $3*3$ strided convolution layer with the rate of 2, a channel-wise attention block, and a residual block, which is a concatenation of three parallel convolution layers with spatial sizes $3*3$, $5*5$, and $7*7$ of the input feature map.

The responsibilities of EB are to extract feature maps of the hazy input video frame and to re-calibrate learned weights of channel feature responses by using a channel wise attention model and scaling operation. Feature extraction of the hazy video frame is done using a $3*3$ convolution layer which extracts the features and forwards it to a $1*1$ convolution layer followed by a $3*3$ convolution layer to acquire further high-level features. Then, the learned feature maps are fed to the residual block as well as CAB. To acquire residuals of these learned feature maps we scaled them using three convolution layers with spatial sizes $3*3$, $5*5$, and $7*7$ parallelly and fused them using concatenation operation. To improve the representational capability of this model we adopted scaling (element-wise multiplication) of generated features from CAB and learned features coming from residual block. To retain low-level information of video frame at the output learned features of first $3*3$ convolution layer are directly fed to element-wise summation with the output features of scaling using skip connection. Then, we use a $3*3$ strided convolution layer with a rate of 2, which helps to compress and downsample the extracted feature maps from the hazy input video frame.

Channel Attention Block (CAB): The use of channel-wise attention block (CAB) in a simple residual module helps to re-calibrate learned weights of channel feature responses by modeling inter-dependencies between channels. We develop CAB motivated by [29,31], which is used in the EB to detect and give more attention to learned haze feature maps by re-calibrating the weights of learned feature maps automatically.

The CAB starts with two streams, one has global average pooling (AvgP) followed by depth-wise and point-wise convolution layer, and the second has maximal pooling (MaxP) followed by depth-wise and point-wise convolution layer. The global average pooling is used to capture common features and maximal pooling for capturing distinctive features. Depth-wise convolution is used both streams after AvgP and MaxP to learn and acquire the weights of each channel separately by applying a convolution filter to each channel independently. The operation of channel attention is based on squeeze and excitation for weight re-calibration of the feature map. The squeeze operates on spatial information with the help of both global average pooling and max pooling and transforms into channel-wise information. On the other hand, excitation operates using a depth-wise convolution layer which enables channel separable mechanism, point-wise convolution layer is used to project output channels of the depth-wise convolution layer onto new channel space. These features are then aggregated and applied to sigmoid activation. Then we adopted scaling (element-wise multiplication) for feature response re-calibration between inputted features and learned output features.

Residual Block: For widening the learning capabilities of deep networks by using a parallel combination of convolution layers, Residual block increases the capability of the proposed network to extract the haze-relevant feature maps and helps to recover the haze-free video frame. it is developed using a parallel connection of three convolution layers with spatial sizes $3 * 3$, $5 * 5$, and $7 * 7$, and integrated their individual learned feature outputs using a concatenation layer.

Dilated Residual Module (DRM): We used DRM motivated by [30] to gain more contextual information by widening the receptive field of incoming feature maps. The figure shows the block diagram of DB, two dilated convolution layers with the same dilated rate of 2, which helps to achieve a large receptive field and one ReLU activation function.

As dilated convolutions are useful for increasing the receptive field by retaining the spatial resolutions of inputted feature maps we provided output feature maps of each encoder to DRM. Then we integrated generated output with output feature maps of each encoder and given as input to the next EB or DB. Dilated convolution layer is used to generate the deep convolution feature maps to acquire contextual information. We added these learned feature map outputs from dilated convolution layer with the initially inputted feature maps from an EB to retain low-level information of video frame at the output.

3.2 Decoder Block (DB)

DB restores the haze free video frames from learned feature maps of the encoder block. DB integrates inputted feature maps with the learned features of each encoder coming via skip connection and reduces the number of feature channels to half. We develop DB motivated by [28]. The block diagram of DB is shown in the Fig. 1, which consists of two $3 * 3$ convolution layers, a $1 * 1$ convolution layer, a $3 * 3$ de-convolution layer with the rate of 2, a channel-wise attention block.

The responsibilities of DB are to restore the hazy free video frames from learned feature maps of the hazy input video frame and to re-calibrate learned weights of channel feature responses by using a channel-wise attention model and scaling operation. extracted features of the hazy video frame are further learned using a $3 * 3$ convolution layer which extracts the features and forwards them to a $1 * 1$

convolution layer followed by a $3*3$ convolution layer to acquire further high-level features. Then, the learned feature maps are fed to the CAB. To improve the representational capability of this model we adopted scaling (element-wise multiplication) of generated features from CAB and learned features coming via short skip connection from preceding convolution layers. To retain low-level information of video frame at the output, learned features of the first $3*3$ convolution layer are directly fed to element-wise summation with the output features of scaling using skip connection. Then, we use a $3*3$ de-convolution layer with a rate of 2, which helps to upsample the extracted feature maps from the hazy input video frame.

3.3 Loss Function

The optimization of the network parameters is done using adversarial loss and then added edge loss, ssim loss and Vgg loss in the overall loss function.

Adversaial Loss: Equation 3 is the equation of adversarial loss [37] used to map G: H to C.

$$\begin{aligned} l_{GAN}(G, D, H, C) = & E_{c \sim p_{data}}(c)\left[\log\left(D_C\left(C\left(c\right)\right)\right)\right] \\ & + E_{h \sim p_{data}}(h)\left[\log\left(1 - D_Y\left(G\left(h\right)\right)\right)\right] \end{aligned} \tag{3}$$

where, G is the generator and D_C is the discriminator. To convert hazy frame to haze-free frame we developed G, where it tries to generate frame G(h) same as that of haze-free frame. D_C is used to compare G_h with ground truth frame and responds by answering generated frame G_h is real or fake.

Edge Loss: We use sobel edge detector for edge map computation and enhancement. Following is the edge loss equation [4].

$$l_{Edge}(G) = \| E_{g(x)} - E_{y(x)} \| \tag{4}$$

where, $E_{g(x)}$ signifies generated scene edge map and $E_{y(x)}$ signifies ground truth scene edge map.

SSIM Loss: The Structural Similarity Index (SSIM) loss [41] is mostly helpful to preserve contrast and high-frequency information than other losses. Hence, we used it in our model. SSIM loss is defined as,

$$l_{SSIM}(G) = 1 - SSIM\left(G(h), C(c)\right) \tag{5}$$

where, $G(h)$ is generated frame from proposed generator and $C(c)$ is ground truth frame.

Vgg Loss: The vgg loss is used to guide the network by gaining texture and structure-related information, and calculation is carried out by considering generated and ground truth frames. Then it is passed through the pre-trained model of VGG19 [38].

$$l_p = \sum_{i=1}^{X}\sum_{j=1}^{Y}\sum_{k=1}^{Z} \frac{1}{XYZ} \parallel \Theta_l(G(h))_{i,j,k} - \Theta_l(C(c))_{i,j,k} \parallel \tag{6}$$

where, C_c is ground truth frame, G_h is the generated frame by proposed generator network. X,Y,Z are the feature map dimensions.

For training of the network, following is the overall loss equation

$$l_{Total} = \lambda_{Edge} l_{Edge} + \lambda_{SSIM} l_{SSIM} + \lambda_p l_p + \lambda_{GAN} l_{GAN} \tag{7}$$

where, λ_{Edge}, λ_{SSIM}, λ_p and λ_{GAN} are the assigned loss function weights for l_{Edge}, l_{SSIM}, l_P and l_{GAN} respectively.

4 Experimental Results

We used synthetic and real-world hazy video frames and haze-free video frames for training the proposed model. The existing benchmark datasets: synthetic videos generated using DAVIS 2016 [13] and NYU depth [15] datasets are used training and testing. The video frame haze removal model is trained on NVIDIA DGX station having the configuration NVIDIA Tesla V100 4 × 32 GB GPU.

In this section, qualitative and quantitative evaluations of the proposed method are analyzed experimentally. The evaluation of the proposed method is carried out by comparing it with the existing SOTA methods. Peak signal-to-noise ratio (PSNR) and structural similarity index (SSIM) parameters are used for the performance evaluation. We compare experimental evaluations of the proposed method against the existing SOTA methods by dividing them into two subparts (1) Quantitative Analysis (2) Qualitative Analysis.

4.1 Quantitative Analysis

The quantitative analysis is carried out by using two existing benchmarks video datasets: DAVIS 2016 [13] and NYU depth [15]. The Davis dataset contains a total of 45 videos which are divided into 25 training having 2079 video frames and 20 testings having 1376 video frames. The NYU-Depth dataset consists of a total of 45 videos which are divided into 25 training having 28,222 video frames and 20 testings having 7528 video frames.

The SOTA methods used for comparative evaluations are TCN [32], FFANet [33], MSBDN [34], GCANet [35], RRO [36], CANCB [39] and FME [40]. We did training and testing of the proposed model on the Davis 2016 [13] and NYU-Depth [15] datasets separately. Then we tested all the trained SOTA models using DAVIS 2016 and NYU depth. The evaluated results obtained by testing the proposed method are compared with the testing results of SOTA methods (Tables 1 and 2).

Table 1. Quantitative analysis of proposed method and SOTA for single video frame Haze removal on DAVIS 2016 [13] database

Method	TCN [32]	FFANet [33]	MSBDN [34]	GCANet [35]	RRO [36]	FMENet [40]	Proposed
PSNR	16.608	12.19	14.64	17.66	15.09	16.16	22.0795
SSIM	0.619	0.6501	0.7319	0.7185	0.7604	0.8297	0.8107

Table 2. Quantitative analysis of proposed method and SOTA for single video frame Haze removal on NYU depth [15] database

Method	TCN [32]	FMENet [40]	GCANet [35]	RRO [36]	CANCB [39]	Proposed
PSNR	18.837	19.81	22.55	19.47	20.87	23.79
SSIM	0.6142	0.8429	0.9013	0.8422	0.8903	0.9096

4.2 Qualitative Analysis

The qualitative analysis of the proposed method against SOTA methods using DAVIS 2016 [13] and NYU depth [15] video datasets is as shown in Figs. 2 and 3 respectively. We observed that qualitative analysis of the proposed method efficiently reduces the effect of haze in the single hazy video frame.

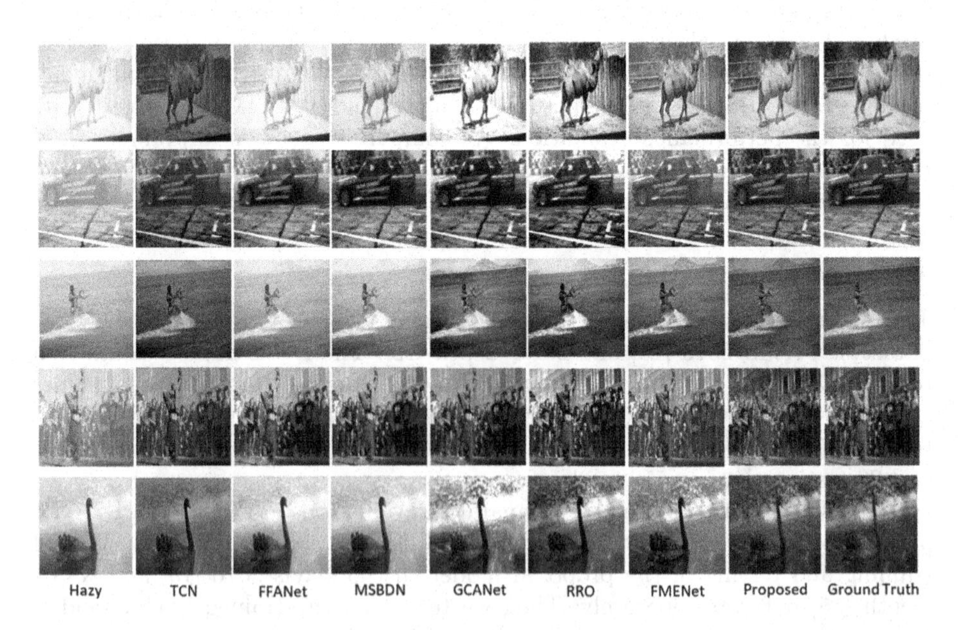

Fig. 2. Analysis of proposed method and SOTA for video dehazing on DAVIS-16 [13] Database. (TCN [32], FFANet [33], MSBDN [34], GCANet [35], RRO [36], and FME [40]

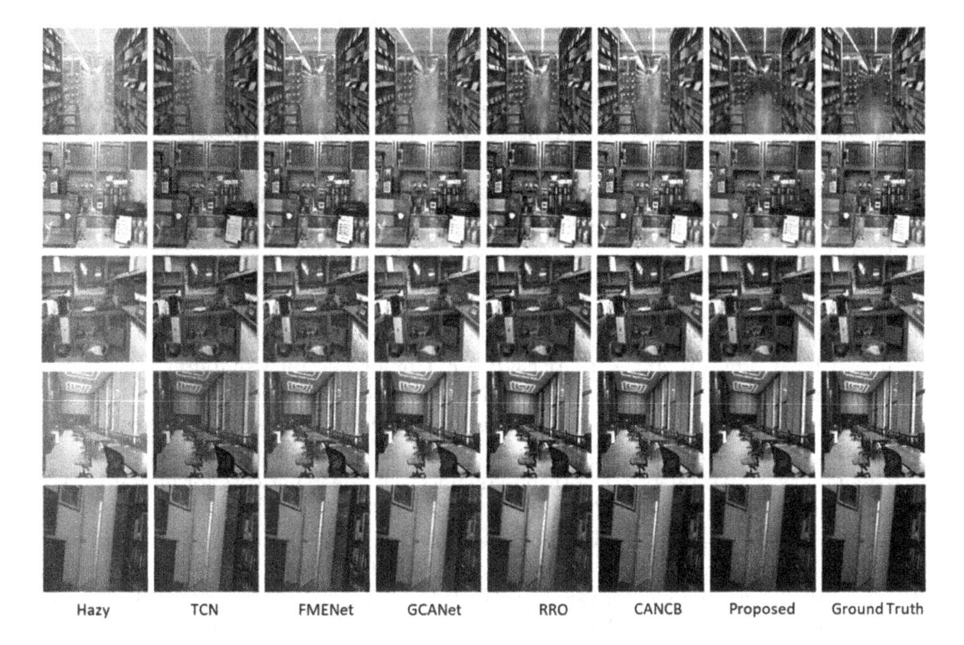

Fig. 3. Analysis of proposed method and SOTA for video dehazing on NYU Depth [15] Database. (TCN [32], FME [40], GCANet [35], RRO [36], and CANCB [39])

5 Conclusion

The proposed encoder-decoder based video enhancement framework is an end-to-end learning approach for single video frame haze removal. The EB is developed to extract feature maps of the hazy video frame and to re-calibrate learned weights of channel feature responses using CAB. We then adopted scaling operation between generated features from CAB and learned features coming from residual block to improve the representational capability of this model. The use of DRM help to improve output results and gain more contextual information by widening the receptive field. The DB restores the hazy free video frames from incoming extracted feature maps of EB. The use of CAB in DB to re-calibrate learned weights of channel feature responses to gain a haze free scene. We used skip connections to provide low-level learned features generated at initial convolution layers of the encoder to the decoder, which helps to generate the noticeable edge information. We used two benchmark datasets: DAVIS 2016 and NYU Depth for performance evaluation of the proposed method, then analyzed and compared the qualitative results of the proposed method with the state-of-the-art video frame dehazing methods. Experimentally we analyzed that the results of the proposed method out performs other dehazing methods.

References

1. Berman, D., Treibitz, T., Avidan, S.: Non-local image dehazing. In: Proceedings of the IEEE Conference on Computer Vision and Pattern Recognition (CVPR) (2016)
2. Choi, L.K., You, J., Bovik, A.C.: Referenceless prediction of perceptual fog density and perceptual image defogging. IEEE Trans. Image Processing **24**(11), 3888–3901 (2015)
3. Zhang, J., et al.: Hierarchical density-aware dehazing network. IEEE Trans. Cybernet. 1–13 (2021)
4. Dudhane, A., Aulakh, H.S., Murala, S.: Ri-gan: an end-to-end network for single image haze removal. In: 2019 IEEE/CVF Conference on Computer Vision and Pattern Recognition Workshops (CVPRW), pp. 2014–2023 (2019)
5. Dudhane, A., Murala, S.: Cdnet: Single Image Dehazing Using Unpaired Adversarial Training, pp. 1147–1155 (2019)
6. Chaudhary, S., Murala, S.: Depth-based end-to-end deep network for human action recognition. IET Comput. Vis. **13**(1), 15–22 (2019)
7. Dudhane, A., Murala, S.: Ryf-net: deep fusion network for single image haze removal. IEEE Trans. Image Processing **29**, 628–640 (2020)
8. Fattal, R.: Single image dehazing. ACM Trans. Graph. **27**(3), 1–9 (2008)
9. Chaudhary, S., Murala, S.: Deep network for human action recognition using Weber motion. Neurocomputing **367**, 207–216 (2019)
10. He, K., Sun, J., Tang, X.: Single image haze removal using dark channel prior. IEEE Trans. Pattern Anal. Mach. Intell. **33**(12), 2341–2353 (2011)
11. Phutke Shruti, S., Murala, S.: Diverse receptive field based adversarial concurrent encoder network for image inpainting. IEEE Signal Process. Lett. **28**, 1873–1877 (2021)
12. Q., Yu, Chen, Y., Huang, J., Xie, Y.: Enhanced pix2pix dehazing network. In: Proceedings of the IEEE/CVF Conference on Computer Vision and Pattern Recognition (CVPR) (2019)
13. Perazzi, F., et al.: A benchmark dataset and evaluation methodology for video object segmentation. In: Proceedings of the IEEE Conference on Computer Vision and Pattern Recognition (2016)
14. Chaudhary, S., Murala, S.: TSNet: deep network for human action recognition in Hazy Videos. In: 2018 IEEE International Conference on Systems, Man, and Cybernetics (SMC), pp. 3981–3986 (2018). https://doi.org/10.1109/SMC.2018.00675
15. Silberman, N., Fergus, R.: Indoor scene segmentation using a structured light sensor. In: 2011 IEEE International Conference on Computer Vision Workshops (ICCV Workshops), pp. 601–608. IEEE (2011)
16. Hambarde, P., Dudhane, A., Patil, P.W., Murala, S., Dhall, A.: Depth estimation from single image and semantic prior. In: 2020 IEEE International Conference on Image Processing (ICIP), pp. 1441–1445. IEEE (2020)
17. Patil Prashant, W., Dudhane, A., Kulkarni, A., Murala, S., Gonde, A.B., Gupta, S.: An unified recurrent video object segmentation framework for various surveillance environments. IEEE Trans. Image Processing **30**, 7889–7902 (2021)
18. Patil Prashant, W., Biradar, K.M., Dudhane, A., Murala, S. An end-to-end edge aggregation network for moving object segmentation. In: Proceedings of the IEEE/CVF Conference on Computer Vision and Pattern Recognition, pp. 8149–8158 (2020)

19. Chaudhary, S., Dudhane, A., Patil, P., Murala, S.: Pose guided dynamic image network for human action recognition in person centric videos. In: 2019 16th IEEE International Conference on Advanced Video and Signal Based Surveillance (AVSS), pp. 1–8 (2019). https://doi.org/10.1109/AVSS.2019.8909835
20. Yang, X., Li, H., Fan, Y.-L., Chen, R.: Single image haze removal via region detection network. IEEE Trans. Multim. **21**(10), 2545–2560 (2019)
21. Zhang, H., Patel, V.M.: Densely connected pyramid dehazing network. In: Proceedings of the IEEE Conference on Computer Vision and Pattern Recognition (CVPR) (2018)
22. Chaudhary, S.: Deep learning approaches to tackle the challenges of human action recognition in videos. Diss. (2019)
23. Zhang, H., Sindagi, V., Patel, V.M.: Multi-scale single image dehazing using perceptual pyramid deep network. In: 2018 IEEE/CVF Conference on Computer Vision and Pattern Recognition Workshops (CVPRW), pp. 1015–101509 (2018)
24. Kulkarni, A., Patil, P.W., Murala, S.: Progressive subtractive recurrent lightweight network for video deraining. IEEE Signal Process. Lett. **29**, 229–233 (2022). https://doi.org/10.1109/LSP.2021.3134171
25. Zhang, J., Li, L., Zhang, Y., Yang, G., Cao, X., Sun, J.: Video dehazing with spatial and temporal coherence. Visual Comput. **27**(6), 749–757 (2011)
26. Zhang, X., Dong, H., Hu, Z., Lai, W.-S., Wang, F., Yang, M.-H.: Gated fusion network for joint image deblurring and super-resolution. arXiv preprint arXiv:1807.10806 (2018)
27. Zhu, Q., Mai, J., Shao, L.: Single image dehazing using color attenuation prior. In: BMVC, Citeseer (2014)
28. Ren, W., Tian, J., Wang, Q., Tang, Y.: Dually connected deraining net using pixel-wise attention. IEEE Signal Process. Lett. **27**, 316–320 (2020). https://doi.org/10.1109/LSP.2020.2970345
29. Wang, S., Wu, H., Zhang, L.: AFDN: attention-based feedback dehazing network for UAV remote sensing image Haze removal. IEEE Int. Conf. Image Process. **2021**, 3822–3826 (2021). https://doi.org/10.1109/ICIP42928.2021.9506604
30. Zhu, L., et al.: Learning gated non-local residual for single-image rain streak removal. IEEE Trans. Circuits Syst. Video Technol. **31**(6), 2147–2159 (2021). https://doi.org/10.1109/TCSVT.2020.3022707
31. Li, P., Tian, J., Tang, Y., Wang, G., Wu, C.: Model-based deep network for single image deraining. IEEE Access **8**, 14036–14047 (2020). https://doi.org/10.1109/ACCESS.2020.2965545
32. Shin, J., Park, H., Paik, J.: Region-based dehazing via dual-supervised triple-convolutional network. In: IEEE Trans. Multim. https://doi.org/10.1109/TMM.2021.3050053
33. Qin, X., Wang, Z., Bai, Y., Xie, X., Jia, H.: FFA-net: feature fusion attention network for single image dehazing. Proc. AAAI Conf. Artif. Intell. **34**, 11908–11915 (2020). https://doi.org/10.1609/aaai.v34i07.6865
34. Dong, H., et al.: Multi-scale boosted dehazing network with dense feature fusion. 2154–2164 (2020). https://doi.org/10.1109/CVPR42600.2020.00223
35. Chen, D., et al.: Gated context aggregation network for image dehazing and deraining. 1375–1383 (2019). https://doi.org/10.1109/WACV.2019.00151
36. Shin, J., Kim, M., Paik, J., Lee, S.: Radiance-reflectance combined optimization and structure-guided ℓ_0-norm for single image dehazing. IEEE Trans. Multim. **22**(1), 30–44 (2020). https://doi.org/10.1109/TMM.2019.2922127

37. Isola, P., et al.: Image-to-image translation with conditional adversarial networks. In: 2017 IEEE Conference on Computer Vision and Pattern Recognition (CVPR), pp. 5967–5976 (2017)
38. Simonyan, K., Zisserman, A.: Very Deep Convolutional Networks for Large-Scale Image Recognition. arXiv preprint arXiv:1409.1556 (2015)
39. Kanti Dhara, S., Roy, M., Sen, D., Kumar Biswas, P.: Color cast dependent image dehazing via adaptive airlight refinement and non-linear color balancing. IEEE Trans. Circuits Syst. Video Technol. 31(5), 2076–2081 (2021). https://doi.org/10.1109/TCSVT.2020.3007850
40. Zhu, Z., Wei, H., Hu, G., Li, Y., Qi, G., Mazur, N.: A novel fast single image dehazing algorithm based on artificial multiexposure image fusion. IEEE Trans. Instrument. Measur. **70**, 5001523 (2021). https://doi.org/10.1109/TIM.2020.3024335
41. Que, Y., Li, S., Lee, H.J.: Attentive composite residual network for robust rain removal from single images. IEEE Trans. Multim. **23**, 3059–3072 (2021). https://doi.org/10.1109/TMM.2020.3019680

Blind Video Quality Assessment Using Fusion of Novel Structural Features and Deep Features

Anish Kumar Vishwakarma(✉) and Kishor M. Bhurchandi

Vivesvaraya National Institute of Technology, Nagpur, India
anish.vishwakarma10@gmail.com

Abstract. We propose a robust and efficient blind video quality assessment model using fusion of novel structural features and deep semantic features. As the human visual system (HVS) is very sensitive to the structural contents in a visual scene, we come up with a novel structural feature extractor that uses a two-level encoding scheme. In addition, we employ a pre-trained Convolutional Neural Network (CNN) model Inception-v3 that extracts semantic features from the sampled video frames. Further, structural and deep semantic features are concatenated and applied to a support vector regression (SVR) that predicts the final visual quality scores of the videos. The performance of the proposed method is validated on three popular and widely used authentic distortions datasets, LIVE-VQC, KoNViD-1k, and LIVE Qualcomm. Results show excellent performance of the proposed model compared with other state-of-the-art methods with significantly reduced computational burden.

Keywords: Structural features · Support vector regression · Convolutional Neural Network

1 Introduction

In recent past, the swift advancement in communication technologies and availability of fast internet have increased the consumption of videos drastically. Many social media platforms and video streaming providers give opportunity to users to upload and access videos at a very high speed. As the communication resources are limited, it is critical to maintain the quality of video services by utilizing the resources in an optimized manner. Here, Video Quality Assessment (VQA) models play an important role. VQA techniques are generally classified as Full-reference (FR), Reduced-reference (RR), and No-reference (NR)or blind based on the availability of reference or original videos. FR and RR methods use complete and partial (statistics of reference videos) reference videos, respectively. However, in many practical cases, reference videos are not available. Only NR-VQA methods are suitable for practical applications where access to reference videos is not possible. However, due to the unavailability of reference videos, NR-VQA models are much more complicated than FR and RR.

B. Raman et al. (Eds.): CVIP 2021, CCIS 1568, pp. 219–229, 2022.
https://doi.org/10.1007/978-3-031-11349-9_19

Here, we present a brief description of the published NR or blind VQA models. The first general-purpose blind VQA model [13] was developed using the 2D Discrete Cosine Transform (DCT) coefficients of consecutive frame differences and motion coherency. However, the computation of motion coherency makes the model computationally very heavy. Li et al. [7] extracted various spatiotemporal features from the 3D DCT coefficients of the 3D video blocks followed by a linear support vector regression. Mittal et al. [10] developed a completely blind VQA model without using true quality scores or subjective score values, but the performance of the model is poor.

Recently researchers mainly focus on the development of VQA models for user-generated content (UGC). UGC videos are captured by professional as well as non-professional persons. In addition, UGC videos are distorted with authentic or in-capture distortions such as in/out focus, under/overexposure, camera shake, etc. Quality assessments of UGC videos are much complex due to these authentic distortions. Korohonen [5] developed an efficient and robust blind VQA model using various spatial and transform-based features. The feature extraction is divided into low-complexity and high-complexity feature extraction. Low-complexity features were extracted from every alternate video frame, and for high-complexity features, only a few video frames were selected based on low-complexity features. The model in [5] is among the best VQA model for authentic distortion datasets.

Dendi et al. [1] presented a blind VQA model using statistics of response of Gabor bandpass filters and three-dimensional mean subtracted contrast normalized (3D MSCN) coefficients of 3D video blocks. The model in [1] exhibit moderate performance with very high computational complexity. Vishwakarma and bhurchandi [19] developed a computationally efficient and robust blind VQA model based on the statistics of 3D discrete wavelet transform (DWT) of local video blocks. Only a few video blocks were selected based on the energy of the 3D gray-level co-occurrence matrix (GLCM) of video blocks. Excellent performance on synthetic distortion datasets and moderate performance on authentic distortion datasets was achieved in [19]. VIDEVAL [18] used features extracted from some top-performing Image Quality Assessment (IQA)/VQA models. They applied feature selection on the extracted features and selected very few features for final quality prediction. The model [18] yields excellent performance. ChipQA-0 [2] is a blind VQA model based on the natural scene statistics (NSS) of local space-time chips sliced in the direction of local motion flow, exhibits good performance for authentic distortion datasets.

Apart from the hand-crafted features based blind VQA models, there are some VQA models available in literature that are developed using CNN. The V-MEON [8] model used 3D CNN and developed a blind VQA model that predict the visual quality of the videos distorted with compression distortions. Using a pre-trained ImageNet followed by a gated current unit (GRU), Li et al. [6] designed an effective blind video quality prediction model. Recently, in [20] an NR-VQA model was developed using 3D CNN followed by a long short-term memory (LSTM) network. An excellent performance was observed by the model [20]. Among blind VQA models VIDEVAL [18] and [5] exhibit best performance.

In this paper, we present a blind VQA model using the fusion of novel structural features and deep semantic features. We proposed a novel structural feature extraction technique that uses a two-level encoding scheme with four directions (0°, 90°, 180°, and 270°). We also employ a pre-trained Inception-v3 [16] model that uses kernels of different sizes at each layer and extracts deep semantic features from the sampled video frames. In the regression stage, we use a support vector regressor that predicts the visual quality scores followed by a polynomial curve fitting that fine-tuned the predicted scores and further improves the final performance of the proposed VQA model.

The major contributions of this paper are as follows:

(i) We propose a novel structural feature extractor that uses a two-level encoding scheme to code the dominant structures in sampled video frames using four directions.
(ii) We extract automatically learned high-level semantic features from the sampled video frames using the pre-trained Inception-v3 model.
(iii) A novel hybrid regression scheme, SVR followed by polynomial fitting, is proposed to predict the final visual quality scores of the distorted videos.
(iv) Exhaustive experiments, including ablation study and time complexity analysis, are conducted using three publicly available authentic distortion databases.

The rest of this paper is organized as follows. In Sect. 2, we present the detailed description of the proposed features extraction and regression scheme. Section 3 provides the dataset description followed by the experimental procedure and performance analysis of the proposed VQA model. Finally, we present conclusion in Section 4.

2 Proposed Blind VQA Model

This section presents the proposed feature extraction process in detail. We also discuss the regression scheme adapted for the quality score predictions and overall view of the proposed model.

2.1 Structural Feature Extraction

Since HVS exhibit strong sensitivity for the structural contents in a natural scene and distortion deteriorate these structural contents. Here, we extract the dominant structural information from the sampled video frames and further use it to estimate the visual quality scores. In this proposed work, we modify the conventional local tetra patterns (LTrPs) [12] by considering four direction instead of two direction. Noise exhibits the point discontinues, and the increased number of direction detect the point discontinuities very effectively.

Unlike conventional LTrPS, the modified LTrPs use two-level encoding scheme. First, we compute the first derivative at center pixel (p_c) of sampled video frame (F) along horizontal directions (0° & 180°) followed by the first-level encoding $F_h^1(p_c)$ as in (1), (2), and (3).

$$F_{0^o}^1(p_c) = F(p_{h,0^o}) - F(p_c) \tag{1}$$

$$F_{180^o}^1(p_c) = F(p_{h,180^o}) - F(p_c) \tag{2}$$

$$F_h^1(p_c) = \begin{cases} 1, & if\ F_{0^o}^1(p_c) \times F_{180^0}^1(p_c) > 0 \\ -1, & otherwise \end{cases} \tag{3}$$

where, $F(p_{h,0^o})$ and $F(p_{h,180^o})$ represent the pixel intensity value at 0° and 180° neighborhood of p_c. Superscript '1' indicates first-derivative. Similarly, we compute the first-derivative at p_c along vertical direction (90° & 270°) followed by the firs-level encoding $F_v^1(p_c)$ as in (4), (5), and (6).

$$F_{90^o}^1(p_c) = F(p_{h,90^o}) - F(p_c) \tag{4}$$

$$F_{270^o}^1(p_c) = F(p_{h,270^o}) - F(p_c) \tag{5}$$

$$F_v^1(p_c) = \begin{cases} 1, & if\ F_{90^o}^1(p_c) \times F_{270^0}^1(p_c) > 0 \\ -1, & otherwise \end{cases} \tag{6}$$

Further (3) and (6) are used for second-level encoding that generates coded values '1', '2', '3', and '4' as in (7).

$$F^1(p_c) = \begin{cases} 1 & ; F_h^1(p_c) > 0\ and\ F_v^1(p_c) > 0 \\ 2 & ; F_h^1(p_c) < 0\ and\ F_v^1(p_c) > 0 \\ 3 & ; F_h^1(p_c) < 0\ and\ F_v^1(p_c) < 0 \\ 4 & ; F_h^1(p_c) > 0\ and\ F_v^1(p_c) < 0 \end{cases} \tag{7}$$

The same procedure is repeated to generate tetra codes for all eight neighbor pixels (p_n) of center pixel (p_c) with $n = 1, 2, 3......8$. Next, we generate 8 digit modified LTrPs (MLTrPs) code at p_c by using the tetra codes of eight neighboring pixels p_n as in (8).

$$MLTrPs(p_c) = \{f_1(F^1(p_c), F^1(p_1)), \\ f_1(F^1(p_c), F^1(p_2)), ..., f_1(F^1(p_c), F^1(p_8))\} \tag{8}$$

where $f_1(.)$ is a function that compares the tetra codes of p_c and p_n defined as in (9).

$$f_1(F^1(p_c), F^1(p_n)) = \begin{cases} 0, & if\ F^1(p_c) = F^1(p_n) \\ F^1(p_n), & otherwise \end{cases} \tag{9}$$

After generating the 8 digit MLTrPs code, we convert it into binary forms. Let the tetra code corresponding to p_c is '1' and the 8 digit MLTrPs code is "0 2 0 3 4 2 4 3". Now for each tetra code of p_n, i.e., '2', '3', and '4', we can generate three 8-bit binary codes by simply replacing tetra code corresponding to p_n by binary

1 and remaining tetra codes by binary 0. For tetra code of $p_n = 2$, the binary pattern is obtained as "0 1 0 0 1 0 0". Similarly, for tetra codes of $p_n = 3, 4$, binary patterns are "0 0 0 1 0 0 0 1" and "0 0 0 0 1 0 1 0", respectively. We can observe that for each center pixel tetra code, three binary codes are generated.

Next we convert these binary codes into decimal by employing binary to decimal conversion. The 8 bit binary code can generate $2^8 = (256)$ possible decimal values. Besides, there are $4 \times 3 = 12$ patterns generated, 4 possible tetra code for the center pixel (p_c) and 3 patterns for each tetra code of neighboring pixel (p_n). Now for each pattern the length of the histogram is 256, for 12 patterns the the length of histogram is $12 \times 256 = 3072$. To reduce the length of the histogram or features, we use only uniform codes. Uniform codes are the codes, where at most only one transition either from (0 to 1) or (1 to 0) occurs, for example, "1 1 1 0 0 0 0 0", "1 0 0 0 0 0 0 0", etc. From literature we observe that uniform codes consists of dominant structural information [14]. Out of 256 codes total 58 uniform codes are possible and the remaining codes are converted into one code, so total 59 codes instead of 256 codes are used. It reduces the length of feature vector significantly, from 3072 to $12 \times 59 = 708$.

2.2 Deep Semantic Feature Extraction

Deep semantic features are extracted from the sampled video frames using pre-trained Inception-v3 [16] CNN model. The reason to select Inception-v3 model is that it uses multiple size kernel filters at the same layer. Large size filters capture the visual information distributed globally, while small size filters capture local visual information in a visual scene. As distortion deteriorate the visual content, a pre-trained Inception-v3 model can effectively extracts the useful high-level semantic features such as edges, corners, blobs, etc. Also, the size of the VQA datasets are limited, a pre-trained CNN model avoid the overfitting problem.

The main difficulty to use the pre-trained CNN model is the mismatch between the input size of CNN model and the size of the sampled video frames that we use as input to the pre-trained model. Our main aim to extract the high-level semantic features from pre-trained CNN model. We downscale the sampled video frames to fit to the input size of Inception-v3 module. Frame selection rate of 1 frame/s is used to extract the features from the pre-trained CNN model. Inception-v3 model provides a feature vector of size 2048.

2.3 Regression Scheme

After extracting both structural and deep semantic features from the sampled video frames, we use average pooling to convert frame-level features into video-level features, and then concatenated to form the final feature vector. Further, to predict the visual quality scores of distorted videos, we use support vector regression model (SVR). To optimize the SVR hyperparameters (γ & C), we have employed a grid of predetermined values of γ and C of size 10×10 [18]. Following the standard convention, we split the video dataset into 80% training set (further 80% training and 20% validation set) and 20% testing set. Different combinations of γ and C are used to train the SVR model and further to test on validation set. We track Root Mean Squared Error (RMSE) between true quality scores and predicted quality scores for each combination of hyperparameters. Finally, the hyperparameters corresponding to minimum RMSE are selected.

2.4 Proposed Methodology

The overall view of the proposed blind VQA model is presented in Fig. 1. Dominant structural features were extracted from the sampled video frames using proposed MLTrPs. We select frame selection rate of 1 frame/s as it is sufficient to design a robust VQA model [5]. High level semantic features were extracted using pre-trained Inception-v3 CNN model for each sampled frame. Again we sampled the video frames with 1 frame/s sampling rate for deep features extraction. Extracted features were further concatenated to form the final feature vector. For visual score prediction, we train an SVR model with optimized hyperparameters. Predicted scores obtained from trained SVR model are then passed through the polynomial fitting that further fine-tuned the predicted score values and improves the video quality prediction accuracy of the proposed blind VQA model.

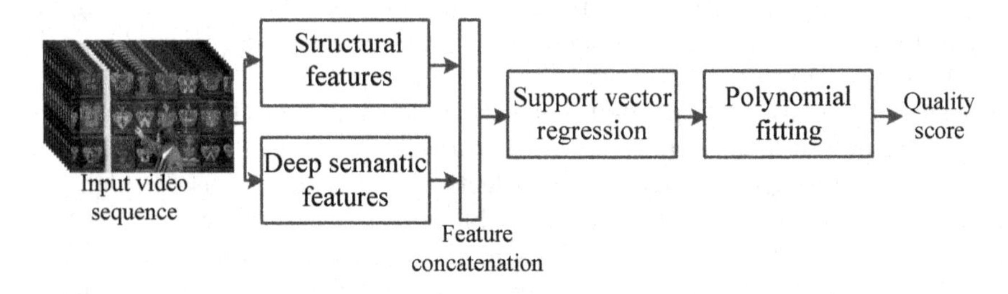

Fig. 1. Block diagram of the proposed blind VQA model.

3 Experimental Results

This section presents the description of the three popular and publicly available video quality datasets followed by the experimental procedures adapted for robust video quality estimation. Finally, we present the performance of the proposed method and compared it with other state-of-the-art techniques.

3.1 VQA Datasets

We analyze the performance of the proposed model using three complex and publicly available video quality datasets. The LIVE-VQC [15] dataset consists of 585 videos distorted with various in-capture distortions such as in/out focus, under/overexposure, sharpness, etc. The resolution and frame rate of the videos varies from 320 × 240 to 1920 × 1080 and 19 to 30 frames per second, respectively. The subjective scores of the videos range from 0–100 and are collected through crowd-sourcing.

The KoNViD-1k [4] dataset consists of 1200 videos distorted with in-capture distortions. The resolution of each video is 960 × 540, and frame rate varies from 23 to 29 frames per second. videos of the KoNViD-1k dataset are sampled from the large-scale YFCC-100M [17] video dataset. Subjective scores are collected through crowd-sourcing and ranges from 1 to 5.

The LIVE Qualcomm [3] dataset consists of 208 video sequences, each with a resolution of 1920 × 1080 and a frame rate of 30 frames per second that have been distorted with various authentic distortions. Subjective scores range from 0–100 and are obtained using a lab-based technique.

3.2 Experiments

The proposed method extracts two types of features, structural features using modified local tetra patterns and semantic features from the sampled video frames by employing a pre-trained Inception-v3 CNN model. These extracted features are further concatenated and applied to support vector regression followed by polynomial curve fitting for video quality score prediction. The performance of the proposed model is compared with various state-of-the-art VQA models. As a performance metric, we have used Pearson's Linear Correlation Coefficient (PLCC) and Spearman Rank-ordered Correlation Coefficient (SRCC). PLCC gives the quality score prediction while SRCC provides monotonicity. Higher values of both the metrics indicate better performance and range between 0 to 1.

3.3 Performance Analysis

We have performed experiments using various state-of-the-art VQA models NIQE [11], BRISQUE [9], V-BLIINDS [13], VIIDEO [10], TLVQM [5], Li et al. [6], Dendi et al. [1], VIDEVAL [18], ChipQA-0 [2], and the proposed method. We have used publicly available authentic source codes provided by the authors of different VQA models for performance comparison. For learning-based models, including the proposed method, the SVR hyperparameters are optimized using grid search algorithm. With optimized hyperparameters, we train the SVR model on the training set and test on the testing set. After quality scores prediction, we further employ polynomial curve fitting of order 16 on predicted quality scores. The polynomial fitting further fine-tuned the predicted score values and reduces the spread in quality scores. We repeat the whole process for 100 iterations and record the PLCC and SRCC values obtained in every iteration. As a final performance metric, we have used the median of the PLCC and SRCC values.

Table 1 depicts the performance comparison of the proposed method and other state-ot-the-art techniques.

Table 1. Median SRCC and PLCC results of the proposed work and other existing NR-VQA techniques on LIVE-VQC, KoNViD-1k, and LIVE Qualcomm datasets. Results in boldface and underlined indicate the top two performances.

Method	LIVE-VQC [15]		KoNViD-1k [4]		LIVE Qualcomm [3]	
	SRCC	PLCC	SRCC	PLCC	SRCC	PLCC
NIQE [11]	0.564	0.583	0.340	0.340	0.545	0.680
BRISQUE [9]	0.611	0.630	0.676	0.668	0.558	0.578
V-BLIINDS [13]	0.725	0.710	0.703	0.682	0.617	0.665
VIIDEO [10]	−0.029	0.137	0.031	−0.015	−0.141	0.098
TLVQM [5]	**0.797**	**0.798**	0.775	<u>0.768</u>	<u>0.780</u>	<u>0.810</u>
Li et al. [6]	–	–	0.755	0.744	–	–
Dendi et al. [1]	0.626	0.625	0.641	0.652	0.589	0.628
VIDEVAL [18]	<u>0.750</u>	0.748	<u>0.783</u>	0.767	0.704	0.735
ChipQA-0 [2]	0.669	0.696	0.697	0.694	–	–
Proposed	0.705	<u>0.774</u>	**0.786**	**0.810**	**0.872**	**0.887**

We can observe that, our proposed model yields best performance for the KoNViD-1k [4] and LIVE Qualcomm [3] datasets, while for the LIVE-VQC [15] dataset the proposed method gives competitive performance. Also, the generalization capability of the proposed model is demonstrated using cross-dataset experiments. The regression model is trained on one dataset and tested it on a different dataset. The cross-dataset performance of the proposed model is depicted in Table 2. We can observe from Table 2 that the proposed blind VQA model shows good generalization capability with PLCC and SRCC values of more than 0.5. We also demonstrate the individual contribution of both structural features and deep semantic features. Figure 2 depicts that both types of features have a significant contribution to the overall video quality prediction of the proposed VQA model.

Table 2. Cross-dataset experiment results of the proposed model.

Train dataset	Test dataset	SRCC	PLCC
LIVE-VQC	KoNViD-1k	0.536	0.561
KoNViD-1k	LIVE-VQC	0.540	0.576

Computational time requirement is a very important performance measure for the blind VQA models. We have compared the computational time of the proposed method and other benchmark VQA models. We have used a Dell Desktop computer with Intel (R) Core (TM) i7-4770 CPU, 3.4 GHz, and 24 GB RAM.

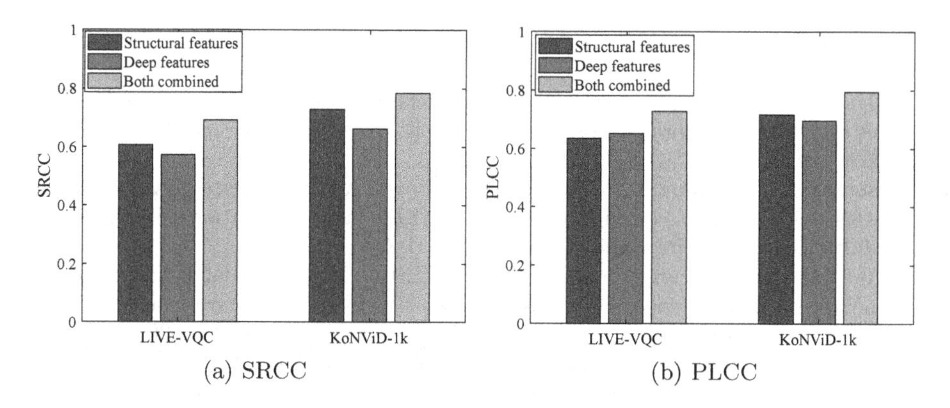

Fig. 2. Ablation study results on LIVE-VQC and KoNViD-1k datasets.

We measure the time required for the proposed method and other benchmark techniques on the same system. We have selected the videos from the LIVE-VQC dataset with a spatial resolution of 1920 × 1080. The time complexity performance of all the compared models is reported in Table 3. It can be observed that the proposed model is very fast and outperforms all the other state-of-the-art methods with a significant margin except the BRISQUE model that yields moderate SRCC and PLCC performance.

Table 3. Average time requirement in seconds on per frame basis for videos with spatial resolution of 1920 × 1080 on LIVE-VQC [15] dataset.

Method	NIQE [11]	BRISQUE [9]	V-BLLINDS [13]	VIIDEO [10]	TLVQM [5]	Dendi et al. [1]	VIDEVAL [18]	Proposed
Time per frame	0.951	0.396	8.429	2.882	1.097	6.618	2.939	**0.498**

4 Conclusion

In this paper, we have proposed a robust and computationally efficient blind VQA model using novel structural features and deep semantic features. We proposed MLTrPs that extract dominant structural information from sampled video frames. A pre-trained Inception-v3 model was used to extract deep semantic features. Both features have shown significant contribution towards the final quality prediction performance of the proposed model. We have conducted experiments on three popular authentic distortion video datasets LIVE-VQC [15], KoNViD-1k [4], and LIVE Qualcomm [3], and compared the performance of the proposed model with various state-of-the-art VQA techniques. The proposed method shows excellent performance with significantly reduced computational overhead. Also, the ablation study results show the robustness of the proposed features.

References

1. Dendi, S.V.R., Channappayya, S.S.: No-reference video quality assessment using natural spatiotemporal scene statistics. IEEE Trans. Image Processing **29**, 5612–5624 (2020)
2. Ebenezer, J.P., Shang, Z., Wu, Y., Wei, H., Bovik, A.C.: No-reference video quality assessment using space-time chips. In: 2020 IEEE 22nd International Workshop on Multimedia Signal Processing (MMSP), pp. 1–6. IEEE (2020)
3. Ghadiyaram, D., Pan, J., Bovik, A.C., Moorthy, A.K., Panda, P., Yang, K.C.: In-capture mobile video distortions: a study of subjective behavior and objective algorithms. IEEE Trans. Circuits Syst. Video Technol. **28**(9), 2061–2077 (2017)
4. Hosu, V., et al.: The Konstanz natural video database (konvid-1k). In: 2017 Ninth International Conference on Quality of Multimedia Experience (QoMEX), pp. 1–6. IEEE (2017)
5. Korhonen, J.: Two-level approach for no-reference consumer video quality assessment. IEEE Trans. Image Process. **28**(12), 5923–5938 (2019)
6. Li, D., Jiang, T., Jiang, M.: Quality assessment of in-the-wild videos. In: Proceedings of the 27th ACM International Conference on Multimedia, pp. 2351–2359 (2019)
7. Li, X., Guo, Q., Lu, X.: Spatiotemporal statistics for video quality assessment. IEEE Trans. Image Processing **25**(7), 3329–3342 (2016)
8. Liu, W., Duanmu, Z., Wang, Z.: End-to-end blind quality assessment of compressed videos using deep neural networks. In: ACM Multimedia, pp. 546–554 (2018)
9. Mittal, A., Moorthy, A.K., Bovik, A.C.: No-reference image quality assessment in the spatial domain. IEEE Trans. Image Processing **21**(12), 4695–4708 (2012)
10. Mittal, A., Saad, M.A., Bovik, A.C.: A completely blind video integrity oracle. IEEE Trans. Image Processing **25**(1), 289–300 (2015)
11. Mittal, A., Soundararajan, R., Bovik, A.C.: Making a "completely blind" image quality analyzer. IEEE Signal Process. Lett. **20**(3), 209–212 (2012)
12. Murala, S., Maheshwari, R., Balasubramanian, R.: Local tetra patterns: a new feature descriptor for content-based image retrieval. IEEE Trans. Image Processing **21**(5), 2874–2886 (2012)
13. Saad, M.A., Bovik, A.C., Charrier, C.: Blind prediction of natural video quality. IEEE Trans. Image Processing **23**(3), 1352–1365 (2014)
14. Singh, R., Aggarwal, N.: A distortion-agnostic video quality metric based on multi-scale spatio-temporal structural information. Signal Process. Image Commun. **74**, 299–308 (2019)
15. Sinno, Z., Bovik, A.C.: Large-scale study of perceptual video quality. IEEE Trans. Image Processing **28**(2), 612–627 (2018)
16. Szegedy, C., Vanhoucke, V., Ioffe, S., Shlens, J., Wojna, Z.: Rethinking the inception architecture for computer vision. In: Proceedings of the IEEE Conference on Computer Vision and Pattern Rrecognition, pp. 2818–2826 (2016)
17. Thomee, B., et al.: Yfcc100m: the new data in multimedia research. Commun. ACM **59**(2), 64–73 (2016)
18. Tu, Z., Wang, Y., Birkbeck, N., Adsumilli, B., Bovik, A.C.: UGC-VGA: benchmarking blind video quality assessment for user generated content. IEEE Trans. Image Processing **30**, 4449–4464 (2021)

19. Vishwakarma, A.K., Bhurchandi, K.M.: 3d-dwt cross-band statistics and features for no-reference video quality assessment (NR-VQA). Optik 167774 (2021)
20. You, J., Korhonen, J.: Deep neural networks for no-reference video quality assessment. In: 2019 IEEE International Conference on Image Processing (ICIP), pp. 2349–2353. IEEE (2019)

On-Device Spatial Attention Based Sequence Learning Approach for Scene Text Script Identification

Rutika Moharir(✉), Arun D. Prabhu, Sukumar Moharana, Gopi Ramena, and Rachit S. Munjal

On-Device AI, Samsung R&D Institute, Bangalore, India
{r.moharir,arun.prabhu,msukumar,gopi.ramena,rachit.m}@samsung.com

Abstract. Automatic identification of script is an essential component of a multilingual OCR engine. In this paper, we present an efficient, lightweight, real-time and on-device spatial attention based CNN-LSTM network for scene text script identification, feasible for deployment on resource constrained mobile devices. Our network consists of a CNN, equipped with a spatial attention module which helps reduce the spatial distortions present in natural images. This allows the feature extractor to generate rich image representations while ignoring the deformities and thereby, enhancing the performance of this fine grained classification task. The network also employs residue convolutional blocks to build a deep network to focus on the discriminative features of a script. The CNN learns the text feature representation by identifying each character as belonging to a particular script and the long term spatial dependencies within the text are captured using the sequence learning capabilities of the LSTM layers. Combining the spatial attention mechanism with the residue convolutional blocks, we are able to enhance the performance of the baseline CNN to build an end-to-end trainable network for script identification. The experimental results on several standard benchmarks demonstrate the effectiveness of our method. The network achieves competitive accuracy with state-of-the-art methods and is superior in terms of network size, with a total of just 1.1 million parameters and inference time of 2.7 ms.

1 Introduction

Most multilingual OCR systems require knowledge of the underlying script for each text instance present in any natural image. Hence, script identification acts as an essential step in increasing the overall efficiency of any text recognition pipeline. Many languages share the same script family, which makes script identification a fine-grained classification task.

Earlier works in script identification have been proposed for document [2,16], handwritten text [11,13] and video-overlaid text [6,27] where the texts are fairly

B. Raman et al. (Eds.): CVIP 2021, CCIS 1568, pp. 230–242, 2022.
https://doi.org/10.1007/978-3-031-11349-9_20

simple with non-complex backgrounds and have been able to achieve great performance. However, scene text images pose additional challenges such as varied appearances and distortions, various text styles and degradation like background noise and light conditions, etc. Our work on scene text script identification focuses on two challenges - first, handling spatial distortions in scene text and exploiting sequential dependencies within the text which can help focus on discriminative features for efficient classification of scripts with relatively subtle differences, for example Chinese and Japanese which share a common subset of characters. Second, be suitable, have low latency and memory footprint, for on-device deployment.

While recent works have been trying to handle script identification in complex backgrounds [1,3–5,26], they have their own challenges. [4] tries to enhance discriminative features using intermediate supervision on patch-level predictions. While, [1] proposed an attention-based Convolutional-LSTM network, analyzing features globally and locally. [3] uses an end-to-end trainable single fully convolutional network (FCN) while [7] adopted a patch-based method containing an ensemble of identical nets to learn discriminative strokepart representations. In [26] they used BLCT to enhance the discriminative features and imposed inverse document frequency (idf) to get codewords occurrences. [5] use Encoder and Summarizer to get local features and fuse them to a single summary by attention mechanism to reflect the importance of different patches. However, few works have been reported to address an important factor that often leads to script identification failures - the degradation of scene text such as blur, tilt, low contrast to the background, etc. which are not rare in natural images. Instead of using computationally expensive attention-based patch weights [1] or optimizing patch level predictions [4], we tackle this problem using a spatial attention mechanism that is capable of directly learning the region of interest for each character in the text for a superior foreground-background separation. Figure 3 shows the effect of using our spatial attention mechanism. The distorted text images are transformed and made suitable for script identification.

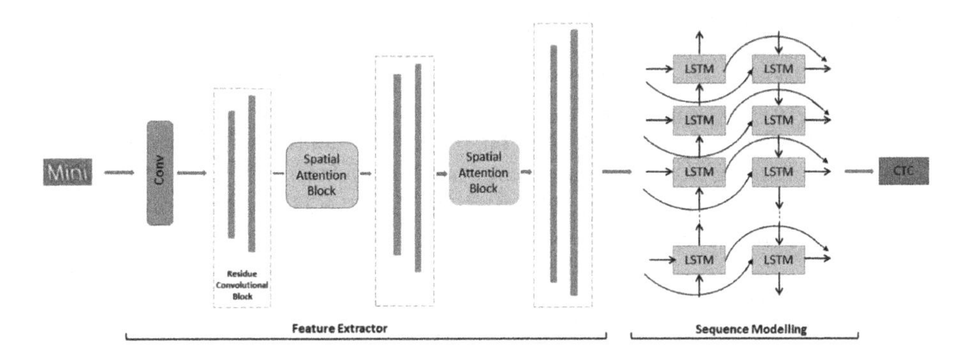

Fig. 1. The architecture of the proposed script identification model.

As for the problem of failing to capture discriminative features, [21] presented a method based on hand-crafted features, a LBP variant, and a deep multi-Layer perceptron. Pre-defined image classification algorithms, such as [17] who proposed a combination of CNN and LSTM model, the robustly tested CNN [24] or the Single-Layer Networks (SLN) [23] normally consider holistic representation of images. Hence they perform poorly in distinguishing similarly structured script categories as they fail to take into account the discriminative features of the scripts. For example, a few languages belonging to the Chinese and Japanese script share a common subset of characters. A word is more likely to be Japanese if both Chinese and Japanese characters occur in an image. But when no Japanese character occurs, the word has to be Chinese. We propose a sequence to sequence mapping of individual characters' shape to their script, from which we can know what scripts the character of a given image could be. After that, the majority of the predicted probability distributions is calculated, and the dominant script is determined.

In this paper, we propose a novel scene text script identification method which uses spatial attention based sequence learning approach for script identification in scene text images. We use spatial attention blocks which extract rich text features while ignoring the disruptions present in the scene text. Inspired by the successful application of residual connections in image classification tasks [12,25], we employ hierarchical residual connections to build a deeper network that helps in learning representative image features. In training phase, instead of learning a single representation of the text and failing to capture the discriminating chars of each script, a segmentation free, sequence to sequence mapping is done using Connectionist Temporal Classification (CTC) to supervise the predictions. The key contributions of our scene text script identification method are:

1. We propose a novel end-to-end trainable scene text script identification network which integrates spatial attention mechanism and residue convolutional blocks with the sequence learning capabilities of the LSTM to generate sequence to sequence mapping for discriminative learning of the script. The module has low latency and low memory footprint and thus is feasible for deployment in on-device text recognition systems.
2. A deep feature extractor with residue convolutional blocks generates a rich image representation for the text.
3. The proposed sequence learning based script identification method does supervision using Connectionist Temporal Classification (CTC). To the extent of our knowledge, this is the very first time that CTC demonstrates strong capabilities for scene text script identification.

2 Methodology

In the following sections, we describe in detail the three key components of our network, namely, the spatial attention based feature extractor (Sect. 2.1), sequence modelling (Sect. 2.2) and prediction (Sect. 2.3). Figure 1 illustrates the overall architecture of our network.

2.1 Spatial Attention Based Feature Extractor

The feature extractor consists of a combination of two modules - Spatial Attention Block and the Residue Convolutional Block (see Fig. 2).

Spatial Attention Block. Spatial attention mechanism in a convolutional neural network (CNN) ignores the information provided by the channels and treats all the channels in a feature map equally. In the CNN, the low level feature maps having fewer channels, are responsible for extracting the spatial characteristics such as edges and contours. Hence in our network, the spatial attention block is introduced for the low level feature maps to learn the interaction of spatial points and focus on the key informative areas while ignoring the irrelevant distortions.

As can be seen from Fig. 2, to compute the spatial attention, firstly, we pass the feature map $F \in R^{H \times W \times C}$ to the channel aggregation operation which generates a global distribution of spatial features, the spatial descriptor $S \in R^{H \times W}$. Global average pooling G_{avg} in the channel dimension (C) is used as the aggregation function.

$$s_{hw} = G_{avg}(f_{hw}) = \frac{1}{C}\sum_{i=l}^{C} f_{hw}(i) \quad (1)$$

where $f_{hw} \in R^{C}$ refers to the local feature at spatial position (h, w). On the generated spatial descriptor S, weight learning is implemented using a convolutional layer to generate the spatial attention map $T \in R^{H \times W}$ which is computed as:

$$T = \sigma(\delta(f_1 p)) \quad (2)$$

where p is spatial position, f_1 represents a convolution operation with filter size 3×3, δ refers to activation function ReLU [18] and σ is a sigmoid activation function used to generate spatial weight $t_{h,w} \in (0, 1)$ at position (h, w). T is finally applied to the feature map F. The feature values at different positions in

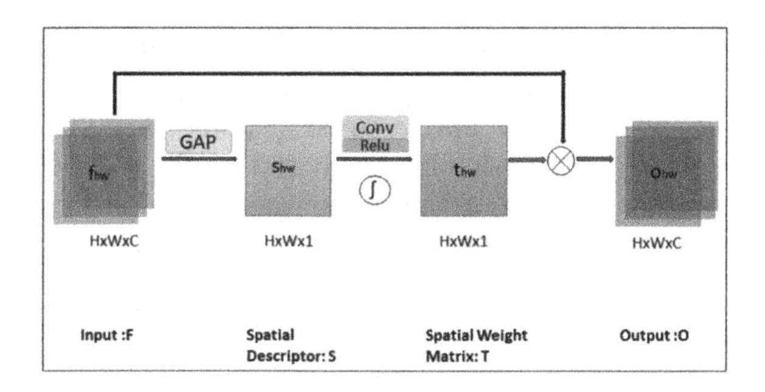

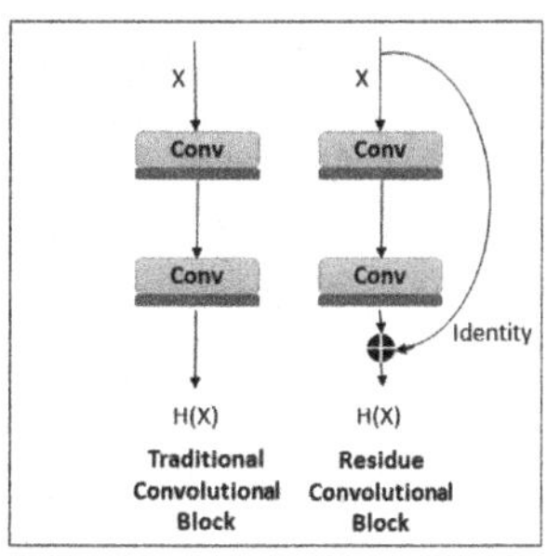

Fig. 2. Illustration of the Spatial Attention Block (left) and the Residue Convolutional Block (right) which are the components of the feature extractor.

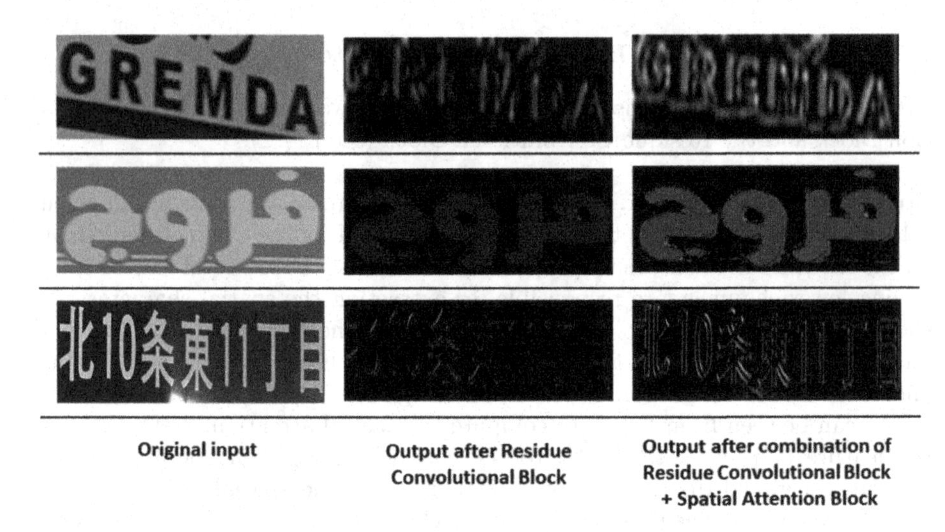

Fig. 3. Illustration of the effect of the Spatial Attention Block during feature extraction. The 2nd column represents feature maps generated in the absence of the attention blocks while the 3rd column shows enhanced foreground-background separation with the removal of distortions on the application of the attention blocks.

F are multiplied by different weights in T to generate the output O of the spatial attention block as:

$$O_{hw} = f_{hw} \cdot t_{hw} \tag{3}$$

Residue Convolutional Block. The residue block (see Fig. 2(b)) consists of two convolutional layers, each followed by ReLU activation and a skip-connection between the input of the first and the output of the second convolutional layer. If the input x is mapped to the true distribution H(x), then the difference can be stated as:

$$R(x) = H(x) - x \tag{4}$$

The layers in a traditional network learn the true output H(X) while the residue convolutional block layers learn the residue R(x), since they have an identity connection reference due to x. Other than the fact that it is easier to optimise the residual mapping than the original unreferenced mapping, an added advantage is that it can learn the identity function by simply setting the residual as zero. This enables to address the problem of vanishing gradients, giving us the ability to build deep feature extractor consisting of 7 convolutional layers.

2.2 Sequence Modelling

Previous works fail to take the spatial dependencies within text word into consideration, which may be discriminative for script identification as mentioned in Sect. 1.

Long Short-Term Memory (LSTM) [14] solves this problem by introducing three gates: input gate, forget gate, output gate. They are integrated together to capture spatial dependencies for a long time by keeping the gradient in the same order of magnitude during back-propagation. The hidden state h_t depends not only on the current input x_t, but also previous time step hidden state h_{t-1}.

In this work, we use one Bidirectional-LSTM (Bi-LSTM) layer with 256 hidden states which calculates h_t using the memory block from both forward and backward directions. For faster inference, we use LSTM with recurrent projection layer [22] to decrease the computational complexity. The projection layer projects the hidden state to a lower dimension of 96 at every time step.

2.3 Prediction

We concatenate the features generated by the Bi-LSTM layer and the per-frame predictions are fed into a fully connected layer, followed by a softmax layer which outputs the normalized probabilities for each script class.

We use Connectionist Temporal Classification (CTC) [9] to help the model associate the shape of the character to a script class.

The formulation of the conditional probability is described as follows. Let L represent the set of all script classes in our task and $L^{'} = L \cup \{blank\}$. The output from the sequential modelling block is transformed into per-frame probability distribution $y = (y_1, y_2 ... y_T)$ where T is the sequence length. Each y_t is the probability distribution over the set L'. A sequence-to-sequence mapping function β is defined on the sequence $\pi \in L'^T$ which deletes repeated characters and the blank symbol. The conditional probability is calculated by summing the probabilities of all paths mapped onto the label sequence l by β:

$$p(l|y) = \sum_{\beta(\pi)=l} p(\pi|y) \tag{5}$$

As directly computing the above equation is computationally expensive, forward-backward algorithm [9] is used to compute it efficiently. While predicting the sequence of labels, we pick the label with the highest probability at each time step and then apply β to the entire output path.

During inference, we calculate the number of times each script is predicted for the given word crop. The script having the highest count is considered as the script for the word crop.

3 Experiments

3.1 Dataset

We evaluate our scene text script identification methods on following benchmark datasets:

Table 1. Script Identification accuracy on benchmark datasets. The best performing result for each column is set in bold font. The number of parameters are in millions (M) and time is denoted in milliseconds per image (ms).

Method	MLT17	CVSI 15	MLe2e	Parameters	Time
Bhunia [1]	**90.23**	97.75	96.70	12	85
ECN [8]	86.46	97.20	94.40	24	13
Cheng [4]	89.42	**98.60**	–	26.7	**2.5**
Mei [17]	–	94.20	–	5.2	92
Zdenek [26]	–	97.11	–	–	60
Ours	89.59	97.44	**97.33**	**1.1**	2.7

ICDAR 2019 [19] (IC19) consists of 8 script classes - Latin, Korean, Chinese, Japanese, Arabic, Hindi, Bangla and Symbols. The test set consists of approximately 102462 cropped word images and is highly biased towards Latin.

RRC-MLT 2017 [20] (MLT17) consists of 97,619 test cropped images. This dataset holds 7 scripts: Arabic, Latin, Chinese, Korean, Japanese, Bangla, Symbols. There exist some multioriented, curved and perspectively distorted texts which make it challenging.

CVSI 2015 [23] (CVSI 15) dataset contains scene text images of ten different scripts: Latin, Hindi, Bengali, Oriya, Gujrati, Punjabi, Kannada, Tamil, Telegu, and Arabic. Each script has at least 1,000 text images collected from different sources (i.e. news, sports etc.)

MLe2e [7] is a dataset for the evaluation of scene text end-to-end reading systems. The dataset contains 643 word images for testing covering four different scripts: Latin, Chinese, Kannada and Hangul many of which are severely corrupted by noise and blur and may have very low resolutions.

3.2 Implementation Details

For feature extraction, the input text crops are scaled to fixed height of 24 while maintaining the aspect ratio. The residue convolutional blocks are stacked with the spatial attention block being introduced after the 1st and 2nd block as shown in Fig. 1. The filter size, stride and padding size of all the convolutional layers is 3, 1 and 1 respectively and each convolutional layers is followed by a max-pooling layer. The output filter size for the convolutional layers are 32, 64, 96, 128, 164, 196, 256 respectively and each layer is followed by the ReLU activation function. Batch normalization [15] is adopted after the 3rd and the 5th convolutional layer, speeding up the training significantly. The input layer is followed by a 3×3-stride max-pool layer. This is followed by three residual blocks, with each unit consisting of two 3×3 convolution operations. The first residual block further downsample the feature maps with a 2×2-stride max-pool layer, while the last two residual blocks each employ a 2×1-stride max-pool layer to reduce the sizes

Table 2. (a) Comparison with different variations of the proposed network on MLT17 validation set. Parameters are mentioned in millions. (b) Script Identification accuracy on MLT17 with respect to the position of spatial attention blocks, where RB stands for Residual Block and SA stands for Spatial Attention Block

Module	Accuracy	Params
CNN+LSTM	84.79	0.65
ResCNN+LSTM	84.66	1.0
Atten+CNN+LSTM	86.75	1.03
Atten+ResCNN+LSTM	88.60	1.1
Atten+CNN+LSTM+CTC	87.85	1.03
Atten+ResCNN+LSTM+CTC	90.17	1.1

Position of SA blocks	Accuracy
After conv and 1st RB	83.81
After conv and 2nd RB	84.59
After 1st and 2nd RB	85.63
After 2nd and 3rd RB	84.81

of feature maps only in the height dimension. The two LSTMs of a BiLSTM layer in the encoder each have 256 hidden units for capturing dependencies in each direction of the feature sequence.

We initialize the network by first training on a synthetic dataset generated using the method followed by [10]. The training set consists of 1.5 million word crops with equal distribution for all the scripts, which are randomly sampled to form minibatches of size 96. Further, the network is fine-tuned on the benchmark datasets and evaluated. We adopt some data augmentation techniques such as varying contrast, adding noise and perspective transformations to make the dataset robust. The Adam optimizer is used with the initial learning rate being set to 0.001 and is automatically adjusted by the optimizer during training. The training of the network on Nvidia Geforce GTX 1080ti GPU, 16 GB RAM takes about 36 h for 10 epochs. The network has approximately 1.1 million parameters and takes 2.7 ms per image for identification.

3.3 Effectiveness of Spatial Attention and Residue Convolutional Blocks for Script Identification

To validate the effectiveness of the proposed spatial attention based residue convolutional network for scene text script identification, in Table 2(a), we compare the performance with that of a baseline CNN network without the application of the spatial attention mechanism or residue convolutional blocks but with the same sequence modelling backbone. Specifically, in the 'CNN+LSTM' model we skip the spatial attention block and residual connections, feeding the input text image into the 'CNN+LSTM' backbone and using cross entropy classification as the loss function. The model 'ResCNN+LSTM' omits the spatial attention block and makes use of the residue convolutional blocks, enhancing the quality of image feature representation. On the other hand, the model 'Atten+CNN+LSTM' skips using the residual connections and employs the spatial attention block, using cross entropy as the optimization function while 'Atten+CNN+LSTM+CTC' makes use of the CTC loss function. The model 'Atten+ResCNN+LSTM' combines the spatial attention blocks with the residue

Table 3. (a) Comparison of our results on MLT 19 with other methods. (b) Script wise results of our method on MLT19 dataset.

Method	Accuracy (%)
Tencent - DPPR team	94.03
SOT: CNN based classifier	91.66
GSPA_HUST	91.02
SCUT-DLVC-Lab	90.97
TH-ML	88.85
Multiscale_HUST	88.64
Conv_Attention	88.41
ELE-MLT based method	82.86
Ours	89.82

Script	Accuracy
Latin	98.30
Korean	94.28
Chinese	90.28
Japanese	84.15
Arabic	92.61
Hindi	86.28
Bangla	87.28
Symbols	85.44
Average	89.82

convolutional blocks, and the model 'Atten+Res+CNN+LSTM+CTC' instead uses CTC loss function to optimise the prediction.

Contribution of Spatial Attention Blocks and Residue Convolutional Blocks. Compared to the baseline model, the introduction of the proposed spatial attention module and the residue convolutional module effectively increases the accuracy of the script identification model. Combining both modules helps further enhance the performance on all the datasets. 'Atten+ResCNN+LSTM' attains more accurate residual enhancement representations as compared to 'Atten+CNN+LSTM', thus capturing low level features, which help improve the identification accuracy. The overall enhanced identification accuracy of the proposed spatial attention based sequence learning approach for script identification relative to the baseline, demonstrates the effectiveness of spatial attention for scene text script identification.

Arrangement of Spatial Attention Blocks. To ensure low latency and computational complexity we limit to using two spatial attention blocks. Given an input image, the two attention blocks compute attention, focusing on exploiting image features while ignoring distortions. Considering this, the two blocks can be placed at either the low level or at the higher level in the network. In this experiment we compare four different ways of arranging the two spatial attention blocks: After conv and the 1st residual block, after conv and 2nd residual block, after the 1st and 2nd residual block and finally, after the 2nd and 3rd residual block. Since the spatial attention works locally, the order may affect the overall performance. Table 2(b) summarises the experimental results on different combinations. From the results we can find that spatial attention blocks after the 1st residual block and the 2nd residual block is slightly better than the other arrangements.

Effect of CTC Loss. To verify the superiority of the proposed sequence to sequence mapping method for script identification, which performs per character script classification, we replace the proposed CTC module in our script identification model 'Atten+Res+CNN+LSTM+CTC', with the Cross entropy classification technique, the model 'Atten+Res+CNN+LSTM', which generates a single representation for the input image and the spatial attention based residual network is trained. The experimental results show that CTC module yields 1.57% higher script identification accuracy on the benchmark dataset compared to the cross entropy module whose output features may be ignoring the discriminative features between similarly structured scripts.

3.4 Comparison with State-of-the-Art Script Identification Methods

Our method achieves the highest script identification accuracy on one of the three benchmark datasets and a highly competitive accuracy with the last two datasets. Note that our network is significantly lighter and faster as compared to other state-of-the-art methods which qualifies it for real-time on-device pre-ocr applications.

For the scene text images in Mle2e, a great improvement has been made. Since in this dataset, many of the text regions are severely distorted and few of them have very low resolutions, the performance of our network shows the ability and the need for a spatial attention mechanism in handling disruptions occurring in natural images. As can been seen from the Table 1, our method also performs well on the MLT17 dataset.Table 4 compares our method with others on CVSI 2015 dataset. The result demonstrates that our method performs well on both document text images and complex background images.

Table 3 mentions script wise performance accuracy of our method ICDAR19 RRC (IC19). Table 3 also compares the performance of our method with other models described in [19]. The baseline methods have based their methods on famous deep nets for text recognition such as ResNet, VGG16, Seq2Seq with CTC etc. and use the recognised characters to identify the scripts while adopting improvements such as multiscale techniques, voting strategy for combining results from multiple nets, training statistics of the scripts etc. Even though the proposed network is slightly less accurate the benchmarks, it is superior in terms of latency and model size and thus, in contrast, is feasible for on-device deployment.

Given the relatively standard feature extraction backbone employed in our method, the results of the experiment demonstrates the significant effect of the proposed spatial attention mechanism coupled with the sequence learning capabilities of LSTM on improving the accuracy of the script identification model. Particularly, compared to the existing script identification methods, Sect. 1, which either use complex self-attention mechanisms or heavy feature extractors, our method with a simple convolution based attention block and a small CNN backbone achieves overall enhanced accuracy, with the additional benefit of being computationally lighter and faster.

Table 4. Script wise performance comparison with various methods on CVSI 2015 dataset.

Script	C-DAC	HUST	CVC-2	CIJK	Cheng	**Ours**
Eng	68.33	93.55	88.86	65.69	94.20	98.54
Hin	71.47	96.31	96.01	61.66	96.50	97.24
Ben	91.61	95.81	92.58	68.71	95.60	97.74
Ori	88.04	98.47	98.16	79.14	98.30	98.16
Guj	88.99	97.55	98.17	73.39	98.70	96.94
Pun	90.51	97.15	96.52	92.09	99.10	97.47
Kan	68.47	92.68	97.13	71.66	98.60	97.45
Tam	91.90	97.82	99.69	82.55	99.20	96.88
Tel	91.33	97.83	93.80	57.89	97.70	96.60
Ara	97.69	100.00	99.67	89.44	99.60	97.36
Average	84.66	96.69	96.00	74.06	97.75	97.44

4 Conclusion

In this paper, we present a novel lightweight real-time on-device spatial attention based sequence learning approach for scene text script identification. The spatial attention mechanism helps removing spatial disruptions in natural images for improving script identification accuracy in complex scenarios. Our network employs residue convolutional blocks to generate rich feature representations of the input image thus capturing discriminative features of a script. Sequence learning using the Bidirectional LSTM layer, optimised using the Connectionist Temporal Classification enables the model to associate the characters shape to its script, thereby enhancing script identification accuracy. While achieving competitive accuracy with state-of-the-art methods, our network is superior, as it achieves 4 times reduction in model size and has low latency, making it suitable for on-device inference.

References

1. Bhunia, A.K., Konwer, A., Bhunia, A.K., Bhowmick, A., Roy, P.P., Pal, U.: Script identification in natural scene image and video frames using an attention based convolutional-LSTM network. Pattern Recogn. **85**, 172–184 (2019)
2. Busch, A., Boles, W.W., Sridharan, S.: Texture for script identification. IEEE Trans. Pattern Anal. Mach. Intell. **27**, 1720–1732 (2005)
3. Bušta, M., Patel, Y., Matas, J.: E2E-MLT - an unconstrained end-to-end method for multi-language scene text. In: Carneiro, G., You, S. (eds.) ACCV 2018. LNCS, vol. 11367, pp. 127–143. Springer, Cham (2019). https://doi.org/10.1007/978-3-030-21074-8_11
4. Cheng, C., Huang, Q., Bai, X., Feng, B., Liu, W.: Patch aggregator for scene text script identification. In: 2019 International Conference on Document Analysis and Recognition (ICDAR). IEEE (2019)

5. Fujii, Y., Driesen, K., Baccash, J., Hurst, A., Popat, A.C.: Sequence-to-label script identification for multilingual OCR. In: 2017 14th IAPR International Conference on Document Analysis and Recognition (ICDAR), vol. 1. IEEE (2017)
6. Gllavata, J., Freisleben, B.: Script recognition in images with complex backgrounds. In: 2005 Proceedings of the 5th IEEE International Symposium on Signal Processing and Information Technology, pp. 589–594. IEEE (2005)
7. Gomez, L., Karatzas, D.: A fine-grained approach to scene text script identification. In: 2016 12th IAPR Workshop on Document Analysis Systems (DAS). IEEE (2016)
8. Gomez, L., Nicolaou, A., Karatzas, D.: Boosting patch-based scene text script identification with ensembles of conjoined networks. arXiv preprint arXiv:1602.07480 (2016)
9. Graves, A., Fernández, S., Gomez, F., Schmidhuber, J.: Connectionist temporal classification: labelling unsegmented sequence data with recurrent neural networks. In: Proceedings of the 23rd International Conference on Machine learning (2006)
10. Gupta, A., Vedaldi, A., Zisserman, A.: Synthetic data for text localisation in natural images. In: Proceedings of the IEEE Conference on Computer Vision and Pattern Recognition (2016)
11. Hangarge, M., Dhandra, B.: Offline handwritten script identification in document images. Int. J. Comput. Appl. **4**, 6–10 (2010)
12. He, K., Zhang, X., Ren, S., Sun, J.: Deep residual learning for image recognition. In: Proceedings of the IEEE Conference on Computer Vision and Pattern Recognition (2016)
13. Hochberg, J., Bowers, K., Cannon, M., Kelly, P.: Script and language identification for handwritten document images. Int. J. Doc. Anal. Recogn. **2**, 45–52 (1999)
14. Hochreiter, S., Schmidhuber, J.: Long short-term memory. Neural Comput. **9**, 1735–1780 (1997)
15. Ioffe, S., Szegedy, C.: Batch normalization: accelerating deep network training by reducing internal covariate shift. In: International Conference on Machine Learning. PMLR (2015)
16. Joshi, G.D., Garg, S., Sivaswamy, J.: A generalised framework for script identification. Int. J. Doc. Anal. Recogn. (IJDAR) **10**(2), 55–68 (2007)
17. Mei, J., Dai, L., Shi, B., Bai, X.: Scene text script identification with convolutional recurrent neural networks. In: 2016 23rd International Conference on Pattern Recognition (ICPR). IEEE (2016)
18. Nair, V., Hinton, G.E.: Rectified linear units improve restricted Boltzmann machines. In: ICML (2010)
19. Nayef, N., et al.: ICDAR 2019 robust reading challenge on multi-lingual scene text detection and recognition-RRC-MLT-2019. In: 2019 International Conference on Document Analysis and Recognition (ICDAR). IEEE (2019)
20. Nayef, N., et al.: ICDAR 2017 robust reading challenge on multi-lingual scene text detection and script identification-RRC-MLT. In: 2017 14th IAPR International Conference on Document Analysis and Recognition (ICDAR). IEEE (2017)
21. Nicolaou, A., Bagdanov, A.D., Gomez, L., Karatzas, D.: Visual script and language identification. In: 2016 12th IAPR Workshop on Document Analysis Systems (DAS). IEEE (2016)
22. Sak, H., Senior, A.W., Beaufays, F.: Long short-term memory recurrent neural network architectures for large scale acoustic modeling (2014)
23. Sharma, N., Mandal, R., Sharma, R., Pal, U., Blumenstein, M.: ICDAR 2015 competition on video script identification (CVSI 2015). In: 2015 13th International Conference on Document Analysis and Recognition (ICDAR). IEEE (2015)

24. Shi, B., Bai, X., Yao, C.: Script identification in the wild via discriminative convolutional neural network. Pattern Recogn. **52**, 448–458 (2016)
25. Szegedy, C., Ioffe, S., Vanhoucke, V., Alemi, A.: Inception-v4, inception-ResNet and the impact of residual connections on learning. In: Proceedings of the AAAI Conference on Artificial Intelligence (2017)
26. Zdenek, J., Nakayama, H.: Bag of local convolutional triplets for script identification in scene text. In: 2017 14th IAPR International Conference on Document Analysis and Recognition (ICDAR), vol. 1. IEEE (2017)
27. Zhao, D., Shivakumara, P., Lu, S., Tan, C.L.: New spatial-gradient-features for video script identification. In: 2012 10th IAPR International Workshop on Document Analysis Systems. IEEE (2012)

Post-harvest Handling of Mangoes: An Integrated Solution Using Machine Learning Approach

D. S. Guru[1], Anitha Raghavendra[2,3](✉), and Mahesh K. Rao[3]

[1] Department of Studies in Computer Science, University of Mysore, Manasagangotri, Mysore 570 006, India
dsg@compsci.uni-mysore.ac.in

[2] Infosys Limited, Infosys Rd, Hebbal Industrial Estate, Hebbal, Mysuru 570027, India
anithbg@gmail.com

[3] Maharaja Research Foundation, Maharaja Institute of Technology Mysore Campus, Belawadi, S R Patna, Mandya 571 477, India

Abstract. In this paper, different steps involved in harvesting and postharvest handling of mangoes are detailed and the possibilities of exploring Artificial Intelligence based solutions for each step are highlighted. A schematic diagram of overall post-harvesting is given. Various suitable methods of classical machine learning such as feature extraction methods, feature reduction methods, and classifiers are presented. Problems of classifying mangoes into mature, non- mature and matured class further into ripened and under-ripened are addressed with a proposal of machine learning approaches. Nevertheless, a hierarchical multi-classifiers fusion approach is designed for classification of ripened mangoes into perfect ripened, over ripened, over ripened with black spots on skin and without black spots on skin. Moreover, method of selecting the best wavelengths of NIR spectroscopy towards proposal of non-destructive method of finding internal defects in mangoes is also introduced. In addition, the most difficult issue of classifying ripened mangoes into naturally ripened and artificially ripened mangoes is also attempted, and a suitable classifier is presented. For each problem being addressed using classical machine learning methods, the corresponding counterparts from deep learning architectures are also highlighted. Results of extensive experimentation conducted to demonstrate the success of our approaches are presented on reasonably sized datasets of mango images created during the course of our research. A description on datasets along with difficulties faced while creating datasets are detailed. A comparative study on different approaches including deep learning based approaches is presented. Scope for future research in the similar directions is also explored. Overall, this paper is an attempt towards an integrated solution for postharvest handling of mangoes especially for automation of sorting and grading processes.

Keywords: Postharvest handling · Mango sorting and grading · Internal defect detection in mangoes · Classification of mangoes · Machine learning · Deep learning

B. Raman et al. (Eds.): CVIP 2021, CCIS 1568, pp. 243–253, 2022.
https://doi.org/10.1007/978-3-031-11349-9_21

1 Introduction

Horticulture plays a significant role in Indian agriculture, as fruit industry contributes a major part in economic development of the nation which is due to the export of quality fruit and to satisfy the consumer's demand. In order to strengthen the export competitiveness, adoption of post-harvest technology is essential. Post-harvest management system includes washing, sorting, grading, packing, transporting and storage of fruits. Amongst all these steps, sorting and grading are the major processing tasks associated for preserving the quality of fresh-market stuff. The main objectives of applying postharvest technology are to maintain quality (appearance, texture, flavor and nutritive value), to protect food safety and reduce losses between harvest and consumption. Hence to have an efficient post-harvest technology, precision agriculture is being used from past two decades. Precision Agriculture (PA) is a management of variability in time and space using information and communication technologies (Goovaerts, 2000). PA in horticulture aims to have effective utilization of resources for achieving targeted production of fruit. The applications of PA in post-harvest process management by the fruit industries use sensors to monitor conditions for storage and grading of fruit to achieve the quality (Erick Saldana et al., 2001). In recent years, machine vision based technology has become more popular in many fields of agriculture and food industry. Machine vision based grading of fruits and vegetables is considered efficient as it avoids inconsistency, inaccuracy, time consuming and labour intensive.

Mango is one of the most famous and delicious fruits which is produced and consumed throughout the world. India is one of the most prolific mango growers. As India constitutes about 57% of the worldwide production, there is a demand for it both in local and international markets (Khoje and Bodhe, 2012). But India is still being poor in exporting mango fruit, as there is lack of knowledge for farmers and also lack of adopting technology during the quarantine measure. Mango's commercial value makes gradation and inspection of it necessary according to APEDA (Agricultural and Processed Food Products Export Development Authority). An automatic grading of mangoes is inevitable, not only for export, even for food processing industries and also in consumers to buy it. By incorporating automation in grading and sorting of mangoes, human intervention can be brought down while dealing with large quantities especially during the export. Such an environment for post-harvesting management of mangoes is recommended by proposing an integrated solution using machine learning approaches as in Fig. 1.

In Fig. 1, three steps are adapted for the purpose of serving the better quality of mangoes for consumption. Initially, mango fruits can be sorted/graded based on their ripeness levels. Once the mangoes are sorted into different ripening stages, later these mangoes can be segregated as naturally and chemically ripened mangoes. Further, these mangoes can be classified as healthy and internally defected ones using NIR spectroscopy. Here complete hardware in the environment is driven by software and the software inside the camera becomes an intelligent vision system. Classification task at each stage is very challenging as interclass variation is less and intra-class variation is more. In order to make the computer vision system more efficacious, deep learning techniques can be inculcated. As Deep Convolutional Neural Networks (DCNN) have become an excellent technique for classification tasks in recent years (Edmar Rezende et al., 2017). Deep

Learning (DL) extends the classical machine learning by appending more "depth" (convolution) to the model and also transforming the data using different functions that permit representation of data in a pyramid form through several abstraction stages (Salunkhe and Patial, 2015), (LeCun and Bengio, 1995). An important advantage of DL is learning the features, i.e. the extraction of features from the original data, the features which are being created by the combination of lower level features (LeCun et al., 2015). The transition from conventional approaches (manually extracted features combined with classifiers) to DCNNs is because of the immense success of DCNNs on classification challenges such as the ILSVRC-ImageNet Large Scale Visual Recognition Challenge (Edmar Rezende et al., 2017). Over the years, there has been a phenomenon that the deeper the model, the more successful and efficient the model can be on the ImageNet challenge. Kamilaris and Prenafeta-Boldú, 2018 have made an extensive survey on deep learning in agriculture and they have mentioned the applications of deep learning in the field agriculture.

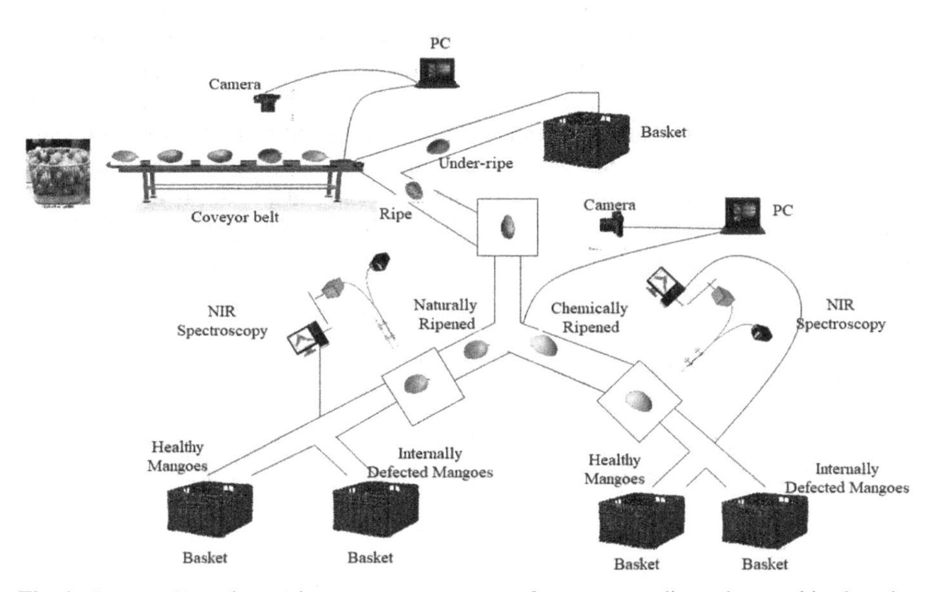

Fig. 1. Proposed post-harvesting management system for mango grading using machine learning approaches

Mango fruits can be evaluated based on their shape, size, ripeness levels, chemical ripening agent used and also on defects (internal and external). In most of the earlier research works, assessing the mango fruit has been done based on their shape, size, color and external defects (Roomi et al., 2012; Pauly and Sankar, 2015; Khoje and Bodhe, 2012; Khoje et al., 2013; Nanna et al., 2014; Musale and Patil, 2014). But some have addressed on classification of mango fruits based on their level of ripeness such as unripe, ripe and overripe (Vyas et al. 2014; Rivera et al; Salunkhe and Patial, 2015; Chhabra et al., 2012; Mansor et al., 2014; Nandi C.S et al., 2014). But very few have addressed on categorizing the different ripening stages of mangoes especially under-ripe and overripe mangoes. Hence there is a requirement to develop a model which needs to categorize mango classes as under ripe, ripe, overripe with internal defects and overripe without internal defects. Developing such a kind of model may contribute in exporting the mango fruit to different countries based on the time taken to obtain the best price for the fruit, fruit product manufacturing industries and also consumer's purchase decision. However, black spotted over-ripened mangoes without any internal defects can still be put to use in food product manufacturing industries. Alongside, method of ripening is terrifying the consumers because the mangoes are being ripened with artificial ripening agent. Consuming such mangoes leads to many health issues and majorly leads to cancer (www.midday.com). Hence the decline in export market for Indian mangoes because of rampant use of artificial ripeness. The presence of artificial ripening agent can be usually identified by the experts based on the skin of the fruit which may be in the form of color or texture. Chemically matured mangoes also have distorted wrinkles on the skin. Mangoes which are ripened with calcium carbide or ethrel have uniformity in color and they are clear without any black spots. Those ripened with calcium carbide are soft; however, they have good peel/skin color but are poor in flavor. Since identification of chemically ripened mango fruits has not been attempted in earlier works, this problem needs to be attempted in further research study. Most of the countries have banned Indian mangoes, because of the internal quality issues such as spongy tissue and seed weevil. Automation in evaluating the mango fruit is a necessary factor during quarantine measure. The same problem has been addressed in the earlier study using only X-ray method. As a warning label needs to be displayed during the quarantine measure and also due to the ionization radiation, X-ray cannot be considered for sorting/grading especially during export level. Hence detecting the internal defects in mango fruits need to be addressed using efficient non-destructive techniques.

2 Proposed Model for Grading

The architecture of the mango grading is shown in Fig. 2. Initially, mango dataset can be considered as images or spectroscopy data based on the problem analysis.

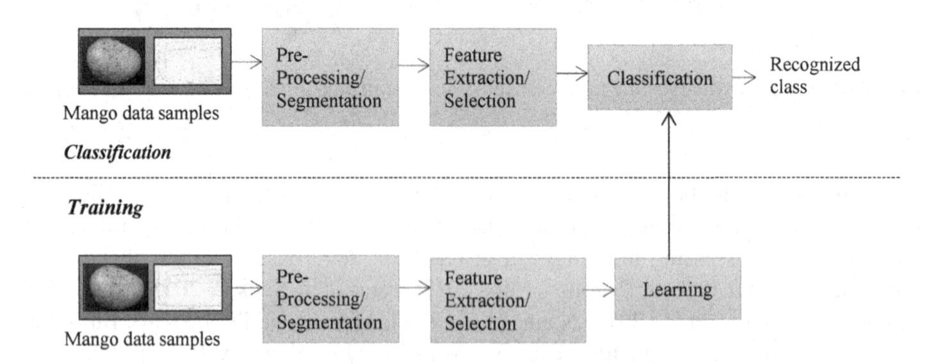

Fig. 2. A typical architecture of a mango grading model using machine learning approaches

2.1 Pre-processing/Segmentation

Initially, any data sample need to be preprocessed before feeding into further recognition. Preprocessing includes image cropping, image resizing, image normalization and data normalization. In general, segmentation was the most difficult part in image processing and it played a very important role in further classification process. Mango region in images should be segmented properly so that, it would make further classification easier. We could find several segmentation techniques such as threshold based methods, color based method (Moureen Ahmed et al., 2015), K-means clustering and C-means algorithm segmentation technique (Ashok and Vinod, 2014).

2.2 Feature Extraction/Selection

Feature extraction is the major step in pattern recognition system. For any further effective classification of samples, suitable features are to be extracted from the samples. In mango sorting and grading system, different features like size, shape, color and texture have been extracted (Roomi et al., 2012; Pauly and Sankar, 2015; Khoje and Bodhe, 2012; Khoje et al., 2013; Nanna et al., 2014; Musale and Patil, 2014). Further feature selection can be used to select the best subset from the input space using filter and wrapper method. Its ultimate goal is to select the optimal features subset that can achieve the highest accuracy results.

2.3 Classification

After optimal feature subset is selected, a classifier can be designed using various approaches. Roughly speaking, there are three different approaches. The first approach is the simplest and the most intuitive approach which is based on the concept of similarity. Template matching is an example. The second one is a probabilistic approach. It includes methods based on Bayes decision rule, the maximum likelihood or density estimator. Three well-known methods are k-Nearnest Neighbour (k-NN), Parzen window classifier and branch-and bound methods (BnB).The third approach is to construct decision boundaries directly by optimizing certain error criterion. Examples are fisher's linear discriminant, multilayer perceptrons, decision tree and support vector machine. There are many kinds of classification technique aiming at reducing the computational burden for pattern recognition (Zheng et al.; Duda et al.). For sorting and grading of mangoes following models have been used to classify the samples at each level.

2.3.1 Ripeness Evaluation

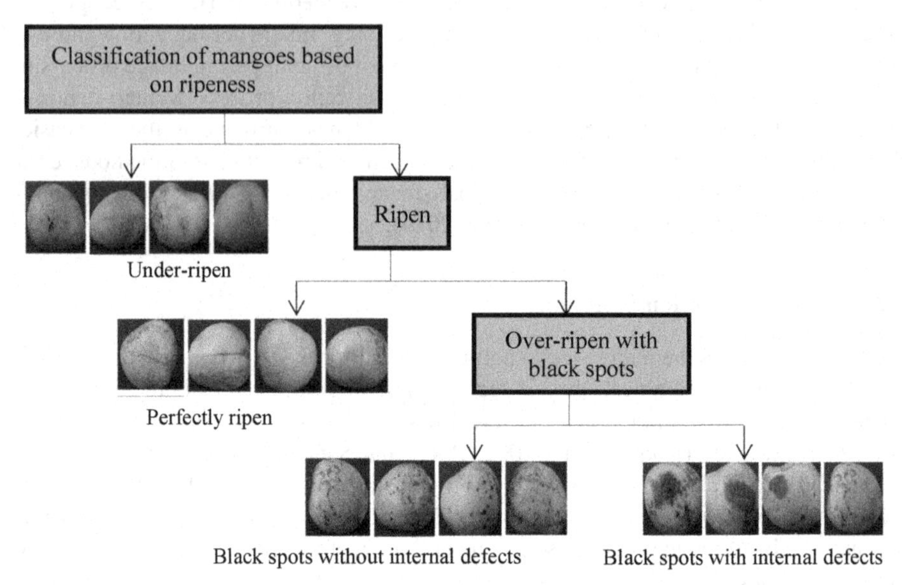

Fig. 3. Hierarchical method for classification of mangoes based on ripeness (Raghavendra et al., 2020b)

In this study, a hierarchical approach was adopted to classify the mangoes into the four classes viz., under-ripe, perfectly ripe, over-ripened with internal defects and over-ripened without internal defects as shown in Fig. 3. At each stage of classification, L*a*b color space features were extracted. For the purpose of classification at each stage, a number of classifiers and their possible combinations were tried out. The study revealed that, the Support Vector Machine (SVM) classifier works better for classifying mangoes into under- ripe, perfectly ripe and over-ripened while; the thresholding classifier has superior classification performance on over-ripened with internal defects and over-ripened without internal defects. Further, to bring out the superiority of the hierarchical approach, a conventional single shot multi-class classification approach with SVM was also studied.

2.3.2 Sorting of Natural and Artificial Ripened Mangoes

Further, to sort out natural and artificial ripened mangoes using machine learning model, the best of color and texture features are chosen based on the highest match of results which were obtained from five well known classifiers like LDA, Naïve Bayesian, k-NN, SVM and PNN. The obtained accuracy was between 69.89–85.72% in all categories by choosing the best of color features. Similarly, best of texture features gave an accuracy between 75- 79%. It has also been observed that accuracy decreases as the fusion of color and texture features are chosen for classification (Raghavendra et al., 2020a)

2.4 Internal Defect Detection

To measure the goodness of the proposed model (Fig. 1) at the last stage, dataset was created using the NIR (Near Infrared) spectroscopy; wavelength ranging from 673 nm-1900 nm. In order to perform classification task between defected and healthy mango fruits, wavelength selection methods were proposed to identify the range of wavelength. Further to express the effectiveness of the model, different feature selection techniques are investigated and found that Fisher's criterion-based technique appeared to be the best method for effective wavelength selection useful for the classification of defected and healthy mango fruits. Classification is performed using Euclidean distance measure. The optimal wavelengths were found in the range of 702.72 nm to 752.34 nm using fisher's criterion with a classification accuracy of 84.5% (Raghavendra et al., 2021).

3 Experimentation

3.1 Dataset

Experiments were conducted on Alphonso variety of mangoes. This variety has four classes such as under-ripe, perfect ripe, over-ripe black spots without internal defects and over-ripe black spots with internal defects at the first level. A total of 230 mangoes were selected from local farm, Mysore, Karnataka, India. Later these were kept for ripening process followed under different ripening procedures. Images were acquired from day 5 after they were kept for ripening, irrespective of the ripeness treatments which were applied for mangoes. Due to the type of ripeness treatment, some mangoes were ripened on day 5 and some were not ripened. These mangoes were first classified manually by empirical method. Mango images were captured with 16 MP resolution camera.

From Fig. 1, the dataset which were used in the first level are being carried out to the second level, this variety constituted into 2 classes based on method of ripening, one class of mangoes were ripened naturally and remaining one class was artificially ripened, with artificially ripening agent (Sachet kind-China Product). Naturally ripened mangoes were ripened with paddy straw bed for a week. From this dataset, 1050 mango images have been considered for experimentation which covers all the views of the mango.

In continuation with second level of dataset, further experiments were conducted at the last level of proposed model shown in Fig. 1 using the Ocean Optics NIR Spectroscopic instrument. 76 ripened mangoes were used at this level, as these mangoes had to be transported to the laboratory within 3 h at arbitrary temperature. In this study, measurement setup was made in reflection mode. In reflection mode, light source and detector were mounted under a specific angle, e.g., 45 degree to avoid specular reflection (Bart M. Nicolai et al., 2007). The spectral reflectance was being measured for each sample. Spectral range of the optical fibre was 673 nm to 1900 nm. Two optical fibre cables were used; one cable generates the spectral range between 673 nm to 1100 nm and another cable provides 1100 to 1900 nm range of wavelength. 1024 wavelengths were obtained from the range of 673 nm to 1100 nm with the approximate period of 0.5 nm and 512 wavelengths were obtained from the range of 1100 to 1900 nm with the approximate period of 0.56 nm. There were no clues of internal defect on the mango fruit surfaces detected as inspected visually. Hence after obtaining the reflectance values, each of the mangoes had to be cut individually to confirm the internal defects. Based on the above procedure, samples are being segregated as 43 defective samples and 33 healthy samples.

3.2 Proposed Model in This Study

The work presented in this study is an extension of our previous work (Raghavendra et al., 2020a) with additional experiments for the same dataset. Irrespective of the type of chemical treatment which is being used for mango ripeness, all the mango samples are pooled and considered as single class which is being named as chemically ripened mangoes. Different deep learning architectures have been adapted for this dataset. Since Convolutional neural networks have proven to be very successful in representation learning, as they extract features through convolutional filters and train the parameters through backpropagation. Single layer convolutional neural network consists of convolutional layer, ReLu layer, max pooling layer and fully connected layer. Single layer convolutional neural network has been used and their result is shown in Fig. 4.

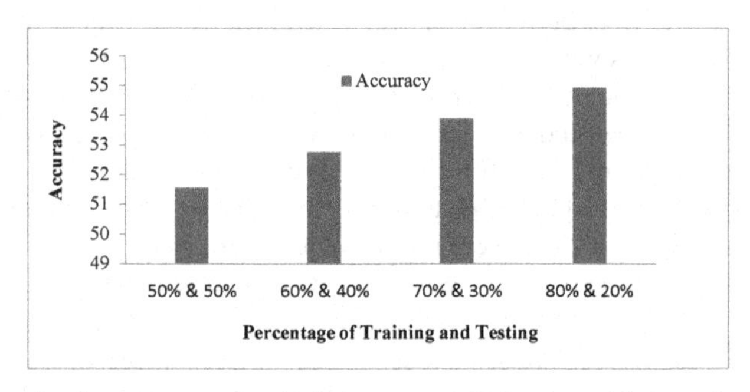

Fig. 4. Accuracy of a single layer convolutional neural network.

In order to improve the efficacy of the same model, different pertained Deep-CNN architecture is used for mango feature extraction. For the same dataset, proposed model is described in two step processes, which includes extracting features from the images using pre-trained deep networks such as VGG16, VGG19, Googlenet, ResNet50, Resnet101 and InceptionV3: and further these features are used to train a Support Vector Machine (SVM) in order to perform the task of classification. Initially mango images are fed as input to the system and then images are resized into 224 × 224 to fit the input requirements of the pre- trained networks (Suhita Ray et al., 2018). Feature vector was obtained by removing the fully connected layers from the network. These feature vectors were obtained easily without the use of much computational power. The feature extracted from the pre-trained deep networks is a minimum of 1000 and maximum of 2048 dimensional feature vector. SVM classification is highly efficient and cost-effective in different applications. The important advantage of SVM is that it offers a possibility to train generalizable, nonlinear classifiers in high dimensional spaces using a small training set. SVMs generalization error is not related to the input dimensionality of the problem but to the margin with which it separates the data. SVMs can have good performance even with a large number of inputs. Hence, obtained features are then used in the training of a SVM classifier. Accuracy has been performed with 50%, 60%, 70% and 80% of the training dataset. The accuracy results for the pooled dataset are shown in Fig. 4 (Fig. 5).

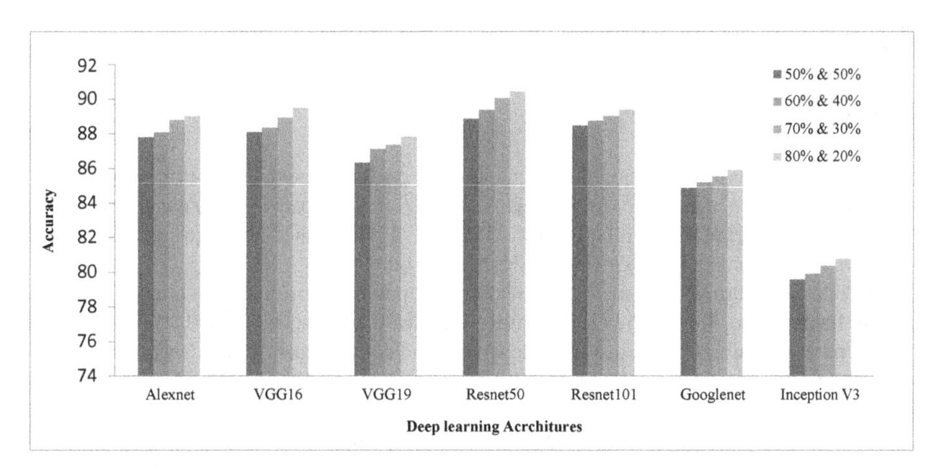

Fig. 5. Accuracy of the pre-trained deep learning architectures

In this model, ResNet-50 is able to achieve higher level of accuracy, indicating that the convolutional pools within ResNet-50 are extracting discriminative features in pooled data. ResNet50 has the potential to provide powerful features for most images. Resnet50 pre- trained deep network is shown to be the effective one which provides approximately 90% of accuracy. Three different models in the proposed post-harvesting management system shown in Fig. 1 have been discussed here (Anitha Raghavendra., Ph.D Dissertation 2021).

4 Further Works

According to fruit seller's knowledge, physiological disorder such as spongy tissue can be detected based on the shape of the mango. With this scenario, shape of the mango looks bulge, hence videos of the mango can be collected in further study to identify the internal defects. Along with the aforementioned problem, the android app can be designed further to identify the artificially ripened mangoes.The success of this study on post-harvest management of mangoes motivated for further studies of other varieties of mangoes in addition to consideration of other parameters such as store time, temperature and moisture.

5 Conclusion

This research focuses on the extraction and analysis of various image and spectroscopy features that contribute to the evaluation of quality of mango fruits. The performance comparison of classifier and feature selection techniques is brought out and finally, the best one is suggested for typical mango ripeness grading, naturally and artificially ripened mangoes sorting and finally to understand the internal quality of mangoes. In this study, a detailed analysis of these methods has been done and has introduced novel methods and techniques, as appropriate. At food industry level, an integrated solution for postharvest handling of mangoes especially for automation of sorting and grading processes can be proposed for economic development of the nation.

References

Raghavendra, A., Guru, D.S., Rao, M.K.: An automatic predictive model for sorting of artificially and naturally ripened mangoes. In: Bhateja, V., Peng, S.-L., Satapathy, S.C., Zhang, Y.-D. (eds.) Evolution in Computational Intelligence. AISC, vol. 1176, pp. 633–646. Springer, Singapore (2021). https://doi.org/10.1007/978-981-15-5788-0_60

Raghavendra, A., Guru, D.S., Rao, M.K.: Hierarchical approach for ripeness grading of mangoes. Artif. Intell. Agric. **4**, 243–252 (2020). https://doi.org/10.1016/j.aiia.2020.10.003

Raghavendra, A., Guru, D.S., Rao, M.K.: Mango internal defect detection based on optimal wavelength selection method using NIR spectroscopy. Artif. Intell. Agric. **5**, 43–51 (2021). https://doi.org/10.1016/j.aiia.2021.01.005

Raghavendra, A.: Machine learning approaches for quality evaluation of mangoes. Ph. D Dissertation, Electronics, University of Mysore, Karnataka. Accessed 18 Aug 2021

Nicolaï, B.M., Beullens, K., Bobelyn, E., Peirs, A., Saeys, W., Theron, K.I., Lammertyn, J.: Nondestructive measurement of fruit and vegetable quality by means of NIR spectroscopy: a review. Postharvest Biol. Technol. **46**(2), 99–118 (2007). https://doi.org/10.1016/j.postharvbio.2007.06.024

Nandi, C.S., Tudu, B., Koley, C.: Machine vision based techniques for automatic mango fruit sorting and grading based on maturity level and size. In: Mason, A., Mukhopadhyay, S.C., Jayasundera, K.P., Bhattacharyya, N. (eds.) Sensing Technology: Current Status and Future Trends II. SSMI, vol. 8, pp. 27–46. Springer, Cham (2014). https://doi.org/10.1007/978-3-319-02315-1_2

Chhabra, M., Gaur, R., Reel, P.S.: Detection of fully and partially RIPED mango by machine vision. ICRTITCS Proceedings Published in IJCA **5**, 25–31 (2012)

Duda, R.O., Hart, P.E., Stork. D.G.: Pattern Classification. A Wiley-Interscience Publication John Wiley & Sons, Inc. (2012)

Rezende, E., Ruppert, G., Carvalho, T., Ramos, F., de Geus, P.: Malicious software classification using transfer learning of ResNet-50 deep neural network. In: 16th IEEE International Conference on Machine Learning and Applications. 0-7695-6321-X/17/31.00 (2017)

Effect of ethrel spray on the ripening behaviour of mango (Mangiferaindica L.) variety 'Dashehari'. J. Appl. Nat. Sci.

Saldana, E., Siche, R., Lujan, M., Quevedo, R.: Computers and electronics in postharvest technology - a review. Baz. J. Food Technol. **30**(1–3), 109–124 (2001)

Goovaerts, P.: Estimation or simulation of soil properties? an optimization problem with conflicting criteria. Geoderma **3**, 165–186 (2000)

https://www.mid-day.com/articles/cac2-may-cause-cancer-blindness-seizures/15346168

Inter Institutional Inclusive Innovations Centre. www.i4c.co.in/idea/getIdeaProfile/idea_id/2969

Nutritional Talk. https://nwg-works.blogspot.in/2013/04/how-to-identify-banana-ripened-using.html

LiveChennai.com. http://www.livechennai.com/healthnews.asp?newsid=10973

Kamilaris, A., Prenafeta-Boldú, F.X.: Deep learning in agriculture: a survey. Comput. Electron. Agric. (2018). https://doi.org/10.1016/j.compag.2018.02.016
Khoje, A.S., Bodhe, S.K.: Comparative performance evaluation of fast discrete curvelet transform and colour texture moments as texture features for fruit skin damage detection" Springer. J. Food Sci. Technol. (2015). https://doi.org/10.1007/s13197-015-1794-3
Khoje, A.S., Bodhe, S.K.: Application of colour texture moments to detect external skin damages in guavas (Psidium guajava L). World Appl. Sci. J. **27**(5), 590–596 (2013)
Khoje, S., Bodhe, S.: Performance comparison of Fourier transform and its derivatives as shape descriptors for mango grading. Int. J. Comput. Appl. **53**(3), 17–22 (2012)
Khoje, S., Bodhe, S.: Comparative performance evaluation of size metrics and classifiers in computer vision based automatic mango grading. Int. J. Comput. Appl. **61**(9), 1–7 (2013)
LeCun, Y., Bengio, Y.: Convolutional networks for images, speech, and time series. The Handbook of Brain Theory and Neural Networks, **3361**(10) (1995)
LeCun, Y., Bengio, Y., Hinton, G.: Deep learning. Nature, Intell. Control Autom. **7**(4), 521(7553), 436–444. LiveChennai.com http://www.livechennai.com/healthnews.asp?newsid=10973 (2015)
Mansor, A.R., et al.: Fuzzy ripening mango index using RGB color sensor model. Res. World-J. Arts Sci. Commer. **5**(2) (2014). E-ISSN2229-4686 ISSN2231-4172
Moureen, A., Rao, M., Raghavendra, A.: An image segmentation comparison approach for lesion detection and area calculation in mangoes. Int. Res. J. Eng. Technol. (IRJET). **2**(5) (2015)
Musale, S.S., Patil, P.M.: Identification of defective mangoes using Gabor wavelets: A non-destructive technique based on texture analysis. Int. J. Agric. Innov. Res. **2**(6) (2014). ISSN 2319-1473
Nanna, K., et al.: Detecting mango fruits by using randomized Hough transform and back-propagation neural network. In: IEEE Conference (2014). https://doi.org/10.1109/IV.2014.54
Pauly, L., Sankar, D.: A New method for sorting and grading of mangoes based on computer vision system. IEEE Conf. (2015). https://doi.org/10.1109/IADCC.2015.715489
Rivea, N.V., Perez, J.J.C.: Description of maturity stages of mango 'Manila' by image analysis and ripening index
Roomi, M.M., et al.: Classification of mangoes by object features and contour modeling. IEEE (2012). https://doi.org/10.1109/MVIP.2012.6428786
Salunkhe, R.P., Patil, A.A.: Image processing for Mango ripening stage detection: RGB and HSV method. IEEE Conf. (2015). https://doi.org/10.1109/ICIIP.2015.7414796
Schmidhuber, J.: Deep learning in neural networks: an overview. Neural Netw. **61**, 85–117 (2015)
Ray, S.: Disease classification within dermascopic images using features extracted by ResNet50 and classification through deep forest. Comput. Vis. Pattern Recogn. (2018). arXiv:1807.05711
Ashok, V., Vinod, D.S.: Using K-means cluster and fuzzy C means for defect segmentation in fruits. Int. J. Comput. Eng. Technol. (2014)
MVyas, A., Talati, B., Naik, S.: Quality inspection and classification of mangoes using color and size features. Int. J. Comput. Appl. **98**(1), 1–5 (2014)
Zheng, W., Zhao, L., Zou, C.: Locally nearest neighbor classifiers for pattern classification. Pattern Regonit. **37**(6), 1307–1309 (2004)

Morphological Gradient Analysis and Contour Feature Learning for Locating Text in Natural Scene Images

B. H. Shekar[1] and S. Raveeshwara[2(✉)]

[1] Department of Computer Science, Mangalore University, Mangalore, Karnataka 574199, India
[2] Government First Grade College Uppinangady, Uppinangady, Karnataka 574241, India
raveeshwara@gmail.com

Abstract. We use text as the primary medium for providing precise information. Text could be major source of information for understanding of a scene imagery or video, once it is recognized. Although identifying and understanding textual information is fairly simple for us, it can be extremely complex task for machines. Variations such as color, orientation, scale, flow, lighting, noise, occlusion, language features and font can make this task of computer vision challenging. Detecting presence of text and precisely locating regions of text are vital for faster and precise recognition. Due to the complexity of the task, most of the popular techniques of today require an intense training phase and powerful computation infrastructure. In the proposed method, we have tried to minimize the amount of training required to achieve a decent text localization result. We have observed that, morphological gradient analysis enhances textual regions and contour feature analysis can help to eliminate non-textual components. Combination of these techniques produces promising results with small dataset, minimal training and limited computational ability. Also, the proposed detector can detect text across multiple languages and is fairly robust against the variations such as orientation and scale. The proposed method achieves an F-measure of 0.77 on MSRA-TD500 after the training with 300 images.

Keywords: Efficient text spotting · Reducing training · Contours of morphological gradient analysis

1 Introduction

Text provides simple, concise and accurate platform for human communication. Text is used in variety of ways. It could provide a simple clue or it may share knowledge across generations. Text reading is vital for both document and scene understanding. Text is a quick and reliable source of contextual information in natural scenes. Hence, the researchers have been working since decades to extract the text from the natural scene images. Yet, detecting and recognizing text is still a challenge. Text reading is a complex task due to presence of variety of distortions caused by factors such as noise, poor illumination, low contrast, presence of loss due to occlusion, other distortion and

B. Raman et al. (Eds.): CVIP 2021, CCIS 1568, pp. 254–261, 2022.
https://doi.org/10.1007/978-3-031-11349-9_22

variation [33]. Though there are fairly accurate modern methods for locating text, these techniques generally require a very large dataset for training. These methods also require a long training process and powerful computational infrastructure. Even with decent computational facility, the process of training may range from days to weeks - which is extreme. Fine tuning of the training is highly complex and a good result might rely upon a luck factor.

We are proposing a semi-supervised novel method to combine benefits of image enhancement using morphological gradient analysis, extraction of contour features and using a nonlinear multi-layer perceptron classifier to detect text regions, in this work. This would reduce training time to a great extent and produce precise results. Also, the proposed method is capable of detecting text belonging to multiple languages, different orientations and of various scale.

This work has also explored benefits and limits of some of the prominent works. In early days, generally a list of rules and some defined calculations were used in text detection. Classification of text region was decided upon a set of extracted features from the image [3, 10, 23, 31]. Later, supervised learning replaced the set of rules. Many of the modern techniques employ unsupervised learning for the prediction of text regions [6, 16, 27]. There is a surge in text reading techniques in the recent years. This surge has also showcased considerable developments in text detection and recognition in natural scene imagery and video. In the recent years, inception of neural network based classifiers, especially deep learning models [14] in text localization and text recognition has promised a great leap in results. All though these techniques significantly improve the results, they demand long hours of training and considerably huge amount of computational ability. These techniques are generally well beyond the limits of standalone and embedded systems.

2 Related Work

The research community has spent decades of effort for enabling machines to detect presence of text, locate coordinates and recognize the content of text regions in scene images. Detection and localization of text regions are an elementary step in understanding the text. Most of the earlier methods were supervised. Some of the important early techniques are discussed in the references [4–6, 10, 18, 23, 26, 31]. In the past decade, the new techniques are centered upon neural network classifiers. Especially, methods started to get more unsupervised. Deeper neural networks are utilized for improving learning in the works [9, 11–13, 21, 22, 24, 32]. The survey works [10, 14, 28, 33] compare these different methods and critically evaluate their efficacy.

Specific character set detection was the focus of many of the early works [10]. Most of the times, effort was limited to detect the presence of text belonging to only a specified language or a subset of it, in the early works. This condition was essential to reduce the extreme complexity caused by the variations in text. As the computational facility and the techniques improved, more capable, accurate and multilingual methods for text reading in the scene imagery gained prominence. For example, MSER [4] and SWT [6] techniques utilize estimated edge based features for determining text regions. These methods are supervised and rely on specific defined features. They try to estimate the stroke features such as width and other allied parameters for eliminating non-text regions. Exploring text reading from different perspective such as the methods based on the frequency domain [18] has also gained attention. Frequency domain, presents entirely different view of the problem of text reading and may ease the complexity. Training with reduced supervision [5] is also explored by the researchers.

Inception of convolutional neural networks introduced possibilities with unsupervised learning. Substantial improvements in the results of text detection and recognition can be found in the works [9, 11–14, 21, 22, 24, 32]. Observed improvement is also a result of emerging deep learning based architectures in text reading domain. In general, these methods offer better accuracy and work better in the challenging scene images. But, this improvement is at the cost of considerable amount of training time, high-performance computing infrastructure and a large set of relevant training images. Researchers are also exploring ways to reduce such demanding requirements. An example for such an effort is discussed here [7]. However, the gain in efficacy costs in reduction of accuracy or limits capability of detecting variations in text such as the text in different orientation or the text in a different scale.

Morphological gradient is utilized for locating horizontal text in the works [18, 20]. Convolutional neural networks such as text attentional networks [9] showcased idea of enhancing contrast of the image and using composite images along with Text CNN. With the achievement of good results in standard datasets, EAST [32] was considered as an efficient approach. EAST also featured multi-orientation text line detection. An oriented bounding box regression method called as Rotation-Sensitive Regression [12] has used Rotation-sensitive Regression Detector (RRD) and shown good results across orientations. Inceptext [24] is known for faster multi-orientation text detection. Open-source implementation of this method with text detection, localization and recognition capabilities is available online. Approach of TextScanner [21] is motivated by the idea of context modelling and semantic segmentation. This method has tested across datasets and attained good results. Text might appear in arbitrary shapes in scene imagery. This variety of text localization with adaptive text region representation has been discussed in the work referenced in [21]. Some signboards, shirts feature curved text. A technique for detection of curved text via transverse and longitudinal sequence connection is explored in [13]. The results of this method is compared with a new dataset of curved scene text and against standard datasets. Text belonging to multiple languages can be mixed. Or, different scene images might content text of different languages. Variations in language specific features is tough challenge to address. It has been attempted by many researchers with considerable results. Multilingual text detection is also explored in the work [1] with fixed brightness and contrast adjustments. Classification on higher level annotations and assembling text components has attempted in the work [11]. This method was able to manage different orientations and also worked across multiple languages.

However, most of the above techniques cannot be implemented on an average computing device. There is a requirement of sophisticated computational facility. Also, these techniques demand a prolonged training process. There is no doubt that we need to improve the process of training. Hence, we have focused on improvement of training process in this work and attained a promising results while reducing a substantial amount of training effort.

3 Dataset Description

Natural scenes may contain text with perspective distortion due to variations in the orientation. Scene text can be composed of multiple languages, such as bilingual sign boards. Text regions may contain words from different languages. For instance, two text lines each in two different languages. Microsoft Research Asia Text Detection 500 dataset (MSRA-TD500) [25] is a small dataset consisting of 300 images for training and 200 images for testing. The images are could be considered of natural scenes and vary in resolution. The dataset fairly demonstrates real world challenges to be faced by a text detection system. Hence, we are evaluating our method against this dataset.

4 Proposed Method

Our focus is to minimize the amount of training required to achieve a decent text localization result. We have observed in our experiments that the morphological gradient analysis (MGA) enhances textual regions. Contour feature learning (CFL) [19] can help to eliminate non-textual components. Combination of Morphological Gradient Analysis and Contour Feature Learning can effectively locate text in natural scene images. This technique produces promising results with small dataset, minimal training and limited computational ability. Also, the proposed text detector can detect text across multiple languages and is proven robust across the normal variations in the text such as scale, orientation and color.

The methodology can be split into two usual phases - training phase and testing phase. In the phase of training, morphological gradient is computed and contour features are extracted from morphological gradient image. Ground truths along with the extracted contour features are used to train our text classifier. The text classifier is designed using a multi-layer perceptron neural network. During the testing phase, this trained classifier is applied on the test images to determine text contours. For the recognition phase it is ideal if the characters are grouped together into words or text lines. To attain this, contours are put together based on their neighborhood.

4.1 Phase 1: Training

During the phase of training, two major activities are performed. First, image is preprocessed and morphological gradient [17] operation is applied on it. Morphological gradient operation would enhance variations of pixel intensity. Second, contours [15] and contour features are extracted from the enhanced image. Contour features are balanced to reduce bias and are used to train our NN classifier. Sequence of these operations are shown in Fig. 1.

We have used contour perimeter, contour area, contour dimensions, enclosing minimum rectangle, number of points, convexity and direction for contour feature learning.

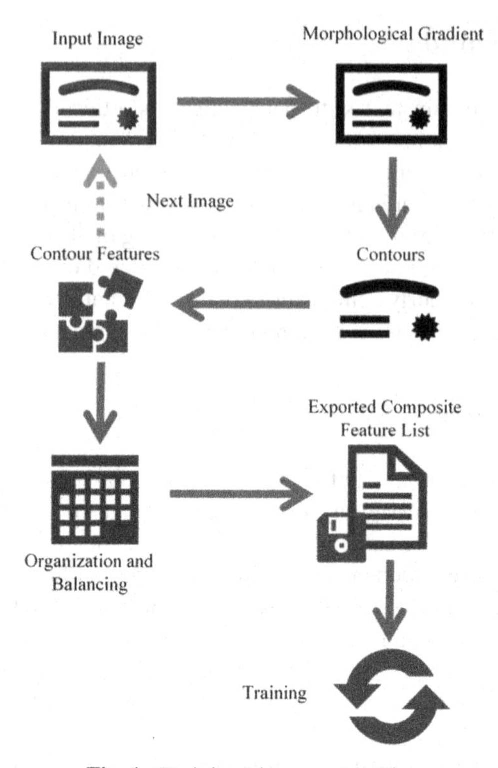

Fig. 1. Training phase composition

4.2 Phase 2: Testing

During testing, an image under goes similar phases as that of training. However, instead of updating the training model, a classification decision is produced to identify text contours and locate text in the input image. Text regions such as words and lines are formed by neighborhood analysis and text contour grouping. The steps involved are shown in Fig. 2.

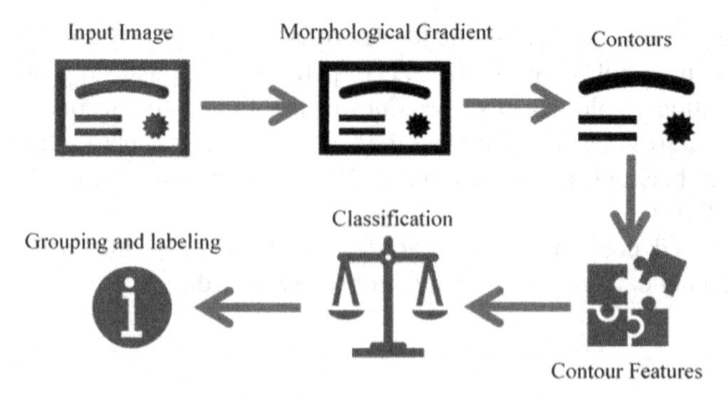

Fig. 2. Steps in the phase of testing

5 Experimental Results

Our proposed method is trained with training images of MSRA-TD500 dataset and evaluated against the testing set. The method has shown promising results. The obtained result and comparison with the standard result can be seen in the Table 1. We have conducted our experiments using Python OpenCV [2], NumPy [8] array image data and tools available with Anaconda Scikit Learn.

Table 1. Experiment results comparison with MSRA-TD500 dataset

Method	F-score	Precision	Recall
SWT Epshtein et al. [6, 11]	0.25	0.25	0.25
Text Detection-ICDAR [25]	0.50	0.53	0.52
Text Detection-Mixture [25]	0.60	0.63	0.63
AT-Text Li et al. [11]	0.73	0.77	0.69
Zhang et al. [30]	0.74	0.83	0.67
Text-Attentional CNN He et al. [9]	0.74	0.77	0.70
Zhang et al. [29]	0.76	0.81	**0.72**
MGA	0.69	0.71	0.69
CFL [19]	0.75	0.84	0.67
MGA + CFL (Proposed)	**0.77**	**0.85**	0.71

The proposed method works best with large text regions on contrasting background. The method has produced promising results against variations in color, font, size, input dimension, orientation of text, blur introduced in natural scene images with limited training of 300 images. This can be seen in Fig. 3. However, we have observed that the method may not always yield good results in large images consisting small text and when text is not clearly distinguishable from the background. Some of the such instances are shown in Fig. 4.

Fig. 3. Some successful detection results of the proposed method

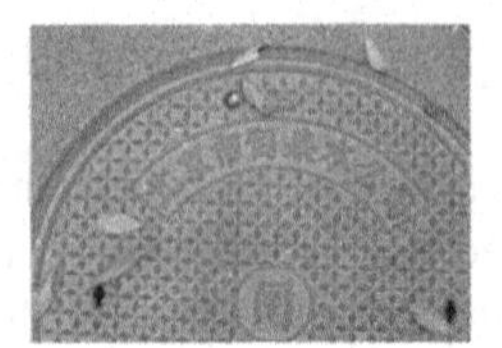

Fig. 4. Some failure cases

6 Conclusion

The proposed method has shown promising results, considering that it is trained with only 300 training images of MSRA-TD500 dataset and has performed well on the testing set of images. The method has successful results in multiple languages, different scales, different input image size and in multiple orientations. The results of the method are comparable with works of recent literature. The method stands out with the ability to learn detecting text with limited training.

References

1. Basu, S., et al.: Multilingual scene text detection using gradient morphology. Int. J. Comput. Vis. Image Process. **10**(3), 31–43 (2020). https://doi.org/10.4018/IJCVIP.2020070103
2. Bradski, G.: The OpenCV library. Dr. Dobb's J. Softw. Tools (2000)
3. Chen, D., Luettin, J.: A survey of text detection and recognition in images and videos (2000)
4. Chen, H., et al.: Robust text detection in natural images with edge-enhanced maximally stable Extremal regions. In: 2011 18th IEEE International Conference on Image Processing, pp. 2609–2612 (2011). https://doi.org/10.1109/ICIP.2011.6116200
5. Coates, A., et al.: Text detection and character recognition in scene images with unsupervised feature learning. In: 2011 International Conference on Document Analysis and Recognition, pp. 440–445 (2011). https://doi.org/10.1109/ICDAR.2011.95
6. Epshtein, B., et al.: Detecting text in natural scenes with stroke width transform. In: 2010 IEEE Computer Society Conference on Computer Vision and Pattern Recognition, pp. 2963–2970 (2010). https://doi.org/10.1109/CVPR.2010.5540041
7. Fu, K., et al.: Text detection for natural scene based on MobileNet V2 and U-Net. In: 2019 IEEE International Conference on Mechatronics and Automation (ICMA), pp. 1560–1564 (2019). https://doi.org/10.1109/ICMA.2019.8816384
8. Harris, C.R., et al.: Array programming with {NumPy}. Nature **585**(7825), 357–362 (2020). https://doi.org/10.1038/s41586-020-2649-2
9. He, T., et al.: Text-attentional convolutional neural network for scene text detection. IEEE Trans. Image Process. **25**(6), 2529–2541 (2016). https://doi.org/10.1109/TIP.2016.2547588
10. Jung, K., et al.: Text information extraction in images and video: a survey. Pattern Recogn. **37**(5), 977–997 (2004). https://doi.org/10.1016/j.patcog.2003.10.012
11. Li, H., Lu, H.: AT-text: assembling text components for efficient dense scene text detection. Future Internet. **12**(11), 1–14 (2020). https://doi.org/10.3390/fi12110200
12. Liao, M., et al.: Rotation-sensitive regression for oriented scene text detection. Proceedings of the IEEE Computer Society Conference on Computer Vision and Pattern Recognition, pp. 5909–5918 (2018). https://doi.org/10.1109/CVPR.2018.00619
13. Liu, Y., et al.: Curved scene text detection via transverse and longitudinal sequence connection. Pattern Recogn. **90**, 337–345 (2019). https://doi.org/10.1016/j.patcog.2019.02.002

14. Long, S., He, X., Yao, C.: Scene text detection and recognition: the deep learning era. Int. J. Comput. Vision **129**(1), 161–184 (2020). https://doi.org/10.1007/s11263-020-01369-0
15. Malik, J., et al.: Contour and texture analysis for image segmentation. Int. J. Comput. Vision **43**(1), 7–27 (2001). https://doi.org/10.1023/A:1011174803800
16. Matas, J., et al.: Robust wide-baseline stereo from maximally stable extremal regions. In: Image and Vision Computing (2004). https://doi.org/10.1016/j.imavis.2004.02.006
17. Rivest, J.-F., et al.: Morphological gradients. J. Electron. Imaging **2**(4), 326–336 (1993). https://doi.org/10.1117/12.159642
18. Shekar, B.H., et al.: Discrete wavelet transform and gradient difference based approach for text localization in videos. In: Proceedings - 2014 5th International Conference on Signal and Image Processing, ICSIP 2014, pp. 280–284 (2014). https://doi.org/10.1109/ICSIP.2014.50
19. Shekar, B.H., Raveeshwara, S.: Contour feature learning for locating text in natural scene images. Int. J. Inf. Technol. (2022). https://doi.org/10.1007/s41870-021-00851-3
20. Shekar, B.H., Smitha M.L.: Morphological gradient based approach for text localization in video/scene images. In: 2014 International Conference on Advances in Computing, Communications and Informatics (ICACCI), pp. 2426–2431 (2014). https://doi.org/10.1109/ICACCI.2014.6968426
21. Wan, Z., et al.: TextScanner: Reading characters in order for robust scene text recognition. arXiv. (2019). https://doi.org/10.1609/aaai.v34i07.6891
22. Wang, X., et al.: Arbitrary shape scene text detection with adaptive text region representation. In: Proceedings of the IEEE Computer Society Conference on Computer Vision and Pattern Recognition. 2019-June, pp. 6442–6451 (2019). https://doi.org/10.1109/CVPR.2019.00661
23. Wu, V., et al.: Textfinder: an automatic system to detect and recognize text in images. IEEE Trans. Pattern Anal. Mach. Intell. **21**(11), 1224–1229 (1999). https://doi.org/10.1109/34.809116
24. Yang, Q., et al.: Inceptext: A new inception-text module with deformable PSROI pooling for multi-oriented scene text detection. In: IJCAI International Joint Conference on Artificial Intelligence, pp. 1071–1077 (2018). https://doi.org/10.24963/ijcai.2018/149
25. Yao, C., et al.: Detecting texts of arbitrary orientations in natural images. Proc. IEEE Comput. Soc. Conf. Comput. Vis. Pattern Recognit. **8**, 1083–1090 (2012). https://doi.org/10.1109/CVPR.2012.6247787
26. Yao, C., et al.: Scene text detection via holistic, multi-channel prediction 1–10 (2016)
27. Ye, Q., Doermann, D.: Text detection and recognition in imagery: a survey. IEEE Trans. Pattern Anal. Mach. Intell. **37**(7), 1480–1500 (2014)
28. Yin, X.C., et al.: Text detection, tracking and recognition in video: a comprehensive survey. IEEE Trans. Image Process. **25**(6), 2752–2773 (2016). https://doi.org/10.1109/TIP.2016.2554321
29. Zhang, Y., Huang, Y., Zhao, D., Wu, C.H., Ip, W.H., Yung, K.L.: A scene text detector based on deep feature merging. Multimed. Tools Appl. **80**(19), 29005–29016 (2021). https://doi.org/10.1007/s11042-021-11101-w
30. Zhang, Z., et al.: Multi-oriented text detection with fully convolutional networks. In: 2016 IEEE Conference on Computer Vision and Pattern Recognition (CVPR), pp. 4159–4167 (2016). https://doi.org/10.1109/CVPR.2016.451
31. Zhong, Y., et al.: Locating text in complex color images. Pattern Recogn. **28**(10), 1523–1535 (1995). https://doi.org/10.1016/0031-3203(95)00030-4
32. Zhou, X., et al.: EAST: an efficient and accurate scene text detector. In: Proceedings - 30th IEEE Conference on Computer Vision and Pattern Recognition, CVPR 2017, pp. 2642–2651 (2017). https://doi.org/10.1109/CVPR.2017.283
33. Zhu, A.: Scene text detection and recognition. Front. Comp. Sci. **10**(1), 19–36 (2017)

Introspecting Local Binary Feature Vectors for Classification of Radiographic Weld Images

Jayendra Kumar[1], Pratul Arvind[2](✉), Prashant Singh[2], and Yamini Sarada[2]

[1] National Institute of Technology Jamshedpur, Jamshedpur, India
[2] ADGITM, New Delhi, India
pratul.arvind@gmail.com

Abstract. With the advent of technology and industrialization, welding has emerged as a potential tool. The process of welding gives a permanent joint to the material which is joined together, but it also affects the properties of the constituents. Any kind of deficiency in the welding process gives rise to weld defects. There should be a mechanism for accurate inspection of welded materials in order to maintain the superiority of design of the welded material and smooth operation. A proper welded material assures proper protection and trustworthiness. Non – Destructive Inspection is one of the significant features for accurate identification of the weld defects. Nowadays, this technique is extensively used for weld flaw detection as it doesn't alter with the property of the welded objects. As computer technology has paved its way in all domain of engineering, the current research work focusses on discovery of a scientific solution for exact identification and classification of imperfections in welding. In the present work, the image database has been created from Welding research laboratory, Department of Mechanical and Industrial Engineering, Indian Institute of Technology Roorkee, Roorkee and consists of 79 radiographic weld images with 8 types of flaws and one without flaws. The present work explores the various local binary pattern feature vectors for the classification of flaws in radiographic weld images. It gives a concise description of the various existing feature extraction techniques for the proposed database.

Keywords: Non – destructive testing · Weld defects · Local binary pattern feature vectors

1 Introduction

Welding is an effective and economical process to join metals together. It is accomplished by adding some additional molten joining material on the melting part of the materials to be joined. In the welding procedure, a several dissimilar types of flaws can be present. It arises due to irregularities in the welding material, the weld operators' fault and several other aspects that is beyond the operators' operational limit. Regardless of the cause of error, the recognition of flaws is important. An improper welding enormously decreases the bond that exists among the materials and may result in welding defect. With the Introduction of the fast development of computer technology, new methods of extracting in vivo functional and dimensional information, for establishing localized

B. Raman et al. (Eds.): CVIP 2021, CCIS 1568, pp. 262–272, 2022.
https://doi.org/10.1007/978-3-031-11349-9_23

flaws and development of defects in the weldment is the need of hour. The manual interpretation of radiographic NDE weld images depends upon the level of expertise of the specialist and is usually a time-consuming process as well as subjective. Image processing plays a vital role to interpret these images with an aim to spot the flaws in the weldment. The present work has been an effort in the direction of automating the inspection process [1, 2].

The principal welding defects or flaws may be classified. The different types of flaws that exists in welding are Gas cavity, Porosity, Lack of fusion, Lack of penetration, Wormhole, Slag inclusion, Undercut, and Cracks. There are two types of testing of materials such as Destructive Testing of Materials and Non- Destructive Testing of Material. In the destructive testing of material, the material is tested under the load till failure is traced out. In the process of Non -Destructive Testing of Materials (NDT), the material is inspected without any harm or everlasting damage [3]. This inspection aids to establish the several properties of the material such as hardness, strength, toughness, etc. The properties stated have a substantial influence on the material usage. So, the material needs to be inspected and evaluated at a priority before utilizing the materials for the application. The examination has the advantages of improved speed during inspection and dependability, sensitivity to flaws of any orientation, suited to high operating temperatures.

The application of image processing in the field of radiographic images has started from 1990 and still, it's a challenging work for computer vision applications. In the beginning, Gayer A. et al. [4] developed a technique for the automatic examination of welding defects from real-time radiographic weld images which involved two-stage. Automatic inspection of the weld image flaw detection by machine vision has been done by Ker J. et al. [4]. Nockeman C. et al. [5] has analyzed the reliability of radiographic weld flaw detection using relative operating characteristics. Kato Y. et al. [6] proposed a computer-aided radiographic inspection expert system for recognizing welding flaws. The assimilation rules were framed on the views received from experts. Further, the authors in [7] have employed fuzzy classifiers, specifically fuzzy K-NN and fuzzy c-means, for classifying the weld patterns. It was used for the segmentation of curved welds and was capable of detecting varying weld flaws in one radiographic image. In the same year, the author in [8] has developed a human visual inspection method that can identify 60–75% of the signification defects. Mery D. et al. [9] used a texture feature for the automatic recognition of welding defects. The features vectors provided by edge detection algorithm, in this case, were based on two features of texture i.e. occurrence matrix and 2D Gabor function. Zapata J. et al. [10] has developed a system to notice, identify and classify welding defects in radiographic images. Further, the process was evaluated for two neuro-classifiers based on an artificial neural network (ANN) and an adaptive-network based fuzzy inference system (ANFIS). The accuracy was 78.9% for the ANN and 82.6% for the ANFIS. This methodology was tested on 86 radiograph images that consist of 375 defects having five types of flaws. Till date, several research works are underway for, feature extraction, identification and classification of the appropriate information contained in weld image. The proposed work aims to improve the process of automated information extraction systems.

In the present work feature extraction using different texture operators such as Local binary pattern, Uniform local binary pattern, Rotation invariant local binary pattern, Local binary pattern histogram Fourier features, Complete local binary pattern, Adaptive local binary pattern, Uniform Adaptive local binary pattern, Rotational Invariant Adaptive local binary pattern, Rotational Invariant Uniform Adaptive local binary pattern have been discussed respectively. Feature extraction has also been carried by means of complete feature vector data of the above-mentioned techniques and reduced feature vector data using Principal Component Analysis.

2 Database for Present Research Work

Appropriate identification and classification of imperfections in welding is a major thrust area of research which still needs attention. The dataset of weld image is developed by the Weld Testing and NDT laboratory, Department of Mechanical and Industrial Engineering, Indian Institute of Technology Roorkee. The database is built up of radiographic images and proper identification of the weld defect was a cumbersome process. In addition, the weld images are of poor quality which makes it inappropriate for exact recognition (Table 1).

Table 1. Description of image dataset

S. No.	Name of the weld defect	Number of images
1	Gas cavity	08
2	Lack of penetration	20
3	Porosity	07
4	Slag inclusion	16
5	Cracks	11
6	Lack of fusion	07
7	Wormhole	02
8	Undercut	03
9	No defect	05
10	Total no of images	**79**

The dataset consists of 79 radiographic weld images. It consists of eight different types of weld defects and five images are flawless.

3 Local Binary Feature Vectors

Texture features [11, 12] has been instrumental in the classification of the image database. In the present work, some widely used texture feature descriptors [13, 14] are employed and their effectiveness for the classification of weld flaws images has been studied.:

3.1 Local Binary Pattern (LBP)

The LBP is an overwhelming texture descriptor strategy for image investigation because of its capability of representation of discriminative information (Ojala et al., 1994 [15, 16]). A circular neighborhood of flexible size is depicted in [17] to conquer the deficiency of the original LBP operator of neighborhood measures that can't catch the predominant texture features in huge scale structures.

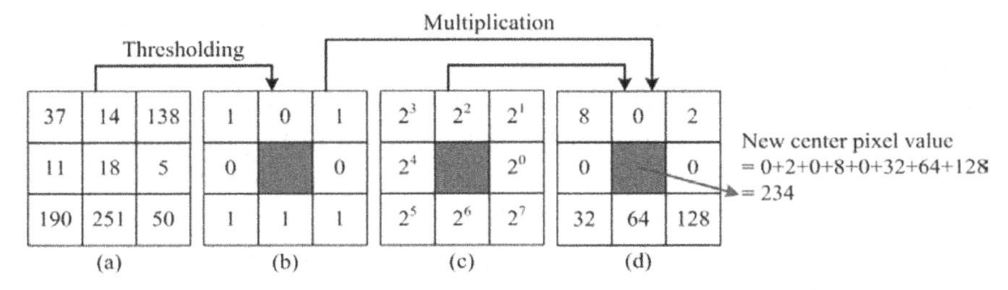

Fig. 1. LBP Computation Process. a) 3 3× local window image, (b) thresholding, (c) weight and d) new center pixel value (decimal)

The LBP of an image with center pixel coordinates (x, y) is given by [88]:

$$LBP_{P,R}(x, y) = \sum_{p=0}^{P-1} s(g_p - g_c)2^p \tag{1}$$

where, g_c and g_p, are the gray value of the center pixel and its p neighbors, respectively. Further, $\mathrm{s(z)} = \begin{Bmatrix} 0, \ z < 0 \\ 1, \ z \geq 0 \end{Bmatrix}$ it signifies a thresholding function.

In LBP_{PR}, P represents amount of sampling points on the roundabout neighborhood, whereas R is the resolution of the spatial goal in the concerned area. Bilinear interpolation is connected to pixel values if the sampling points excludes the integer coordinates. The LBP operator produces a 256-dimensional texture descriptor for a given image. The pictorial representation of the calculation of new center pixel value for LBP is shown in Fig. 1.

3.2 Uniform Local Binary Pattern ($\mathbf{LBP^{u2}}$)

The LBP designs are supposed to be unvarying examples if at maximum 2-bit wise transition (1 to 0 or 0 to 1) is accounted for the circular binary pattern of LBP [18]. The histogram includes a distinct bin for uniform patterns and just a solo bin is assigned to all the non-uniform patterns. For a given pattern of P bits, P(P-1) +3 bits are generated. The decrease in the non-uniform pattern is because of the natural images since, the LBP patterns are usually uniform. Further, uniform patterns of texture represent around 90% of the whole pattern with (8, 1) neighborhood and near 70% for (16, 2) neighborhood [19]. The LBPu2 produces 59-dimensional surface descriptors.

3.3 Rotational Invariant Local Binary Pattern (LBP^{ri})

The image rotation creates various LBP codes. To address, LBP^{ri} has been proposed [17–19] to influence the matter of the image rotation. In this way, to make every one of the adaptations of parallel codes the equivalent, the LBP codes are turned back to the referred position of the pixel to invalidate the result of interpretation for location of the pixel. The $LBP^{ri}_{P,R}$ is created by circularly turning the fundamental LBP code and considering the pattern which has a base an incentive as represented by [17–19]

$$LBP^{ri}_{P,R} = \min_{i}\{ROR(LBP_{P,R}, i)\} \tag{2}$$

where, i = 0, 1, 2, … –, P 1. The circular bit – by - bit right shift activity is performed on x (a P-bit number) for i times by the function ROR(x,i). The $LBP^{ri}_{P,R}$ descriptor produces in general 36-bin histograms for each image due to 36 different, 8-bit rotation invariant codes [17–19].

3.4 Local Binary Pattern Histogram Fourier Features (LBP-HF)

A rotation invariant LBP-HF protects the utmost discriminative qualities which are attained by taking the discrete Fourier transform (DFT) of [18, 19]. It is developed comprehensively for the whole image contrasted with other histogram-based invariant texture descriptor strategies that leads to standardization of rotation in the neighborhood region. The LBPHF's are invariant to cyclic moves along the rows of input histogram, and are said to be invariant to the rotational movement of image under investigation [19]. The DFT is utilized to build the features as given by [18, 19]:

$$H(n, u) = \sum_{r=0}^{P-1} h_I(U_P(n, r))e^{-i2\pi ur/P} \tag{3}$$

where H(n, u) corresponds to the DFT of the nth row of LBP^{u2} histogram $h_I(U_P(n, r))$. It produces 38-bin histograms for a given texture image.

3.5 Adaptive Local Binary Pattern (ALBP)

In the year 2010, an adaptive local binary pattern which improves the effectiveness of images classification by minimizing the variations of the oriented mean and standard deviation of absolute local difference $|g_c - g_p|$ was introduced by Guo et al. [20]. A weight parameter (w_p) as given below is introduced to minimize the overall directional differences $|g_c - w_p * g_p|$ along diverse orientations. The objective function for ALBP [21] is expressed by:

$$w_p = \arg\min_{w}\left\{\sum_{i=1}^{N}\sum_{j=1}^{M}\left|g_c(i, j) - w.g_p(i, j)\right|^2\right\} \tag{4}$$

where, N and M resembles the number of rows and columns present in the weld image. For each of the orientations 2πp P/ of entire image, a weight factor wp is approximated.

The expression for ALBP is then given by

$$ALBP_{P,R} = \sum_{p=0}^{p-1} s\left(g_p * w_p - g_c\right)2^p, \quad s(x) = \begin{cases} 1, & x \geq 0 \\ 0, & x < 0 \end{cases} \tag{5}$$

3.6 Complete Local Binary Pattern (CLBP)

The completed local binary pattern (CLBP) enhances the capability of prominent texture feature extraction of LBP as projected by Guo et al. [22]. Figure 2(a) shows 3 3× block of an image having center pixel value 34. Figure 2 (b), (c) and (d) depicts the local difference, sign component and magnitude components.

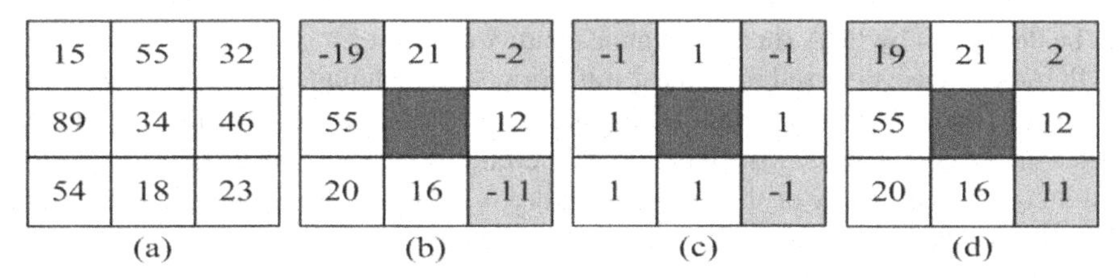

Fig. 2. CBLP Computation Process (a) 3 3× block of image, (b) local difference (g gp – c) (c) sign component, and (d) magnitude component.

The structure of CLBP is depicted in Fig. 3. In CLBP, the local difference and center gray level are found from the gray scale image. The sign (*S*) and magnitude (*M*) components of local difference is shaped by evaluating the local difference sign-magnitude transform (LDSMT) represented by [95]:

$$d_p = g_p - g_c = s_p * m_p, \quad \text{and} \begin{cases} s_p = \text{sign}(d_p) \\ m_p = |d_p| \end{cases} \tag{6}$$

where, m_p and $s_p = \begin{cases} 1, d_p \geq 0 \\ -1, d_p < 0 \end{cases}$ are magnitude and sign of d_p, respectively. The CLBP Sign (*CLBP_S*) and CLBP Magnitude (*CLBP_M*) operator portrays the complementary components of local structure of weld image. Further, the CLBP Centre (*CLBP_C*) operator is created by altering center pixel into binary code using global thresholding [20]. Here, *CLBP_S*, and *CLBP_M* operator are sequenced to form the CLBP histogram.

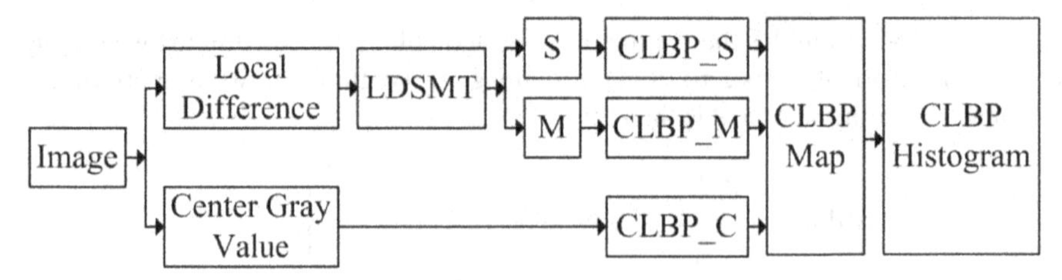

Fig. 3. Structure of CBLP

4 Feature Reduction by PCA

The high-dimensional features obtained from the weld images raises the computational requirements of the classifier due to which the classification accuracy may not be enhanced as required. Along these lines, there is a need for reduction in the dimension of the features thus obtained [23, 24]. Also, it is required to change the information content from high-dimension to low-dimension. The dimension of feature vector information can be decreased by PCA (include dimensionality decrease).

PCA is extensively used as a linear transformation technique [25]. It decreases data dimensions by registering a couple of symmetrical straight mixes of the first dataset features with greatest change. Also, it includes ascertaining the eigenvalues and eigenvectors of the covariance matrix of the original feature matrix. The first principal component (PC) is described by the eigenvectors as described by largest eigenvalue contained in the eigenvectors while, the second largest difference is termed as second PC and it is orthogonal to the first PC.

5 Methodology

The methodology adopted for the present algorithm is demonstrated below as the flowchart in Fig. 4.

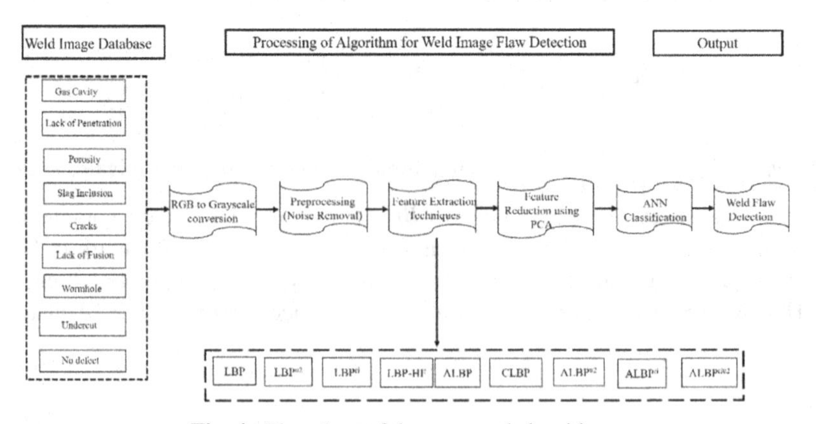

Fig. 4. Flowchart of the proposed algorithm.

5.1 Assessment of Different Texture Feature Extraction Techniques

The performance of various texture feature extraction technique as mentioned in Sect. 3 has been examined by dividing the data randomly to achieve the accurateness in classification of weld images.

In the present research, the performance has been executed by combining two different proportion of testing and training as mentioned below (Table 2):

Table 2. Performance evaluation using different combination of ANN

S. No.	Training data	Testing data
1	80	20
2	70	30

6 Results and Discussion

The investigational work carried out in the present research work examines the effectiveness of the various Local Binary Texture features extraction techniques as discussed in Sect. 3 for the accurate classification of radiographic weld images from the dataset considered into 9 different types of weld flaws and one without flaw classes by employing the use of ANN classifiers.

6.1 Performance Evaluation of Feature Extraction Techniques

The P and R parameter values are 8 and 1 that are used for the present work. It is applied for all the variants of LBP (LBP, LBP^{u2}, LBP^{ri}, LBP^{riu2}, LBP-HF, ALBP, CLBP, $ALBP^{u2}$, $ALBP^{ri}$ and $ALBP^{riu2}$). It is due to the fact that these particular values have generated fast and correct feature extraction [120]. Also, in case of texture images, 90% of uniform patterns is obtained for all patterns when (8, 1) neighborhood is considered, while 70% uniform pattern accounts for (16, 2) neighborhood is obtained. Moreover, the *R* parameter is generally selected as a small value since with increase in distance, the correlation among the pixels decreases. From local neighborhoods, ample amount of the texture information can be gained.

6.2 Performance Evaluation Using ANN

In the case of ANN classifier [26], again the ALBP texture features with 216 features are obtained to give the optimal classification accuracy of 91.14% for 80/20 training and testing ratios and 89.87% for 70/30 training and testing ratios is obtained for the present dataset. Also, the performance of LBP and its variants are also good.

As evident from Fig. 5, after applying PCA technique, the compact feature vector of $ALBP^{ri}$ technique yields the finest classification accuracy of 91.28% (120 features) for 80/20 and 90.54% for 70/30 training and testing respectively, whereas the second

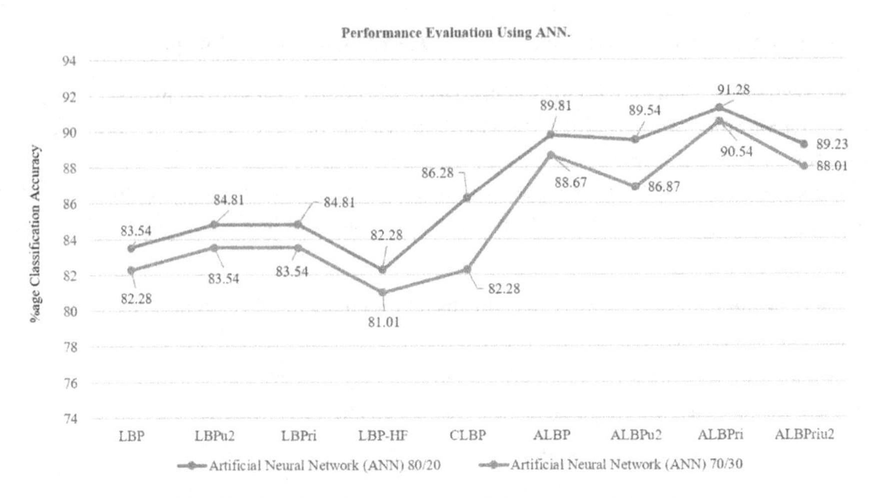

Fig. 5. Classification accuracy of the proposed algorithm

highest accuracy of 89.81% obtained by ANN classifier for 80/20 and 88.67 for 70/30 ratio of randomly divided data set.

From the above figure, it is evident that 09 texture features performance was evaluated and inference can be drawn that the features collected after PCA for ALBP^{ri} outperforms other features. The minimal result was obtained by LBP -HF features. The motive behind introducing so many feature extraction techniques are that an extensive experiment was carried out to find the optimal feature for the selected weld image database.

7 Conclusion

It is obvious from the above experiment that the texture feature obtained from ABLP^{ri} has the qualities of accurate and prompt classification, as well as minimal size of feature. Also, its classification accuracy is remarkably better. It is also evident that the PCA reduced feature vector does not affect the accuracy in most of the cases. However, the advantage of PCA reduced database is that it takes less time for classification and increases the results by very little amount. Also, ANN [27] is skilled operator that is able to detect the nonlinear classification boundaries present in the image. The better performance is due to its capability to locate the nonlinear classification boundaries.

References

1. Basak, D.: Non-destructive evaluation of drive ropes: a case study. Nondest. Test. Eval. **20**(4), 221–229 (2005)
2. Anouncia, S.M., Saravana, R.: A knowledge model for gray image interpretation with emphasis on welding defect classification. Comput. Ind. **61**, 742–749 (2010)
3. Gayer, A., Saya, A., Shiloh, A.: Automatic recognition of welding defects in real-time radiography. NDT E Int. **23**(3), 131–136 (1990)

4. Ker, J., Kengshool, K.: An efficient method for inspecting machined part by a fixtureless machine vision system. Proc SME Vision, Detroit **2**, 45–51 (1990)
5. Nockeman, C., Heidt, H., Tomsen, N.: Reliability in NDT: ROC study of radiographic weld inspection. NDT E Int. **24**(5), 235–245 (1991)
6. Kato, Y., et al.: Development of an automatic weld defect identification system for radiographic testing. Weld World **30**(7/8), 182–188 (1992)
7. Liao, T.W., Li, D., Li, Y.: Extraction of welds from radiographic images using fuzzy classifiers. Inf. Sci. **126**(1–4), 21–40 (2000). https://doi.org/10.1016/S0020-0255(00)00016-5
8. Chan, C., Grantham, K.H.: Fabric defect detection by Fourier analysis. IEEE Trans. Ind. Appl. **36**(5), 12671275 (2000)
9. Zapata, J., Vilar, R., Ruiz, R.: Performance Evaluation of an automatic inspection system of weld defects in radiographic images based on neuro – classifiers **38**(7), 8812–8824 (2011)
10. Li, W., You, J., Zhang, D.: Texture-based palmprint retrieval using a layered search scheme for personal identification. IEEE Trans. Multimedia **7**(5), 891–898 (2005)
11. Zhang, D., Kong, W.-K., You, J., Wong, M.: Online palmprint identification. IEEE Trans. Pattern Anal. Mach. Intell. **25**(9), 1041–1050 (2003)
12. Ojala, T., Pietikainen, M., Harwood, D.: Performance evaluation of texture measures with classification based on Kullback discrimination of distributions. In: Proceedings of the 12th IAPR International Conference on Pattern Recognition, Vol. 1-Conference A: Computer Vision & Image Processing, pp. 582–585 (1994)
13. Ojala, T., Pietikäinen, M., Harwood, D.: A comparative study of texture measures with classification based on featured distributions. Pattern Recogn. **29**(1), 51–59 (1996)
14. Ojala, T., Pietikainen, M., Maenpaa, T.: Multiresolution gray-scale and rotation invariant texture classification with local binary patterns. IEEE Trans. Pattern Anal. Mach. Intell. **24**(7), 971–987 (2002)
15. Pietikäinen, M., Hadid, A., Zhao, G., Ahonen, T.: Computer vision using local binary patterns. Computational Imaging and Vision (2011)
16. Mäenpää, T.: The Local Binary Pattern Approach to Texture Analysis: Extenxions and Applications. Oulun Yliopisto (2003)
17. Pietikäinen, M., Ojala, T., Xu, Z.: Rotation-invariant texture classification using feature distributions. Pattern Recogn. **33**(1), 43–52 (2000)
18. Ahonen, T., Matas, J., He, C., Pietikäinen, M.: Rotation invariant image description with local binary pattern histogram fourier features. Image Analysis, Springer. pp. 61-70 (2009)
19. Zhao, G., Ahonen, T., Matas, J., Pietikainen, M.: Rotation-invariant image and video description with local binary pattern features. IEEE Trans. Image Process. **21**(4), 1465–1477 (2012)
20. Guo, Z., Zhang, D., Zhang, S.: Rotation invariant texture classification using adaptive LBP with directional statistical features. In: 17th IEEE International Conference on Image Processing (ICIP), pp. 285–288 (2010)
21. Guo, Z., Zhang, D.: A completed modeling of local binary pattern operator for texture classification. IEEE Trans. Image Process. **19**(6), 1657–1663 (2010)
22. Zhang, L., Zhou, Z., Li, H.: Binary Gabor pattern: An efficient and robust descriptor for texture classification. In: 19th IEEE International Conference on Image Processing (ICIP), pp. 81–84 (2012)
23. Canuto, A.M., Vale, K.M., Feitos, A., Signoretti, A.: ReinSel: a classbased mechanism for feature selection in ensemble of classifiers. Appl. Soft Comput. **12**(8), 2517–2529 (2012)
24. Yaghouby, F., Ayatollahi, A., Soleimani, R.: Classification of cardiac abnormalities using reduced features of heart rate variability signal. World Appl. Sci. J. **6**(11), 1547–1554 (2009)
25. Camastra, F.: Data dimensionality estimation methods: a survey. Pattern Recogn. **36**(12), 2945–2954 (2003)

26. Pratul, A., M. R., P.: A wavelet packet transform approach for locating faults in distribution system. In: 2012 IEEE Symposium on Computers & Informatics (ISCI), pp. 113–118 (2012). https://doi.org/10.1109/ISCI.2012.6222677
27. Kumar, J., Arvind, P., Singh, P., Sarada, Y., Kumar, N., Bhardwaj, S.: LBPriu2 features for classification of radiographic weld images. International Conference on Innovative Trends and Advances in Engineering and Technology (ICITAET) **2019**, 160–165 (2019). https://doi.org/10.1109/ICITAET47105.2019.9170146

Performance Evaluation of Deep Learning Models for Ship Detection

Rahul Sharma, Harshit Sharma, Tamanna Meena(✉), Padmavati Khandnor, Palak Bansal, and Paras Sharma

Punjab Engineering College, Chandigarh 160012, India
tamanna.meena95@gmail.com

Abstract. Although enormous success is achieved in the field of object detection, yet ship detection in high-resolution images is still an crucial task. Ship detection from optical remote sensing images plays a significant role in military and civil applications. For maritime surveillance, monitoring and traffic supervision ship detection deserves optimal solutions to identify objects accurately with faster speed. In this work, various object detection methods such as You Only Look Once (YOLO) v3, YOLO v4, RetinaNet152, EfficientDet-D2 and Faster-RCNN have been implemented to improve efficiency, speed and accuracy. Numerous experiments were conducted to evaluate the efficiency of object detection methods. The YOLO v4 with custom selection of anchor boxes using K-means++ clustering algorithm outperformed as compared to other detection methods in terms of accuracy, which is evaluated using COCO metrics, training and detection time. All the experiments are performed on the Airbus detection dataset from https://www.kaggle.com/c/airbus-ship-detection.

Keywords: Ship detection · Convolution neural network · YOLO · Satellite imagery

1 Introduction

Transport has always been an important factor for development throughout history, as a result, the marine transportation business is currently evolving at a breakneck speed to deliver convenience, luxury, or better ways for importing-exporting goods. The maritime surveillance, monitoring and traffic supervision has grown as an important research subject in the field of remote sensing and computer science to deal with many security applications. Therefore, Earth Observation satellites represent a potentially important source for near real-time detection of small objects.

All the objects on Earth absorb or reflect a particular range of frequencies called as spectral signatures. Satellite instruments such as camera and scanners record these signatures and give significant information about each frequency band that passed through the atmosphere to satellite instrument and generate satellite images. These images can be used to map or investigate different surfaces and objects. But sometimes these images need some processing by manual efforts. To make the task faster and more automated,

B. Raman et al. (Eds.): CVIP 2021, CCIS 1568, pp. 273–287, 2022.
https://doi.org/10.1007/978-3-031-11349-9_24

machine learning and computer vision algorithms have been used in this study. In this work, we have implemented following object detection models. YOLO v3 [45] and v4 [43] because of the growing popularity as they are pretty fast and accurate and have achieved and improved upon the state-of-the-art benchmarks. Also implemented the RetinaNet [51] because it has proven to work well with small and dense scale objects and our dataset has 60.062% small ship objects. Also implemented the Faster R-CNN [52] because it is a two-stage detector that has region proposal network that improves its detection performance and has also achieved state-of-the art benchmarks. EfficientDet D2 [50] created by Google Brain team that is considered highly efficient and accurate. Thus, we selected these models for the performance evaluation on Airbus ship dataset based on the COCO metrics, training and detection time. First set of experiment is carried out using the baseline models. Based on the performance evaluation YOLOv4 model is selected and further set of experiments are done. Clustering algorithms like spectral [53], agglomerative [54] and k-means++ [48] are implemented to generate anchor boxes that further improved the performance of YOLOv4. Based on the experiments YOLOv4 with k-means++ clustering outperformed rest of the methods.

2 Related Work

Machine learning and artificial intelligence have recently gained a lot of attention and have had a lot of success in a variety of fields, including computer vision, image segmentation, natural language processing, and object detection. On the other hand, due to increase in launch of satellites and availability of high-resolution data, aerial remote sensing technology have also developed rapidly and provided detailed information about the targets. Ship detection for maritime surveillance is important to detect any kind of movements on the water surface and warn against potential threats and disasters. In recent years, there have been many technologies developed for ship detection which included a combination of deep learning methods and satellite imagery and proved to be helpful for controlling water movements.

There are chances of high success rate while performing deep learning-based methods for object detection, if the dataset is labelled which is usually the important part, as collecting such datasets either are expensive and difficult to obtain or they are available with low labelling accuracy. The Airbus dataset [1] used in this paper was uploaded to a Kaggle competition in 2018 is the largest sample set in the ship detection research community and contains JPEG images captured by SPOT [2] satellite imagery of 1.5 resolution in RGB bands. It also provides the annotation for ship location with masks for rotated bounding boxes.

The goal of Kaggle challenge 2018 was to analyze satellite photos of container ships and create ship segmentation masks. Among traditional methods for ship segmentation, random forest gives better results [3] but in recent years, with rapid progress in machine learning, especially in deep learning a notable effort has been made in the identification of targets using Convolutional Neural Networks (CNNs).

Along with CNNs, lots of state-of-the-art models based on deep CNNs have also developed with time such as AlexNet [4], ResNet [5] and DenseNet [6] which have achieved good results for classification and detection. One of the excellent models is

Mask Region based Convolutional Neural Network (R-CNN) [7, 9–11] which aim to solve instance segmentation problem with Feature Pyramid Network (FPN) [8–12] as feature extractor. FPN [8] is a feature extractor which uses bottom up and top down pathway designed without compromising accuracy and speed of the model.

Nie M. et al. [9] adopted a voting mechanism to ensure the more accuracy of Mask R-CNN with FPN. Štepec, D. et al. [10] modified the faster RCNN model using FPN and a residual learning framework and ResNet-50 as backbone architecture.

Li, L. et al. [13] used a CNN based domain adaption method based on image level as well as instance level to deal with large domain shifts. Experiments showed that the proposed method works on both supervised and unsupervised domain adaption scenarios effectively. A new cascaded CNN model CCNet proposed by Zhang, Z. X. et al. [14] contains two independent models, an ROI extraction model (REM) and an ROI detection model (RDM) and achieves high detection speed and precision compared to the Mask R-CNN. Huang et al. [15] proposed Fast Single-shot ship instance segmentation called SSS-Net to predict the position of the target and its mask.

In remote sensing images, to separate small objects such as ships from its background covering large portion of the image, image segmentation is needed which is basically an image processing approach to break down into segments of small regions. Polat et al. [16] performed semantic segmentation with transfer learning to detect ships. Xia, X., et al. [17] combined U-Net [18] for segmentation with ResNets which performed better than U-Net34. Hordiiuk et al. [19] used a Xception [20] classifier and baseline U-Net model with ResNet for segmentation. De Vieilleville, F. et al. [21] proposed a distillation method to reduce the size of the network. Smith, B. et. al. [22] used U-Net model with additional layer to enhance system performance. Talon, P. et al. [23] chose semantic segmentation using a custom U-Net architecture with EfficentNet-B0 [24] as backbone over deeper CNNs. Experiments by Karki, S. & Kulkarni, S. [25] suggested that the U-Net with EfficientNet encoder is good choice for detection of big ships but the Fastai [26] U-Net model with GoogleNet [27] encoder model generates better ship segmentation masks, even for smaller and closer ships.

A new anchor-free method was proposed by Chen et al. [28] to identify arbitrary oriented ships in the natural images. Rogers, C. [29] proposed a method to learn the effects of perturbations on image classifications and its relationship with robustness of the deep learning models. Hu J. et al. [30] proposed a novel and stable approach derived from visual saliency to detect multiple targets with more accuracy and speed. The method included saliency maps based on background prior and foreground constraint. Xu, W. et al. [31] used a multi-scale deep residual network to solve the problem.

The maritime objects are small and cover a very small fraction of the entire image. Considering this, Duan, Y. et al. [32] also proposed EEG signals [33] based object detection algorithm using IoT [34] devices. The method used fewer samples which reduces human subjective bias related to the cognitive abilities. Zhong Z. et al. [35] proposed a ship detection method based on local features like shape and texture. The method considers Local Binary Pattern (LBP) [36] algorithm as a feature vector for ship classification and LightGBM [37] classifier to identify and confirm the candidate region. Inspired from CenterNet [38] Wang J. et al. [39] proposed a multiple center points-based learning network M-CenterNet using Deep Layer Aggregation (DLW) [40] for feature extraction to perform better localization for tiny objects.

Cordova, A. W. A. et al. [41] proposed a novel technique for ship detection using You Only Look Twice (YOLT) [42] and YOLOv4 [43] where transfer learning was used to speed up various training sessions. Lisbon, P. T. [44] used a two-phased approach that applies image detection using You Only Look Once (YOLO) [45] network to identify possible locations of the ship, and then, semantic segmentation using U-Net for real-time detection. Mohamed, E. et. al. [46] used YoloV3 model with pre-trained ResNet-50 encoder. Ramesh, S. S. et. al. [47] combined YOLT with inception model and Li, X., & Cai, K. [12] proposed an improved YOLO V3 algorithm using FPN optimization and sliding segmentation with k mean algorithm to deal with the problem of missing ships in remote sensing images.

3 Training Data

The Airbus Dataset [1] from the Kaggle Challenge (2018) consisted of 1,92,556 labeled images, each having a resolution of 768 × 768 pixels with labels present in Run Length Encoding (RLE). The data was studied and visualised using python tools such as matplotlib and seaborn. It was found that approximately 79% of images don't contain any ships and rest of the images have number of ships varying between 1 and 15. Thus each image was categorized into a class depending upon the number of ships in that image. There are total 16 classes over which the images are distributed. Since the distribution is skewed (shown in Fig. 1 (a)) and volume of data is also very high therefore a maximum of 1000 images from each class (shown in Fig. 1 (b)) was selected. Class 0 images are ignored for training as it has no ship. Images having area distribution of ship less than 1 percentile are also ignored because they are very tiny extending up to 1 or 2 pixels and

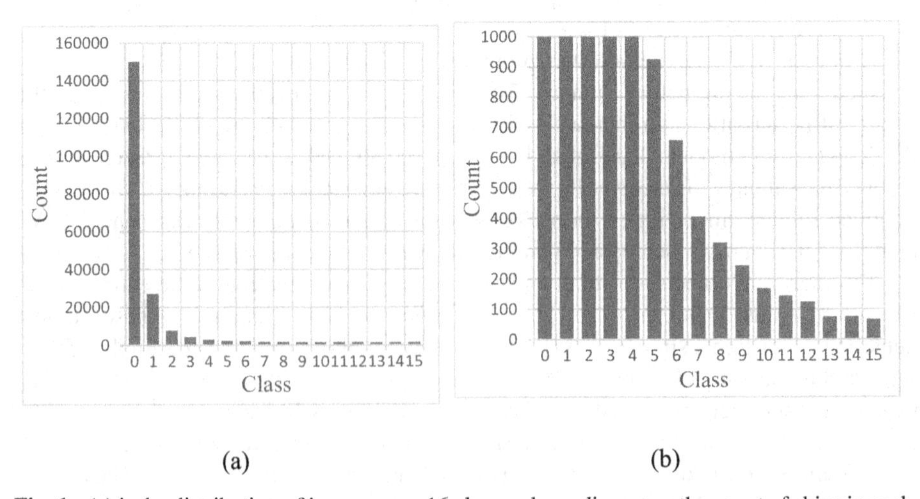

Fig. 1. (a) is the distribution of images over 16 classes depending upon the count of ships in each image and (b) is the balanced distribution by putting an upper bound of 1000 count of images per class

thus may not be helpful during training of the models. We finally used 7202 images out of which 75% (5402 images) are used for training.

4 Methods

Based on object detection (Fig. 2) studies, YOLOv4 [43] is considered fast with high accuracy results. YOLOv4's backbone comprises of bag of freebies, bag of specials and CSPDarknet53, its neck has blocks of Spatial Pyramid Pooling layer (SPP) and Path Aggregation Network (PANet) and the head is same as that of YOLOv3 [45]. YOLOv4 uses 9 anchor boxes as initial bounding boxes for prediction and fine-tune its center co-ordinates' offset, width and height while training. Thus, choosing the anchor boxes plays an important role in object detection. The selection of anchor boxes can be done by considering it as a clustering problem. We have used agglomerative clustering, spectral clustering, and K-means++ [48] clustering on the training data for anchor generation.

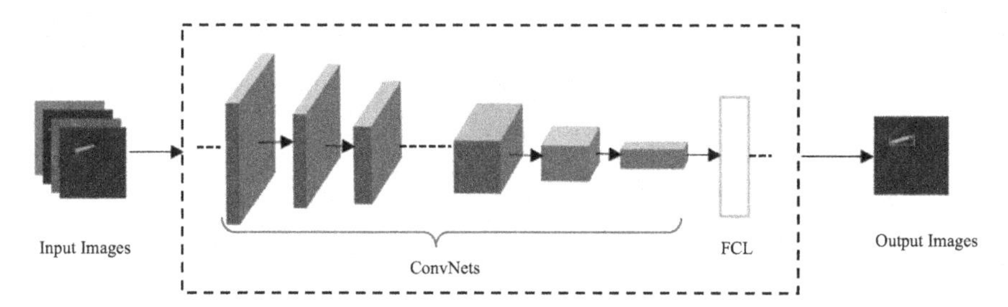

Fig. 2. General architecture for object detection

We have compared various state of the art models namely EfficientDet-D2 [50], published by Google Brain Team, build over EfficientNet with feature fusion using Bidirectional Feature Pyramid Network (BiFPN); RetinaNet152 which has the same architecture of RetinaNet50 or RetinaNet101 [51] but instead has ResNet152 as feed-forward network; Faster R-CNN [52] with backbone as ResNet152; YOLOv3 that uses Darknet53 model as backbone and YOLOv4, YOLOv4 with agglomerative clustering, YOLOv4 with spectral clustering and YOLOv4 with k-means++ clustering.

5 Experiments

All the experiments have been performed on Ubuntu 18.04, 64-bit operating system having Intel Xeon W-21155 CPU, NVIDIA Quadro P4000 with GPU memory of 8192 MB and CUDA toolkit.

Evaluation is done using Common Objects in Context (COCO) metrics based on various mean average precision (mAP) scores (refer Table 2). A higher mAP is an indication of better detection performance. For all models, pre-trained weights on MS COCO dataset are used as initial weights.

For YOLO models, the labeling of dataset was converted from RLE to YOLO format, which is a tuple of 5 values namely object class id, object center coordinate x, object center coordinate y, object width and object height for each ground truth bounding box. In the first experiment, we used YOLOv3 and the results were acceptable. Next, the experiment performed with YOLOv4 with anchors used for pre-trained weights and results improved significantly as compared to that of YOLOv3.

Next, the experiments are performed using EfficientDet-D2, RetinaNet152 and Faster R-CNN. Tensorflow object detection API was used. The labeling of dataset was converted from RLE to tensorflow records. During initial iterations of experimentation of each model, a sudden explosion in loss was noted, which was due to very high learning rate. Thus, for EfficientDet-D2, RetinaNet152 and Faster R-CNN, learning rates are set as 8e-3, 5e-4, 10e-4 respectively. The Backbone, hyperparameters and training time of models are given in Table 1.

Table 1. Backbone, hyperparameters and training time of models

Models	Backbone	Network resolution	Batch size	Sub divisions	Learning rate	Warm up steps	Epochs	Training time (hrs.)
Yolo_v3	Darknet53	416 × 416	64	16	0.001	1000	47	12
RetinaNet152	ResNet152+FPN	1024 × 1024	1	–	0.0005	2000	29	61
EfficientDet-D2	EfficientNet+BiFPN	768 × 768	2	–	0.008	2500	92	74
Faster R-CNN	ResNet152	1024 × 1024	2	–	0.001	2000	18	28
Yolo_v4	CSPDarknet53	608 × 608	64	64	0.001	1000	47	22

From the above experimentations, it can be observed that the performance of Faster-RCNN and YOLOv4 are good and comparable, with Faster-RCNN being slightly better (refer mAP@[0.5:0.95] in Table 2). But in terms of training speed, Faster-RCNN is very slow as compared to YOLOv4. Faster R-CNN took around 28 h to run 18 epochs, whereas, YOLOv4 took around 22 h to run 47 epochs (refer Table 1). Also, it can be observed that the detection time of YOLOv4 (1390 ms) is approximately 24 times faster than Faster R-CNN (58 ms). Thus, remaining experiments were performed with YOLOv4 in order to improve its accuracy.

YOLOv4 uses 3 anchors in each of the three YOLO layers for prediction, therefore, 9 anchor boxes are required in total. Each anchor is defined by its width and height. To generate 9 anchors, we used clustering algorithms on the data points that are defined by the width and height parameters of the labeled training dataset in YOLO format. Clustering algorithms use either distance or affinity score between two data points on the basis of which clustering is performed. Each data point corresponds to a rectangular box. For all data points, top left co-ordinates are kept as 0,0 because we are only interested in the width and height dimensions. For any two data points, say α and β, the distance formula is defined through in Eq. 1

$$dist(\alpha, \beta) = 1 - IoU(\alpha, \beta) \quad (1)$$

where IoU is called intersection over union and is calculated as a ratio between the area overlapping and total area between two data points. It is given through Eq. 2.

$$IoU(\alpha, \beta) = \frac{\alpha \cap \beta}{\alpha \cup \beta} \quad (2)$$

We compared 3 different clustering techniques for anchor generation for YOLOv4. Previous YOLOv4 experiment's best trained weights are used as initial weights for the experiment with each clustering technique to fine-tune the weights which eventually improve the results.

First, we used K-means++ [48] clustering which is the standard K-means algorithm [49] coupled with smarter initialization of centroids that improves the quality of clusters. The centroids of 9 clusters are selected as 9 anchors. It uses distance score (refer Eq. 1) between the data points for clustering. Nine anchors generated using K-means++ clustering are [(7, 8), (14, 15), (34, 19), (22, 44), (61, 28), (50, 54), (130, 56), (68, 113), (145, 132)]. The visualization of clusters and anchors obtained through K-means++ clustering, is shown in Fig. 3.

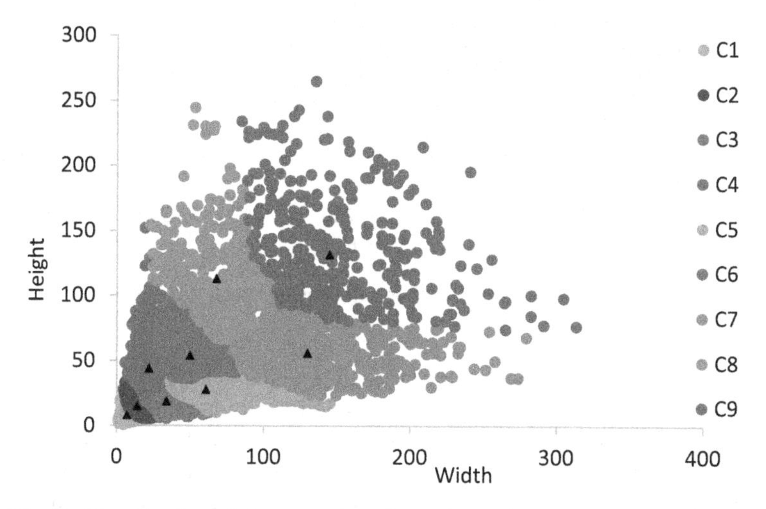

Fig. 3. Visualization of clusters and anchor points using K-means++ clustering

Next, we used Spectral clustering [53] which considers each data point as graph node and transforms the problem as graph partitioning problem. The weights between each node corresponds to affinity score. IoU between data points is used as the affinity score (refer Eq. 2.). The mean values of width and height of the 9 clusters each (refer Eq. 3 and Eq. 4), are chosen as anchor points. The 9 anchor points generated by Spectral clustering are [(4, 5), (7, 11), (12, 7), (14, 16), (37, 18), (24, 40), (78, 37), (50, 71), (134, 121)].

$$\forall C, anchor_w^C = \frac{1}{N_C} \sum\nolimits_{i=1}^{N_C} w_i \tag{3}$$

$$\forall C, anchor_h^C = \frac{1}{N_C} \sum\nolimits_{i=1}^{N_C} h_i \tag{4}$$

where $C \in \{c_1, c_2, c_3, c_4, c_5, c_6, c_7, c_8, c_9\}$ i.e., 9 clusters, N_C is the number of data points in cluster C, w_i and h_i are width and height of ith data point in cluster C. The visualization of clusters and anchors obtained through spectral clustering is shown in Fig. 4.

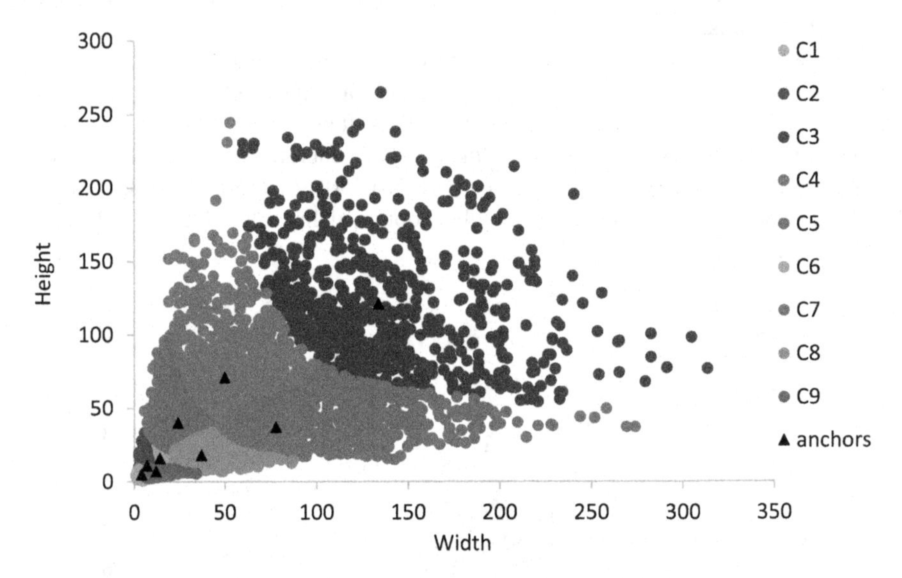

Fig. 4. Visualization of anchor points using spectral clustering

Finally, we used agglomerative clustering [54] which is a type of hierarchical clustering that performs clustering in bottom-up fashion by grouping the data points. It also uses distance score (refer Eq. 1) between the data points to perform clustering. We used average linkage for the grouping of two clusters to form a new cluster. Once 9 clusters are formed, the mean values of width and height of each cluster are chosen as anchor points (refer Eq. 3 and Eq. 4). The 9 anchor points generated by agglomerative clustering are [(1, 5), (4, 2), (5, 6), (20, 48), (47, 36), (132, 33), (120, 109), (58, 235)]. The visualization of clusters and anchors obtained through agglomerative clustering is shown in Fig. 5.

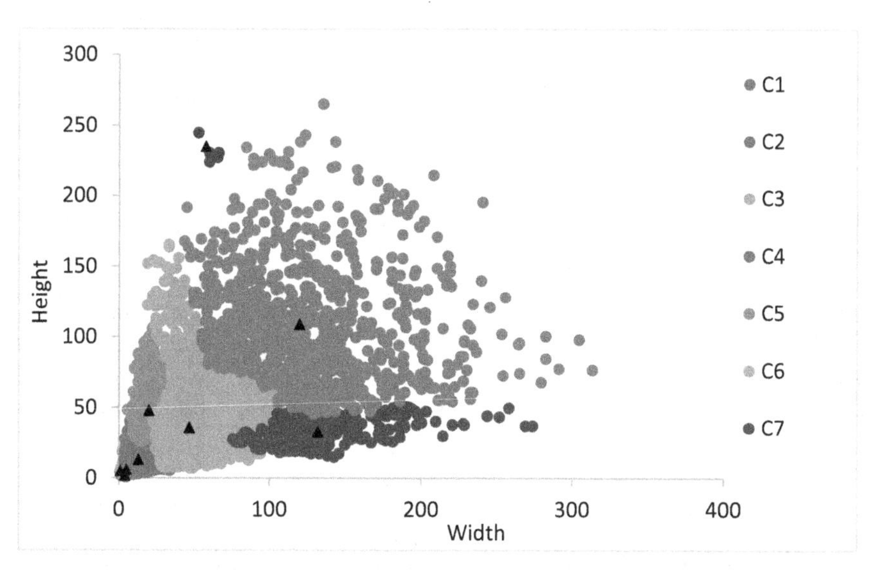

Fig. 5. Visualization of anchor points using agglomerative clustering

6 Evaluation and Results

The COCO metrics [55] are used for evaluation on 25% test data (1800 images). The Mean average precision (mAP) is calculated over 10 IoU thresholds starting from 0.5 till 0.95 with step size of 0.05. The mAP small is calculated for ships having bounding box area less than 32 sq. pixels. mAP medium is calculated for ships having bounding box area between 32 sq. pixels and 96 sq. pixels and mAP large is calculated for ships having bounding box area greater than 96 sq. pixels. mAP@0.5 and mAP@0.75 are mAP values at IoU threshold 0.5 and 0.75 respectively. The average precision of a single class object is defined in Eq. 5.

$$AP = \frac{1}{101} \sum\nolimits_{Recall_i} Precision\,(Recall_i) \tag{5}$$

where $Recall_i \in \{0, 0.01, 0.02, 0.03\ldots\ldots 1\}$ i.e., 101 values ranging from 0 to 1.

If there are N classes of objects, then mean average precision (mAP) threshold is defined by Eq. 6.

$$mAP = \frac{1}{N} \sum\nolimits_{i=0}^{N-1} AP_i \tag{6}$$

where AP_i is average precision of ith class. Since in our case, $N = 1$ thus refer Eq. 7.

$$mAP = AP \tag{7}$$

The comparisons of stated methods are done on the basis of COCO metrics and detection time (in ms) which measures how much time has elapsed to process the given input and generate the output image (Refer Table 2). The best results for each comparison parameter are highlighted in bold.

Table 2. Model comparison on the basis of COCO metrics and Detection time

Models	Clustering	mAP@[0.5:0.95] (%)	mAP@0.5 (%)	mAP@0.75 (%)	mAP small (%)	mAP medium (%)	mAP large (%)	Detection time (ms)
Yolov3	–	30.23	66.57	24.69	14.172	50	57.24	**30**
RetinaNet152	–	35.1	59.2	35.9	16.8	56.6	69.6	962
EfficientDet-D2	–	39.48	68.09	40.37	16.96	64.95	77.36	800
Yolov4	–	48.87	81.92	51.86	28.58	66.28	71.86	58
Faster R-CNN	–	49.99	81.15	53.11	30.96	**71.37**	**83.36**	1390
Yolov4	Agglomerative	50.24	85.02	52.36	31.14	65.62	72.76	58
Yolov4	Spectral	50.55	86.01	53.19	33.72	62.86	70.42	58
Yolov4	K-means++	**53.39**	**86.22**	**57.53**	**34.02**	67.86	72.57	58

Based on the results, YOLOv4 with K-means++ clustering performs the best in terms of mAP@[0.5:0.95], mAP@0.5, mAP@0.75 and mAP small with performance values of 53.39%, 86.22%, 57.53% and 34.02% respectively. Faster R-CNN gave the best results for mAP of medium size ship detection with a value of 71.37%. Faster R-CNN also gave the best results for large ship detection with a value of 83.36% followed by second best EfficientDet-D2 with a value of 77.36%. YOLOv4 with K-means++ clustering has a score of 72.57% which is less than Faster R-CNN for mAP large. It is because of high false positives as compared to Faster R-CNN but it still depicts significantly good performance. In terms of detection time, YOLOv3 is the fastest of all with detection time of 30 ms but has poor performance among all. YOLOv4 is the second best in terms of detection time which is 58 ms and very fast as compared to RetinaNet152, EfficientDet-D2 and Faster-RCNN. 60.062% of data consists of small ship objects, 34.766% of medium and only 5.172% consists of large ship objects. Majority of data has small ships and YOLOv4 with K-means++ clustering has the highest value of mAP small and a good score for medium and large thus, YOLOv4 with K-means++ clustering performs significantly better than the rest of the stated methods in terms of speed and accuracy. K-means++ clustering is a centroid based clustering. Thus, grouping done by K-means++ seems to be a good choice because we will get clustering around the centroids. Selecting these centroids as anchors will help the model to better learn and fine tune them to generate output bounding box that could cover maximum data points belonging within a cluster during training thus improving the performance of model. This is also proven in the experimentation results.

The sample detections from eight different methods are shown in Fig. 6. The first two rows are Ground Truth Bounding Boxes and Ground Truth Masks respectively. From 3rd row till last are the detections using eight methods in the increasing order of their mAP@[0.5:0.95] (refer Table 2). There are seven different comparison images. First two columns are large images as the ship object bounding box area is greater than 96 sq. pixels. Next two columns are medium images as the ship object bounding box area is between 32 sq. pixels and 96 sq. pixels. The last three columns are small images as the ship object bounding box area is less than 32 sq. pixels.

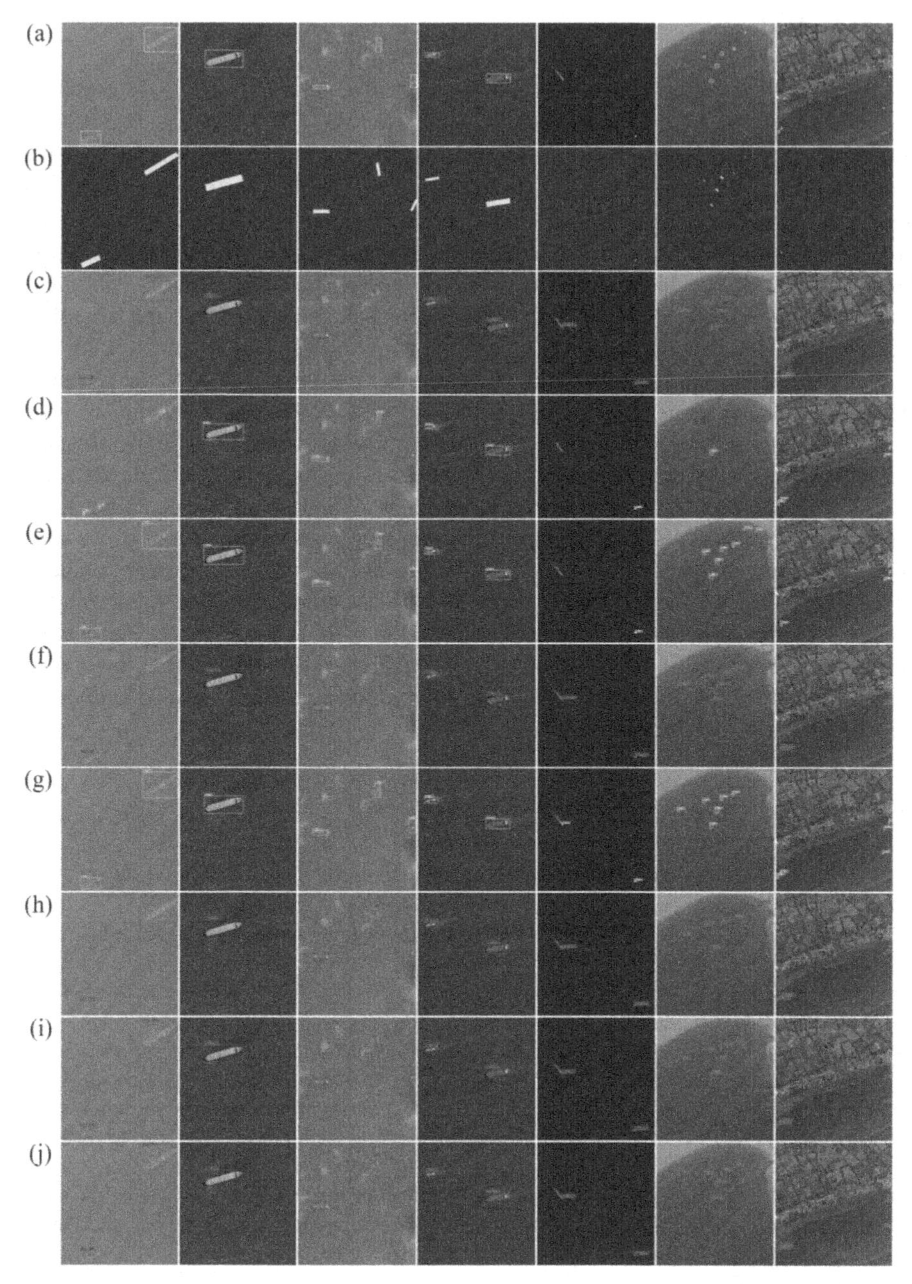

Fig. 6. Seven detection instances (columns) using eight different methods(rows) from 3rd row till last in increasing order of their mAP@[0.5:0.95]. Ground Truth Bounding Boxes (a), Ground Truth Masks (b), YOLOv3 (c), RetinaNet152 (d), EfficientDet-D2 (e), YOLOv4 (f), Faster R-CNN (g), YOLOv4 + agglomerative (h), YOLOv4 + spectral (i), YOLOv4 + k-means + + (j).

7 Conclusion

RetinaNet152 performed poor in detecting large ships. It was more inclined towards localizing smaller areas for small object detection (refer fourth row & first column of Fig. 6). It also missed out the detections for medium and small ships and has low confidence score. EfficientDet-D2 also missed out detections for medium and small ships and has a low confidence score for localized ship area. Faster R-CNN has a decent confidence score for small, detected objects while it performed very well for medium and large objects. YOLOv3 has a decent confidence score but has an overall poor performance on the validation set. YOLOv4 made good detections for small, medium, and large ships with a good confidence score. Faster R-CNN seems slightly better than YOLOv4 but it is very slow in terms of training as well as detection time. Thus, for the choice of object detection in terms of speed and accuracy, YOLOv4 seems a good choice. There is an improvement in the performance of YOLOv4 with the use of agglomerative and spectral clustering for anchor selection. Both the clustering techniques performed equally good. Spectral clustering shows better improvement in mAP small. Finally, K-means++ clustering for anchors selection significantly improves the results and confidence score of YOLOv4 detection. It shows good percentage increase in mAP scores compared to baseline YOLOv4 method. Thus, YOLOv4 with custom anchors generated using K-means++ clustering performs best as compared to rest of the stated methods. This method seems the best choice in terms of speed and accuracy, given the fact that majority of data consists of small ship objects on which this method performs best and significantly good on medium and large ships.

References

1. Airbus Ship Detection Challenge.: https://www.kaggle.com/c/airbus-ship-detection. Last accessed 22 June 2021
2. European Space Agency: SPOT. https://earth.esa.int/eogateway/missions/spot. Last accessed 29 June 2021
3. Chen, Y., Zheng, J., Zhou, Z.: Airbus Ship Detection-Traditional vs Convolutional Neural Network Approach (n.d.)
4. Krizhevsky, A., Sutskever, I., Hinton, G.E.: Imagenet classification with deep convolutional neural networks. Adv. Neural. Inf. Process. Syst. **25**, 1097–1105 (2012)
5. He, K., Zhang, X., Ren, S., Sun, J.: Deep residual learning for image recognition. In: Proceedings of the IEEE Conference on Computer Vision and Pattern Recognition, pp. 770–778 (2016)
6. Huang, G., Liu, Z., Van Der Maaten, L., Weinberger, K.Q.: Densely connected convolutional networks. In: Proceedings of the IEEE Conference on Computer Vision and Pattern Recognition, pp. 4700–4708 (2017)
7. Girshick, R., Donahue, J., Darrell, T., Malik, J.: Rich feature hierarchies for accurate object detection and semantic segmentation. In: Proceedings of the IEEE Conference on Computer Vision and Pattern Recognition, pp. 580–587 (2014)
8. Lin, T.Y., Dollár, P., Girshick, R., He, K., Hariharan, B., Belongie, S.: Feature pyramid networks for object detection. In: Proceedings of the IEEE Conference on Computer Vision and Pattern Recognition, pp. 2117–2125 (2017)

9. Nie, M., Zhang, J., Zhang, X.: Ship segmentation and orientation estimation using keypoints detection and voting mechanism in remote sensing images. In: International Symposium on Neural Networks, pp. 402–413. Springer, Cham (2019)
10. Štepec, D., Martinčič, T., Skočaj, D.: Automated system for ship detection from medium resolution satellite optical imagery. In: Oceans 2019 MTS/IEEE, Seattle, pp. 1–10, IEEE (2019)
11. Nie, X., Duan, M., Ding, H., Hu, B., Wong, E.K.: Attention mask R-CNN for ship detection and segmentation from remote sensing images. IEEE Access **8**, 9325–9334 (2020)
12. Li, X., Cai, K.: Method research on ship detection in remote sensing image based on Yolo algorithm. In: 2020 International Conference on Information Science, Parallel and Distributed Systems (ISPDS), pp. 104–108, IEEE (2020)
13. Li, L., Zhou, Z., Wang, B., Miao, L., An, Z., Xiao, X.: Domain adaptive ship detection in optical remote sensing images. Remote Sens. **13**(16), 3168 (2021)
14. Zhang, Z.X., et al.: CCNet: a high-speed cascaded convolutional neural network for ship detection with multispectral images. Infrared. Millim. Waves **38**(3), 290–295 (2019)
15. Huang, Z., Sun, S., Li, R.: Fast single-shot ship instance segmentation based on polar template mask in remote sensing images. In: IGARSS 2020–2020 IEEE International Geoscience and Remote Sensing Symposium, pp. 1236–1239, IEEE (2020)
16. Polat, M., Mohammed, H.M.A., Oral, E.A.: Ship detection in satellite images. In: ISASE2018, p. 200 (2018)
17. Xia, X., Lu, Q., Gu, X.: Exploring an easy way for imbalanced data sets in semantic image segmentation. J. Phys. Conf. Ser. **1213**(2) 022003 (2019)
18. Ronneberger, O., Fischer, P., Brox, T.: U-net: convolutional networks for biomedical image segmentation. In: International Conference on Medical Image Computing and Computer-Assisted Intervention, pp. 234–241. Springer, Cham (2015)
19. Hordiiuk, D., Oliinyk, I., Hnatushenko, V., Maksymov, K.: Semantic segmentation for ships detection from satellite imagery. In: 2019 IEEE 39th International Conference on Electronics and Nanotechnology (ELNANO), pp. 454–457, IEEE (2019)
20. Chollet, F.: Xception: deep learning with depthwise separable convolutions. In: Proceedings of the IEEE Conference on Computer Vision and Pattern Recognition, pp. 1251–1258 (2017)
21. De Vieilleville, F., May, S., Lagrange, A., Dupuis, A., Ruiloba, R.: Simplification of deep neural networks for image analysis at the edge. In: Actes de la Conférence CAID 2020, p. 4 (2020)
22. Smith, B., Chester, S., Coady, Y.: Ship detection in satellite optical imagery. In: 2020 3rd Artificial Intelligence and Cloud Computing Conference, pp. 11–18 (2020)
23. Talon, P., Pérez-Villar, J.I B., Hadland, A., Wyniawskyj, N.S., Petit, D., Wilson, M.: Ship detection on single-band grayscale imagery using deep learning and AIS signal matching using non-rigid transformations. In: IGARSS 2020–2020 IEEE International Geoscience and Remote Sensing Symposium, pp. 248–251, IEEE (2020)
24. Tan, M., Le, Q.: Efficientnet: rethinking model scaling for convolutional neural networks. In: International Conference on Machine Learning, pp. 6105–6114, PMLR (2019)
25. Karki, S., Kulkarni, S.: Ship detection and segmentation using Unet. In: 2021 International Conference on Advances in Electrical, Computing, Communication and Sustainable Technologies (ICAECT), pp. 1–7, IEEE (2021)
26. Howard, J., Gugger, S.: Fastai: a layered API for deep learning. Information **11**(2), 108 (2020)
27. Szegedy, C., et al.: Going deeper with convolutions. In: Proceedings of the IEEE Conference on Computer Vision and Pattern Recognition, pp. 1–9 (2015)
28. Chen, J., Xie, F., Lu, Y., Jiang, Z.: Finding arbitrary-oriented ships from remote sensing images using corner detection. IEEE Geosci. Remote Sens. Lett. **17**(10), 1712–1716 (2019)

29. Rogers, C., et al.: Adversarial artificial intelligence for overhead imagery classification models. In: 2019 Systems and Information Engineering Design Symposium (SIEDS), pp. 1–6, IEEE (2019)
30. Hu, J., Zhi, X., Zhang, W., Ren, L., Bruzzone, L.: Salient ship detection via background prior and foreground constraint in remote sensing images. Remote Sens. **12**(20), 3370 (2020)
31. Xu, W., Zhang, C., Wu, M.: Multi-scale deep residual network for satellite image super-resolution reconstruction. In: Chinese Conference on Pattern Recognition and Computer Vision (PRCV), pp. 332–340. Springer, Cham (2019)
32. Duan, Y., Li, Z., Tao, X., Li, Q., Hu, S., Lu, J.: EEG-based maritime object detection for iot-driven surveillance systems in smart ocean. IEEE Internet Things J. **7**(10), 9678–9687 (2020)
33. Haas, L.F.: Hans berger (1873–1941), richard caton (1842–1926), and electroencephalography. J. Neurol. Neurosurg. Psychiatry **74**(1), 9 (2003)
34. Ashton, K.: That 'internet of things' thing. RFID J. **22**(7), 97–114 (2009)
35. Zhong, Z., Li, Y., Han, Z., Yang, Z.: Ship target detection based on Lightgbm algorithm. In: 2020 International Conference on Computer Information and Big Data Applications (CIBDA), pp. 425–429, IEEE (2020)
36. Ojala, T., Pietikäinen, M., Harwood, D.: A comparative study of texture measures with classification based on featured distributions. Pattern Recogn. **29**(1), 51–59 (1996)
37. Ke, G., et al.: Lightgbm: a highly efficient gradient boosting decision tree. Adv. Neural. Inf. Process. Syst. **30**, 3146–3154 (2017)
38. Zhou, X., Wang, D., Krähenbühl, P.: Objects as points (2019)
39. Wang, J., Yang, W., Guo, H., Zhang, R., Xia, G.S.: Tiny object detection in aerial images. In: 2020 25th International Conference on Pattern Recognition (ICPR), pp. 3791–3798, IEEE (2021)
40. Yu, F., Wang, D., Shelhamer, E., & Darrell, T.: Deep layer aggregation. In: Proceedings of the IEEE Conference on Computer Vision and Pattern Recognition, pp. 2403–2412 (2018)
41. Cordova, A.W.A., Quispe, W.C., Inca, R.J.C., Choquehuayta, W.N., Gutierrez, E.C.: New approaches and tools for ship detection in optical satellite imagery. J. Phys. Conf. Ser. **1642**(1), 012003 (2020)
42. Van Etten, A.: You only look twice: rapid multi-scale object detection in satellite imagery. arXiv preprint arXiv:1805.09512 (2018)
43. Bochkovskiy, A., Wang, C.-Y., Liao, H.-Y. M.: YOLOv4: optimal speed and accuracy of object detection (2020)
44. Lisbon, P.T.: Ship Segmentation in Areal Images for Maritime Surveillance (n.d.)
45. Redmon, J., Farhadi, A.: Yolov3: an incremental improvement (2018)
46. Mohamed, E., Shaker, A., Rashed, H., El-Sallab, A., Hadhoud, M.: INSTA-YOLO: Real-time instance segmentation (2021)
47. Ramesh, S.S., Kimtani, M.Y., Talukdar, M.Y., Shah, M.A.K.: Ship detection and classification of satellite images using deep learning (n.d.)
48. Arthur, D., Vassilvitskii, S.: k-means++: the advantages of careful seeding. Stanford (2006)
49. Jin, X., Han, J.: K-means clustering. In: Sammut, C., Webb, G.I., (eds.), Encyclopedia of Machine Learning, pp. 563–564 (2010)
50. Tan, M., Pang, R., Le, Q.V.: Efficientdet: scalable and efficient object detection. In: Proceedings of the IEEE/CVF Conference on Computer Vision and Pattern Recognition, pp. 10781–10790 (2020)
51. Lin, T.-Y., Goyal, P., Girshick, R., He, K., Dollar, P.: Focal loss for dense object detection. In: Proceedings of the IEEE International Conference on Computer Vision (ICCV), pp. 2980–2988 (2017)
52. Ren, S., He, K., Girshick, R., Sun, J.: Faster R-CNN: towards real-time object detection with region proposal networks. IEEE Trans. Pattern Anal. Mach. Intell. **39**(6), 1137–1149 (2016)

53. Ng, A.Y., Jordan, M.I., Weiss, Y.: On spectral clustering: analysis and an algorithm. In: Advances in Neural Information Processing Systems, pp. 849–856 (2002)
54. Zepeda-Mendoza, M.L., Resendis-Antonio, O.: Hierarchical agglomerative clustering. Ency. Syst. Biol. **43**(1), 886–887 (2013)
55. Lin, T.-Y., et al.: Detection evaluation. https://cocodataset.org/#detection-eval. Last accessed 2 July 2021

COVID-19 Social Distance Surveillance Using Deep Learning

Praveen Nair, Uttam Kumar(✉), and Sowmith Nandan

Spatial Computing Laboratory, Center for Data Sciences, International Institute of Information Technology Bangalore (IIITB), 26/C, Electronics City Phase-1, Bangalore 560100, India
{praveen.nair,uttam,sowmith.nandan}@iiitb.ac.in

Abstract. COVID-19 disease discovered from the novel corona virus can spread through close contact with a COVID-19 infected person. One of the measures advised to contain the spread of the virus is to maintain social distancing by minimizing contact between potentially infected individuals and healthy individuals or between population groups with high rates of transmission and population groups with no or low-levels of transmission. Motivated by this practice, we propose a deep learning framework for social distance detection and monitoring using surveillance video that can aid in reducing the impact of COVID-19 pandemic. This work utilizes YOLO, Detectron2 and DETR pre-trained models for detecting humans in a video frame to obtain bounding boxes and their coordinates. Bottom-centre points of the boxes were determined and were then transformed to top-down view for accurate measurement of distances between the detected humans. Based on the depth of each bottom-centre point estimated using monodepth2, dynamic distance between pairs of bounding boxes and corresponding distance threshold (safe distance) to prevent violation of social distancing norm were computed. Bounding boxes which violate the distance threshold were categorized as unsafe. All the experiments were conducted on publicly available Oxford Town Center, PETS2009 and VIRAT dataset. Results showed that Detectron2 with top-down view transformation and distance thresholding using pixel depth estimation outperformed other state-of-the-art models. The major contribution of this work is the estimation and integration of variable depth information in obtaining the distance threshold for evaluating social distances between humans in videos.

Keywords: COVID-19 · Social distance · Object detection · Perspective transformation · Depth estimation · Detectron2 · Monodepth2

1 Introduction

Coronavirus disease (COVID-19) is an infectious disease caused by a newly discovered coronavirus. COVID-19 has affected many countries worldwide from December 2019 since its origin from Wuhan, China. World Health Organization (WHO) announced it a pandemic as the virus spread through many countries causing many fatalities. COVID-19 virus spreads primarily when an infected person coughs or sneezes, and the droplets from their nose or mouth disperse through the air, affecting nearby people. Individuals

B. Raman et al. (Eds.): CVIP 2021, CCIS 1568, pp. 288–298, 2022.
https://doi.org/10.1007/978-3-031-11349-9_25

who are infected can be a carrier of the virus and may spread them without showing any visible symptoms. Therefore, it is necessary to maintain social distancing from others, even if people do not have any symptoms of COVID-19.

Social distancing is suggested as the best spread curtailer in the present scenario along with wearing masks, and all affected countries went under complete lockdown to implement social distancing. It aims to decrease or interrupt transmission of COVID-19 in a population by minimizing contact between potentially infected individuals and healthy individuals, or between population groups with high rates of transmission and population groups with no or low levels of transmission. It has been suggested that maintaining a distance of approximately 2 m from another individual results in a marked reduction in transmission of most flu virus strains, including COVID-19. But there are many people who ignore or neglect public health advisories such as social distancing in these difficult times. Hence, there is a need for continuous effort on development of methods that facilitate automated detection of social distance violation.

Harith et al. [1] proposed a social distance evaluation system to reduce physical contact between people in restricted areas through triggered warnings. A pre-trained CNN (convolutional neural network) based SSD (single shot detector) MobileNet v1 COCO model was selected for object detection and bounding box estimation, and Euclidean distances between centroid of bounding boxes were compared against fixed threshold. The approach showed 100% accuracy for self-taken video, 62.5% for Town Centre, 68% for PETS2009 and 56.5% for VIRAT dataset. Person detection in restricted area accuracy was close to 95% for CamNet dataset. A social distance evaluation framework [2] based on a deep learning platform included an overhead perspective that explored pre-trained CNN model. YOLOv3 and transfer learning models were selected for object detection and class probabilities estimation that achieved detection accuracy of 92% and 95% with and without transfer learning respectively. Houl et al. [3] proposed a social distancing detection system using deep learning to evaluate the distance between people in order to reduce the impact of the coronavirus effect. Pre-trained YOLOv3 model trained on COCO dataset was used for object detection with top-down view transformation and distance measurement. However, it lacks visual representation of the results in terms of precision, accuracy/speed and social distance detection. A social distancing monitoring system [4] via fine-tuned YOLOv3 and deep sort techniques were proposed through a pre-trained object detection model to segregate humans from the background. A tracking algorithm was used to trace the identified people using bounding boxes to generate a social distancing violation index using YOLOv3 with deep sort algorithm. Pairwise vectorized L2 norm was computed for distance calculation. Shorfuzzaman et al. [5] proposed a deep learning-based framework for social distance video surveillance in real time towards the sustainable development of smart cities through mass video surveillance and a response to the COVID-19 pandemic. Pedestrian detection using pre-trained object detection models such as YOLOv3, Faster R-CNN and SSD were used. Birds eye view transformation of detected bounding box of pedestrians and distance between every pair of detected bounding boxes of pedestrians below a minimum fixed threshold was highlighted as a social distance violation that triggered non-intrusive alerts. To summarize, previous studies have suggested that YOLOv3 is the best model with a fair trade-off between accuracy and the speed of detection. However, most literature suggest the selection of optimized and improved models for detection and tracking of objects such as mask detection and human body temperature detection for increasing the speed and accuracy along with top-down view transformation. They also lack implementation and validation of models in varying environments such restaurant, school and offices. Therefore, in this work, we propose an automated method using a deep learning model for detection of humans who violate social distancing norm. The objectives of this work are

i) To detect humans from surveillance video/images using pre-trained object detection models.
ii) To assess social distancing between detected humans.
iii) To analyze the effect of depth information in measuring social distancing.

The paper is organized as follows: Sect. 2 details the methodology with respect to (1) human detection, (2) top-down view transformation, (3) distance calculation, and (4) pixel depth-based distance thresholding and social distance evaluation. Section 3 details the dataset and evaluation metrics, while results are presented and discussed in Sect. 4 with concluding remarks in Sect. 5.

2 Methodology

The major steps involved in social distance evaluation experiment are listed below. A brief overview of the system is shown in Fig. 1 followed by a detailed description.

1. Human detection in the form of bounding boxes with various object detection models in an image frame.
2. Top-down view transformation of the detected bounding box ground points.
3. Computation of distance between every pair of detected humans in the frame.
4. Pixel depth estimation of the ground points to set the distance threshold.
5. Finding and marking the detected bounding boxes as unsafe which are below the distance threshold or marking them safe otherwise.
6. Evaluation of the model's accuracy.

2.1 Human Detection

In the first step, pedestrians from the video frame are detected with YOLO [7–9], Detectron2 [10] and DETR [11] object detection models trained on real-world datasets. Multiple detection of a pedestrian results in multiple overlapping bounding boxes, where the unwanted boxes were removed using non-max suppression. Class based filtering was used to ensure that only humans are detected. It is to be noted that each detection model returns bounding box coordinates, confidence and class of the prediction in different format. For example, YOLO returns bounding box centre coordinates along with width and height from which top left corner and bottom right corner coordinates can be derived whereas Detectron2 model returns top left corner and bottom right corner coordinates of the bounding box. Figure 2 shows the output of an object detection model highlighting only humans.

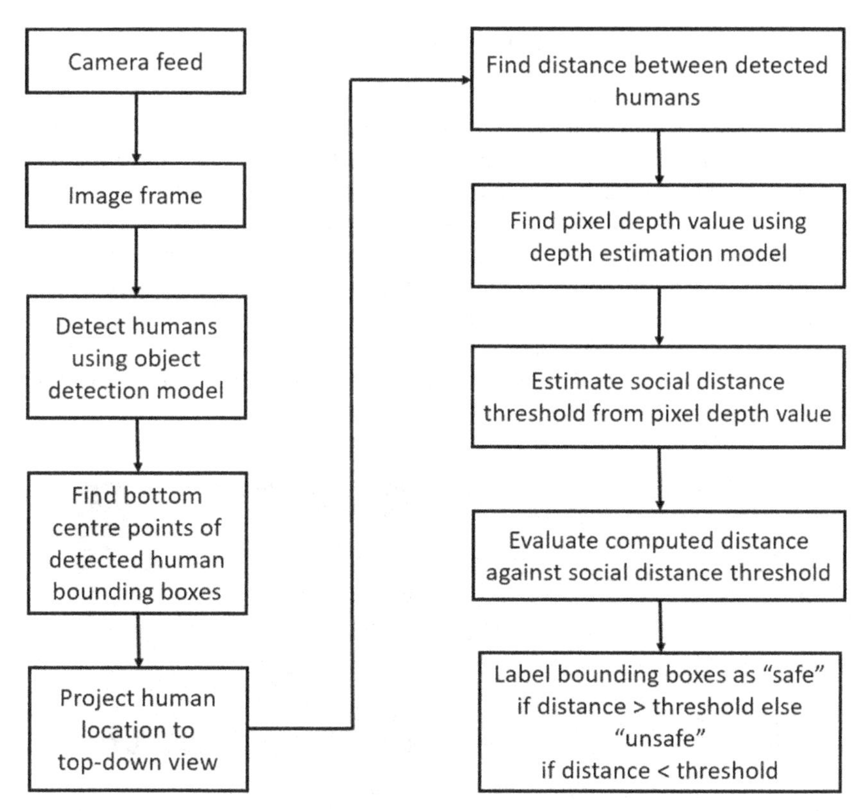

Fig. 1. Overall methodology of social distance evaluation.

Fig. 2. Human detection shown as bounding boxes in sample video frame (image).

2.2 Top-Down View Transformation

A top view or perspective transformation [6] of the scene is used to transform an image taken from a perspective to a top view. This allows to define the constraints of the person's location while comparing them with the next nearest person in the frame (performed using OpenCV). Four corner points on the original source image and on the destination image that form a rectangle with at least two opposite parallel sides were used in the transformation that were wrapped with the original image. The coordinates for each ground points were transformed using a transformation matrix. In perspective view, the distance is not the same when people are in different planes, at least not at the same distance from the camera. So, top view transformed points are preferred over the original ground points, which also improves social distancing evaluation. Figure 3 shows the top-down view warped image of an input frame.

Fig. 3. Perspective transformation of an image.

2.3 Distance Calculation

Once the bounding box for the humans were obtained in the image, bird's eye view location of the bottom centre coordinates was calculated from top left and bottom right corner coordinates. If (x_1,y_1) and (x_2,y_2) represents the top left and bottom right corner points, then bottom ground centre points $D(x,y)$ is

$$D(x,y) = \left(\frac{x_1 + x_2}{2}, y_2\right). \tag{1}$$

Next, the bird's eye view distance for every bounding box ground point to all the other detected bounding box points in the frame were computed. Given the ground centre points of two pedestrians in an image, Euclidean distance between them were computed. The pair of detected bounding boxes for which the inter-distance was below the minimum fixed distance (or threshold) was considered a violation of the social distancing norm. This minimum acceptable distance threshold was set based on the pixel depth model as explained in the next section.

2.4 Pixel Depth-Based Distance Thresholding and Social Distance Evaluation

In most of the work related to social distancing evaluation using deep learning models, the minimum acceptable distance threshold is a fixed value in terms of pixels. However, the camera is placed at a particular angle and the inter-pedestrian distance varies as per the initial pedestrian's location in the video. The minimum acceptable distance threshold can be improved by extracting the pixel level depth information as per the pixel depth map that contains information relating to the distance of the surfaces. Depth can also be acquired via dedicated hardware that directly sense depth but these sensors are often expensive, power-consuming and limited to indoor environments. In some cases, depth can be inferred from multiple cameras through the use of multi-view geometry, but building a stereo camera requires significant complexity in the form of calibration, rectification and synchronization. Supervised machine learning methods such as neural networks have also been trained on pairs of images and their depth maps. However, improving the performance of supervised methods beyond certain level of accuracy is difficult due to the challenges associated with acquisition of per-pixel ground-truth depth data at scale. To overcome these limitations, self-supervised machine learning technique like monocular depth estimation was used to estimate depth value of each pixel in the video frame/image. Monodepth2 [12], a supervised model is used here to extract pixel depth value for ground centre points to be used for variable depth-based distance thresholding.

Based on the top-down view transformation, distance calculation and pixel depth-based thresholding, bounding boxes that were below distance threshold were considered unsafe and were marked in red rectangle (labelled as "1") and bounding boxes that were above distance threshold were considered as safe and were marked in blue rectangle (labelled as "0"). Note that green bounding boxes were created manually for safe cases as ground truth as shown in Fig. 4.

3 Dataset and Evaluation Metrics

Dataset used for the experiment were the Oxford Town Centre dataset [14], PETS2009 dataset [15] and VIRAT dataset [16]. The Oxford Town Centre dataset is a CCTV video of pedestrians in a busy down-town area in Oxford used for research and development, and face recognition systems released by Oxford University as part of the visual surveillance projects. It contains video data with a resolution of 1920 x 1080 at 25 FPS with 7500 frames from a semi-crowded town center where people were walking. The PETS (Performance Evaluation of Tracking and Surveillance) 2009 are multisensory sequences containing different crowd activities called S2.L1 walking datasets. This video data has a resolution of 768 $\times$ 576 sampled at 25 FPS and is of a one-minute 20 s duration with 795 frames. The Video and Image Retrieval and Analysis Tool (VIRAT) program is a video surveillance project funded by the Information Processing Technology Office (IPTO) of the Defense Advanced Research Projects Agency (DARPA). Videos were captured by stationary HD cameras (1080p or 720p) and have slight jitter due to wind. The video data used for social distancing evaluation were obtained from the VIRAT Public Video Dataset Release 2.0 (VIRAT Ground Dataset) which is a 24 s video with 584 frames and a resolution of 1280 $\times$ 720 sampled at 25 FPS.

The dataset used in this work do not have ground truth label for social distance evaluation. Therefore, the videos were converted to frames and were then manually labelled as safe or unsafe using bounding boxes with a freeware tool called Yolo Label [13]. It provides mouse click control to draw bounding boxes over objects and allows assignment of label with predefined colors for each bounding box. Table 1 shows the number of frames manually labelled for ground truth for each dataset.

Table 1. Details of ground truth labels.

Dataset	Number of frames labelled for ground truth
Oxford Town Centre	149
PETS2009	228
VIRAT	146

Average precision (AP) was used for calculating the detection accuracy. Intersection over Union (IoU also known as Jaccard Index) was used to distinguish correct and incorrect detections. IoU (with a range between 0 and 1) is defined as the area of the intersection divided by the area of the union of a predicted bounding box. The probability of an anchor box containing an object predicted by a detection model is the confidence score. A detection is a true positive or a false positive based on confidence score and IoU thresholds. If an object detection has a confidence score greater than threshold or the predicted class matches the class of a ground truth or the predicted bounding box has an IoU greater than a threshold (say 0.5) with the ground truth, then the detection is considered as true positive (TP). If any of the latter two conditions are violated, then the detection is considered as a false positive (FP). If the confidence score of a detection which should detect a ground truth is lower than the threshold, then the detection is counted as a false negative (FN) and if the confidence score of a detection that should not detect anything is lower than the threshold, then the detection is considered as a true negative (TN).

Different values of precision and recall are obtained by setting different threshold values for IoU and confidence scores and its relation is shown via a curve with recall on the X-axis and precision on the Y-axis. Average precision is the precision averaged across all unique recall levels from the precision-recall curve. An 11-point interpolation of the precision-recall curve is taken by averaging the precision at a set of eleven equally spaced recall levels. The interpolated precision p_{interp} at some recall level r is defined as the maximum precision found for a y recall level $r' \geq r$

$$p_{interp}(r) = \max_{r' \geq r} p(r'). \tag{2}$$

The average of maximum precision value for all the 11 recall values was computed for a fixed IoU. AP is calculated at various IoU's typically from 0.5 to 0.95 at a step size of 0.05. Mean average precision or MAP is the mean of all AP's computed for various IoU's.

$$AP = \frac{1}{11} \sum\nolimits_{r \in \{0,0.1,0..,0.9,1\}} p_{\text{interp}}(r). \tag{3}$$

Training and testing of the algorithms were performed using TensorFlow library in Python. The models were trained in Google Colab platform with GPUs.

4 Results and Discussion

Tables 2, 3 and 4 show accuracy assessment for the Oxford Town Centre, PETS2009 and VIRAT dataset with various fixed thresholds and depth-based thresholds. Figure 4 shows ground truth, detection using fixed threshold and detection using depth-based threshold for a sample video frame from the three dataset respectively. Notice that in all the three datasets, fixed threshold has wrongly identified and depth based threshold has correctly classified humans with safe and unsafe distances by comparing the corresponding green, red and blue bounding boxes (as highlighted by white arrow and yellow oval shapes in Fig. 4). Tables 2–4 highlight the results (in bold font) obtained from Detectron2 which is the best performing model in terms of accuracy with depth-based distance threshold. The major contribution of the present work is the incorporation of the depth information in estimating the distance threshold, which is an improvement over the fixed thresholding method while evaluating the social distances between people in a public place. One limitation of this method is the pairwise distance computation overhead for crowded scenes. Moreover, the time aspect needs further analysis as violating the social distance for a fraction of second when two individuals cross paths is not the same as violating the social distance for an extended period of time.

Future work will involve improvement in the data labelling process and distance measurement (which was done through visual judgement in this work) for ground truth generation to perform rigorous accuracy measurement. Fine tuning of the depth for distance evaluation, speedy assessment of human detection, and exploration of the benefits of transfer learning are topics of extended research. Social distance evaluation using video surveillance and monitoring at public places will also help in identifying behavior and patterns of people to trigger non-intrusive alerts by authorities.

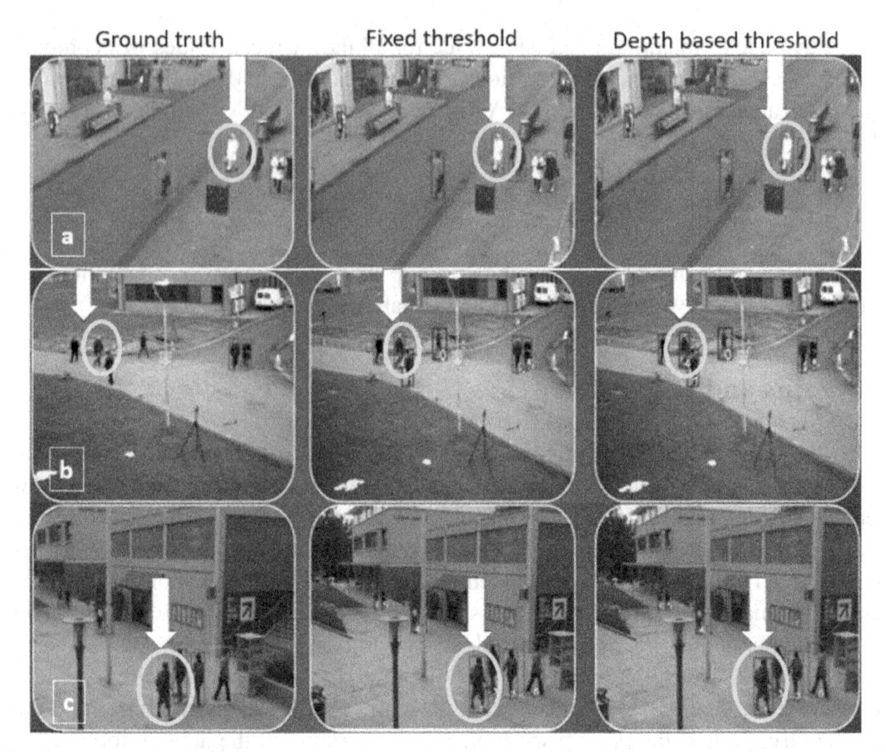

Fig. 4. Ground truth (in green), detection using fixed and depth-based threshold for a sample video frame from the (a) Oxford Town Centre dataset, (b) PETS2009 dataset and (c) VIRAT dataset. (Color figure online)

Table 2. Accuracy results for Oxford Town Centre dataset with various fixed social distance threshold (35, 60 and 90 pixels) against depth based distance threshold. AP is average precision and MAP is mean average precision.

Model	35 pixels		60 pixels		90 pixels		Depth threshold	
	AP	MAP	AP	MAP	AP	MAP	AP	MAP
Detectron2	63.41	36.49	60.23	34.55	44.36	25.99	**71.75**	**39.46**
DETR	60.93	34.54	52.36	31.36	39.73	26.26	62.2	35.97
Yolov3	51.56	29.83	60.31	31.45	40.91	24.48	61.7	32.13
Yolov4	52.41	32.05	61.49	34.44	61.49	34.44	62.99	35.02
Yolov5	52.95	29.98	55.74	33.54	42.52	26.96	56.41	33.4

Table 3. Accuracy results for PETS2009 dataset with various fixed social distance threshold (35, 60 and 90 pixels) against depth based distance threshold. AP is average precision and MAP is mean average precision.

Model	35 pixels		60 pixels		90 pixels		Depth threshold	
	AP	MAP	AP	MAP	AP	MAP	AP	MAP
Detectron2	78.73	55.1	66.9	48.2	54.42	40.44	**78.96**	**55.15**
DETR	78.37	54.91	66.4	48.03	52.42	38.95	78.44	54.91
Yolov3	77.76	51.64	66.11	46.87	51.51	34.94	77.4	51.43
Yolov4	78.37	53.37	65.89	46.31	52.33	38.32	78.39	53.33
Yolov5	78.53	54.43	65.91	47.59	51.79	38.39	78.84	54.98

Table 4. Accuracy results for VIRAT dataset with various fixed social distance threshold (35, 60 and 90 pixels) against depth based distance threshold. AP is average precision and MAP is mean average precision.

Model	35 pixels		60 pixels		90 pixels		Depth threshold	
	AP	MAP	AP	MAP	AP	MAP	AP	MAP
Detectron2	61.96	41.21	60.93	40.2	60.2	39.62	**62.18**	**41.34**
DETR	57.64	34.88	58.23	35.58	57.12	34.55	56.79	34.58
Yolov3	43.85	26.7	53.07	32.46	53.1	32.46	42.12	25.23
Yolov4	44.68	32.34	61.11	40	61.26	40.45	43.21	30.19
Yolov5	50.76	34.75	61.95	40.56	61.98	40.89	42.81	30.9

5 Conclusion

Object detection is an emerging and challenging field in the area of deep learning and computer vision. Recent developments in models such as Detectron2 is a state-of-the-art object detection algorithm developed by Facebook AI Research. DETR or Detection transformer is the most recent end-to-end object detection model using transformer architecture. During social distance evaluation experiments, these pre-trained models were used and compared against state-of-the-art object detection models such as YOLO and the results showed that Detectron2 accuracy exceeded YOLO and DETR for various datasets. Pixel depth were estimated using monocular depth estimation model to obtain distance thresholds based on human's location in the image, which is an improvement over the fixed thresholding method.

Acknowledgements. We are grateful to the International Institute of Information Technology Bangalore (IIITB), India for the infrastructure support. We are thankful to Infosys Foundation for the financial assistance and project grant through the Infosys Foundation Career Development Chair Professor.

References

1. Afiq, A., Norliza, Z., Mohd, F.: Person detection for social distancing and safety violation alert based on segmented ROI. In: 2020 10th IEEE International Conference on Control System, Computing and Engineering (ICCSCE), pp. 113–118 (2020). https://doi.org/10.1109/ICCSCE50387.2020.9204934
2. Imran, A., Misbah, A., Joel, R., Gwanggil, J., Sadia, D.: A deep learning-based social distance monitoring framework for COVID-19. Sustain. Cities Soc. **65**, 102571 (2021). ISSN 2210-6707
3. Yew, C., Mohd, Z., Salman, Y., Sumayyah, D.: Social distancing detection with deep learning model. In: 2020 8th International Conference on Information Technology and Multimedia (ICIMU), pp. 334–338 (2020). https://doi.org/10.1109/ICIMU49871.2020.9243478
4. Narinder, P., Sanjay, S., Sonali, A.: Monitoring COVID-19 social distancing with person detection and tracking via fine-tuned YOLOv3 and deep sort techniques (2020). http://arxiv.org/abs/2005.01385
5. Mohammad, S., Shamim, H., Mohammed, A.: Towards the sustainable development of smart cities through mass video surveillance: A response to the COVID-19 pandemic. Sustain. Cities Soc. **64**, 102582 (2021). ISSN 2210-6707
6. Lin-Bo, K., In-Sung, K., Kyeong-yuk, M., Jun, W., Jongwha, C.: Low-cost implementation of bird's-eye view system for camera-on-vehicle. In: ICCE 2010 - 2010 Digest of Technical Papers International Conference on Consumer Electronics, pp 311–312 (2010). https://doi.org/10.1109/ICCE.2010.5418845
7. Joseph, R., Ali, F.: Yolov3: An incremental improvement. arXiv preprint arXiv:1804.02767 (2018)
8. Alexey, B., Chein-Yao, W., Hong-Yang, L.: Yolov4: Optimal speed and accuracy of object detection, arXiv preprint arXiv:2004.10934 (2020)
9. Glen, J., Alex S., Jirka, B., Ayush, C., Liu, C., Abhiram, V., Jan, H., Laurentiu, D., Yonghye, K.: Ultralytics/yolov5(v5.0), https://doi.org/10.5281/zenodo.4679653Detectron2 (2021). Accessed 16 Oct 2020
10. https://github.com/facebookresearch/detectron2/blob/master/MODELZOO.md. Accessed 16 Oct 2020
11. Nicolas, C., Fransisco, M., Gabriel, S., Nicolas, U., Alexander, K., Sergey, Z.: End-to-End Object Detection with Transformers (2020)
12. Clement, G., Oisin, A., Micheal, F., Gabriel, B.: Digging into self-supervised monocular depth estimation. arXiv preprint arXiv:1806.01260 (2018)
13. Yolo_Label (2020). https://github.com/developer0hye/Yolo_Label. Accessed 16 Oct 2020
14. Oxford Town Centre video data: http://www.robots.ox.ac.uk/ActiveVision/Research/Projects/2009bbenfold_headpose/project.html. Accessed 16 Oct 2020
15. PETS2009 dataset: http://www.cvg.reading.ac.uk/PETS2009/a.html. Accessed 17 Oct 2020
16. VIRAT Video Dataset: https://viratdata.org/. Accessed 18 Oct 2020

Towards Semi-supervised Tree Canopy Detection and Extraction from UAV Images

Uttam Kumar(✉), Anindita Dasgupta, Lingamallu S. N. Venkata Vamsi Krishna, and Pranav Kumar Chintakunta

Spatial Computing Laboratory, Centre for Data Sciences, International Institute of Information Technology Bangalore (IIITB), 26/C, Electronics City Phase-1, Bangalore 560100, India
{uttam,anindita.dasgupta,pranav.kumar}@iiitb.ac.in

Abstract. With unmanned aerial vehicle (UAV) becoming more accessible, remote sensing using UAVs have garnered a lot of attention. UAVs have applications in traffic management, weather monitoring, precision agriculture, orchard management, etc. Now, it is possible to detect and monitor trees from their canopy with the availability of high spatial resolution images acquired from cameras mounted on UAV. Tree canopy detection and counting has been important in orchard management, forest surveys and inventory, monitoring tree health, tree counting, and so on. Previous studies have focused on usage of deep neural networks for detecting tree canopy and in a few cases, they have delineated the tree canopy masks. However, creating training samples of masks by annotation is an extremely challenging task for two important reasons. Firstly, due to the sheer volume of data required for deep neural networks and the effort required for creating labelled masks through bounding boxes can be manifold. Secondly, resolution of the UAV images and irregular shapes of the tree canopies make it a difficult process to hand draw the masks around the canopies. In this work, a two stage semi-supervised approach for detecting the tree canopy is proposed. The first stage comprises of detecting tree canopy through bounding boxes using RetinaNet, and the second stage finds the tree canopy masks using a combination of thresholded ExGI (excess green index) values, neural networks with back propagation and SLIC (simple linear iterative clustering). The results showed a mean average precision of 90% for tree canopy detection and 65% accuracy for the tree canopy extraction.

Keywords: Tree canopy · UAV images · Object detection · Clustering · Image segmentation

1 Introduction

Data acquired by unmanned aerial vehicle (UAV) are fast rising and less expensive alternatives to space-borne remote sensing satellite images. UAV based remote sensing have applications in traffic management, weather monitoring, precision agriculture, forest surveys, biodiversity studies, monitoring tree health, and so on. In agricultural and ecology domain, UAV is used for inventory management of orchards, for detecting tree

B. Raman et al. (Eds.): CVIP 2021, CCIS 1568, pp. 299–307, 2022.
https://doi.org/10.1007/978-3-031-11349-9_26

or plant health, in plant/tree growth monitoring, forest management, etc. Individual tree canopy detection and delineation serves as a primary step for identifying tree species, finding the distance/gap between trees on the ground, for understanding tree distribution patterns and for biomass estimation.

Advancements in deep learning and computer vision based techniques have drawn a lot of attention in detecting objects from UAV data. State-of-the-art techniques for object detection include YOLO [1], YOLOv3 [2], RetinaNet [3], Faster R-CNN [4] and U-Net [5]. While the techniques mentioned so far detect objects by determining bounding boxes around them, there also exist methods that aim at extraction and delineation of objects at pixel level such as Mask R-CNN [6]. These techniques have also been found to be useful in tree canopy delineation from UAV images. For example, there have been studies to assess the performance of convolutional neural network (CNN) based techniques such as YOLOv3, RetinaNet and Faster R-CNN for individual tree crown detection [7]. RetinaNet performed best in terms of average precision on the studied dataset while YOLOv3 was the fastest amongst the three. The study concluded that the techniques yielded satisfactory results but the process of creating annotations for the purpose of training the CNN models was expensive, time consuming and often the annotations were non-reproducible, which affect the efficiency of these techniques. In another study, a semi-supervised approach using a combination of LiDAR and RGB images [8] achieved a precision of 0.53. Noisy labels generated from LiDAR data were used in an unsupervised technique [9], which were overlaid on the RGB images and fed as input to a pre-trained RetinaNet model. This model was enhanced by re-training on a small set of hand annotated images, which led to an improvement of precision to 0.69. Further, possibility of using a model that was trained at a certain geographical site and was used for prediction at a different spatial site was also explored [10]. It was observed that a cross-site model performed well than the individual models trained for each local site. There are also a few studies that focus on reducing the cost of data acquisition, for example, an approach for individual tree crown detection from low resolution RGB (Red-Green-Blue) images using a combination of GANs (Generative Adversarial Networks) and RetinaNet was proposed [11].

While the literature deliberated so far emphasize on tree canopy detection by producing bounding boxes surrounding them, extracting tree canopies is also important for many applications. As such, Adhikari et al. [12] proposed combination of super pixel segmentation with Random Forest along with few textural measures for extracting tree canopies for monitoring and delineation. A combination of textural and spectral features for monitoring the tree growth from UAV platform was also suggested [13]. UAV images often have varying illuminations and shadow related problem, for which a solution was proposed using vegetation parameters and spatial information [14]. Most studies discussed here focused on usage of machine learning and deep learning techniques for detecting tree canopies. However, supervised approaches for tree canopy extraction is challenging for two reasons. First, due to the sheer volume of data required for deep neural networks, the effort in creating labeled masks could be manifold. Second, resolution of the UAV images and irregular shapes of the tree canopies make it a difficult process to handcraft the masks around the canopies. Irregular shapes of the tree canopies, which cannot be identified and demarcated effectively for creating training masks for extraction, make it a laborious process and often lead to inconsistencies among the created masks. Therefore this study attempts to address these problems using a 2-stage semi-supervised approach for detecting and extracting the pixels corresponding to the tree canopies. The objectives of this study are:

(i) To detect trees from UAV images, and
(ii) To extract individual tree canopies.

The paper is organized as follows: Sect. 2 details the methodology in two stages: (1) supervised tree canopy detection, and (2) unsupervised tree canopy segmentation along with the UAV data details and preprocessing steps. Section 3 discusses the results with concluding remarks in Sect. 4.

2 Methodology and Data

2.1 Methodology

The process of extracting tree canopies from the UAV images was divided into two stages. Stage-I is supervised which deals with predicting tree canopies through bounding boxes. Stage-II is unsupervised that extracts the tree canopies from the tree canopy region of interest (ROI) obtained from stage-I. The overall methodology is shown in Fig. 1.

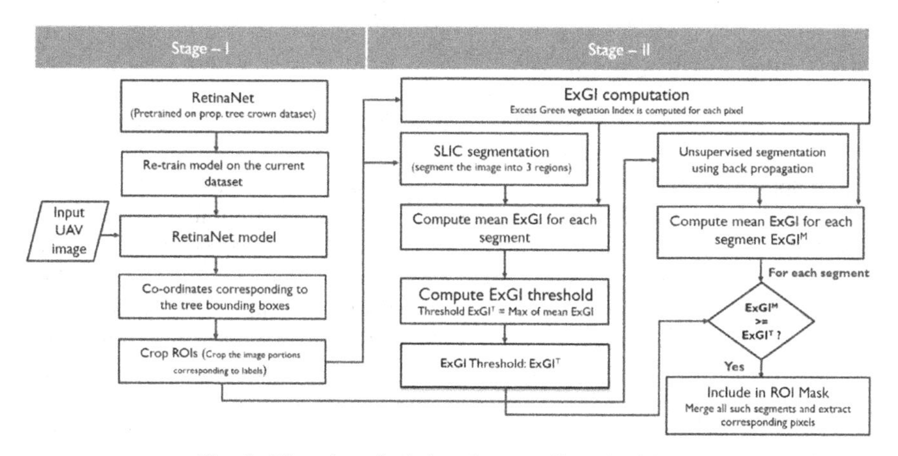

Fig. 1. Flowchart depicting the overall methodology.

Stage-I: Supervised Tree Canopy Detection using RetinaNet – This stage detects the tree canopies from the individual image tiles obtained after preprocessing the original UAV data. RetinaNet and Deep Forest package were used as base model which was then re-trained on the AgTrC dataset, followed by evaluation and bounding box detection on the evaluation dataset. Tiles from evaluation dataset were fed to the final model to predict labels/tree crowns. ROIs were obtained by cropping the regions corresponding to the predicted labels, where each ROI is expected to contain a single tree crown. These ROIs were further processed in stage-II for tree canopy extraction.

Stage-II: Unsupervised Tree Canopy Segmentation – This stage deals with extracting tree canopies from the ROIs obtained from stage-I using unsupervised segmentation by extracting green pixels from the image using Excess Green Index (ExGI) [15] which was used earlier to distinguish weed from non-plant background [16]. Assuming R-G-B channels in an 8-bit image, chromatic coordinates for each pixel in the R, G and B channels are computed using Eqs. (1)–(3):

$$r = \frac{R^*}{R^* + G^* + B^*}, \quad g = \frac{G^*}{R^* + G^* + B^*}, \quad b = \frac{B^*}{R^* + G^* + B^*} \tag{1}$$

where, R*, G* and B* for a given pixel are normalized channel values, defined in (2).

$$R^* = \frac{R}{R_m + G_m + B_m}, \quad G^* = \frac{G}{R_m + G_m + B_m}, \quad B^* = \frac{B}{R_m + G_m + B_m} \tag{2}$$

where, R, G and B are the integer values of the corresponding channels for the given pixel, and R_m, G_m and B_m indicate the maximum integer values possible. For a 8-bit channel representation, $R_m = G_m = B_m = 255$. Hence, from (1) and (2), Eq. (3) is obtained.

$$r = \frac{R}{R+G+B}, \quad g = \frac{G}{R+G+B}, \quad b = \frac{B}{R+G+B} \tag{3}$$

ExGI is then calculated as in (4):

$$ExGI = 2g - r - b \tag{4}$$

ExGI applied on an image with a zero binary threshold eliminates non-green pixels from the image. It is to be noted that non-green pixels were eliminated and marked as black and green pixels remained in the resultant image. However, due to the image capturing dynamics of the UAV such as its altitude, illumination conditions and resolution of the image, the same technique when applied on a ROI obtained from stage-I did not render promising results. Therefore, alternative methods of extracting tree crowns using super pixel clustering and CNN based segmentation were pursued as outlined in the following three steps:

Step 1: ExGI Threshold Computation – For the obtained ROI, it was observed that the image consists of three broad regions: tree canopy, background and the shadow. To identify these broad regions, SLIC (simple linear iterative clustering) [17] was used to find the three segments. For each of these segments, mean of ExGI for all the pixels in the segment was calculated. Since, ExGI is a measure of greenness, the segment corresponding to tree canopy will have the highest mean ExGI. The maximum among the mean ExGI was taken as the threshold identified by $ExGI^T$.

Step 2: ROI Segmentation Using Back-Propagation – ROI was subjected to unsupervised segmentation using back propagation [18] which used a combination of CNN and super pixel refinement for finding the segments in an image. It is to be noted that no inputs corresponding to the number and location of segments are required here. Finally, mean ExGI ($ExGI^M$) for each segment was calculated.

Step 3: Tree Canopy Extraction – Let n represent the number of segments, ith segment be represented as S_i, and mean of ExGI of all the pixels in that segment be represented as $ExGI_i^M$. For each segment S_i, the pixels of the segment are retained if $ExGI_i^M \geq ExGI^T$, else the pixels are discarded.

2.2 Data

Data for this work were obtained from a UAV flown over an agricultural area near Bangalore City, Karnataka state in India. A frame from the UAV video at a point is referred to as a scene and this work used a total of 3 such scenes. Each scene was further cropped into tiles of size 400 × 400 pixels. An overlap of 0.5% was allowed so as to accommodate the trees being cut at the edges of the tiles. About 57 such tiles were obtained for this study consisting of more than 120 tree canopies, where 80% of the data were used for training and the rest for evaluation. Bounding boxes were drawn on the images using LabelImg [19], a graphical image annotation tool. Ground truth masks used for evaluation were generated using VGG Image Annotator (VIA) [20].

3 Results and Discussion

Figure 2 (a) shows the predicted bounding boxes enclosing tree canopies obtained by the RetinaNet model on a tile from the evaluation dataset. Boxes in red indicate the predicted outputs, while the boxes in black are the ground truth labels. Predictions of the model on a test tile is shown in Fig. 2 (b). Stage I (discussed in Sect. 2 earlier), which was used to get bounding boxes around tree canopies was evaluated using a different

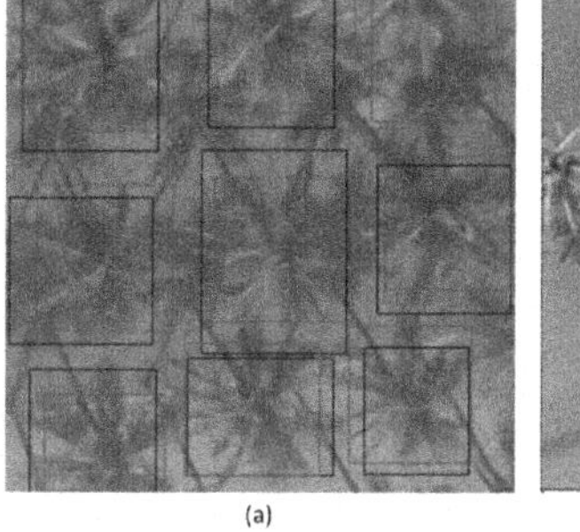

Fig. 2. Predictions of the RetinaNet model. (a) A sample evaluation image, and (b) A sample test image.

test dataset that showed a mean average precision of 0.897 under an IoU (Intersection over Union) threshold of 0.5.

Figure 3 (a) is an image of a plant and Fig. 3 (b) is the output obtained after ExGI filtering. SLIC segmentation of a sample ROI is shown in Fig. 4. Figure 5 depicts the output of the unsupervised segmentation on the tree canopy ROI using back propagation, where each segment is identified by a distinct number and unique colour representation.

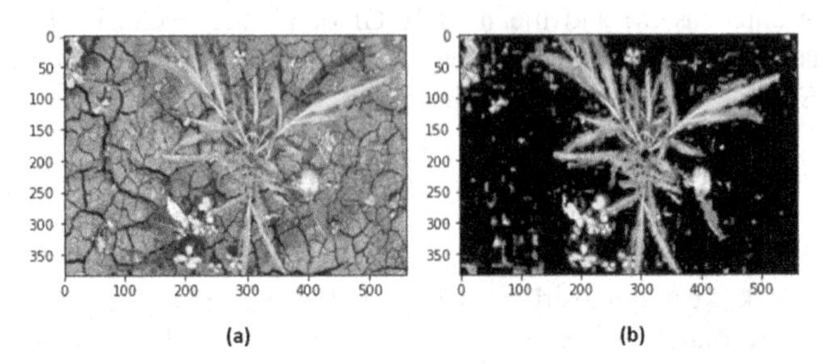

Fig. 3. Effect of ExGI based filtering. (a) An original image, and (b) ExGI filtered image.

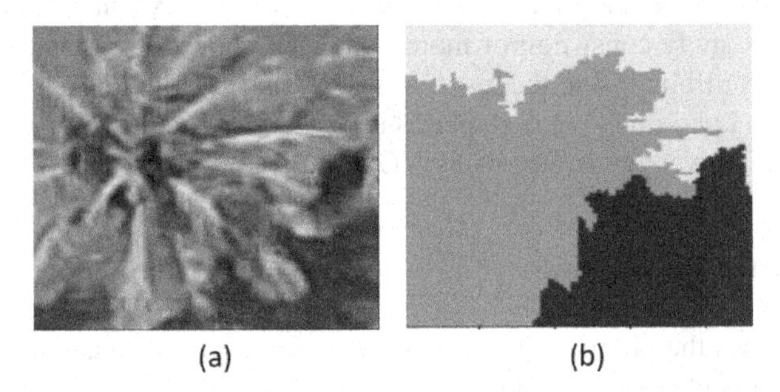

Fig. 4. SLIC segmentation on tree canopy ROI.

Fig. 5. (a) Tree canopy ROI subjected to segmentation, and (b) Output of segmentation using back propagation.

A binary mask was hence obtained as shown in Fig. 6 (a) which when applied on the ROI gave the extracted tree canopy shown in Fig. 6 (d). Thus, a combination of stage-I and II resulted in tree canopy extraction from UAV images as depicted in Fig. 7. Pixels corresponding to the tree canopy were evaluated against handcrafted ground truth masks that gave an accuracy of 65% under an IoU threshold of 0.4 which is comparable to the existing state-of-the-art approaches for tree canopy detection as reported in [12, 21].

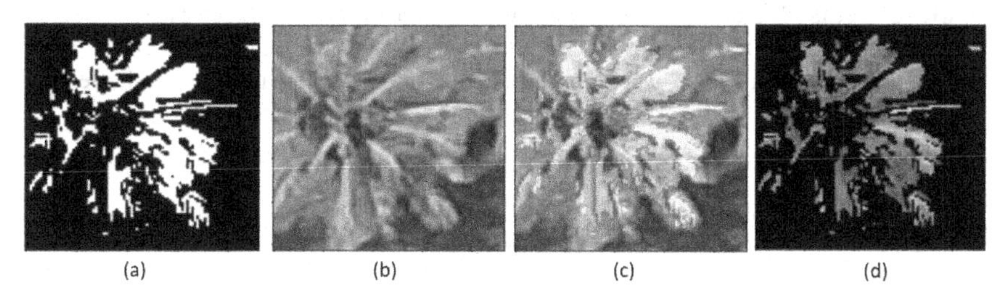

Fig. 6. (a) Binary mask obtained as a result of the applied ExGI threshold, (b) Original ROI, (c) Mask overlaid on the ROI, and (d) The extracted tree canopy.

Fig. 7. Final output of tree canopy extraction on ROI.

4 Conclusion

A semi-supervised approach for extracting tree canopies from UAV images was proposed involving supervised stage for bounding box detection and an unsupervised stage for mask detection. This approach helped to deal with the small and irregular shapes of the tree canopies in the UAV images. The method employed in this work avoids laborious process of generating ground truth masks for the trees as required for supervised learning. The unsupervised segmentation approach does not require a large input data for segmentation and uses only RGB images which can be easily obtained as compared to LiDAR or multispectral data. Future work will focus on re-purposing tree canopies extracted from this approach as training data for other supervised image segmentation or instance segmentation algorithms like U-Net, Mask R-CNN etc. Algorithms can also be developed to analyze tree health from the extracted tree canopies using the above proposed method.

Acknowledgements. We are grateful to the International Institute of Information Technology Bangalore (IIITB), India for the infrastructure support. We are thankful to Infosys Foundation for the financial assistance and project grant through the Infosys Foundation Career Development Chair Professor.

References

1. Redmon, J., Divvala, S., Girshick, R., Farhadi, A.: You only look once: Unified, real-time object detection. In: Proceedings of the IEEE Conference on Computer Vision and Pattern Recognition, pp. 779–788 (2016)
2. Redmon, J., Farhadi, A.: Yolov3: An incremental improvement. arXiv preprint arXiv:1804. 02767 (2018)
3. Lin, T.Y., Goyal, P., Girshick, R., He, K., Dollár, P.: Focal loss for dense object detection. In: Proceedings of the IEEE International Conference on Computer Vision, pp. 2980–2988 (2017)
4. Ren, S., He, K., Girshick, R., Sun, J.: Faster R-CNN: Towards real-time object detection with region proposal networks. In: Proceedings of the 28th International Conference on Neural Information Processing Systems – NIPS'15, vol. 1, pp. 91–99. Cambridge, MA, USA, MIT Press (2015)
5. Ronneberger, O., Fischer, P., Brox, T.: U-Net: Convolutional Networks for Biomedical Image Segmentation. In: Navab, N., Hornegger, J., Wells, W., Frangi, A. (eds) Medical Image Computing and Computer-Assisted Intervention – MICCAI 2015. MICCAI 2015. Lecture Notes in Computer Science, vol. 9351. Springer, Cham (2015). https://doi.org/10.1007/978-3-319-24574-4_28
6. He, K., Gkioxari, G., Dollár, P., Girshick, R.: Mask r-cnn. In: 2017 IEEE International Conference on Computer Vision (ICCV), pp. 2980–2988 (2017)
7. Santos, A.A.D., et al.: Assessment of cnn-based methods for individual tree detection on images captured by rgb cameras attached to UAVs. Sensors **19**(16), 3595 (2019)
8. Weinstein, B.G., Marconi, S., Bohlman, S., Zare, A., White, E.: Individual tree-crown detection in RGB imagery using semi-supervised deep learning neural networks. Remote Sensing **11**(11), 1309 (2019)
9. Silva, C.A., et al.: Imputation of individual longleaf pine (Pinus palustris mill.) tree attributes from field and lidar data. Can. J. Remote Sens. **42**(5), 554–573 (2016)
10. Weinstein, B.G., Marconi, S., Bohlman, S.A., Zare, A., White, E.P.: Cross-site learning in deep learning RGB tree crown detection. EcologicalInformatics **56**, 101061 (2020)
11. Roslan, Z., Long, Z.A., Ismail, R.: Individual tree crown detectionusing GAN and RetinaNet on tropical forest. In: 2021 15th International Conference on Ubiquitous Information Management and Communication (IMCOM), pp. 1–7. IEEE (2021)
12. Adhikari, A., Kumar, M., Agrawal, S., Raghavendra, S.: An integrated object and machine learning approach for tree canopy extraction from UAV datasets. J. Indian Soc. Remote Sens. **49**(3), 471–478 (2021)

13. Guo, Y., et al.: Integrating spectral and textural information for monitoring the growth of pear trees using optical images from the UAV platform. Remote Sens. **13**(9), 1795 (2021)
14. Agarwal, A., Kumar, S., Singh, D.: An adaptive techniqueto detect and remove shadow from drone data. J. Indian Soc. Remote Sens. **49**(3), 491–498 (2021)
15. Woebbecke, D.M., Meyer, G.E., Von Bargen, K., Mortensen, D.A.: Color indices for weed identification under various soil, residue, and lighting conditions. Trans. ASAE **38**(1), 259–269 (1995)
16. Meyer, G.E., Neto, J.C.: Verification of color vegetation indices for automated crop imaging applications. Comput. Electron. Agric. **63**(2), 282–293 (2008)
17. Achanta, R., Shaji, A., Smith, K., Lucchi, A., Fua, P., & Süsstrunk, S.: SLIC superpixels. Tech. Rep. 149300, EcolePolytechnique Fédéral de Lausssanne (EPFL) (2010)
18. Kanezaki, A.: Unsupervised image segmentation by backpropagation. In: 2018 IEEE international conference on acoustics, speech and signal processing (ICASSP), pp. 1543–1547. IEEE (2018)
19. Tzutalin. Labelimg. https://github.com/tzutalin/labelImg (2015)
20. Dutta, A., Zisserman, A.: The via annotation software for images, audio and video. In: Proceedings of the 27th ACM International Conference on Multimedia, pp. 2276–2279 (2019)
21. Micheal, A.A., Vani, K., Sanjeevi, S., Lin, C.-H.: Object detection and tracking with UAV data using deep learning. J. Indian Soc. Remote Sens. **49**(3), 463–469 (2020)

Pose Guided Controllable Gesture to Gesture Translation

Mallika, Debashis Ghosh(✉), and Pyari Mohan Pradhan

Department of Electronics and Communication Engineering,
Indian Institute of Technology Roorkee, Roorkee 247667, Uttarakhand, India
mallika@ec.iitr.ac.in, {debashis.ghosh,pmpradhan}@ece.iitr.ac.in

Abstract. Gesture-to-gesture (G2G) translation is a very difficult task due to various poses, size of hands, location of fingers and other important factors. This task serves various applications, such as gesture synthesis, gesture recognition, etc. We propose a pose guided controllable Generative Adversarial Network (GAN). We have taken into consideration the structural similarity based gesture synthesis for conditional image to image translation. Along with this, controllable structures are also used to generate image. The proposed model learns the translational mapping from a combination of losses which are used for training the proposed GAN. The generated images are evaluated based on Peak Signal-to-Noise Ratio (PSNR), Inception Score (IS) and Frechet Inception Distance (FID) on NTU hand digit dataset and Senz3D dataset.

Keywords: Hand gesture-to-gesture translation · Pose guidance · Conditional generative adversarial network · Controllable structures

1 Introduction

Recently, GANs have gained much attention in various face generation [11], style transfer [13], human pose generation [26], image restoration [29], image-to-image translation [1,5], semantic label into realistic photos [25] and image denoising task. One such task is gesture-to-gesture translation which is a challenging task. Training a GAN is also a difficult task as a lot of computation is required for the generator to learn the target image mapping from the input distribution. Also, since minmax optimization is used for training the GAN, it is very difficult to converge for both generator and discriminator. So, to encourage convergence, feature matching, minibatch discrimination, historical averaging, etc. were the solutions presented in [18].

Tang et al. [22] proposed GestureGAN which translated from input to target gesture using target hand skeleton image as input to the GAN. Tang also proposed a unified GAN [20] for controllable image translation that also uses controllable image generation similar to GestureGAN. The proposed GAN uses controllable translation conditioning on the reference image. Conditioning helps

B. Raman et al. (Eds.): CVIP 2021, CCIS 1568, pp. 308–318, 2022.
https://doi.org/10.1007/978-3-031-11349-9_27

GAN to learn the target distribution more easily than the normal GAN while controllable structure provides structural information for generating the target image. We have considered perceptual loss and structural similarity index loss as the learning criteria along with cyclic loss, identity loss for training the proposed controllable GAN.

The main objective is to perform gesture translation using paired conditional controllable GAN. The contributions are as follows:

(i) A controllable GAN for paired gesture translation has been proposed. The gestures are generated in both directions using cycle consistency.
(ii) Color, identity, cyclic, perceptual, structural similarity loss functions are used along with adversarial loss to train the proposed GAN and get better results. We have used Structural Similarity index measure as objective function for controllable GAN for the first time in the present state of art.
(iii) Experiments are conducted on the NTU hand gesture dataset and Creative Senz3D dataset which are publicly available datasets and results show that the proposed GAN model is capable of generating realistic images, which can further be used for other purposes like gesture recognition.

2 Review of Literature

Initially, GAN was presented by Isola et al. [7] which was developed on the concept of Game Theory. In [18], an enhancement in the training process of both the generator and discriminator has been presented. Class level and image level scene can also be generated using sematic guided GANs [23]. Enormous development has been seen in the GAN frameworks in all generation fields, either text synthesis, image synthesis, or speech synthesis etc. Attention-Guided Generative Adversarial Network (AGGAN) [21] uses unsupervised image to image translation which can produce attention mask, content mask and target image at the same time.

Self-attention GAN (SAGAN) [28] produces self-attention mask to generate the realistic target image. Also, CycleGAN [30] carries the translation process in the absence of paired data. Later, paired cycleGAN has also been proposed [3] that was developed for applying and removing makeup from one to many transforms problem. Also, conditional cyclic GAN [11] also used for face generation application uses attribute guided generation like gender, hair, color, emotions etc.

Along with image generation, GANs are also used for 3D model generation. Model aware [6] 3D hand g2g translation is also based on GAN and MANO hand model [17]. Disentangling a GANs is one major issue, especially for high resolution image generation. Disentanglement can be done in unsupervised manner [4], semi-supervised manner [15] and self-supervised manner [9]. With an advancement in transformer model [24], transformer based GAN model has also come into focus [8,10].

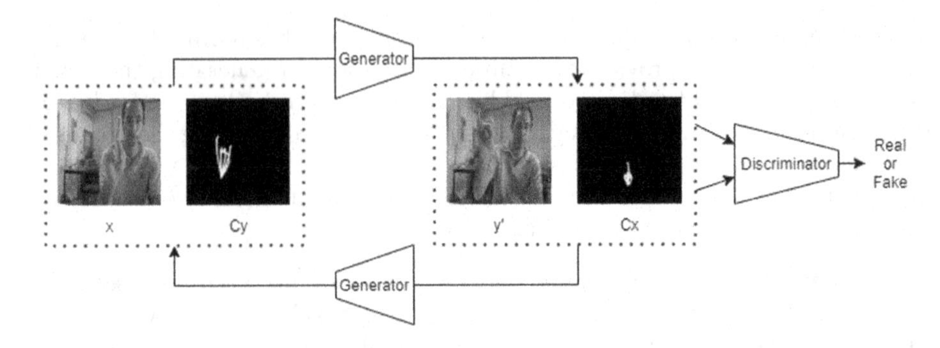

Fig. 1. The proposed GAN model for gesture to gesture translation.

3 Proposed Method

Our model includes the conditional generation process with controllable structures. In this paper, we will train the GAN from input domain to output domain with some target information as input for controlling the training process. The keypoints are extracted from Openpose [2] for each target gesture. For each target gesture, Openpose extracts 21 keypoints for hand pose. Paired image translation is used to train the proposed GAN with cyclic arrangement. The proposed GAN model is shown in Fig. 1. For translating, the input gesture along with target hand skeleton is given as input to the generator for training. The forward cycle can be,

$$x'' = G(y', C_x) = G(G(x, C_y), C_x) \approx x, \tag{1}$$

where x is the given conditional input gesture, y' is the generated or fake image, x'' is the reconstructed image for forward cycle, and C_y & C_y are the structural details of the target gesture for forward and backward cycle, respectively.

Similarly, for another cycle, we have,

$$y'' = G(x', C_y) = G(G(y, C_x), C_y) \approx y, \tag{2}$$

where y is the given conditional input gesture, x' is the generated or fake image, y'' is the reconstructed image for backward cycle.

The training of proposed GAN model includes the following loss functions:

Adversarial Generative Loss: The generator G learn the translation from input to target gesture with some loss or objective function. This learning is based on an adversarial loss expressed as

$$\begin{aligned} \mathcal{L}_{adv}(G, D, C_y) &= \mathbb{E}_{[x,C_y],y}[logD([x, C_y], y)] \\ &+ \mathbb{E}_{x,C_y}[log(1 - D([x, C_y], G(x, C_y)))], \end{aligned} \tag{3}$$

Similarly, for backward,

$$\begin{aligned} \mathcal{L}_{adv}(G, D, C_x) &= \mathbb{E}_{[y,C_x],x}[logD([y, C_x], x)] \\ &+ \mathbb{E}_{y,C_x}[log(1 - D([y, C_x], G(y, C_x)))]. \end{aligned} \tag{4}$$

So, the total adversarial loss becomes,

$$\mathcal{L}_{adv}(G, D) = \mathcal{L}_{adv}(G, D, C_y) + \mathcal{L}_{adv}(G, D, C_x). \tag{5}$$

Color Loss: This loss [20] is used to preserve each color component of the target image which is derived from $L1$ loss function and is expressed as

$$\begin{aligned} \mathcal{L}_{Color_c}(G) &= \mathbb{E}_{[x_c, C_y], y_c}[||y_c - G(x_c, C_y)||_1] \\ &+ \mathbb{E}_{[y_c, C_x], x_c}[||x_c - G(y_c, C_x)||_1], \end{aligned} \tag{6}$$

where c takes the value r, g and b for red, green and blue color component, respectively.

Thus the total color loss becomes,

$$\mathcal{L}_{Color}(G) = \mathcal{L}_{Color_r} + \mathcal{L}_{Color_g} + \mathcal{L}_{Color_b}. \tag{7}$$

Identity Loss: Identity loss is initialized to preserve the image quality. It indicates that if x translates to some y' after passing through the generator, then if y is fed to the generator, it should output the real y. Mathematically, it is represented as,

$$\begin{aligned} \mathcal{L}_{iden}(G) &= \mathbb{E}_{[x_c, C_x]}[||x - G(x, C_x)||_1] \\ &+ \mathbb{E}_{[y, C_y]}[||y - G(y, C_y)||_1]. \end{aligned} \tag{8}$$

Cyclic Loss: Cyclic loss guarantees that the overall output image should be close to the original input. It states that if generator translates image x to y', then y' must translate to some $G(y', C_x)$ that must closely resemble with the original x. Therefore, it is defined as

$$\begin{aligned} \mathcal{L}_{cyc}(G) &= \mathbb{E}_{x, C_x, C_y}[||x - G(G(x, C_y), C_x)||_1] \\ &+ \mathbb{E}_{y, C_x, C_y}[||y - G(G(y, C_x), C_y)||_1], \end{aligned} \tag{9}$$

Perceptual Loss: This loss is used to generate the two similar images which are similar in style [26]. Similarity of two images are calculated based on their features from VGG network and is given by

$$\mathcal{L}_{vgg}(y') = \frac{1}{W_{i,j} H_{i,j}} \sum_{w=1}^{W_{i,j}} \sum_{h=1}^{H_{i,j}} [||\mathcal{F}^k(y) - \mathcal{F}^k(G(x, C_y))||_1], \tag{10}$$

where $\mathcal{F}^k$ are the features from k_{th} layer of VGG, $W_{i,j}$ and $H_{i,j}$ are the height and width of the features obtained.

Hence, for backward cycle,

$$\mathcal{L}_{vgg}(x') = \frac{1}{W_{i,j} H_{i,j}} \sum_{w=1}^{W_{i,j}} \sum_{h=1}^{H_{i,j}} [||\mathcal{F}^k(x) - \mathcal{F}^k(G(y, C_x))||_1], \tag{11}$$

Therefore, total perceptual loss is given as,

$$\mathcal{L}_{VGG}(G) = \mathcal{L}_{vgg}(x') + \mathcal{L}_{vgg}(y'). \tag{12}$$

Structural Similarity Loss. Structural Similarity Loss is used to produce visually pleasing images [29]. As in the proposed GAN, we need to generate images of same shape and structure as that of the target image, with SSIM, the structural similarity loss for both the cycles are given as,

$$\mathcal{L}_{SSIM}(y') = 1 - E(\frac{1}{K}\sum_{j=1}^{K} SSIM(G_j(x, C_y), y_j)), \tag{13}$$

$$\mathcal{L}_{SSIM}(x') = 1 - E(\frac{1}{K}\sum_{j=1}^{K} SSIM(G_j(y, C_x), x_j)), \tag{14}$$

where K is the number of patches in an image, and SSIM is given by

$$SSIM(a, b) = \frac{2\mu_a\mu_b + c_1}{2\mu_a^2\mu_b^2 + c_1} \cdot \frac{2\sigma_{ab} + c_2}{2\sigma_a^2\sigma_b^2 + c_2} \tag{15}$$

where a and b represent the two patches, and μ and σ are the mean and variance, respectively. c_1 and c_2 are small constants incorporated in the denominators for avoiding division by zero.

Therefore, structural similarity loss for complete cycle is,

$$\mathcal{L}_{SSIM}(G) = \mathcal{L}_{SSIM}(x') + \mathcal{L}_{SSIM}(y'). \tag{16}$$

Overall Loss: The overall loss for the proposed GAN is,

$$\begin{aligned} G* = \arg \min_{G} \max_{D}(&\mathcal{L}_{cGAN}(G, D) + \lambda\mathcal{L}_{Color}(G) + \lambda_{iden}\mathcal{L}_{iden}(G) \\ &+ \lambda_{cyc}\mathcal{L}_{cyc}(G) + \lambda_{SSIM}\mathcal{L}_{SSIM}(G) + \lambda_{VGG}\mathcal{L}_{VGG}(G)) \end{aligned} \tag{17}$$

where λ, λ_{SSIM}, λ_{VGG}, λ_{iden}, λ_{cyc} are the hyperparameters for controlling the quality of the generated image.

4 Experimental Set-Up

Nvidia GeForce GTX 1080 Ti GPU with CUDA 10.2 and cuDNN 8.1.1 is used for training the proposed GAN. The model is implementated on Tensorflow-gpu = 1.14 and is optimised with Adam optimize with $\beta 1 = 0.9$ and $\beta 2 = 0.999$. 200 epochs are used for training the model on training datasets with a learning rate of 0.0001. Empirically, the values of λ, λ_{SSIM}, λ_{VGG}, λ_{iden}, λ_{cyc} are chosen to be 800, 1000, 1000, 0.01, and 0.1 respectively.

4.1 Datasets

We have used publically available NTU Hand Digit [16] and Creative Senz3D [14]. The details of all the datasets used in the implementation and analysis are provided below:

(i) **NTU Hand Digit**: It contains 10 hand gestures from 0 to 9. The dataset is collected using kinect sensor. There are total 10 subjects. Each gesture is performed 10 times by each subject. So, a total of 1000 images are present in the dataset. The size of each image is 640 × 480. The dataset consists of depth maps along with RGB images.

(ii) **Creative Senz3D**: Creative Senz3D dataset uses Creative Senz3D camera to capture static hand gestures. 4 subjects are used to perform the gestures. Each subject performs 11 gestures repeated 30 times each. So, the complete dataset contains overall 1320 images. All images are of resolution 640 × 480.

Table 1. Comparison results on NTU hand digit and Senz3D. For all except FID, higher is better. (*) are reported in [20].

Method	NTU dataset			Senz3D		
	PSNR↑	IS↑	FID↓	PSNR↑	IS↑	FID↓
PG^2 [26]	28.2403*	2.4152*	24.2093*	26.5138*	3.3699*	31.7333*
SAMG [27]	28.0185*	2.4919*	31.2841*	26.9545*	3.3285*	38.1758*
DPIG [12]	30.6487*	2.4547*	6.7661*	26.9451*	3.3874*	26.2713*
PoseGAN [19]	29.5471*	2.4017*	9.6725*	27.3014*	3.2147*	24.6712*
GestureGAN [22]	32.6091*	2.5532*	7.5860*	27.9749*	3.4107*	18.4595*
UnifiedGAN [20]	32.6574*	2.3783*	6.7493*	31.5420*	2.2159*	18.4595*
Ours	32.03	2.4815	6.5	29.54	2.7	11.536

4.2 Results

The implementation has been performed similar to as mentioned in [22]. Only RGB images are used for training the GAN and skeleton images be given as controlling structures. This section presents the results obtained for the proposed GAN model. Table 1 presents the comparative analysis obtained from experiments on NTU and Senz3D dataset with other methods. PSNR, IS, and FID metrices are used for evaluating the proposed model. From the table, this could be concluded that our model performs better in terms of FID for both the datasets. Similar comparisons are shown in Fig. 2 and Fig. 3 for both the datasets.

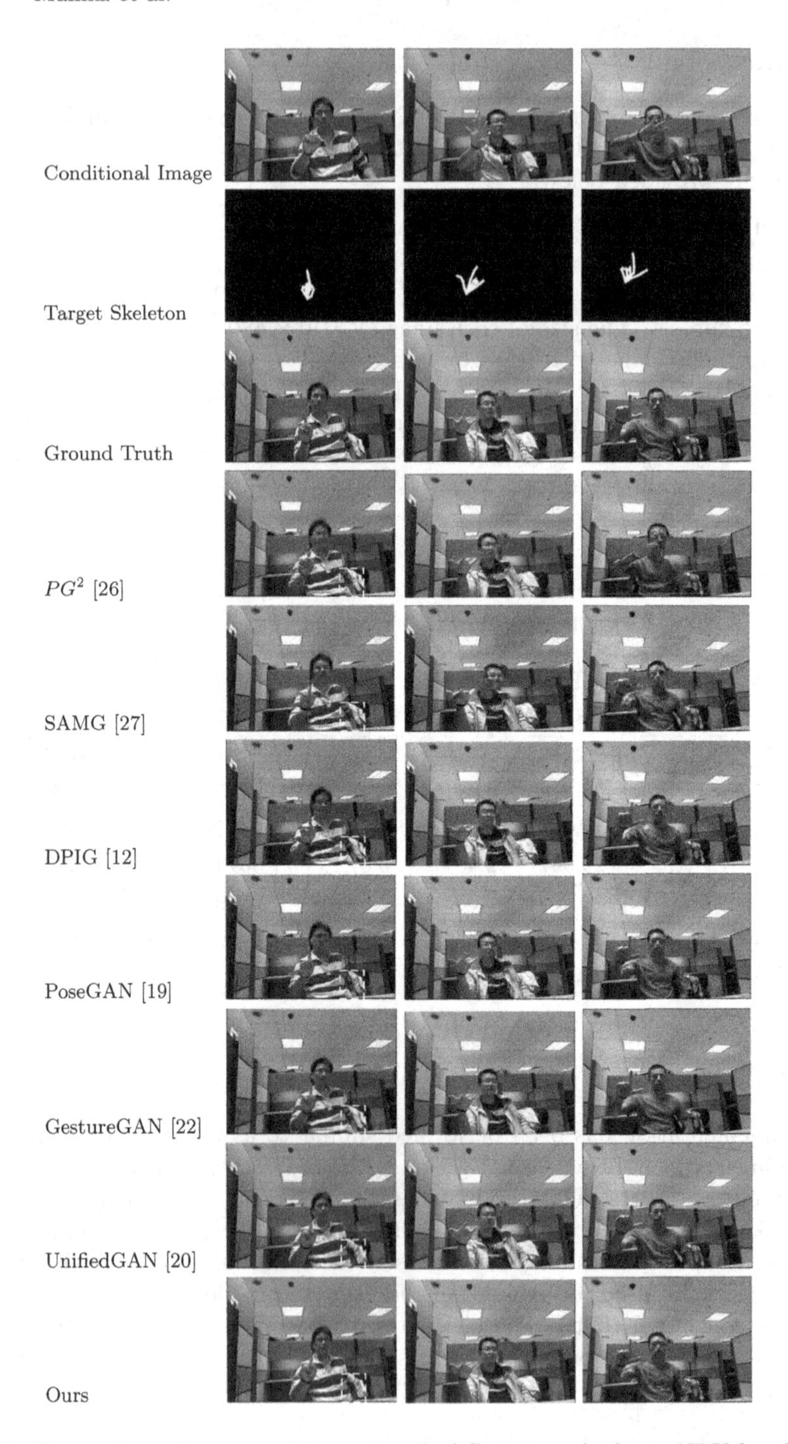

Fig. 2. Comparison of generated images with different methods on NTU hand digit dataset.

Fig. 3. Comparison of generated images with different methods on Senz3D dataset.

5 Conclusion

A novel g2g method has been proposed in this work which takes as input, the given conditional gesture, skeleton and the RGB image of the target gesture and translates to target gesture using controllable cGAN. Since, gesture to gesture translation is a challenging task, so, conditional cyclic GAN are used for the translation process that considers various losses including color, perceptual, structural similarity, cyclic, identity and adversarial loss. Results shows that proposed approach has convincing results. Therefore, in future, the translated hand gestures obtained as output from the proposed GAN can be utilized for data augmentation, gesture recognition purposes etc.

References

1. AlBahar, B., Huang, J.B.: Guided image-to-image translation with bi-directional feature transformation. In: Proceedings of the IEEE/CVF International Conference on Computer Vision, pp. 9016–9025 (2019)
2. Cao, Z., Hidalgo, G., Simon, T., Wei, S.E., Sheikh, Y.: OpenPose: realtime multi-person 2D pose estimation using part affinity fields. IEEE Trans. Pattern Anal. Mach. Intell. **43**(1), 172–186 (2019)
3. Chang, H., Lu, J., Yu, F., Finkelstein, A.: PairedCycleGAN: asymmetric style transfer for applying and removing makeup. In: Proceedings of the IEEE Conference on Computer Vision and Pattern Recognition, pp. 40–48 (2018)
4. Chen, X., Duan, Y., Houthooft, R., Schulman, J., Sutskever, I., Abbeel, P.: InfoGAN: interpretable representation learning by information maximizing generative adversarial nets. arXiv preprint arXiv:1606.03657 (2016)
5. Garg, M., Ghosh, D., Pradhan, P.M.: Generating multiview hand gestures with conditional adversarial network. In: 2021 IEEE 18th India Council International Conference (INDICON), pp. 1–6 (2021)
6. Hu, H., Wang, W., Zhou, W., Zhao, W., Li, H.: Model-aware gesture-to-gesture translation. In: Proceedings of the IEEE/CVF Conference on Computer Vision and Pattern Recognition, pp. 16428–16437 (2021)
7. Isola, P., Zhu, J.Y., Zhou, T., Efros, A.A.: Image-to-image translation with conditional adversarial networks. In: Proceedings of the IEEE Conference on Computer Vision and Pattern Recognition, pp. 1125–1134 (2017)
8. Jiang, Y., Chang, S., Wang, Z.: TransGAN: two pure transformers can make one strong GAN, and that can scale up. In: Advances in Neural Information Processing Systems 34 (2021)
9. Li, Y., Zeng, J., Shan, S., Chen, X.: Self-supervised representation learning from videos for facial action unit detection. In: Proceedings of the IEEE/CVF Conference on Computer Vision and Pattern Recognition, pp. 10924–10933 (2019)
10. Lin, C.H., Yumer, E., Wang, O., Shechtman, E., Lucey, S.: ST-GAN: spatial transformer generative adversarial networks for image compositing. In: Proceedings of the IEEE Conference on Computer Vision and Pattern Recognition, pp. 9455–9464 (2018)

11. Lu, Y., Tai, Y.-W., Tang, C.-K.: Attribute-guided face generation using conditional CycleGAN. In: Ferrari, V., Hebert, M., Sminchisescu, C., Weiss, Y. (eds.) ECCV 2018. LNCS, vol. 11216, pp. 293–308. Springer, Cham (2018). https://doi.org/10.1007/978-3-030-01258-8_18
12. Ma, L., Sun, Q., Georgoulis, S., Van Gool, L., Schiele, B., Fritz, M.: Disentangled person image generation. In: Proceedings of the IEEE Conference on Computer Vision and Pattern Recognition, pp. 99–108 (2018)
13. Mallika, Ubhi, J.S., Aggarwal, A.K.: Neural style transfer for image within images and conditional GANs for destylization. J. Vis. Commun. Image Represent. **85**, 103483 (2022)
14. Memo, A., Zanuttigh, P.: Head-mounted gesture controlled interface for human-computer interaction. Multimedia Tools Appl. **77**(1), 27–53 (2016). https://doi.org/10.1007/s11042-016-4223-3
15. Nie, W., et al.: Semi-supervised styleGAN for disentanglement learning. In: International Conference on Machine Learning, pp. 7360–7369. PMLR (2020)
16. Ren, Z., Yuan, J., Meng, J., Zhang, Z.: Robust part-based hand gesture recognition using kinect sensor. IEEE Trans. multimedia **15**(5), 1110–1120 (2013)
17. Romero, J., Tzionas, D., Black, M.J.: Embodied hands: modeling and capturing hands and bodies together (2017)
18. Salimans, T., Goodfellow, I., Zaremba, W., Cheung, V., Radford, A., Chen, X.: Improved techniques for training GANs. arXiv preprint arXiv:1606.03498 (2016)
19. Siarohin, A., Sangineto, E., Lathuiliere, S., Sebe, N.: Deformable GANs for pose-based human image generation. In: Proceedings of the IEEE Conference on Computer Vision and Pattern Recognition, pp. 3408–3416 (2018)
20. Tang, H., Liu, H., Sebe, N.: Unified generative adversarial networks for controllable image-to-image translation. IEEE Trans. Image Process. **29**, 8916–8929 (2020)
21. Tang, H., Liu, H., Xu, D., Torr, P.H., Sebe, N.: AttentionGAN: unpaired image-to-image translation using attention-guided generative adversarial networks. arXiv preprint arXiv:1911.11897 (2019)
22. Tang, H., Wang, W., Xu, D., Yan, Y., Sebe, N.: GestureGAN for hand gesture-to-gesture translation in the wild. In: Proceedings of the 26th ACM International Conference on Multimedia, pp. 774–782 (2018)
23. Tang, H., Xu, D., Yan, Y., Torr, P.H., Sebe, N.: Local class-specific and global image-level generative adversarial networks for semantic-guided scene generation. In: Proceedings of the IEEE/CVF Conference on Computer Vision and Pattern Recognition, pp. 7870–7879 (2020)
24. Vaswani, A., et al.: Attention is all you need. In: Advances in Neural Information Processing Systems 30 (2017)
25. Wang, T.C., Liu, M.Y., Zhu, J.Y., Tao, A., Kautz, J., Catanzaro, B.: High-resolution image synthesis and semantic manipulation with conditional GANs. In: Proceedings of the IEEE Conference on Computer Vision and Pattern Recognition, pp. 8798–8807 (2018)
26. Xu, C., Fu, Y., Wen, C., Pan, Y., Jiang, Y.G., Xue, X.: Pose-guided person image synthesis in the non-iconic views. IEEE Trans. Image Process. **29**, 9060–9072 (2020)
27. Yan, Y., Xu, J., Ni, B., Zhang, W., Yang, X.: Skeleton-aided articulated motion generation. In: Proceedings of the 25th ACM International Conference on Multimedia, pp. 199–207 (2017)
28. Zhang, H., Goodfellow, I., Metaxas, D., Odena, A.: Self-attention generative adversarial networks. In: International Conference on Machine Learning, pp. 7354–7363. PMLR (2019)

29. Zhao, H., Gallo, O., Frosio, I., Kautz, J.: Loss functions for image restoration with neural networks. IEEE Trans. Comput. Imaging **3**(1), 47–57 (2016)
30. Zhu, J.Y., Park, T., Isola, P., Efros, A.A.: Unpaired image-to-image translation using cycle-consistent adversarial networks. In: Proceedings of the IEEE International Conference on Computer Vision, pp. 2223–2232 (2017)

EDR: Enriched Deep Residual Framework with Image Reconstruction for Medical Image Retrieval

Rohini Pinapatruni(✉) and Shoba Bindu Chigarapalle

JNTUA, Anatapuramu, India
rohinipinapatruni@gmail.com

Abstract. In recent times, the advancement of Artificial Intelligence (AI) attracted many researchers in medical image analysis. Analyzing the vast medical data through traditional approaches is a bit tedious and time-consuming in designing feature descriptors. Therefore, we presented an EDR: Enriched Deep Residual Framework for robust medical image retrieval in this paper. The proposed EDR framework consists of an image reconstruction module using a residual encoder and sequential decoder. Also, the image matching module is followed by retrieval to retrieve similar images from the database. The encoder module of the EDR framework consists of series of residual connections that encode the features from a given image and are forwarded to the reconstruction decoder module. The extracted encoded features provide the latent representation for the robust reconstruction of the input image. Further, this latent information is used in the image matching and retrieve similar images from the database. The performance of the proposed EDR framework is analyzed on benchmark medical image databases such as VIA/ELCAP-CT, ILD for image retrieval tasks. The proposed EDR framework is compared with the state-of-the-art approaches for average precision and recall over two datasets. The experiments and results show that the proposed framework outperformed existing works in medical image retrieval.

Keywords: Medical image retrieval · Res-Unet framework · Image reconstruction · Index matching

1 Introduction

Content-Based Image Retrieval (CBIR) is an emerging research field of science, law enforcement, education, entertainment, medical, etc. with the ease availability of big visual data through the internet, youtube, medical hospitals, patient diagnosis, treatment planning, and response assessment gain attention to use robust for CBIR system. Existing algorithms make use of traditional/hand-crafted approaches to retrieve similar medical images from the database.

B. Raman et al. (Eds.): CVIP 2021, CCIS 1568, pp. 319–328, 2022.
https://doi.org/10.1007/978-3-031-11349-9_28

Feature extraction is one of the essential tasks in the development of a robust CBIR system. The feature extraction highly depending on the selected features from a given medical data. Visual texture, color, shape features, etc., are regularly utilized in the literature. The blur, rotation, illumination, etc., highly affect the performance of the CBIR system. The texture is one of the most noticeable features in the image. Handcrafted features in the literature [5,14,15,29,31]. The Local binary pattern (LBP) [3] and scale-invariant features (SIFT) [1] proven breakthrough guidance in the extraction of texture classification, image retrieval, motion recognition, etc. Also, many varients of LBP [16,18–20,23,25] have proven the descriptor's capabilities in these fields. Most of these features have utilized the relationship between neighboring pixels with reference pixels. However, feature descriptor design is one of the complex and time-consuming tasks. Therefore, many researchers have utilized deep learning-based approaches for simultaneous feature extraction and learning in recent times.

The field of deep learning is started by [32] for the Imagenet dataset, and after many researchers follow the network for multiple applications like CBIR, MIR, Classification,Image2Image generation [2,6–13,36] etc. After that, many deep depth networks like VGG [34], ResNet [33] proposed for the same application. In the field of medical imaging also CNN getting boosted, initially U-Net [41] is proposed to segment medical images. There are many applications in the computer vision area like image to image transform, haze removal, moving object segmentation, anomaly detection, etc., utilizing U-Net's strength [4,35,37,38, 40]. Hambarde et al. [27] have proposed a prostate gland and prostate lesion segmentation algorithm.

Generative Adversarial Networks are boosted for image segmentation and other similar application. So inspired by the pix2pix [30] application and developed an identity mapping-based generative adversarial network for image reconstruction. In adversarial training, the generator generates input image and discriminator differentiate input and output image. Likewise, both generator and discriminator models are trained. The significant contributions of the proposed approach are:

- An End-to-end robust enriched deep residual framework is proposed for the medical image retrieval
- The latent encoded features are estimated with series of residual blocks in the encoder module of EDR framework. These latent features are fed to the decoder for the robust reconstruction of a given input medical image.
- The achievement of the stated strategy is empirically verified on two benchmark medical image databases.

Brief introduction of MIR and existing state-of-the-art methods like handcrafted, deep learning, and GAN-based approaches explained in a 1st section. In the 2nd section, we will discuss the details of the proposed architecture. Training details and architecture discussion results are discussed in Sect. 3. The final conclusion of the proposed work is described in Sect. 4.

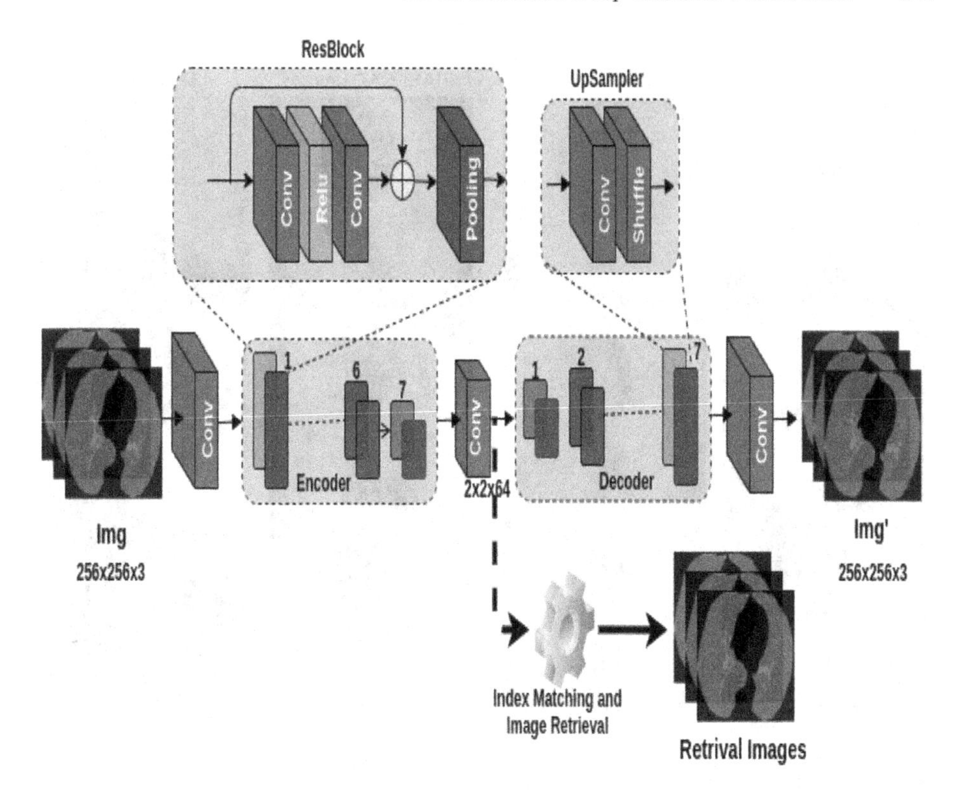

Fig. 1. Enriched Deep Residual Framework for MCBIR. Stage1: Reconstruct image using Enriched Deep Residual Network having input and output image of size $256 \times 256 \times 3$, Stage2: Encoder Feature extraction from stage 1 and index matching.

2 Proposed EDR: Enriched Deep Residual Framework For MCBIR

In this section, the proposed EDR framework for medical image retrieval is discussed. The EDR framework consists of an encoder-decoder module for latent feature extraction and reconstruction of given input medical images. Further, the encoded latent features are fed to the similarity index and retrieval module. The latent features are estimated using series of residual blocks, as shown in Fig. 1. The residual connections are employed for abstract representation of the given medical image and ease the reconstruction capability of the decoder network. Further, these latent abstract input features are used to robust index matching and retrieve similar images from the medical image database. The EDR framework of medical image retrieval is divided into Enriched Deep Residual Framework and Image indexing and retrieval modules, respectively.

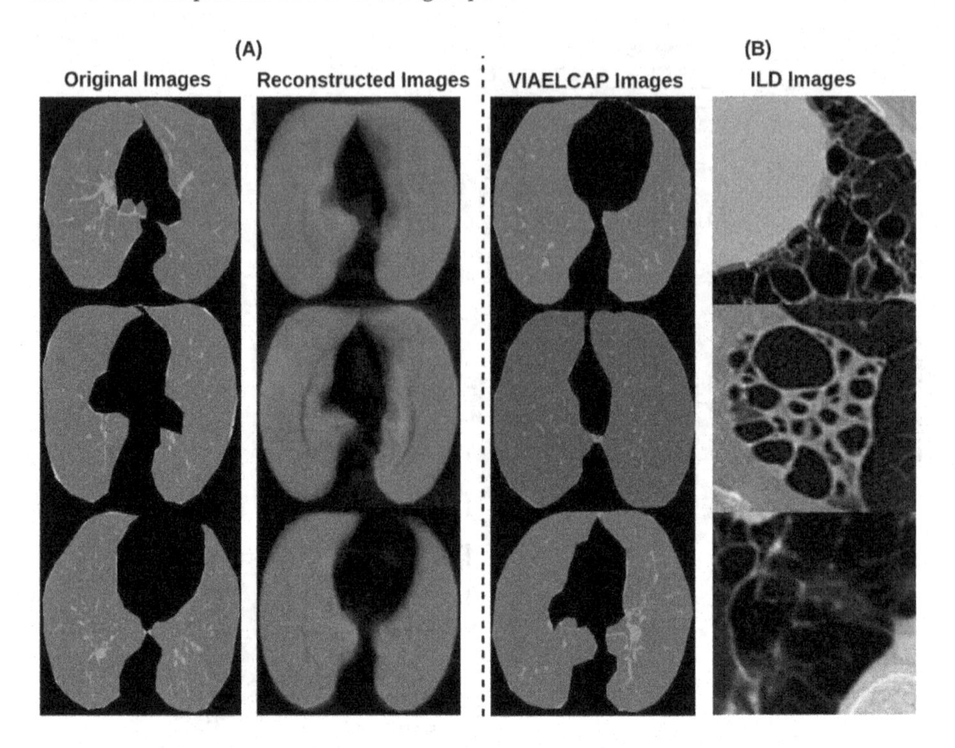

Fig. 2. Left two rows shows EDR Net reconstructed image, 1st column depicts original images and 2nd column reconstructs original image. 3rd and 4th column are sample images from VIAELCAP and ILD images.

2.1 Enriched Deep Residual Framework For Medical Image Reconstruction

The auto encoder-decoder networks have gained much attention in image enhancement, reconnection, image super-resolution, object/moving object segmentation, depth estimation, etc. Inspired by the technical functionality of the network, we proposed an end-to-end enriched deep residual framework (EDR) for medical image reconstruction and retrieval. The EDR framework aims to learn salient input features using a stack of seven residual blocks as shown in Fig 1. The residual block consists of a pooling operation followed by sequential convolutions and activation functions with residuality.

In the proposed EDR framework, The encoder layers are designed using res Block $RB^f_{3\times3}$ is a combination of two convolutions and relu with input features shown in fig and extracted using Eq. 1 and

$$RB^f_{3\times3} = MP_{2\times2}(C^f_{3\times3}(\Re(C^f_{3\times3}(x)))) \tag{1}$$

where, x is a input feature to a res block, $C^f_{3\times3}$ is convolution with 3×3 filter size and f filters, $MP_{2\times2}$ is a maxpooling with stride 2.

The proposed network architecture is characterized by seven sequential residual blocks to represent the input image by its abstract version in the encoder module. Using seven residual blocks on the input image of size 256×256 image reduces to the $2 \times 2 \times 64$ features as shown in Fig. 1 and Encoder features are extracted using Eq. 2. This abstract representation is further used to recover the input medical image with the help of decoder architecture, as shown in Fig. 1.

$$latentfeat = c_{3x3}^{64}((RB_{3x3}^{64}(c_{3x3}^{64}(I)) * 7) \tag{2}$$

The decoder architecture of the proposed framework is designed using convolution and shuffle with upsample shown in Fig. 1. Input for the decoder is a latent space produced from encoder, and output for this is reconstructed input image which can be done by Eq. 3

$$I^{'} = c_{3\times3}^{3}((Shuff(c_{3\times3}^{f}(latentfeat))) * 7) \tag{3}$$

2.2 Effective Feature Extraction and Similarity Indexing

Robust feature extraction is one of the most prominent steps in the CBIR framework to achieve significant performance. The traditional approaches in the literature like LBP [3], ANTIC [14], MEPP [5], LTrP [17], etc. are following the rule-based strategies in designing the descriptors. However, these descriptors fail to handle the various challenges in big medical data processing. therefore, the learning-based feature descriptors, effectively extract the features from input image (AlexNet [32], ResNet [33], VGG-16[34]).

Figure 2(A) shows the qualitative responses of the proposed EDR network for image reconstruction from the abstract features of an encoder. This indicates the efficiency of the encoder in EDR network for the generation of abstract feature representation for the input medical image. Thus, the abstract representation from the encoder of the proposed EDR network is used as the robust features ($latentfeat$) of the respective input image. In this work, we use these powerful features for index matching and retrieval tasks. Figure. 1 shows the feature extraction from the query medical image and the index matching and retrieval process.

Let, $fq = [f_{q1}, f_{q2}, f_{q3}, ..., f_{qN}]$ are feature vector of input query image. $f_{Di} = [f_{D1}, f_{D2}, f_{D3}, ..., f_{DN}]$ are the feature vectors of N images from Database i.e. $1, 2, 3, ..., N$.

The similarity between the input query image fq and the images in the database fDi is estimated using d1 distance measure. Based on the d1 similarity, n top similar images are retrieved from the database. The d1 distance is given as in Eq. 4

$$d(fq, f_D) = \sum_{i=1}^{N} \left| \frac{f_{Di} - f_q}{1 + f_{Di} + f_q} \right| \tag{4}$$

where, f_q is the query image, N is the length of feature vector, f_D is database images.

Table 1. Average retrieval precision (ARP) of the proposed EDR network over descriptor and learning-based techniques on ILD database for top 10 image retrieval.

Method	1	2	3	4	5	6	7	8	9	10
SS3D [39]	100	83.21	76.85	72.34	69.27	66.44	64.48	62.71	61.43	60.40
LTCoP [24]	100	84.95	76.29	71.77	67.44	64.79	62.57	60.83	59.14	58.13
LTrP [17]	100	85.26	77.56	71.96	68.54	65.02	62.48	60.75	59.08	57.61
MDMEP [28]	100	86.17	79.43	75.87	72.58	70.09	68.32	67.21	66.18	65.4
AlexNet [32]	100	88.91	83.99	79.6	76.66	74.54	72.38	71.12	69.79	68.45
ResNet [33]	100	89.67	84.75	81.31	78.84	76.55	75.51	74.34	73.44	72.67
VGG-16 [34]	100	89.21	83.94	80.24	77.96	75.94	74.4	73.21	72.26	71.34
EDR Net	**100**	**87.50**	**84.66**	**83.00**	**82.00**	**79.50**	**79.00**	**78.62**	**78.66**	**78.70**

Table 2. Group-wise ARP comparison of the proposed EDR network over descriptor and learning based techniques on ILD database

Method	Group 1	Group 2	Group 3	Group 4	Group 5	Average
LTCoP [24]	62.64	69.95	53.96	48.72	78.97	62.85
SS3D [39]	47.17	72.62	53.21	52.82	75.68	60.30
LTrP [17]	54.72	72.41	50.94	53.85	77.51	61.88
MDMEP [28]	75.09	70.37	68.11	43.08	79.56	67.24
AlexNet [32]	65.28	83.10	65.85	48.72	82.64	69.12
ResNet [33]	71.70	88.24	61.51	46.67	85.13	70.65
VGG-16 [34]	73.96	87.49	57.17	57.44	83.22	71.86
EDR Net	**88.62**	**84.6**	**76.15**	**79.29**	**83.10**	**82.35**

3 Experimental Results and Discussion

The details of the training procedure for Enriched deep residual framework is elaborated in the section. Specifically, the EDR-NET is trained in ad-versarial fashion for medical image reconstruction task. BraTS-2015 database [26] is considered for the training of the proposed EDR-NET, which comprises 54 LGG and 220 HGG patient scans.

This section will describe the effectiveness of the proposed network over the state of the art hand-crafted and learning-based methods. The proposed EDR framework is trained on BraTS-2015 database which comprises 54 LGG and 220 HGG patient scans and done inference on integrated interstitial lung disease (ILD) [22], and VIA/I-ELCAP-CT [21] databases with CT scans images. The performance of the approaches is measured in terms of average retrieval precision (ARP), average retrieval rate (ARR). The sample images of the ILD and VIA/I-ELCAP-CT datasets are shown in Fig. 2(B).

3.1 Retrieval Accuracy on ILD Database

The interstitial lung disease (ILD) dataset [22] is considered to validate the effect of the proposed EDR framework and other existing state-of-the-art approaches. The ILD diagnosis can be efficiently enhanced using a relatively less invasive X-ray method since X-ray and CT of the same patient are stored in the same resource. This dataset includes lung CT scans from 130 patients in which every ILD has highlighted disease regions by the experienced Radiologists [4]. Also, a small dataset (from ILD) is available consisting of 658 regions with marking of patches/regions for different classes such as fibrosis (187 regions), ground glass (106 regions), emphysema (53 regions), micronodules (173 regions), and healthy (139 regions).

Table 3. Average retrieval precision (ARP) of the proposed EDR network over descriptor and learning-based techniques on VIA/I-ELCAP database for top 10 image retrieval.

Method	1	2	3	4	5	6	7	8	9	10
SS3D [39]	100	74.25	64.27	58.70	54.78	52.55	50.47	48.68	47.44	46.23
LTCoP [24]	100	93.80	91.10	88.75	86.80	85.35	84.13	82.60	81.31	80.28
LTrP [17]	100	81.60	74.73	70.55	67.70	65.32	63.49	61.96	60.88	59.62
MDMEP [28]	100	81.60	74.73	70.55	67.70	65.32	63.49	61.96	60.88	59.62
AlexNet [32]	100	99.60	99.13	98.63	98.06	97.43	96.56	95.85	94.89	93.88
ResNet [33]	100	98.35	96.60	94.88	93.38	91.98	90.27	88.68	87.22	85.78
VGG-16 [34]	100	94.55	90.33	86.88	84.16	81.85	79.76	77.78	75.96	74.23
EDRNet	**100**	**99.5**	**98.00**	**98.25**	**98.00**	**98.00**	**97.71**	**98.00**	**98.00**	**97.60**

Table 4. Average retrieval rate (ARR) of the proposed EDR network over descriptor and learning based techniques on VIA/I-ELCAP database for top 10 image retrieval.

Method	1	2	3	4	5	6	7	8	9	10
SS3D [39]	10	14.85	19.28	23.48	27.39	31.53	35.33	38.94	42.70	46.24
LTCoP [24]	10	18.76	27.33	35.50	43.40	51.21	58.89	66.08	73.18	80.28
LTrP [17]	10	16.32	22.42	28.22	33.85	39.19	44.44	49.57	54.79	59.62
MDMEP [28]	10	16.32	22.42	28.22	33.85	39.19	44.44	49.57	54.79	59.62
AlexNet [32]	10	19.92	29.74	39.45	49.03	58.46	67.59	76.68	85.40	93.88
ResNet [33]	10	19.67	28.98	37.95	46.69	55.19	63.19	70.94	78.50	85.78
VGG-16 [34]	10	18.91	27.10	34.75	42.08	49.11	55.83	62.22	68.36	74.23
EDR Net	**10**	**19.9**	**29.4**	**39.3**	**49.0**	**58.8**	**68.4**	**78.4**	**88.2**	**97.6**

The comparison of the proposed method with existing methods in terms of the ARP on top 10 images with the ILD dataset is given in Table 1. Table 2 shows the comparison of the proposed method and existing methods in terms of group-wise average retrieval precision (GARP) on the ILD dataset. From Table 1 and Table 2, it is clear that the proposed EDR framework outperforms all the existing methods for image retrieval.

3.2 Retrieval Accuracy on VIA/I-ELCAP-CT Dataset

VIA/I-ELCAP-CT dataset [21] is a benchmark dataset used in CBMIR. This dataset has 1000 images from 10 classes, each class having 100 images. Table 3 shows the average retrieval precision (ARP) of the proposed EDR network over descriptor and learning-based techniques on VIA/I-ELCAP database for top 10 image retrieval. Similarly, Table 4 shows the average retrieval rate (ARR) of the proposed EDR network over descriptor and learning-based techniques on VIA/I-ELCAP database for top 10 image retrieval. From the above Table 3, 4, it is concluded proposed method outperforms the different existing approaches in MIR.

4 Conclusion

This paper proposed an end-to-end deep residual framework utilizing latent encoded features with a series of residual blocks.The proposed enriched deep residual framework (EDR) performs the robust retrieval of images from the medical database. Encoding is performed by the residual block which generates the latent features. These features fed to the decoder for reconstruction of the a given input image. EDR performance is verified on the two benchmark medical datasets. The results validates the improved performance of the proposed EDR framework.

References

1. Lowe, D.G.: Object recognition from local scale-invariant features. In: Proceedings of the Seventh IEEE International Conference on Computer Vision, vol. 2. IEEE (1999)
2. Kulkarni, A., Patil, P.W., Murala, S.: Progressive subtractive recurrent lightweight network for video deraining. IEEE Signal Process. Lett. **29**, 229–233 (2021)
3. Ojala, T., Pietikainen, M., Maenpaa, T.: Multiresolution gray-scale and rotation invariant texture classification with local binary patterns. IEEE Trans. Pattern Anal. Mach. Intell. **24**(7), 971–987 (2002)
4. Patil, P., Murala, S.: FgGAN: a cascaded unpaired learning for background estimation and foreground segmentation. In: 2019 IEEE Winter Conference on Applications of Computer Vision (WACV), pp. 1770–1778. IEEE (2019)
5. Mandal, M., et al.: ANTIC: ANTithetic isomeric cluster patterns for medical image retrieval and change detection. IET Comput. Vis. **13**(1), 31–43 (2019)

6. Phutke, Sh.S., Murala, S.: Diverse receptive field based adversarial concurrent encoder network for image inpainting. IEEE Signal Process. Lett. **28**, 1873–1877 (2021)
7. Nancy, M., Murala, S.: MSAR-Net: multi-scale attention based light-weight image super-resolution. Pattern Recognit. Lett. **151**, 215–221 (2021)
8. Patil, P.W., et al.: An unified recurrent video object segmentation framework for various surveillance environments. IEEE Trans. Image Process. **30**, 7889–7902 (2021)
9. Akshay, D., Hambarde, P., Patil, P., Murala, S.: Deep underwater image restoration and beyond. IEEE Signal Process. Lett. **27**, 675–679 (2020)
10. Hambarde, P., Dudhane, A., Murala, S.: Single image depth estimation using deep adversarial training. In: 2019 IEEE International Conference on Image Processing (ICIP), pp. 989–993. IEEE (2019)
11. Hambarde, P., Dudhane, A., Patil, P.W., Murala, S., Dhall, A.: Depth estimation from single image and semantic prior. In: 2020 IEEE International Conference on Image Processing (ICIP), pp. 1441–1445. IEEE (2020)
12. Praful, H., Murala, S.: S2DNet: depth estimation from single image and sparse samples. IEEE Trans. Comput. Imaging **6**, 806–817 (2020)
13. Patil, P.W., Dudhane, A., Chaudhary, S., Murala, S.: Multi-frame based adversarial learning approach for video surveillance. Pattern Recognit. **122**, 108350 (2022)
14. Vipparthi, S.K., et al.: Local directional mask maximum edge patterns for image retrieval and face recognition. IET Comput. Vis. **10**(3), 182–192 (2016)
15. Vipparthi, S.K., et al.: Local Gabor maximum edge position octal patterns for image retrieval. Neurocomputing **167**, 336–345 (2015)
16. Vipparthi, S.K., Murala, S., Nagar, S.K.: Dual directional multi-motif XOR patterns: a new feature descriptor for image indexing and retrieval. Optik **126**(15-16), 1467–1473 (2015)
17. Murala, S., Maheshwari, R.P., Balasubramanian, R.: Local tetra patterns: a new feature descriptor for content-based image retrieval. IEEE Trans. Image Process. **21**(5), 2874–2886 (2012)
18. Vipparthi, S.K., Nagar, S.K.: Directional local ternary patterns for multimedia image indexing and retrieval. Int. J. Signal Imaging Syst. Eng. **8**(3), 137–145 (2015)
19. Mohite, N., et al.: 3D local circular difference patterns for biomedical image retrieval. Int. J. Multimedia Inf. Retr. **8**(2), 115–125 (2019)
20. Vipparthi, S.K., Nagar, S.K.: Integration of color and local derivative pattern features for content-based image indexing and retrieval. J. Inst. Eng. (India) Ser. B **96**(3), 251–263 (2015)
21. Via/i-elcap database. http://www.via.cornell.edu/lungdb.html. Accessed 10 Mar 2019
22. Depeursinge, A., et al.: Building a reference multimedia database for interstitial lung diseases. Comput. Med. Imaging Graph. **36**(3), 227–238 (2012)
23. Vipparthi, S.K., Nagar, S.K.: Local extreme complete trio pattern for multimedia image retrieval system. Int. J. Autom. Comput. **13**(5), 457–467 (2016). https://doi.org/10.1007/s11633-016-0978-2
24. Murala, S., Wu, Q.M.J.: Local ternary co-occurrence patterns: a new feature descriptor for MRI and CT image retrieval. Neurocomputing **119**, 399–412 (2013)
25. Biradar, K.M., et al.: Local Gaussian difference extrema pattern: a new feature extractor for face recognition. In: 2017 Fourth International Conference on Image Information Processing (ICIIP). IEEE (2017)

26. Isensee, F., Kickingereder, P., Wick, W., Bendszus, M., Maier-Hein, K.H.: Brain tumor segmentation and radiomics survival prediction: contribution to the BRATS 2017 challenge. In: Crimi, A., Bakas, S., Kuijf, H., Menze, B., Reyes, M. (eds.) BrainLes 2017. LNCS, vol. 10670, pp. 287–297. Springer, Cham (2018). https://doi.org/10.1007/978-3-319-75238-9_25
27. Hambarde, P., et al.: Prostate lesion segmentation in MR images using radiomics based deeply supervised U-Net. Biocybern. Biomed. Eng. **40**(4), 1421–1435 (2020)
28. Galshetwar, G.M., Waghmare, L.M., Gonde, A.B., Murala, S.: Multi-dimensional multi-directional mask maximum edge pattern for bio-medical image retrieval. Int. J. Multimedia Inf. Retr. **7**(4), 231–239 (2018). https://doi.org/10.1007/s13735-018-0156-0
29. Vipparthi, S.K., Nagar, S.K.: Expert image retrieval system using directional local motif XoR patterns. Expert Syst. Appl. **41**(17), 8016–8026 (2014)
30. Isola, P., et al.: Image-to-image translation with conditional adversarial networks. In: Proceedings of the IEEE Conference on Computer Vision and Pattern Recognition (2017)
31. Vipparthi, S.K., Nagar, S.K.: Multi-joint histogram based modelling for image indexing and retrieval. Comput. Electr. Eng. **40**(8), 163–173 (2014)
32. Krizhevsky, A., Sutskever, I., Hinton, G.E.: ImageNet classification with deep convolutional neural networks. Adv. Neural Inf. Process. Syst. **25**, 1097–1105 (2012)
33. He, K., et al.: Deep residual learning for image recognition. In: Proceedings of the IEEE Conference on Computer Vision and Pattern Recognition (2016)
34. Simonyan, K., Zisserman, A.: Very deep convolutional networks for large-scale image recognition. arXiv preprint arXiv:1409.1556 (2014)
35. Dudhane, A., et al.: Varicolored image de-hazing. In: Proceedings of the IEEE/CVF Conference on Computer Vision and Pattern Recognition (2020)
36. Patil, P.W., et al.: MSFgNet: a novel compact end-to-end deep network for moving object detection. IEEE Trans. Intell. Transp. Syst. **20**(11), 4066–4077 (2018)
37. Patil, P.W., et al.: An end-to-end edge aggregation network for moving object segmentation. In: Proceedings of the IEEE/CVF Conference on Computer Vision and Pattern Recognition (2020)
38. Biradar, K.M., et al.: Challenges in time-stamp aware anomaly detection in traffic videos. arXiv preprint arXiv:1906.04574 (2019)
39. Murala, S., Wu, Q.M.J.: Spherical symmetric 3D local ternary patterns for natural, texture and biomedical image indexing and retrieval. Neurocomputing **149**, 1502–1514 (2015)
40. Biradar, K., Dube, S., Vipparthi, S.K.: DEARESt: deep convolutional aberrant behavior detection in real-world scenarios. In: 2018 IEEE 13th International Conference on Industrial and Information Systems (ICIIS). IEEE (2018)
41. Ronneberger, O., Fischer, P., Brox, T.: U-Net: convolutional networks for biomedical image segmentation. In: Navab, N., Hornegger, J., Wells, W.M., Frangi, A.F. (eds.) MICCAI 2015. LNCS, vol. 9351, pp. 234–241. Springer, Cham (2015). https://doi.org/10.1007/978-3-319-24574-4_28

Depth Estimation Using Sparse Depth and Transformer

Roopak Malik, Praful Hambarde(✉), and Subrahmanyam Murala

CVPR Lab, IIT Ropar, Ropar, India
{2019eem1012,2018eez0001,subbumurala}@iitrpr.ac.in

Abstract. Depth prediction from single image is a challenging task due to the intra scale ambiguity and unavailability of prior information. The prediction of an unambiguous depth from single RGB image is very important aspect for computer vision applications. In this paper, an end-to-end sparse-to-dense network using transformers is proposed for depth estimation. The proposed network processes single images along with the additional sparse depth samples which have been generated for depth estimation. The additional sparse depth sample are acquired either with a low-resolution depth sensor or calculated by visual simultaneous localization. Here, we have proposed a model that utilises both sparse samples and transformers and along with a encoder-decoder structure that helps us in giving great depth results that are comparable to other state-of-the-art results.

1 Introduction

Depth estimation is a very crucial topic required in many different engineering applications such as 3D mapping, robotic vision, measurements and length predictions, autonomous driving *etc.* An accurate depth map is very important for navigation, scene understanding in robotics. Any inaccuracy would lead to poor performance of the entire operation. Therefore, depth prediction makes up one of the most crucial aspects of the aforementioned fields [1]. Although this can be done using existing depth sensors such as LiDAR (Light Detection and Ranging), RGB-D cameras or stereo cameras, they each have their own shortcomings. For example, even though most 3D LiDAR sensors are extremely expensive, they often only provide sparse measurements of depth for distant objects. Others are sunlight sensitive and heavy on power consumption, and most face the problem of short ranging distances.

The main method proposed here is a deep learning based network which takes as its input a pair of RGB images as well as a set of sparse samples and generates a proper depth map from them. We pair this approach with the use of visual transformers, an emerging new method inspired by transformers from the domain of Natural Language Processing (NLP) [2]. This provides much better global self-attention mechanisms to capture long-range dependencies as compared to

B. Raman et al. (Eds.): CVIP 2021, CCIS 1568, pp. 329–337, 2022.
https://doi.org/10.1007/978-3-031-11349-9_29

standard convolution-based architectures. Transformers were originally used for sequence-to-sequence modelling of NLP tasks but have recently attracted a lot of attention from different fields such as computer vision [3]. We show that our approach offers a different perspective in depth estimation tasks and achieves comparable results to most other methods.

2 Literature Survey

Depth Estimation Using Only RGB Images: Nowadays, deep-learning-based architecture are used in different application such as object segmentation [19,24,25,27,28], image de-hazing [20,21,30], depth estimation [22,23,26], image super-resolution [18], image-inpainting [17], Object detection [29,31] *etc.*. Earlier work done in the field of depth map generation by using RGB images was based on hand-crafted methods. Others also used probabilistic graphical methods, Markov Random field models *etc.* Saxena *et al.* [4] is an example of a work of depth estimation done using Markov Random field models. Non-parametric techniques such as combination of image depths with other photometric data were also used to estimate depth.

In Ma *et al.* [4], they introduce additional sparse depth samples, which are either acquired with a low-resolution depth sensor or computed via visual Simultaneous Localization and Mapping (SLAM) algorithms. They also propose the use of a single deep regression network to learn directly from the RGB-D raw data, and explore the impact of number of depth samples on prediction accuracy.

Similarly, in Hambarde *et al.* [5], they have explored the idea of using a two-stream deep adversarial network for single image depth estimation in RGB images. The first stream utilised a novel encoder-decoder architecture using residual concepts to extract course-level depth features, whereas the second stream was used to purely process the information through the residual architecture for fine-level depth estimation. They had also designed a feature map sharing architecture to share the learned feature maps of the decoder module of first stream. This would strengthen the residual learning to estimate the scene depth and increase the robustness of the proposed network.

Works on sparse samples have mostly been focused on using sparse depth representations for generating depth signal values. As an example, Hawe *et al.* [6] worked on the assumption that Wavelet basis has sparse maps of disparity and generate these dense image with a complex gradient technique. Some other works make use of the sparse values defining the 2nd order derivatives created from depth values, which are shown to perform better than most others in re-construction accuracy and speed

3 Proposed Method

The overall architecture of the proposed method is illustrated in Fig. 1. In this section, we mainly introduce the visual transformers and sparse samples based network.

3.1 Sparse Depth

Sparse depth samples are essentially a few randomly selected pixel points from the ground truth depth map of an image that are taken and concatenated along with their respective RGB input images to aid in the process of depth map prediction [7]. The number of sparse samples taken are represented by N, and are mostly taken around 200. We have performed testing with different N values such as 100, 200, 500 etc., and have found N = 200 to give the best results. The advantage of using sparse depth is that it is readily available in many forms such as low resolution depth sensors (e.g., a low-cost LiDARs), or can be generated artificially. This helps us to use sparse samples taken from the ground truth during the training phase and then using sparse depth generated from sources such as low cost LiDARs during testing and real-life implementation phases. We generate sparse depth from the ground truth and concatenate it with RGB input. This causes our input to go from 3 channel (from RGB) to a 4 channel input image and the consequent network is designed by keeping this in mind.

Sparse Sample Generation. For our training phase, the depth image from the ground truth (denoted as D*) is sampled randomly to generate the required input sparse depth map (denoted as D) [4]. Therefore, for a specified number of depth samples N (decided before training), we compute a Bernoulli probability given by $\boldsymbol{p} = \boldsymbol{N/n}$, where n is the total number of valid pixels in D* given by the image dimensions of D* (width * height). So, for any pixel denoted by (i, j) in the image D*, its corresponding value in D is given by:

$$D(i,j) = \begin{cases} D^*(i,j) & With\, probability\, p \\ 0 & Otherwise \end{cases} \tag{1}$$

With this method, the number of non-zero values from depth pixels varies for each training sample around the expectation N. This is different from dropout since in dropout we scale the output up by 1/p during training. This is done to compensate for the deactivated neurons.

Our strategy helps in not only creating more data in the form of data augmentation, but also helps to increase the robustness of our model against a different number of inputs.

3.2 Proposed Network

Our proposed model consists of an encoder-decoder structure that incorporates a transformer in the middle. The proposed structure contains a total of 40 million trainable parameters. It takes in a 4 channel (RGB + sparse) image as its input and outputs an image with the same height and width dimensions but with a single channel. This configuration is called the 'rgbd' modality. It also supports a 'rgb' modality with only 3 input channels.

The encoder structure is based on the Resnet18 backbone architecture. It incorporates the encoder structure similar to that of Resnet18 but with a few

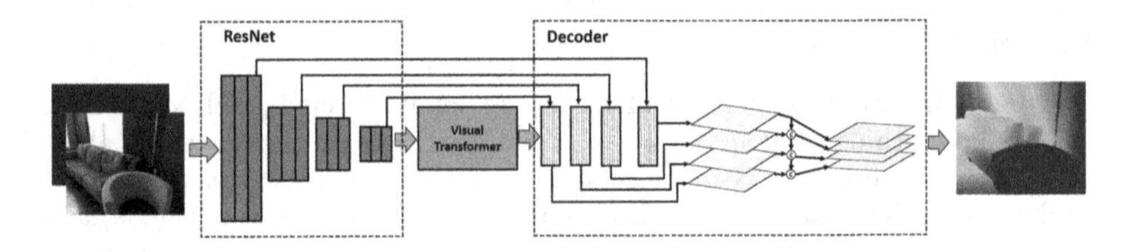

Fig. 1. Proposed transformer model with sparse samples

notable differences. Unlike a conventional Resnet model, our encoder also implements a few features of Unet structures such as skip connections. Therefore, along with the final output layer, the encoder also returns 3 skip connections after each layer of the encoder.

The encoder structure is followed by a visual transformer. In vision transformers, the input sequence is given in the form of flattened vector of pixel values extracted from a patch of size PxP [8]. The flattened elements are then given to a linear projection layer to produce their respective patch embeddings. This embedding helps to give a positional value to the patches generated. A Multi Layer Perceptron (MLP) head is added to the top of the transformer [9].

The final output of the encoder is first passed to the positional embedding block followed by the transformer [10]. The embedding block performs the necessary positional embedding on to the input given which has been described in the transformer architecture. The result obtained after embedding is then fed to the transformer to produce intermediate results. This encoder takes in the input as the last layer and results the intermediate layers generated as the output along with the input passed, after generating the final Z values. This final result is then passed to the decoder structure along with the skip connections.

We have designed a multi-level feature aggregation decoder structure as shown in the figure. The decoder structure proposed are can be described as (Fig. 2):

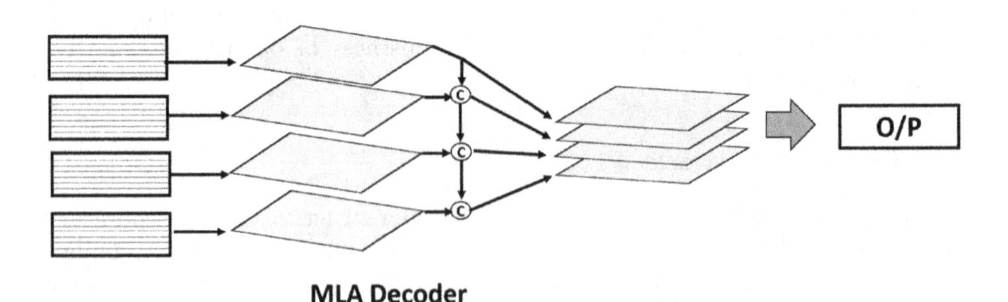

Fig. 2. Proposed decoder architecture

MLA (Multi-level Feature Aggregation) Decoder is similar to a feature pyramid network but unlike the pyramid network, it contains similar dimensions for all the inputs. We obtain the fused feature from all the streams via channel-wise concatenation which is then bilinearly upsampled 4× to the full resolution [12].

We use proposed decoder_mla as our base decoder since the results from this have proven to beat those of other standard decoder structures.

3.3 Losses

We use $L1$ loss as the error metric in our proposed model. The $L1$ loss is generate by calculating the sum of all the absolute values of the difference between the predicted image and the target image.

$$L1Loss = \sum_{i=1}^{n} |y_{true} - y_{predicted}| \tag{2}$$

Apart from this loss, we also added a different loss function in the form of edge loss. Edge loss is calculated in the form of gradients and is useful for calculating the difference between the target image edge gradients and the predicted ones [13].

$$EdgeLoss = \sum (\|g_1' - g_2'\|)^2 \tag{3}$$

where, $g'_1 = \sqrt{g_1(x)^2 + g_1(y)^2}$ and $g'_2 = \sqrt{g_2(x)^2 + g_2(y)^2}$ *i.e.* $g_1(x)$ and $g_1(y)$ are gradient edges in x and y direction for predicted image. Similarly, $g_2(x)$ and $g_2(y)$ are gradient edges in x and y direction for the target image.

We give the L1 loss a weightage of 1 while the edge loss is given a much smaller weightage of 0.2 since these values were tried and found to be the most effective and the L1 loss is the primary loss function used. The overall loss function can be then summarised by using the formula:

$$loss = L1Loss + 0.1 * EdgeLoss \tag{4}$$

4 Experimental Analysis

4.1 Training Details

We have implemented the proposed sparse and transformer model using PyTorch deep learning framework. The proposed network is trained on the NYU-Depth-V2 dataset utilizing stochastic gradient descent (SGD) back propagation algorithm. Training hyperparameters are updated on NVIDIA DGX station with processor 2.2 GHz, Intel Xeon E5-2698, NVIDIA Tesla V100 1 × 16 GB GPU. We have used a batch size of 4 and epoch 30 for NYU-Depth-V2 dataset. During the training of our model, learning rate started at 0.01 and decreased successively after every 5 epochs. For regularization purpose we use 0.0001 as the weight decay.

4.2 Dataset Description

NYU Dataset. In this work, we have used the NYU Depth v2 dataset as a benchmark dataset. This dataset has 1449 image pairs of input images along with their ground truth. Out of these, 795 images were taken for training dataset whereas the remaining 654 images made up the testing dataset. Although this dataset is available in both raw images along with their dump files as well as matlab formatted data, we have extracted the images from the mat file due to the comparably smaller size of the mat file as compared to the raw images. We have worked with both img and hdf5 file formats to obtain the results. Preprocessing of the dataset is done in the form of rotation, random jittering and horizontal flip with a probability of 0.5. The images are all then resized via the centercrop transform to obtain the final size of (640, 480) [14] (Fig. 3).

Fig. 3. Example of images from the NYU depth v2 dataset [8]

5 Results

The proposed algorithm is evaluated on the publicly available dataset from NYU Depth v2 Dataset. Our results the comparison of the proposed deep learning based network both qualitatively and quantitatively with the state-of-the-art methods on NYU Depth-v2 dataset.

The following are the results we obtained after trying various different configurations with this transformer structure. We have used a jet color map here instead of the grayscale map as used before. The jet colormap gives us a much better representation of the depths of the image and is more frequently used for depth estimation problems. As can be seen here, we have used a jet color map here instead of the grayscale map as used before. The jet colormap gives us a much better representation of the depths of the image and is more frequently used for depth estimation problems. We have performed the comparison of our network- in both quantitative as well as qualitative manner, on the described NYU depth v2. The proposed deep network has been evaluated by examining error metrics such as Root Mean Square Error (RMSE), log 10, REL and δ_i values (Fig. 4 and Table 1).

Fig. 4. Proposed model results

Table 1. Comparison and evaluation of the metrics observed on the datasets.

Methods	RMSE	MAE	log10	Absrel	δ_1	δ_2	δ_3
Ma *et al.* [4]	0.230	0.120	0.022	0.044	97.1	99.4	99.8
Hambarde and Murala [5]	0.251	0.125	0.079	0.141	97.8	99.5	99.9
Ours (Proposed)	**0.219**	**0.121**	**0.019**	**0.044**	**97.9**	**99.5**	**99.9**

6 Conclusion

The deep learning model proposed consists of the self-attention based model that learns and extracts important features from the images and performs the estimation of a accurate depth prediction. Many thorough experiments have been carried out to prove the robustness of the proposed Network to estimate the scene depth map. To analyse the performance of the proposed model, one benchmark dataset namely: NYU Depth v2 Dataset has been utilized. We have carried out the evaluation both qualitative as well as quantitative, of our network proposed on our benchmark data set and compared the results of our proposed model with our other methods. Also, our proposed method of depth estimation of monocular images which incorporates transformers can also be used for various other applications such as robotic vision, if sufficient modifications can be made.

References

1. Bousmalis, K., Silberman, N., Dohan, D., Erhan, D., Krishnan, D.: Unsupervised pixel-level domain adaptation with generative adversarial networks. In: Proceedings of the IEEE Conference on Computer Vision and Pattern Recognition (2017)
2. Chen, J., et al.: Transformers make strong encoders for medical image segmentation. arXiv preprint arXiv:2102.13645 (2021)
3. Soleymani, A.A.M.-M., Deep Learning: Transformer Networks (2019)
4. Ma, F., Karaman, S.: Sparse-to-dense: depth prediction from sparse depth samples and a single image. In: 2018 IEEE International Conference on Robotics and Automation (ICRA) (2018)
5. Hambarde, P., Murala, S.: S2DNet: depth estimation from single image and sparse samples. IEEE Trans. Comput. Imaging **6**, 806–817 (2020)
6. Vaswani, A., et al.: Attention is all you need. arXiv preprint arXiv:1706.03762 (2017)
7. Zheng, S., et al.: Rethinking semantic segmentation from a sequence-to-sequence perspective with transformers. arXiv preprint arXiv:2012.15840 (2020)
8. Geiger, A., Lenz, P., Stiller, C., Urtasun, R.: Vision meets robotics: the KITTI dataset. Int. J. Robot. Res. (IJRR) **32**, 1231–1237 (2013)
9. Koch, T., Liebel, L., Fraundorfer, F., Körner, M.: Evaluation of CNN-based single-image depth estimation methods. In: Leal-Taixé, L., Roth, S. (eds.) ECCV 2018. LNCS, vol. 11131, pp. 331–348. Springer, Cham (2019). https://doi.org/10.1007/978-3-030-11015-4_25
10. Roy, A., Todorovic, S.: Monocular depth estimation using neural regression forest. In: Proceedings of the IEEE Conference on Computer Vision and Pattern Recognition (2016)
11. Wang, W., Chen, C., Ding, M., Li, J., Yu, H., Zha, S.: TransBTS: multimodal brain tumor segmentation using transformer. arXiv preprint arXiv:2103.04430 (2021)
12. Han, K., et al.: A survey on visual transformer. arXiv preprint arXiv:2012.12556 (2020)
13. Karimi, D., Vasylechko, S., Gholipour, A.: Convolution-free medical image segmentation using transformers. arXiv preprint arXiv:2102.13645 (2021)
14. Silberman, N., Hoiem, D., Kohli, P., Fergus, R.: Indoor segmentation and support inference from RGBD images. In: Fitzgibbon, A., Lazebnik, S., Perona, P., Sato, Y., Schmid, C. (eds.) ECCV 2012. LNCS, vol. 7576, pp. 746–760. Springer, Heidelberg (2012). https://doi.org/10.1007/978-3-642-33715-4_54

15. Dosovitskiy, A., et al.: An image is worth 16x16 words: transformers for image recognition at scale. arXiv preprint arXiv:2010.11929 (2020)
16. Yang, G., Tang, H., Ding, M., Sebe, N., Ricci, E.: Transformers solve the limited receptive field for monocular depth prediction. arXiv preprint arXiv:2103.12091 (2021)
17. Phutke, S.S., Murala, S.: Diverse receptive field based adversarial concurrent encoder network for image inpainting. IEEE Signal Process. Lett. **28**, 1873–1877 (2021)
18. Mehta, N., Murala, S.: MSAR-Net: multi-scale attention based light-weight image super-resolution. Pattern Recognit. Lett. **151**, 215–221 (2021)
19. Patil, P.W., et al.: An unified recurrent video object segmentation framework for various surveillance environments. IEEE Trans. Image Process. **30**, 7889–7902 (2021)
20. Dudhane, A., Hambarde, P., Patil, P., Murala, S.: Deep underwater image restoration and beyond. IEEE Signal Process. Lett. **27**, 675–679 (2020)
21. Dudhane, A., Biradar, K.M., Patil, P.W., Hambarde, P., Murala, S.: Varicolored image de-hazing. In: Proceedings of the IEEE/CVF Conference on Computer Vision and Pattern Recognition, pp. 4564–4573 (2020)
22. Hambarde, P., Dudhane, A., Murala, S.: Single image depth estimation using deep adversarial training. In: 2019 IEEE International Conference on Image Processing (ICIP), pp. 989–993. IEEE (2019)
23. Hambarde, P., Dudhane, A., Patil, P.W., Murala, S., Dhall, A.: Depth estimation from single image and semantic prior. In: 2020 IEEE International Conference on Image Processing (ICIP), pp. 1441–1445. IEEE (2020)
24. Patil, P.W., Biradar, K.M., Dudhane, A., Murala, S.: An end-to-end edge aggregation network for moving object segmentation. In: Proceedings of the IEEE/CVF Conference on Computer Vision and Pattern Recognition, pp. 8149–8158 (2020)
25. Patil, P.W., Dudhane, A., Chaudhary, S., Murala, S.: Multi-frame based adversarial learning approach for video surveillance. Pattern Recognit. **122**, 108350 (2022)
26. Hambarde, P., Murala, S., Dhall, A.: UW-GAN: single image DepthEstimation and image enhancement for underwater images. IEEE Trans. Instrum. Meas. **70**, 1–12(2021)
27. Hambarde, P., Talbar, S.N., Sable, N., Mahajan, A., Chavan, S.S., Thakur, M.: Radiomics for peripheral zone and intra-prostatic urethra segmentation in MR imaging. Biomed. Signal Process. Control **51**, 19–29 (2019)
28. Hambarde, P., Talbar, S., Mahajan, A., Chavan, S., Thakur, M., Sable, N.: Prostate lesion segmentation in MR images using radiomics based deeply supervised U-Net. Biocybern. Biomed. Eng. **40**(4), 1421–1435 (2020)
29. Bhagat, S., Kokare, M., Haswani, V., Hambarde, P., Kamble, R.: WheatNet-lite: a novel light weight network for wheat head detection. In: Proceedings of the IEEE/CVF International Conference on Computer Vision, pp. 1332–1341 (2021)
30. Alaspure, P., Hambarde, P., Dudhane, A., Murala, S.: DarkGAN: night image enhancement using generative adversarial networks. In: Singh, S.K., Roy, P., Raman, B., Nagabhushan, P. (eds.) CVIP 2020. CCIS, vol. 1376, pp. 293–302. Springer, Singapore (2021). https://doi.org/10.1007/978-981-16-1086-8_26
31. Bhagat, S., Kokare, M., Haswani, V., Hambarde, P., Kamble, R.: Eff-UNet++: a novel architecture for plant leaf segmentation and counting. Ecol. Inform. **68**, 101583 (2022)

Analysis of Loss Functions for Image Reconstruction Using Convolutional Autoencoder

Nishant Khare, Poornima Singh Thakur, Pritee Khanna(✉), and Aparajita Ojha

Computer Science and Engineering, PDPM IIITDM, Jabalpur, India
{nishantk,poornima,pkhanna,aojha}@iiitdmj.ac.in

Abstract. In recent years, several loss functions have been proposed for the image reconstruction task of convolutional autoencoders (CAEs). In this paper, a performance analysis of a CAE with respect to different loss functions is presented. Quality of reconstruction is analyzed using the mean Square error (MSE), binary cross-entropy (BCE), Sobel, Laplacian, and Focal binary loss functions. To evaluate the performance of different loss functions, a vanilla autoencoder is trained on eight datasets having diversity in terms of application domains, image dimension, color space, and the number of images in the dataset. MSE, peak signal to noise ratio (PSNR), and structural similarity index (SSIM) metrics have been used as the performance measures on all eight datasets. The assessment shows that the MSE loss function outperforms two datasets with a small image dimension and a large number of images. At the same time, BCE excels on six datasets with high image dimensions and a small number of training samples in datasets compared with the Sobel and Laplacian loss functions.

Keywords: Image reconstruction · Autoencoder · Loss functions

1 Introduction

Autoencoders are an unsupervised learning technique, where neural networks are used for representation learning. Recently, autoencoders are being used for learning features [1,2], removing noise (Denoising Autoencoder) [3], image enhancement (deblurring or dehazing images) [4,5], feature retrieval, image coloring, and compression. Autoencoders can reconstruct an image similar to the input image by feature learning and representation and found to be useful for image reconstruction in fields like medical science and forensic science, where the decoded and noise-free images are required from the available incomplete or noisy images [6].

An autoencoder works with encoder and decoder modules. The encoder module learns important features from a given image. The decoder module attempts to reconstruct the matching input image from this representation. The CAE learns useful features to reproduce the input image as the output. Due to the loss of some features, the reconstructed image may not match exactly with the

B. Raman et al. (Eds.): CVIP 2021, CCIS 1568, pp. 338–349, 2022.
https://doi.org/10.1007/978-3-031-11349-9_30

input image. Loss functions play an important role in achieving the desired reconstructed image. The performance of autoencoder depends on input data and the loss function. The motivation behind this work is to explore the performance of a convolutional autoencoder (CAE) using various existing loss functions on datasets from various domains. The objective is to analyze the reconstruction quality of images generated through an autoencoder with various loss functions. A variety of datasets including natural images, sketches, and medical images, are considered to fulfill this objective. The work is organized as follows. Section 2 presents related works. The CAE architecture used in the present analysis is discussed in Sect. 3. Loss functions used for the experimentation are explained in Sect. 4. Experimental setup, including the details of datasets, are discussed in Sect. 5. The experimental results on different datasets in terms of quantitative evaluation and qualitative comparisons are discussed in Sect. 6. Finally, the work is concluded in Sect. 7.

2 Related Work

Recent works on image reconstruction are focused on the use of autoencoders [3, 6, 7] (see also, [8, 9]). Autoencoders are primarily used for image reconstruction, but have also been utilized in a variety of different tasks like image denoising [3], anomaly detection [7], learning sparse representation and generative modeling. Although convolutional autoencoders can reconstruct images and have noise removal capability, there are specific issues about overall predicted image quality, edges, small artifacts, and blurriness. As autoencoders learn from raw data, they require large number of samples and clean representative data [10]. Also, autoencoders are data-specific and can lead to lossy reconstruction. The amount of acceptable reconstruction loss is application-specific. Considering the impact of the loss layer of neural networks, Zhao et al. [11] compared the performance of several loss functions (MS-SSIM, L1, and SSIM) for image restoration. They showed the importance of perceptually motivated losses when the resulting image is evaluated by a human observer and proposed a novel, differentiable error function. The new metric combines the advantages of L1 and MS-SSIM.

Li et al. [12] proposed steer image synthesis with Laplacian loss. They used neural style transfer based on CNN to synthesizes a new image that retains the high-level structure of a content image, rendered in the low-level texture of a style image. The Laplacian matrix produced by the Laplacian operator helped to detect edges and contours. A new optimization objective for neural style transfer named Lapstyle is achieved using Laplacian loss along with CNN. The minimization of this objective produced a stylized image that better preserves the detail structures of the content image and eliminates the artifacts.

Bai et al. [8] proposed an autoencoder like GAN. The proposed network predicts the missing projections and improves the reconstructed CT (X-Ray) images. Reconstruction loss and adversarial loss are jointly implemented to produce more realistic images. In the generator network, an encoder-decoder is trained to predict missing projections. The output of the decoder and ground truth for missing projections are fed to the discriminator network for image generation.

Hong et al. [5] proposed an effective loss function that uses the Sobel operator to improve image deblurring methods based on deep learning. They reported that loss functions can affects the network's learning. Compared to conventional deep learning-based image deblurring methods based on the mean square error (MSE) loss function, the Sobel loss function proposed in this work helps the network recover the high-frequency components of resultant images. Zhu et al. [13] proposed a wavelet loss function for the autoencoder to generate the reconstructed image. The wavelet loss function overcomes the problem of blurriness for large dimension images. A new image quality index called wavelet high-frequency signal-to-noise ratio is also proposed to showcase the effectiveness of the wavelet loss function. Most of the previous works using the autoencoder with loss function are focused on the specific type of datasets. The quality of the results may improve significantly with a better loss function with the same network architecture Zhao et al. [11]. A detailed comparison between loss functions is required to differentiate their performance over different datasets. In this paper, we focus on the performance of some frequently used loss functions of different datasets. The contribution in this paper can be summarized as follows:

- Reconstruction performance of a CAE is analyzed with respect to five different loss functions on eight color and grey-scale image datasets from different domains.
- Sobel and Laplacian loss functions are analyzed for their edge reconstruction capabilities in the autoencoder.

3 Convolutional Autoencoder (CAE)

Loss functions plays a crucial role when image reconstruction is performed using a convolutional autoencoder. In this study to analyze the performance of various loss functions in image reconstruction, a simple CAE architecture is chosen. There are total eight blocks in the CAE containing different combinations of convolution layers, downsampling, upsampling, batch normalization, and transpose convolution layers as illustrated in Fig. 1. The convolutional layer performs convolution operations over the images and generates feature maps. These maps store useful features of a given image. The downsampling layers (MaxPool) and the upsampling layers (UpSample) are applied in the convolutional neural network to reduce the dimension and restore the original size of the image, respectively. The batch normalization layer improves training speed by optimizing the model parameters. The transpose convolution layer performs the operations of upsampling and convolution as a single layer. The encoder part performs image feature extraction and the decoder is used to reconstruct input images with the help of encoder features.

The encoder consists of four blocks. Each block has two convolutional layers with ReLU activation function followed by a batch normalization layer. The

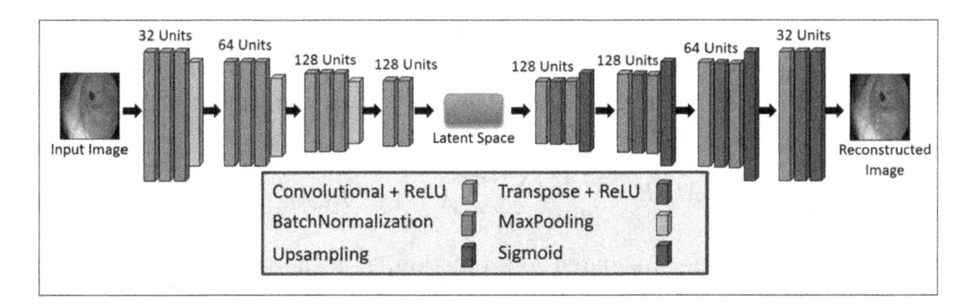

Fig. 1. The proposed CAE network

initial three blocks of the encoder have a max-pooling layer for dimension reduction. In the first block, 32 filters are used in the convolutional layer for feature extraction. In the following blocks, the number of filters is increased by two times, i.e., the second block has 64 filters, the third block has 128 filters, and the fourth block has 128 filters in the convolutional layers. The filter size in the convolutional layers is 3×3, and the maximum pooling size is 2×2. The value of stride is set as 1 in all the blocks. The decoder comprises four blocks in the opposite setup to the encoder part. Each block has a convolutional layer with ReLU activation function and a transpose layer with ReLU activation function followed by the batch normalization layer. The filter size is set as 3×3 in convolution and transpose layer. The decoder part uses transpose convolution to reconstruct the spatial resolution. Transpose convolutions change the order of dimension during convolutional operation. Along with the transpose layer, an upsampling layer is used to retrieve similar-sized feature maps as encoder blocks. An upsampling layer of size 2×2 is used to increase the feature dimensions in the initial three blocks. The first convolutional layer of the decoder takes the input from the last convolutional layer of the encoder, which has 128 feature maps. The decoder's first, second, third, and fourth blocks have 128, 128, 64, and 32 filters, respectively. In the last block, a convolutional layer with three filters is applied with the sigmoid activation function to generate the input image as output. This architecture is selected after multiple rounds of hyperparameter tuning in terms of the number of layers and filters in each layer.

4 Loss Functions

The loss function is used to optimize model performance. It tells us the prediction performance of the model. In the present context, reconstruction loss is a measure to check how good reconstructed images are produced by the proposed autoencoder model. The following loss functions are investigated with the CAE to analyze their suitability for datasets from different domains.

Mean square error (MSE) is the most common regression function which compares the pixels of the input image to the pixels of the reconstructed image. The mean of the corresponding pixel's difference is taken and squared as shown in Eq. (1) [14]. The advantage of using MSE is to get only one global minimum. There are no local minima. MSE is very easy to compute but very sensitive to outliers. MSE makes error values large by computing the square of them, and the model tries to minimize that cost.

$$MSE = (\frac{1}{mn})\sum_{i=1}^{n}\sum_{i=1}^{m}(f_{(i,j)} - \hat{f}_{(i,j)})^2 \tag{1}$$

where MSE is the mean squared loss function, f is the true input image fed to the autoencoder, $\hat{f}$ is the reconstructed image by the autoencoder, (i, j) are image pixel location, and $m \times n$ is image dimension.

Binary cross-entropy (BCE) loss compares pixel probabilities of the reconstructed and input image and produces output in terms of 0 or 1. It then calculates the score that penalizes the probabilities based on how close or far the expected value is from the actual value. BCE is the negative average of the log of corrected predicted probabilities as given in Eq. (2) [15]. Sigmoid is the only activation function that is suitable for BCE.

$$\text{BCE} = -(f_i \cdot \log \hat{f}_i + (1 - f_i) \cdot \log (1 - \hat{f}_i)) \tag{2}$$

where $\hat{f}_i$ is the i-th scalar value of the model output, and f_i is the corresponding target value.

Sobel loss function [5,16] is specially designed to highlight the edges of any objects in the input image using the Sobel operator. It gives high-frequency reconstruction guidance with clear and sharp images. Sobel is an edge detection filter that has two kernels, Horizontal and vertical. Convolution of the image with Sobel kernels gives edge maps. The distance between the reconstructed image and input image to the autoencoder can be minimized in terms of minimizing the distance between the two in Sobel edge space. The idea is to calculate the mean squared loss of a Sobel filtered prediction and a sobel filtered ground truth image as shown in Eq. (3) [16,17].

$$SL = (\frac{1}{nm})\sum_{i=1}^{n}\sum_{j=1}^{m}(F(i,j) - \hat{F}(i,j))^2 \tag{3}$$

where SL is Sobel Loss, F is the Sobel filtered input image fed to the autoencoder, $\hat{F}$ is the Sobel filtered reconstructed image by the autoencoder, (i, j) are image pixel location, and m and n are image dimensions.

Laplacian loss function [12,16] is designed with the Laplacian operator, widely used in image processing and computer vision to detect edges and contours. The Laplacian operator generates images that have gray edge lines and dark background. The Laplacian operator highlights gray level discontinuities in an image and tries to deemphasize regions with slowly varying gray levels. This produces inward and outward edges in an image. Mean squared loss of a Laplacian filtered prediction and a Laplacian filtered ground truth image as given in Eq. (4).

$$LL = (\frac{1}{nm})\sum_{i=1}^{n}\sum_{j=1}^{m}(G(i,j) - \hat{G}(i,j))^2 \tag{4}$$

where LL is the Laplacian Loss, G is the Laplacian filtered true input image fed to the autoencoder, $\hat{G}$ is the Laplacian filtered reconstructed image by the autoencoder, (i, j) are image pixel location, and m and n are image dimensions.

Focal binary loss function [18,19] generalizes the binary cross-entropy loss by introducing a hyperparameter γ which is a focusing parameter for hard to classify cases. The focal loss [20] is more robust to noisy labels and is defined as in Eq. (5).

$$FL(y, \hat{p}) = -\alpha y(1-\hat{y})^{\gamma} log(\hat{y}) - (1-y)\hat{y}^{\gamma} log(1-\hat{y}) \tag{5}$$

where y is the class label and $\hat{y}$ is the predicted probability of the positive class.

In some works, Structure Similarity measure is also used as the loss function. In this paper, we have used SSIM and peak signal to noise ratio for performance evaluation.

5 Experimental Setup

5.1 Datasets Used for Experimentation and Pre-processing

Experiments have been carried out on 9 datasets selected from three distinct fields. It includes medical images (Kvasir [21], MRI Brain [22], and COVID chest X-Ray [23] datasets), natural images (CIFAR-10 [24], SBU Shadow [25], COIL-100 [26] datasets), and handmade digits and sketches images (MNIST [27] and TU-Berlin [28] datasets) as shown in Table 1. The purpose of selecting multiple datasets from 3 different domains is to analyze the performance of different loss functions on heterogeneous images. The datasets chosen for the experiment have varying image dimensions, and color channels.

Table 1. Datasets for experiment with description

Dataset	Description	Images	Size	Type
TU-Berlin [28]	Handmade sketches	1600	1111×1111	Binary
Kvasir [21]	Medical Images (Endoscopic)	1500	Varying size	RGB
MRI Brain [22]	Brain MRI	253	630×630	Grayscale
SBU Shadow [25]	Shadow images	5938	Varying size	RGB
X-Ray [23]	Covid Chest X-Ray	3000	299×299	Grayscale
COIL-100 [26]	Natural Images	7200	128×128	RGB
CIFAR-10 [24]	Natural Images	60000	32×32	RGB
MNIST [27]	Handwritten digits	70000	28×28	Grayscale

As seen in Table 1, images in various datasets are of different sizes. Moreover, images in SBU Shadow and Kvasir datasets vary in size. For further processing, images of TU-Berlin (1111×1111), Kvasir, MRI Brain (630×630), and SBU

Shadow datasets are resized into 512 × 512 using interpolation method. Similarly, images of the X-Ray (299 × 299) dataset are resized into 256 × 256. In the case of datasets with images of small dimensions, like COIL-100 (128 × 128), CIFAR-10 (32 × 32), and MNIST (28 × 28), the autoencoder with only six blocks is used. For these datasets, the encoder has three blocks with 32, 64, and 128 convolutional filters, and the decoder is mirror of the encoder.

5.2 Evaluation Metrics

The CAE used in this work reduces the size of input images into a smaller representation. There is some information loss while reconstructing the image using the latent representation of the image data. Structural similarity and noise generation during reconstruction by autoencoder are to be observed to find out reconstruction error. Image similarity metrics are used to generate quantitative evaluation between two images. MSE (as discussed in Sect. 4), PSNR, and SSIM are computed to measure the reconstructed image quality.

PSNR (Peak Signal to Noise Ratio) is useful in estimating the efficiency of the autoencoder for representing features. PSNR represents the ratio between the maximum intensity of the image and noise that affects the quality of the image, as shown in Eq. (6) [29].

$$PSNR = 20 \log_{10} \frac{(L-1)}{RMSE} \tag{6}$$

where L is the highest intensity value of the image, and RMSE is the root mean square error.

The SSIM metric compares the structural similarity between two images f the original and the reconstructed images $\hat{f}$. It is a perception-based metric that evaluates image degradation as the error in structural information during reconstruction. Structural information is based on pixel's inter-dependencies that are spatially closed to each other. It is calculated using Eq. (7) [30].

$$SSIM(f, \hat{f}) = \frac{(2\mu_f \mu_{\hat{f}} + t_1) + (2\sigma_{f\hat{f}} + t_2)}{(\mu_f^2 + \mu_{\hat{f}}^2 + t_1)(\sigma_f^2 + \sigma_{\hat{f}}^2 + t_2)} \tag{7}$$

where μ_f, $\mu_{\hat{f}}$, σ_f^2, $\sigma_{\hat{f}}^2$ are the average and variance of f and $\hat{f}$ respectively and $\sigma_{f\hat{f}}$ is the co-variance of f and $\hat{f}$; t_1 and t_2 are constants.

6 Results and Discussion

Both quantitative and qualitative results have been reported in this section. As per the quantitative results shown in Table 2 for all the eight datasets, the MSE and BCE loss functions provide better reconstruction capability to the CAE as compared to Sobel and Laplacian loss functions. The CAE with MSE loss function gives better results for CIFAR-10 and MNIST datasets. However, CAE with BCE loss outperforms for TU-Berlin sketches, Covid X-Ray, MRI Brain, SBU Shadow, COIL-100, and Kvasir dataets. It can be seen that CAE with MSE loss function performs better for low-dimensional images but with a large number of training samples, whereas CAE with BCE is able to converge well on high-dimension images even with smaller number of images.

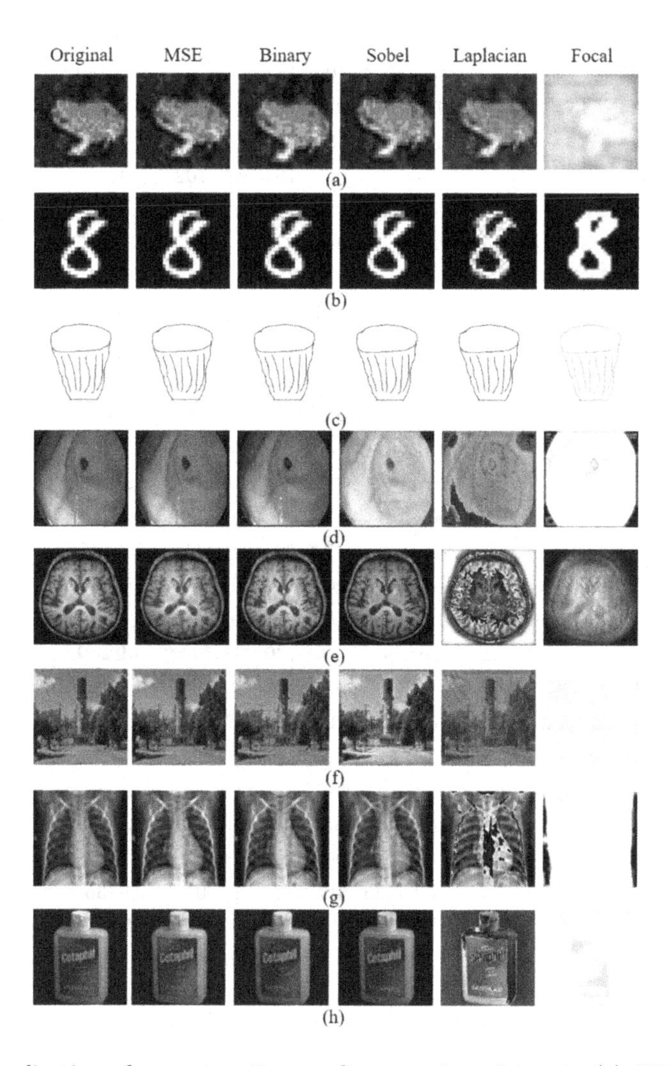

Fig. 2. Visualization of reconstruction results on various datasets: (a) CIFAR-10, (b) MNIST, (c) TU-Berlin, (d) Kvasir, (e) MRI Brain, (f) SBU Shadow, (g) X-Ray, and (h) COIL-100. Loss functions used with CAE are mentioned at the top of the column.

The qualitative results are generated to analyze the reconstruction of edges and lines in terms of image quality. Figure 2 shows the qualitative results for all the datasets. The MSE loss function squares the difference between predicted

Table 2. Quantitative results for all the datasets

Dataset	Loss type	MSE	PSNR	SSIM
CIFAR-10	**MSE**	**30.7353**	**33.5410**	**0.9377**
	Binary	34.8772	32.9695	0.9259
	Sobel	38.3657	32.5157	0.9268
	Laplacian	96.1099	28.3799	0.8819
	Focal	102.9514	0.2604	28.0189
MNIST	**MSE**	**0.0002**	**38.7742**	**0.9983**
	Binary	15.2759	36.5869	0.9380
	Sobel	0.000194	37.8702	0.9985
	Laplacian	32.6411	33.0845	0.7842
	Focal	0.0407	0.7554	62.2592
TU-Berlin	MSE	2.5797	44.4779	0.9900
	Binary	**2.4129**	**44.8066**	**0.9907**
	Sobel	2.5151	44.6088	0.9903
	Laplacian	5.7002	40.7972	0.9778
	Focal	5.0341	0.9603	41.3340
X-Ray	MSE	74.3817	28.6159	0.8735
	Binary	**70.2714**	**30.0993**	**0.8941**
	Sobel	67.7070	30.0734	0.8769
	Laplacian	97.7320	28.2346	0.6269
	Focal	104.0770	0.6093	27.9662
MRI Brain	MSE	25.5430	34.6040	0.9040
	Binary	**22.8570**	**34.7756**	**0.9209**
	Sobel	59.3236	30.7709	0.7860
	Laplacian	108.2985	27.8012	0.4402
	Focal	109.9161	0.2692	27.7253
SBU Shadow	MSE	58.8917	30.5607	0.7249
	Binary	**58.0454**	**30.6898**	**0.7439**
	Sobel	96.8310	28.2956	0.7154
	Laplacian	105.3909	27.9058	0.3136
	Focal	106.2414	0.3950	27.8794
COIL-100	MSE	14.1881	36.9310	0.9523
	Binary	**11.2689**	**38.0655**	**0.9556**
	Sobel	89.4440	28.7077	0.9200
	Laplacian	107.4096	27.8260	0.4455
	Focal	104.5412	0.2859	27.9466
Kvasir	MSE	46.3198	31.6111	0.8534
	Binary	**22.5377**	**34.8506**	**0.8706**
	Sobel	114.2045	27.5561	0.6733
	Laplacian	94.9930	28.3767	0.5526
	Focal	100.3878	0.5715	28.1283

and input images, which makes the error prominent and results in blurry images. In addition, it removes small artifacts, i.e., thin lines and edges are faded from images or are blurred. On the other hand, image reconstruction with BCE shows better performance. The Sobel loss function also works well for producing fine lines and edges. Laplacian loss function generates sharp edges. However, the overall image quality is not good. Also, in the case of the SBU Shadow dataset shown in Fig. 2 (g), it can be seen that the MSE loss function reconstructs blurry electricity cables in the image, whereas BCE and Sobel loss functions perform better than MSE. The predicted image with the Laplacian function is not similar to the original image, as it reconstructs dark and thick electricity cables. The edges and lines in the image generated by the MSE loss function are not as sharp and smooth as the original image in X-Ray and SBU Shadow datasets. The performance of focal loss function is not upto the mark as can be seen from its reconstructed images in Fig. 2 and the quantitative results given in Table 2.

7 Conclusion

In the present work, five different loss functions are evaluated for their strength in enabling a CAE in image reconstruction. The experiments are performed on eight diverse image datasets with MSE, BCE, Sobel, Laplacian and focal loss functions. The results show that the MSE loss function outperforms on datasets with sample low-dimension and large number of training samples. The CAE can adapt well with BCE loss function on high-dimensional images even with a relatively small number of training samples. The CAE with BCE loss function reconstructs good quality images. Sharp and smooth edges of the objects are achieved with CAE using the Sobel loss function. However, slight blurriness is noted in some cases, similar to the MSE loss function results. Overall, the CAE model with binary cross-entropy performs well in most datasets, whereas it is close to the MSE loss function in some experiments.

References

1. Odaibo, S.: Tutorial: deriving the standard variational autoencoder (VAE) loss function. arXiv preprint arXiv:1907.08956 (2019)
2. Kingma, D.P., Welling, M.: An introduction to variational autoencoders. arXiv preprint arXiv:1906.02691 (2019)
3. Gondara, L.: Medical image denoising using convolutional denoising autoencoders. In: 2016 IEEE 16th International Conference on Data Mining Workshops (ICDMW), pp. 241–246. IEEE (2016)
4. Fazlali, H., Shirani, S., McDonald, M., Brown, D., Kirubarajan, T.: Aerial image dehazing using a deep convolutional autoencoder. Multimedia Tools Appl. **79**(39), 29493–29511 (2020)
5. Hong, J.-P., Cho, S.-J., Lee, J., Ji, S.-W., Ko, S.-J.: Single image deblurring based on auxiliary Sobel loss function. In: 2020 IEEE International Conference on Consumer Electronics-Asia (ICCE-Asia), pp. 1–3. IEEE (2020)

6. Chen, M., Shi, X., Zhang, Y., Wu, D., Guizani, M.: Deep features learning for medical image analysis with convolutional autoencoder neural network. IEEE Trans. Big Data **7**, 750–758 (2017)
7. Chow, J.K., Su, Z., Wu, J., Tan, P.S., Mao, X., Wang, Y.-H.: Anomaly detection of defects on concrete structures with the convolutional autoencoder. Adv. Eng. Inform. **45**, 101105 (2020)
8. Bai, J., Dai, X., Wu, Q., Xie, L.: Limited-view CT reconstruction based on autoencoder-like generative adversarial networks with joint loss. In: 2018 40th Annual International Conference of the IEEE Engineering in Medicine and Biology Society (EMBC), pp. 5570–5574. IEEE (2018)
9. Liu, X., Gherbi, A., Wei, Z., Li, W., Cheriet, M.: Multispectral image reconstruction from color images using enhanced variational autoencoder and generative adversarial network. IEEE Access **9**, 1666–1679 (2020)
10. Pandey, R.K., Saha, N., Karmakar, S., Ramakrishnan, A.G.: MSCE: an edge-preserving robust loss function for improving super-resolution algorithms. In: Cheng, L., Leung, A.C.S., Ozawa, S. (eds.) ICONIP 2018. LNCS, vol. 11306, pp. 566–575. Springer, Cham (2018). https://doi.org/10.1007/978-3-030-04224-0_49
11. Zhao, H., Gallo, O., Frosio, I., Kautz, J.: Loss functions for image restoration with neural networks. IEEE Trans. Comput. imaging **3**(1), 47–57 (2016)
12. Li, S., Xu, X., Nie, L., Chua, T.-S.: Laplacian-steered neural style transfer. In: Proceedings of the 25th ACM International Conference on Multimedia, pp. 1716–1724 (2017)
13. Zhu, Q., Wang, H., Zhang, R.: Wavelet loss function for auto-encoder. IEEE Access **9**, 27101–27108 (2021)
14. Ephraim, Y., Malah, D.: Speech enhancement using a minimum mean-square error log-spectral amplitude estimator. IEEE Trans. Acoust. Speech Signal Process. **33**(2), 443–445 (1985)
15. Creswell, A., Arulkumaran, K., Bharath, A.A.: On denoising autoencoders trained to minimise binary cross-entropy. arXiv preprint arXiv:1708.08487 (2017)
16. Lu, Z., Chen, Y.: Single image super resolution based on a modified U-net with mixed gradient loss. arXiv preprint arXiv:1911.09428 (2019)
17. Kanopoulos, N., Vasanthavada, N., Baker, R.L.: Design of an image edge detection filter using the Sobel operator. IEEE J. Solid State Circuits **23**(2), 358–367 (1988)
18. Lin, T.-Y., Goyal, P., Girshick, R., He, K., Dollar, P.: Focal loss for dense object detection. In: Proceedings of the IEEE International Conference on Computer Vision (ICCV), October 2017
19. Wang, C., Deng, C., Wang, S.: Imbalance-XGBoost: leveraging weighted and focal losses for binary label-imbalanced classification with XGBoost. Pattern Recognit. Lett. **136**, 190–197 (2020)
20. Ma, X., Huang, H., Wang, Y., Romano, S., Erfani, S., Bailey, J.: Normalized loss functions for deep learning with noisy labels. In: International Conference on Machine Learning, pp. 6543–6553. PMLR (2020)
21. Pogorelov, K., et al.: KVASIR: a multi-class image dataset for computer aided gastrointestinal disease detection. In: Proceedings of the 8th ACM on Multimedia Systems Conference, pp. 164–169 (2017)
22. Chakrabarty, N.: Brain MRI images for brain tumor detection. https://www.kaggle.com/navoneel/brain-mri-images-for-brain-tumor-detection (2019)
23. Rahman, T., et al.: Exploring the effect of image enhancement techniques on COVID-19 detection using chest X-ray images. Comput. Biol. Med. **132**, 104319 (2021)

24. Krizhevsky, A., Nair, V., Hinton, G.: CIFAR-10 (Canadian institute for advanced research), vol. 5, vol. 4 (2010). http://www.cs.toronto.edu/kriz/cifar.html
25. Vicente, T.F.Y., Hou, L., Yu, C.-P., Hoai, M., Samaras, D.: Large-scale training of shadow detectors with noisily-annotated shadow examples. In: Leibe, B., Matas, J., Sebe, N., Welling, M. (eds.) ECCV 2016. LNCS, vol. 9910, pp. 816–832. Springer, Cham (2016). https://doi.org/10.1007/978-3-319-46466-4_49
26. Nene, S.A., Nayar, S.K., Murase, H., et al.: Columbia object image library (COIL-100) (1996)
27. LeCun, Y., Cortes, C.: MNIST handwritten digit database (2010). http://yann.lecun.com/exdb/mnist/
28. Eitz, M., Hays, J., Alexa, M.: How do humans sketch objects? ACM Trans. Graph. (Proc. SIGGRAPH) **31**(4), 44:1–44:10 (2012)
29. Silva, E.A., Panetta, K., Agaian, S.S.: Quantifying image similarity using measure of enhancement by entropy. In: Mobile Multimedia/Image Processing for Military and Security Applications 2007, vol. 6579, p. 65790U. International Society for Optics and Photonics (2007)
30. Wang, Z., Bovik, A.C., Sheikh, H.R., Simoncelli, E.P.: Image quality assessment: from error visibility to structural similarity. IEEE Trans. Image Process. **13**(4), 600–612 (2004)

Fuzzy Entropy k-Plane Clustering Method and Its Application to Medical Image Segmentation

Puneet Kumar[1(✉)], Dhirendra Kumar[2], and Ramesh Kumar Agrawal[1]

[1] School of Computer and Systems Sciences, Jawaharlal Nehru University, Delhi, India
me.puneetkumar94@gmail.com

[2] Department of Applied Mathematics, Delhi Technological University, Delhi, India

Abstract. MRI images are complex, and the data distribution of tissues in MRI are non-spherical and overlapping in nature. Plane-based clustering methods are more efficient in comparison to centroid based clustering for non-spherical data, and soft clustering methods can efficiently handle the overlapping nature by representing clusters in terms of fuzzy sets. In this paper, we propose fuzzy entropy k-plane clustering (FEkPC), which incorporates the fuzzy partition entropy term with a fuzzy entropy parameter in the optimization problem of the conventional kPC method. The fuzzy entropy parameter controls the degree of fuzziness, the same as the fuzzifier parameter in the fuzzy clustering method. The fuzzy entropy terms try to minimize averaged non-membership degrees in the cluster. The performance of the proposed method has been evaluated over three publicly available MRI datasets: one simulated and two real human brain MRI datasets. The experimental results show that the proposed FEkPC method outperforms other state-of-the-art methods in terms of ASA and Dice Score.

Keywords: Fuzzy sets · Fuzzy entropy measure · k-plane clustering · Fuzzy clustering · MRI image segmentation

1 Introduction

Clustering is a popular unsupervised learning technique in the field of machine learning. The aim of clustering methods is to obtain the groups containing similar objects by identifying the data's dominant structure. Clustering has wide application in many research area, including data mining, pattern recognition, computer vision, bioinformatics and medical image analysis. There are two categories of clustering methods such as hard clustering and soft clustering [4]. The hard clustering assigns each object to only one cluster at a time which is not reasonable in some situation. In many real-world scenarios [5], the objects may

B. Raman et al. (Eds.): CVIP 2021, CCIS 1568, pp. 350–361, 2022.
https://doi.org/10.1007/978-3-031-11349-9_31

belong to more than one cluster with a degree of belongingness. For such a situation, soft clustering method performs well as the objects can be assigned to more than one cluster.

One of the popular application of clustering is image segmentation, where the pixel intensity values are clustered to obtain the segmented image [8]. Medical image analysis for diagnosis of diseases using clustering techniques may help the expert radiologists and provide a diagnosis of a large number of medical images in less time [10]. The MRI is a popular imaging modality for the diagnosis of many brain diseases as it provides good contrast of brain tissues compared to other imaging modalities. The brain MRI effectively delineate the soft tissues such as Grey Matter (GM), White Matter (WM), and Cerebrospinal Fluid (CSF). Fuzzy logic-based clustering methods [2] are widely used for segmentation of brain MRI images as the voxel intensity values on boundary regions may contain more than one tissue. The distribution of intensity values corresponding to different brain tissue captured by MRI machine is non-spherical and overlapping in nature [1]. Due to overlapping nature, the fuzzy c-means (FCM) [3] clustering method is suitable for brain MRI image segmentation. The FCM clustering method assigns a membership value for a data point to each cluster, i.e., a given data point may belong to a different cluster with a membership value. In FCM, a fuzzier exponent parameter controls the degree of fuzziness. Another clustering method that will produce the fuzzy partitions with entropy principle is known as fuzzy entropy clustering (FEC) [15]. The FEC clustering method aims to minimize the partition entropy and generate the cluster centroid. Both FCM and FEC clustering techniques utilize the centroid to obtain the fuzzy partition to represent different clusters. Both of them provide appropriate clusters when data are distributed around centroid or spherical, and all such approaches are categorized as a centroid-based clustering method. The performance of centroid based clustering methods is poor when data points for a given cluster distributed around multiple points or non-spherical or elliptical. To obtain the clusters in such situation, multiple centroids may be used to handle the non-spherical or elliptical nature of data. One such clustering method that makes clustering decision with multiple centroids (means) is known as K-Multiple-Means (KMM) [12]. The KMM clustering solves the clustering problem as an optimization problem by converting the original clustering problem into a bipartite graph partitioning problem. However, the KMM works well for smaller size datasets but requires huge computation time for large data. Another approach to address the above problem is plane based clustering techniques where the cluster prototype is represented as a plane. In this way, the plane based clustering can be viewed as a generalization of clustering techniques where the cluster is represented using multiple centroid or means.

Bradley et al. [6] proposed the first plane based clustering by representing the plane as a cluster prototype which is known as k-plane clustering [6] (kPC). The kPC clustering method assigns each data point to only one cluster. Motivated from the advantage of the fuzzy set theory based clustering, Zhu et al. [13] proposed a variant of kPC method termed as fuzzy k-plane clustering [13] (FkPC). The fuzzy partition obtained by FkPC method efficiently deals with overlapping structure and uncertainty in non-spherical data. The solution of the optimization problem of FkPC method can be obtained using Lagrange's multiplier method for obtaining the fuzzy partition matrix. Further, for calculating the plane parameter for each cluster, an eigenvalue problem is obtained using the Lagrangian. To take advantage of both FCM and FkPC method, the research work [11] proposed a hybrid method termed as fuzzy mix-prototype clustering (FMP). The FMP method handles both spherical and non-spherical geometric structures present in a dataset by representing centroid and plane as a cluster prototype using the FCM and the FkPC respectively. The optimization problem of the FMP clustering method includes the individual optimization problem of FCM and FkPC with a trade-off parameter. The trade-off parameter controls the cluster prototypes according to the distribution of data. Another research work [9], cluster the data points by obtaining weight for each dimension corresponding to each cluster termed as fuzzy subspace clustering (FSC). The weight in FSC method is known as fuzzy dimension weights, which indicated the degree of importance of the dimension to the cluster. The clusters represented in terms of fuzzy sets are associated with uncertainty in defining their membership value which is difficult to measure. Entropy can be considered as an alternative way to quantify the uncertainty involved in defining the fuzzy sets. It is calculated as the average value of the non-membership of elements in fuzzy sets[15].

Motivated by the concept of entropy in fuzzy sets, in this research work, we explore the plane based clustering method with fuzzy entropy information named as fuzzy entropy k-plane clustering (FEkPC). The proposed FEkPC method incorporates the fuzzy partition entropy term with a fuzzy entropy parameter in the optimization problem of the conventional kPC method [6]. The fuzzy entropy parameter controls the degree of fuzziness, the same as the fuzzifier parameter in the fuzzy clustering method. The fuzzy entropy terms try to minimize averaged non-membership degrees in the cluster. The optimization problem of the proposed FEkPC method can be solved by utilizing the underlined constraints with the help of the Lagrange's method. This research work is the first attempt to explore fuzzy entropy information in the k-plane clustering method to the best of our knowledge. Thus, the proposed FEkPC clustering method produces an optimized fuzzy partition matrix and cluster plane by solving a series of eigenvalue problems.

The contribution of the proposed approach can be summarized as:

- We propose a variant of the kPC method, termed as the FEkPC method.
- The proposed FEkPC method minimizes the uncertainty of the fuzzy partition matrix by the fuzzy entropy term, which gives more confidence in the defuzzification decision process.
- The proposed FEkPC method produces better performance in lesser time in comparison to the fuzzy variant of kPC (FkPC).
- The proposed FEkPC method outperforms the state-of-the-art methods on publicly available datasets in terms of average segmentation accuracy and Dice Score.

To check the effectiveness of the proposed FEkPC method, experiments have been carried over the three publicly available MRI dataset one simulated and two real human brain MRI dataset. The average segmentation accuracy and Dice score is used for comparing the performance of the proposed FEkPC method with related methods FCM [3], FEC [15], FkPC [13], FSC [9], FMP [11]. The experimental results indicate that the performance of the proposed FEkPC method is better than the state-of-the-art methods for the MRI image segmentation

The rest part of the article is organized as follows. In Sect. 2, we presented preliminaries and related work. The optimization formulation of the proposed FEkPC method is discussed in Sect. 3. Section 4 presents brief details of the datasets and experimental results. Finally, Sect. 5 includes the conclusion and future direction.

2 Preliminaries and Related Work

In this section, we presented some of the preliminary related work to the proposed method.

2.1 Concept of Entropy for Fuzzy Sets

Fuzzy set F, over a universal set X can be represented as [16]

$$F = \{(x, \mu_F(x)) : x \in X\}\mu_F(x) \in [0, 1] \tag{1}$$

where $\mu_F(x)$ is membership function for an element $x \in X$. This membership function quantifies the degree of belongingness of the element x in fuzzy set F.

Entropy of Fuzzy Set: For a fuzzy set F, the fuzzy entropy measures the uncertainty involved in defining the fuzzy set F. The fuzzy entropy measure $E(F)$ can be defined as [15]:

$$E(F) = -\sum_{i=1}^{N} \mu_F(x_i) \log \mu_F(x_i) \tag{2}$$

where $\mu_F(x_i)$ is the membership value of the element $x_i \in X, i = 1, 2, \ldots, N$ in fuzzy set F.

2.2 FEC

The research work [15] proposed fuzzy entropy clustering (FEC) method by utilizing the concept of entropy of fuzzy sets. The objective function of FEC method is formulated which minimizes the overall difference of the sum of square error of ith datapoints from j^{th} cluster prototype and the negative of fuzzy entropy function multiplied by a fuzzy entropy factor. The optimization problem of the FEC method is defined as [15]

$$\min_{\{\mathbf{U},\mathbf{V}\}} J_{FEC} = \sum_{i=1}^{N}\sum_{j=1}^{k} \mu_{ij}\|x_i - v_j\|^2 + \lambda \sum_{i=1}^{N}\sum_{j=1}^{k} \mu_{ij}\log \mu_{ij} \tag{3}$$

$$\text{s.t. } \mu_{ij} \in [0,1], \textstyle\sum_{j=1}^{k} \mu_{ij} = 1$$

where $\lambda > 0$ is fuzzy entropy factor, x_i denotes a datapoints in $\mathbf{X}$, v_j is cluster prototype and $\mathbf{U} = [u_{ij}]_{N\times k}$ represents the fuzzy partition matrix corresponding to the membership value for i^{th} datapoint to j^{th} cluster.

The fuzzy entropy factor controls the degree of fuzziness of each cluster. The FEC performs well when data points are distributed around a given centroid or spherical structure as it utilizes the cluster centroid to compute the fuzzy partition matrix. The performance of the FEC method is poor when data points are distributed around multiple points or non-spherical in nature.

2.3 kPC

Bradley et al. [6] suggested a k-plane clustering (kPC) where a cluster prototype is represented with the help of a plane rather than point. This strategy can efficiently cluster the data points which are non-spherical in nature. The objective function of the kPC method is formulated, which minimizes the overall sum of the squares of distances of ith datapoints from j^{th} cluster plane prototype. The optimization problem of the kPC method is expressed as:

$$\min_{\{\mathbf{W},b\}} J_{kPC} = \sum_{i=1}^{N}\sum_{j=1}^{k} \|x_i^T \mathbf{w_j} + b_j\|^2 \tag{4}$$

$$\text{s.t.} \|\mathbf{w_j}\|^2 = 1,\ 1 \leq j \leq k$$

where x_i denotes a datapoints in $\mathbf{X}$, $\mathbf{w_j}, b_j$ are the plane parameters to represent the cluster prototype.

3 Proposed Fuzzy Entropy k-Plane Clustering Approach

We formulated a new objecting function for clustering the data points to include the advantage of fuzzy entropy measure and the k-plane clustering (kPC) method. The proposed plane based clustering method is termed as fuzzy entropy k-plane clustering (FEkPC) method. The optimization problem of the proposed FEkPC method is comprised of two terms. The first term minimizes the overall sum of squares distances between each ith data point to j^{th} cluster prototype defined by a plane. The second term is the negative of the entropy function $E(\mathbf{U})$ multiplied by weight factor, which reduces the uncertainty of clusters represented using fuzzy sets. The optimization problem of the proposed FEkPC method is given as:

$$\min_{\{\mathbf{U},\mathbf{W},\mathbf{b}\}} J_{FEkPC} = \sum_{i=1}^{N}\sum_{j=1}^{k} \mu_{ij}\|x_i^T\mathbf{w_j} + b_j\|^2 + \alpha\sum_{i=1}^{N}\sum_{j=1}^{k}\mu_{ij}\log\mu_{ij} \tag{5}$$

$$\text{s.t.}\mu_{ij} \in [0,1], \textstyle\sum_{j=1}^{k}\mu_{ij} = 1, \|\mathbf{w_j}\|^2 = 1,$$

where x_i denotes a datapoints in $\mathbf{X}$, $\alpha > 0$ is fuzzy entropy factor, $\mathbf{w_j}, b_j$ are the parameters of the cluster prototype and $\mathbf{U} = [u_{ij}]_{N\times k}$ represents the fuzzy partition matrix corresponding to the membership value for i^{th} datapoint to j^{th} cluster. The above optimization problem (5) is reduced to eigenvalue problems using Lagrange's method of the undetermined multiplier. After solving the Lagrangian function corresponding to (5), for each jth cluster, we can obtain plane parameters from a $d \times d$ characteristic matrix $\mathbf{D_j}$. The characteristic equation is represented by the following equation:

$$\mathbf{D_j}\mathbf{w_j} = \xi_\mathbf{j}\mathbf{w_j} \tag{6}$$

where $\mathbf{w_j}$, $\xi_\mathbf{j}$ are eigenvector and eigenvalue corresponding to the jth cluster, respectively. The expression of characteristic matrix $\mathbf{D_j}$ can be expressed as:

$$\mathbf{D_j} = \frac{\left(\sum_{i=1}^{N}\mu_{ij}x_i\right)\left(\sum_{i=1}^{N}\mu_{ij}x_i^T\right)}{\sum_{i=1}^{N}\mu_{ij}} - \sum_{i=1}^{N}\mu_{ij}x_ix_i^T \tag{7}$$

The plane parameter $\mathbf{w_j}$ is an eigenvector corresponding to the smallest eigenvalue. After obtaining $\mathbf{w_j}$ from Eq. (6), we can compute b_j as follows:

$$b_j = \frac{-\sum_{i=1}^{N}\mu_{ij}x_i^T\mathbf{w_j}}{\sum_{i=1}^{N}\mu_{ij}} \tag{8}$$

Using the values of $\mathbf{w_j}$ and b_j, we can compute the fuzzy partition membership matrix μ_{ij} as follows:

$$\mu_{ij} = \left\{\sum_{p=1}^{k}\left(\frac{exp(\|x_i^T\mathbf{w_j} + b_j\|^2)}{exp(\|x_i^T\mathbf{w_p} + b_p\|^2)}\right)^{\frac{1}{\alpha}}\right\}^{-1} \tag{9}$$

The proposed FEkPC method computes the cluster plane prototype and fuzzy partition matrix using alternatively optimization procedure until convergence. The steps involved for clustering using the proposed method is summarized in Algorithm 1.

4 Dataset and Experimental Results

This section describes the datasets, evaluation metrics and their corresponding results.

Algorithm 1. The proposed FEkPC algorithm

Input: Set values for the number of clusters k, the fuzzy entropy factor $\alpha > 0$, and the stopping criterion value ϵ.
1.Randomly initialise fuzzy membership matrix $\mathbf{U}^1$
2.t$\leftarrow$ 1
3.**Repeat**
4.Update the plane parameters $\mathbf{w}_j^t$ using equation (6)$\forall j \in \{1, 2, \cdots k\}$
5.Update the plane parameters b_j^t using equation (8) $\forall j \in \{1, 2, \cdots k\}$
6.Update the fuzzy membership matrix $\mathbf{U}^{t+1} = \{\mu_{ij}^{t+1}\}_{N \times k}$ using equation (9)
7. t $\leftarrow$ t+1;
8.**Until** $||\mathbf{U}^{t+1} - \mathbf{U}^t|| < \epsilon$
9.**Return** : $\mathbf{U} = \{\mu_{ij}\}_{N\times k}$, $\mathbf{W} = \{\mathbf{w_j}\}_{d\times k}$ and $\mathbf{b} = \{b_j\}_{1\times k}$

4.1 Datasets

To validate and compare the performance of the proposed FEkPC method with state of-the-art methods, we have utilized three publicly available T1-weighted MRI imaging datasets with given ground truth. The first one is the simulated Brainweb dataset [7], the second one is real brain MRI dataset acquired from the Internet Brain Segmentation Repository (IBSR) (available: https://www.nitrc.org/projects/ibsr), and the third one is real brain MRI dataset acquired from the grand challenge on MR Brain Segmentation at MICCAI 2018 (MRBrainS18) (available: https://mrbrains18.isi.uu.nl/). We used the brain extraction tool [14] for scull striping. The detailed description of the dataset is given in Table 1.

4.2 Evaluation Metrics

To evaluate the performance of the proposed FEkPC clustering algorithm, we have converted fuzzy partition into crisp set with the maximum membership aka defuzzification process. We used average segmentation accuracy (ASA) and Dice score (DS) as a performance measure for imaging data which are mathematically defined as:

$$ASA = \sum_{i=1}^{k} \frac{|X_i \cap Y_i|}{\sum_{j=1}^{k} |X_j|} \tag{10}$$

$$DS = \frac{2|X_i \cap Y_i|}{|X_i| + |Y_i|} \tag{11}$$

where X_i denotes the set of pixels belonging to the ith class of ground truth image, Y_i denotes the set of pixels belonging to the ith cluster of the segmented image. $|X_i|$ denotes the cardinality of pixels in X_i corresponding to i^{th} region of ground truth image, similarly $|Y_i|$ denotes the cardinality of pixels in Y_i corresponding to i^{th} region of segmented image. The high value of DS and ASA for a given method indicates the better performance of the method.

Table 1. Dataset description

3D MRI dataset	Volume dimension	Voxel size
Simulated Brainweb MRI dataset	181 × 217 × 181	1 mm × 1 mm × 1 mm
Real brain IBSR dataset	256 × 256 × [49–59]	1 mm × 1 mm × 1 mm
Real brain MRBrainS18 dataset	240 × 240 × 56	0.958 mm × 0.958 mm × 3.0 mm

4.3 Results on Simulated Brainweb MRI Dataset

To check the effectiveness and applicability of the proposed FEkPC method on MRI image segmentation. We perform experiment on simulated Brainweb MRI dataset. We present quantitative results on Brainweb MRI dataset. The following observation can be drawn from Table 2 and Fig. 1(a):

- The proposed FEkPC method outperforms other related methods in terms of ASA.
- The proposed FEkPC method outperforms other related methods in terms of average DS of GM on the simulated dataset and performs better in terms of average DS of WM.

Table 2. Average segmentation accuracy and Dice score for BrainWeb dataset

Performance measure\Method	FCM	FEC	FkPC	FSC	FMP	FEkPC
ASA	0.9567	0.9562	0.9562	0.9561	0.9567	**0.9571**
DICE (GM)	0.9609	0.9603	0.9603	0.9601	0.9609	**0.9615**
DICE (WM)	**0.9904**	0.9901	0.9901	0.9902	**0.9904**	**0.9904**
Avg. performance	0.9693	0.9689	0.9689	0.9688	0.9693	**0.9697**

Table 3. Comparison of average computation time.

Clustering methods	FCM	FEC	FKPC	FSC	FMP	FEkPC
Computation time (in Sec.)	0.1740	0.1258	0.1887	0.0530	0.1811	0.1481

We present a comparison of the proposed FEkPC method and FCM, FEC, FkPC, FSC, FMP methods in terms of average computation time of 100 runs on a simulated brain MRI of size 181 × 217 in Table 3. It can be observed from Table 3 that the proposed FEkPC Method takes less computation time in comparison to the three methods (FCM, FkPC, FMP) and more compared to two methods (FEC and FSC).

4.4 Results on Real IBSR Brain MRI Dataset

To check the effectiveness and applicability of the proposed FEkPC method on a real MRI dataset, we consider MRI data with case no. 100_23, 110_3, 112_3,

191_3, 205_3, 6_10 and 8_4. We present quantitative results on the IBSR Brain MRI dataset. The following observation can be drawn from Table 4, Table 5 and Fig. 1(b):

- The proposed FEkPC method outperforms other related methods in terms of ASA over all MRI considered from the real IBSR Brain dataset.
- The proposed FEkPC method outperforms other related methods in terms of DS for GM over all MRI considered from the real IBSR Brain dataset.
- The proposed FEkPC method outperforms other related methods in terms of average DS for WM

Table 4. Average segmentation accuracy for real brain IBSR dataset

Images\Methods	FCM	FEC	FkPC	FSC	FMP	FEkPC
100_23	0.6022	0.6080	0.6081	0.6044	0.6082	**0.6220**
110_3	0.6507	0.6817	0.6817	0.6433	0.6828	**0.6903**
112_2	0.6394	0.6661	0.6661	0.6401	0.6666	**0.6743**
191_3	0.6504	0.6782	0.6782	0.6395	0.6794	**0.6901**
205_3	0.6241	0.6446	0.6503	0.6100	0.6506	**0.6632**
6_10	0.5161	0.5545	0.5545	0.4334	0.5545	**0.5685**
8_4	0.6775	0.6923	0.6923	0.6741	0.6923	**0.6950**
Avg. performance	0.6229	0.6465	0.6473	0.6064	0.6478	**0.6576**

Table 5. Brain tissues dice score for real brain IBSR dataset

Images\Methods		FCM	FEC	FkPC	FSC	FMP	FEkPC
GM	100_23	0.5996	0.6058	0.6058	0.6116	0.6081	**0.6255**
	110_3	0.6710	0.7007	0.7007	0.6671	0.7049	**0.7112**
	112_2	0.6577	0.6792	0.6792	0.6561	0.6829	**0.6925**
	191_3	0.6385	0.6697	0.6697	0.6319	0.6740	**0.6860**
	205_3	0.6078	0.6252	0.6338	0.5980	0.6338	**0.6496**
	6_10	0.5440	0.5402	0.5402	0.5216	0.5402	**0.5615**
	8_4	0.6779	0.6919	0.6919	0.6740	0.6919	**0.6955**
WM	100_23	0.8327	**0.8337**	**0.8337**	0.8023	0.8327	**0.8337**
	110_3	0.8142	**0.8194**	**0.8194**	0.7826	0.8182	**0.8194**
	112_2	0.8246	**0.8383**	**0.8383**	0.8074	0.8368	0.8368
	191_3	0.8223	**0.8229**	**0.8229**	0.8058	0.8224	**0.8229**
	205_3	0.7856	0.7939	0.7939	0.7629	0.7939	**0.7948**
	6_10	0.7115	**0.7270**	**0.7270**	0.5722	**0.7270**	**0.7270**
	8_4	**0.8038**	0.8014	0.8014	0.7828	0.8014	0.8014
Avg. performance		0.7137	0.7250	0.7256	0.6912	0.7263	**0.7327**

4.5 Results on Real MRBrainS18 MRI Dataset

To check the effectiveness and applicability of the proposed FEkPC method on a real MRI dataset, we consider MRI data with case no. 070, 1, 4, 5 and 7. We present quantitative results on the MRBrainS18 Brain MRI dataset. The following observation can be drawn from Table 6, Table 7 and Fig. 1(c):

- The proposed FEkPC method outperforms other related methods in terms of ASA over all MRI considered from the real brain MRBrainS18 dataset.
- The proposed FEkPC method outperforms other related methods in terms of DS for GM and WM over all MRI considered from the real Brain MRBrainS18 dataset.

Table 6. Average segmentation accuracy for real MRbrainS18 dataset

Images\Method	FCM	FEC	FkPC	FSC	FMP	FEkPC
070	0.7165	0.7164	0.7164	0.7161	0.7252	**0.7350**
1	0.7457	0.7458	0.7444	0.7438	0.7486	**0.7509**
4	0.6962	0.6966	0.6967	0.6964	0.7029	**0.7070**
5	0.6814	0.6814	0.6817	0.6816	0.6831	**0.6848**
7	0.7229	0.7230	0.7240	0.7227	0.7275	**0.7323**
Avg. performance	0.7126	0.7126	0.7127	0.7121	0.7174	**0.7220**

Table 7. Dice score for real MRbrainS18 dataset

Images\Method		FCM	FEC	FkPC	FSC	FMP	FEkPC
GM	070	0.7004	0.6998	0.7026	0.6996	0.7166	**0.7279**
	1	0.7319	0.7320	0.7332	0.7317	0.7446	**0.7535**
	4	0.6897	0.6897	0.6910	0.6893	0.7027	**0.7102**
	5	0.6515	0.6511	0.6518	0.6519	0.6594	**0.6644**
	7	0.6906	0.6907	0.6919	0.6901	0.7012	**0.7149**
WM	070	0.7919	0.7915	0.7923	0.7915	0.7985	**0.8028**
	1	0.8374	0.8374	0.8386	0.8377	0.8434	**0.8470**
	4	0.7156	0.7156	0.7164	0.7153	0.7235	**0.7265**
	5	0.7496	0.7495	0.7499	0.7497	0.7553	**0.7594**
	7	0.7934	0.7935	0.7938	0.7932	0.8004	**0.8090**
Avg. performance		0.7352	0.7351	0.7362	0.7350	0.7446	**0.7516**

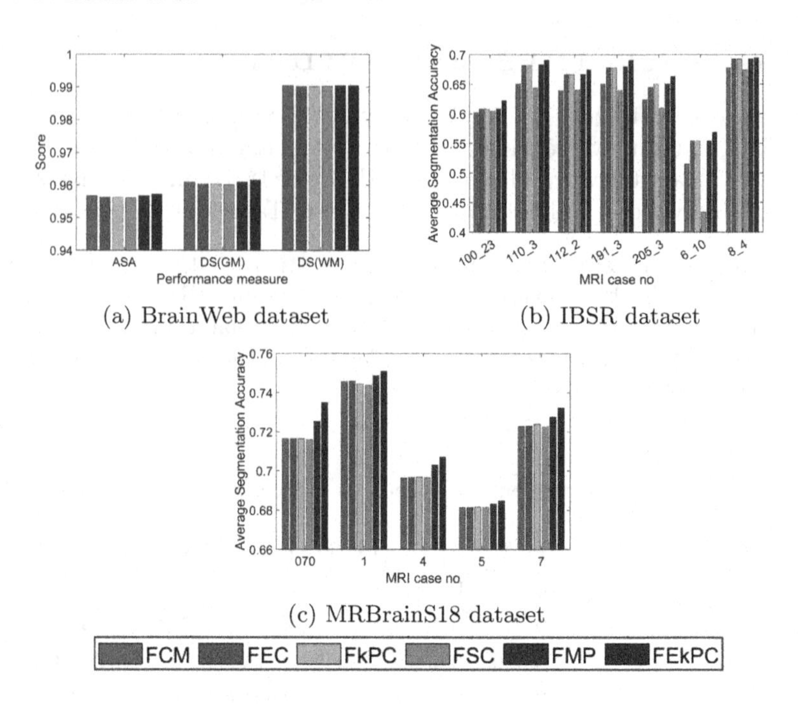

Fig. 1. Comparision of segmentation performance on (a) BrainWeb dataset (b) real brain IBSR dataset (c) MRBrainS18 dataset in terms of ASA and DS

5 Conclusion and Future Works

In this research work, a fuzzy entropy k-plane clustering (FEkPC) method is proposed. The method is formulated by a new objective function with fuzzy entropy information and a k-plane clustering method to minimize uncertainty in the fuzzy partition matrix in the segmentation procedure. The proposed FEkPC method takes advantage of the fuzzy entropy term to retain more information and produce better segmentation performance. Experimental results are presented in term of two quantitative measures on the three publicly available MRI datasets. Experimental results demonstrate superior performance of the proposed method over other related methods. In future, we will also focus on extending the proposed method in intuitionistic fuzzy metric space to handle the problem of fuzzy set.

References

1. Bai, X., Zhang, Y., Liu, H., Chen, Z.: Similarity measure-based possibilistic FCM with label information for brain MRI segmentation. IEEE Trans. Cybern. **49**(7), 2618–2630 (2018)
2. Balafar, M.: Fuzzy C-mean based brain MRI segmentation algorithms. Artif. Intell. Rev. **41**(3), 441–449 (2014)

3. Bezdek, J.C., Ehrlich, R., Full, W.: FCM: the fuzzy C-means clustering algorithm. Comput. Geosci. **10**(2–3), 191–203 (1984)
4. Bora, D.J., Gupta, D., Kumar, A.: A comparative study between fuzzy clustering algorithm and hard clustering algorithm. arXiv preprint arXiv:1404.6059 (2014)
5. Bouchon-Meunier, B., Detyniecki, M., Lesot, M.J., Marsala, C., Rifqi, M.: Real-world fuzzy logic applications in data mining and information retrieval. In: Wang, P.P., Ruan, D., Kerre, E.E. (eds.) Fuzzy Logic. STUDFUZZ, vol. 215, pp. 219–247. Springer, Heidelberg (2007). https://doi.org/10.1007/978-3-540-71258-9_11
6. Bradley, P.S., Mangasarian, O.L.: K-plane clustering. J. Glob. Optim. **16**(1), 23–32 (2000)
7. Cocosco, C.A., Kollokian, V., Kwan, R.K.S., Pike, G.B., Evans, A.C.: BrainWeb: online interface to a 3D MRI simulated brain database. In: NeuroImage. Citeseer (1997)
8. Dhanachandra, N., Chanu, Y.J.: A survey on image segmentation methods using clustering techniques. Eur. J. Eng. Technol. Res. **2**(1), 15–20 (2017)
9. Gan, G., Wu, J., Yang, Z.: A fuzzy subspace algorithm for clustering high dimensional data. In: Li, X., Zaïane, O.R., Li, Z. (eds.) ADMA 2006. LNCS (LNAI), vol. 4093, pp. 271–278. Springer, Heidelberg (2006). https://doi.org/10.1007/11811305_30
10. Ghosal, A., Nandy, A., Das, A.K., Goswami, S., Panday, M.: A short review on different clustering techniques and their applications. Emerg. Technol. Model. Graph., 69–83 (2020)
11. Liu, J., Pham, T.D., Yan, H., Liang, Z.: Fuzzy mixed-prototype clustering algorithm for microarray data analysis. Neurocomputing **276**, 42–54 (2018)
12. Nie, F., Wang, C.L., Li, X.: K-multiple-means: a multiple-means clustering method with specified k clusters. In: Proceedings of the 25th ACM SIGKDD International Conference on Knowledge Discovery & Data Mining, pp. 959–967 (2019)
13. Zhu, L., Wang, S,, Pan, Y., Han, B.: Improved fuzzy partitions for K-plane clustering algorithm and its robustness research. J. Electron. Inf. Technol. **30**(8), 1923–1927 (2008)
14. Smith, S.M.: BET: brain extraction tool (2000)
15. Tran, D., Wagner, M.: Fuzzy entropy clustering. In: Ninth IEEE International Conference on Fuzzy Systems. FUZZ-IEEE 2000 (Cat. No. 00CH37063), vol. 1, pp. 152–157. IEEE (2000)
16. Zadeh, L.A.: Fuzzy sets. In: Zadeh, L.A. (ed.) Fuzzy Sets, Fuzzy Logic, and Fuzzy Systems: Selected Papers, pp. 394–432. World Scientific (1996)

FMD-cGAN: Fast Motion Deblurring Using Conditional Generative Adversarial Networks

Jatin Kumar[1(✉)], Indra Deep Mastan[2], and Shanmuganathan Raman[1]

[1] IIT Gandhinagar, Gandhinagar, India
kumar_jatin@alumni.iitgn.ac.in, shanmuga@iitgn.ac.in
[2] LNMIIT Jaipur, Jaipur, India
indradeep.mastan@lnmiit.ac.in

Abstract. In this paper, we present a Fast Motion Deblurring-Conditional Generative Adversarial Network (FMD-cGAN) that helps in blind motion deblurring of a single image. FMD-cGAN delivers impressive structural similarity and visual appearance after deblurring an image. Like other deep neural network architectures, GANs also suffer from large model size (parameters) and computations. It is not easy to deploy the model on resource constraint devices such as mobile and robotics. With the help of MobileNet [1] based architecture that consists of depthwise separable convolution, we reduce the model size and inference time, without losing the quality of the images. More specifically, we reduce the model size by 3–60x compare to the nearest competitor. The resulting compressed Deblurring cGAN faster than its closest competitors and even qualitative and quantitative results outperform various recently proposed state-of-the-art blind motion deblurring models. We can also use our model for real-time image deblurring tasks. The current experiment on the standard datasets shows the effectiveness of the proposed method.

Keywords: Fast deblurring · Generative adversarial networks · Depthwise separable convolution · Hinge loss

1 Introduction

Image degradation by motion blur generally occurs due to movement during the capture process from the camera or capturing using lightweight devices such as mobile phones and low intensity during camera exposure. Blur in the images degrades the perceptual quality. For example, blur distorts the object's structure (Fig. 1).

Supplementary Information The online version contains supplementary material available at https://doi.org/10.1007/978-3-031-11349-9_32.

B. Raman et al. (Eds.): CVIP 2021, CCIS 1568, pp. 362–377, 2022.
https://doi.org/10.1007/978-3-031-11349-9_32

(a) Blur Image

(b) Sharp Image

Fig. 1. The figure shows object detection on images becomes easy after deblurring using FMD-cGAN.YOLO [2] object detection on the (a) blurry picture and on the (b) sharp picture from the GoPro dataset [3]

Image Deblurring is a method to remove the blurring artifacts and distortion from a blurry image. Human vision can easily understand the blur in the image. However, it is challenging to create metrics that can estimate the blur present in the image. Image degradation model using non-uniform blur kernel [4,5] is given in Eq. 1.

$$I_B = K(M) * I_S + N \tag{1}$$

where, I_B denotes a blurred image, K(M) denotes unknown blur kernels depending on M's motion field. I_S denotes a latent sharp image, $*$ denotes a convolution operation, and N denotes the noise. As an inverse problem, we retrieve sharp image I_S from blur image I_B during the deblurring process. The deblurring problem generally classified as non-blind deblurring [9] and blind deblurring [10,11], according to knowledge of blur kernel $K(M)$ is known or not.

Our work aims at a single image blind motion deblurring task using deep-learning. The deep-learning methods are effective in performing various computer vision tasks such as object removal [15,16], style transfer [17], and image restoration [3,19,20]. More specifically, convolution neural networks (CNNs) based approaches for image restoration tasks are increasing, e.g., image denoising [18], super-resolution [19], and deblurring [3,20].

The applications of Generative Adversarial Networks (GANs) [30] are increasing immensely, particularly image-to-image conversion GANs [7] have been successfully used on image enhancement, image synthesis, image editing and style transfer. Image deblurring could be formulated as an image-to-image translation task. Generally, applications that interact with humans (e.g., Object Detection) require to be faster and lightweight for a better experience. Image deblurring could be useful pre-processing steps of other computer vision tasks such as Object Detection (Fig. 1 and Fig. 2).

In this paper, we propose a Fast Motion Deblurring conditional Generative Adversarial Network architecture (FMD-cGAN). Our FMD-cGAN architecture is based on conditional GANs [40] and the resnet network architecture [6] (Fig. 5). We also used depthwise separable convolution (Fig. 3) inspired from MobileNet

(a) Corrupted Image (b) FMD-cGAN (ours) (c) Original Image

Fig. 2. First-row images are from the GoPro dataset [22], and second-row images are from the REDS dataset [38] processed by Fast Deblurring cGAN.

to improve efficiency. A MobileNet network [1] has fewer Multiplications and Additions (smaller complexity) operations, and fewer parameters (smaller model size) compare to the same network with regular convolution operation.

Unlike other GAN frameworks, where we give the sharp image (real example) and output image from generator network (fake example) as the inputs into Discriminator network [7,40], we train our Discriminator (Fig. 4) by providing input as combining blurred image with the output image from the generator network (or blurred image with sharp image).

Different from previous work, we propose to use Hinge loss [31] and Perceptual loss [8] to improve the quality of the output image. Hinge loss improves the fidelity and diversity of the generated images [49]. Using the Hinge loss in our FMD-cGAN allows building lightweight neural network architectures for the single image motion deblurring task compared to standard Deep ResNet architectures. The Perceptual loss [8] is used as content loss to generate photo-realistic images in our GAN framework.

Contributions: The major contributions are summarized as below.

- We propose a faster and light-weight conditional GAN architecture (FMD-cGAN) for blind motion deblurring tasks. We show that FMD-cGAN (ours) is efficient with lesser inference time than DeblurGAN [3], DeblurGANv2 [27], and DeepDeblur [22] models (Table 1).
- We have performed extensive experiments on GoPro dataset and REDS dataset (Sect. 6). The results shows that our FMD-cGAN outputs images with good visual quality and structure similarity (Fig. 6, Fig. 7, and Table 2).
- We also provide two variants (WILD and Comb) of FMD-cGAN to show that image deblurring task could be improved by pre-training network (Table 1 and Sect. 5).
- We have also performed ablation study to illustrate that our network design choices improves the deblurring performance (Sect. 7).

2 Background

2.1 Image Deblurring

Images can have different types of blur problems, such as motion blur, defocus blur, and handshake blur. We have described that image deblurring is classified into two types: Non-blind image deblurring and Blind image deblurring (Sect. 1).

Non-blind deblurring is an ill-posed problem. The noise inverse process is unstable; a small quantity of noise can cause critical distortions. Most of the earlier works [12–14] aims to perform non-blind deblurring task by assuming that blur kernels $K(M)$ are known. Blind deblurring techniques for a single image, which use Deep-learning based approaches, are observed to be effective in single image deblurring tasks [22,39] because most of the kernel-based methods are not sufficient to model the real world blur [37]. The task is to estimates both the sharp image I_S and the blur kernel $K(M)$ for image restoration. There are also classical approaches such as low-rank prior [46] and dark channel prior [47] that are useful for deblurring, but they also have shortcomings.

2.2 Generative Adversarial Networks

Generative Adversarial Network (GAN) was initially developed and introduced by Ian Goodfellow and his fellow workers in 2014 [30]. GAN framework includes two competing network architectures: a generator network G and a discriminator network D. Generator (G) task is to generate fake samples similar to input by capturing the input data distribution, and on the opposite side, the Discriminator (D) aims to differentiate between the fake and real samples; and pass this information to the G so that G can learn. Generator G and Discriminator D follows the minimax objective defined as follows.

$$\min_G \max_D V(D,G) = E_{x \sim p_{data}(x)}[log(D(x))] + E_{z \sim p_z(z)}[log(1 - D(G(z))] \quad (2)$$

Here, in Eq. 2, the generator G aims to minimize the value function V, and the discriminator D tries to maximize the value function V. Moreover, the generator G faces problems such as mode collapse and gradient diminishing (e.g., Vanilla GAN).

WGAN and WGAN-GP: To deal with mode collapse and gradient diminishing, WGAN method [25] uses Earth-Mover (Wasserstein-1) distance in the loss function. In this implementation, the discriminator output layer is a linear one, not sigmoid (discriminator output's a real value). WGAN [25] performs weight clipping $[-c,\ c]$ to enforce the Lipschitz constraint on the critic (i.e., discriminator). This method faces the issue of gradient explosion/vanishing without proper value of weight clipping parameter c. WGAN with Gradient penalty (WGAN-GP) [26] resolve above issues with WGAN [25]. WGAN-GP enforces a penalty on the gradient norm for random samples $\tilde{x} \sim P_{\tilde{x}}$. The objective function of WGAN-GP is as below.

$$V(D,G) = \min_G \max_D E_{\tilde{x}\sim p_g}[D(\tilde{x})] - E_{x\sim p_r}[D(x)] + \lambda E_{\tilde{x}\sim P_{\tilde{x}}}[(||\nabla_{\tilde{x}} D(\tilde{x})||_2 - 1)^2] \tag{3}$$

WGAN-GP [26] makes the WGAN [25] training more stable and does not require hyperparameter tuning. The DeblurGAN [3] used WGAN-GP method (Eq. 3) for single image blind motion deblurring.

Hinge Loss: In our method, we used Hinge loss [31,32] which is giving better result as compared to WGAN-GP [26] based deblurring method. Hinge loss output also a real value. Generator loss L_G and Discriminator loss L_D in the presence of Hinge loss is defined as follows.

$$L_D = -E_{(x,y)\sim p_{data}}[min(0, -1 + D(x,y))] - E_{z\sim p_z, y\sim p_{data}}[min(0, -1 - D(G(z),y))] \tag{4}$$

$$L_G = -E_{z\sim p_z, y\sim p_{data}} D(G(z), y) \tag{5}$$

Here, D tries that a real image will get a large value, and a fake or generated image will get a small value.

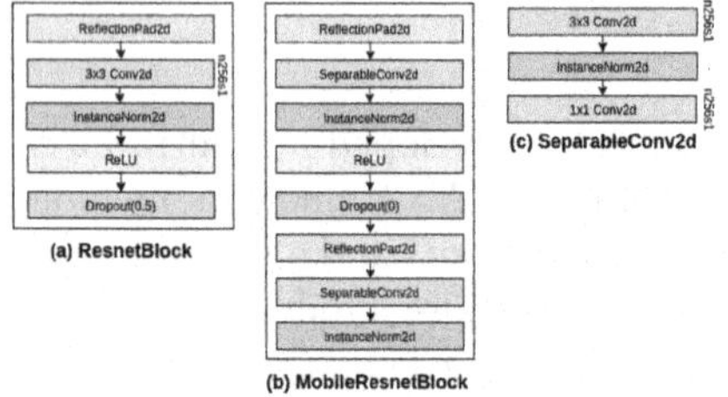

Fig. 3. Modified resnet block

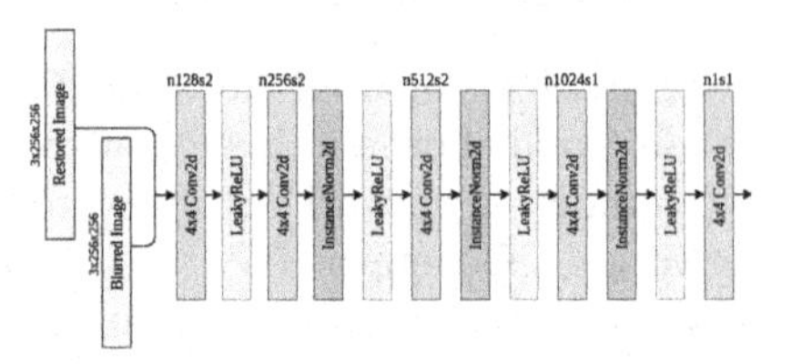

Fig. 4. The figure shows the architecture of the critic network (Discriminator).

3 Related Works

The deep learning-based methods attempt to estimate the motion blur in the degraded image and use this blurring information to restore the sharp image [21]. The methods which use the multi-scale framework [22] to recover the deblurred image are computationally expensive. The use of GANs also increasing in blind kernel free single image deblurring tasks such as Ramakrishnan et al. [24] used image translation framework [7] and densely connected convolution network [23]. The methods above performs image-deblurring task, when input image may have blur due to multiple sources. Kupyn et al. [3] proposed the DeblurGAN method, which uses the Wasserstein GAN [25] with gradient penalty [26] and the Perceptual loss [8]. Kupyn et al. [27] proposed a new method DeblurGAN-v2, which is faster and has better results than the previously proposed method; this method uses the feature pyramid network [28] in the generator. A study of various single image blind deblurring methods is provided in [29].

4 Our Method

In our proposed method, the blur kernel knowledge is not present, and from a given blur image I_B as an input, our purpose is to develop a sharp image I_S from I_B. For the deblurring task, we train a Generator network denoted by G_{θ_G}. During the training period, along with Generator, there is one another CNN also present D_{θ_D} referred to as the critic network (i.e., Discriminator). The Generator G_{θ_G} and the Discriminator D_{θ_D} are trained in an adversarial manner. In what follows, we describe the network architecture and the loss functions for our method.

4.1 Network Architecture

The generator network, a chief component of proposed model, is a transformed version of residual network architecture [6] (Sect. 4.1). The discriminator architecture, which helps to learn the Generator, is a transformed version of Markovian Discriminator (PatchGAN) [7] (Sect. 4.1). The residual network architecture helps us to build deeper CNN architectures. Also, this architecture is effective because we want our network to learn only the difference between pairs of sharp and blur images as they are almost alike in values.

We used the depthwise separable convolution in place of the standard convolution layer to reduce the inference time and model size [1]. Generator aims to generate sharp images given the blurred images as input. Note that generated images need to be realistic so that the Discriminator thinks that generated images are from the real data distribution. In this way, the Generator helps to generate a visually attractive sharp image from an input blurred image. Discriminator goal is to classify if the input is from the real data distribution or output from the generator. Discriminator accomplish this by analyzing the patches in the input image for making a decision. The changes which we made in the resnet block displayed in Fig. 3, we convert structure (a) into structure (b).

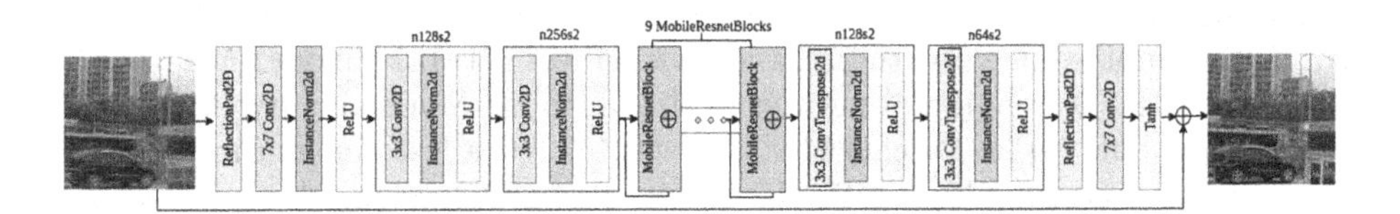

Fig. 5. The figure shows the generator architecture of our Fast Motion Deblurring-cGAN. Given a blurred image as input, Generator outputs a realistic-looking sharp image as output.

Generator Architecture. The Generator's CNN architecture is displayed in Fig. 5. This architecture is alike to the style transfer architecture which is proposed by Johnson et al. [8]. The generator network has two strided convolution blocks in begining with stride 2, nine depthwise separable convolutions based residual blocks (MobileResnet Block) [1,6], and two transposed convolution blocks, and the global skip connection. In our architecture, most of the computation is done by MobileResNet-Block. Therefore, we use depthwise separable convolution here to reduce computation cost without affecting accuracy.

Every convolution and transposed convolution layer have an instance normalization layer [33] and a ReLU activation layer [34] behind it. Each Mobile Resnet block consists of two depthwise separable convolutions [1], a dropout layer [35], two instance normalization layers after each separable convolution block, and a ReLU activation layer. In each mobile resnet block, after the first depth-wise separable convolution layer, a dropout regularization layer with a probability of zero is added. Furthermore, we add a global skip connection in the model, also referred to as ResOut.

When we use many convolution layers, it will become difficult to generalize over first-level features, deep generative CNNs often unintentionally memorize high-level representations of edges. The network will be unable to retrieve sharp boundaries at proper positions from the blur photos as a result of this. We combine the head and tail of the network. Since the gradients value now can reach from the tail straight to the beginning and affect the update in the lower layers, generation efficiency improves significantly [36]. In the blurred image I_B, CNN learns residual correction I_R, so the resulting sharp image is $I_S = I_B + I_R$. From experiments, we come to know that such formulation improves the training time, and generalizes the resulting model better.

Discriminator Architecture. In our model, we create a critic network D_{θ_D} also refer to as Discriminator. D_{θ_D} guides Generator network G_{θ_G} to generate sharp images by giving feedback on the input is from real data distribution or generator output. The architecture of Discriminator network is shown in Fig. 4. We avoid high-depth Discriminator network as it's goal is to perform the classification task unlike image synthesis task of Generator network. In our FMD-cGAN framework, the Discriminator network is similar to the Markovian patch discriminator, also refer to as PatchGAN [7]. Except for the last convolutional layer, InstanceNorm layer and LeakyReLU with a value of 0.2, follow all convolutional layers of the network. This architecture looks for explicit structural characteristics at many local patches. It also ensures that the generated raw images have a rich color.

Table 1. The table shows the results on GoPro test dataset. Here, FMD-cGAN$_{WILD}$ and FMD-cGAN$_{Comb}$ are our methods (Sect. 5). It could be observed that our frameworks achieves good quantitative performance.

Method	PSNR	SSIM	Time (GPU)	Time (CPU)	#Parameters	MACs
Sun et al. [5]	24.64	0.842	N/A	N/A	N/A	N/A
Xu et al. [41]	25.1	0.89	N/A	N/A	N/A	N/A
DeepFilter [24]	28.94	0.922	0.3 s	3.09 s	3.20M	N/A
$DeblurGAN_{WILD}$ [3]	27.2	0.954	0.45 s	3.36 s	6.06M	35.07G
$DeblurGAN_{Comb}$	28.7	0.958				
$DeblurGANv2_{Resnetv2}$ [27]	29.55	0.934	0.14 s	3.67 s	66.594M	274.20G
$DeblurGANv2_{Mobnetv2}$	28.17	0.925	0.04 s	1.23 s	3.12M	39.05G
SRN [39]	30.10	0.932	1.6 s	28.85 s	6.95M	N/A
DeepDeblur [22,48]	**30.40**	0.901	2.93 s	56.76 s	11.72M	4727.22G
FMD-cGAN$_{WILD}$	28.33	0.962	**0.01 s**	**0.28 s**	**1.98M**	**18.36G**
FMD-cGAN$_{Comb}$	29.675	**0.971**				

4.2 Loss Functions

The total loss function for FMD-cGAN deblurring framework is the mixture of adversarial loss and content loss.

$$L_{total} = L_{GAN} + \lambda \cdot L_X \tag{6}$$

In Eq. 6, L_{GAN} represents the advesarial loss (Sect. 4.2), L_X represents the content loss (Sect. 4.2) and λ represents the hyperparameter which controls the effect of L_X. The value of λ is equal to 100 in the current experiment.

Adversarial Loss. To train a learning-based image restoration network, we need to compare the difference between the restored and the original images during the training stage. Many image restoration works are using an adversarial-based network to generate sharp images [19,20]. During the training stage, the adversarial loss after pooling with other losses helps to determine how good the Generator is working against the Discriminator [22]. Initial works based on conditional GANs use the objective function of the vanilla GAN as the loss function [19]. Lately, least-square GAN [42] was observed to be better balanced and produce the good quality desired outputs. We apply Hinge loss [31] (Eq. 4 and Eq. 5) in our model to provide good results with the generator architecture [49]. Generator loss (L_G) and Discriminator loss (L_D) are computed as follows (Eq. 7 and Eq. 8).

$$L_G = -\sum_{n=1}^{N} D_{\theta_D}(G_{\theta_G}(I^B)) \tag{7}$$

$$L_D = -\sum_{n=1}^{N} min(0, D_{\theta_D}(I^S) - 1) - \sum_{n=1}^{N} min(0, -D_{\theta_D}(G_{\theta_G}(I^B)) - 1) \tag{8}$$

If we do not use adversarial loss in our network, it still converges. However, the output images will be dull with not many sharp edges, and these output images are still blurry because the blur at edges and corners is still intact. If we only use adversarial loss in our network, edges are retained in images, and more practical color assignment happens. However, it has two issues: still, it has no idea about the structure, and Generator is working according to the guidance provided by Discriminator based on the generated image. We remove these issues with the adversarial loss by combining adding with the Perceptual loss.

Content Loss. Generally, there are two choices for the pixel-based content loss: (a) L1 or MAE loss and (b) L2 or MSE loss. Moreover, above loss functions may produce blurry artifacts on the generated image due to the average of pixels [19]. Due to this issue, we used Perceptual loss [8] function for content loss. Unlike L2 Loss, Perceptual compares the difference between CNN feature maps of the restored image and the original image. This loss function puts structural knowledge into the Generator, which helps it against the patch-wise decision of the Markovian Discriminator. The equation of the Perceptual loss is as follows:

$$L_X = \frac{1}{W_{i,j}H_{i,j}} \sum_{x=1}^{W_{i,j}} \sum_{y=1}^{H_{i,j}} (\phi_{i,j}(I^S)_{x,y} - \phi_{i,j}(G_{\theta_G}(I^B))_{x,y})^2 \tag{9}$$

where $W_{i,j}$ and $H_{i,j}$ are the width and height of the $(i,j)^{th}$ ReLU layer of the **VGG-16** network [43], here i and j denote j^{th} convolution (after activation) before the ith max-pooling layer. $\phi_{i,j}$ denotes the feature map. In our current method, we use the output of activations from $VGG_{3,3}$ convolutional layer. The output from activations of the end layers of the network represents more features information [19,44]. The Perceptual loss helps to restore the general content [7,19]; on the other side adversarial loss helps to restore texture details. If we do not use the Perceptual loss in our network or use simple MSE based loss on pixels, the network will not converge to a good state.

4.3 Training Datasets

GoPro Dataset. The images of the GoPro dataset [22] are generated using the GoPro Hero 4 camera. The camera captures 240 frames per second video sequences. The blurred images are captured by averaging consecutive short-exposure frames. It is the most commonly used benchmark dataset in motion deblurring tasks, containing 3214 pairs of blur and sharp images. We use 1111 pairs of images for testing purposes and the remaining 2103 pairs of images for training [22].

REDS Dataset. The Realistic and Dynamic Scenes dataset [38] was designed for video deblurring and super-resolution, but it is also helpful in the image deblurring. The dataset comprises 300 video sequences having a resolution of 720×1280. Here, the training set contains 240 videos, the validation set contains 30 videos, and the testing set contains 30 videos. Each video has 100 frames. REDS dataset is generated from 120 fps videos, synthesizing blurry frames by merging subsequent frames. We have 240*100 pairs of blur and sharp images for training, 30*100 pairs of blur and sharp images for testing.

5 Training Details

The Pytorch[1] deep learning library is used to implement our model. The training of the model is accomplished on a single Nvidia Quadro RTX 5000 GPU using different datasets. The model takes image patches as input and fully convolutional to be used on images of arbitrary size. There is no change in the learning rate for the first 150 epochs; after it, we decrease the learning rate linearly to zero for the subsequent 150 epochs. We used Adam [45] optimizers for loss functions in both the Generator and the Discriminator with a learning rate of 0.0001. During the training time, we kept the batch size of 1, which gives a better result. Furthermore, we used the dropout layer (rate = 0) and the Instancenormalization layer instead of the batch-normalization layer concept both for the Generator and the Discriminator [7]. The training time of the network is approximately 2.5 days, which is significantly less than its competitive network. We have provided training details in Table 3. We discuss the two variants of FMD-cGAN as follows.

(1) FMD-cGAN$_{wild}$: our first trained model is **WILD**, which represents that the model is trained only on a single dataset such as GoPro and REDS dataset on which we are going to evaluate it. For example, in the case of the GoPro dataset model is trained on 2103 pairs of blur and sharp images of the GoPro dataset.

(2) FMD-cGAN$_{comb}$: The second trained model is **Comb**, which is first trained on the REDS training dataset; after training, we evaluate its performance on the REDS testing dataset. Now we train this pre-trained model on the GoPro dataset. We test both trained models **Comb** and **WILD** final performance on the GoPro dataset's 1111 test images.

[1] https://pytorch.org/.

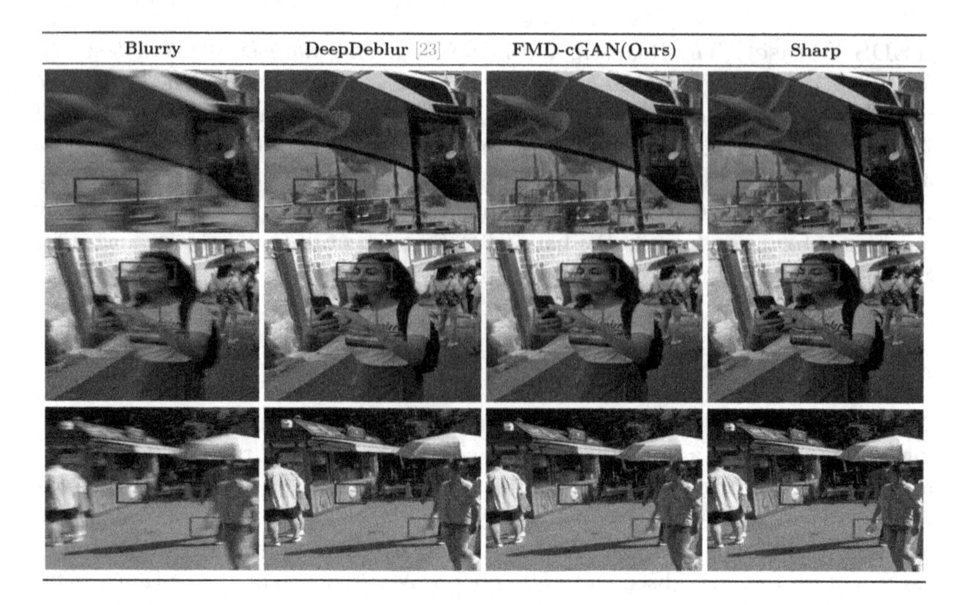

Fig. 6. The figure shows visual comparison on the REDS dataset (images are best viewed after zooming).

Table 2. The table shows the PSNR and SSIM comparison between FMD-cGAN (ours) and DeepDeblur [22,48] on the REDS test dataset.

Method	PSNR	SSIM
DeepDeblur [22,48]	32.89	0.9207
FMD-cGAN (ours)	31.79	**0.9804**

Table 3. The table summarises training details of our methods.

Model	Dataset	#Train images	#Test images
FMD-cGAN$_{WILD}$	GoPro	2103	1111
FMD-cGAN$_{Comb}$	1. REDS	24000	3000
	2. GoPro	2103	1111

6 Experimental Results

We compare the results of our FMD-cGAN with relevant models using the standard performance metrics (PSNR, SSIM). We also show inference time of each model (i.e., average running time per image) on a single **GPU (Nvidia RTX 5000)** and **CPU (2 X Intel Xeon 4216 (16C))**. To calculate Number of parameters and Number of MACs operations in PyTorch based model, we use pytorch-summary[2] and torchprofile[3] libraries.

[2] https://github.com/sksq96/pytorch-summary.
[3] https://github.com/zhijian-liu/torchprofile.

Table 4. Performance and efficiency comparison on the different no. of generator filters (#ngf)

#ngf	PSNR	SSIM	Time (CPU)	#Param	MACs
48	27.95	0.960	0.20 s	**1.13M**	**10.60G**
64	28.33	0.963	0.28 s	1.98M	18.36G
96	**28.52**	**0.964**	0.5 s	4.41M	40.23G

Table 5. Performance comparison after applying convolution decomposition in different parts of network and #ngf=64

Model	PSNR	SSIM	#Param	MACs
Only ResNetBlock	**28.33**	0.963	1.98M	18.36G
Downsample + ResNetBlock	28.24	0.962	**1.661M**	16.81G
Upsample + ResNetBlock	28.19	0.961	1.663M	**11.79G**

6.1 Quantitative Evaluation on GoPro Dataset

Here, we discuss the performance of our method on GoPro Dataset. We used 1111 pairs of blur and sharp images from GoPro test dataset for evaluation. We compare our model's results with other state-of-the-art model's results: where Sun et al. [5] is a traditional method, while others are deep learning-based methods: Xu et al. [41], DeepDeblur [22], DeepFilter [24], *DeblurGAN* [3], *DeblurGANv*2 [27] and SRN [39]. We use PSNR and SSIM value of other methods from their respective papers.

We show the results in Table 1. It could be observed that FMD-cGAN (ours) has high efficiency in terms of performance and inference time. FMD-cGAN also has the lowest inference time, and in terms of no. of parameters and macs operations also has the lowest value. Furthermore, FMD-cGAN output PSNR and SSIM values comparable to the other models in comparison.

6.2 Quantitative Evaluation on REDS Dataset

We also show the performance of our framework on the REDS dataset. We used 3000 pairs of blur and sharp images from REDS test dataset for evaluation. We compare the performance of FMD-cGAN (ours) with the DeepDeblur model [22]. We used the results of DeepDeblur from official GitHub repository - DeepDeblur-PyTorch[4].

We show the results in Table 2. It could be observed that our method achieves high SSIM and PSNR values which are comparable to DeepDeblur [22]. We emphasise that our network has a significantly lesser size as compared to DeepDeblur [22]. Currently, only the DeepDeblur model used the REDS dataset for training and performance evaluation.

[4] https://github.com/SeungjunNah/DeepDeblur-PyTorch.

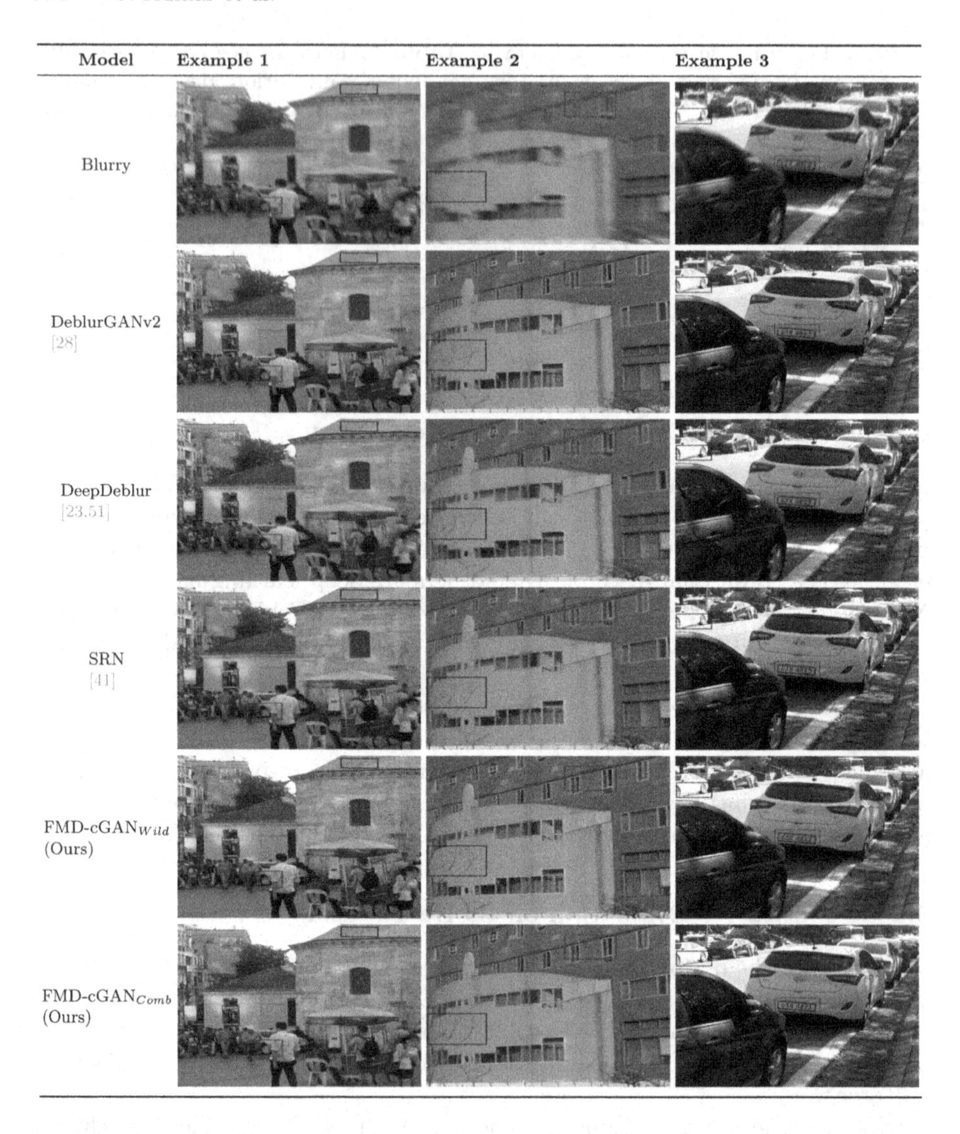

Fig. 7. The figure shows visual comparison on the GoPro dataset (images are best viewed after zooming).

6.3 Visual Comparison

Figure 6 shows the visual comparison on the REDS dataset. It could be observed that FMD-cGAN (ours) restore images comparable to the relevant top-performing works such as DeepDeblur [22] and SRN [39]. For example, row 1 of Fig. 6 shows that our method preserves the fine object structure details (i.e., building) which are missing in the blurry image.

Figure 7 shows the visual comparison results on the GoPro dataset. It could be observed that the output of our method is visually appealing in the presence of motion blur in the input image (see Example 3 of Fig. 7). To provide more clarity, we show the results for both FMD-cGAN$_{Wild}$ and FMD-cGAN$_{Comb}$ (Sect. 5). FMD-cGAN (ours) is faster and output better reconstruction than other motion deblurring methods even though our model has fewer parameters (Table 1). We have provided the extended versions of Fig. 6 and Fig. 7 in the supplementary material for better visual comparisons.

7 Ablation Study

Table 4 shows an ablation study on the generator network architecture for different design choices. Here, we train and test our network's performance only on the GoPro dataset. Suppose #ngf denotes the initial layer's filters count in the generator network, affecting filters count of subsequent layers. Table 4 demonstrates how #ngf affects model performance. It could be observed that if we increase the #ngf then image quality (PSNR) will increase. However, it increases #parameters and MACs operations also, affecting inference time and model size.

We divide our generator network into three parts according to its structure: Downsample (two 3×3 convolutions), ResnetBlocks (9 blocks), and Upsample (two 3×3 deconvolutions). To check the network performance, we put separable convolution into different parts. Table 5 demonstrates model performance after applying convolution decomposition in different parts of the generator network. ResNet blocks do most of the computation in the network; from Table 5, we can see applying convolution decomposition in this part giving better performance.

8 Conclusion

We proposed a Fast Motion Deblurring method (FMD-cGAN) for a single image. FMD-cGAN does not require knowledge of the blur kernel. Our method uses the conditional generative adversarial network for this task and is optimized using the multi-part loss function. Our method shows that using MobileNetv1 architecture consists of depthwise separable convolution to reduce computational cost and memory requirement without losing accuracy. We also proposed that using Hinge loss in the network gives good results. Our method produces better blur-free images, as confirmed by the quantitative and visual comparisons. FMD-cGAN is faster with low inference time and memory requirements, and it outperforms various state-of-the-art models for blind motion deblurring of a single image (Table 1). We propose as future work to deploy our model in lightweight devices for real-time image deblurring tasks.

References

1. Howard, A.G., et al.: MobileNets: Efficient Convolutional Neural Networks for Mobile Vision Applications. arXiv preprint arXiv:1704.04861v1 (2017)
2. Redmon, J., Divvala, S., Girshick, R., Farhadi, A.: You Only Look Once: Unified, Real-Time Object Detection. arXiv e-prints, June 2015
3. Kupyn, O., Budzan, V., Mykhailych, M., Mishkin, D., Matas, J.: DeblurGAN: blind motion deblurring using conditional adversarial networks. In: CVPR (2018)
4. Gong, D., et al.: From motion blur to motion flow: a deep learning solution for removing heterogeneous motion blur. IEEE (2017)
5. Sun, J., Cao, W., Xu, Z., Ponce, J.: Learning a convolutional neural network for non-uniform motion blur removal. In: CVPR (2015)
6. He, K., Hang, X., Ren, S., Sun, J.: Deep residual learning for image recognition. In: CVPR (2016)
7. Isola, P., Zhu, J.Y., Zhou, T., Efros, A.: Image-to-image translation with conditional adversarial networks. In: CVPR (2017)
8. Johnson, J., Alahi, A., Fei-Fei, L.: Perceptual losses for real-time style transfer and super-resolution. In: Leibe, B., Matas, J., Sebe, N., Welling, M. (eds.) ECCV 2016. LNCS, vol. 9906, pp. 694–711. Springer, Cham (2016). https://doi.org/10.1007/978-3-319-46475-6_43
9. Hansen, P.C., Nagy, J.G., O'Leary, D.P.: Deblurring images:matrices, spectra, and filterin. SIAM (2006)
10. Almeida, M.S.C., Almeida, L.B.: Blind and semi-blind deblurring of natural images. IEEE (2010)
11. Levin, A., Weiss, Y., Durand, F., Freeman, W.T.: Understanding blind deconvolution algorithms. IEEE (2011)
12. Szeliski, R.: Computer Vision: Algorithms and Applications. Springer, London (2011). https://doi.org/10.1007/978-1-84882-935-0
13. Richardson, W.H.: Bayesian-based iterative method of image restoration. JoSA **62**(1), 55–59 (1972)
14. Wiener, N.: Extrapolation, interpolation, and smoothing of stationary time series, with engineering applications. Technology Press of the MIT (1950)
15. Cai, X., Song, B.: Semantic object removal with convolutional neural network feature-based inpainting approach. Multimedia Syst. **24**(5), 597–609 (2018)
16. Chen, J., Tan, C.H., Hou, J., Chau, L.P., Li, H.: Robust video content alignment and compensation for rain removal in a CNN framework. In: CVPR (2018)
17. Luan, F., Paris, S., Shechtman, E., Bala, K.: Deep photo style transfer. In: CVPR (2017)
18. Zhang, K., Zuo, W., Che, Y., Meng, D., Zhang, L.: Beyond a gaussian denoiser: residual learning of deep CNN for image denoising. IEEE (2017)
19. Ledig, C., et al.: Photo-realistic single image super-resolution using a generative adversarial network. In: CVPR (2017)
20. Zhang, J., et al.: Dynamic scene deblurring using spatially variant recurrent neural networks. In: CVPR (2018)
21. Sun, J., Cao, W., Xu, Z., Ponce, J.: Learning a convolutional neural network for non-uniform motion blur removal. IEEE (2015)
22. Nah, S., Kim, T.H., Lee, K.M.: Deep multi-scale convolutional neural network for dynamic scene deblurring. In: CVPR (2017)
23. Huang, G., Liu, Z., van der Maaten, L., Weinberger, K.Q.: Densely connected convolutional networks. In: CVPR (2017)

24. Ramakrishnan, S., Pachori, S., Gangopadhyay, A., Raman, S.: Deep generative filter for motion deblurring. In: ICCVW (2017)
25. Arjovsky, M., Chintala, S., Bottou, L.: Wasserstein GAN. arXiv preprint arXiv:1701.07875 (2017)
26. Gulrajani, I., Ahmed, F., Arjovsky, M., Dumoulin, V., Courville, A.C.: Improved training of wasserstein GANs. In: Advances in Neural Information Processing Systems (2017)
27. Kupyn, O., Martyniuk, T., Wu, J., Wang, Z.: DeblurGAN-v2: deblurring (orders-of-magnitude) faster and better. In: ICCV, August 2019
28. Lin, T., Dollár, P., Girshick, R., He, K., Hariharan, B., Belongieg, S.: Feature pyramid networks for object detection. In: CVPR, July 2017
29. Lai, W., Huang, J.B., Hu, Z., Ahuja, N., Yang, M.H.: A comparative study for single image blind deblurring. In: CVPR (2016)
30. Goodfellow, I.J., et al.: Generative adversarial nets. In: NIPS (2014)
31. Lim, J.H., Ye, J.C.: Geometric GAN. arXiv preprint arXiv:1705.02894 (2017)
32. Zhang, H., Goodfellow, I., Metaxas, D., Odena, A.: Self-Attention GANs. arXiv:1805.08318v2 (2019)
33. Ulyanov, D., Vedaldi, A., Lempitsky, V.S.: Instance normalization: the missing ingredient for fast stylization. CoRR, abs/1607.08022 (2016)
34. Fred, A., Agarap, M.: Deep Learning using Rectified Linear Units (ReLU). arXiv:1803.08375v2, February 2019
35. Srivastava, N., Hinton, G., Krizhevsky, A., Sutskever, I., Salakhutdinov, R.: Dropout: a simple way to prevent neural networks from overfitting. J. Mach. Learn. Res. **15**(1), 1929–1958 (2014)
36. He, K., Zhang, X., Ren, S., Sun, J.: Identity mappings in deep residual networks. In: Leibe, B., Matas, J., Sebe, N., Welling, M. (eds.) ECCV 2016. LNCS, vol. 9908, pp. 630–645. Springer, Cham (2016). https://doi.org/10.1007/978-3-319-46493-0_38
37. Liang, C.H., Chen, Y.A., Liu, Y.C., Hsu, W.H.: Raw image deblurring. IEEE Trans. Multimedia (2020)
38. Nah, S., et al.: NTIRE 2019 challenge on video deblurring and super-resolution: dataset and study. In: CVPR Workshops, June 2019
39. Tao, X., Gao, H., Shen, X., Wang, J., Jia, J.: Scale-recurrent network for deep image deblurring. In: CVPR (2018)
40. Mirza, M., Osindero, S.: Conditional Generative Adversarial Nets. arXiv preprint arXiv:1411.1784v1, November 2014
41. Xu, L., Zheng, S., Jia, J.: Unnatural L0 sparse representation for natural image deblurring. In: CVPR (2013)
42. Mao, X., Li, Q., Xie, H., Lau, R.Y.K., Wang, Z.: Least squares generative adversarial networks. arxiv:1611.04076 (2016)
43. Simonyan, K., Zisserman, A.: Very deep convolutional networks for large-scale image recognition. In: ICLR (2015)
44. Zeiler, M.D., Fergus, R.: Visualizing and understanding convolutional networks. CoRR, abs/1311.2901 (2013)
45. Kingma, D.P., Ba, J.: Adam: a method for stochastic optimization. CoRR, abs/1412.6980 (2014)
46. Ren, W., Cao, X., Pan, J., Guo, X., Zuo, W., Yang, M.H.: Image deblurring via enhanced low-rank prior. IEEE Trans. Image Process. **25**(7), 3426–3437 (2016)
47. Pan, J., Sun, D., Pfister, H., Yang, M.H.: Blind image deblurring using dark channel prior. In: CVPR, June 2016
48. Nah, S.: DeepDeblur-PyTorch. https://github.com/SeungjunNah/DeepDeblur-PyTorch
49. Gong, X., Chang, S., Jiang, Y., Wang, Z.: AutoGAN: neural architecture search for GANs. In: ICCV (2019)

Region Extraction Based Approach for Cigarette Usage Classification Using Deep Learning

Anshul Pundhir(✉), Deepak Verma, Puneet Kumar, and Balasubramanian Raman

Computer Science and Engineering Department, Indian Institute of Technology, Roorkee, Roorkee, India
{anshul_p,d_verma,pkumar99,bala}@cs.iitr.ac.in

Abstract. This paper has proposed a novel approach to classify the subjects' smoking behavior by extracting relevant regions from a given image using deep learning. After the classification, we have proposed a conditionally active detection module based on Yolo-v3, which improves the model's performance and reduces its complexity. To the best of our knowledge, we are the first to work on the dataset named "Dataset containing smoking and not-smoking images (smoker vs. non-smoker)". This dataset contains a total of 2,400 images that include smokers and non-smokers equally in various environmental settings. We have evaluated the proposed approach's performance using quantitative and qualitative measures, which confirms its effectiveness in challenging situations. The proposed approach has achieved a classification accuracy of 96.74% on this dataset.

Keywords: Smoking behavior classification · Conditional detection · Small object detection · Region extraction · Bounding box adjustment

1 Introduction

Today's world, which is developing posthaste, has seen various technological innovations and financial advancements that positively serve society in many areas. Despite that, we have many problems such as pollution, an increasing number of road accidents, health issues such as lung cancer, respiratory diseases, and eye-vision problems. These hazards happen due to various factors, out of which daily use of cigarettes is a prominent one. As per the doctors' advice, one should avoid cigarette use since it has adverse effects on our health, environment, and life span. The governments have also established rules to prevent their use in public areas, but some break laws when they find themselves not monitored by any authority. Unfortunately, cigarette use by one person has adverse effects on

This work was supported by the University Grants Commission (UGC) INDIA with grant number: 190510040512.

B. Raman et al. (Eds.): CVIP 2021, CCIS 1568, pp. 378–390, 2022.
https://doi.org/10.1007/978-3-031-11349-9_33

others' lives in the form of pollution, health issues, and car accidents. So, there is an essential need to develop an automated system that can help to find a person's smoking behavior. Such systems have a wide range of applications such as automated smoke monitoring systems, cigarette censoring in videos, and controlling the number of road accidents due to drivers' smoking behavior [6]. The proposed approach consists of a region extraction module, classification module, and conditionally active Yolo-v3 [16] based real-time detection module. It provides a simple yet effective tool to judge the subjects' smoking behavior by analyzing their visual information. The region extraction module refines the visual information by extracting face and hand proposals. The classification module processes these proposals for final classification. Based on the classification result, the detection module will perform cigarette detection. The proposed approach has been evaluated on a recent dataset named "Dataset containing smoking and not-smoking images (smoker vs. non-smoker)" [11], containing 2,400 images with a nearly equal number of smoker and non-smoker images, and achieved an accuracy of 96.74%. This approach's effectiveness is determined quantitatively by measuring accuracy, precision, recall, and qualitatively by visualizing its performance in various challenging situations.

During the evaluation, results show that the proposed approach can handle multiple challenges like variability in hand, face postures, different illumination conditions, and little difference between smoker and non-smoker in the larger scene.

The contributions of the paper are summarised as follows.

- A novel region extraction-based deep-learning approach has been proposed for cigarette usage classification. It is capable of handling challenging situations such as low brightness, little visibility of cigarettes, and various gestures of hands.
- The architecture of the proposed approach has been determined after performing an extensive ablation study to analyze the effect of region proposal and transfer learning. The performance of the proposed approach has been compared with the state-of-the-art methods for a similar problem and found effective in handling challenging situations.
- Conditionally active detection module has been incorporated to save the computational cost and improve the detection performance by reducing the false positives.

The rest of the paper is organized as follows. The related work in this area is discussed in Sect. 2, the proposed methodology is discussed in Sect. 3. We have discussed the procedures used in experiments and results in Sect. 4. Finally, in Sect. 5, we have concluded our findings with future directions for further research.

2 Related Work

Various research attempts have been made for smoking behaviour detection and classification [2–4, 9, 12–14, 17, 19]. We have found that researchers have applied image-based and sensor-based approaches to overcome the underlying challenges.

- **Image based approaches**: The image-based methods process the image-related information such as the presence of smoke and the color of smoking object [4,6,12,13,19,20]. Wu *et al.* [20] have proposed the framework to detect smokers' behavior in videos based on visual cues. For tracking the cigarette, they have used the color histogram projection and Gaussian Mixture Models. Chien *et al.* [4] have used the custom dataset to detect the usage of cigarettes using a deep learning-based approach. This work was mainly useful to increase road safety by detecting the smoking behavior of the driver. Dhanwal *et al.* [6] have proposed the convolution-based model that processes the information obtained from multiple online resources by segmentation, cleaning, and augmentation to recognize the smoking behavior. Stitt *et al.* [19] have used the video camera feeds to process the information with the help of image processing algorithms that can quantify the parameters related to the smoking activity. For detecting the distraction behavior of the driver in the car, which includes using a cigarette, a phone call during driving, etc., Mao *et al.* have proposed a deep learning-based model to detect the track the region around the driver's face [13].
- **Sensor based approaches**: These approaches work on the application of sensors to measure the related parameters instead of direct processing of visual information. Sensor-based methods deploy the sensors to detect the smoking behavior and process the data collected by them [1,7–9,17]. Senyurek *et al.* [17] have proposed the CNN and LSTM based approach that uses the signals generated from the wearable sensors to identify the smoking puffs of cigarettes in free-living environmental conditions. Imtiaz *et al.* [9] have used the support vector machine classifier that could perform feature analysis from the information related to the body parameters like heart rate, breathing rate, etc., to find the smoking behavior. They also evaluate the results in free-living and laboratory conditions and analyze which features are more decidable to identify smoking behavior. Imtiaz *et al.* [7] have proposed an approach making use of a camera module, an inertial sensor, to detect smoking behavior. Añazco *et al.* [1] have proposed a two-stage approach that can use the smartphone app, which can operate with the wrist band sensor for relevant features and hence helps to detect the smoking puffs. Imtiaz *et al.* [8] finds the smoking behavior in free-living using a wearable sensors-based approach.

It has been found that the image-based approaches are comparatively more preferred in comparison to the sensor-based approaches. The primary reason to incline towards the image-based methods is the extra setup cost involvement and less convenience in the sensor-based methods for the users under consideration. Also, image-based approaches are more readily scalable for large-scale deployment as compared to sensor-based approaches. The literature survey suggests that this problem needs further exploration since most researchers have focussed on controlled surroundings. Moreover, the issue includes challenges due to the tiny shape of a cigarette. The robust and convincing cigarette usage detection and classification methods to address these challenges will contribute to a safer and greener world. It motivated us to develop an approach to effectively overcome the challenges mentioned above and accurately detect smoking behavior.

3 Proposed Methodology

This section elaborates on the proposed methodology. The proposed method's architecture has been shown in Fig. 1, and various components are discussed in the following sections.

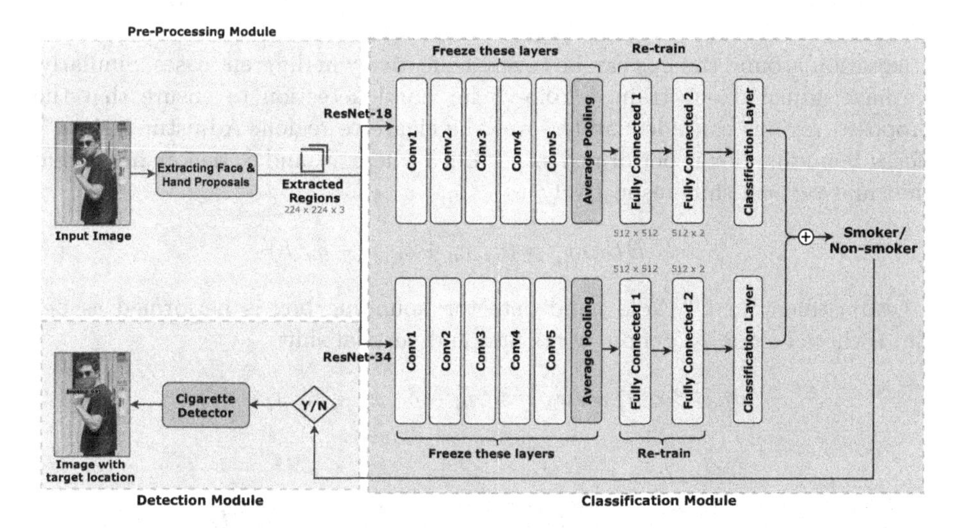

Fig. 1. Schematic architecture of the proposed approach.

3.1 Region Extraction Module

Due to the cigarette's small size, smoker and non-smoker objects look very similar in the broader view. This module performs data preprocessing, which solves challenges due to the small size of a cigarette. As shown in Fig. 2, the input images are preprocessed using *Faced* algorithm (for detecting face regions) [10] and Yolo-v3 (trained by us for detecting hand regions) [16], to extract the probable cigarette regions (i.e., face, hand). It helped us improve the model's performance since it needs to process relevant regions rather than process the whole image. We found *Faced* algorithm more convincing on this dataset than the 'Haar Cascade Classifier' [18] for extracting variable face poses. Further, we have fine-tuned the parameters of *Faced* algorithm to improve the predictions.

For a given image, I, we have extracted face proposals $F_1, F_2, ..., F_i$, and hand proposals $H_1, H_2, ..., H_j$, where i and j are the number of proposals extracted by *Faced* algorithm and trained yolo hand detector. *Faced* algorithm returns the bounding box for a kth face proposal as per Eq. 1 where, c_x, c_y corresponds to coordinates of center and w, h denotes width and height of the bounding box.

$$BBox_{F_k} = (c_x, c_y, w, h) \tag{1}$$

Likewise, the Yolo hand detector returns the bounding box for k^{th} hand proposal as per Eq. 2 where x_1, y_1 denote the coordinates of top left corner and x_2, y_2 denote the coordinate of bottom right corner.

$$BBox_{H_k} = (x_1, y_1, x_2, y_2) \quad (2)$$

To improve the region's coverage for a cigarette, we have adjusted the *Faced* bounding box by shifting them vertically down with wider width so that cigarette orientation around the lips can be covered effectively in different cases. Similarly, we have adjusted our trained Yolo-v3 for hand detection to ensure that the proposed regions can adequately cover the cigarette region. Adjustment in k^{th} *Faced* bounding box is performed as per Eq. 3 where, δ_h and δ_v denote horizontal shift and vertical shift respectively.

$$Adj\_BBox_{F_k} = (c_x, c_y + \delta_v, w + \delta_h, h) \quad (3)$$

Adjustment in k^{th} Yolo hand detector bounding box is performed as per Eq. 4 where δ_h *and* δ_v denote horizontal and vertical shift.

$$Adj\_BBox_{H_k} = (x_1 - \delta_h, y_1 - \delta_v, x_2 + \delta_h, y_2 + \delta_v) \quad (4)$$

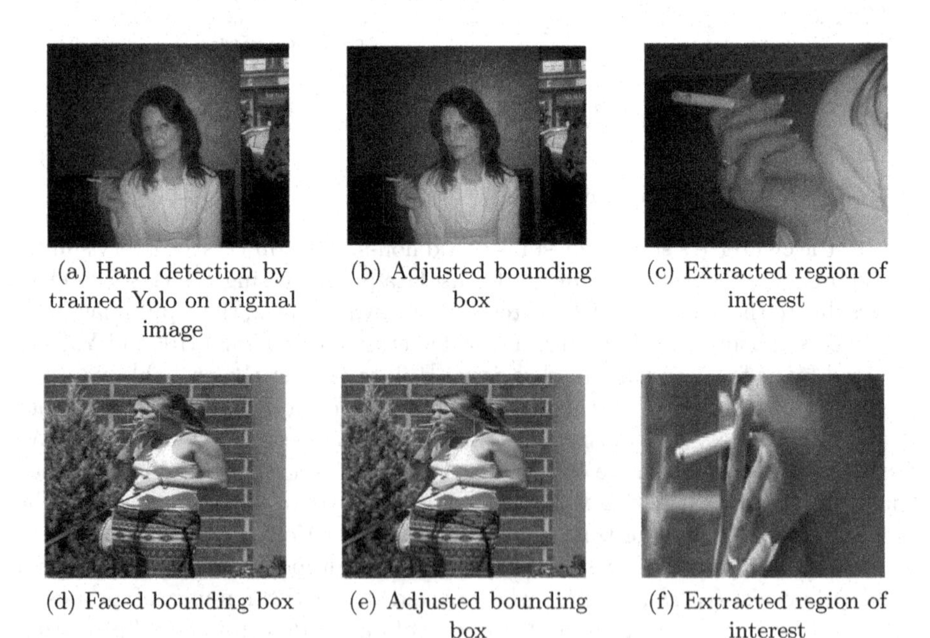

(a) Hand detection by trained Yolo on original image (b) Adjusted bounding box (c) Extracted region of interest

(d) Faced bounding box (e) Adjusted bounding box (f) Extracted region of interest

Fig. 2. Detection and region extraction using Yolo and Faced algorithm.

3.2 Classification Module

From baseline experiments in Sect. 4.1, it was observed that instead of using simple CNN and training from scratch, we could use transfer learning which gives two-fold benefits in this problem by giving low-cost models and solving small dataset size issues. Here, we have used the ensemble of ResNet-18 and ResNet-34 models with their pre-trained weights. We have modified the ResNet model's architecture (in each variant) by adding an extra fully connected layer as shown in Fig. 1. This extra FC layer takes 512 input neurons and outputs 512 output neurons. After this FC layer, we have added the ReLU activation followed by a dropout layer having the probability of 0.5. Finally, the last FC layer classifies the input neurons into two desired classes as smoker and non-smoker.

For any given image I, model performs classification on face and hand proposals as $C_{F_1}, C_{F_2}, ...C_{F_i}$ and $C_{H_1}, C_{H_2}, ...C_{H_j}$ respectively, where any C_F, C_H is either 0 or 1. Finally classified category C_I for the image I is determined as per Eq. 5 where $\oplus$ denotes the operation to takes maximum of all predicted classification categories.

$$C_I = \oplus(C_{F_1}, C_{F_2}, ...C_{F_i}, C_{H_1}, C_{H_2}, ...C_{H_j}) \tag{5}$$

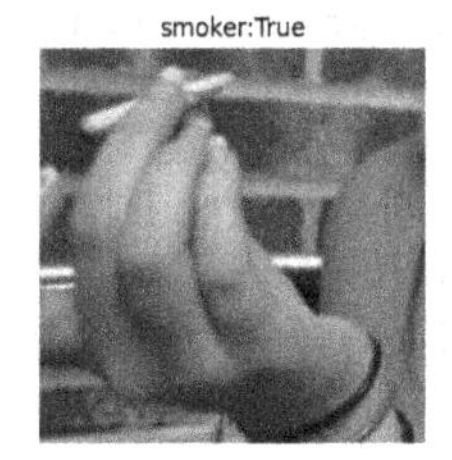

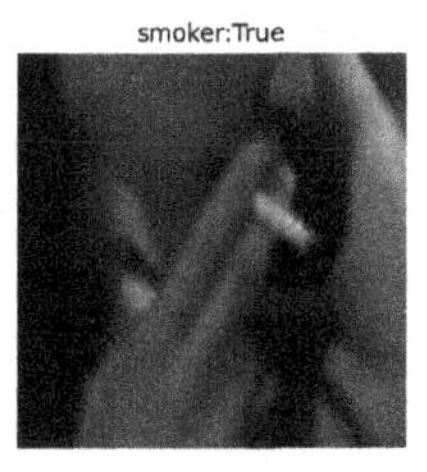

Fig. 3. Classification on sample images.

3.3 Detection Module

This module implements real-time cigarette detection using Yolo-v3 model trained on the cigarette images using LabelImg [5]. This module is conditionally active and performs detection only if the given image is classified with smoking behavior. This idea helps in reducing false positives and improves performance significantly. The detection module gets triggered only when the given image, I is classified as a smoker. Suppose $ind_i, ind_2, ..., ind_l$ denotes the indices of l proposals on which C_F, C_H have classified as a smoker. Then cigarette detector performs l cigarette detections on the input image I. It is robust for various challenging situations (as shown in Fig. 4). In case of failed detection in the first attempt, it performs the detection on l proposals to improve our results (as shown in Fig. 5). It overlays the proposals on the raw image to hold smokers' identities, which helps in the cases where cigarette is in hand. This idea is useful for cigarette monitoring systems.

4 Experiments and Results

This section discusses the implementation details, evaluation metrics, and the results obtained during the experiments.

4.1 Implementation

Experimental Setup. We have trained our proposed model on Nvidia RTX 2060 GPU having 1920 CUDA cores. This proposed model has been tested on Intel(R) Core(TM) i5-9300H, 2.40 GHz, 16 GB RAM CPU machine with 64-bit Ubuntu-20.04 OS machine.

Dataset and Training. The proposed model has been trained and evaluated on the Mendeley smoker dataset [11], which has never been used before to the best of our knowledge. It contains 2,400 images with the smoker and non-smoker images in various poses and environmental settings. The training and evaluation of our model are performed based on the splits provided in the dataset. The classification model is trained with the valid proposals obtained from the pre-processing module. At evaluation time, the model classifies the input class using Eq. 5. We have used the Adam optimizer to fine-tune the last two FC layers only, with a learning rate of 0.001. The final classification is performed using a softmax classifier along with using the cross-entropy loss function. Yolo-v3 based hand detector and cigarette detector are trained using 400 images from the training dataset over 4000 epochs. We had not trained any custom model for detecting face proposals since qualitative visuals obtained by the Faced model on the dataset were compelling. After training, we have found the preprocessing module robust enough to detect the region proposals.

Ablation Study. We have decided on the final experimental setup during the ablation study by investigating the effect of region proposal extraction and transfer learning on the model's performance. These effects are detailed as follows:

- **Effect of region proposals**: For this, we have set up two experimental settings named Baseline 1 and Baseline 2. These baselines are simple convolutional architectures consists of convolutional and pooling layers. These baselines are further explained in Sect. 4.1. In Baseline 1, we have used only raw images, while in Baseline 2, we have used the region proposals instead of raw images. From Table 1, we found significant performance improvement if we focus on region proposals instead of the original raw image. Therefore, the region proposal approach favors improvement in the classification performance of our model.
- **Effect of transfer learning**: After finding the importance of region proposals, we want to test whether the transfer learning approach can improve the performance further. In Baseline 3, we have tested how well the transfer

learning can work on the raw images using the ensemble model of ResNet-18 and ResNet-34 as discussed in Sect. 4.1. With this experimental setup, we have found huge performance gain in classification as shown in Table 1. Hence, we conclude that transfer learning favors the classification capability of our model.

Baselines and State of the Arts. An ablation study to decide the proposed approach's architecture is performed and summarized in Table 1. Here, baselines are designed by considering the challenges due to the small dataset and cigarette size. Baseline 1 and Baseline 2 focus on the challenge due to the small cigarette size while Baseline 3 on the small dataset size. Further details about the baselines are as follows:

- **Baseline 1:** This baseline is designed using a simple convolutional neural network (CNN) based architecture. Here, we are experimenting to test any performance improvement by the proposed region extraction module. Therefore, to build this simple architecture, we designed a custom CNN using two convolution blocks and two fully connected layers. After each convolution block, we have applied the max-pooling layer followed by the ReLU activation function. This model takes the raw input images from the dataset (i.e., without region extraction) to perform classification using softmax and cross-entropy loss function, along with adam optimizer to update the model parameters. This model is trained with these raw images and then yields the classification accuracy of 59% on the test dataset.
- **Baseline 2:** This baseline has similar architectural settings to Baseline 1, but the introduction of the proposed region extraction module. Due to this change, the model is trained with the refined proposals, which contain the features responsible for the desired classification task since these refined regions help to avoid analysis on the unimportant area of the given image. After training, this model performs much better than Baseline 1 and achieves the classification accuracy of 67% (improved from 59%). This confirms the importance of the region extraction module to handle the challenge of the tiny size of cigarettes in the image.
- **Baseline 3:** For solving small dataset issues, it is a common practice to use the transfer learning approach. We tested this idea for our problem using the ensemble of the pre-trained ResNet-18 and ResNet-34 (since the individual model could not significantly improve the performance). Before starting its training, we froze the convolutional layers, and only the fully connected layers were trained using the raw images from the train split of the dataset. This idea works for this problem, and with the limited dataset size, our trained model achieves the classification accuracy of 90%. This confirms the importance of using the pre-trained model for this problem.

After analyzing the results from the above baselines, it has been concluded to use the transfer learning approach along with the region extraction module. With the observations mentioned above, we came up with the proposed approach's architecture, as shown in Fig. 1. We have compared the performance obtained by our model on the dataset with the other state-of-the-arts results in Table 4.

Table 1. Summary of the ablation study.

Model	Processing strategy	Accuracy
Baseline 1	Raw input image	59%
Baseline 2	Extracting ROIs	67%
Baseline 3	Raw input image	90%
Proposed approach	Extracting ROIs	**96.74%**

4.2 Results and Evaluation

The proposed approach has obtained an accuracy of 96.74% for classification. Its performance has been evaluated using the following quantitative and qualitative measures during the classification and detection.

Quantitative Performance Measures. For our model's quantitative measure during classification, we have shown its precision, accuracy, recall, and F1 score in Table 2. We have also shown the quantitative measure for Yolo-v3 based hand detector and cigarette detector in Table 3. Our trained Yolo-v3 based hand detector and cigarette detector have achieved the detection accuracy of 88% and 72.72%, respectively. We have calculated the intersection over union score to calculate the overlap between the predicted proposals concerning their actual ground truth proposals during hand detection and cigarette detection. For hand detection and cigarette proposals, the mean intersection over union score is 0.89 and 0.85, respectively.

Table 2. Quantitative performance measures during classification.

Metric	Obtained
Precision	95%
Recall	98%
Accuracy	96.74%
F1 score	96.47%

Table 3. Quantitative performance measures during detection.

Metric	Hand detection	Cigarette detection
Precision	95.13%	87.86%
Recall	92.14%	80.85%
Accuracy	88%	72.72%
F1 score	93.61%	84.21%

Qualitative Performance Measures. The visuals related to classification results are shown in Fig. 3. The detection capability by our trained Yolo-v3 based cigarette and hand detector in different challenging scenarios are shown in Fig. 4 and Fig. 5 respectively. Also, in Fig. 6, we have shown how accurate our detection is with respect to the actual ground truth by showing the overlap between them. From Fig. 6, we can see that predicted proposals are consistent with respect to actual ground truth proposals.

4.3 Comparison with State-of-the-Art Approaches

As per the literature, no state-of-the-art (SOTA) approaches for cigarette usage analysis are available for the Mendeley smoker dataset [11]. Moreover, few datasets are available for this problem. We have compared the proposed approach to SOTA approaches for other problems with similar objectives and use-cases. The comparison shown in Table 4 affirms the proposed approach's applicability for this problem.

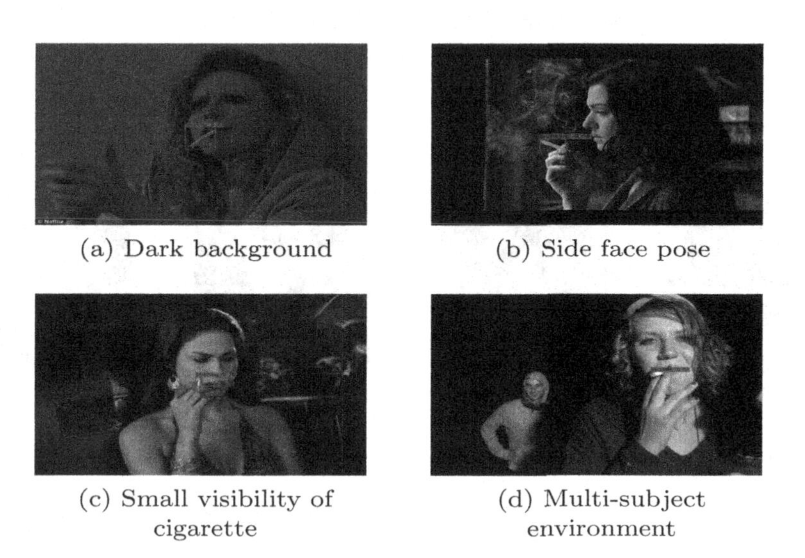

(a) Dark background (b) Side face pose

(c) Small visibility of cigarette (d) Multi-subject environment

Fig. 4. Cigarette detection in different settings.

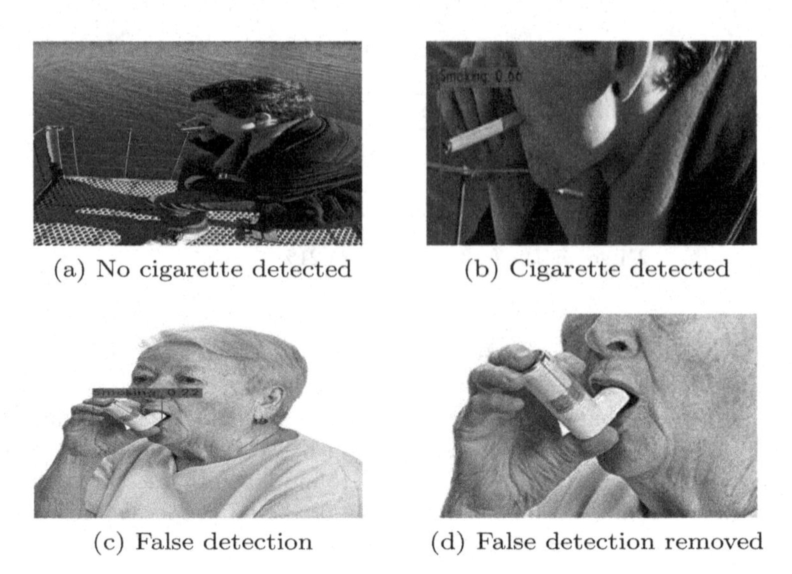

(a) No cigarette detected (b) Cigarette detected

(c) False detection (d) False detection removed

Fig. 5. Improvement in detection using proposed approach.

Table 4. Comparison with state-of-the-art approaches.

Method	Author	Accuracy
CNN based	Ou *et al.* [15]	79.4%
Deep learning	Dhanwal *et al.* [6]	89.9%
Wrist IMU	Añazco *et al.* [1]	91.38%
Faster-RCNN	Lu *et al.* [12]	92.1%
Proposed		**96.74%**

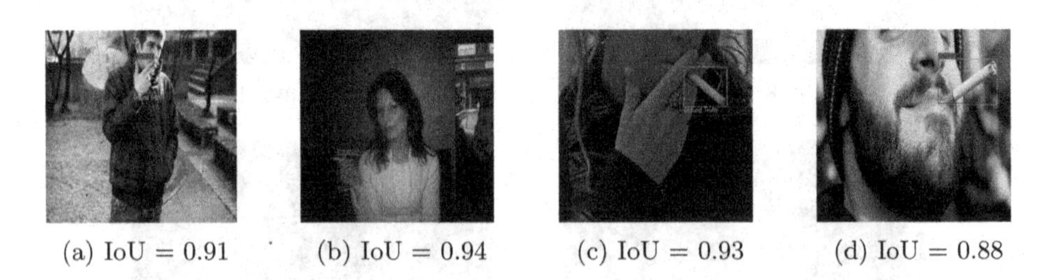

(a) IoU = 0.91 (b) IoU = 0.94 (c) IoU = 0.93 (d) IoU = 0.88

Fig. 6. Detection of hand, cigarette with respect to their ground truth.

4.4 Discussion

In this work, we have found that the region proposals extraction approach is helpful to improve the classification and detection ability of the model. As shown in Eq. 5, the proposed approach assigns the smoker category during classification by considering all probable cigarette regions while discarding unimportant areas of the given input. This way of classification is also effective if any invalid proposal (i.e., region without cigarette) is present in detected proposals. Therefore, problems related to false positive detection do not affect the final classification accuracy. This paper has addressed the challenge related to the small dataset size, the very tiny size of the target object. These proposals help us extract the relevant features by avoiding the less critical regions, and thus robust feature set is obtained. Also, low-level feature computation becomes easy for the model using the pre-trained model, and considerable performance is achieved on the dataset. In experiments, we have found that proposal extraction helps to improve the true positive rate and reduce the false positive rate for detection. Also, the proposal-based classification leads to a conditional active and cost-effective detection module with a reduced false-positive detection rate.

5 Conclusion and Future Work

This paper has proposed a novel approach for smoking behavior classification and detection using deep learning. The proposed approach has obtained significant results in challenging situations for the Mendeley smoker dataset, which has not been used so far. It can be extended for anomalous human activity recognition and detection of other small objects. In the future, we aim to make the proposed approach more effective by including the information from more modalities such as videos and feeds through night-vision cameras.

References

1. Añazco, E.V., Lopez, P.R., Lee, S., Byun, K., Kim, T.S.: Smoking activity recognition using a single wrist IMU and deep learning light. In: International Conference on Digital Signal Processing (ICDSP), pp. 48–51 (2018)
2. Arief, L., Tantowi, A.Z., Novani, N.P., Sundara, T.A.: Implementation of YOLO and smoke sensor for automating public service announcement of cigarette's hazard in public facilities. In: IEEE International Conference on Information Technology Systems and Innovation (ICITSI), vol. 782, no. 2, pp. 101–107 (2020)
3. Chang, Y., Du, Z., Sun, J.: Dangerous behaviors detection based on deep learning. In: International Conference on Artificial Intelligence and Pattern Recognition (AIPR), pp. 24–27 (2019)
4. Chien, T.C., Lin, C.C., Fan, C.P.: Deep learning based driver smoking behavior detection for driving safety. J. Image Graph. **8**(1), 15–20 (2020)
5. Darrenl, T.: Labelimg: A Graphical Image Annotation Tool (2017). https://github.com/tzutalin/labelImg. Accessed 02 Oct 2021

6. Dhanwal, S., Bhaskar, V., Agarwal, T.: Automated censoring of cigarettes in videos using deep learning techniques. In: Kapur, P.K., Singh, G., Klochkov, Y.S., Kumar, U. (eds.) Decision Analytics Applications in Industry. AA, pp. 339–348. Springer, Singapore (2020). https://doi.org/10.1007/978-981-15-3643-4_26
7. Imtiaz, M.H., Hossain, D., Senyurek, V.Y., Belsare, P., Sazonov, E.: PACT CAM: wearable sensor system to capture the details of cigarette smoking in free-living. In: IEEE Sensors Journal, pp. 1–4 (2020)
8. Imtiaz, M.H., Ramos-Garcia, R.I., Wattal, S., Tiffany, S., Sazonov, E.: Wearable sensors for monitoring of cigarette smoking in free-living: a systematic review. MDPI Sens. J. **19**(21), 4678 (2019)
9. Imtiaz, M.H., Senyurek, V.Y., Belsare, P., Tiffany, S., Sazonov, E.: Objective detection of cigarette smoking from physiological sensor signals. In: IEEE Annual International Conference of the Engineering in Medicine and Biology Society (EMBS), pp. 3563–3566 (2019)
10. Itzcovich, I.: Faced: Algorithm for Detecting Face Regions (2018). https://github.com/iitzco/faced. Accessed 02 Oct 2021
11. Khan, A.: Dataset Containing Smoking and Not-Smoking Images (Smoker vs Non-Smoker) (2020). https://data.mendeley.com/datasets/7b52hhzs3r/1. Accessed 02 Oct 2021
12. Lu, M., Hu, Y., Lu, X.: Driver action recognition using deformable and dilated faster R-CNN with optimized region proposals. Appl. Intell. J. **50**(4), 1100–1111 (2020)
13. Mao, P., Zhang, K., Liang, D.: Driver distraction behavior detection method based on deep learning. J. Manag. Sci. Eng. (JMSE) **782**(2), 022012 (2020)
14. Melek, C.G., Sonmez, E.B., Albayrak, S.: Object detection in shelf images with YOLO. In: IEEE EUROCON International Conference on Smart Technologies (SmartTech), pp. 1–5 (2019)
15. Ou, Y.Y., Tsai, A.C., Wang, J.F., Lin, J.: Automatic drug pills detection based on convolution neural network. In: IEEE International Conference on Orange Technologies (ICOT) (2018)
16. Redmon, J., Farhadi, A.: Yolov3: An Incremental Improvement. arXiV (2018). Accessed 02 Oct 2021
17. Senyurek, V.Y., Imtiaz, M.H., Belsare, P., Tiffany, S., Sazonov, E.: A CNN-LSTM neural network for recognition of puffing in smoking episodes using wearable sensors. Biomed. Eng. Lett. **10**(2), 195–203 (2020). https://doi.org/10.1007/s13534-020-00147-8
18. Soo, S.: Object Detection using Haar Cascade Classifier. Institute of Computer Science, University of Tartu, pp. 1–12 (2014)
19. Stitt, J.P., Kozlowski, L.T.: A system for automatic quantification of cigarette smoking behavior. In: IEEE International Conference of Engineering in Medicine and Biology Society (EMBS), pp. 4771–4774 (2006)
20. Wu, P., Hsieh, J.W., Cheng, J.C., Cheng, S.C., Tseng, S.Y.: Human smoking event detection using visual interaction clues. In: IEEE International Conference on Pattern Recognition (ICPR), pp. 4344–4347 (2010)

Fire Detection Model Using Deep Learning Techniques

Arun Singh Pundir(✉) and Balasubramanian Raman

Department of Computer Science and Engineering, Indian Institute of Technology, Roorkee, India
arunpundir2006@gmail.com, balarfma@iitr.ac.in

Abstract. Frame-based and motion-based deep learning framework have been employed in this paper for fire detection. This paper presents a novel method for fire detection that employs different Convolutional Neural Networks (CNN) architectures for fire detection. Firstly, frame based features such as fire-color, fire texture, and analysis of perimeter disorder which are present in the still images were extracted using transfer learning. Secondly, motion-based CNN is used for extracting motion based features of fire such as growing region, moving area and uprising part detection. Optical flow was employed to calculate the motion of frame intensities. These extracted intensity features which being projected as image were fed into Deep CNN for find out the uniqueness in the motion of fire. Features from both the models are combined to feed into different classifiers. Method is tested on different varieties of non-fire videos which are similar to forest fire such as fire-works, sun based videos, traffic at night and flower-valley videos. Accuracy on such similar situations has proven the precision and robustness in the proposed method.

Keywords: Transfer learning · Fire detection · Deep CNN · Optical flow

1 Introduction

All around the world, regions with hot climate and intense vegetation covered areas are facing threat in the form of wildfires. According to Zhang et al. [25], inland wetland in Australia are facing crisis, due to wildfire, not only in the form of human life and economic loss but also wipe out the fire sensitive wetland breed like river red gems. In a report given by [7], it is mentioned that China's 8% of the forest and world's 1% forest is destroyed by wildfire every year.

There are different approaches for detection of fire such as remote monitoring of entire forest using surveillance cameras. Secondly, using optical sensor based fire alarms. Thirdly, detection by non-optical sensors based techniques like surveillance cameras using computer vision techniques for automatic fire

B. Raman et al. (Eds.): CVIP 2021, CCIS 1568, pp. 391–402, 2022.
https://doi.org/10.1007/978-3-031-11349-9_34

detection. Optical detectors consist of LEDs and opening area that receives carbon particles. If there is any occurrence of fire, the burning carbon particles interrupts the constant circuit between LED and photocell within device, this breaking of circuit will raise the alarm.

Remote surveillance cameras and fire alarms show 100% accuracy in fire detection. However, they also faced various shortcomings such as in case of remotely installed cameras, it always demands some persons to get engaged in watching the cameras, it causes visual fatigue and there is no one surveillance during shift change. In case of fire alarms, its limitations are, firstly they should be close to the fire for triggering of alarm while in case of forest fire, no one can predict its point of occurrence. Secondly, their detection capability is well suited to a small or large hall but when it comes to few kilometres, these alarms fail. Thirdly, alarms cannot predict whether fire alarm is raised due to smoke coming from cigarette or some uncontrollable fire.

In case of fire detection by computer vision, the accuracy is not 100% but there are many pros that can make it better approach than previously mentioned approaches. If surveillance cameras, capable of automatic fire detection using computer vision based methods, are placed on a height so that their vision area is few kilometres such as placed on hill top or installed on a tower in forest area, it can detect any occurrence of fire within few kilometres. This approach does not need any carbon particle to be sensed by, hence can be installed anywhere in in-door location or in deep forest area. Easy installation and cheap set up can be considered as advantages over other mentioned approaches.

In our work, our proposed method mainly focuses on extracting all the features of fire using the power of transfer learning. For spatial features such as fire-color, perimeter disorder analysis and fire-texture, we have extracted fire-based frames from our dataset of fire-containing videos and finetune them on different CNN architectures. For motion features of fire, such as unique fire growing motion optical flow [11] is calculated. Then, these optical flow are feed into various deep network architectures for extracted motion features of fire. The computed features are aggregated together and feed into classifiers for detection.

2 Literature Review

Literature review is done in Table No. 1

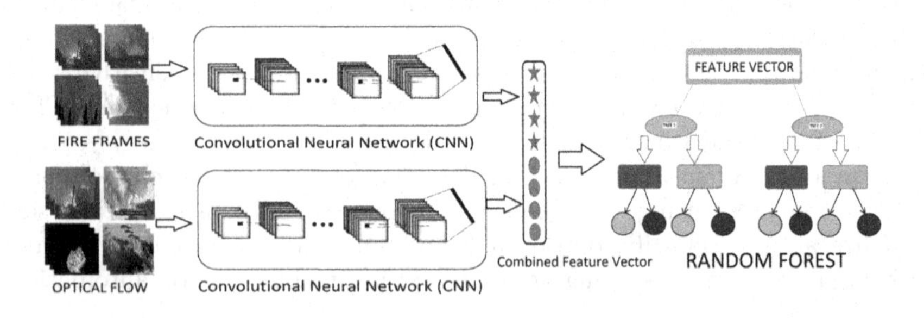

Fig. 1. Block diagram of the proposed method (ResNet with Random Forest).

Table 1. Literature survey

Related work	Authors	Summary
Color based related work	Chen et al. [4]	Incorporated RGB (Red, Green, Blue) color model for extracting fire pixels using chromatic and disorder measurements. Conclusive functions based on red part, had been deduced for fire pixels while verification was done by growth and disorder of smoke
	Qi et al. [17]	Color model based on pigmentation values was given by which include both of RGB and HSV color models. Proposed color space was used to give cumulative fire color matrix to aggregate the fire color present in each image
	Chen et al. [3]	An extended Gaussian Mixture Model was adopted by to extract egions of high motion, then fire based color algorithm utilized to find the fire and non-fire areas
	Gunay et al. [8]	Hidden Markov Model was employed by to detect the irregularities present in the nature of flame boundaries. Spatial wavelet transform was computed to figure out the variations in the different color of flame
Motion based related work	Toyerin et al. [21]	Considered both ordinary motion and fire flickering by processing the videos in the wavelet domain. Quasi-periodic nature of fire boundaries were analyzed by temporal wavelet transform
	Xu et al. [23]	Adaptive background subtraction model using different Gaussian distributions. They verified, whether a foreground object is a fire or not, by extracting the foreground object using adaptive background subtraction and confirm the presence of fire using statistical fire color model
	Celik et al. [2]	Fire pixels were extracted using segmentation technique for the creation of continual dynamic envelops. Minor pre-processed envelops were computed based on complex and transitory motion, finally recognition was based on smoke plumes velocity
	Pundir et al. [16]	Proposed dense trajectories based method to counter the challenge of motion in back-ground scene or motion due to moving camera. Further GMM was computed to aggregate the color based features in the frame

(*continued*)

Table 1. (*continued*)

Related work	Authors	Summary
Texture based related work	Cui et al. [5]	Wavelet analysis and Gray Level Co-occurrence Matrices (GLCM) were utilized by Cui et al. for analysing the texture of fire. Wavelet transform was computed for textural images and GLCM was used to exploit various scales of wavelet transform
	Yuan et al. [24]	Histogram sequence of texture based pyramids was given by to generate three level pyramids. Local Binary Patterns (LBP) and Local Binary Patterns based on variance (LBPV) were extracted at each pyramidal level for the generation of LBP pyramids and LBPV pyramids
	Dimitropoulos et al. [6]	Dynamic texture analysis for regions of high motion in flame was given by and linear dynamical system and bag-of-system approach was employ-ed for extracting various spatio-temporal features of fire which were used to train two-class SVM
Deep learning based related work	Xu et al. [22]	Employed variational autoencoder (VAE) and deep Long-Short Term Memory (LSTM) networks to carry out the monitoring of fire based regions
	Jeon et al. [12]	Used multi scale feature maps by stackingthe convolutional layers in the deep convolutional neural networks. Further they proposed a block based on squeezing the features to extract the information at different scales
	Li et al. [14]	They also employed the extraction of features at multi scale to remove the artifacts of fire like objects. A channel attention method was used to give weightage for various feature maps
	Shamsoshoara et al. [18]	Given the FLAME dataset for images by drones (aerial-images) which further help researchers to build various strategies to fight the forest fire issues. They employed U-Net for creating the segmented parts in the fire images

3 Methodology

Our given method is built on extracting the prominent features of fire using frame-based and optical flow based deep learning frameworks, and classify the non-fire and fire frames by various classifiers. Given method is summarized in Fig. 1.

3.1 About Dataset

Available datasets of fire consists of very few videos which are not sufficient to train a deep CNNs, as training the CNNs requires large number of dataset and huge computation power. Therefore, we decided to create our dataset by handpicking videos from various Internet sources[1,2,3,4,5] and preprocess them. As videos (from various sources) have dissimilar frame per second, we choose the videos with less frame rate per seconds. Challenges that we faced while creation of dataset are, first we need a still camera in videos which are hard to find. Once we found a still camera videos, some obstacle come in between fire region and camera, which leads to dispose of those videos. Sometimes, it requires a lot of time to crop a specific portion from video as it requires the whole searching of the video, ensuring that no obstacle will not ruin the video. In our case, we took many videos from the farm (keeping in mind about forest fires), buildings on fire, etc. First of all still camera videos were very hard to find because in forest fire the video is generally taken through a helicopter or drones which keeps on moving. We got various still videos for buildings on fire but then also many obstacles (fire squad, nearby people) were constantly interrupting the videos. Hence, sometimes we have to sit through hours to crop a 15–20 s video from an hour long video.

A total of 80 videos have been employed in our work, 40 were fire videos and 40 were non-fire videos. Further we discuss about dataset in testing and analysis section.

3.2 Transfer Learning

In our method, we fine-tune our dataset on different CNNs. For fine-tuning we employed pretrained models in TensorFlow such as Alexnet [13], VGG-16 [19], ResNet [10] and GoogleNet [20]. Finetuning was done on fire and non-fire frames.

3.3 Optical Flow

Optical flow technique is used to estimate the approximate local image motion, whose ground root is local derivatives in the consecutive succession of frames. In two-dimension, it estimates the motion of given pixel value between two consecutive frames, while in three-dimension, it determines the motion of volume voxel between two successive volumes. For calculating the three-dimension optical flow, an assumption has been used for intensity changes that is, all time based intensity changes are because of motion part only. In our approach, optical flow is determined, which is projected as a image to be feed into CNN architectures.

---

[1] http://cvpr.kmu.ac.kr/.
[2] http://www.openvisor.org.
[3] http://signal.ee.bilkent.edu.tr/VisiFire/Demo.
[4] https://www.shutterstock.com/video/search/fire.
[5] https://www.youtube.com/results?search_query=forest+fire.

This approach was given by [1], for interpreting the optical flow field in order to nullify the effect of noise and independently moving objects in the flow field.

Let $P(a,b,t)$ represents the central pixel with $m \times m$ neighborhood and shift by ∂a, ∂b in ∂t time to pixel value $P(a+\partial a, b+\partial b, t+\partial t)$. Since, $P(a,b,t)$ and $P(a+\partial a, b+\partial b, t+\partial t)$ represents the same pixel value, hence, we have:

$$P(a,b,t) = P(a+\partial a, b+\partial b, t+\partial t). \tag{1}$$

Two dimensional motion constraint equation is based on assumption represented by Eq. 1, which forms the basis for optical flow. Above mentioned assumption is true for small local translations, with constraints that ∂a, ∂b and ∂t are not too big. Hence, Taylor series expansion can be computed on Eq. 1 to get:

$$P(a+\partial a, b+\partial b, t+\partial t) = P(a,b,t) + \frac{\partial P}{\partial a}\partial a + \frac{\partial P}{\partial b}\partial b + \frac{\partial P}{\partial t}\partial t + \gamma, \tag{2}$$

where γ represents the higher order terms, which are assumed to be very small and can be neglected. Using the above equations, the following equations can be obtained as:

$$\frac{\partial P}{\partial a}\partial a + \frac{\partial P}{\partial b}\partial b + \frac{\partial P}{\partial t}\partial t = 0 \ or \tag{3}$$

$$\frac{\partial P}{\partial a}\frac{\partial a}{\partial t} + \frac{\partial P}{\partial b}\frac{\partial b}{\partial t} + \frac{\partial P}{\partial t} = 0 \ and \tag{4}$$

$$\frac{\partial P}{\partial a}v_a + \frac{\partial P}{\partial b}v_b + \frac{\partial P}{\partial t} = 0. \tag{5}$$

Here, $v_a = \frac{\partial a}{\partial t}$ and $v_b = \frac{\partial b}{\partial t}$ represents the $a-$ and $b-$components of frame velocity and $\frac{\partial P}{\partial a}$, $\frac{\partial P}{\partial b}$ and $\frac{\partial P}{\partial t}$ represents frame intensity derivatives at pixel value $P(a,b,t)$. These can be written as $P_a = \frac{\partial P}{\partial a}$, $P_b = \frac{\partial P}{\partial b}$ and $P_t = \frac{\partial P}{\partial t}$.

Using, difference between $a-$ and $b-$ components of frame velocity or optical flow that is (v_a, v_b) and intensity derivatives (P_a, P_b, P_t). The equation can be mathematically represented as:

$$(P_a, P_b).(v_a, v_b) = -P_t \ or \tag{6}$$

$$\nabla P.\overrightarrow{v} = -P_t, \tag{7}$$

where ∇P represents the spatial intensity gradient and $\overrightarrow{v}$ is the optical flow at pixel value (a,b) at time t. Equation $\nabla P.\overrightarrow{v} = -P_t$ is known as the $2D$ motion constraint equation.

Fig. 2. Few fire videos used as dataset.

Fig. 3. Non-fire videos that are used as a challenging dataset in our method which comprises of flowers, fire-works, sun, traffic-light and water-fountain.

4 Testing and Analysis

As explained in Sect. 3.1, we have downloaded fire-containing videos from different Internet sources. A total of 80 videos have been employed in our work, 40 are non-fire-based videos and 40 are fire containing videos. To quantify the strength of proposed method, the given method is evaluated on very challenging non-fire dataset. If observed for a long duration of time, these non-fire videos are very similar to fire containing videos. These non-fire videos consisting of fire-works, valley of flowers, Sun, traffic lights and water-fountain. Some of the challenging non-fire dataset videos are shown in Fig. 3.

The fire videos contain various scenarios of fire ranging from forest based fire, fire on hill to fire in the burning houses. Non- fire videos have taken such that either they contain background similar to as in fire-videos or some fire-color moving object is present in the videos. Frames are extracted from videos. Hence, around 20000 fire-containing images, along with 20000 non-fire images have been employed in our work.

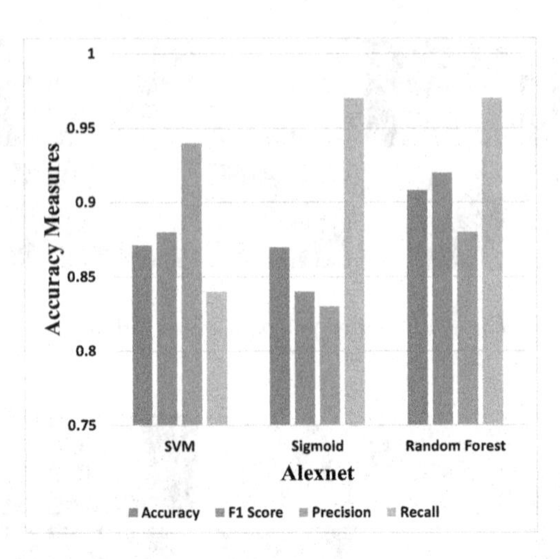

Fig. 4. Accuracy measure for AlexNet architecture on fire conditions.

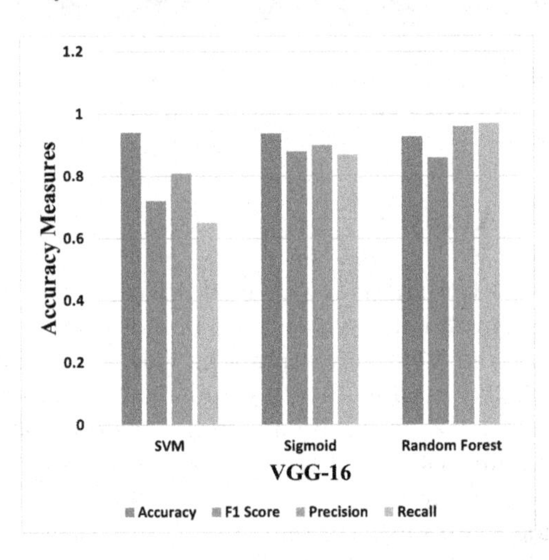

Fig. 5. Accuracy measure for VGG-16 architecture on fire conditions.

In our work, 40000 extracted images are employed with equal distribution of non-fire and fire classes. Similarly, for motion features of fire, optical flow have been calculated on 40 fire and 40 non-fire videos. These retrieved vectors in the form of image have been used as motion-based dataset to be feed into different CNN architectures. Frame-based features and motion based features extracted from deep CNN are combined to train the different classifiers. Figure 2 and Fig. 3 represent few of the fire and non-fire videos. To prove that our system is free from overfitting or biased free, we used 5-fold cross validation method. We have implemented our method using GeForce GTX 1080 Ti Graphics Card, i7 intel processor using TensorFlow libraries.

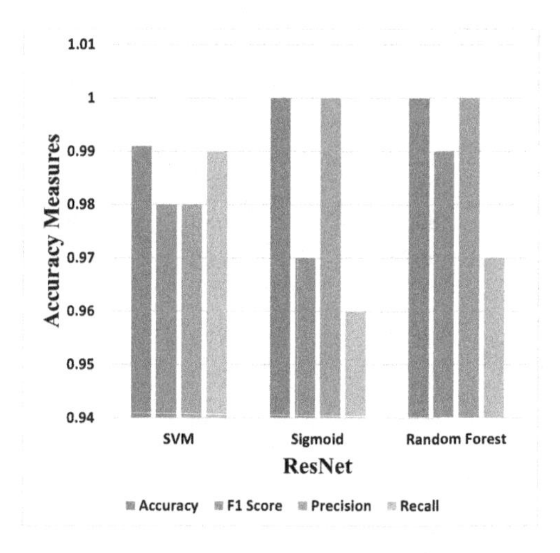

Fig. 6. Accuracy measure for ResNet architecture on fire conditions.

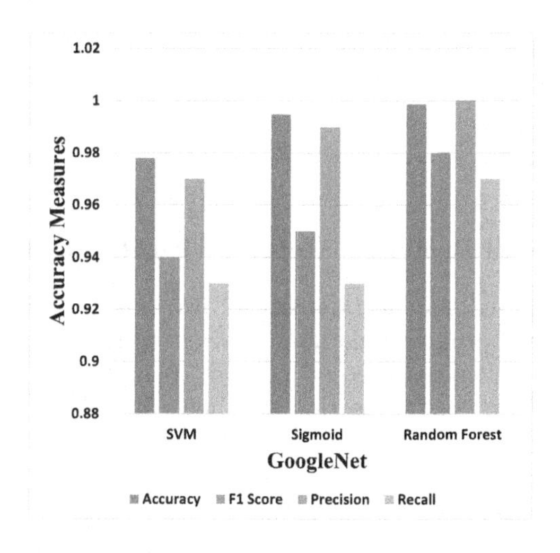

Fig. 7. Accuracy measure for GoogleNet architecture on fire conditions.

4.1 Overall Performance of System on Fire Detection

In this section, we have followed the above mentioned method, like 40000 extracted frames with equal distribution of fire and non-fire classes, have been used to get the spatial features from CNN. And for motion features of fire, optical flow calculated on 40 fire videos and 40 non-fire videos. The results are shown in Table 2 and Alexnet Fig. 4, VGG-16 Fig. 5, ResNet Fig. 6 and GoogleNet Fig. 7. On performing experiments on different architectures we conclude that ResNet with random forest has given the best accuracy and other accuracy measures.

Table 2. Over all evaluation of system performance.

Classifier	Accuracy (in %)	F1 Score	Precision	Recall
AlexNet				
SVM	87.11	0.88	0.94	0.84
Sigmoid	86.98	0.84	0.83	0.86
Random forest	90.79	0.92	0.88	0.97
VGG-16				
SVM	93.98	0.72	0.81	0.65
Sigmoid	93.73	0.88	0.90	0.87
Random forest	92.81	0.86	0.96	0.97
ResNet				
SVM	99.10	0.98	0.98	0.99
Sigmoid	100	0.97	1.0	0.96
Random forest	100	0.99	1.0	0.99
GoogleNet				
SVM	97.81	0.94	0.97	0.93
Sigmoid	99.47	0.95	0.99	0.93
Random Forest	99.87	0.98	1.0	0.98

4.2 Comparison of Proposed Method with Recent Papers

There is no image based standard dataset for fire detection. Hence, to prove the better performance of proposed method (ResNet with random forest) over other methods, we have implemented recent method [16] and two other methods [9,15] on our dataset. Results for this section are shown in Table 3 and Fig. 8.

Table 3. Comparison of proposed method with few recent techniques.

Methods	F1 score	Precision	Recall
Dense trajectories [16]	0.82	0.81	0.83
MHI method [9]	0.55	0.41	0.83
Saliency Method [15]	0.76	0.89	0.67
Proposed Method (ResNet with random forest)	**0.99**	**1.00**	**0.99**

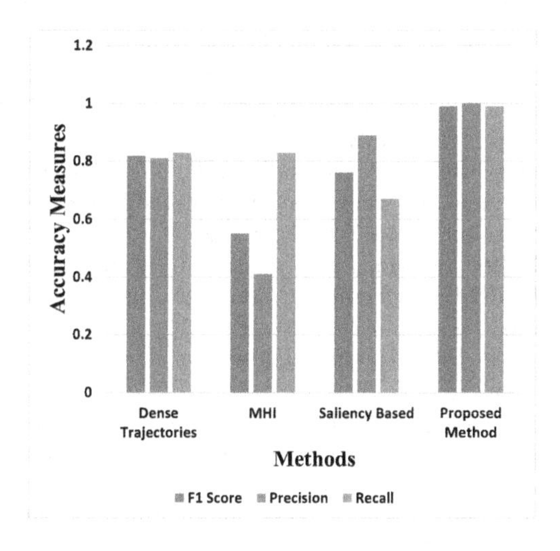

Fig. 8. Comparisons of proposed method with different recent papers.

5 Conclusion

This paper proposes a surveillance camera based fire detection system, which is capable of detecting fire based frames, using deep learning techniques. We exploit the power of transfer learning in our work to train image-based and motion-based deep learning models. The proposed work can also be extended by using various deep learning techniques that works on temporal information. In our work we found out that ResNet with random forest classifier gives the best F1-score. Proposed work has also shown good results when compared with few recent fire detection methods.

References

1. Adiv, G.: Determining three-dimensional motion and structure from optical flow generated by several moving objects. IEEE Trans. Pattern Anal. Mach. Intell. **4**, 384–401 (1985)
2. Celik, T., Demirel, H., Ozkaramanli, H., Uyguroglu, M.: Fire detection using statistical color model in video sequences. J. Vis. Commun. Image Represent. **18**(2), 176–185 (2007)
3. Chen, J., He, Y., Wang, J.: Multi-feature fusion based fast video flame detection. Build. Environ. **45**(5), 1113–1122 (2010)
4. Chen, T.H., Wu, P.H., Chiou, Y.C.: An early fire-detection method based on image processing. In: International Conference on Image Processing, ICIP, vol. 3, pp. 1707–1710. IEEE (2004)
5. Cui, Y., Dong, H., Zhou, E.: An early fire detection method based on smoke texture analysis and discrimination. In: Congress on Image and Signal Processing, CISP, vol. 3, pp. 95–99. IEEE (2008)
6. Dimitropoulos, K., Barmpoutis, P., Grammalidis, N.: Spatio-temporal flame modeling and dynamic texture analysis for automatic video-based fire detection. IEEE Trans. Circuits Syst. Video Technol. **25**(2), 339–351 (2015)

7. Fang, L., Xu, A., Tang, L.: A study of the key technology of forest fire prevention based on a cooperation of video monitor and GIS. In: Fourth International Conference on Natural Computation, (ICNC 2008), vol. 5, pp. 391–396. IEEE (2008)
8. Günay, O., Taşdemir, K., Töreyin, B.U., Çetin, A.E.: Fire detection in video using LMS based active learning. Fire Technol. **46**(3), 551–577 (2010)
9. Han, D., Lee, B.: Flame and smoke detection method for early real-time detection of a tunnel fire. Fire Saf. J. **44**(7), 951–961 (2009)
10. He, K., Zhang, X., Ren, S., Sun, J.: Deep residual learning for image recognition. In: Proceedings of the IEEE Conference on Computer Vision and Pattern Recognition, pp. 770–778 (2016)
11. Horn, B.K., Schunck, B.G.: Determining optical flow. Artif. Intell. **17**(1–3), 185–203 (1981)
12. Jeon, M., Choi, H.S., Lee, J., Kang, M.: Multi-scale prediction for fire detection using convolutional neural network. Fire Technol. 1–19 (2021)
13. Krizhevsky, A., Sutskever, I., Hinton, G.E.: Imagenet classification with deep convolutional neural networks. In: Advances in Neural Information Processing Systems, NIPS, pp. 1097–1105 (2012)
14. Li, S., Yan, Q., Liu, P.: An efficient fire detection method based on multiscale feature extraction, implicit deep supervision and channel attention mechanism. IEEE Trans. Image Process. **29**, 8467–8475 (2020)
15. Liu, Z.G., Yang, Y., Ji, X.H.: Flame detection algorithm based on a saliency detection technique and the uniform local binary pattern in the YCbCr color space. SIViP **10**(2), 277–284 (2016)
16. Pundir, A.S., Buckchash, H., Rajput, A.S., Tanwar, V.K., Raman, B.: Fire detection using dense trajectories. In: The Second International Conference on Computer Vision and Image Processing, CVIP 2017 (2017)
17. Qi, X., Ebert, J.: A computer vision based method for fire detection in color videos. Int. J. Imaging **2**(S09), 22–34 (2009)
18. Shamsoshoara, A., Afghah, F., Razi, A., Zheng, L., Fulé, P.Z., Blasch, E.: Aerial imagery pile burn detection using deep learning: the flame dataset. Comput. Netw. **193**, 108001 (2021)
19. Simonyan, K., Zisserman, A.: Very deep convolutional networks for large-scale image recognition. arXiv preprint arXiv:1409.1556 (2014)
20. Szegedy, C., et al.: Going deeper with convolutions. In: Proceedings of the IEEE Conference on Computer Vision and Pattern Recognition, pp. 1–9 (2015)
21. Töreyin, B.U., Dedeoğlu, Y., Güdükbay, U., Cetin, A.E.: Computer vision based method for real-time fire and flame detection. Pattern Recogn. Lett. **27**(1), 49–58 (2006)
22. Xu, Z., Guo, Y., Saleh, J.H.: Advances toward the next generation fire detection: deep LSTM variational autoencoder for improved sensitivity and reliability. IEEE Access **9**, 30636–30653 (2021)
23. Xu, Z., Xu, J.: Automatic fire smoke detection based on image visual features. In: International Conference on Computational Intelligence and Security Workshops, CISW, pp. 316–319. IEEE (2007)
24. Yuan, F.: Video-based smoke detection with histogram sequence of LBP and LBPV pyramids. Fire Saf. J. **46**(3), 132–139 (2011)
25. Zhang, Y., Lim, S., Sharples, J.: Drivers of wildfire occurrence patterns in wetlands of riverine bioregion in New South Wales, Australia. In: EGU General Assembly Conference Abstracts, vol. 19, p. 231 (2017)

Two Novel Methods for Multiple Kinect v2 Sensor Calibration

Sumit Hazra[1,2](✉), Manasa Pisipati[1], Amrit Puhan[1], Anup Nandy[1], and Rafał Scherer[3]

[1] Machine Intelligence and Bio-Motion Research Lab, Department of Computer Science and Engineering, NIT Rourkela, Rourkela 769008, Odisha, India
[2] Department of Computer Science and Engineering, Koneru Lakshmaiah Education Foundation, Hyderabad 500075, Telangana, India
sumhaz15@gmail.com
[3] Department of Intelligent Computer Systems, Czestochowa University of Technology, al. Armii Krajowej 36, 42-200 Czestochowa, Poland
rafal.scherer@pcz.pl

Abstract. Camera calibration is an essential step for measuring an instrument's accuracy by using its parameters. In this paper, we propose two methods for calibrating eight Kinect v2.0 sensors, namely, pairwise and simultaneous single-camera calibration. In the first method, an experimental setup is managed so that the eight Kinect cameras can view a 3D object on a treadmill in consecutive pairs. The novelty of this method infers pairwise calibration of eight Kinects using six steps, specifically: precalibration process, acquiring images from Kinect, exporting the parameters, generation of independent live point clouds, feeding exported parameters for point cloud matching and ensuring successful calibration through merged point clouds. In the second method, a novel octagonal model is developed to calibrate each of the eight cameras on the same experimental setup. We obtain an overall mean reprojection error of 1.37 pixels and 0.42 pixels for the first method and second method, respectively. The smallest reprojection error for the first method is reported to be 0.63 pixels and 0.27 pixels for the second method. The efficiency of the proposed methods is compared with the state-of-art techniques using the root mean square and the reprojection error metrics. We observed that the proposed methods outperform the existing techniques.

Keywords: Kinect v2 sensors · Iterative Closest Point (ICP) · Multiple cameras · Camera calibration · Checkerboard · Root Mean Square (RMS)

We are extremely thankful to Science and Engineering Research Board (SERB), DST, Govt. of India to support this research work. The Kinect v2.0 sensors used in our research experiment are purchased from the project funded by SERB with FILE NO: ECR/2017/000408. We would also like to extend our sincere thanks to the students of Department of Computer Science and Engineering, NIT Rourkela for their uninterrupted cooperation and participation catering to the data collection.

B. Raman et al. (Eds.): CVIP 2021, CCIS 1568, pp. 403–414, 2022.
https://doi.org/10.1007/978-3-031-11349-9_35

1 Related Works

Calibration using Kinect sensors is required when accuracy & precision are of utmost importance [8,12]. A comparison of several depth-camera calibration methods [9,13,17] was presented to provide an extensive evaluation of the algorithms but no new self-driven methods were proposed. [6] used a seven camera-based Vicon motion capture system as a gold standard to measure the accuracy of Kinect sensor and their work had mean errors of up to 10 cm in the range which covered the distances up to 3 m from the sensor with an effective field of view of 54.0° horizontal and 39.1° vertical. In his work, although different types of sensors were calibrated properly, there was a need to reduce the calibration errors. The drawback of his work was that multiple Kinect sensors were not explored for calibration. The authors in [7] carried out investigations on the accuracy and precisions of Kinect XBOX 360 sensor and Asus Xtion using a specific measurement tool. It was confirmed through their measurements that both the important parameters, accuracy, and precision deteriorated to the order of a second-order polynomial. The RMS error values were of the order of 10 mm at a distance of 2 m. Two Kinect sensors and an Xtion sensor's accuracy was also checked. The calibration or accuracy of multiple Kinects was not addressed in their work. [3] worked on the augmented reality-based 3D reconstruction of objects using multiple Kinect sensors and point clouds. The limitation of their work was if the person moved out from the center of the Kinect view, the result became unpredictable as the angle was calculated with the information of the calibration object. The problem of interference with multiple Kinects was reported in their work. [19] investigated camera calibration using one-dimensional objects with three points of minimal configuration. [15] worked upon multiple camera calibration in virtual environments. A laser-pointer was the only additional hardware, used as the calibration object. The bright spot was generated to pass through the considered volume, whose projections were found with sub-pixel precision verified by the RANSAC analysis. Nonlinear distortions were estimated from their existing dataset. [16] developed a technique for self-calibrating a moving projecting camera from at least five views of a scene that was planar in nature. The method involved the 'rectification' of inter-image homographies. If only the focal length is estimated, it requires at least three images, and five images when all five internal parameters are computed. The flexibility of this technique was good, but the initialisation part was difficult. Some works [4,11] start to utilize deep neural networks for camera calibration what requires long training stage. [14] utilised a spherical calibration object to establish view correspondence. In their work, they used various view transformation functions for determining the most robust mapping. However, as per the results, the reprojection error obtained by using these processes did not seem to be compatible enough for camera calibration.

2 Theoretical Background

In this section, we describe Kinect sensors and camera lens-sensor system errors that need to be corrected. Camera calibration is a pre-requisite in various applications such as tracking people, motion capture, integration of digital information with live video, or the environment of the users in real-time. Sensors are an integral part of the modern scientific era [2,18]. We consider the Kinect v2 sensor, which needs calibration to achieve satisfactory accuracy. Kinect consists of an infrared (IR) laser, an IR camera, and a supplementary RGB camera. Thus, sensor calibration involves refinement of the colour (RGB) images as well as that of correspondences between object coordinates, which are 3D in nature and coordinates in RGB, IR, and depth images, respectively.

Camera calibration takes an assumption for parameter identification of functions, which models transformation of the 3D coordinates of external objects to coordinates with image plane [10]. Thus, calibration involves the identification of intrinsic and extrinsic parameters values.

Intrinsic parameters describe the elements of the camera lens-sensor system and its geometry, such as focal length or camera resolution. They also characterize various forms of distortions. Distortion is a deviation from the rectilinear projection. Two major distorting factors in an image are radial and tangential distortions. In our experiments, we consider two types of distortions:

- Radial distortion: It is mostly seen when pictures of vertical structures having straight lines are taken, which then appear curved. Radial distortion leads to a barrel or fish-eye effect in an image. To remove radial distortion, d_1, d_2 and d_3 were considered as the radial distortion factors, c_1 and c_2 as coefficients of correcting tangential distortion, and m, n as the coordinates in the original image, then the corrected coordinates ($m_{corrected}$, $n_{corrected}$) will be

$$m_{corrected} = m(1 + d_1 r^2 + d_2 r^4 + d_3 r^6) \quad (1)$$

$$n_{corrected} = n(1 + d_1 r^2 + d_2 r^4 + d_3 r^6) \quad (2)$$

- Tangential distortion occurs due to the fact that image taking lenses are not perfectly parallel to the imaging plane. It can be corrected via formulae

$$m_{corrected} = m + [2c_1 mn + c_2(r^2 + 2m^2)] \quad (3)$$

$$n_{corrected} = n + [c_1(r^2 + 2n^2) + 2c_2 mn] \quad (4)$$

Extrinsic parameters describe rotation and translation of a camera.

3 Method I – Calibration Procedure for Pair-wise Eight Cameras

This method deals with calibration of eight Kinect sensors in consecutive pairs. The procedure is depicted in Fig. 1. The experimental setup uses eight Microsoft Kinect v2.0 sensors and a treadmill. Each Kinect device has a viewpoint change of 45° with respect to the reference point (the treadmill). In order to perform pairwise calibration, we capture a checkerboard (oriented at a particular angle) image from camera 1 and camera 2. A similar process is carried out for all the eight cameras, changing the reference sensor each time for a new pair.

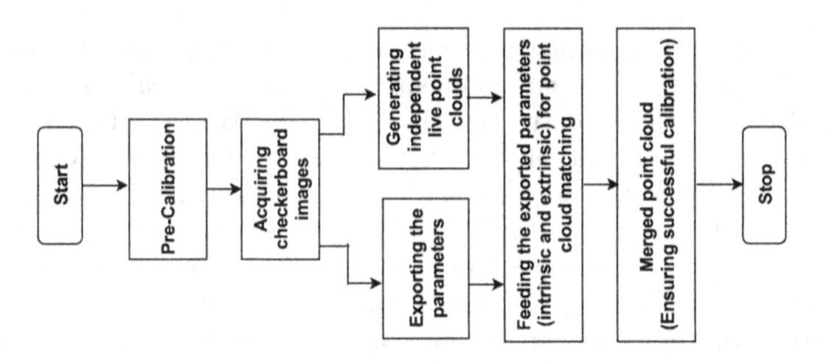

Fig. 1. Calibration process for pairwise eight cameras.

In order to perform Pre-Calibration, our setup consists of two laptops and eight Kinect v2 models. Images of RGBD format are collected from the Colour Basics section of the SDK Browser v2.0 (Kinect for Windows). To calibrate the Kinect camera lens, a 10×7 checkerboard, with each square 105 mm wide, is used. In order to capture RGB and depth images, we utilize the Color Basics D2D (for RGB image capture) module available in Kinect SDK Browser v2.0. A minimum of 10 to 15 images are taken by each Kinect camera, and they are later given as input (adjacent two cameras) for Stereo Camera Calibration. Number of accepted pairs of checkerboard images for each Kinect pair in Method I is 5–20. The images captured from the adjacent camera pairs are thus given as input for Stereo Camera Calibration. The rotational matrix and translational vectors of each camera with respect to the other are obtained. These parameters are individually saved. At this point, a significant amount of camera calibration is achieved.

In order to verify whether the transformation parameters for camera calibration obtained in the previous steps are accurate, live point clouds are captured from each individual Kinect camera by measuring the x, y, z coordinates of the object with respect to the lens of the camera. Both the positive and the negative co-ordinates (upper and lower) are considered for each of the x, y and z axes respectively after proper measurements (considering the Kinect sensor, the subject as well as the position of the subject on the treadmill). They are taken as inputs in each specific case, which is solely camera-specific. An instance of the x, y and z values taken are as follows: $lx = -0.48$, $ux = 0.34$, $ly = -1.22$, $uy = 0.72$, $lz = 2.10$ and $uz = 2.50$, where u and l mean lower and upper signals in x, y and z directions, respectively.

We apply the Iterative Closest Point (ICP) Algorithm [1] for merging two point clouds of adjacent cameras. The point clouds are given as inputs with their respective translation and rotation parameters, explained in the next step of the Calibration Pipeline. If the two point clouds overlap almost exactly, then the transformation matrix obtained are assumed to be correct. In our experiment, the point clouds overlap correctly during the pairwise calibration procedure.

To feed the exported parameters for point cloud matching, first we compute the extrinsic parameters. The parameter values are measured and fed as inputs for the live point cloud matching. An instance of the measured values for the extrinsic parameters is demonstrated for the pair of Kinect v2 sensors 1 and 2 (Method I) and is as follows: translationX = 1.82, translationY = −1.41, translationZ = 6.19 and rotation = $-\pi/4$.

3.1 Merged Point Clouds (Ensuring Successful Calibration)

In this paper, the ICP algorithm is used as a method of verification for the obtained transformation matrix (for Method I). The ICP algorithm is a matching algorithm which finds the minimum distance between two point clouds which we already obtained. By observing the overlapped point clouds, it is inferred that the transformation matrix is an accurate one. The main objective of the ICP can be defined as

$$\tau_A^B = min_\tau(Error(\tau(P_A), Q_B)) \tag{5}$$

where τ_A^B is the transformation matrix, P_A is the reading point cloud obtained in the co-ordinate system A, Q_B is the reference point cloud obtained from coordinate system B. The error is generally measured as the distance between each closest point pairs in different point clouds which was in our case as well. In our work, the implementation of the algorithm is done as follows: the program takes as input two point clouds, say p and q, the parameters namely a rotational matrix R and a translational vector T and iteratively aligns the points of p to q until it reaches a position where minimum error is obtained between two mapped points which is represented in Eq. 6.

$$q = R * p + repmat(T, 1, length(p)) \tag{6}$$

4 Method II – Procedure for Simultaneous Camera Calibration

This method deals with the single camera calibration of eight Kinects at the same time using an octagonal checkerboard model. The steps are described in Fig. 2. For pre-calibration, eight laptops with the amount of RAM ranging from 8 GB to 16 GB and having either Intel i5 or i7 processors are used. A three-dimensional octagonal structure (Fig. 3) is built with each side having a 10 × 7 checkerboard wherein each square is 105 mm wide. For this method, it is not possible to fix position of two cameras to face the same side of the octagon. This is because each adjacent pair of sides of the octagon are aligned at an angle of 45° with respect to each other and hence there are tangential distortions.

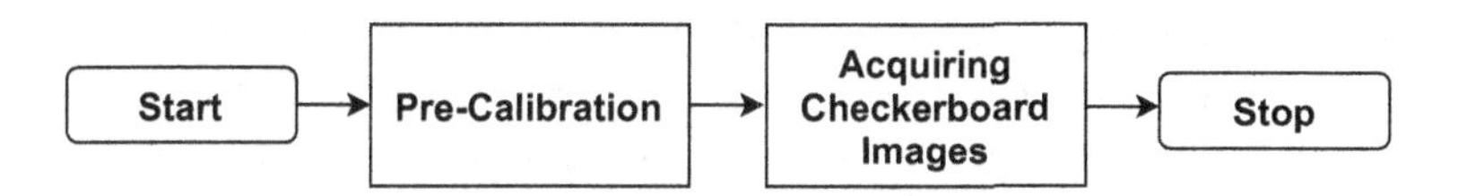

Fig. 2. Simultaneous camera calibration procedure.

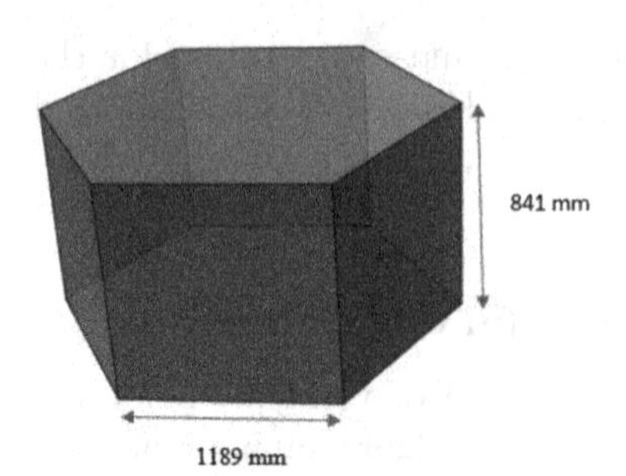

Fig. 3. Proposed three-dimensional octagonal model.

In order to capture RGB images, we utilize the Color Basics D2D (for the RGB image capture) module available in Kinect SDK Browser v2.0. A minimum of ten images is taken from each Kinect camera which is later processed using Single Camera Calibration. Owing to the geometrical features and constraints of the octagon model, the camera can only detect the face which is at a perpendicular position with respect to it. The other two adjacent faces are not detected due to tangential reflections. Thus, each camera is facing a separate side of the octagon. Hence, the Single Calibration App is used instead of the Stereo Calibration App. In order to collect the checkerboard images, two approaches are available. Approach I where the octagon structure can be placed at different orientations and then the images can be captured from all eight Kinects simultaneously. Approach II where the cameras can be moved by equal distances and angles. Following which the images can be captured from the various positions while keeping the octagon structure stationary at the center. In our experiment, Approach II is applied to acquire data. The number of input images taken from each camera which are later processed in the Single Camera Calibration App is about 30–45.

5 Result Analysis and Discussion

Checkerboard images for calibration are concurrently captured using an eight-camera Kinect v2 sensor system in two separate ways for camera calibration, and the results are demonstrated in the following manner.

For Method I: The Kinect sensors are positioned at an angle of 45° to the treadmill for our experiment. Unlike the previous systems that has only one camera and that calculates 2D position with their corresponding angles, the Kinect-based tracking system estimates 3D position. Thus, it allows us to position the camera in such a manner which optimizes the spatial distance with respect to the subject instead of restricting it's position perpendicular to the plane of motion. Therefore, the Kinect is able to track a human figure at 45° angle. This sensor position is optimal to capture gait data at the sagittal plane without any obstruction by treadmill. It ensures a successful calibration based on the merged point cloud results. A subject is stationed on the treadmill at the specified coordinate position with respect to a specific pair of Kinect cameras in the first case. Individual point clouds are generated from the two cameras in the pair. We obtain a transformation matrix and rotational vector for each pair to provide as input for point cloud matching. The individual point cloud generation is demonstrated in Figs. 4a and 4b for Kinect v2 sensors number 2 and 3. The same process is repeated for all the eight pairs of the Kinect v2 sensors. The translation and the rotation parameters for each pair is obtained and given as input for the point cloud matching. The merged point cloud for Kinect sensors number 2 and 3 is demonstrated in Fig. 4c.

We consider 10 to 15 images captured from the pairwise adjacent cameras. Those images are given as inputs to the stereo camera calibration. It discards certain pairs of images and accepts a considerable number of Kinect pairs as mentioned in Table 1. The camera positions for each pairwise calibration process are carefully adjusted to ensure proper results. The position of Kinect v2 sensor pair 1 and 2 along with the different angles of checkerboard images is shown in Fig. 5a. A similar approach is followed for all the eight pairs. The overall reprojection error for Kinect pair 1 and 2 is shown in Fig. 5b.

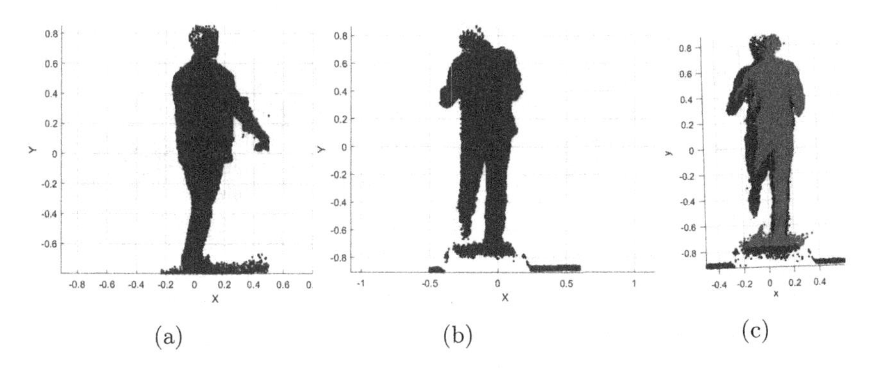

Fig. 4. Example point cloud from Kinect number 2 (a) and 3 (b), and merged point clouds for the pair of Kinects numbered 2 and 3 (c).

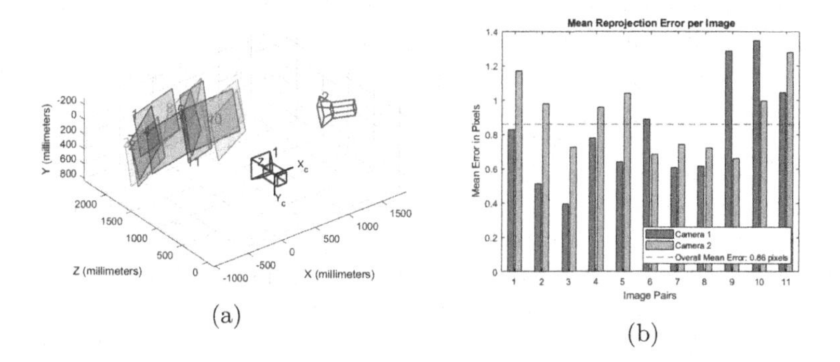

Fig. 5. Camera positions along with the checkerboard angles (a) and the reprojection error (b) for the pair of Kinects numbered 1 and 2.

For Method II: In this step, the proposed three-dimensional octagonal model is used. It is constructed such that each of its sides has a 10×7 checkerboard pattern wherein each square is 105 mm wide. The Kinect cameras are moved by equal distances and angles. The captured images are taken as input for single camera calibration. The errors are found to be less than in the case of Method I and the literature. The position of Kinect v2 sensor number 5 and the different angles of checkerboard images captured through it are shown in Fig. 6a. A similar approach is followed for all the eight Kinects. The reprojection error for this case for sensor number 5 is shown in Fig. 6b. Similarly, the reprojection errors are calculated for all the eight Kinects facing each of the eight sides of the octagon model.

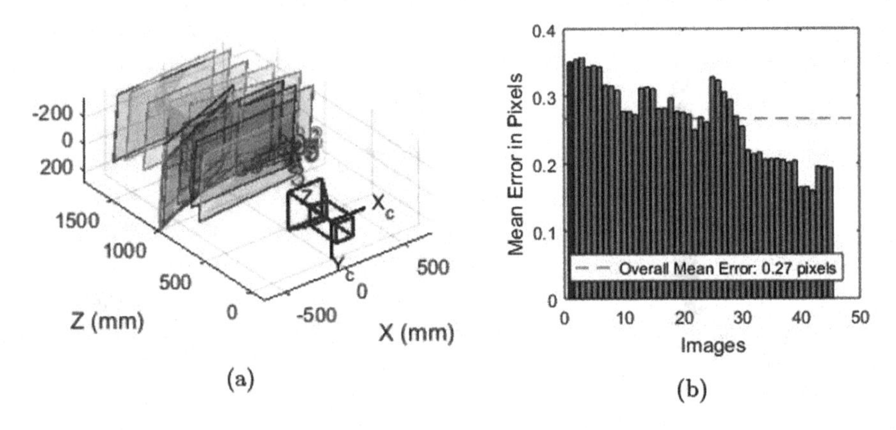

Fig. 6. Camera positions along with the checkerboard angles (a) and the reprojection error (b) for Kinect number 5.

For Method I, the overall result for all pairs, is illustrated in Table 1. It also shows the corresponding rotational angle and the translational vector along with the overall mean reprojection error for each pair. An overall mean reprojection error of 1.37 for all the eight pairs is reported.

We also compute the RMS error for validating the results of the ICP algorithm for this method. The RMS error is calculated by comparing the source coordinates that are known, to the coordinates for the same point in the input coordinate system. They are measured by $\sqrt{(\sum_{i=1}^{n}(\hat{x}_i - x_i)^2)/n}$, where $\hat{x}_i - x_i$ are the differences between the actual values and the predicted values also known as residuals, where x_i is the ith observation value and $\hat{x}_i$ is the predicted value. Performing a square of the residuals followed by taking the average of the squares, and then the square root gives us the RMS error. Thus, minimum difference between the actual and the value after prediction produces lesser RMS error. The RMS convergence curves for Method I for Kinect pair 4 and 5 and Kinect pair 8 and 1 are shown in Fig. 7a and 7b respectively. It is well understood from the graphs that the RMS error decreases exponentially till it approaches 0 for both the cases. Similar results are observed for the remaining six Kinect pairs.

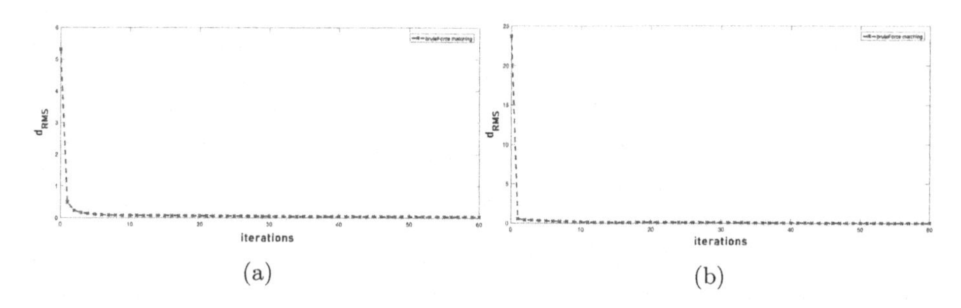

Fig. 7. The RMS error for Kinect sensors number 4 and 5 (a) and 8 and 1 (b).

For Method II, the developed octagonal structure is kept static. The eight Kinect sensors are moved by equal distances and angles. The captured images are then given as inputs for single camera calibration which gives the reprojection errors for Kinect $1 = 1.05$, $2 = 0.43$, $3 = 0.34$, $4 = 0.4$, $5 = 0.27$, $6 = 0.4$, $7 = 0.45$, $8 = 0.43$. We obtained the overall mean reprojection error of 0.42 for all the eight cameras which is much less than in the previous methods. Therefore, the average of Overall Mean Error for Method I is $(0.86 + 0.63 + 3.07 + 0.88 + 0.83 + 0.83 + 3.07 + 0.84)/8 = 1.37$. Similarly, for Method II, the average of overall mean error is $(1.05 + 0.43 + 0.34 + 0.40 + 0.27 + 0.40 + 0.45 + 0.43)/8 = 0.42$.

Table 1. Overall mean reprojection errors for each Kinect pair for Method I

Camera number	Rotational angle	Translational vector (x, y, z)	Overall mean reprojection error
1 and 2	$\pi/4$	1.82, −1.41, 6.19	0.86
2 and 3	$\pi/4$	−1.73, 17.09, 557.94	0.63
3 and 4	$\pi/4$	1.79, −1.68, 1.11	3.07
4 and 5	$\pi/4$	1.82, −1.41, 6.19	0.88
5 and 6	$\pi/4$	1.88, −16.54, 580.15	0.83
6 and 7	$\pi/4$	1.73, −51.94, 5.31	0.83
7 and 8	$\pi/4$	1.78, −1.68, 1.11	3.07
8 and 1	$\pi/4$	1.64, 24.89, 3.62	0.84

6 Comparison with Existing Methods

We compare our results with the existing calibration method proposed by [5] which is similar to our approach for RGBD camera calibration. In [5], calibration was done using 1D and 2D calibration pattern wands. In our paper, both methods used 3D calibration objects (checkerboard and octagon) and Zhang's calibration method. While [5] focused on calibration of four Kinect cameras, both methods proposed in our paper focus on calibration of 8 Kinect cameras. [5] obtained the results for camera 1, 2, 3 and 4 through iterative non-linear optimization technique. The least RMS error reported by them was 0.331. Our first method produces a least RMS error which is close to 0 and a reprojection error of 1.37 pixels for 13 calibration points while our second method produces a reprojection error of 0.42 pixels for 40 calibration points. [14], presented a fast and robust method for calibration of network of RGB-D cameras using a spherical calibration object. They used three regression schemes for their proposed calibration technique. As per their explanation, Regression II and III introduced distortions in the background wall. It was because the data from the training set were overfitted by parameters which were of higher degree and the cross terms, which could no longer preserve the geometry of the far-distance objects outside the calibration area. Regression I produced the best reconstruction results for them but with slight misalignment as well as geometrical distortion. The average re-projection errors obtained by them are more than that obtained from our proposed methods as can be seen in Table 2. Verification using ICP algorithm and point cloud matching is possible for pairwise calibration because the algorithm takes two point clouds as inputs and produces a merged point cloud as output. Hence, Method I allows us to verify whether the calibration is done properly and if it is within error limits for each pair with the help of the RMS error metric value. For Method II however, ICP cannot be applied as eight point clouds are simultaneously obtained from the 8 cameras used. Also, two cameras cannot be adjusted to face the same side of the octagon in Method II due to tangential distortions. Thus, the results of Method II are verified with those of Method I to check that they are within the error limits and are accurate. Hence, re-projection errors are used as metric for comparison for both Methods I & II whereas the RMS error is used only for Method I. An elaborative study of comparison of our work with the state-of-the-art methods, is depicted in Table 2.

Table 2. Comparison of our methodologies with state of the art methods

Paper	Method	No. of kinects	Average no. of calibration points	Reprojection errors
[14]	Pairwise calibration + Regression I	5	–	1.92
	Pairwise calibration + Regression II	5	–	1.80
	Pairwise calibration + Regression III	5	–	1.77
Proposed Method I	Zhang's calibration + Stereo camera calibration	8	13	**1.37**
Proposed Method II	Zhang's calibration + Single camera calibration	8	40	**0.42**

7 Conclusion and Future Work

Working with Kinect sensors itself is a challenging task because of the requirements of its efficiency, accuracy and robustness and hence the calibration. We proposed two methods of calibrating multiple Kinect V2.0 sensors. Compared to the literature, our methods are applicable to multiple Kinect v2.0 sensors. The calibration procedure for multiple Kinects is fast. Moreover, the proposed calibration techniques are advantageous as we use a checkerboard method as the only calibration tool with no requirement of additional calibration objects or external measurement devices like the existing works. Both the proposed methodologies heavily involve software-based techniques for enhancing the accuracy of multiple Kinect cameras. Thus, the model parameters of the sensors obtained for calibration are more precise and accurate. All these cumulatively provide new insights of efficiently using multiple Kinect cameras for clinical gait analysis. The proposed techniques provision a fast and easy to setup solution of calibrating multiple resources for analysing human gait patterns. The calibration methods implemented does not require intensive computational resources. Through the above results and the experiment conducted, the following conclusions are drawn about the methods proposed. Method I requires a lesser amount of resources as compared to Method II and hence is more affordable. The RMS convergence curves clearly show a steep decrease almost about the origin in every Kinect pair with an increase in the number of iterations for the point cloud matching for Method I. This shows that our adopted approach (Method I) is negligibly erroneous. Method II requires a slightly more significant amount of resources certainly, but the time taken is considerably lesser, and the output produced is significantly more accurate. Also, a notable feature about the second method as per our observation is that better results are obtained if Kinects are at a distance greater than 1.5 m. Additionally, it is observed that with the increase in the number of calibration points, there is a significant decrease in reprojection errors. Hence, when there is a limitation of resources or time, and the errors of calibration do not affect the results substantially, Method I is better. Alternatively, when the resources are sufficient and time and accuracy are of utmost importance, Method II is a better option. There is always a need for calibrating a camera without any kind of constraints, also known as automatic camera calibration. Most of the methods used to date required a known and specific object for calibration. It, in turn, caused hindrance in the process of camera calibration. The future goal is obtaining all the camera parameters for calibration from any single image taken. However, it is still a very challenging issue.

References

1. Besl, P.J., McKay, N.D.: Method for registration of 3-d shapes. In: Sensor Fusion IV: Control Paradigms and Data Structures, vol. 1611, pp. 586–606. International Society for Optics and Photonics (1992)
2. Borghese, N.A., Cerveri, P.: Calibrating a video camera pair with a rigid bar. Pattern Recogn. **33**(1), 81–95 (2000)

3. Chang, W.C., Chang, W.C.: Real-time 3D rendering based on multiple cameras and point cloud. In: 2014 7th International Conference on Ubi-Media Computing and Workshops, pp. 121–126. IEEE (2014)
4. Chen, J., Little, J.J.: Sports camera calibration via synthetic data. In: Proceedings of the IEEE/CVF Conference on Computer Vision and Pattern Recognition (CVPR) Workshops, June 2019
5. Córdova-Esparza, D.M., Terven, J.R., Jiménez-Hernández, H., Vázquez-Cervantes, A., Herrera-Navarro, A.M., Ramírez-Pedraza, A.: Multiple kinect v2 calibration. Automatika **57**(3), 810–821 (2016)
6. Dutta, T.: Evaluation of the kinect^{TM} sensor for 3-D kinematic measurement in the workplace. Appl. Ergon. **43**(4), 645–649 (2012)
7. Gonzalez-Jorge, H., Riveiro, B., Vazquez-Fernandez, E., Martínez-Sánchez, J., Arias, P.: Metrological evaluation of Microsoft Kinect and Asus Xtion sensors. Measurement **46**(6), 1800–1806 (2013)
8. Hazra, S., Pratap, A.A., Tripathy, D., Nandy, A.: Novel data fusion strategy for human gait analysis using multiple kinect sensors. Biomed. Signal Process. Control **67**, 102512 (2021)
9. Herrera C., D., Kannala, J., Heikkilä, J.: Accurate and practical calibration of a depth and color camera pair. In: Real, P., Diaz-Pernil, D., Molina-Abril, H., Berciano, A., Kropatsch, W. (eds.) CAIP 2011. LNCS, vol. 6855, pp. 437–445. Springer, Heidelberg (2011). https://doi.org/10.1007/978-3-642-23678-5_52
10. Liebowitz, D., Zisserman, A.: Metric rectification for perspective images of planes. In: Proceedings of 1998 IEEE Computer Society Conference on Computer Vision and Pattern Recognition (Cat. No. 98CB36231), pp. 482–488. IEEE (1998)
11. Lopez, M., Mari, R., Gargallo, P., Kuang, Y., Gonzalez-Jimenez, J., Haro, G.: Deep single image camera calibration with radial distortion. In: Proceedings of the IEEE/CVF Conference on Computer Vision and Pattern Recognition (CVPR), June 2019
12. Nguyen, M.H., Hsiao, C.C., Cheng, W.H., Huang, C.C.: Practical 3D human skeleton tracking based on multi-view and multi-kinect fusion. Multimedia Syst. 1–24 (2021)
13. Staranowicz, A., Morbidi, F., Mariottini, G.: Depth-camera calibration toolbox (DCCT): accurate, robust, and practical calibration of depth cameras. In: Proceedings of the British Machine Vision Conference (BMVC) (2012)
14. Su, P.C., Shen, J., Xu, W., Cheung, S.C.S., Luo, Y.: A fast and robust extrinsic calibration for RGB-D camera networks. Sensors **18**(1), 235 (2018)
15. Svoboda, T., Martinec, D., Pajdla, T.: A convenient multicamera self-calibration for virtual environments. Presence Teleoper. Virtual Environ. **14**(4), 407–422 (2005)
16. Triggs, B.: Autocalibration from planar scenes. In: Burkhardt, H., Neumann, B. (eds.) ECCV 1998. LNCS, vol. 1406, pp. 89–105. Springer, Heidelberg (1998). https://doi.org/10.1007/BFb0055661
17. Zhang, C., Zhang, Z.: Calibration between depth and color sensors for commodity depth cameras. In: Shao, L., Han, J., Kohli, P., Zhang, Z. (eds.) Computer Vision and Machine Learning with RGB-D Sensors. ACVPR, pp. 47–64. Springer, Cham (2014). https://doi.org/10.1007/978-3-319-08651-4_3
18. Zhang, Z.: A flexible new technique for camera calibration. IEEE Trans. Pattern Anal. Mach. Intell. **22**(11), 1330–1334 (2000)
19. Zhang, Z.: Camera calibration with one-dimensional objects. IEEE Trans. Pattern Anal. Mach. Intell. **26**(7), 892–899 (2004)

Residual Inception Cycle-Consistent Adversarial Networks

Ekjot Singh Nanda, Vijay M. Galshetwar, and Sachin Chaudhary(✉)

Punjab Engineering College, Chandigarh, India
{ekjotsingh.be18cse,vijaymadhavraogalshetwar.phdcse20,
sachin.chaudhary}@pec.edu.in

Abstract. Unpaired Image-to-image translation is a problem formulation where our aim is to learn a function which can convert an image of one domain into another different domain without using a paired set of examples. One of the methods to tackle this problem is CycleGAN. Even though it had remarkable success in the recent years, it still have some issues Our method enhances CycleGAN formulation by replacing the Residual block with our proposed Residual-Inception module for multi-scale feature extraction and by adding a cyclic perceptual loss for improving the quality of texture in recovered image and generating visually better results. Qualitative results are presented on horse2zebra dataset and Quantitative results on I-Haze and Rain 1200 datasets. We show both quantitative and qualitative results on 3 datasets and show that our method improves the CycleGAN method.

Keywords: Unpaired translation · Residual-inception module · Cyclic perceptual loss

1 Introduction

Unpaired Image-to-image translation is a problem formulation where our aim is to learn a function which can convert an image of one domain into another different domain without using a paired set of examples. Recently deep learning have been applied to many applications like- human action recognition [37–41], image dehazing [43], depth estimation [44–46], unpaired image to image translation and many more. One of the methods to tackle this problem is CycleGAN. Even though it had remarkable success in the recent years, it still have some issues. One of the flaws in CycleGAN is that it is not able to extract multi-scale features, due to which it is not capable to convert images at different scales. Another issue is that sometimes it is not able to localize the part of the object whose appearance has to be changed. For example- Sometimes while converting horse to zebra, the strips of the converted zebra goes out of the body. To tackle these issues we developed a Residual-Inception module for multi-scale feature extraction which is comprised of parallel convolutions, skip connection and feature concatenation. We also added a cyclic perceptual loss for improving the

B. Raman et al. (Eds.): CVIP 2021, CCIS 1568, pp. 415–425, 2022.
https://doi.org/10.1007/978-3-031-11349-9_36

quality of texture in recovered image, generating visually better results and for better loacalizing the objects. We also increased the number of skip connections in the generator network for faster training.

The key contributions of the proposed method are explained below:

1. We developed a Residual-Inception module which consists of parallel convolutions, skip connection and feature concatenation for multi-scale feature extraction.
2. We enhance CycleGAN [34] architecture by adding cyclic perceptual consistency loss besides cycle-consistency loss.
3. We increased the number of skip connections in the generator model for better training.

The rest of the paper is structured as follows: Literature review is illustrated in Sect. 2. The detail discussion on the proposed method is done in Sect. 3. Experimental results are presented and discussed in Sect. 4. Finally paper is concluded in Sect. 5.

2 Literature Review

Generative Adversarial Networks (GANs). The remarkable results shown by [44–46] in depth estimation, [1,2] in image editing [11,43] in image dehazing, [42] in image super-resolution, image generation [12,13] and representation learning [12,14,16]. Recently image generation methods have adopted the same idea in applications, such as image inpainting [20], future prediction [18], and text2image [15]. Adversarial loss in GAN helps for image generation tasks.

Image-to-Image Translation. Only two levels of Hertzmann et al. [17] have introduced image analogy for image-to-image translation using non-parametric texture model [19], who considered pair of input-output images while training. Recent methods [21] use input-output dataset to learn CNN based parametric translation function. Mapping of input images to output images is achieved using conditional generative adversarial network [1] based pix2pix framework [22]. The areas of sketches [23] and semantic layouts [24] have also considered the similar idea.

Unpaired Image-to-Image Translation. Two domains of data are used in unpaired image-to-image translation. Rosales et al. [25] introduced Bayesian framework, which uses source image and a likelihood obtained by multiple style images for computation of priors using patch-based Markov random field. The weight-sharing strategy is used to learn a common illustration across domains using Cross-modal scene networks [26] and CoGAN [27]. Liu et al. [29] extended the above work by considering generative adversarial networks [1] and variational autoencoders [30]. Other methods [31–33] inspires input and output for sharing content specific features although difference in their styles.

Neural Style Transfer. It is the another type of image-to-image translation [28,34–36], where synthesizing of the image is taken place by integrating content and style of two different images namely content image and style image. The pre-trained deep feature statistics are used as Gram matrix for matching purpose.

3 Proposed Method

Residual Inception CycleGAN is an enhanced version of CycleGan [3]. For extracting multi-scale features, we have incorporated a Residual -Inception block in place of Residual block in the generator of original CycleGAN. Proposed Residual-Inception module comprises of parallel convolution and skip connections. Convolution layers having kernel size $3*3$, $5*5$, and $7*7$ are applied in parallel in proposed RI module. To merge features learned by the convolution layers, we have incorporated feature concatenation followed by a convolution layer. For achieving better results we have incorporated perceptual loss inspired by EnhanceNet [5]. The main motive behind using this loss is rather than comparing images in just pixel space, we should also compare images in feature space. Therefore, Residual Inception CycleGAN compares the reconstructed cyclic-image with the original image at both pixel space and feature space, where cycle-consistency loss gives a high PSNR metric value and perceptual loss preserves the sharpness of the image. We also increased the number of skip connections in the generator for avoiding vanishing gradient problem.

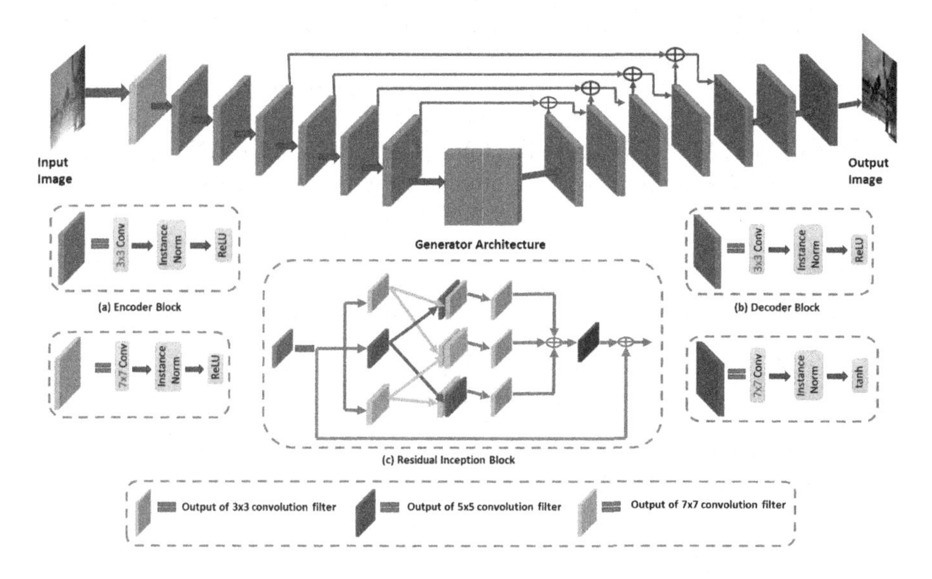

Fig. 1. Generator architecture of proposed residual-inception CycleGAN

3.1 Residual-Inception Module

Proposed Residual-Inception module as shown in Fig. 1(c) comprises of parallel convolution and skip connections. Convolution layers having kernel size $3 * 3$, $5 * 5$, and $7 * 7$ are applied in parallel in proposed RI module. To merge features learned by the convolution layers, we have incorporated feature concatenation followed by a convolution layer. As the integration of feature maps is done in two stages i.e. dense concatenation followed by element-wise summation the proposed feature integration approach differs from original inception module [4]. To match the feature dimension with the input of the RI module, a 1×1 convolution layer is designed. Olaf et al. The main reason behind using skip connections in the generator architecture was to pass on the low-level features learned at initial convolution layers to the deconvolution layers. This merging of low level feature maps with deconvolution layers helps in generating prominent edge information in the output image. These skip connections help to share the learned features across the network which causes better convergence and better training.

3.2 Cyclic Perceptual-Consistency Loss

CycleGAN [3] architecture used the cycle-consistency loss, which calculates L1 norm between the cyclic image and original image. But, this calculated loss between the cyclic and original image is not enough to recover all textural information. To preserve the original image structure we have used a cyclic perceptual loss which uses VGG16 [7] architecture for extracting features. Cyclic perceptual loss basically takes the combination of high and low-level features extracted from 2nd and 5th pooling layers of VGG16 [7] architecture.

As compared to CycleGAN [3] architecture our proposed Residual Inception CycleGAN has one extra cyclic perceptual loss. So, the objective of Residual Inception CycleGAN can be stated as follows, where LCycleGAN (G, F, Dx, Dy) is the full objective of CycleGAN [3] architecture, D stands for the discriminator and controls the effect of cyclic perceptual consistency loss:

$$Ł\left(G, F, D_x, D_y\right) = Ł_{CycleGAN}\left(G, F, D_x, D_y\right) + \gamma * Ł_{Perceptual}(G, F) \quad (1)$$

$$G^*, F^* = \arg\min\max Ł\left(G, F, D_x, D_y\right) \quad (2)$$

Conclusively, Residual Inception CycleGAN optimizes CycleGAN [3] architecture with the additional cyclic perceptual-consistency loss according to Eqs. 1 and 2.

4 Experiments and Results

We have presented the experimental results in this section. We trained our network seperately on 3 different datasets named: horse2zebra [9], I-Haze [8] and

Rain1200 [10]. We show both qualitative and quantitative results on Rain 1200 and I-Haze dataset. However for horse2zebra dataset only qualitative results are shown. The model is trained on NVIDIA DGX station having the configuration NVIDIA Tesla V100 4×32 GB GPU.

4.1 Results on I-haze Dataset

Due to too few images in the training set, we augmented the dataset by random cropping. Random crops were taken from the image by selecting random pixel coordinates and crop sizes. We then resized the crops to 256×256 and then feed them into the network. This augmentation was performed for each image in the dataset. We obtained 200 images per original image in the training set of the I-HAZE [8] from this augmentation. Both quantitative and qualitative results are shown for I-Haze [8] dataset (Table 1).

Table 1. Results on I-HAZE dataset

Approach	SSIM	PSNR
CycleGAN	0.59	15.63
Proposed method	0.64	16.51

4.2 Results on Rain1200 Dataset

We trained our network on 1200 images in Rain1200 [10] training dataset and have shown both qualitative and quantitative results on Rain1200 [10] test dataset (Table 2).

Table 2. Results on Rain1200 dataset

Approach	SSIM	PSNR
CycleGAN	0.75	23.26
Proposed method	0.77	24.08

4.3 Results on Horse2zebra Dataset

We trained the network on training set of horse2zebra dataset [9] and have shown qualitative results on the test dataset. Quantitative results are not shown due to unavailability of ground-truth (Figs. 2, 3 and 4).

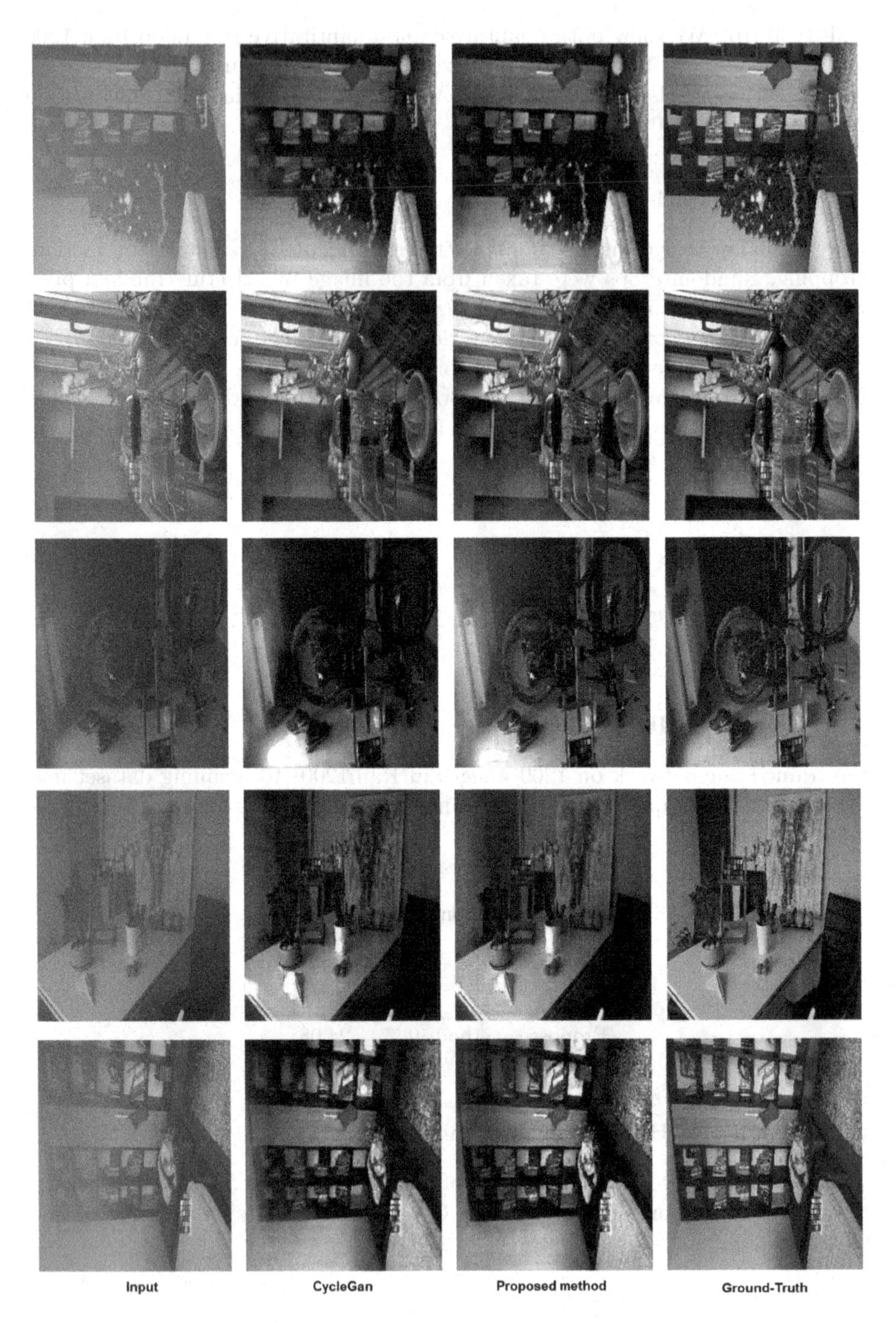

Fig. 2. Qualitative results on I-Haze [8] dataset

Fig. 3. Qualitative results on horse2zebra [9] dataset

Fig. 4. Qualitative results on Rain1200 [10] dataset

5 Conclusion

We proposed an enhanced version of CycleGAN, named as Residual Inception CycleGAN. We replaced the residual block with our proposed Residual Inception module for multi-scale feature extraction. We incorporated cyclic perceptual loss to improve the quality of texture in the recovered image and for better loacalizing the objects. We also increased the number of skip connections for better training. The experimental results shown on 3 datasets demonstrate that our method achieves higher accuracy values and produces visually better images and than CycleGAN architecture.

References

1. Goodfellow, I., et al.: Generative adversarial nets. In: NIPS (2014)
2. Zhao, J, Mathieu, M, LeCun, Y.: Energy-based generative adversarial network. In: ICLR (2017)
3. Zhu, J.-Y, Park, T., Isola, P., Efros, A.A.: Unpaired image-to-image translation using cycle-consistent adversarial networks. IEEE International Conference on Computer Vision (ICCV) (2017)
4. Szegedy, C., et al.: Going deeper with convolutions. In: Proceedings of the IEEE Conference on Computer Vision and Pattern Recognition, pp. 1–9 (2015)
5. Sajjadi, M.S., Scholkopf, and B., Hirsch, M.: Enhancenet: single image super-resolution through automated texture synthesis. In: IEEE International Conference on Computer Vision (ICCV) (2017)
6. Ronneberger, O., Fischer, P., Brox, T.: U-Net: convolutional networks for biomedical image segmentation. In: Navab, N., Hornegger, J., Wells, W.M., Frangi, A.F. (eds.) MICCAI 2015. LNCS, vol. 9351, pp. 234–241. Springer, Cham (2015). https://doi.org/10.1007/978-3-319-24574-4_28
7. Simonyan, K., Zisserman, A.: Very deep convolutional networks for large-scale image recognition. In: International Conference on Learning Representations (ICLR) (2015)
8. Ancuti, C.O, Ancuti, C, Timofte, R, De Vleeschouwer, C.: I-HAZE: a dehazing benchmark with real hazy and haze-free indoor images. arXiv (2018)
9. Deng, J., Dong, W., Socher, R, Li, L.-J, Li, K, Fei-Fei, L.: Imagenet: a large-scale hierarchical image database. In: CVPR (2009)
10. Zhang, H., Patel, V.M.: Density-aware single image de-raining using a multi-stream dense network. In: Proceedings of the IEEE Conference On Computer Vision and Pattern Recognition (2018)
11. Zhu, J.-Y., Krähenbühl, P., Shechtman, E., Efros, A.A.: Generative visual manipulation on the natural image manifold. In: Leibe, B., Matas, J., Sebe, N., Welling, M. (eds.) ECCV 2016. LNCS, vol. 9909, pp. 597–613. Springer, Cham (2016). https://doi.org/10.1007/978-3-319-46454-1_36
12. Radford, A., Metz, L., Chintala S.: Unsupervised representation learning with deep convolutional generative adversarial networks. In: ICLR (2016)
13. Denton, E.L., et al. :Deep generative image models using a Laplacian pyramid of adversarial networks. In: NIPS 2(015)
14. Salimans, T., Goodfellow, I., Zaremba, W., Cheung, V., Radford, A., Chen, X.: Improved techniques for training GANs. In: NIPS (2016)

15. Reed, S., Akata, Z., Yan, X., Logeswaran, L., Schiele, B., Lee, H.: Generative adversarial text to image synthesis. In: ICML (2016)
16. Mathieu, M.F., Zhao, J., Ramesh, A., Sprechmann, P., LeCun, Y.: Disentangling factors of variation in deep representation using adversarial training. In: NIPS (2016)
17. Hertzmann, A., Jacobs, C.E., Oliver, N., Curless, B., Salesin, D.H.: Image analogies. In: SIGGRAPH (2001)
18. Mathieu, M., Couprie, C., LeCun, Y.: Deep multiscale video prediction beyond mean square error. In: ICLR (2016)
19. Efros, A.A., Leung, T.K.: Texture synthesis by non-parametric sampling. In: ICCV (1999)
20. Pathak, D., Krahenbuhl, P., Donahue, J., Darrell, T., Efros, A.A.: Context encoders: feature learning by inpainting. In: CVPR (2016)
21. Long, J., Shelhamer, E., Darrell, T.: Fully convolutional networks for semantic segmentation. In: CVPR (2015)
22. Isola, P., Zhu, J.-Y., Zhou, T., Efros, A.A.: Imageto-image translation with conditional adversarial networks. In: CVPR (2017)
23. Sangkloy, P., Lu, J., Fang, C., Yu, F., Hays, J.: Scribbler: controlling deep image synthesis with sketch and color. In: CVPR (2017)
24. Karacan, L., Akata, Z., Erdem, A., Erdem, E.: Learning to generate images of outdoor scenes from attributes and semantic layouts. arXiv preprint arXiv:1612.00215 (2016)
25. Rosales, R., Achan, K., Frey, B.J.: Unsupervised image translation. In: ICCV (2003)
26. Aytar, Y., Castrejon, L., Vondrick, C., Pirsiavash, H., Torralba, A.: Cross-modal scene networks. In: PAMI (2016)
27. Liu, M.-Y., Tuzel, O.: Coupled generative adversarial networks. In: NIPS (2016)
28. Ulyanov, D., Lebedev, V., Vedaldi, A., Lempitsky, V.: Texture networks: Feed-forward synthesis of textures and stylized images. In: ICML (2016)
29. Liu, M.-Y., Breuel, T., Kautz, J.: Unsupervised image-to-image translation networks. In: NIPS (2017)
30. Kingma, D.P., Welling, M.: Auto-encoding variational bayes. In: ICLR (2014)
31. Shrivastava, A., Pfister, T., Tuzel, O., Susskind, J., Wang, W., Webb, R.: Learning from simulated and unsupervised images through adversarial training. In: CVPR (2017)
32. Taigman, Y., Polyak, A, Wolf, L.: Unsupervised cross-domain image generation. In: ICLR (2017)
33. Bousmalis, K., Silberman, N., Dohan, D., Erhan, D., Krishnan, D.: Unsupervised pixel-level domain adaptation with generative adversarial networks. In: CVPR (2017)
34. Gatys, L.A., Ecker, A.S, and M. Bethge. Image style transfer using convolutional neural networks. In: CVPR (2016)
35. Johnson, J., Alahi, A., Fei-Fei, L.: Perceptual losses for real-time style transfer and super-resolution. In: Leibe, B., Matas, J., Sebe, N., Welling, M. (eds.) ECCV 2016. LNCS, vol. 9906, pp. 694–711. Springer, Cham (2016). https://doi.org/10.1007/978-3-319-46475-6_43
36. Gatys, L.A., Bethge, M., Hertzmann, A., Shechtman, E.: Preserving color in neural artistic style transfer. arXiv preprint arXiv:1606.05897 (2016)
37. Chaudhary, S., Murala, S.: Deep network for human action recognition using Weber motion. Neurocomputing **367**, 207–216 (2019)

38. Chaudhary, S., Murala, S.: Depth-based end-to-end deep network for human action recognition. IET Comput. Vision **13**(1), 15–22 (2019)
39. Chaudhary, S., Murala, S.: TSNet: deep network for human action recognition in hazy videos. In: 2018 IEEE International Conference on Systems, Man, and Cybernetics (SMC), pp. 3981–3986 (2018). https://doi.org/10.1109/SMC.2018.00675
40. Chaudhary, S., Dudhane, A., Patil, P., Murala, S.: Pose guided dynamic image network for human action recognition in person centric videos. In: 2019 16th IEEE International Conference on Advanced Video and Signal Based Surveillance (AVSS), pp. 1–8 (2019) .https://doi.org/10.1109/AVSS.2019.8909835
41. Chaudhary, S.: Deep learning approaches to tackle the challenges of human action recognition in videos. Dissertation (2019)
42. Nancy, M., Murala, S.: MSAR-Net: multi-scale attention based light-weight image super-resolution. Pattern Recogn. Lett. **151**, 215–221 (2021)
43. Akshay, D., Biradar, K.M., Patil, P.W., Hambarde, P., Murala, S.: Varicolored image de-hazing. In: proceedings of the IEEE/CVF Conference on Computer Vision and Pattern Recognition, Pp. 4564–4573 (2020)
44. Praful, H., Dudhane, A., Murala, S.: Single image depth estimation using deep adversarial training. In: 2019 IEEE International Conference on Image Processing (ICIP), pp. 989–993. IEEE (2019)
45. Hambarde, P., Dudhane, A., Patil, P.W., Murala, S., Dhall, A.: Depth estimation from single image and semantic prior. In: 2020 IEEE International Conference on Image Processing (ICIP), pp. 1441–1445. IEEE (2020)
46. Hambarde, P., Murala, S.: S2DNet: depth estimation from single image and sparse samples. IEEE Trans. Comput. Imaging **6**, 806–817 (2020)

MAG-Net: A Memory Augmented Generative Framework for Video Anomaly Detection Using Extrapolation

Sachin Dube, Kuldeep Biradar, Santosh Kumar Vipparthi(✉), and Dinesh Kumar Tyagi

Malviya National Institute of Technology, Jaipur 302017, India
skvipparthi@iitg.ac.in, dktyagi.cse@mnit.ac.in
https://visionintelligence.github.io/

Abstract. Anomaly detection is one of the popular problem in computer vision as it is elementary to several computer vision applications. However, robust detection of spatiotemporal anomalies with dependencies across multiple frames is equally challenging. Most reconstruction error based unsupervised anomaly detection approaches struggle in case of intraclass dissimilarities and interclass similarities. Robust reconstruction capability of CNN often results in misclassification. To this end, a memory augmented generative encoder-decoder network (MAG-Net) is proposed for anomaly detection. A memory element is inserted between encoder and decoder to store diverse representations of normal feature space. A query is compared with multiple memory items representing normality space to ensure low reconstruction loss for normal frames and vice versa. MAG-Net also employs a residual block in the encoder stream to preserve visual features by allowing feedback from a layer to its successors. Residual block contains channel and pixel attention blocks to further accentuate spatiotemporal representations. Evaluation is done by comparing MAG-Net with other methods on benchmark datasets like UCSD ped-2 and Avenue. Experimental results validate the robustness of the proposed framework.

Keywords: GAN · Extrapolation · Anomaly detection

1 Introduction

Visual anomaly detection is one of most sought after problems in computer vision, as it forms the basis of many vision based applications such as: Intelligent transport systems, Automated surveillance systems, Quality control etc. Pervasive use of CCTV surveillance cameras generate huge volumes of data from public spaces; however, manual monitoring of data stream leads to underutilization of such infrastructure. Various anomaly detection techniques [1–4] can be used for automatic detection unwanted and threatening situations without human supervision. Different perspectives have led to numerous definitions of

B. Raman et al. (Eds.): CVIP 2021, CCIS 1568, pp. 426–437, 2022.
https://doi.org/10.1007/978-3-031-11349-9_37

anomaly depending upon the domain but with one commonality: anything that doesn't agree with prevalent normal behaviour can be termed as anomaly or outlier. Frame level anomalies are usually detected by application of a classifier on spatial features extracted from the frame, for example detection of an anomalous dog amongst cats or a defective item is relatively soft as it doesn't need much context building. However videos usually contain anomalous events evolving across multiple frames. Such anomalous sequences often have long dependencies across frames; and require spatio-temporal as well as contextual information for accurate detection. For example: aggressive charge followed by punching seems normal in a boxing ring but not in a mall. A robust anomaly detector has to deal with classic computer vision challenges caused by viewpoint and illumination change, occlusion, motion dynamics, resolution etc.; apart from anomaly specific challenges as pointed below:

1. *Data availability:* Capturing an anomalous event is rarer than the event itself [28] This translates to scarcity of high quality labelled anomalous data. Problem is even more evident in the case of data driven supervised deep learning based approaches. Apart from small size most of the staged datasets are also shot in a controlled environment and lack the variations present in spontaneous/candid scenarios. It is not feasible to capture infinite variations possible in unknown and unbounded events for classification. Framewise labelling and data balancing is also an arduous task.
2. *Contextual/long term dependencies:* Accurate identification of an anomalous event may be achieved by also considering information just before and after the anomaly. For example in case of accident, pre-collision and post collision data improves the accuracy of accident detection [29]. Such a long sequence of frames can be used to build context that plays a crucial role in deciding whether an event is anomalous or not. Extraction and retention of relevant spatiotemporal features also adds to challenges.
3. *Scene density and rate of motion:* Approaches that are effective in detecting anomalies in presence of limited number of subjects sees sharp decline in performance when tested on crowd datasets [9]. This can be attributed to occlusion and diminished subject boundaries. This variation has spawned entire class of anomaly detection approaches.
4. *Interclass similarities and intraclass dissimilarities:* At times there is high spatial or spatiotemporal similarity among two classes leading to a very thin or sometimes fluidic decision boundary [7,8]. Robust modelling in such scenarios require high number of samples to ensure multiple instances of a particular pattern in data. Problem becomes even more prominent in case of wild unseen scenarios.

In literature various approaches have been proposed around above challenges; however, emergence of CNN along with deep learning has completely revolutionized anomaly detection research. Initial spatio-temporal feature based supervised approaches [1,2] modelled the problem as a two class classification problem and were limited to detection of anomalous frames, without any attention to localization, timestamp or type of anomaly. Biradar et al. [3] proposed a time stamp aware anomaly detection approach, however unavailability of large scale labelled dataset shifted the interest towards more intuitive semi-supervised and unsupervised techniques [3–6] so much so that anomaly detection is seen as an unsupervised problem. Generative adversarial network (GAN) based approaches [4,5] were introduced to handle skewed datasets as well as unseen/unknown anomalies. Defining normality solely from normal data samples while ignoring abnormal data completely has become a prevalent practice. Such approaches rely on the fact that unseen abnormal patterns will exhibit high divergence from defined normality. Majority of such methods try to minimize reconstruction error during training and hope that an anomalous event will result in high reconstruction error. However, visual and structural similarities between normal and anomalous samples can lead to very small reconstruction error. For example CNN based Auto encoders leverage the representation capability of CNN to generate a latent representation of normality from training data. However High capacity CNN can reconstruct unseen anomalous frames satisfactorily. Memory based approaches [7,8] were proposed to capture and store explicit representation of diverse normal features. Spatiotemporal features often diminish through successive layers of CNN due to repeated use of pooling layers. This results in poor detection and can be compensated by usage of Feedback/skip connections from a layer to its successive layers [6]. Liu et al. [27] have proposed a future frame prediction based approach that compares test frame with predicted frame for anomaly decision.

Above works and challenges have motivated us to propose a video anomaly detection framework which tries to leverage benefits of state of the art methods while masking their shortcomings. Our contribution is summarized in the following points.

- A memory augmented generative framework termed as MAG-Net is proposed for Video anomaly detection using extrapolation. We try to extrapolate the next frame by using previous frame features and compare it with the groundtruth for anomaly decision.
- We have introduced a customized residual block for better feature retention. Apart from feedback connections it contains a channel and a pixel attention block in succession to accentuate visual features across channels.
- We have also used memory augmented encoder-decoder architecture by inserting a memory module between encoder and decoder. This memory module records diverse latent representations of normality. Memory items complements encoder features to ensure high reconstruction loss in case of anomalous frames and vice versa.
- Performance of the MAG-Net is validated by comparing it with other state-of-the-art methods on UCSD ped-2 and Avenue dataset.

2 Literature Review

Anomaly detection approaches can be broadly divided into two categories: Classic handcrafted feature based methods and learning based methods. Appearance and trajectory based approaches constitute major chunk feature based approaches. Feature extractors like LBP-TOP [9] are applied to extract dynamic texture features from image patches followed by bayesian classification [10]. Histogram of oriented gradient [11,12], optical flow [13,14] are also very popular techniques. Researchers have also tried to leverage benefits offered by Hog and optical flow by combining them and presented Histograms of optical flows [15,16]. Trajectory-based methods [17] leverage semantic information obtained by feature point tracking to capture effects of speed and direction changes. Yuan et al. [18] used a 3D DCT for pedestrian detection and tracking. Lin et al. [19] used multiple hypothesis tracking algorithms. However, the trajectory detection pipeline is highly error prone as it depends on object detector and segmentor. These errors see a spike in crowded or occluded scenarios.

Detection and tracking happens to be computation intensive in comparison to appearance features. However, success of feature based approaches is highly dependent on domain knowledge. In contrast learning based approaches automatically learn to identify relevant features. CNN's robust representation capability is utilized in various computer vision domains. Zhou et al. [20] proposed a 3D convolutional network for anomaly classification. Biradar et al. [2] used a two stream CNN for extracting spatial and temporal motion information separately. Spatial features were extracted through VGG-19 while flownet [21] extracts optical flow. These features were combined before training a Neural network based classifier. Hinami et al. [1] also tried to judge false positives by explaining causality behind an anomalous event along with detection. Furthermore biradar et al. [3] have proposed a timestamp aware anomaly detection algorithm. Malhotra et al. [22] Proposed to use LSTM to model long term dependencies among frames by maintaining previous states. Luo et al. [23] combined LSTM with CNN to leverage spatial as well as temporal information. Anomaly detection research in recent times has shifted in favour of more intuitive unsupervised learning based methods.

Hasan et al. [24] used end to end autoencoders to model temporal regularities in video sequences. Deep autoencoders [25] are also very effective in modelling high-dimensional data in unsupervised settings. These methods try to model normality i.e. latent representation or feature space to represent normality with the use of CNN based encoder decoder duo. Such approaches completely ignore abnormal data during training and hope that anomalous query will yield a high reconstruction error. However, this assumption can not guarantee high reconstruction error for anomalies due to various factors like robust representation capabilities of CNN, Perceptual and structural similarities among two classes, dynamic nature of events etc. To mitigate this shortcoming of AE's, dong et al. [7] proposed a memory augmented autoencoder MemAE. Encoder and decoder are trained together as usual but latent normal feature space is generated and stored within limited number of memory slots. During testing encoder output is not fed directly to decoder; instead it is used as a query to find the most relevant items in the memory. These items are then aggregated with encoder output and forwarded to the decoder. Park et al. [8] further improved upon that by updating memory to store prototypical patterns of normal data. They also took

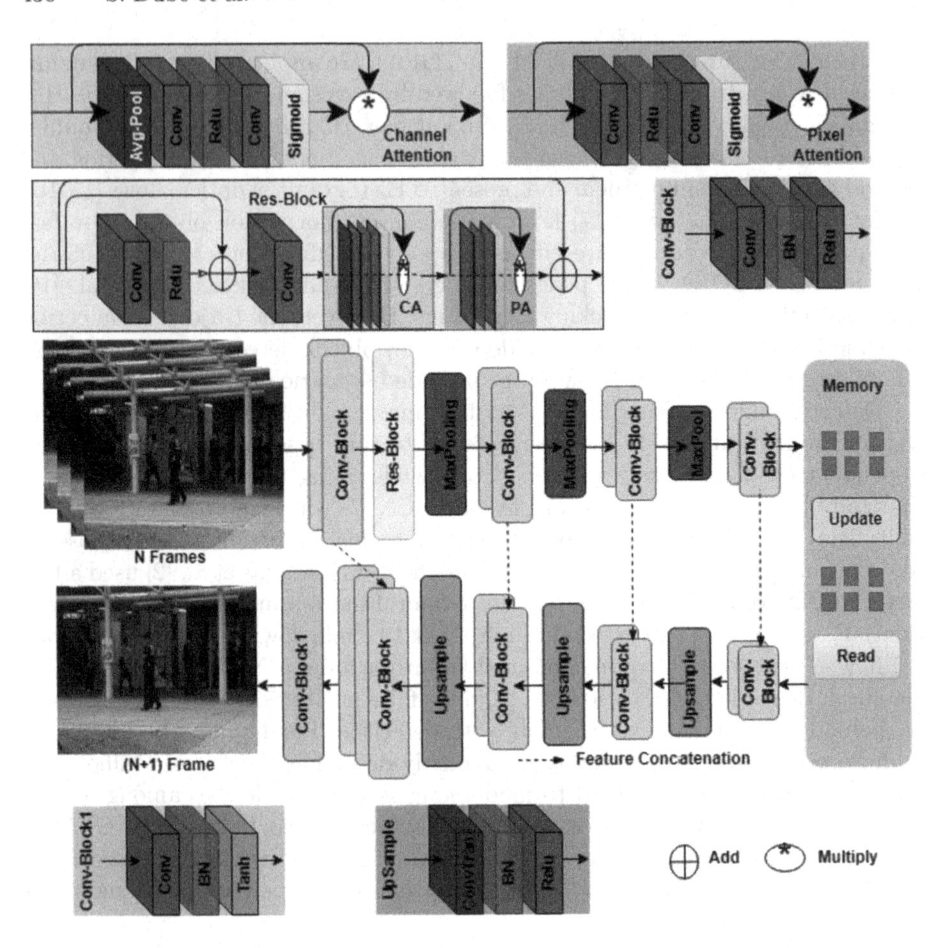

Fig. 1. Schematic diagram of proposed MAG-Net shows three components namely encoder decoder and a memory module. It also depicts layered architecture of conv and residual block while drilling down to conv, conv1, upsample, channel and pixel attention block.

reconstruction error between the query frame and its predicted counterpart. Akcay et al. [4] have proposed "GANomaly": A semi-supervised Conditional-GAN based method for anomaly detection with limited data. They have also used an encoder-decoder architecture in the generator module. Zenati et al. [5] propose an efficient GAN-Based anomaly detector and claim it to be 100 times more efficient than its counterparts. Taking inspiration from above methods we have tried to leverage benefits offered by these approaches while masking their shortcomings in the proposed framework.

3 Proposed Method: Memory Augmented Generative Network

Proposed memory augmented CNN based Generative encoder-decoder network mitigates mis-classification caused by interclass similarities and intraclass dissimilarities. Proposed method can be divided into three major components namely: Encoder, Memory element and decoder as shown in Fig. 1. Memory element is crucial to the performance of MAG-Net as it captures and stores diverse prototypical patterns to represent normality. Majority of unsupervised anomaly detection methods try to model normality by training solely on normal data. These approaches assume that abnormal patterns will give high reconstruction error when passed through a model trained on normal data. However, at times a strong decoder can reconstruct anomalies as well due to high similarity among two classes and diversity within normal class. Proposed MAG-Net tries to overcome this loophole to improve classification accuracy.

MAG-Net Extrapolates $(N+1)^{th}$ frame from a sequence of N input frames. Process starts by feeding N input frames of size $256 \times 256 \times N$ (N = 4 RGB Frames) to the encoder via a convolution block using Eq. 1. Conv block contains serial arrangement of convolution layers followed by Batch normalization BN and Relu $\Re$ activation. Figure 1 depicts actual no. of layers in each block. Output of Conv. block is fed to a residual block RB for better retention of spatiotemporal features extracted using Eq. 2. Channel attention and pixel attention blocks from [26] are highlight of residual block and are discussed below. Features from residual block are further concentrated by using a sequence of maxpooling and convolution blocks to find latent representation q_t of normality using Eq. 3.

$$CB = \Re(BN(Conv(x))) \tag{1}$$

$$RB = \tilde{F}(F_c^*(\Re(Conv(x)) + x)) + x \tag{2}$$

$$q_t = CB(MP(CB(MP(CB(MP(RB(CB(x)))))))) \tag{3}$$

$$US = R(BN(ConvT(x))) \tag{4}$$

$$CBT = T(BN(Conv(x))) \tag{5}$$

where, x is input features for convblock CB and q_t is output of encoder i.e. latent feature. BN, $\Re$,T, RB and MP stands for Batch Normalization, ReLU, Tanh, residual Block and maxPooling respectively.

Encoder and decoder follow Unet architecture with memory augmentation. Process starts with N = 4 frames each with 3 RGB channels making the input size $256 \times 256 \times 4 \times 3$ First conv. block keeps the same resolution while increasing the depth. Residual block accentuates channel and pixel wise features without changing the shape of input i.e. $256 \times 256 \times 64$. Output from Residual block is fed to maxpooling layer that uses stride two hence reshaping the input to $128 \times 128 \times 64$. This sequence of Conv and maxpooling continues till latent feature of size $32 \times 32 \times 512$ is reached. These latent features are mapped to memory or used as key to extract normality features from memory. Now latent feature q_t is reinforced with memory and are fed to decoder. Decoder unfolds by following inverse process of encoder by applying a series of convolution and upsample blocks as shown in Fig. 1 along with Eq. 4, ??; however last conv. block uses Tanh activation instead of ReLU as given in Eq. 5 to give extrapolated image of size $256 \times 256 \times 3$. An anomalous frame is supposed exhibit significant divergence when extrapolated by using proposed framework.

3.1 Channel Attention (CA)

Channel attention module averages features along different channels. Global average pooling is done over spatial information from individual channels.

$$g_c = L_p(F_c) = \frac{1}{H \times W} \sum_{i=1}^{H} \sum_{j=1}^{W} X_c(i,j) \tag{6}$$

Here $X_c(i,j)$ represents value of c_{th} channel Xc at location (i,j), where Lp is the global pooling function. Hence feature maps from previous conv block changes shape from $C \times H \times W$ to $C \times 1 \times 1$. Features from different channels are weighted by passing then through two convolution layers followed by sigmoid and ReLu activation functions.

$$CA_c = \sigma(Conv(\Re(Conv(g_c)))) \tag{7}$$

Here, σ is sigmoid function and, $\Re$ is ReLu. Finally, element-wise multiplication of the input F_c and channel weights CA_c is taken.

$$F_c^* = CA_c \bigotimes F_c \tag{8}$$

3.2 Pixel Attention (PA)

As different channels consist feature maps obtained by application of different kernals they usually contain different distributions. Considering this we also use a pixel attention (PA) module so that network pays explicit attention to informative spatial information. Like CA, convolution layers are fed withe input F^* (the output of the CA) followed by Re-Lu and Sigmoid. The shape changes from $C \times H \times W$ to $1 \times H \times W$.

$$PA = \sigma(Conv(\Re(Conv(F^*)))) \tag{9}$$

Finally element-wise multiplication is performed on input F^* and PA, $\tilde{F}$ is final output of (FA) module.

$$\tilde{F} = F^* \bigotimes PA \tag{10}$$

Table 1. Quantitative comparison of the state of the art anomaly detection methods in terms of average AUC (Ped2 [37] and Avenue [36].

Methods	Ped2	Avenue
MPPCA [30]	69.3	
MPPCA+FFA [30]	61.3	12
MDT [31]	82.9	10
AMDN [32]	90.8	10
Unmasking [33]	82.2	80.6
MT-FRCN [10]	92.2	
ConvAE [24]	85	80
TSC [24]	91	80.6
StackRNN [34]	92.2	81.7
AbnormalGAN [35]	93.5	
MemAE w/o Mem [7]	91.7	81
MemAE w/ Mem. [7]	94.1	83.3
OUR	92.92	84.69

3.3 Memory Module

A memory module is introduced in between encoder and decoder as proposed by Park et al. [8]. Memory is trained along with encoder-decoder network and stores prototypical patterns of normality to emulate a feature space. A memory update operation is executed to ensure only normal patterns are stored in memory. Multiple queries are associated with a memory item to enhance discrimination power of entire model. During testing a query sequence is fed to the encoder which extract latent spatiotemporal query features. These features are used to find appropriate memory item as explained in trailing subsections. Both feature vectors are then combined and fed to decoder for extrapolating next frame in the sequence. Memory is adaptive as it includes new items by using probabilistic matching at later stages. Thus memory contains M items that can be read or updated to maintain protoypical normality.

Memory Read: Output from the encoder i.e. spatiotemporal query feature q_t is mapped to memory by using cosine similarity between each query item q_t^k (size to $1 \times 1 \times c$ and k=$H \times W$.) and individual memory item p_m. This results in correlation map with dimensions to $M \times K$. A softmax function is then applied to correlation map as given by Eq. 11 to give matching probability $w_t^{k,m}$. For every query item q_t^k memory is read by weighted memory items p_m as given by Eq. 14 to give weighted feature p_t^k which is concatenated with q_t and fed to decoder.

Memory Update: Matching probabilities of query feature and memory item is found by using Eq. 11. A memory item has to correspond to multiple query features. The set of query features corresponding to memory item p_m is denoted by U_t^m. queries indexed by are used to update the memory according to Eq. 14, matching probability $v_t^{k,m}$ is calculated similar to memory read however softmax is applied in horizontal direction according to Eq. 14.

$$w_t^{k,m} = \frac{exp((p_m)^T q_t^k)}{\sum_{m'=1}^{M} exp((p_{m'})^T q_t^k)} \tag{11}$$

$$\hat{p}_t^k = \sum_{m'=1}^{M} w_t^{k,m'} p_{m'} \tag{12}$$

$$p^m \leftarrow f(p^m + \sum_{k \epsilon U_t^m} v_t^{'k,m} q_t^k) \tag{13}$$

$$v_t^{k,m} = \frac{exp((p_m)^T q_t^k)}{\sum_{k'=1}^{K} exp((p_m)^T q_t^{k'})} \tag{14}$$

4 Experimental Setup

Datasets: MAG-Net is evaluated on two most popular datasets namely UCSD peds2 [37] and Avenue [37]. **Peds2 dataset** contains 16 training and 12 testing videos shot from a stationary camera in a park. Annotations as well as binary pixels masks are also provided along-with the dataset. Camera is at elevated position and captures spontaneous anomalies. Pedestrians walking on pathway are considered normal while other vehicles like cycle, cart, van etc. plying on pathway is considered anomaly. **Avenue dataset** contains 16 training videos of normality and 21 testing videos containing both normality and abnormality. Videos are shot in CUHK campus with stationary camera with few testing videos containing camera shake. Walking is considered normal while panic running and throwing things is considered abnormal. Dataset consists of 47 abnormal events in total and also contains annotations for the same.

Training: Training is carried out by making chunks of N = 4 frames to be fed to MAG-Net. Each frame is resized have a resolution of 256 × 256. Output of encoder i.e. query feature map or latent representation is of the size $32 \times 32 \times 512$ while memory size is kept at 10 items. Adam optimizer is used with $\beta1 = .9$ and $\beta2 = .999$. Learning rate is kept at .0002. Models are trained on pytorch using Nvidia GTX TITAN XP.

5 Results and Discussion

Experimental results are tabulated in Table 1 which compares Average ROC score of proposed framework with other state-of-the-art methods. It is evident that majority of methods give very good result on ped-2 data-set but performance sees sharp decline in case of Avenue dataset. This disparity is common to majority of approaches but is exceptionally high in case of AMDN [32]. MemAE [7] is our closest contender and performs marginally (1.2) better in case of ped2. However it lags behind in case of more challenging avenue data-set by slightly higher margin (1.4). This improvement can be attributed to memory module as well as channel attention and pixel attention mechanism in residual block. It can also be noted that memory based approaches trump all other state-of-the-art methods by a significant margin and establishes significance of memory element in presence of long term dependencies.

6 Conclusion

Experimental results clearly convey that reconstruction based unsupervised anomaly detection approaches are susceptible to mis-classification. Sole reliance on reconstruction error often fails when put against high representation capability of CNN in presence of thin boundaries among contrasting classes. However, usage of memory to store diverse normality feature can significantly improve classification accuracy. Experimental results also Indicate superiority of proposed framework which follows memory augmented encoder decoder architecture. Performance gain can be attributed to Channel attention and pixel attention modules.

References

1. Hinami, R., Mei, T., Satoh, S.: Joint detection and recounting of abnormal events by learning deep generic knowledge. In: Proceedings of the IEEE International Conference on Computer Vision (2017)
2. Biradar, K., Dube, S., Vipparthi, S.K.: DEARESt: deep convolutional aberrant behavior detection in real-world scenarios. In: 2018 IEEE 13th International Conference on Industrial and Information Systems (ICIIS). IEEE (2018)
3. Biradar, K.M., et al.: Challenges in time-stamp aware anomaly detection in traffic videos. arXiv preprint arXiv:1906.04574 (2019)
4. Akcay, S., Atapour-Abarghouei, A., Breckon, T.P.: GANomaly: semi-supervised anomaly detection via adversarial training. In: Jawahar, C.V., Li, H., Mori, G., Schindler, K. (eds.) ACCV 2018. LNCS, vol. 11363, pp. 622–637. Springer, Cham (2019). https://doi.org/10.1007/978-3-030-20893-6_39
5. Zenati, H., et al.: Efficient GAN-based anomaly detection. arXiv preprint arXiv:1802.06222 (2018)
6. Chouhan, N., Khan, A.: Network anomaly detection using channel boosted and residual learning based deep convolutional neural network. Appl. Soft Comput. **83**, 105612 (2019)

7. Gong, D., et al.: Memorizing normality to detect anomaly: memory-augmented deep autoencoder for unsupervised anomaly detection. In: Proceedings of the IEEE/CVF International Conference on Computer Vision (2019)
8. Park, H., Noh, J., Ham, B.: Learning memory-guided normality for anomaly detection. In: Proceedings of the IEEE/CVF Conference on Computer Vision and Pattern Recognition (2020)
9. Xu, J., Denman, S., Fookes, C., Sridharan, S.: Unusual event detection in crowded scenes using bag of LBPs in spatio-temporal patches. In: Proceedings of DICTA, pp. 549–554 (2011)
10. Wang, X., Ma, X., Grimson, E.: Unsupervised activity perception by hierarchical Bayesian models. In: Proceedings of CVPR, pp. 1–8 (2007)
11. Chen, D.Y., Huang, P.C.: Motion-based unusual event detection in human crowds. J. Vis. Commun. Image Represent. **22**(2), 178–186 (2011)
12. Cheng, K., Chen, Y., Fang, W.: Video anomaly detection and localization using hierarchical feature representation and gaussian process regression. In: Proceedings of CVPR, pp. 2909–2917 (2015)
13. Xu, D., Ricci, E., Yan, Y., Song, J., Sebe, N.: Learning deep representations of appearance and motion for anomalous event detection. In: Proceedings of BMVC (2015)
14. Adam, A., Rivlin, E., Shimshoni, I., Reinitz, D.: Robust real-time unusual event detection using multiple fixed-location monitors. IEEE Trans. Pattern Anal. Mach. Intell. **30**(3), 555–560 (2008)
15. Sabokrou, M., Fathy, M., Hoseini, M., Klette, R.: Real-time anomaly detection and localization in crowded scenes. In: Proceedings of CVPR Workshops, pp. 56–62 (2015)
16. Wu, S., Wong, H.S., Yu, Z.: A Bayesian model for crowd escape behavior detection. IEEE Trans. Cir. Syst. Video Technol. **24**(1), 85–98 (2014)
17. Jiang, F., Wu, Y., Katsaggelos, A.K.: A dynamic hierarchical clustering method for trajectory-based unusual video event detection. IEEE Trans. Image Process. **18**(4), 907–913 (2009)
18. Yuan, Y., Fang, J., Wang, Q.: Online anomaly detection in crowd scenes via structure analysis. IEEE Trans. Cybern. **45**(3), 548–561 (2015)
19. Lin, H., Deng, J.D., Woodford, B.J., Shahi, A.: Online weighted clustering or real-time abnormal event detection in video surveillance. In: Proceeding of ACMMM, pp. 536–540 (2016)
20. Zhou, S., Shen, W., Zeng, D., Fang, M., Wei, Y., Zhang, Z.: Spatial-temporal convolutional neural networks for anomaly detection and localization in crowded scenes. Signal Process. Image Communicat. **47**, 358–368 (2016)
21. Dosovitskiy, A., et al.: FlowNet: learning optical flow with convolutional networks. In: Proceedings of the IEEE International Conference on Computer Vision (2015)
22. Malhotra, P., et al.: Long short term memory networks for anomaly detection in time series. In: Proceedings, vol. 89 (2015)
23. Luo, W., Liu, W., Gao, S.: Remembering history with convolutional LSTM for anomaly detection. In: 2017 IEEE International Conference on Multimedia and Expo (ICME). IEEE (2017)
24. Hasan, M., Choi, J., Neumann, J., Roy-Chowdhury, A.K., Davis, L.S.: Learning temporal regularity in video sequences. In: Proceedings of CVPR, pp. 733–742 (2016)
25. Kingma, D.P., Welling, M.: Auto-encoding variational Bayes. In: International Conference on Learning Representations (ICLR) (2014)

26. Qin, X., et al.: FFA-Net: feature fusion attention network for single image dehazing. In: Proceedings of the AAAI Conference on Artificial Intelligence, vol. 34, no. 07 (2020)
27. Liu, W., et al.: Future frame prediction for anomaly detection-a new baseline. In: Proceedings of the IEEE Conference on Computer Vision and Pattern Recognition (2018)
28. Sultani, W., Chen, C., Shah, M.: Real-world anomaly detection in surveillance videos. In: Proceedings of the IEEE Conference on Computer Vision and Pattern Recognition (2018)
29. Singh, D., Mohan, D.K.: Deep spatio-temporal representation for detection of road accidents using stacked autoencoder. IEEE Trans. Intell. Transp. Syst. **20**(3), 879–887 (2018)
30. Kim, J., Grauman, K.: Observe locally, infer globally: a space-time MRF for detecting abnormal activities with incremental updates. In: 2009 IEEE Conference on Computer Vision and Pattern Recognition. IEEE (2009)
31. Mahadevan, V., Li, W., Bhalodia, V., Vasconcelos, N.: Anomaly detection in crowded scenes. In: CVPR (2010)
32. Xu, D., Yan, Y., Ricci, E., Sebe, N.: Detecting anomalous events in videos by learning deep representations of appearance and motion. In: CVIU (2017)
33. Tudor Ionescu, R., Smeureanu, S., Alexe, B., Popescu, M.: Unmasking the abnormal events in video. In: ICCV (2017)
34. Luo, W., Liu, W., Gao, S.: A revisit of sparse coding based anomaly detection in stacked RNN framework. In: ICCV (2017)
35. Ravanbakhsh, M., Nabi, M., Sangineto, E., Marcenaro, L., Regazzoni, C., Sebe, N.: Abnormal event detection in videos using generative adversarial nets. In: ICIP (2017)
36. Lu, C., Shi, J., Jia, J.: Abnormal event detection at 150 FPS in MATLAB. In: ICCV (2013)
37. Li, W., Mahadevan, V., Vasconcelos, N.: Anomaly detection and localization in crowded scenes. IEEE TPAMI **36**, 18–32 (2013)

Hand Gesture Recognition Using CBAM-RetinaNet

Kota Yamini Suguna[1], H Pallab Jyoti Dutta[1], M. K. Bhuyan[1(✉)], and R. H. Laskar[2]

[1] Department of Electronics and Electrical Engineering, Indian Institute of Technology, Guwahati 781039, Assam, India
{k.yamini,h18,mkb}@iitg.ac.in
[2] Department of Electronics and Communication Engineering, National Institute of Technology, Silchar 788010, Assam, India
rhlaskar@ece.nits.ac.in

Abstract. Hand gesture recognition has become very popular with the advancements in computer vision, and efficient detection of hand gestures is certainly the talk of the hour. It finds numerous worthy real-world applications such as human computer interaction, sign language interpretation, immersive gaming experience, robotics control systems, etc. This paper presents an end-to-end system for the recognition of hand gestures, based on an efficient object detection architecture-RetinaNet. Convolutional Block Attention Module is used to improve performance. The method has been tested on Ouhands dataset and achieved better recognition accuracies and real-time application. Results show that our CBAM-RetinaNet model is robust and efficient in recognition of hand gestures in complex backgrounds and different illumination variations.

Keywords: Attention mechanism · Bounding box · Deep learning · Hand gesture recognition

1 Introduction

Hand gesture recognition is one of the more explored problems in Computer Vision. Hand Gestures play an important role in many fields like communication with specially abled people (deaf and mute), Human-Computer-Interaction (HCI), robotics, etc. Simple gestures can be used to control and communicate with devices without any physical contact. Hand gesture recognition involves detection of the hand in the image or video, followed by recognizing meaning conveyed by the gesture. Having a good hand gesture recognition system can not only make HCI more natural and better, but it can also be used to bridge the communication gap between people who don't understand sign-language and the deaf.

For hand gesture recognition, many methods have been adopted. Papers like [1] and [2] involve electronics and gloves and are intrusive in nature for

B. Raman et al. (Eds.): CVIP 2021, CCIS 1568, pp. 438–449, 2022.
https://doi.org/10.1007/978-3-031-11349-9_38

hand gesture estimation. In [3], a 3D hand-based model is proposed that uses geometrical aspects of the hand, such as joints, to represent hand pose. A CNN based network is shown in [4] for black and white hand pose images. HGRNet [5] is a two-stage fusion network, where pixel-level semantic segmentation is used to detect hand regions and then classify the hand gestures by a two-stream CNN. Even in [6], the authors utilized a UNET architecture to obtain the segmentated masks of the gestures and fed them to a pre-trained network to get the desired classification results. CrossInfoNet [7] uses 2D and 3D heatmaps for feature extraction. The methods used for hand gesture estimation involve a lot of preprocessing and obtaining depth maps or heatmaps. For this, a significant amount of time and resources are required. This motivates the object detection based methods like [8–10], etc. for hand gesture recognition.

Object Detection (one of the cutting-edge topics in Computer Vision) is widely explored for hand gesture recognition. It recognizes "what" the object is and "where" it is located. Object Detection involves "Object Localization", i.e., identifying the position of one or more objects in the image and "Image Classification", i.e., prediction of the class of the object in the image. Many algorithms and methods have been proposed and have gained huge popularity for Object Detection like Faster-RCNN [19], SSD (Single Shot MultiBox Detector) [12], YOLO [13], etc. The deep learning-based object detection methods involve 2 stage methods for feature extraction and region proposals, followed by the classification networks. Though these methods have higher accuracy, the inference time required is higher, making it infeasible for real-time applications. The single stage object detection methods like SSD, YOLO, etc. have very less inference time, but they compromise on the accuracy of detection. The advantages of RetinaNet [14], like lower inference time, higher accuracy, multi scale detection, etc., give a strong motivation to select RetinaNet for our Hand Gesture Recognition system. Further, an attention mechanism has been employed which shows where to focus. An attention based method is shown in [15] for hand pose estimation against complex backgrounds. We propose a RetinaNet based end-to-end hand gesture recognition system which employs a Convolutional Block Attention Module(CBAM) [16] for better focus on the essential features.

2 Proposed Network

A RetinaNet-based network has been designed for the task of hand gesture recognition. Figure 1 shows the architecture of the CBAM-RetinaNet model in detail. The following sections describe the details of each module in the proposed network.

2.1 ResNet-18 Backbone

We use ResNet-18 [17] as the backbone of our CBAM-RetinaNet architecture. Residual Network uses shortcut connections (or skip connections). When the input and the output are of the same dimension, we can simply add them (i.e.,

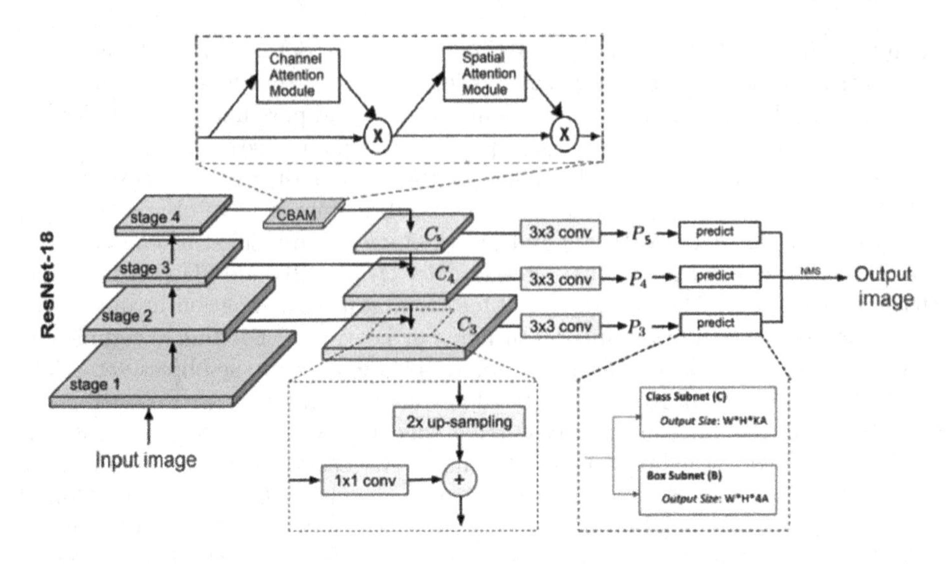

Fig. 1. Complete Architecture of the Proposed model

identity shortcuts). However, when the dimensions vary, either zero padding can be done to match the dimensions before adding, or 1×1 convolutions can be utilized to match the dimensions (i.e., projection shortcuts). When the sizes are different, shortcuts are done with a stride 2.

The last layer of each stage (the feature activation output from each stage's last residual block of ResNet-18) is taken as a reference set of feature maps, because the deepest layer of each stage is expected to have the strongest and the most vital features. Further, the output from the last stage is fed to the CBAM before passing it as a reference to the Feature Pyramid Network, as it is the feature map richest in information.

2.2 Convolutional Block Attention Module

We employ a Convolutional Block Attention Module (CBAM) to the last layer of the fourth stage of ResNet-18. CBAM has two sub-modules: Channel Attention Module and Spatial Attention module.

The feature map F obtained from the fourth stage of ResNet-18 is first passed through the Channel Attention Module, where 'what' to focus on is obtained in the form of a 1D attention map M_c. This is then multiplied, element-wise, with F to get F'. Then, F' is passed through the Spatial Attention module, which produces a 2D attention map M_s, which shows 'where' to focus. This is again multiplied, element-wise, with F' to obtain F''.

$$F' = M_c(F) \bigotimes F \tag{1}$$

$$F'' = M_s(F') \bigotimes F' \tag{2}$$

$\otimes$ denotes element-wise multiplication. M_c and M_s are obtained as:

$$M_c(F) = \sigma(MLP(AvgPool(F)) + MLP(MaxPool(F))) \tag{3}$$

Here, MaxPooling and AvgPooling are done along the spatial dimension.

$$M_s(F) = \sigma(f^{7\times7}([AvgPool(F); MaxPool(F)])) \tag{4}$$

Here, MaxPooling and AvgPooling are done along the channel dimension.
σ represents Sigmoid activation function. F'' is of the same dimension as the original F, the feature map obtained from the last stage of ResNet-18 backbone.

2.3 Feature Pyramid Network

Feature Pyramid Network (FPN) [18] is built on top of ResNet-18 architecture. FPN comprises of a bottom-up and a top-down pathway with lateral connections. The bottom-up pathway is the standard convolutional network for feature extraction. The spatial resolution decreases, as we go up. As we go down the path, the top-down pathway restores resolution with rich semantic information by upsampling the previous layer by 2 using nearest neighbor upsampling. Lateral connection performs element-wise addition between feature maps of the same spatial size from the bottom-up pathway (after going through 1×1 convolutions to match channel dimensions) and the top-down pathway. To reduce aliasing effects due to upsampling, a 3×3 convolution is performed and the final feature map is obtained.

The feature pyramid constructed has 5 layers called P_3 to P_7. Layer P_l has the resolution 2^l lower than the input resolution (where l is the level in the pyramid). P_3 to P_4 are computed from the corresponding layer outputs from ResNet-18. P5 is computed from the output of CBAM. P_6 is computed by 3×3 convolutions with stride 2 and P_7 is computed by taking ReLU, followed by 3×3 convolutions on P_6 with stride 2. The number of output channels is kept 256 for all the layers of the pyramid. A rich, multi-scale feature pyramid, obtained from a single input image, improves the performance significantly.

Anchor Boxes. Anchor boxes [11] of different heights and widths are defined to capture the scales and aspect ratios of different classes in the training data. The use of anchor boxes enables a network to detect objects of different scales, multiple objects, and overlapping objects all at once, thus increasing the overall speed.

The anchors used at levels P_2 to P_6 have areas 32^2 to 512^2, respectively. For each level, anchors of 3 aspect ratios are used: 1:2, 1:1, 2:1. For better scale coverage, additional anchors of size $2^1/3$, $2^2/3$ times the original sizes are added, making a total of 9 anchors for each level. For each anchor box, two things are to be predicted:

1. The 4 location offsets against the anchor box (a vector of length 4 containing box coordinate offsets obtained from the Box Regression Subnetwork).

2. The K number of class probabilities to tell us which class the bounding box belongs to (a one-hot vector of length K obtained from the Classification Subnetwork).

In total, we are predicting $(4 + K)$ values for one anchor box. A threshold of 0.5 is taken for the Intersection-over-union (IoU) for the foreground objects; and IoU of [0, 0.4) for the background. If the overlap is in [0.4, 0.5), the anchor box is unassigned and it is ignored during training.

2.4 Classification Subnetwork

The Classification Subnetwork is a small Fully Connected Network attached to each level of the feature pyramid. It is used to predict the probability of the hand being present at each spatial position for each of the $A = 9$ anchors and K classes. The parameters are shared across all pyramid levels. Four 3×3 convolutional layers, each having $C = 256$ channels, are applied to each pyramid level, followed by ReLU activations, followed by a 3×3 conv layer with $K * A$ filters. Then, Sigmoid Activations are applied to get an output of KA binary predictions per spatial location.

2.5 Box Regression Subnetwork

The Box Regression Subnetwork is used to obtain the relative offsets between the anchor and the ground truth box, if any. The design of this subnetwork is same as the classification subnetwork except that it outputs a $(4{*}A)$ linear vector per spatial location for each of the anchor boxes. The parameters for the Classification Subnetwork and the Box Regression Subnetwork are not shared and are different from each other.

Focal Loss. We use Focal loss, used in RetinaNet, to cater to the extreme imbalance of a large number of easy background samples vs the sparse foreground samples and the hard background samples. It reduces the loss for the well-trained class such that, whenever the model is good at detecting background, it will reduce its loss and reemphasize the training on the class and the hard background samples. Focal loss can effectively discount the effect of easy negatives, focusing all attention on the hard negative examples.

Focal Loss is taken as the loss on the Classification Subnetwork, applied to all the anchors in each sampled image. The total Focal Loss is the sum of the focal losses over all the anchors, normalized by the number of anchors assigned to a ground-truth box, not the total number of anchors.

If $p \in [0, 1]$ is the estimated probability for the label $y = 1$. Let's define p_t as:

$$p_t = \begin{cases} p, & \text{if } y = 1 \\ 1 - p, & \text{otherwise} \end{cases} \tag{5}$$

Focal Loss is defined as:

$$FL(p_t) = -\alpha_t * (1 - p_t)^\gamma * log(p_t) \tag{6}$$

where $(1 - p_t)^\gamma$ is the modulating factor to the Cross Entropy Loss and $\gamma \geq 0$ is tunable focusing parameter (default: 2). And

$$\alpha_t = \begin{cases} \alpha, & \text{if } y = 1 \\ 1 - \alpha, & \text{otherwise} \end{cases} \tag{7}$$

where $\alpha \in [0, 1]$ is weighting factor for addressing class imbalance.

Decoding is done on at most 1k top predictions for each level in the feature pyramid after thresholding the detector confidence at 0.05. Non-maximal suppression (threshold: 0.5) is done on the merged predictions from all the levels to get the final results.

3 Experimentation

We use pre-trained ResNet-18 (on ImageNet-1K dataset) for the experimentation and transfer learning is applied to fine-tune the network. The hyperparameters of the Focal loss are kept the same as in RetinaNet [14]. Adam optimizer is used for parameter optimization with a mini-batch size of 2. The learning rate is kept $1e{-}5$. Since transfer learning is used, the network shows significantly good results without the need for data augmentation. Furthermore, batch normalization in the ResNet-18 architecture ensures we are not over-fitting the network on the training dataset.

3.1 Dataset

We use OUHands Dataset [20] for our task of hand gesture recognition. Only colour images are used for training and testing purposes. There are 10 classes of gestures, 10 sample images are collected from each of the 20 subjects for the training data. The training data has a total of 2000 images, which is split into a training set of 1600 images from 16 subjects and a validation set of 400 images from the remaining 4 subjects. The testing is done on 1000 images, 10 sample images from each of the 10 subjects. These images are taken in different illumination setting and against complex backgrounds.

3.2 Position of CBAM in the Network

We experimented with the position of the CBAM module in the architecture for optimal performance. We trained each of the networks for 30 epochs. The positions explored are:

1. after the FPN layer
2. after all the stages of ResNet-18 (x2-x3-x4)

3. after combinations of the outputs from stages of ResNet-18 (x2-x3, x2-x4, x3-x4)
4. after individual outputs from stages of ResNet-18 (x2, x3 and x4)

Fig. 2. Hand Gesture Recognition Outputs

Fig. 3. Grad-CAM visualization outputs

Table 1 shows F1 score and recognition accuracy comparison of networks with CBAM in different positions in the network. Table 2 shows the class-wise comparison of the mAP scores.

After experimentation, we chose the architecture with CBAM module after the output of the last stage of ResNet-18 (x4) before passing it to the FPN. The confidence score and IOU threshold for the detection taken is 0.5. The detection results, confusion matrix, and attention maps are shown in the following section.

Table 1. Precision, Recall, F1 score, and Accuracy comparison for different positions of CBAM in the network

CBAM position	F1 score(%)	Accuracy(%)
Without CBAM	86.15	81.07
After FPN	83.58	78.32
Before FPN: all (x2–x3–x4)	76.32	74.27
Before FPN: x2–x3	83.54	80.09
Before FPN: x2–x4	80.2	73.88
Before FPN: x3–x4	83.19	76.75
Before FPN: x2	85.93	81.16
Before FPN: x3	81.56	74.71
Before FPN: x4	**86.7**	**82.05**

4 Results

This section shows different results obtained for the hand gesture recognition with the RetinaNet based model with Convolutional Block Attention Module. Figure 2 shows a few examples of the detected hand along with the predicted class. We can see CBAM-RetinaNet model predicts the hand gestures accurately in varying illumination and contrast conditions as well as against complex backgrounds.

Table 2. mAP score comparison for different positions of CBAM in the network

CBAM position	A	B	C	D	E	F	H	I	J	K
Without CBAM	0.99	0.98	0.90	0.87	**0.96**	0.76	0.89	0.91	0.88	0.93
After FPN	0.99	0.98	0.89	0.86	0.76	**0.83**	0.86	0.84	**0.90**	0.92
Before FPN: all	0.91	0.92	0.84	0.65	0.74	0.69	0.59	0.63	0.70	0.83
Before FPN: x2-x3	0.97	0.98	0.90	0.71	0.82	0.78	0.76	0.86	0.85	0.89
Before FPN: x2–x4	0.99	0.98	0.80	0.83	0.77	0.76	0.79	0.86	0.88	0.89
Before FPN: x3–x4	**1.00**	**0.99**	0.79	0.85	0.92	0.82	0.86	0.89	0.87	0.91
Before FPN: x2	0.99	0.98	0.90	0.87	0.93	0.78	0.86	0.87	0.88	0.93
Before FPN: x3	0.97	0.96	**0.91**	0.80	0.89	0.76	0.81	0.93	0.86	0.89
Before FPN: x4	0.98	0.98	0.88	**0.88**	0.92	0.73	**0.90**	**0.94**	0.89	**0.94**

(a) Without CBAM.

(b) With CBAM.

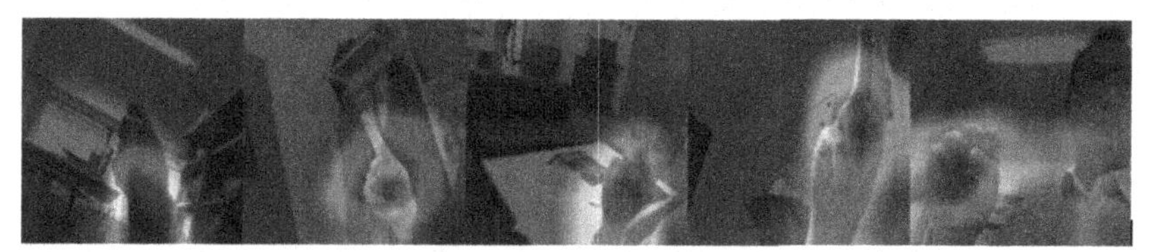

(c) Grad-CAM visualization for without CBAM case.

(d) Grad-CAM visualization for with CBAM case.

Fig. 4. Effects of CBAM on the Network

We use Grad-CAM [21] for visualizing the network. Grad-CAM uses gradients to calculate the importance of spatial locations in convolutional layers. Grad-CAM results show attended regions because it calculates the gradients with respect to each class.

The Grad-CAM visualization is computed for the last layer of the fourth stage of ResNet-18. Figure 3 shows the Grad-CAM visualization outputs for the images in Fig. 2.

We observed that without the attention module, the network outputs multiple detections for a single gesture and that too with incorrect classification. Figure 4(a) shows a few examples of images where the model detects false positives and multiple gestures and Fig. 4(b) shows their corresponding output from CBAM-RetinaNet model with the attention module. Figure 4(c) and Fig. 4(d) show the Grad-CAM visualization of the impact of the attention module for detection.

		Predicted									
		A	B	C	D	E	F	H	I	J	K
Actual	A	99	0	0	0	0	0	0	12	0	0
	B	0	98	0	1	0	2	0	0	0	0
	C	6	0	89	5	1	0	2	19	0	2
	D	1	1	5	93	6	0	24	8	0	0
	E	0	0	0	8	94	0	12	0	0	1
	F	0	7	0	4	5	73	1	0	0	0
	H	0	0	0	10	3	0	92	15	0	0
	I	3	0	0	6	0	0	2	98	1	0
	J	7	1	0	1	0	0	0	4	89	3
	K	1	0	2	2	0	0	0	0	7	94

Fig. 5. Confusion Matrix

We also computed the confusion matrix for the testing data. The Fig. 5 shows the obtained confusion matrix. The columns represent the predicted class and the rows represent the ground truth.

5 HCI Application

We have developed a Human-Computer-Interaction Application for real-time hand gesture recognition based on our CBAM-RetinaNet model. The application allows capturing snapshots from live feed with the click of a button. The snapshot is processed in the backend to detect the hand gesture posed in the image. The recognised gesture is displayed in the message box along with the confidence score of the same. If no gesture is detected, then the message "No label detected" is printed in the message box. The application allows the user to take snapshots

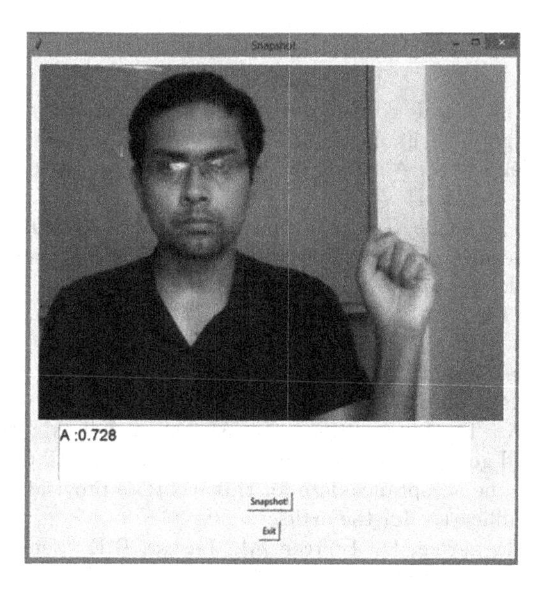

Fig. 6. HCI Application

continuously to detect gestures from multiple images from the live feed. Figure 6 shows an instance of our HCI Application for hand gesture recognition using CBAM-RetinaNet.

6 Conclusion

The hand gesture recognition system can be used in many real-world applications like improving Human-Computer-Interaction, robot control, and even facilitating communication using Sign Language. We proposed an end-to-end system for the recognition of hand gestures, based on an efficient object detection architecture-RetinaNet with Convolutional Block Attention Module to improve performance. We tested our CBAM-RetinaNet model on OUHands dataset and achieved better recognition accuracies. Results show that our method is robust and efficient in recognition of hand gestures in complex backgrounds and different illumination variations. Although, we focused on static hand gestures in the paper, CBAM-RetinaNet model can be used in real-time dynamic hand gesture recognition because of low inference time. Further, we intend to study the effectiveness of CBAM-RetinaNet model on different datasets and develop a real-time hand gesture-to-text system.

Acknowledgement. We acknowledge the Department of Biotechnology, Government of India for the financial support for the Project BT/COE/34/SP28408/2018.

References

1. Santos, L., et al.: Dynamic gesture recognition using a smart glove in hand-assisted laparoscopic surgery. Technologies **6**, 8 (2018)
2. Sturman, D., Zeltzer, D.: A survey of glove-based input. IEEE Comput. Graphics Appl. **14**(1), 30–39 (1994)
3. Rehg, James M., Kanade, Takeo: Visual tracking of high DOF articulated structures: an application to human hand tracking. In: Eklundh, Jan-Olof. (ed.) ECCV 1994. LNCS, vol. 801, pp. 35–46. Springer, Heidelberg (1994). https://doi.org/10.1007/BFb0028333
4. Zhan, F.: Hand gesture recognition with convolution neural networks. In: 2019 IEEE 20th International Conference on Information Reuse and Integration for Data Science (IRI), pp. 295–298 (2019)
5. Dadashzadeh, A., Targhi, A., Tahmasbi, M., Mirmehdi, M.: HGRNet: a fusion network for hand gesture segmentation and recognition. IET Comput. Vis. **13**(8), 700–707 (2019), the acceptance date for this record is provisional and based upon the month of publication for the article
6. Jyoti Dutta, H.P. Sarma, D., Bhuyan, M., Laskar, R.H.: Semantic segmentation based hand gesture recognition using deep neural networks. In: 2020 National Conference on Communications (NCC), pp. 1–6, (2020)
7. Du, K., Lin, X., Sun, Y., Ma, X.: CrossInfoNet: multi-task information sharing based hand pose estimation. In: Proceedings of the IEEE/CVF Conference on Computer Vision and Pattern Recognition (CVPR), June 2019
8. Zhang, Q., Zhang, Y., Liu, Z.: A dynamic hand gesture recognition algorithm based on CSI and YOLOv3. J. Phys. Conf. Ser. **1267**, 012055 (2019)
9. Liu, P., Li, X., Cui, H., Li, S., Yafei, Y.: Hand gesture recognition based on single-shot multibox detector deep learning. Mobile Inf. Syst. **2019**, 1–7 (2019)
10. Sharma, S., Pallab Jyoti Dutta, H., Bhuyan, M., Laskar, R.: Hand gesture localization and classification by deep neural network for online text entry. In: 2020 IEEE Applied Signal Processing Conference (ASPCON), pp. 298–302 (2020)
11. Girshick, R.: Fast R-CNN. In: IEEE International Conference on Computer Vision (ICCV), vol. 2015, pp. 1440–1448 (2015)
12. Liu, W., et al.: SSD: single shot multibox detector. In: Leibe, Bastian, Matas, Jiri, Sebe, Nicu, Welling, Max (eds.) ECCV 2016. LNCS, vol. 9905, pp. 21–37. Springer, Cham (2016). https://doi.org/10.1007/978-3-319-46448-0_2
13. Redmon, J., Divvala, S., Girshick, R., Farhadi, A.: You only look once: unified, real-time object detection. In: IEEE Conference on Computer Vision and Pattern Recognition (CVPR), vol. 2016, pp. 779–788 (2016)
14. Lin, T.Y., Goyal, P., Girshick, R., He, K., Dollar, P.: Focal loss for dense object detection. IEEE Trans. Pattern Anal. Mach. Int. **PP**, 1 (2018)
15. Pisharady, P., Vadakkepat, P., Loh, A.: Attention based detection and recognition of hand postures against complex backgrounds (2013)
16. Woo, Sanghyun, Park, Jongchan, Lee, Joon-Young., Kweon, In So.: CBAM: convolutional block attention module. In: Ferrari, Vittorio, Hebert, Martial, Sminchisescu, Cristian, Weiss, Yair (eds.) ECCV 2018. LNCS, vol. 11211, pp. 3–19. Springer, Cham (2018). https://doi.org/10.1007/978-3-030-01234-2_1
17. He, K., Zhang, X., Ren, S., Sun, J.: Deep Residual Learning for Image Recognition In: IEEE Conference on Computer Vision and Pattern Recognition (CVPR), vol. 2016, pp. 770–778 (2016)

18. Lin, T.Y., Dollr, P., Girshick, R., He, K., Hariharan, B., Belongie, S.: Feature pyramid networks for object detection. In: IEEE Conference on Computer Vision and Pattern Recognition (CVPR). vol. 2017, pp. 936–944 (2017)
19. Ren, S., He, K., Girshick, R., Sun, J.: Faster R-CNN: towards real-time object detection with region proposal networks. IEEE Trans. Pattern Anal. Mach. Intell. **39**, 06 (2015)
20. Matilainen, M., Sangi, P., Holappa, J., Silvn, O.: OUHANDS database for hand detection and pose recognition. In: 2016 Sixth International Conference on Image Processing Theory, Tools and Applications (IPTA), pp. 1–5 (2016)
21. Selvaraju, R.R., Cogswell, M., Das, A., Vedantam, R., Parikh, D., Batra, D.: Grad-CAM: visual explanations from deep networks via gradient-based localization. In: IEEE International Conference on Computer Vision (ICCV), vol. 2017, pp. 618–626 (2017)

Elderly Patient Fall Detection Using Video Surveillance

Amartya Raghav(✉) and Sachin Chaudhary

Punjab Engineering College, Chandigarh, India
amartyaraghav@gmail.com, sachin.chaudhary@pec.edu.in

Abstract. The important goal of this work is to design a system for elderly people who are at higher risk of falling due to some illness. We aim to do this without burdening them with the different number of devices around them including wearable devices. We aim to continuously monitor them in a closed environment for any kind of falls or injuries that might occur to them. This paper aims to achieve this with the help of only video surveillance [13, 16]. We also aim to propose a method where the model is very cost-friendly and easy to implement in any closed environment. Our model removes any use of wearable sensors and proposes an approach where we use inputs from only RGB camera-based sensors and then uses different computer vision approaches [2, 6] to decide if the person has fallen or not. We are using a two-stream network to attain more accuracy. Estimating the motion of a person on regular basis is the main step in the proposed method. There are two most popular publically available methods for motion detection namely Optical Flow (OF) [26] and Motion History Image (MHI) [17]. In the proposed method, we are using OF for the motion estimation. With the help of OF [26] and VGG16 [25] we have proposed a method for elderly patient monitoring system using live camera to detect a falling person. Despite not using multiple wearable and non-wearable sensors our proposed method outperforms the existing fall detection models. We have worked on the publically available UR-Fall Dataset [10].

Keywords: Fall detection · Computer vision · VGG16 · Optical flow

1 Introduction

With the increase of the elderly population, the phenomenon of the elderly falling at home or out is more and more common. It is reported that unexpected falling is a major problem and about one-third of people over 65 fall unexpectedly [1]. Figure 1 shows that fall death rates has increased by 30% from 2007 to 2016 for older adults in the U.S. [14]. The risk of falling is higher among older people, individuals with Parkinson's disease, or patients in rehabilitation units [29]. The main reasons for the occurrence of falls are physical factors like muscle weakness, posture, GAIT balance, vision, old age, and physiological or environmental factors [29]. Falling is a major cause of injuries and hip fractures.

B. Raman et al. (Eds.): CVIP 2021, CCIS 1568, pp. 450–459, 2022.
https://doi.org/10.1007/978-3-031-11349-9_39

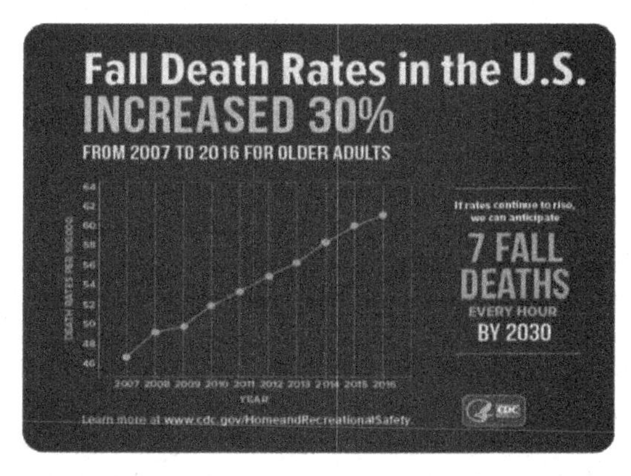

Fig. 1. Fall death rates in the US [14]

Before we begin we must understand the environment of the video stream that we are going to take as our first input. Various methods are available for fall detection but none of them were found to be very efficient without the use of wearable sensors. The use of Convolutional Neural Net-work (CNN) for classification is one of the best in the market as CNN can easily handle lots and lots of data at once and we have used VGG16 [25] architecture of CNN.

To improve the accuracy of the model, we have used motion detection algorithms. The major algorithms present for motion detection [27] are namely Motion History Image (MHI) [17] and Optical Flow (OF) [26]. We have used the method of OF here as it can detect and describe the velocity among consecutive video frames. OF helps in estimating the motion by calculating the motion of image intensities in subsequent frames [3] which is described in the Fig. 2. While the optical flow field appears to be similar to a dense motion field derived from motion estimation techniques, it is the study of not only the OF determination but also its application in estimating the three-dimensional nature and structure of the scene, as well as the 3D motion of objects and the observer relative to the scene.

Fig. 2. Optical flow motion detection between consecutive frames

While on the other hand, the MHI is a static visual template that aids in determining the motion's location and path as it moves forward. The temporal motion information is condensed into a single image template in MHI, and intensity is a function of motion recency. The enormous amount of MHIs per video, the selection of optimal temporal window size, and the training of a model without explicitly specifying MHIs are all major limitations of this approach.

One of the major aspects that need to be taken care of in the region of interest [12, 32], the video frames may contain an excessive background which in our case may be termed as noise. Hence, we needed to detect the area of interest which was the part of the frame with the person in it. To overcome this, we needed an object detection approach as in our case the object is an elderly person. We majorly came across 8 different types of object detection algorithms namely- Spatial Pyramid Pooling (SPP-net) [18], Region-based Convolutional Neural Networks (R-CNN) [19], Fast R-CNN [20], Faster R-CNN [21], Histogram of Oriented Gradients (HOR) [22], Regin based Fully Convolutional Network (R-FCN) [23], Single Shot Detector (SSD) [24] and You Only Look Once (YOLO) [11].

After comparison of all these algorithms and based on our requirement of real-time object detection we proposed an algorithm based on YOLO [11]. Since it is considered as one of the most effective algorithms which work on the principle of CNN. In YOLO [11] as shown in Fig. 3, a single neural network is applied to the complete image which is then divided into different regions and a bounding box is created among each region with objects. These bounding boxes are weighted by the probabilities of prediction.

Fig. 3 Working of YOLO

The next major part of our proposed methodology needs the use of estimation and classification of the frames with a person falling or not falling which is then being performed using CNN and we have used the VGG16 [25] architecture to do this.

VGG16 [25] uses a convolution layer of 3×3 filter with the stride of 1 and used the same padding and max pool layer of 2×2 filter with the stride of 2. This arrangement is followed consistently in the whole architecture. In the end, it adds 2 fully connected layers which are then followed by a softmax layer which gives the output. Figure 4 described this structure in detail. As the name suggests it has 16 weighted layers. It has a very large network with approximately 138 million parameters.

In this paper, we are proposing a two-stream method for the detection of falls in the case of elderly people.

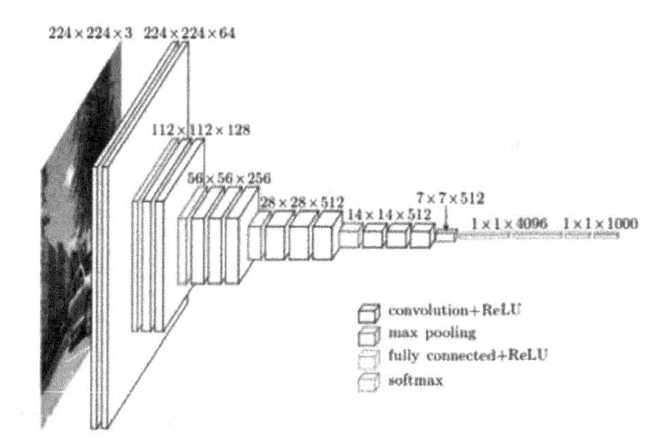

Fig. 4. VGG16 architecture [15]

2 Related Work

Previous literature on human fall detection mostly comprises different methods of detection using wearable sensors. These approaches commonly use the accelerometer sensor and less commonly uses gyroscope sensors to re-confirm their findings. Yazar, Erden, and Cetin et al. [4] worked using one vibration and two passive infrared sensors. They used the PIR sensors to detect any kind of motion and then reconfirmed it using the vibration sensors by detecting human footsteps on the floor. Finally, they analyzed the motion activities from both and combined their results to conclude whether the person has fallen or not. The PIR sensors were used by them to calculate the variance and when the variance surpassed a decided threshold it was used to measure how long there was no motion. Agarwal et al. [5] proposed a method that extracted foreground objects using background subtraction and then used contour-based template matching methods for the detection of sudden change, height-width ratio, and distance between the top and mid-center point of the rectangle to decide human fall. This was proposed for indoor videos only. Adrian et al. [28] presented the application of transfer learning from action recognition to fall detection to create a vision-based fall detection method. They further tested the generality of their algorithm in different conditions of lights and worked on three different datasets. Astriani et al. [7] proposed a threshold-based low-power consumption fall detection system, wherein they used a combination of accelerometer-based technique with four critical characteristics. In their experiment, they used accelerometer and gyroscope sensors to analyze if there has been a fall or not. If the fall is being detected they also used the GPS to communicate with the person and in case of no communication, they used the SMS and/or Voice Call alert for notification. Wang et al. [8] designed a method to detect falls using YOLOv3 based on deep learning. They used the K-means algorithm to optimize the anchored parameters of YOLOv3 [25] and based on that they trained and tested the model on GPU servers. Badgujar et al. [29] proposed a machine learning-based approach over threshold-based approaches. They experimented using Support Vector Machine (SVM) and Decision Tree and concluded that decision tree gives better accuracy and is faster since decision tree can define and classify each attribute to each class precisely and also prediction time of SVM is greater which leads to slower system. Kamble et al. [9] worked on the detection of falls using the YOLO width-height algorithm. An inspired and modified version of which is being used in one stream of our proposed model. They detected the person using state-of-the-art method of YOLO [25] and created a bounding box around them and then calculated the ratio of width and height of that bounding box in equal intervals of times. Whenever the ratio crossed a well-calculated threshold the fall was detected.

3 Proposed Approach

3.1 Dataset

We have used the UR Fall Detection Dataset from Michal Kepski from Interdisciplinary Center of Computational Modelling, University of Rzeszow [10]. They have collected this dataset using 2 Microsoft Kinect cameras. It contains 70 (30 falls and 40 activities of daily life) sequences. This dataset also contains data from accelerometer sensors but we are going to use only the RGB camera data since we are working to eliminate the use of wearable sensors and work only on vision-based approaches (Fig. 5).

Fig. 5 Dataset sample

3.2 Surveillance Video

This is going to serve as the input to our model. It is the continuous video stream being captured by the RGB cameras (Fig. 6).

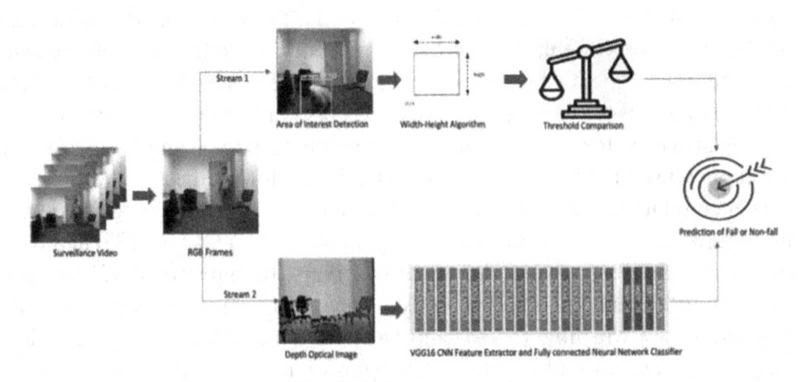

Fig. 6 Proposed model

3.3 RGB Frames

Next, we have divided the input videos into multiple frames of RGB images.

3.4 Stream 1

Area of Interest Detection

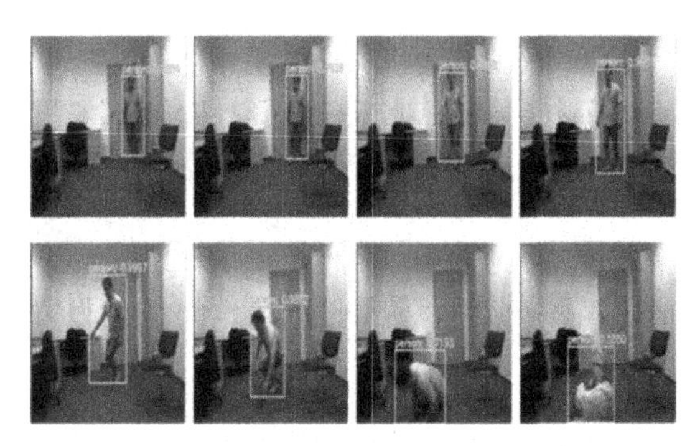

Fig. 7 YOLO person detection

From the RGB frames of the previous step, we are detecting the region of interest, which in this experiment is the person in the video. This is being done because we need to focus only on that area from the complete video since other background things do not interest us in this experiment (Fig. 7).

This detection is done using the approach of YOLO [11]. It is a CNN based approach for object detection in real-time. It uses the approach in which the image is divided into regions by applying a single neural network to the full image. It creates a bounding box around the object with its confidence value.

Width-Height Algorithm. When a person is in a standing or sitting position the height of the person is always more than the width of the person [9]. This algorithm calculates the ratio of objects, in this case, a person's width and height using the dimensions of the bounding box created by the YOLO object detector.

Threshold Comparison. After experimentation on the whole dataset we came across a threshold value of 0.3, and whenever the ratio of the previous algorithm is more than the threshold value the Stream 1 considers it as a fall but our model does not output it directly as fall. But the model then passes this to Stream 2 for secondary validation.

3.5 Stream 2

Once stream 1 detects it as fall, to reconfirm we then pass on that particular RGB frame to the 2nd stream of our model.

Depth Optical Flow Image. The frames from the RGB Frame output are now converted into Depth Optical Images [30, 31]. This is done because optical flow images help in estimating the motion by calculating the motion of image intensities in subsequent frames.

Optical flow is the vector field between two subsequent images which shows how the second image can be formed by moving pixels from the first image frame (Fig. 8).

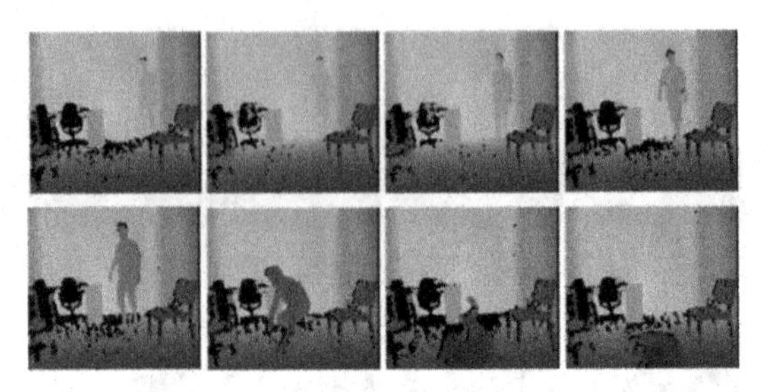

Fig. 8 Depth optical flow image

VGG-16 CNN Feature Extractor and Fully Connected Neural Network Classifier. The Depth optical flow frames are then sent as an input to the VGG16 [25] CNN feature extractor. We are using VGG16 because for this project we have a very large-scale image as the video surveillance needs to be continuously working and this model is considered the best [25].

Prediction of Fall or Non-fall. We used 5-Fold cross validation to check for accuracy. In this method the dataset is shuffled randomly then 5 different groups are made and for each model, we have a different set of testing and training data and finally the summarised model is given out as result.

3.6 Results

We use 5-Fold cross validation to check for accuracy. In this method, the dataset is shuffled randomly then 5 different groups are made and for each model, we have a different set of testing and training data and finally the summarised model is given out as result. For each set, we then build our confusion matrix.

	False Alarm Rate	Missed Detection Rate	Accuracy	Specificity	Sensitivity
Harrou[26]	7.54%	2.00%	-	-	-
Marcos and Azkune[29]	9.00%	0.00%	95.00%	92.00%	100.00%
Zerrouki[27]	-	-	-	89.40%	98.00%
Zerrouki and Houacine[28]	-	-	96.88%	-	-
Proposed in this work	3.28%	0.33%	96.88%	96.72%	99.67%

Fig. 9 Comparative result from pre existing approaches

After 5-cross validation we attained an accuracy of 96.88% while running the model for 20 epochs, the sensitivity attained is 99.67%, specificity at 96.72% (Fig. 10).

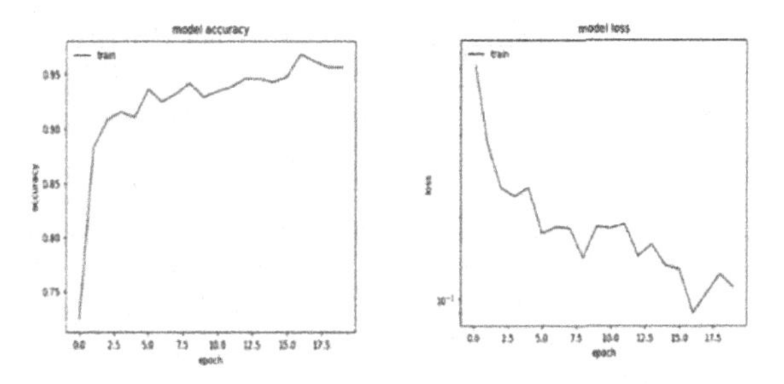

Fig. 10 Model accuracy and loss vs epochs

3.7 Conclusion and Future Work

In this work, we first analyzed all previous work done for the monitoring of elderly people at their homes and came across that most of the approaches consist of the need to wear one or more sensors on them continuously. After analyzing we here propose a model to detect the fall with no use of wearable sensors. We used a vision-based approach and collected data from RGB cameras which are also cost-friendly and easily available.

Based on the data collected from them we built a two-stream network, the first one based on the YOLO detector and width-height algorithm [9], and to reconfirm we built another stream based on the neural network using VGG16 [25] and OF [26] images. We have analyzed that using only the second stream would have increased the time complexity since we would have to use it for every frame. In our proposed model we are only sending our frames to the VGG16 model when the width height model has detected a fall, which in turn improves the time complexity of the model.

In the future, this model could be trained to use in outdoor surveillance methods or could be tested on different datasets for more improved results.

References

1. Liu, J., Lockhart, T.E.: Development and evaluation of a prior-to-impact fall event detection algorithm. IEEE Trans. Biomed. Eng. **61**(7), 2135–2140 (2014). https://doi.org/10.1109/TBME.2014.2315784
2. Chaudhary, S. Deep learning approaches to tackle the challenges of human action recognition in videos. Dissertation (2019)
3. Patil, P.W., Dudhane, A., Chaudhary, S., Murala, S.: Multi-frame based adversarial learning approach for video surveillance. Pattern Recognit. **122**, 108350 (2022)
4. Yazar, A., Erden, F., Cetin, A.: Multi-sensor ambient assisted living system for fall detection (2014)
5. Agrawal, S.C., Tripathi, R.K., Jalal, A.S.: Human-fall detection from an in-door video surveillance. In: 2017 8th International Conference on Computing, Communication and Networking Technologies (ICCCNT), Delhi, pp. 1–5 (2017). https://doi.org/10.1109/ICCCNT.2017.8203923.

6. Chaudhary, S., Dudhane, A., Patil, P., Murala, S.: Pose guided dynamic image network for human action recognition in person centric videos. In: 2019 16th IEEE International Conference on Advanced Video and Signal Based Surveillance (AVSS), pp. 1–8 (2019). https://doi.org/10.1109/AVSS.2019.8909835.
7. Astriani, M.S., Bahana, R., Kurniawan, A., Yi, L.H.: Threshold-based low power consumption human fall detection for health care and monitoring system. In: 2020 International Conference on Information Management and Technology (ICIMTech), Bandung, Indonesia, pp. 853–857 (2020). https://doi.org/10.1109/ICIMTech50083.2020.9211233.
8. Wang, X., Jia, K.: Human fall detection algorithm based on YOLOv3. In: 2020 IEEE 5th International Conference on Image, Vision and Computing (ICIVC), Beijing, China, pp. 50–54 (2020). https://doi.org/10.1109/ICIVC50857.2020.9177447.
9. Kamble, K.P., Sontakke, S.S., Donadkar, H., Poshattiwar, R., Ananth, A.: Fall alert: a novel approach to detect fall using base as a YOLO object detection. In: Hassanien, A., Bhatnagar, R., Darwish, A. (eds.) Advanced Machine Learning Technologies and Applications, AMLTA 2020. Advances in Intelligent Systems and Computing, vol. 1141, pp. 15–24. Springer, Singapore (2021). https://doi.org/10.1007/978-981-15-3383-9_2
10. UR Fall Detection Dataset (n.d.). http://fe-nix.univ.rzeszow.pl/~mkepski/ds/uf.html. 28 Sept 2021
11. Redmon, J., Divvala, S., Girshick, R., Farhadi, A.: You only look once: unified, real-time object detection. In: IEEE Conference on Computer Vision and Pattern Recognition (CVPR), pp. 779–788 (2016). https://doi.org/10.1109/CVPR.2016.91
12. Patil, P.W., Biradar, K.M., Dudhane, A., Murala, S.: An end-to-end edge aggregation network for moving object segmentation. In: Proceedings of the IEEE/CVF Conference on Computer Vision and Pattern Recognition, pp. 8149–8158 (2020)
13. Chaudhary, S., Murala, S.: Deep network for human action recognition using Weber motion. Neurocomputing **367**, 207–216 (2019)
14. Center for Disease Control and Prevention. www.cdc.gov/HomeandRecre-ationalSafety
15. a/l Kanawathi, J., Mokri, S.S., Ibrahim, N., Hussain, A., Mustafa, M.M.: Motion detection using Horn Schunck algorithm and implementation. In: 2009 International Conference on Electrical Engineering and Informatics, pp. 83–87 (2009). https://doi.org/10.1109/ICEEI.2009.5254812.
16. Chaudhary, S., Murala, S.: Depth-based end-to-end deep network for human action recognition. IET Comput. Vis. **13**(1), 15–22 (2019)
17. Ahad, M., Rahman, A., Jie, T., Kim, H., Ishikawa, S.: Motion history image: Its variants and applications. Mach. Vis. Appl. **23**, 255–281 (2010). https://doi.org/10.1007/s00138-010-0298-4
18. He, K., Zhang, X., Ren, S., Sun, J.: Spatial pyramid pooling in deep convolutional networks for visual recognition. In: Fleet, D., Pajdla, T., Schiele, B., Tuytelaars, T. (eds.) Computer Vision – ECCV 2014, ECCV 2014. LNCS, vol. 8691, pp. 346–361. Springer, Cham (2014). https://doi.org/10.1007/978-3-319-10578-9_23
19. Girshick, R., Donahue, J., Darrell, T., Malik, J.: Region-based convolutional networks for accurate object detection and segmentation. IEEE Trans. Pattern Anal. Mach. Intell. **38**(1), 142–158 (2016). https://doi.org/10.1109/TPAMI.2015.2437384
20. Girshick, R.: Fast R-CNN (2015)
21. Ren, S., He, K., Girshick, R., Sun, J.: Faster R-CNN: towards real-time object detection with region proposal networks (2016)
22. Dalal, N., Triggs, B.: Histograms of oriented gradients for human detection. In: 2005 IEEE Computer Society Conference on Computer Vision and Pattern Recognition (CVPR 2005), vol. 1, pp. 886–893 (2005). https://doi.org/10.1109/CVPR.2005.177
23. Dai, J., Li, Y., He, K., Sun, J.: R-FCN: object detection via region-based fully convolutional networks (2016).

24. Liu, W., et al.: SSD: single shot multibox detector. In: Leibe, B., Matas, J., Sebe, N., Welling, M. (eds.) Computer Vision, ECCV 2016. LNCS, vol. 9905, pp. 21–37. Springer, Cham (2016). https://doi.org/10.1007/978-3-319-46448-0_2
25. Simonyan, K., Zisserman, A.: Very deep convolutional networks for large-scale image recognition (2015)
26. Horn, B., Schunck, B.: Determining optical flow. Artif. Intell. **17**, 185–203 (1981). https://doi.org/10.1016/0004-3702(81)90024-2
27. Chaudhary, S., Murala, S.: TSNet: deep network for human action recognition in hazy videos. In: 2018 IEEE International Conference on Systems, Man, and Cybernetics (SMC), pp. 3981–3986 2018. https://doi.org/10.1109/SMC.2018.00675
28. Núñez-Marcos, A., Azkune, G., Arganda-Carreras, I.: Vision-based fall detection with convolutional neural networks. Wirel. Commun. Mob. Comput. 1–16 (2017). https://doi.org/10.1155/2017/9474806
29. Badgujar, S., Pillai, A.S.: Fall detection for elderly people using machine learning. In: 2020 11th International Conference on Computing, Communication and Networking Technologies (ICCCNT), Kharagpur, India, pp. 1–4 (2020). https://doi.org/10.1109/ICCCNT49239.2020.9225494
30. Hambarde, P., Dudhane, A., Patil, P.W., Murala, S., Dhall, A.: Depth estimation from single image and semantic prior. In: 2020 IEEE International Conference on Image Processing (ICIP), pp. 1441–1445. IEEE (2020)
31. Hambarde, P., Dudhane, A., Murala, S.: Single image depth estimation using deep adversarial training. In: 2019 IEEE International Conference on Image Processing (ICIP), pp. 989–993. IEEE (2019)
32. Patil, P.W., Dudhane, A., Kulkarni, A., Murala, S., Gonde, A.B., Gupta, S.: An unified recurrent video object segmentation framework for various surveillance environments. IEEE Trans. Image Process. **30**, 7889–7902 (2021)

OGGN: A Novel Generalized Oracle Guided Generative Architecture for Modelling Inverse Function of Artificial Neural Networks

V. Mohammad Aaftab[1(✉)] and Mansi Sharma[2]

[1] Department of Mechanical Engineering, Indian Institute of Technology Madras, Chennai, India
aaftaabv@gmail.com

[2] Department of Electrical Engineering, Indian Institute of Technology Madras, Chennai, India
mansisharma@ee.iitm.ac.in

Abstract. This paper presents a novel Generative Neural Network Architecture for modelling the inverse function of an Artificial Neural Network (ANN) either completely or partially. Modelling the complete inverse function of an ANN involves generating the values of all features that corresponds to a desired output. On the other hand, partially modelling the inverse function means generating the values of a subset of features and fixing the remaining feature values. The feature set generation is a critical step for artificial neural networks, useful in several practical applications in engineering and science. The proposed Oracle Guided Generative Neural Network, dubbed as OGGN, is flexible to handle a variety of feature generation problems. In general, an ANN is able to predict the target values based on given feature vectors. The OGGN architecture enables to generate feature vectors given the predetermined target values of an ANN. When generated feature vectors are fed to the forward ANN, the target value predicted by ANN will be close to the predetermined target values. Therefore, the OGGN architecture is able to map, inverse function of the function represented by forward ANN. Besides, there is another important contribution of this work. This paper also introduces a new class of functions, defined as constraint functions. The constraint functions enable a neural network to investigate a given local space for a longer period of time. Thus, enabling to find a local optimum of the loss function apart from just being able to find the global optimum. OGGN can also be adapted to solve a system of polynomial equations in many variables. The experiments on synthetic datasets validate the effectiveness of OGGN on various use cases. Our code is available at https://github.com/mohammadaaftabv/OGGN.

Keywords: Inverse of neural networks · Solving polynomial equations in many variables · Oracle Guided Generative Neural Networks

B. Raman et al. (Eds.): CVIP 2021, CCIS 1568, pp. 460–471, 2022.
https://doi.org/10.1007/978-3-031-11349-9_40

1 Introduction

Neural networks are known to be great function approximators for a given data, *i.e.*, for a given dataset with features and targets, neural networks are very useful in modelling the function which maps features to targets [1,2]. The inverse function of a dataset, *i.e.*, the function that maps the targets to corresponding features is also very important. Neural networks have been used to solve the inverse problem, especially in design and manufacturing applications [3–8], where the ability to predict features corresponding to a given target value is vital. The modelling of the inverse function can be accomplished by modelling the inverse function of a neural network that maps features to targets. The problem of finding the inverse of a neural network is also crucial in various engineering problems, especially the optimisation ones [9–11]. Inverse problems in imaging and computer vision are being solved with the help of deep convolution neural networks [12–15]. Generative Adversarial Networks[16] have also been used extensively in solving inverse problems [17–19].

Our paper introduces a novel neural network architecture called Oracle Guided Generative Neural Networks (OGGN). The proposed OGGN model the inverse function (*i.e.*, the function from targets to features) using the forward function (*i.e.*, the function from features to targets). We define oracle as the function mapping features to targets. Theoretically, an oracle can either be a mathematical function or a neural network. The generative neural network is responsible for predicting features that correspond to a desired target, *i.e.*, when the predicted features are modelled using oracle, the output must be close to desired target value. The generator's loss function is dependent on the oracle. Generator tries to minimize loss using gradient descent. Thus, the proposed architecture is called Oracle Guided Generative Neural Network.

The primary contribution of the proposed architecture is solving the problem of modelling the inverse of a given function or neural network. OGGN architecture can generate feature vector based on a given oracle and a given target vector value, such that, when the generated feature vector is fed into the oracle, it outputs the target vector close to the predetermined target value. The proposed OGGN architecture can also find feature vectors subject to constraints like fixed range for a few or all of the feature values. In addition, OGGN can also generate

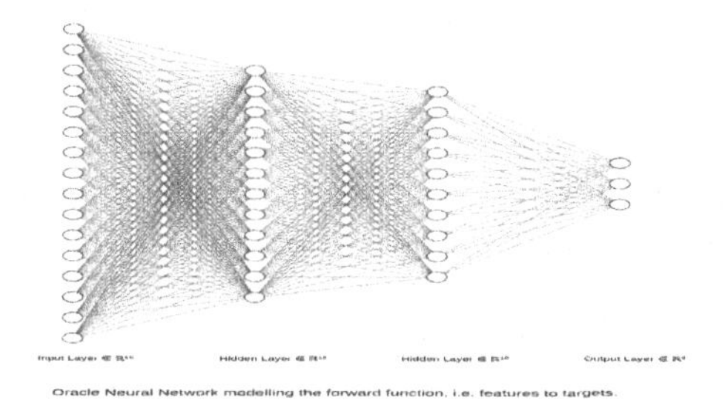

Fig. 1. Oracle neural network: maps features to targets

feature vectors corresponding to a required target with one or more feature values fixed as constants. It can also be modified to tackle the problem of solving a system of polynomial equations in many variables. All use cases of OGGN are analyzed on synthetic datasets in various experiments.

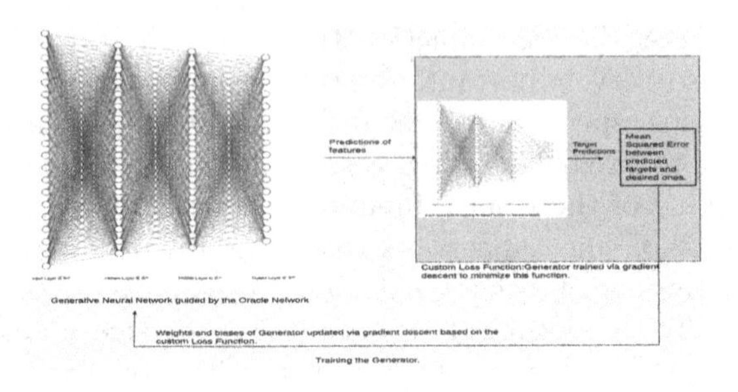

Fig. 2. Generative network training methodology

2 Related Work

The problem of developing generative neural networks for finding features has been explored well, especially in the recent years. In this section, we explore some latest methods.

Chen et al. [20] describes a general-purpose inverse design approach using generative inverse design networks along with active learning. They introduce a framework named Generative Inverse Design Networks (GIDNs). Their framework has two DNNs namely, the predictor and the generator. The weights and biases of the predictor are learned using gradient decent and backpropagation, while training on a given dataset. In the designer, the weights and biases are adapted from the predictor and set as constants. Initial Gaussian distribution features are fed into the designer as inputs and the optimized design features are generated as outputs. The designer inverse model optimizes the initial inputs or designs based on an objective function via gradient descent with weights and biases of all layers kept as constants and only the input layer values are learned. Active learning takes place in the feedback loop, where the optimized designs or features are verified and added to the dataset. We differ from their approach, in using an additional generator neural network to generate features from random data whose size is tunable according to the complexity of the inverse neural network.

Yang et al. [21] describe neural network inversion in adversarial settings. They present an approach to generate the training data that were used in training a forward neural network using only black box access to it. Our approach also uses a black box access to the forward neural network, but the use cases, training procedure, and architecture are different. While their approach mainly focuses on generating the exact training data (features and targets) that were used at the time of training the forward neural network, our approach shows how to generate features corresponding to any desired target.

Tahersima et al. [22] present a method to design integrated photonic power splitters using deep neural networks to model the inverse relation. Their approach deals along the same lines as ours, i.e., using neural networks to find feature vector that corresponds to a given target value. However, their network is different from our approach as it is built for a very specific application of designing integrated photonic power splitters. We proposed an architecture that can be generalized and used for any data.

Zhang et al. [23] describe a multi-valued neural network for inverse modeling and application to microwave filters. The paper attempts to find the feature vector corresponding to a required target value. The focus is mainly on the existence of non-unique feature vectors corresponding to a given target value. Their approach overcomes the above mentioned non-uniqueness problem by generating multiple sets of feature vectors at a time for a given target vector.

Our proposed methodology uses a separate generator neural network to convert random data into features using oracle neural network. Hence, our approach is different from the previous works.

3 Proposed Oracle Guided Generative Neural Network

We introduce a novel neural network architecture known as OGGN, for modelling the inverse of a given neural network. Modelling the inverse of a neural network involves finding the values of features that correspond to a desired target value. The most straight forward way to accomplish this task is to train a new neural network with targets as inputs and features as outputs. The data required for training the new inverse neural network can be obtained by swapping the input and output data used to train the forward network. The straight forward method of modelling the inverse function can often suffer from the multi-valued relationship between inputs and outputs. The multi-valued relationship refers to the fact that different feature vectors can correspond to a same target vector. It means that same input values to the inverse network (targets) can have different outputs (features). This causes significant problems while training the inverse neural network directly from the swapped data. Humayun et al. [24] discusses this problem in detail.

The proposed OGGN architecture provides a better alternative method for mapping the inverse of a neural network. Oracle network is the name of the forward neural network whose inverse function we are trying to model (Fig. 1). The workflow of our architecture is depicted in (Fig. 2). A generator neural network takes random data as input (the dimensionality of the random data is a hyper-parameter and can be increased or decreased to fit the complexity of the inverse function) and outputs the predicted feature vector value. The training procedure of OGGN is summarized in Algorithm 1.

Once predicted features from generator are obtained, they are modelled into target vectors using the oracle neural network. The loss value of the generator neural network is the mean squared error between the predicted target vectors and desired target vectors. The training of generator neural network is done

Algorithm 1: OGGN Methodology

```
Result: features corresponding to a desired target value
random data (size tunable according to the complexity of system of equations);
Oracle Neural Network (maps the features to targets);
e is the acceptable error between predicted targets and desired targets.;
while loss > e do
    features = generator(random data);
    predicted target = oracle(features) ;
    loss = (predicted target−desired target)^2.0;
    optimizer adjusts weights and biases of generator neural network based on loss;
end
```

via gradient descent to minimize the obtained loss value. Minimizing this loss means that the predicted features are close to the features corresponding to the desired target vector. Thus, as the generator trains, it predicts features closer and closer to the required features (*i.e.*, those that correspond the desired target vector). By training the generator, the values of input features corresponding to a desired value of target vector have been derived. It is possible to generate multiple such input feature vectors at a time by passing many rows into the generator neural network and optimising all of them at the same time. Thus, multiple input feature vectors that all correspond to a same or similar value of a output vector can be found.

OGGN can solve the following problems:

- It is possible to model the inverse function to a given neural network by trying to produce the values of feature vectors that correspond to a required output in the forward neural network.
- OGGN can also model only a part of the inverse function of a forward neural network, *i.e.*, it is possible to fix one or more input feature values and train the generator to produce only the remaining feature values such that the whole feature vector corresponds to a desired target value.
- OGGN can constraint the values of one or more generated feature values. This is accomplished by using a family of functions introduced in our paper called *constraint functions.*
- OGGN can be adapted to solve a system of polynomial equations.

All these use cases are validated using synthetic datasets.

3.1 Constraint Functions

Activation functions [25] in neural networks such as ReLu, Tanh serve various purposes ranging from introducing non-linearity to the modelling process or avoiding gradient vanishing.

We propose one such family of functions named, *constraint functions*, which serve the purpose of constraining the search/exploration space in a neural network. Constraint functions, like activation functions are applied to each output from a neuron. Constraint functions are conditional and they help the neural network minimize the loss while still exploring in a prescribed space. Constraint functions help in maintaining the outputs in a certain range. One such example function is described in the experiments section below.

4 Experiments and Results

4.1 Dataset Preparation

To test the performance of the proposed model, we generated a synthetic dataset. We consider an arbitrary polynomial exponential function in four variables $x_i, i = 1, ..., 4$. The function is defined as:

$$y = 9x_1^{0.87} + 8.97x_2^{0.02} + 0.876x_3^{0.12} + 2.9876x_4^{0.987} \quad (1)$$

The training data has been prepared by randomly choosing values of $x_i, i = 1, ..., 4$ variables in (1). The target variable y is calculated from polynomial function by substituting the randomly selected x values. We consider 10,000 random sample points of this function (1) for training our model. For testing, we generated a new set of 1000 random samples of the function (1).

4.2 Experimental Settings

The generated training data is used to train the proposed oracle neural network, which maps the forward function between features and target outputs, *i.e.*, the function mapping from x values to y value. To validate the proposed idea, all experiments have been carried out using the same oracle network. Experiments are designed to validate all use cases of the proposed OGGN architecture. The use cases include:

- Generating values of all features corresponding to a desired target value.
- Generating values of some features corresponding to a desired target value, while other feature values are set as constants.
- Generating feature values corresponding to a desired target value, such that, the feature values are restricted within a customizable range.
- Solving a system of polynomial equations.

Each use case requires a different generator neural network architecture.

4.3 Use Cases of the Proposed OGGN Architecture

Here, different use cases of proposed OGGN architecture are described and experiments have been performed to validate the proposed model.

Modelling the Inverse Function. The main purpose of this experiment is to generate values of the features corresponding to a desired output. For example, given a target value y, our objective is to generate the values of $x_i, i = 1, ..., 4$, such that, when they are substituted in the reference function (1), we get the output very near to the previously given target value y. In this experiment, we aim to find feature vectors $x_i, i = 1, ..., 4$ corresponding to a required output $y = 1900$. So, the generated $x_i, i = 1, ..., 4$ values from this experiment when modelled using the polynomial function (1), the output is very close to 1900.

Algorithm 2: Model inverse function using OGGN

```
Result: features corresponding to a desired target value
random data (size tunable according to the complexity of system of equations);
Oracle Neural Network (maps the features to targets);
while loss > e do
    x1, x2, x3, x4 = generator(random data);
    predicted target = oracle(x1, x2, x3, x4) ;
    loss = (predicted target−desired target)^2.0;
    optimizer adjusts weights and biases of generator neural network based on loss;
end
```

Generalized steps of this use case is summarized in Algorithm 1. Pseudo code of this experiment is given in Algorithm 2. It is not guaranteed to get the exact desired value always, while modelling the generated features using the forward neural network. It may be either because of the error in modelling the polynomial function by the Oracle network, or due to the polynomial function itself never reach the desired value y for any value of $x_i, i = 1, ..., 4$. Hence, the generator network only tries to minimize the error between the desired output and the predicted target.

After training the generator network for 2000 epochs, we get the feature values $x_1 = 224.6277$, $x_2 = 0.000$, $x_3 = 283.2135$, $x_4 = 328.2939$. The target value corresponding to the predicted features is 1911.4099. It is clear that this predicted target, 1911.4099 is very close to the desired target value of 1900.00. With this experiment, we conclude that we can compute feature vectors that correspond to an output of our choice. Thus, this experiment demonstrates the feasibility of the proposed model and its principle. It verifies the proof of concept or theoretical underpinnings that the proposed OGGN has potential to solve a number of inverse problems. The feature vectors can be further optimized so that the target can be achieved closer to 1900. We can explore different approaches like using more data for training of the oracle network or using extended architecture with more layers for oracle and generator neural networks. Training the networks for a larger number of epochs could also improve the accuracy.

Finding the Feature Vector with Some Features in a Desired Range and Some Features Specified to Be Constant. The objective of this experiment is to find the optimal values of feature vectors $x_i, i \neq j$, such that they are in a desired range and one or more variables x_j can be specified as a constant. For instance, the value of x_4 is set to 100.0 in the polynomial function (1). The objective is to compute the values of x_1, x_2, x_3 using proposed model, such that, the target output should be close to the desired value $y = 788.0$ and the computed values of x_1, x_2, x_3 are within the desired range of 1 to 100. The generator predicts the values of x_1, x_2, x_3. The training methodology used in this experiment is summarized in Algorithm 3.

We introduce a family of functions dubbed as constraint functions. These constraint functions decide the range of the output of a neural network. Thus, the feature vectors predicted by generator network are within the predetermined range specified by the constraint functions. The constraint functions are defined as,

Algorithm 3: Model constrained subset of inverse function using OGGN

```
Result: some features constrained to a given space corresponding to a desired target value while other
        features are fixed
random data (size tunable according to the complexity of system of equations);
Oracle neural network (maps the features to targets);
x4=100.00 (this is constant);
constraints on range of features is given;
constrainedgenerator is the neural network with constraint functions applied to its output layer;
while loss > e do
    x1, x2, x3 = constrainedgenerator(random data);
    predicted target = oracle(x1, x2, x3, x4) ;
    loss = (predicted target-desired target)^2.0;
    optimizer adjusts weights and biases of generator neural network based on loss;
end
```

$$\begin{cases} x/c_1 \;\; UpperBound \leq x \\ x * c_2 \; x \leq LowerBound \end{cases} \tag{2}$$

where, c_1 and c_2 are constants. The values of c_1 and c_2 are decided after checking the range of outputs from the generator network, trained without using constraint functions. If the outputs of unconstrained generator network are higher than the required upper bound, it is necessary to divide those outputs with an appropriate number in order to bring outputs into the desired range. This appropriate value is set as constant c_1. Similarly, if the outputs of unconstrained generator network are lower than the required lower bound, it is necessary to multiply those outputs with an appropriate number in order to bring outputs into the desired range. This suitable value is set as constant c_2.

For example, in this experiment, the generator before using constraint functions produce the outputs in the range of 200 – 500. To let the feature values lie between 1 and 100, initially c_1 is set to 2. To further increase the exploration time of generator in the range of 1 and 100, experimentally, the value of c_1 is set to 20.0. Similarly, the value of c_2 has been chosen as 10.0. Thus, the constraint function is defined as

$$\begin{cases} x/20 \;\; 100 \leq x \\ x * 10 \;\;\; x \leq 1 \end{cases} \tag{3}$$

Like an activation function, the constraint function can be applied to neurons in any layer. However in this experiment, constraint function has been applied on neurons in the output layer only. The constrained generator network predicts x_1, x_2, x_3 in the required range. The value of x_4 is set as constant, $x_4 = 100$. The oracle network models the generated features x_1, x_2, x_3 and x_4 and predicts the corresponding target. Final loss is calculated as a mean squared error between the predicted target and the required target. The generator neural network then tries to minimize the final loss via gradient descent.

The constraint function enables a neural network to search in a prescribed space for a longer period of time. It helps neural network explore the prescribed space thoroughly. If the global minimum of the loss function lies outside the prescribed range, even a constrained neural network will eventually optimize itself to produce outputs that correspond to global minima of loss function. We can arrive at the local optima by stopping training of the constrained neural network once the outputs go beyond the prescribed range.

We obtained output $y = 777.5806$ with the predicted feature vector $x_1 = 100.0423$, $x_2 = 0.0000$, $x_3 = 53.7249$, $x_4 = 100.00$. The values of x_1, x_2, x_3 were generated by the generator after training for 200 epochs. The theoretical maximum value achievable for the function (1) is 787.34, if all the features x_1, x_2, x_3 and x_4 are constrained to lie within the range of 1 and 100. It is apparent that the predicted value is pretty close to the theoretical maximum value. This experiment is designed to show that proposed OGGN architecture can explore different projections of the inverse function in a subspace. Thus, proposed OGGN architecture can impose constraints on one or more features and find the local optimum of a function.

Solving a Simultaneous System of Polynomial Equations. In this experiment, we describe another use case of proposed OGGN architecture. OGGN can be used to solve a system of polynomial equations. It is highly versatile and adaptable. Solving a system of polynomial equations involve finding the values of the variables that satisfy all equations. We consider an arbitrary system of polynomial equations to demonstrate the usefulness of the proposed OGGN architecture.

$$\begin{aligned} 9x^2 + 8.97y^{7.8} + 0.876z - 32.0 = 0, \\ 12x^3 + 9.97y^8 + 10.876z^3 - 43.0 = 0 \end{aligned} \tag{4}$$

To solve (4), we consider the following functions

$$\begin{aligned} f(x, y, z) &= 9x^2 + 8.97y^{7.8} + 0.876z - 32.0, \\ g(x, y, z) &= 12x^3 + 9.97y^8 + 10.876z^3 - 43.0 \end{aligned} \tag{5}$$

Solving the simultaneous system of equations involve finding the values of x, y, z such that both $f(x, y, z)$ and $g(x, y, z)$ would come close to zero. To this end, we consider the oracle function for generator neural network, given as.

$$oracleoutput = (f(x, y, z)^2 + g(x, y, z)^2)^{0.5} \tag{6}$$

The main purpose of oracle is to convert the features into targets. In most real life datasets, a mathematical equation is not available to model the features into targets. The only way to model features into targets in real life datasets is to use a neural network. In the previous experiments, a neural network was used as oracle to demonstrate the ability of the OGGN to work with practical datasets. However, in this example, the objective is to solve a system of equations. A mathematical equation converting variables into targets is readily available. Hence, here the oracle is not a neural network, rather it is a function (6).

The value of the oracle function (6) will be close to zero if and only if both functions $f(x, y, z)$ and $g(x, y, z)$ are close to zero. Hence, the solutions of system of Eq. (4) are the values of x, y, z such that upon modelling them using the oracle function (6), the output is close to zero. To this end, we task the generator neural network to find the values of x, y, z such that the oracle function value (6) will be close to zero. The training process is summarized in Algorithm 4.

Algorithm 4: Solving System of Eq. (4): the OGGN way

```
Result: solutions of system of equations, x, y, z
random data (size tunable according to the complexity of system of equations);
while loss > e do
    x, y, z = generator(random data);
    predicted target = oracle(x, y, z) (6);
    loss = (predicted target−0.00)^2.0;
    optimizer adjusts weights and biases of generator neural network based on loss;
end
```

The error in Algorithm 4 is a small number close to zero. The variables x, y, z predicted by the generator are modelled using the oracle (6). The corresponding loss is calculated as a mean square error between predicted oracle function value and zero. Generator neural network now tries to minimize the loss via gradient descent and back propagation.

After training generator neural network, it predicts values of x, y, z such that when modelled using oracle (6), we get an output of 0.0589. The values generated are $x = 1.2279$, $y = 1.0952$, $z = 0.2624$. For these variables, $f(x, y, z) = 0.03179$ and $g(x, y, z) = 0.04953$.

If the given system of equations has a solution, we can be certain that the generated values $x = 1.2279$, $y = 1.0952$, $z = 0.2624$ are close to the actual solution. If the system does not have an exact solution, we can conclude that the obtained values of variables x, y, z take both $f(x, y, z)$ and $g(x, y, z)$ as close to zero as possible. Hence, we conclude in this example that OGGN is capable of solving a system of polynomial equations.

5 Conclusions and Future Work

This paper introduces a novel Oracle Guided Generative Neural Networks (OGGN) to solve the problem of finding the inverse of a given function either fully or partially. The problem of predicting the features corresponding to a desired target is very important and useful in several design and manufacturing fields [3–8]. It is also very useful in various engineering problems, especially optimization ones [9–11]. OGGN can also generate features subject to constraints, *i.e.*, it can generate features that lie within a given range. Our paper also introduces a class of functions known as *Constraint Functions.* Constraint functions enable a given neural network to explore a given space thoroughly and arrive at a local optima. The proposed concept of OGGN architecture coupled with constraint functions have huge potential in several practical applications in design and manufacturing industries. OGGNs are flexible and can be adapted to solve a large variety of research problems.

This paper has described a way to generate feature vectors that are able to model a particular fixed target value using OGGN. In the future, we can further extend the proposed idea for generating feature vectors corresponding to different target values. This aids in the process of extending a given dataset by creating new synthetic data. The synthetic data can build a better oracle neural network. This improved oracle can guide the generator more effectively and can lead to the creation of better synthetic data. Thus, OGGNs can potentially help in active learning of both forward mapping and inverse mapping neural networks for a given dataset. Synthetic dataset generation using the described architecture could helpful in classification tasks [26]. Especially, if there is a lot of class imbalance [27–29]. OGGNs can create synthetic data corresponding to the lesser represented class and potentially overcome the problem of class data imbalance.

References

1. Hornik, K., Stinchcombe, M., White, H.: Multilayer feedforward networks are universal approximators. Neural Netw. **2**(5), 359–366 (1989)
2. Wu, S., Er, M.J.: Dynamic fuzzy neural networks-a novel approach to function approximation. IEEE Trans. Syst. Man Cybern. Part B (cybern.) **30**(2), 358–364 (2000)
3. Liu, D., Tan, Y., Khoram, E., Yu, Z.: Training deep neural networks for the inverse design of nanophotonic structures. ACS Photonics **5**(4), 1365–1369 (2018)
4. Xu, X., Sun, C., Li, Y., Zhao, J., Han, J., Huang, W.: An improved tandem neural network for the inverse design of nanophotonics devices. Optics Commun. **481**, 126513 (2021)
5. Qu, Y., et al.: Inverse design of an integrated-nanophotonics optical neural network. Sci. Bull. **65**(14), 1177–1183 (2020)
6. Sekar, V., Zhang, M., Shu, C., Khoo, B.C.: Inverse design of airfoil using a deep convolutional neural network. AIAA J. **57**(3), 993–1003 (2019)
7. So, S., Mun, J., Rho, J.: Simultaneous inverse design of materials and structures via deep learning: demonstration of dipole resonance engineering using core-shell nanoparticles. ACS Appli. Mater. Interf. **11**(27), 24 264–24 268 (2019)
8. Kim, K., et al.: Deep-learning-based inverse design model for intelligent discovery of organic molecules. NPJ Comput. Mater. **4**(1), 1–7 (2018)
9. Rajesh, R., Preethi, R., Mehata, P., Pandian, B.J.: Artificial neural network based inverse model control of a nonlinear process. In: 2015 International Conference on Computer, Communication and Control (IC4), pp. 1–6. IEEE (2015)
10. Hattab, N., Motelica-Heino, M.: Application of an inverse neural network model for the identification of optimal amendment to reduce copper toxicity in phytoremediated contaminated soils. J. Geochem. Explor. **136**, 14–23 (2014)
11. Krasnopolsky, V.M.: Neural network applications to solve forward and inverse problems in atmospheric and oceanic satellite remote sensing. In: Artificial Intelligence Methods in the Environmental Sciences, pp. 191–205. Springer (2009). https://doi.org/10.1007/978-1-4020-9119-3_9
12. McCann, M.T., Jin, K.H., Unser, M.: A review of convolutional neural networks for inverse problems in imaging (2017). arXiv preprint arXiv:1710.04011
13. Ongie, G., Jalal, A., Metzler, C.A., Baraniuk, R.G., Dimakis, A.G., Willett, R.: Deep learning techniques for inverse problems in imaging. IEEE J. Sel. Areas Inf. Theory **1**(1), 39–56 (2020)
14. Li, H., Schwab, J., Antholzer, S., Haltmeier, M.: Nett: Solving inverse problems with deep neural networks. Inverse Prob. **36**(6), 065005 (2020)
15. Wang, F., Eljarrat, A., Müller, J., Henninen, T.R., Erni, R., Koch, C.T.: Multiresolution convolutional neural networks for inverse problems. Sci. Rep. **10**(1), 1–11 (2020)

16. Goodfellow, I.J.: Generative adversarial networks (2014). arXiv preprint arXiv:1406.2661
17. Lenninger, M.: Generative adversarial networks as integrated forward and inverse model for motor control (2017)
18. Creswell, A., Bharath, A.A.: Inverting the generator of a generative adversarial network. IEEE Trans. Neural Netw. Learn. Syst. **30**(7), 1967–1974 (2018)
19. Asim, M., Daniels, M., Leong, U., Ahmed, A., Hand, P.: Invertible generative models for inverse problems: mitigating representation error and dataset bias. In: International Conference on Machine Learning, PMLR, pp. 399–409 (2020)
20. Chen, C.-T., Gu, G.X.: Generative deep neural networks for inverse materials design using backpropagation and active learning. Adv. Sci. **7**(5), 1902607 (2020)
21. Yang, Z., Chang, E.-C., Liang, Z.: Adversarial neural network inversion via auxiliary knowledge alignment (2019) arXiv preprint arXiv:1902.08552
22. Tahersima, M.H., et al.: Deep neural network inverse design of integrated photonic power splitters. Sci. Rep. **9**(1), 1–9 (2019)
23. Zhang, C., Jin, J., Na, W., Zhang, Q.-J., Yu, M.: Multivalued neural network inverse modeling and applications to microwave filters. IEEE Trans. Microw. Theory Tech. **66**(8), 3781–3797 (2018)
24. Kabir, H., Wang, Y., Yu, M., Zhang, Q.-J.: Neural network inverse modeling and applications to microwave filter design. IEEE Trans. Microw. Theory Tech. **56**(4), 867–879 (2008)
25. Nwankpa, C., Ijomah, W., Gachagan, A., Marshall, S.: Activation functions: Comparison of trends in practice and research for deep learning (2018). arXiv preprint arXiv:1811.03378
26. Fawaz, H.I., Forestier, G., Weber, J., Idoumghar, L., Muller, P.-A.: Deep learning for time series classification: a review. Data Min. Knowl. Disc. **33**(4), 917–963 (2019)
27. Japkowicz, N., Stephen, S.: The class imbalance problem: a systematic study. Intell. Data Anal. **6**(5), 429–449 (2002)
28. Liu, X.-Y., Wu, J., Zhou, Z.-H.: Exploratory undersampling for class-imbalance learning. IEEE Trans. Syst. Man Cybern. Part B (Cybern.) **39**(2), 539–550 (2008)
29. Johnson, J.M., Khoshgoftaar, T.M.: Survey on deep learning with class imbalance. J. Big Data **6**(1), 1–54 (2019)

Deep Learning Based DR Medical Image Classification

Preeti Deshmukh[1](✉) and Arun N. Gaikwad[2]

[1] Vidya Vikas Pratisthan Institute of Engineering and Technology, Solapur, Maharashtra, India
preetikadam110@gmail.com

[2] Zeal College of Engineering and Research Center, Pune, Maharashtra, India

Abstract. Diabetic retinopathy (DR) is a prevailing disease that causes blindness among diabetic patients. The timely intervention with the regular fundus photography screening is the efficient way to cope with this disease. Screening of a large number of diabetic patients urges to the computer-aided and fully automatic DR diagnosis. Deep neural networks are gaining more attention due to their effectiveness in various tasks. The diagnosis of the can be made automatic and accurate suggestions can be provided to DR patients. The classification of DR images is challenging and important step. Therefore, in this paper, we have proposed a learning-based DR image reconstruction followed by a classification approach for DR image classification. Initially, the learning-based image reconstruction approach is proposed with a multi-encoder decoder network and residual inception block to delve into the features of input medical image and convert it into a set of abstract features followed by the reconstruction of the input medical image. The features from encoder are the abstract version of actual image representation which can be used for robust reconstruction for the input image. Therefore, these encoded features are used for DR image classification. The results analysis of the proposed framework with existing SOTA feature extraction algorithms is conducted on the MESSIDOR database for DR image classification. From the results' analysis, it is evident that the proposed reconstruction-based classification framework outperforms the existing SOTA feature extraction algorithms.

Keywords: Image reconstruction · Residual inception block · Diabetic retinopathy · Classification

1 Introduction

Traditionally, the determination of DR grade is based on the combined evaluation of different structural features observed in the color fundus images, for instance, existence of exudates, microaneurysms, hemorrhages, and neovascularization [25]. Thus, from the last two decades, image classification is accomplished

B. Raman et al. (Eds.): CVIP 2021, CCIS 1568, pp. 472–482, 2022.
https://doi.org/10.1007/978-3-031-11349-9_41

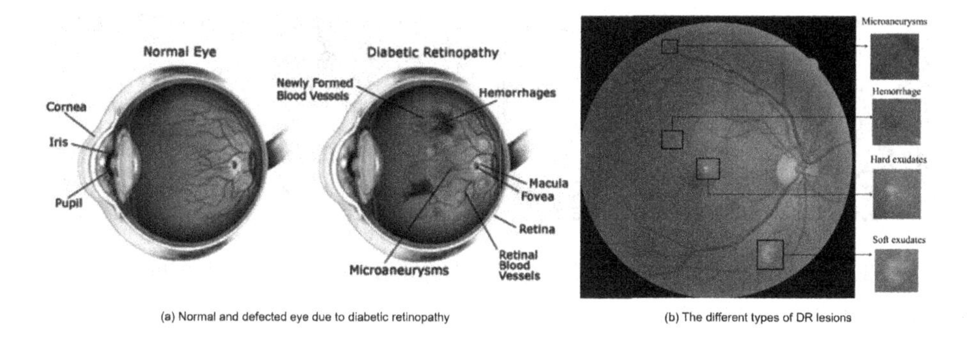

(a) Normal and defected eye due to diabetic retinopathy

(b) The different types of DR lesions

Fig. 1. Normal and defected eye due to diabetic retinopathy with different types of DR lesions.

and has high demand in the area of computer vision applications, autonomous medical applications *etc.* The increasing demand of images and videos leads to increase in capturing devices requirement such as the cameras, smart-phones. This also seeks for the high internet speed for efficient and fast sharing of the data. This increase in the multi-media resource in turn creates ample repositories or databases. In the field of medical screening applications, day-to-day, there is increase in the image and video repositories which are used for diagnosis of the patients. According to the World Health Organization (WHO), in 2014, 422 million diabetic patients were detected, 35% of whom had the damage of small blood vessels in the retina due to retinopathy [15]. Analysis in laboratory animals of diabetic neovascularization and macular edema was problematic because most laboratory species lacked a macula and wont display the characteristics of advanced diabetic retinopathy in patients for retinal neovascularization and thickening. Figure 1 shows the diabetic retinopathy in normal and defected eye with different types of DR lesions. To effectively access this medical data it is required to classify these datasets for which image/video classification plays a vital role. The state-of-the-art methods for image classification generally utilize the hand-crafted approaches. These approaches lack in producing efficient outcomes when considered for large databases. Therefore, the classification and searching of the images or videos from the medical or educational repositories becomes a hectic task. Image classification comprise two major steps: feature extraction either via rule-based (hand-crafted) features descriptor or learning-based algorithms followed by classification task. The sample images from MESSIDOOR [23] database is shown in the Fig. 2. The detailed literature survey on medical image analysis and classification is discussed on the next subsection.

2 Literature Survey

An increase in the inhabitants of diabetic patients and the generality of DR amid them urge automatic DR diagnosing systems. So far, umpteen research with fruitful outcomes is done in many sub problems like lesion detection, vessel

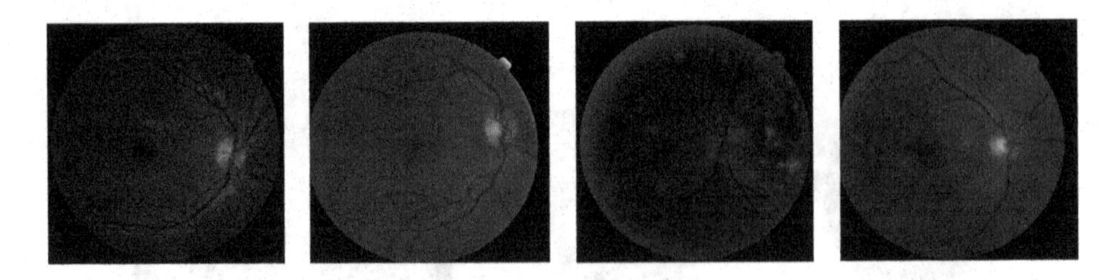

Fig. 2. Sample image from MESSIDOR [23] database used for classification.

segmentation, *etc.* Although, these reported results are provided with very limited or minimal database for DR. Also, these provided results are not suitable for real-world health-care applications. There is variation in the treatment options among patients with different stages. The person with no DR and delicate NPDR required a regular screening. Further, the treatment from medical experts vary from scattering leaser treatment to viterctomy for the persons having moderate NPDR. Therefore, the DR assessment with different grading analysis is in high demand to guide a DR person with suitable treatment [34].

Based on the different characteristics, DR can be defined as four different stages, namely Micro aneurysms (Mas), Hemorrhage (HMs), Exudates (EXs), and Inter-Retinal Microvascular Irregularities (IRMAs), which are determined by Different stages of DR, and the summary is given below. Seepthi *et al.* [32] suggested that morphological operations and segmentation procedures should be used to detect blood vessels, exudates, and microaneurysms. The retinal fundus representation is broken down into four subframes. Various features are obtained from the retinal fundus graphic. It imposes hair wavelet transformations on the extracted elements. Then, the main component analysis approach is applied to improve function quality. Neural network backpropagation and one rule classifier methods are used to classify the photos as diabetic or non-diabetic. Amin *et al.* [4], a DR model was developed to automatically distinguish retinal photos in regions of both exudates and non-exudates. The technique was based on pre-processing, beginning with lesion extraction, extraction of features, and classification.

Local Binary Patterns (LBP) was proposed for face recognition and texture classification by Ojala *et al.* in [33]. Further, a novel edge-texture features based object recognition descriptor is proposed in [30]. To overcome LBP's limitations, Tan *et al.*. [2] proposed Local Ternary Pattern (LTP). In LBP, the single threshold in used to divide the pixels are into values *i.e.,* zero and one. In LTP, unlike LBP, the pixels intensity values are divided into three values (-1, 0, $+1$) using constant threshold (th). Spherical symmetric three dimensional local ternary patterns for biomedical, natural, and texture image feature extraction was proposed in [22]. Similar approaches are proposed by many researchers for feature extraction in [12,21,31].

Rubini and Kunthavai [17] applied hessian-based candidate selection before the feature extraction and the classification is done using a support vector machine (SVM) classifier. The hybrid features like entropy for exudate area and texture based features are combined effectively for DR classification in [13]. The authors in [20] proposed a neural network architecture with multilayer perception and the effective 64 features extracted from the discrete cosine transform. Also, these authors have included the statistical features like entropy, mean, standard deviation, *etc.*. Even-though the superior performance is achieved by state-of-the-art traditional approaches, the limitations are obvious as compared to learning based approaches.

On one hand, if straightforward features are already utilized, the generation of the new efficient feature maps with hand crafted methods becomes more difficult. On the other hand, it is harder to improve the results as the performance of these methods plateaus. The existing rule-based *i.e.,* hand-crafted feature extraction descriptors have the limitations such as: (a) they depend on the rule-based assumptions (b) does not update according to the variations in the data from repositories, *etc.* So, they fail at producing robust systems which will give faithful outcomes even with the complex data. The convolution neural networks (CNN) are gaining more attention due to their reliable learning ability for variety of data. CNNs are highly effective and widely used in medical image analysis as compared to other methods [10]. The CNN architecture mainly consists of the convolution to extract features, pooling to change the feature size, fully connected layers., *etc.* The different parameters of these layers like kernel size, number of filters vary according to the applications. Each of these layers plays unique role. The CONV layers are used for feature extraction form the inputs. The pooling layer is used for extracting the relevant information from input feature maps ans also for dimension reduction. The pooling can be done in different ways but among all the methods, mostly average pooling and max pooling is adopted [10].

A very first and efficient deep learning-based network was proposed by Le *et al.* [1] for character classification. The detailed review of the existing methods used for diabetic retinopathy detection through deep learning is discussed in [18]. With the motivation of stupendous performance by [1,8] come up with an efficinet architecture for object recognition consisting 100 various categories. Also, from the remarkable outcome of [8], the researchers utilized it for various computer vision applications such as object detection, segmentation, and image classification *etc.* Similarly, Szegedy *et al.* [5] proposed a VGGNet with more computation complexity in terms of depth of the network as comapred to AlexNet for the application of object recognition. Also, researchers utilize this effective feature learning ability of the CNN for analysing the medical images. Quellec *et al.* [26] generated heat-maps of the image representing the pixel-wise importance for classification, and the heat-maps are generated by utilizing the generalization of backpropagation for training the CNNs. Yang *et al.* [27] proposed DCNN-based algorithm a two-stage architecture to detect the damaged part of fundus images and effective DR grading respectively. Similar to this approach, Chandore and Asati [29] proposed learning based approach for DR classification of larger dataset.

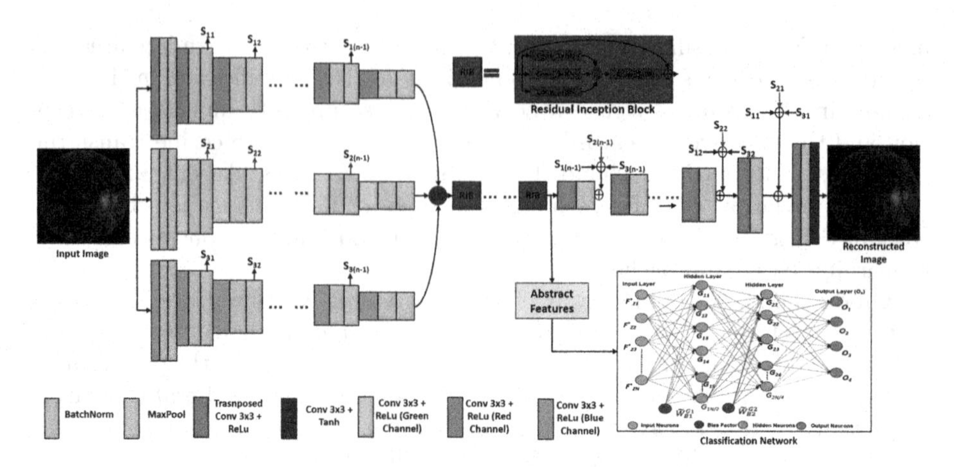

Fig. 3. Proposed learning based framework for DR medical image classification.

3 Proposed Framework

Earlier image classification architectures are generally divided into three key steps: image preprocessing to remove any noise in the image or apply some transformations if the dataset has less number of images, feature extraction by different layers in the network, and feature classification *i.e.,* the last layer. The manual diagnosis of DR by ophthalmologists is requires more efforts as it is time consuming task, and also there is a chance of disease misdiagnosis. With the efficiency of a computer-aided diagnosis system, there is a less chance of misdiagnosis with less the overall cost, time and effort. In earlier era, the deep learning (DL) approaches have emerged and same are adopted in various fields which includes medical image analysis [6], object segmentation [3,7,11,19], depth estimation [14], image dehazing [16], image deraining [28], image inpainting [24], *etc.* The DL helps the classification and segmentation task by identifying the accurate features from the input data. Utilizing the DL for these tasks generally outperform all traditional image analysis techniques. There exists numerous methods for automation of DR image classification using DL which help ophthalmologists for detecting the disease in its early stages [9]. However, most of the existing methods are mainly focused on detecting the DR rather than detection of DR stages. Further, there are limited works for classification and localization of different types of DR lesions. Detection of these stages and localization of lesions will be helpful to the ophthalmologists for evaluating the severity and monitoring the progression based lesions' appearance.

Initially, the input RGB image is proviede as input to the proposed image reconstruction Network.

This network is mainly designed for extraction of the features from the input image. TO process each channel of input image, the parallel path encoders are designed for each of the channels. These parallel path encoder features are then merged and forwarded to the decoder for reconstruction. The overall architecture will reconstruct the same medical image. This in turn helps the network to learn the effective features for each of the input image. The outcome of the learned encoder merging will then be considered as the abstract version of input. Therefore, we extended these encoded features for classification task. The detailed architecture of the proposed framework is depicted in Fig. 3. The proposed approach for classification is divided into two sub-parts as given below:

3.1 Image Reconstruction

For image reconstruction task, we have considered the processing of three channels (Red, Green and Blue) separately. This separate encoding of channel-wise features help to extract the relevant information separately unlike the other methods where all the three channels are encoded in one path. Here, to encode the input image into abstract version, we have used $n = 7$ encoder layers where each encoder layer is represented as `Convolution` $\rightarrow$ `BatchNormalization` $\rightarrow$ `Maxpooling`. The single channel input image with $256 \times 256 \times 1$ dimensions is encoded to feature maps with $2 \times 2 \times 128$ size. These encoded feature maps for each of the channels (Red, Green and Blue) are then combined and then forwarded to five consecutive Residual Inception Block (RIB) to merge the relevant features from each of the channels. The output of RIB can be formulated as:

$$RIB_{out} = f_{in} + Conv_{3\times3}\{Conv_{1\times1}(f_{in}) \otimes Conv_{3\times3}(f_{in}) \otimes Conv_{5\times5}(f_{in})\} \quad (1)$$

where, f_{in} are the input feature maps, $Conv_{m\times m}(x)$ is the convolution on x with filter size $m \times m$ followed by ReLu activation function. This RIB produces the feature maps by efficiently selecting the appropriate scale features from the incoming channel-wise feature map inputs. These feature maps are considered as abstract features which are then given to the classification network.

To reconstruct the input image, the output feature maps of last RIB are then fed to consecutive decoder blocks. Each decoder layer is represented as `Deconvolution` $\rightarrow$ `BatchNormalization`. Similar to encoder, the decoder consists of $n = 7$ decoder layers, to generate output reconstructed image same as that of input image.

3.2 Image Classification

To classify the input image in four different categories, we have considered the abstract features from the reconstruction network as input features to classification network (see Fig. 3). The abstract features are taken from the output of RIB_l where, $l = 5$ having dimension of $2 \times 2 \times 128$. These abstract features are then flattened to a size of 1 which are then fed to the classification network with 512 neurons in input layers. The output neurons of the neural network is kept as 4 (normal, microanaurysms, Hemorrhage amd Exudates). We have trained the neural network for 250 epochs. For comparison purpose, the feature from existing state-of-the-art feature descriptors (LBP, EdgeLBP, LTP, SS3D, LTrP) are extracted for complete database. With similar settings, the neural network is trained on these extracted feature maps and classification accuracies are examined.

Table 1. Specification of each class present in the MESSIDOR database.

Label	Specification
Group 1	Normal
Group 2	(0< μA <= 5) and (H=0)
Group 3	(5< μA < 15) or (0< H <5) and (NV = 0)
Group 4	(μA > = 15) or (H>=5) and (NV = 1)

4 Result Analysis

To compare the proposed method with existing state-of-the-art methods in terms of classification accuracy, we use the MESSIDOR dataset. This is a publicly available dataset with 1200 retinal images from diabetic retinopathy patients. These 1200 images are then segregated in four groups according to the severity of disease. There are 546, 153, 247 and 254 images in the four groups respectively. The resolution of these images are in different size like 1440×960, 2240×1488, 2304×1536. The grading of the retinopathy is determined depending on the number of micro-aneurysms, hemorrhages, signs of neovascularization, *etc.* The images with no abnormalities are then considered as the normal images. The specifications for different stages of retinopathy are given in Table 1 and sample images are shown in the Fig. 2. The earliest symptom occur as small red coloured dots on retina are the Microaneurysms (μA) which cause due to the weakness in the vessel's walls. Whereas the larger spots on retina with the size greater than 125 μm are the Haemorrhages (HM) which have irregular margin. Also, the hard exudates are with sharp margins and found in the outer layer of retina which are caused by leakage of plasma. The color of hard exudates appear as bright-yellow. The stages are segregated according to these specifications found on the retina. The proposed reconstruction based classification network accuracy analysis is compared with existing state-of-the-art feature descriptors and illustrated in Fig. 4. We have trained the neural network for 250 epochs. For comparison purpose, the feature from existing state-of-the-art feature descriptors (LBP, EdgeLBP, LTP, SS3D, LTrP) are extracted for complete database. With similar settings, the neural network is trained on these extracted feature maps and classification accuracies are examined on MESSIDOR database.

The classification accuracy for various feature descriptors is EdgeLBP (46.7%), LBP (60.8%), LTP(53.6%), SS3D (59.2%) and Tetra (46.7%). Whereas, the proposed algorithm achieved improved classification accuracy of 63.8% on MESSIDOR database. From Fig. 4, it is evident that the proposed reconstruction based learning framework is achieved significantly improved performance as compared to existing state-of-the-art feature descriptors while used for classification.

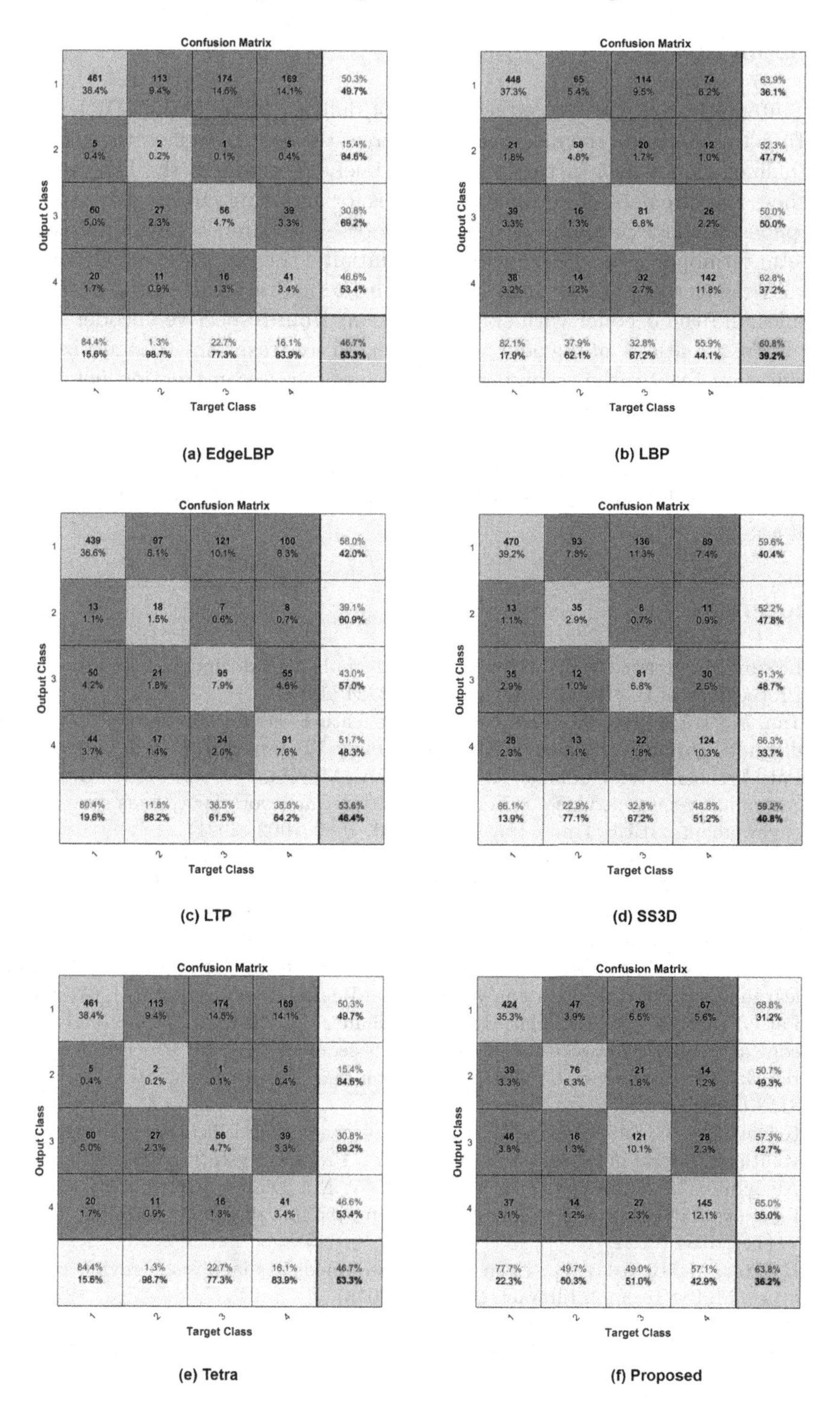

Fig. 4. Classification accuracy analysis of proposed framework with existing state-of-the-art feature descriptor on MESSIDOR database.

5 Conclusion

The automatic computer aided screening of diabetic retinopathy (DR) reduce the time to determine the diagnosis. This saves the cost as well as the efforts of ophthalmologists which in turn results in timely diagnosis of the DR patients. So, there is a dire need of automated systems for detecting DR at early stage. In this paper, we have proposed reconstruction based feature learning approach for diabetic retinopathy image classification. Initially, the proposed reconstruction network regenerates the medical image with the multi-stream encoder-decoder modules and the decoder with skip connections from respective encoder layers. The learned features of encoder while regeneration task, are then utilized for classification task in the second stage. The proposed reconstruction network is composed of residual inception block which helps to delve into the feature maps with different kernel sizes and reduces the feature redundancy. The proposed framework gives significantly improved performance as compared to existing state-of-the-art feature descriptors for DR image classification on MESSIDOR database.

References

1. LeCun, Y., et al.: Gradient-based learning applied to document recognition. In: Proceedings of the IEEE 86.11, pp. 2278–2324 (1998)
2. Tan, X., Triggs, B.: Enhanced local texture feature sets for face recognition under difficult lighting conditions. IEEE Trans. Image Process. **19**(6), 1635–1650 (2010)
3. Patil Prashant, W., Dudhane, A., Kulkarni, A., Murala, S., Gonde, A.B., Gupta, S.: An unified recurrent video object segmentation framework for various surveillance environments. IEEE Trans. Image Proc. **30**, 7889–7902 (2021)
4. Amin, J., et al.: A method for the detection and classification of diabetic retinopathy using structural predictors of bright lesions. J. Comput. Sci. **19**, 153–164 (2017)
5. Simonyan, K., Zisserman, A.: Very deep convolutional networks for large-scale image recognition (2014). arXiv preprint arXiv:1409.1556
6. Sikder, N., et al.: Severity classification of diabetic Retinopathy using an ensemble learning algorithm through analyzing retinal Images. Symmetry **13**(4), 670 (2021)
7. Patil Prashant, W., Biradar, K.M., Dudhane, A., Murala, S.: An end-to-end edge aggregation network for moving object segmentation. In: Proceedings of the IEEE/CVF Conference on Computer Vision and Pattern Recognition, pp. 8149–8158 (2020)
8. Krizhevsky, A., Sutskever, I., Hinton, G.E.: Imagenet classification with deep convolutional neural networks. Adv. Neural. Inf. Process. Syst. **25**, 1097–1105 (2012)
9. Alyoubi, W.L., Abulkhair, M.F., Shalash, W.M.: Diabetic Retinopathy fundus Image classification and lesions localization system using deep learning. Sensors **21**(11), 3704 (2021)
10. Bakator, M., Radosav, D.: Deep learning and medical diagnosis: a review of literature. Multi. Technol. Interact. **2**(3), 47 (2018)
11. Patil, P.W., Murala, S.: Msfgnet: A novel compact end-to-end deep network for moving object detection. IEEE Trans. Intell. Transp. Syst. **20**(11), 4066–4077 (2018)

12. Galshetwar, G.M., et al.: Local directional gradient based feature learning for image retrieval. In: 2018 IEEE 13th International Conference on Industrial and Information Systems (ICIIS). IEEE (2018)
13. Mookiah, M.R.K., et al.: Evolutionary algorithm based classifier parameter tuning for automatic diabetic retinopathy grading: a hybrid feature extraction approach. Knowl. Based Syst. **39**, 9–22 (2013)
14. Hambarde, P., Dudhane, A., Patil, P.W., Murala, S., Dhall, A.: Depth estimation from single image and semantic prior. In: 2020 IEEE International Conference on Image Processing (ICIP), pp. 1441–1445. IEEE (2020)
15. Ophthalmoscopy, Dilated, ETDRS Levels: International clinical diabetic retinopathy disease severity scale detailed table (2002)
16. Akshay, D., Hambarde, P., Patil, P., Murala, S.: Deep underwater image restoration and beyond. IEEE Signal Process. Lett. **27**, 675–679 (2020)
17. Rubini, S.S., Kunthavai, A.: Diabetic retinopathy detection based on eigenvalues of the hessian matrix. Proc. Comput. Sci. **47**, 311–318 (2015)
18. Alyoubi, W.L., Shalash, W.M., Abulkhair, M.F.: Diabetic retinopathy detection through deep learning techniques: a review. Informatics in Medicine Unlocked **20**, 100377 (2020)
19. Patil, P.W., Dudhane, A., Chaudhary, S., Murala, S.: Multi-frame based adversarial learning approach for video surveillance. Pattern Recogn. **122**, 108350 (2022)
20. Bhatkar, A.P., Kharat, G.U.: Detection of diabetic retinopathy in retinal images using MLP classifier. In: 2015 IEEE International Symposium on Nanoelectronic and Information Systems. IEEE (2015)
21. Gonde, A.B., et al.: Volumetric local directional triplet patterns for biomedical image retrieval. In: 2017 Fourth International Conference on Image Information Processing (ICIIP). IEEE (2017)
22. Murala, S., Jonathan Wu, Q.M.: Spherical symmetric 3D local ternary patterns for natural, texture and biomedical image indexing and retrieval. Neurocomputing **149**, 1502–1514 (2015)
23. Decencière, E., et al.: Feedback on a publicly distributed image database: the Messidor database. Image Anal. Stereology **33**(3), 231–234 (2014)
24. Phutke, S.S., Murala, S.: Diverse receptive field based adversarial concurrent encoder network for image inpainting. IEEE Signal Process. Lett. **28**, 1873–1877 (2021)
25. Hua, C.-H., et al.: Convolutional network with twofold feature augmentation for diabetic retinopathy recognition from multi-modal images. IEEE J. Biomed. Health Inf. **25**(7), 2686–2697 (2020)
26. Quellec, G., et al.: Deep image mining for diabetic retinopathy screening. Med. Image Anal. **39**, 178–193 (2017)
27. Yang, Y., Li, T., Li, W., Wu, H., Fan, W., Zhang, W.: Lesion detection and grading of diabetic retinopathy via two-stages deep convolutional neural networks. In: Descoteaux, M., Maier-Hein, L., Franz, A., Jannin, P., Collins, D.L., Duchesne, S. (eds.) MICCAI 2017. LNCS, vol. 10435, pp. 533–540. Springer, Cham (2017). https://doi.org/10.1007/978-3-319-66179-7_61
28. Kulkarni, A., Patil, P.W., Murala, S.: Progressive subtractive recurrent lightweight network for video deraining. IEEE Signal Process. Lett. **29**, 229–233 (2022). https://doi.org/10.1109/LSP.2021.3134171
29. Chandore, V., Asati, S.: Automatic detection of diabetic retinopathy using deep convolutional neural network. Int. J. Adv. Res. Ideas Innov. Technol. **3**, 633–641 (2017)

30. Satpathy, A., Jiang, X., Eng, H.-L.: LBP-based edge-texture features for object recognition. IEEE Trans. Image Process. **23**(5), 1953–1964 (2014)
31. Murala, S., Maheshwari, R.P., Balasubramanian, R.: Local tetra patterns: a new feature descriptor for content-based image retrieval. IEEE Trans. Image Proc. **21**(5), 2874–2886 (2012)
32. Prasad, D.K., Vibha, L., Venugopal, K.R.: Early detection of diabetic retinopathy from digital retinal fundus images. In: 2015 IEEE Recent Advances in Intelligent Computational Systems (RAICS). IEEE (2015)
33. Ojala, T., Pietikäinen, M., Harwood, D.: A comparative study of texture measures with classification based on featured distributions. Pattern Recogn. **29**(1), 51–59 (1996)
34. Resnikoff, S., et al.: The number of ophthalmologists in practice and training worldwide: a growing gap despite more than 200 000 practitioners. British J. Ophthalmol. **96**(6), 783–787 (2012)

Human Action Recognition in Still Images

Palak(✉) and Sachin Chaudhary

Punjab Engineering College, Chandigarh, India
palakgirdhar99@gmail.com, sachin.chaudhary@pec.edu.in

Abstract. In this research work, we are addressing the problem of action recognition in still images, in which the model focuses on recognizing the person's action from a single image. We are using the dataset published by V. Jacquot, Z. Ying, and G. Kreiman in CVPR 2020. The dataset consists of the 3 action classes: Drinking, Reading, and Sitting. The images are not classified into these 3 classes. Instead, binary image classification is used on each class i.e., whether the person is performing that particular action or not. To classify the images, we started with the Detectron2 Object detection model for detecting the person performing the activity and the object related to it (foreground) and then we remove everything else (background) from the image. And then, these images without the background are used for the classification task. The classification is done by using various deep learning models with the help of transfer learning. And as a result, the classification accuracy of HAR in still images increases by 10% on VGG16, 7% on InceptionV3, 1% on Xception, and 4% on the Inception-Resnet model.

Keywords: Action recognition · Still images · Detectron2 · Binary image classification · Background removal

1 Introduction

1.1 Human Action Recognition

Human Action Recognition (HAR) is an important problem in the field of computer vision research. The main goal of an action recognition model is to analyze the image/video for identifying the action in the provided data. HAR [7, 15, 37, 40] is used in a variety of applications which includes video storage and retrieval, video surveillance systems [34, 36], human–machine interface, healthcare, image annotation, identity recognition system, elderly patient monitoring system, and a lot more. Although, HAR is used in many applications, it is still a challenging problem [12] in computer vision due to less accuracy and efficiency. Humans can easily understand the action going on in the given image/video through their senses (eyes in this case). To monitor human actions in some real-world situations, a lot of manpower is required which is quite expensive. Thus, we require a machine that can identify the actions with good accuracy. Earlier, hand-crafted approaches were used for recognizing the actions, and these days we are using learning-based approaches. In this research work, we explored the field of HAR in still images by using deep learning-based approaches.

B. Raman et al. (Eds.): CVIP 2021, CCIS 1568, pp. 483–493, 2022.
https://doi.org/10.1007/978-3-031-11349-9_42

1.2 Human Action Recognition in Still Images

Action Recognition in still images has recently become an active research topic in the field of computer vision. In this technique, we try to identify the action performed by the person through a single image. As the image does not contain any temporal information, it makes this problem of HAR in images more difficult than the HAR in videos. There is a huge amount of still images present over the internet. Therefore, it is important to develop an efficient model for recognizing action in still images which will help in better understanding and retrieval of the images. Although, a lot of work has been done in HAR based on videos, HAR in still images is not explored enough. Due to increase in the usage of digital cameras in everyday life, more and more image content is generated and uploaded to the Internet or stored in a large image dataset. Categorizing rich image content based on the actions appearing in the image is a good way to reach the initial goal of organizing these images. Images are used for performing some other important tasks as well like image inpainting [42], image dehazing [44], single image depth estimation [35] etc. Fig. 1 shows some of the activities which can be recognized from a single image only.

Fig. 1. Activities that can be recognized by single images

According to the survey in [1], the state-of-the-art methods available for recognizing actions in the images are categorized based on the low-level and high-level features available in the images.

Low-Level Features. These are basically the small details of the image like corners or edges. Some of the popular techniques for low-level feature extraction are Dense Sampling of Invariant Feature Transform (DSIFT), Histogram of Oriented Gradient (HOG), Shape Context (SC), and Global Image Descriptor (GIST).

High-Level Features. These features are the complete objects which can be detected by the different object detection models for recognizing the actions in the still images. The various high-level features are: The Human Body, Body Parts, Objects, Human-Object Interaction, Context or Scene. These features are extracted by using machine learning techniques.

2 Related Work

Previous literature on human action recognition in still images comprises the methods of extracting features from the images. The dense sampling of the grayscale images is used to extract low-level features for action analysis, using the Scale Invariant Feature Transform (SIFT) method [2]. DSIFT based feature is used to recognize action classes as discussed in the methods in [3–6]. Still image-based action recognition in [8–11] uses HOG feature proposed by [38]. Approaches in [10, 13, 14] use the shape context feature proposed by [41] to extract the shape features for object matching, to detect and segment the human contour. GIST technique [39] is used for extracting the spatial properties of the scene as an abstract representation of the scene. This feature is used to integrate background or scene information. GIST has been used in [10, 16, 17] for recognizing actions in still images.

High-level features like the human body, body parts, objects, etc. can be extracted using the Object Detection [46] models. These models draw boundaries along the object which helps us in identifying the object. Some of the popular object detection models are R-CNN [29], Fast R-CNN [28], Faster R-CNN [27], YOLO [19], and SSD [30]. R-CNN algorithm [29] puts some boxes randomly in the image and then checks for the objects in those boxes. After choosing the region of the object from the image, the model extracts the features of the object and then combines the regions. Training is a multi-step process and therefore this method is very slow and needs almost 1 min for an image. Fast R-CNN [28] puts a huge number of frames in the image, which results in the duplication of the features extracted by the model. Training speed is increased. Space needed for training is increased. This algorithm puts no constraints on the input data. Better than the R-CNN but still has issues. In Faster R-CNN [27], Deep CNN is used to determine candidate regions. RPN gives the output of the regions with a detection score. This is the fastest of all of its versions and can perform at the rate of 7 images/second. YOLO [19] is different from all the models discussed above. The networks for training and testing are completely different from each other. In this, a single neural network is applied to the complete image which is then divided into different regions, and a bounding box is created among each region containing objects. In SSD [30], a complete image is needed for the input, with the description of the bounding boxes. The default boxes are compared to the ground truth while training the model. It generates a sequence of same-sized boxes and scores for them. It is a newer version of YOLO [19]. Speed is 58 frames/sec with 72% accuracy.

3 Proposed Approach

3.1 Dataset Used

The dataset that we are using in our research work is published by V. Jacquot, Z. Ying, and G. Kreiman in [18]. The dataset consists of 3 action classes: Drinking, Reading, and Sitting where Drinking is defined as liquid in mouth, reading as gaze towards text and sitting as buttocks on support. Every action class is further divided into yes and no category and each category contain images for Training, Testing and Validation. The dataset contains a total of 6804 images, out of which Drinking has 2164 images, Reading has 2524 images and sitting has 2116 images. The complete dataset details are shown in the Table 1.

The images in the dataset are gathered from two different sources:

i) Clicked by the investigators in their lab.
ii) From the internet.

Table 1. Dataset details

	Training	Testing	Validation	Total
Drinking	862	110	110	1082
Not-Drinking	862	110	110	1082
Reading	1002	130	130	1262
Not-Reading	1002	130	130	1262
Sitting	826	110	122	1058
Not-Sitting	826	110	122	1058
				6804

3.2 Methodology

The major tasks associated with our work includes: object detection, background removal (removing noise), Fine-tuning the pre-trained model, Feature extraction and classification. We worked on the drinking dataset and we are interested in classifying whether the person in the image is drinking or not where drinking is defined as the liquid in the mouth.

Our proposed method is divided into 2 modules: Training and Testing. So, we will be treating both training and testing data differently. Training data is used in the training module as shown in Fig. 2 and Testing data is used in the testing module as shown in Fig. 3.

Training Module. We used training data of both the categories i.e., drinking and not-drinking. Firstly, we separate the google images from the images clicked by them i.e., we are dividing the training data into 2 parts:

1. Images clicked by the investigators are named Training Data 1.
2. Images downloaded from the Internet are named Training Data 2.

After that, we perform object detection and background removal on Training Data 1, combines it with Training Data 2, and then using the entire training data for fine-tuning the CNN model pre-trained on the ImageNet Dataset [21]. This whole process is represented in Fig. 2.

Object Detection Using Detectron2. Object Detection is done with the help of the Detectron2 [25] model which is developed by Facebook AI Research. It is the replacement for Detectron and Mask R-CNN [26] benchmark. Detectron2 [25] includes models like RetinaNet [32], DensePose [33], Mask R-CNN [26], Faster R-CNN [27], TensorMask [43], Panoptic FPN [45] and Cascade R-CNN [47]. Detectron2 [25] can perform a variety of tasks like detecting objects using a bounding box, estimating the pose of the humans present in the image, it can also perform panoptic segmentation which is a combination of semantic and instance segmentation. Detectron2 [25] can be used with the help of PyTorch which was a major drawback for the Detectron model. We have detected the human present in the image along with the object it is related to. Image masks are created by Detectron2 [25], these image masks are used to manipulate the image. The output from the object detection model (Detectron2) consists of these three things:

i) predicted class
ii) bounding box coordinates
iii) mask arrays

We used mask outlining and class ID for selecting the correct mask when there are multiple objects present of same class.

Background Removal. After detecting the relevant objects from the image, we removed the background of the image which is not contributing to the classification task. The motive behind removing the background is that our CNN model will not extract the features which are not contributing to the classification task. For removing the background [31], we generate the binary mask from the mask array that we received from Detectron2 [25] and these pixels with the value = 1 in the binary mask are extracted from the original image and then copied to a new image having blank background.

After background removal [31], all these images (Training Data 1) along with Training Data 2 are used for fine-tuning the CNN models pre-trained on the ImageNet Dataset [21]. We received fine-tuned model as an output of the Training module and this fine-tuned model is then used in Testing module for binary action classification.

Testing Module. We will be using testing data of both the categories i.e., Drinking and Not-Drinking. Firstly, we are dividing the testing data into 2 categories:

1. Images clicked by the investigators are named Testing Data 1.
2. Images downloaded from the Internet are named Testing Data 2.

Then, we perform object detection and background removal on Testing Data 1 by using the same method explained in the Training module used for Training Data. Then feeds Testing Data 1 and Testing Data 2 to the fine-tuned model, which we have received as an output of the training module. This fine-tuned CNN model is used to extract features from the images and then classify them as Drinking or Not Drinking. The block diagram of the Testing module is shown below in Fig. 3.

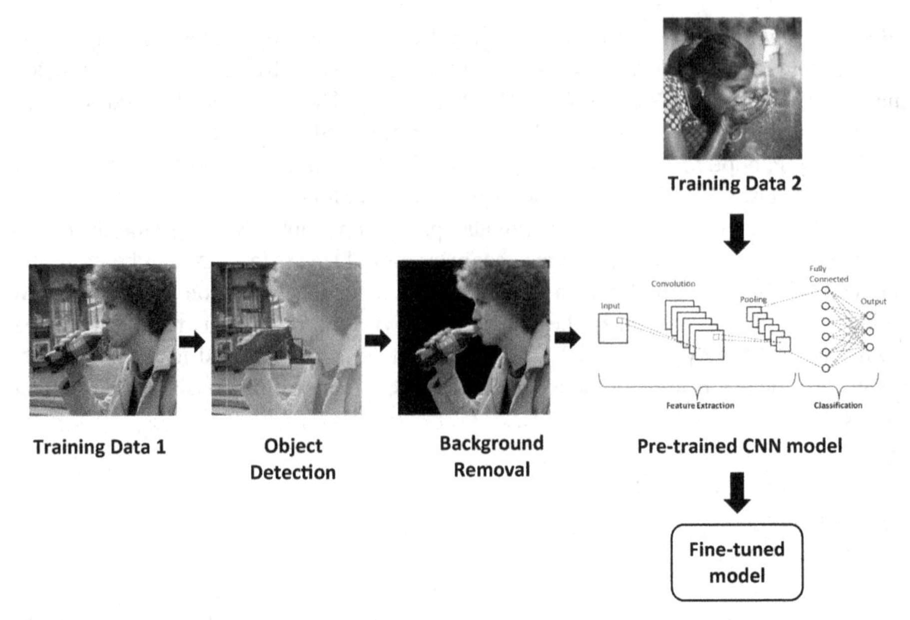

Fig. 2. Training module

We used transfer learning in our research because the dataset we are using is not large enough to train the model. We have also used data augmentation on our dataset before fine-tuning our model to prevent overfitting. We have done all of our work on the GPU.

To evaluate our proposed method, we have used 4 different CNN models pre-trained on the ImageNet Dataset [21]:

VGG16 [20]. It is also known as OxfordNet. It is named after the name of the group (Visual Geometry Group) who developed it from Oxford. This neural network is 16 layers deep. The model uses a set of weights that have been pre-trained on ImageNet [21]. The ImageNet dataset [21] consists of over 14 million images divided into 1000 classes. The top-5 accuracy of the model is 92.7%. The input to the VGG16 model is an RGB image of the size 224×224. The image is then processed through a stack of convolutional layers having the filters of size 3×3 with stride = 1, max pool layer with stride = 2, and a filter of size 2×2.

Inception V3 [23]. It is an Image Recognition model. It consists of symmetric and asymmetric building blocks which include convolution layers, pooling layers (average pooling and max pooling), concats, dropouts, and the fully connected layers. Batch normalization is used in the model. This model has shown an accuracy > 78% on the ImageNet Dataset [21]. As the deep networks are more likely to be overfitted, in this model, rather than going deep into the network, they have widened the network on different levels. So, there is more than 1 filter present on some levels of this model. So, several filters are applied parallelly on the same level along with the pooling operation. Then, all the outputs from a single level are concatenated and sent to the next level.

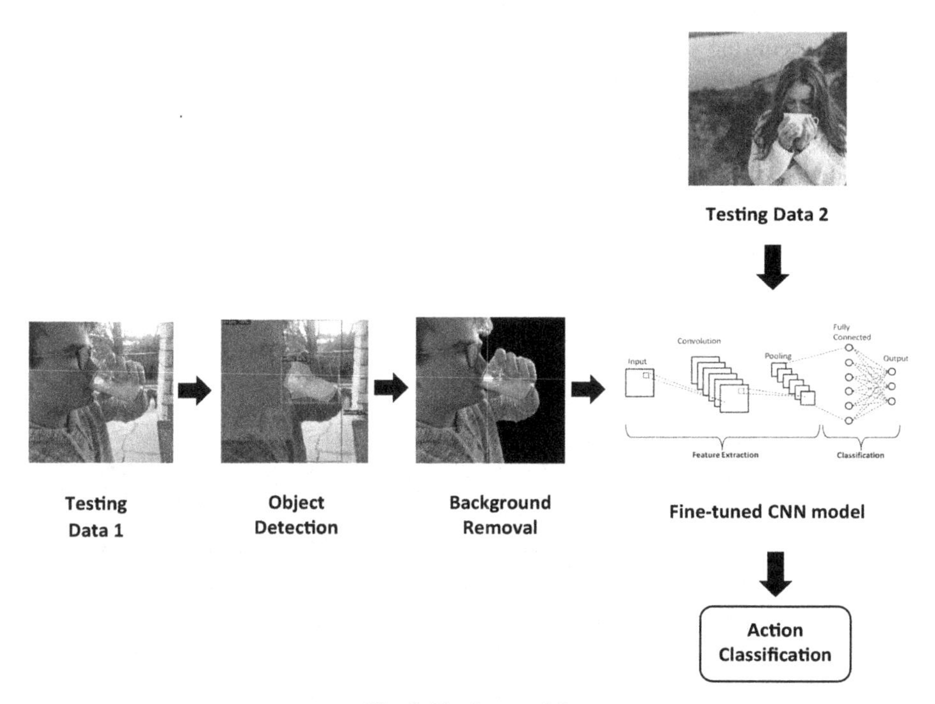

Fig. 3 Testing module

Xception [24]. It is 71 layers deep CNN. This neural network is based on the depth-wise separation of the convolution layers. It stands for Extreme Inception. The concepts of the Inception model are used to an extreme in this model. The depth-wise convolution is followed by pointwise convolution in this model. Its results are somewhat better than the inception v3 model on the ImageNet data [21] but their number of parameters are the same. Thus, this model is efficient as compared to Inception.

Inception-ResNet [22]. This model is a combination of the Inception and the ResNet model because both the model's performances individually are quite appreciable. This model has two versions, the computational cost of the first model is similar to the InceptionV3 [23] and the second's cost is similar to the InceptionV4. Both have different stems. The structure is the same for all the modules and the reduction block. Hyperparameters are different for both. These models can achieve higher accuracies even with the fewer number of iterations.

4 Results

The accuracies of different state-of-the-art models by using our proposed method are calculated and shown in Fig. 4. The accuracies achieved on VGG16 [20], InceptionV3 [23], Xception [24] and Inception-ResNet V2 [22] are 59.09%, 65.91%, 62.75% and 63.18% respectively. These accuracies are then compared with the accuracies of the same dataset on the same deep learning models without using our method and the comparison is shown in Fig. 4.

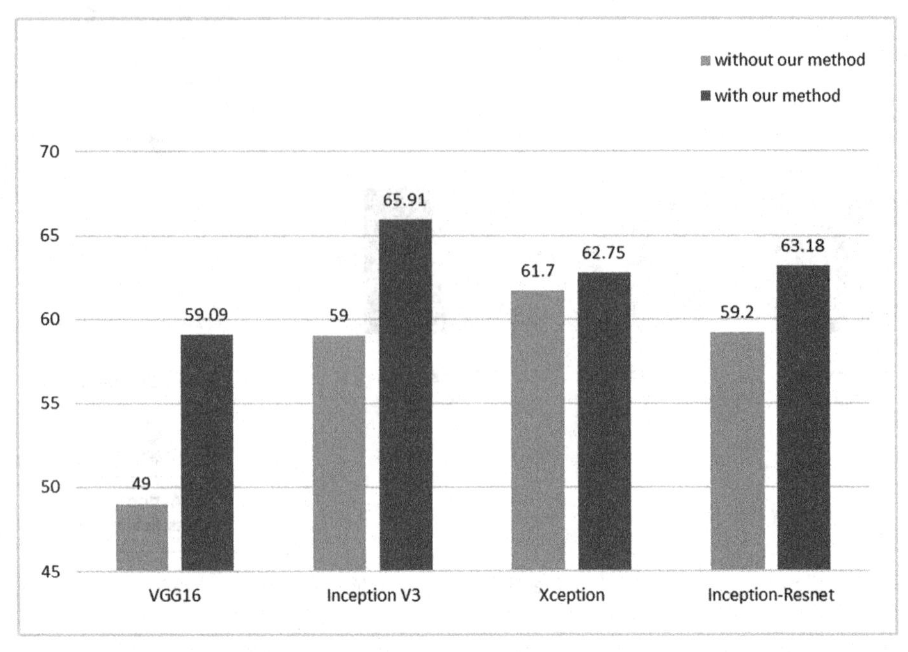

Fig. 4. Comparison of the classification accuracies of the models with and without using our proposed method on the drinking dataset

5 Conclusion and Future Work

In this work, we first analyzed all previous work done on HAR in still images and came across a dataset for basic human activities: Drinking, Reading, and Sitting. The dataset contains fewer biases and thus the state-of-the-art methods are showing less accuracy on this dataset.

So, we proposed a classification method, in which we first perform object detection on the images and then we remove the background of images. After that, we used the same state-of-the-art methods to classify those images. Our method performs well and the accuracy is increased by 10% on VGG16 model, 7% on InceptionV3, 1% on Xception and 4% on Inception-Resnet V2 model.

In the future, this model can be applied for action classification tasks having more than 2 action classes to check if it performs better there as well or not. This model can also be used for binary action classification like in medical sciences, where we need to determine whether the person is suffering from some disease or not.

References

1. Guo, G., Lai, A.: A survey on still image-based human action recognition. Pattern Recognit. **47**(10), 3343–3361 (2014)
2. Lowe, D.G.: Object recognition from local scale-invariant features. In: Proceedings of the Seventh IEEE International Conference on Computer Vision, vol. 2, pp. 1150–1157. IEEE (1999)
3. Li, L.-J., Li, F.-F.: What, where and who? classifying events by scene and object recognition. In: ICCV, vol. 2, no. 5, p. 6 (2007)
4. Delaitre, V., Laptev, I., Sivic, J.: Recognizing human actions in still images: a study of bag-of-features and part-based representations. In BMVC 2010 (2010)
5. Shapovalova, N., Gong, W., Pedersoli, M., Roca, F.X., Gonzàlez, J.: On importance of interactions and context in human action recognition. In: Vitrià, J., Sanches, J.M., Hernández, M. (eds.) Pattern Recognition and Image Analysis, IbPRIA 2011. LNCS, vol. 6669, pp. 58–66. Springer, Heidelberg (2011). https://doi.org/10.1007/978-3-642-21257-4_8
6. Yao, B., Fei-Fei, L.: Grouplet: a structured image representation for recognizing human and object interactions. In: 2010 IEEE Conference on Computer Vision and Pattern Recognition (CVPR), pp. 9–16. IEEE (2010)
7. Chaudhary, S., Murala, S.: Depth-based end-to-end deep network for human action recognition. IET Comput. Vis. **13**(1), 15–22 (2019)
8. Desai, C., Ramanan, D.: Detecting actions, poses, and objects with relational phraselets. In: Fitzgibbon, A., Lazebnik, S., Perona, P., Sato, Y., Schmid, C. (eds.) Computer Vision – ECCV 2012. LNCS, vol. 7575, pp. 158–172. Springer, Heidelberg (2012). https://doi.org/10.1007/978-3-642-33765-9_12
9. Thurau, C., Hlavác, V.: Pose primitive based human action recognition in videos or still images. In: 2008 IEEE Conference on Computer Vision and Pattern Recognition, pp. 1–8. IEEE (2008)
10. Gupta, A., Kembhavi, A., Davis, L.S.: Observing human-object interactions: using spatial and functional compatibility for recognition. IEEE Trans. Pattern Anal. Mach. Intell. **31**(10), 1775–1789 (2009)
11. Desai, C., Ramanan, D., Fowlkes, C.: Discriminative models for static human-object interactions. In: 2010 IEEE Computer Society Conference on Computer Vision and Pattern Recognition-Workshops, pp. 9–16. IEEE (2010)
12. Chaudhary, S.: Deep learning approaches to tackle the challenges of human action recognition in videos. Dissertation (2019)
13. Wang, Y., Jiang, H., Drew, M.S., Li, Z.-N., Mori, G.: Unsupervised discovery of action classes. In: 2006 IEEE Computer Society Conference on Computer Vision and Pattern Recognition (CVPR2006), vol. 2, pp. 1654–1661. IEEE (2006)
14. Yao, B., Fei-Fei, L.: Modeling mutual context of object and human pose in human-object interaction activities. In: 2010 IEEE Conference on Computer Vision and Pattern Recognition, pp. 17–24 (2010)
15. Chaudhary, S., Murala, S.: TSNet: deep network for human action recognition in hazy videos. In: 2018 IEEE International Conference on Systems, Man, and Cybernetics (SMC), pp. 3981–3986 (2018). https://doi.org/10.1109/SMC.2018.00675
16. Prest, A., Schmid, C., Ferrari, V.: Weakly supervised learning of interactions between humans and objects. IEEE Trans. Pattern Anal. Mach. Intell. **34**(3), 601–614 (2011)
17. Li, P., Ma, J.: What is happening in a still picture? In: The First Asian Conference on Pattern Recognition, pp. 32–36. IEEE (2011)
18. Jacquot, V., Ying, Z., Kreiman, G.: Can deep learning recognize subtle human activities? In: 2020 IEEE/CVF Conference on Computer Vision and Pattern Recognition (CVPR), Seattle, WA, USA, pp. 14232–14241 (2020). https://doi.org/10.1109/CVPR42600.2020.01425

19. Redmon, J., Divvala, S., Girshick, R., Farhadi, A.: You only look once: unified, real-time object detection. In: 2016 IEEE Conference on Computer Vision and Pattern Recognition (CVPR), pp. 779–788 (2016). https://doi.org/10.1109/CVPR.2016.91
20. Simonyan, K., Zisserman, A.: Very Deep Convolutional Networks for Large-Scale Image Recognition (2015)
21. Russakovsky, O., et al.: ImageNet large scale visual recognition challenge. Int. J. Comput. Vis. (IJCV), **115**(3), 211–252 (2015)
22. Szegedy, C., Ioffe, S., Vanhoucke, V., Alemi, A.: Inception-v4, inception-resnet and the impact of residual connections on learning (2016)
23. Szegedy, C., Vanhoucke, V., Ioffe, S., Shlens, J., Wojna, Z.: Rethinking the inception architecture for computer vision. CoRR, abs/1512.00567 (2015)
24. Chollet, F.: Xception: deep learning with depthwise separable convolutions. CoRR abs/1610.02357 (2016)
25. Girshick, R., Radosavovic, I., Gkioxari, G., Dollár, P., He, K.: Detectron (2018). https://github.com/facebookresearch/detectron
26. He, K., Gkioxari, G., Dollár, P., Girshick, R.: Mask R-CNN. In: 2017 IEEE International Conference on Computer Vision (ICCV), pp. 2980–2988 (2017). https://doi.org/10.1109/ICCV.2017.322
27. Ren, S., He, K., Girshick, R., Sun, J.: Faster R-CNN: towards real-time object detection with region proposal networks. IEEE Trans. Pattern Anal. Mach. Intell. **39**(6), 1137–1149 (2017). https://doi.org/10.1109/TPAMI.2016.2577031.
28. Girshick, R.: Fast R-CNN. In: 2015 IEEE International Conference on Computer Vision (ICCV), pp. 1440–1448 (2015). https://doi.org/10.1109/ICCV.2015.169
29. Girshick, R., Donahue, J., Darrell, T., Malik, J.: Rich feature hierarchies for accurate object detection and semantic segmentation. In: Proceedings of the IEEE Conference on Computer Vision and Pattern Recognition, pp. 580–587 (2014)
30. Liu, W., et al.: SSD: single shot multibox detector. In: Leibe, B., Matas, J., Sebe, N., Welling, M. (eds.) Computer Vision – ECCV 2016. LNCS, vol. 9905, pp. 21–37. Springer, Cham (2016). https://doi.org/10.1007/978-3-319-46448-0_2
31. Reaper, T.: Automated image background removal with python. tobias.fyi (2020). https://tobias.fyi/blog/remove-bg-python
32. Lin, T., Goyal, P., Girshick, R., He, K., Dollár, P: Focal loss for dense object detection. In: 2017 IEEE International Conference on Computer Vision (ICCV), pp. 2999–3007 (2017). https://doi.org/10.1109/ICCV.2017.324
33. Güler, R.A., Neverova, N., Kokkinos, I.: DensePose: dense human pose estimation in the wild. In: 2018 IEEE/CVF Conference on Computer Vision and Pattern Recognition, pp. 7297–7306 (2018). https://doi.org/10.1109/CVPR.2018.00762
34. Patil, P.W., Dudhane, A., Kulkarni, A., Murala, S., Gonde, A.B., Gupta, S.: An unified recurrent video object segmentation framework for various surveillance environments. IEEE Trans. Image Process. **30**, 7889–7902 (2021)
35. Praful, H., Dudhane, A., Murala, S.: Single image depth estimation using deep adversarial training. In: 2019 IEEE International Conference on Image Processing (ICIP), pp. 989–993. IEEE I(2019)
36. Patil, P.W., Dudhane, A., Chaudhary, S., Murala, S.: Multi-frame based adversarial learning approach for video surveillance. Pattern Recogn. **122**, 108350 (2022)
37. Chaudhary, S., Murala, S.: Deep network for human action recognition using Weber motion. Neurocomputing **367**, 207–216 (2019)
38. Dalal, N., Triggs, B.: Histograms of oriented gradients for human detection. In: Computer Vision and Pattern Recognition CVPR, vol. 1, pp. 886–893 (2005)
39. Oliva, A., Torralba, A.: Modeling the shape of the scene: a holistic representation of the spatial envelope. Int. J. Comput. Vis. **42**(3), 145–175 (2001)

40. Chaudhary, S., Dudhane, A., Patil, P., Murala, S.: Pose guided dynamic image network for human action recognition in person centric videos. In: 2019 16th IEEE International Conference on Advanced Video and Signal Based Surveillance (AVSS), pp. 1–8 (2019). https://doi.org/10.1109/AVSS.2019.8909835
41. Belongie, S., Mori, G., Malik, J.: Matching with shape contexts. In: Krim, H., Yezzi, A. (eds) Statistics and Analysis of Shapes. Modeling and Simulation in Science, Engineering and Technology, pp. 81–105. Birkhäuser Boston, Boston (2006). https://doi.org/10.1007/0-8176-4481-4_4
42. Phutke, S.S., Murala, S.: Diverse receptive field based adversarial concurrent encoder network for image inpainting. IEEE Signal Process. Lett. **28**, 1873–1877 (2021)
43. Chen, X., Girshick, R., He, K., Dollár, P.: Tensormask: a foundation for dense object segmentation. In: Proceedings of the IEEE/CVF International Conference on Computer Vision, pp. 2061–2069 (2019)
44. Akshay, D., Biradar, K.M., Patil, P.W., Hambarde, P., Murala, S.: Varicolored image de-hazing. In: proceedings of the IEEE/CVF Conference on Computer Vision and Pattern Recognition, pp. 4564–4573 (2020)
45. Kirillov, A., Girshick, R., He, K., Dollár, P.: Panoptic feature pyramid networks. In: Proceedings of the IEEE/CVF Conference on Computer Vision and Pattern Recognition, pp. 6399–6408 (2019)
46. Patil, P.W., Biradar, K.M., Dudhane, A., Murala, S.: An end-to-end edge aggregation network for moving object segmentation. In: proceedings of the IEEE/CVF Conference on Computer Vision and Pattern Recognition, pp. 8149–8158 (2020)
47. Cai, Z., Vasconcelos, N.: Cascade R-CNN: high quality object detection and instance segmentation. IEEE Trans. Pattern Anal. Mach. Intell. **43**, 1483–1498 (2019).

Enhancing Unsupervised Video Representation Learning by Temporal Contrastive Modelling Using 2D CNN

Vidit Kumar(✉), Vikas Tripathi, and Bhaskar Pant

Graphic Era Deemed to be University, Dehradun, India
viditkumaruit@gmail.com

Abstract. Using video sequence order as a supervised signal has proven to be effective in initializing 2d convnets for downstream tasks like video retrieval and action recognition. Earlier works used it as sequence sorting task, odd-one out task and sequence order prediction task. In this work, we propose an enhanced unsupervised video representation learning method by solving order prediction and contrastive learning jointly using 2d-CNN (as backbone). With contrastive learning we aim to pull different temporally transformed versions of same video sequence closer while pushing the other sequences away in the latent space. In addition, instead of pair wise feature extraction, the features are learned with 1-d temporal convolutions. Experiments conducted on UCF-101 and HMDB-51 datasets show that our proposal outperforms the other methods on both downstream tasks (video retrieval and action recognition) with 2d-CNN and, achieves satisfactory results compared to 3d-CNN based methods.

Keywords: Deep learning · Self-supervised learning · Unsupervised representation learning · Video analysis · Video representation

1 Introduction

Visual representation learning, especially the task of representing high-level concepts in video or image data, has come a long way since the success of supervised learning [1]. Despite having strong quantitative results, the success is depend on the large scale labeled dataset like ImageNet [2] for training deep networks. Even though labeled video samples can be obtained with a large numbers of annotators, the process is expensive and time-consuming. Also, manual annotations do not scale well to more problem domains, because they can be expensive and hard to obtain (e.g., labeling medical images requires specific expertise). Instead, a large quantity of untagged images and videos that are freely available can be employed for representation learning.

B. Raman et al. (Eds.): CVIP 2021, CCIS 1568, pp. 494–503, 2022.
https://doi.org/10.1007/978-3-031-11349-9_43

Like in image domain, 2-d Convolutional Neural Networks (2d-CNNs) has also shown to be effective in capturing the underlying video representations by transfer learning [3] and large scale supervised training [4]. Now the problem is how to learn these transferable underlying visual representations in a cost-effective way. The solution is to build a pretext-task that builds free training labels and train the network by solving it. This kind of technique is called Self-supervised learning in which the supervisory signal can be easily obtained from the data itself. For e.g., in image domain the self-supervised methods are rotation prediction [5], solving jigsaw puzzles [6], and predicting relative positions of image patches [7], etc. For video domain, such task involves temporal relations such as order prediction [8] and order verification [9, 10] etc. There are also tasks existed [11–17] which are based on 3D-CNN. However we focus on learning with 2d-CNN, as 2d-CNN is computationally efficient and also capable of capturing temporal relations [3, 4]. Existing 2d-CNN based methods are focused on single pretext-tasks which are either concatenation based [9, 10] or pair-wise based [8]. However, we learn the temporal relations between frames with 1-d convolutions. Moreover, we tend to enhance unsupervised video representation learning by joint optimization of two tasks using 2d (CNN) as a backbone.

Key contributions are:

- We propose an enhanced unsupervised video representation learning method by joint learning of order prediction and contrastive learning.
- We use the 1-d temporal convolutions on top of 2d-CNN to capture the temporal relationships between frames.
- We used a slight variant of AlexNet and ResNet as our backbone network and tested our method on UCF101 and HMDB51 datasets and, validate the learned features with state-of-the-art experiments in nearest neighbour retrieval and action recognition tasks.

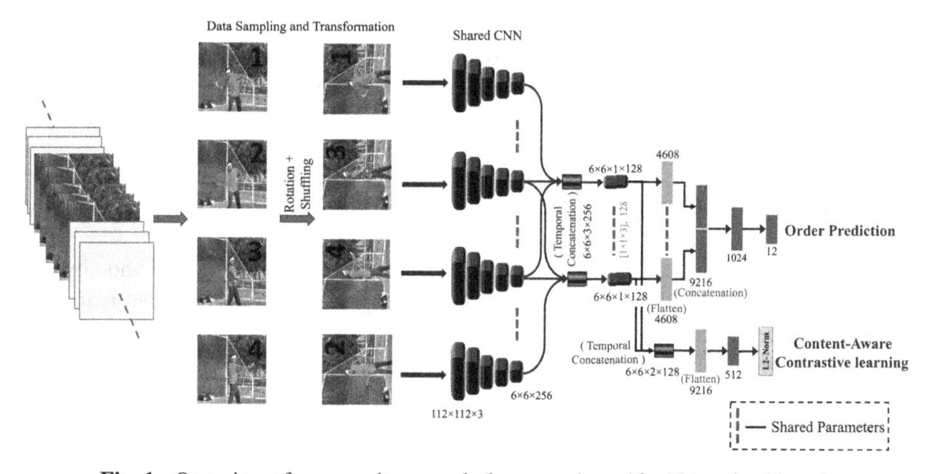

Fig. 1. Overview of proposed approach (here we show AlexNet as backbone).

2 Literature Review

There have been decades of research into video understanding, particularly action recognition, with the goal of solving other video-related problems such as action temporal localization [18], complex action recognition [19], automatic video caption [20] generation, surveillance [21, 22] and video retrieval [23–25], etc. In the literature, prior local spatio-temporal descriptors are hand-crafted, such as STIP [26] and HOG-3D [27] are initially proposed for video representations. Subsequently, the improved dense trajectory (iDT) descriptor has been proposed by Wang et al. [28], which combined the dominant HOF [29], MBH [30] and HOG [31] to produce the finest outcomes out of all the hand-crafted ones. Thereafter, deep learning based methods have been proposed in [3, 32–35], which have proved to be effective than handcrafted features. However, with human-annotated labels, the video representation learning is money-intensive and time consuming, despite the impressive results that have been achieved so far.

Recently, in computer vision community, learning representations from unlabeled data has emerged as a promising research area. The main objective is to design a pretext task that generates labels from data itself for training without human manual labels. In this regard, lots of methods are proposed in recent time such as colorization [36], rotation prediction [5], clustering [37], inpainting [38]. Recently, contrastive learning [39–41] has emerged as a powerful self-supervised representation learning method in image domain. The goal of the contrastive learning is to pull positives closer and push the negatives farther apart. The key problem is to extract the positives and negatives from the data. In recent, unsupervised learning methods are also explored in video domain. For e.g. in [9] author proposed sequence verification task using 2d-CNN for video representation learning. In [8] sorting the sequence task is explored to learn the representations. In [10], odd-one-out task is explored. Further, in [42], a deep reinforcement learning method is presented for designing sampling permutations policy for order prediction tasks. Some 3d CNN based methods are also been proposed for video representations learning like rotation prediction [11], pace prediction [12], space-time cubic puzzles [13], order prediction [14] and speednet [15]. Despite 3d-CNN based methods have shown great performance, it is computationally expensive. In contrast, in this paper, we propose to enhance self-supervised learning to improve the video representation capability using 2d-CNN as a backbone.

3 Proposed Method

Consider the N training videos $V = [V_1, V_2, \ldots, V_N]$ of width M, height N and channel C with frame F. The objective is to learn underlying video representations using the proposed enhanced unsupervised learning approach (see Fig. 1). For this we train a 2d-CNN a slight variant of alexnet. Next we discuss the tasks used for learning.

3.1 Sequence Order Prediction

As proposed in [8–10], we exploit the sequence order prediction task as it provides the strong supervisory signals to the network to learn underlying statistical temporal structure of video sequences. The order prediction task can be define as: given a tuple consists of n randomly shuffled frames. Let f(.) be the CNN to train to learn the representations. The goal is to predict the order of the sequence fed to the network, which can be one of the n!. However some of the sequences of n! are similar i.e. grouped in forward and backward permutations. Hence, only n!/2 is needed. We consider the $n = 4$ frames a tuple which makes $n!/2 = 24/2 = 12$ classes. First, 4 frames a video sequence is fed to the network which outputs $6 \times 6 \times 256$ from the conv5 layer and concatenated to $6 \times 6 \times 4 \times 256$. Thereafter $1 \times 1 \times 3$ convolution with filter size 128 with stride 1 is applied and outputs $[6 \times 6 \times 1 \times 128] \times 2$ followed by concatenation to 9216. Then two fully connected layers of 1024 and 12 units are added to the end. The cross entropy loss is used for the order prediction which is given as:

$$L_{ord} = -\log f^{j}(V_j) \tag{1}$$

3.2 Content-Aware Contrastive Learning

We further include the contrastive learning approach to regularize the learning process and enhance the representation power. Contrastive learning deals with the positives and negatives created using some transformations applied to the data. Then the network is train to learn representations by making positives closer against the negatives. We create the positives by sampling the frame sequences with different pace. Let $Pace(V_i, p)$ be the pace sampling transformation function, which samples the video frames with speed p. Here, we choose three pace candidates {normal, fast and super-fast}. For normal pace ($p = 1$), we sample the frames consecutively from the video. For fast pace ($p \geq 1$ & $p \leq 2$), we sample the frames by skipping the $p = [1, 2]$ frames. For super-fast pace ($p > 2$), we sample the frames by skipping the $p > 2$ frames. Let V_i and V_i^k be the ith video clip with its temporal transformed video using $Pace(V_i, p)$, where k represents k[th] pace candidate. We consider the positive pairs from the $\{V_i, V_i^k\}$ and negative pairs from the $\{V_i, V_j\}$ and $\{V_i, V_j^k\}$. Let F_i and F_i^k be the L2 normalized feature embeddings of video V_i and V_i^k. Now the objective is to make the positive pairs $\{F_i, F_i^k\}$ to closer and the negative pairs $\{F_i, F_j\}$ and $\{F_i, F_j^k\}$ to farther apart in the embedding space. Let $Sim(F_i, F_j)$ denotes the cosine similarity between F_i and F_j. The contrastive loss is defined as:

$$L_c = \sum_i \sum_k \left[-\log \frac{\exp(Sim(F_i, F_i^k)/\tau)}{\sum_j \exp(Sim(F_i, F_j^k)/\tau) + \sum_{i \neq j} \exp(Sim(F_i, F_j)/\tau} \right] \tag{2}$$

3.3 Implementation

We used the AlexNet [1] architecture with the slight modification as reported in Table 1. The input size is set to $112 \times 112 \times 3$. We also used ResNet50 [43] with 112×112

× 3 sized input. We choose the minibatch size as: first we select 24 videos and for each videos, 4 frames a clip is sampled using pace sampling transformation Pace(V_i, p) followed by rotation transformation. All clips randomly sampled from 128 × 170 × 3 sized frames. We set the learning rate to 1e-3 for 120k iterations. Also, we utilized the channel splitting, random horizontal flipping as augmentation technique as the network can find shortcut to learn representations.

Table 1. Modified AlexNet network

Layer	{Kernel size}	Output size
Input	-	112 × 112 × 3
Conv1	{3, 3}	55 × 55 × 96
Pool1 (max)	{2, 2}	27 × 27 × 96
Conv2	{3, 3}	27 × 27 × 256
Pool2 (max)	{2, 2}	13 × 13 × 256
Conv3	{3, 3}	13 × 13 × 384
Conv4	{2, 2}	13 × 13 × 384
Conv5	{3, 3}	13 × 13 × 256
Pool5 (max)	{2, 2}	6 × 6 × 256

4 Experimental Settings

The first dataset used is UCF-101 [44], which has 13 thousand video clips of human actions. The second dataset used is HMDB-51 [45], which has 7 thousand video clips divided in 51 actions. The network is trained using the UCF-101's training split-1 without labels. For video retrieval task, we used the evaluation protocol of [14], where 10 clips (each of 4 frames with skipping factor 1) per video are taken. The activations used from the flatten layer comprising 9216 units. We used the clips in the testing set as queries and retrieves the clips from the training set. We used the cosine distance for matching videos. We used the top-k accuracy for measure the retrieval performance. And for the action recognition task, first the network is initialize with our unsupervised approach and then fine-tune it and, used protocol of [14] for evaluation. To conduct the experiments, we used the Matlab tool with Nvidia tesla k40c gpu.

5 Results

5.1 Influence of Joint Learning on Retrieval Performance

Now we analyze the effect of our joint learning approach. For this we report the results obtain through the experiments on UCF-101 in Table 2. We can see that with contrastive learning, the network's learning ability increases as reflected in the retrieval performance.

Table 2. Effect of contrastive learning on retrieval performance (at Clip Level) for UCF-101

Methods	k = 1	k = 5	k = 10	k = 20	k = 50
Random	18.8	25.7	30.0	35.0	43.3
OPN [8]	19.9	28.7	34.0	40.6	51.6
Büchler et al. [42]	25.7	36.2	42.2	49.2	59.5
Ours (Order Prediction + Contrastive Learning)	**25.67**	**36.50**	**43.36**	**50.01**	**60.74**

5.2 Comparison to Existing Methods in Context of Retrieval Performance

Next, we analyze our results with other methods in Table 3, where we can observe that our method outperforms other 2d-CNN based methods. Numerically, with AlexNet, we achieve 25.67% top-1, 36.50% top-5, 43.36% top-10, 50.01% top-20 and 60.74% top-50. And with ResNet-50, we achieve 26.51% top-1, 37.87% top-5, 44.58% top-10, 52.79% top-20 and 65.17% top-50. Also, our method performs consistently with 3d-CNN based methods.

Table 3. Top-k retrieval accuracy (%) on UCF-101 (at Clip Level)

Network	Methods	k = 1	k = 5	k = 10	k = 20	k = 50
AlexNet	Jigsaw [6]	19.7	28.5	33.5	40.0	49.4
	OPN [8]	19.9	28.7	34.0	40.6	51.6
	Buchler et al. [42]	25.7	36.2	42.2	49.2	59.5
AlexNet (input-112 × 112)	**Ours**	**25.67**	**36.50**	**43.36**	**50.01**	**60.74**
ResNet-50 (input-112 × 112)	**Ours**	**26.51**	**37.87**	**44.58**	**52.79**	**65.17**
S3D-G (3d-CNN)	SpeedNet [15]	13.0	28.1	37.5	49.5	65.0
C3D (3d-CNN)	Clip Order [14]	12.5	29.0	39.0	50.6	66.9
C3D (3d-CNN)	Kumar et al. [17]	28.17	37.92	43.24	51.41	62.93

We visualize the conv6's activations of unsupervised trained model using [12] for 5 actions selected randomly in Fig. 2, where we can see that the model able to learn to pay attention to the object motion's areas.

Fig. 2. Visualization of conv6 layer of unsupervised trained model (AlexNet) using approach as [12]

5.3 Action Recognition

Next, we study the effect of unsupervised learning approach in initializing the network, we transfer the weights of the network trained from the unsupervised model and randomly initialize the fully connected layer and fine-tuned. The learning rate is set to 1e-4 and run for 15k iterations. The results are reported in Table 4, where we consistently achieve better results than other 2d-CNN based methods and satisfactory performance compare to 3d-CNN based methods for both datasets.

Table 4. Average action recognition accuracy (%) over 3 splits

	Method	Network	UCF-101	HMDB-51
2d-CNN	Random	Alexnet	46.5	13.3
	Imagenet	Alexnet	68.1	28.5
	Jigsaw [6]	Alexnet	51.5	22.5
	Shuffle&Learn [9]	Alexnet	50.9	18.1
	OPN [8]	Alexnet	56.3	22.1
	Buchler et al. [42]	Alexnet	58.6	25.0
	Ours	**Alexnet**	**60.02**	**25.11**
	Ours	**ResNet-18**	**62.11**	**25.68**
	Ours	**ResNet-50**	**66.84**	**28.42**
3d-CNN	CubicPuzzle [13]	3D-ResNet18	65.8	28.3
	3D RotNet [11]	3D-ResNet18	66.0	37.1
	Speednet [15]	I3D	66.7	43.7
	Clip Order [14]	C3D	65.6	28.4
	Kumar et al. [17]	C3D	66.8	25.6

6 Conclusion

This work deals with enhanced unsupervised video representation learning approach, which is accomplished by joint learning of sequence sorting task and contrastive learning.

For sorting task, we sample 4 frames a sequence and shuffle in one of the 12 permutations. For contrastive learning, first we sample 3 pace candidates and make these candidates of same video closer against the different videos. Then, we test our approach on two datasets. We report the results on two downstream tasks: video retrieval and action recognition, where we achieve better results. In future, we will explore more ways to sample the positives and negatives and explore more self-supervised learning techniques to enhance representation learning. In addition, we will explore self-supervised learning techniques in medical domain [46] as well.

References

1. Krizhevsky, A., Sutskever, I., Hinton, G.E.: Imagenet classification with deep convolutional neural networks. In: Advances in Neural Information Processing Systems, pp. 1097–1105 (2012)
2. Deng, J., Dong, W., Socher, R., Li, L.-J., Li, K., Fei-Fei, L.: ImageNet: a large-scale hierarchical image database. In: 2009 IEEE Conference on Computer Vision and Pattern Recognition, pp. 248–255. IEEE (2009)
3. Simonyan, K., Zisserman, A.: Two-stream convolutional networks for action recognition in videos. In: Advances in Neural Information Processing Systems, pp. 568–576 (2014)
4. Karpathy, A., Toderici, G., Shetty, S., Leung, T., Sukthankar, R., Fei-Fei, L.: Large-scale video classification with convolutional neural networks. In: Proceedings of the IEEE Conference on Computer Vision and Pattern Recognition, pp. 1725–1732. IEEE (2014)
5. Gidaris, S., Singh, P., Komodakis, N.: Unsupervised representation learning by predicting image rotations. In: International Conference on Learning Representations (2018)
6. Noroozi, M., Favaro, P.: Unsupervised learning of visual representations by solving Jigsaw puzzles. In: Leibe, B., Matas, J., Sebe, N., Welling, M. (eds.) ECCV 2016. LNCS, vol. 9910, pp. 69–84. Springer, Cham (2016). https://doi.org/10.1007/978-3-319-46466-4_5
7. Doersch, C., Gupta, A., Efros, A.A.: Unsupervised visual representation learning by context prediction. In: 2015 IEEE International Conference on Computer Vision (ICCV), pp. 1422–1430. IEEE (2015)
8. Lee, H.-Y., Huang, J.-B., Singh, M., Yang, M.-H.: Unsupervised representation learning by sorting sequences. In: 2017 IEEE International Conference on Computer Vision (ICCV), pp. 667–676. IEEE (2017)
9. Misra, I., Zitnick, C.L., Hebert, M.: Shuffle and learn: unsupervised learning using temporal order verification. In: Leibe, B., Matas, J., Sebe, N., Welling, M. (eds.) Computer Vision – ECCV 2016. LNCS, vol. 9905, pp. 527–544. Springer, Cham (2016). https://doi.org/10.1007/978-3-319-46448-0_32
10. Fernando, B., Bilen, H., Gavves, E., Gould, S.: Self-supervised video representation learning with odd-one-out networks. In: Proceedings of the IEEE Conference on Computer Vision and Pattern Recognition, pp. 3636–3645. IEEE (2017)
11. Jing, L., Yang, X., Liu, J., Tian, Y.: Self-supervised spatiotemporal feature learning via video rotation prediction. arXiv preprint arXiv:1811.11387 (2018)
12. Wang, J., Jiao, J., Liu, Y.-H.: Self-supervised video representation learning by pace prediction. In: Vedaldi, A., Bischof, H., Brox, T., Frahm, J.-M. (eds.) ECCV 2020. LNCS, vol. 12362, pp. 504–521. Springer, Cham (2020). https://doi.org/10.1007/978-3-030-58520-4_30
13. Kim, D., Cho, D., Kweon, I. S.: Self-supervised video representation learning with space-time cubic puzzles. In: Proceedings of the AAAI Conference on Artificial Intelligence, vol. 33, pp. 8545–8552 (2019)

14. Xu, D., Xiao, J., Zhao, Z., Shao, J., Xie, D., Zhuang, Y.: Self-Supervised Spatiotemporal Learning via Video Clip Order Prediction. In: 2019 IEEE/CVF Conference on Computer Vision and Pattern Recognition (CVPR), pp. 10326–10335. IEEE (2019)
15. Benaim, S., et al.: SpeedNet: learning the speediness in videos. In: 2020 IEEE/CVF Conference on Computer Vision and Pattern Recognition (CVPR), pp. 9922–9931. IEEE (2020)
16. Kumar, V., Tripathi, V., Pant, B.: Learning spatio-temporal features for movie scene retrieval using 3D convolutional autoencoder. In: International Conference on Computational Intelligence in Analytics and Information System (CIAIS) (2021)
17. Kumar, V., Tripathi, V., Pant, B.: Unsupervised learning of visual representations via rotation and future frame prediction for video retrieval. In: Singh, M., Tyagi, V., Gupta, P.K., Flusser, J., Ören, T., Sonawane, V.R. (eds.) ICACDS 2021. CCIS, vol. 1440, pp. 701–710. Springer, Cham (2021). https://doi.org/10.1007/978-3-030-81462-5_61
18. Chao, Y.-W., Vijayanarasimhan, S., Seybold, B., Ross, D.A., Deng, J., Sukthankar, R.: Rethinking the faster R-CNN architecture for temporal action localization. In: 2018 IEEE Conference on Computer Vision and Pattern Recognition (CVPR), pp. 1130–1139. IEEE (2018). https://doi.org/10.1109/CVPR.2018.00124
19. Hussein, N., Gavves, E., Smeulders, A.W.M.: Timeception for complex action recognition. In: 2019 IEEE Conference on Computer Vision and Pattern Recognition (CVPR), pp. 254–263. IEEE (2019). https://doi.org/10.1109/CVPR.2019.00034
20. Wang, B., Ma, L., Zhang, W., Liu, W.: Reconstruction network for video captioning. In: 2018 IEEE Conference on Computer Vision and Pattern Recognition (CVPR), pp. 7622–7631. IEEE (2018). https://doi.org/10.1109/CVPR.2018.00795
21. Kumar, V.: A Multi-face recognition framework for real time monitoring. In: 2021 Sixth International Conference on Image Information Processing (ICIIP). IEEE (2021)
22. Hu, X., Peng, S., Wang, L., Yang, Z., Li, Z.: Surveillance video face recognition with single sample per person based on 3D modeling. Neurocomputing **235**, 46–58 (2017)
23. Kumar, V., Tripathi, V., Pant, B.: Learning compact spatio-temporal features for fast content based video retrieval. IJITEE **9**, 2404–2409 (2019)
24. Mühling, M., et al.: Deep learning for content-based video retrieval in film and television production. Multimed. Tools Appl. **76**, 22169–22194 (2017)
25. Kumar, V., Tripathi, V., Pant, B.: Content based movie scene retrieval using spatio-temporal features. IJEAT **9**, 1492–1496 (2019)
26. Laptev, I.: On space-time interest points. IJCV **64**(2–3), 107–123 (2005)
27. Klaser, A., Marsza lek, M., Schmid, C.: A spatio-temporal descriptor based on 3D-gradients. In: BMVC (2008)
28. Wang, H., Schmid, C.: Action recognition with improved trajectories. In: 2013 IEEE International Conference on Computer Vision (ICCV), pp. 3551–3558. IEEE (2013)
29. Laptev, I., Marszalek, M., Schmid, C., Rozenfeld, B.: Learning realistic human actions from movies. In: 2008 IEEE Conference on Computer Vision and Pattern Recognition (CVPR), pp. 1–8. IEEE (2008)
30. Dalal, N., Triggs, B., Schmid, C.: Human detection using oriented histograms of flow and appearance. In: Leonardis, A., Bischof, H., Pinz, A. (eds.) ECCV 2006. LNCS, vol. 3952, pp. 428–441. Springer, Heidelberg (2006). https://doi.org/10.1007/11744047_33
31. Dalal, N., Triggs, B.: Histograms of oriented gradients for human detection. In: 2005 IEEE Conference on Computer Vision and Pattern Recognition (CVPR), pp. 886–893. IEEE (2005). https://doi.org/10.1109/CVPR.2005.177
32. Tran, D., Bourdev, L., Fergus, R., Torresani, L., Paluri, M.: Learning spatiotemporal features with 3D convolutional networks. In: 2015 IEEE International Conference on Computer Vision (ICCV), pp. 4489–4497. IEEE (2015)

33. Kumar, V., Tripathi, V., Pant, B.: Exploring the strengths of neural codes for video retrieval. In: Tomar, A., Malik, H., Kumar, P., Iqbal, A. (eds.) Machine Learning, Advances in Computing, Renewable Energy and Communication. LNEE, vol. 768, pp. 519–531. Springer, Singapore (2022). https://doi.org/10.1007/978-981-16-2354-7_46
34. Kumar, V., Tripathi, V., Pant, B.: Content based surgical video retrieval via multideep features fusion. In: 2021 IEEE International Conference on Electronics, Computing and Communication Technologies (CONECCT). IEEE (2021)
35. Kumar, V., Tripathi, V., Pant, B.: Content based fine-grained image retrieval using convolutional neural network. In: 2020 7th International Conference on Signal Processing and Integrated Networks (SPIN), pp. 1120–1125. IEEE (2020)
36. Zhang, R., Isola, P., Efros, A.A.: Colorful image colorization. In: Leibe, B., Matas, J., Sebe, N., Welling, M. (eds.) ECCV 2016. LNCS, vol. 9907, pp. 649–666. Springer, Cham (2016). https://doi.org/10.1007/978-3-319-46487-9_40
37. Caron, M., Bojanowski, P., Joulin, A., Douze, M.: Deep clustering for unsupervised learning of visual features. In: Ferrari, V., Hebert, M., Sminchisescu, C., Weiss, Y. (eds.) Computer Vision – ECCV 2018. LNCS, vol. 11218, pp. 139–156. Springer, Cham (2018). https://doi.org/10.1007/978-3-030-01264-9_9
38. Pathak, D., Krahenbuhl, P., Donahue, J., Darrell, T., Efros, A.A.: Context encoders: feature learning by inpainting. In: IEEE Conference on Computer Vision and Pattern Recognition (CVPR), pp. 2536–2544. IEEE (2016)
39. Wu, Z., Xiong, Y., Stella, X.Y., Lin, D.: Unsupervised feature learning via non-parametric instance discrimination. In: 2018 IEEE Conference on Computer Vision and Pattern Recognition (CVPR), pp. 3733–3742. IEEE (2018)
40. Chen, T., Kornblith, S., Norouzi, M., Hinton, G.: A simple framework for contrastive learning of visual representations. In: International Conference on Machine Learning, pp. 1597–1607. PMLR (2020)
41. He, K., Fan, H., Wu, Y., Xie, S., Girshick, R.: Momentum contrast for unsupervised visual representation learning. In: 2020 IEEE/CVF Conference on Computer Vision and Pattern Recognition (CVPR), pp. 9729–9738. IEEE (2020)
42. Büchler, U., Brattoli, B., Ommer, B.: Improving spatiotemporal self-supervision by deep reinforcement learning. In: Ferrari, V., Hebert, M., Sminchisescu, C., Weiss, Y. (eds.) ECCV 2018. LNCS, vol. 11219, pp. 797–814. Springer, Cham (2018). https://doi.org/10.1007/978-3-030-01267-0_47
43. He, K., Zhang, X., Ren, S., Sun, J.: Deep residual learning for image recognition. In: 2016 IEEE/CVF Conference on Computer Vision and Pattern Recognition (CVPR), pp. 770–778. IEEE (2016). https://doi.org/10.1109/CVPR.2016.90
44. Soomro, K., Zamir, A.R., Shah, M.: UCF101: a dataset of 101 human actions classes from videos in the wild. arXiv preprint arXiv:1212.0402 (2012)
45. Kuehne, H., Jhuang, H., Garrote, E., Poggio, T., Serre, T.: HMDB: a large video database for human motion recognition. In: 2011 International Conference on Computer Vision ICCV, pp. 2556–2563. IEEE (2011)
46. Kumar, V., et al.: Hybrid spatiotemporal contrastive representation learning for content-based surgical video retrieval. Electron. **11**, 1353 (2022). https://doi.org/10.3390/electronics11091353

Deep Two-Stage LiDAR Depth Completion

Moushumi Medhi(✉) and Rajiv Ranjan Sahay

Indian Institute of Technology, Kharagpur, Kharagpur, India
medhi.moushumi@iitkgp.ac.in, rajiv@ee.iitkgp.ac.in

Abstract. LiDAR depth completion aims at accurately estimating dense depth maps from sparse and noisy LiDAR depth scans, often with the aid of the color image. However, most of the existing deep learning-based LiDAR depth completion approaches focus on learning one-stage networks with computationally intensive RGB-D fusion strategies to compensate for the prediction errors. To eliminate such drawbacks, we have explored a simple yet effective two-stage learning framework where the former stage generates a coarse dense output which is processed in the latter stage to produce a fine dense depth map. The refined dense depth map is obtained at the output of the second stage by employing iterative feedback mechanism that removes any ambiguity associated with a single feed-forward network. Our two-stage learning mechanism allows for simple RGB-D fusion operations devoid of high computational overload. Experiments conducted on the KITTI depth completion benchmark validate the efficacy of our proposed method.

Keywords: LiDAR depth completion · Two-stage depth completion · Coarse depth generation · Depth refinement

1 Introduction

Depth completion of LiDAR (Light Detection and Ranging) data aims at generating dense depth map from sparse LiDAR samples obtained from the Lidar sensor. In the recent decade, deep learning-based methods [5,11,20] have significantly improved the quality of sparse depth completion results over traditional methods [7]. However, most methods have tackled the problem of sparse depth completion using a one-stage network where dense depth maps are recovered in a single shot. In such cases, the reliability of the completion results lies solely on the single-stage network. This necessitates more complex and computationally intensive network modules and fusion schemes for dense depth recovery. In the proposed work, we have considered a two-stage deep convolution neural network (DCNN) (Fig. 1) consisting of two sub-networks, namely, coarse depth completion network ($CDCN$) and fine depth completion network ($FDCN$). $CDCN$ learns to efficiently interpolate the sparse depth data to form a coarse depth

B. Raman et al. (Eds.): CVIP 2021, CCIS 1568, pp. 504–515, 2022.
https://doi.org/10.1007/978-3-031-11349-9_44

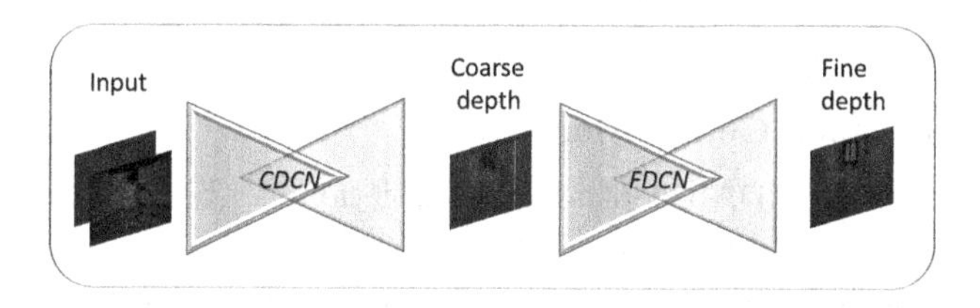

Fig. 1. Overall pipeline of the proposed two-stage LiDAR depth completion.

map in the first stage. *FDCN* subsequently refines the coarse depth map in the second stage using an iterative feedback mechanism. The contextual cues from the corresponding color image is leveraged in both the stages to guide the completion process. Our work is closely related to the recent work in [18] where final dense depth map is recovered from a coarse-to-fine (*FCFR*) network. The authors in [18] have used the outputs of [19] to obtain the coarse dense depth maps. *FCFR*-Net highlights RGB-D feature fusion strategy by using channel shuffle operations and computationally intensive energy functions. However, the major improvement has been brought about by the use of coarse dense depth map at the input and residual learning. On the other hand, we develop an end-to-end DCNN where we fully automate the generation of coarse dense depth map at an intermediate stage. Our work differs substantially from the approach in [18], considering that with the inclusion of several residual and dense blocks in *CDCN*, we have kept *FDCN* model structure relatively simple to facilitate easier fine-tuning and convergence. Furthermore, *FCFR*-Net [18] learns the residuals between the coarse depth maps obtained from [19] and the ground truth in a single feed-forward fashion. Whereas, we aspired to directly assess the dense depth map at first feed-forward pass through *FDCN* and then repeatedly compute the residuals in the subsequent iterations through a feedback mechanism. Once *FDCN* is trained along with *CDCN*, it can also be applied as an easy-to-plug-in module to existing LiDAR depth completion methods to enhance further the completion results. We conduct experiments on KITTI depth completion benchmark [24] to demonstrate the effectiveness of the proposed network, which was followed by an ablation study on the KITTI dataset to investigate the impact of different components of our network.

To summarize, we proposed the following contributions:

- We propose a novel two-stage DCNN architecture that consists of two sub-networks, namely, coarse depth completion network (*CDCN*) and fine depth completion network (*FDCN*) that tackles the problem of depth completion in two stages. *CDCN* generates coarse dense depth output, which acts as a prior for *FDCN* to synthesize the final dense depth map.
- We employ residual dense blocks (RDBs) for the first time, to the best of our knowledge, to tackle the problem of depth completion. RDB allows the preservation of relevant information contained in the fused multi-modal features of the preceding layers.
- An iterative feedback mechanism is employed in *FDCN* network to refine the output of *CDCN* iteratively.

2 Literature Survey

2.1 LiDAR Depth Completion

Existing methods on LiDAR depth map completion can be broadly divided into depth-only methods and image-guided methods.

Depth-Only Methods: Due to lack of guidance from color image for depth completion, these methods usually emphasize the use of auxiliary operators, e.g., sparsity observation masks [24], input confidence masks [15], to deal with the input sparse LiDAR data. [24] proposed sparse convolution modules that propagate observation masks across all layers of a CNN. [5] adopted normalized convolution layers [15] to propagate confidence values through consecutive layers in a CNN. On the other hand, a succession of basic image processing-based traditional hand-crafted operators were also successfully employed to generate a dense depth map. The authors in [4] employed compressed sensing in a deep recurrent auto-encoder for LiDAR depth completion. The authors in [23] trained a CNN in an adversarial framework for dense depth recovery.

Image-Guided Methods: Most of the existing methods [8,9] reasonably exploit the available color information, which usually guarantees better performance due to the presence of powerful information related to the overall contexts, semantics, occlusions or edges, etc. procured from the color image. The authors in [27] first trained two additional deep networks to estimate surface normal and occlusion boundary, which were later used to assist depth completion of indoor scenes. Following the same line of idea, [21] extended the use of surface normal to guide the completion of outdoor sparse LiDAR depth maps. The authors in [19] (STD) minimized a photo-consistency loss between the actual RGB image and the warped synthesized RGB image in a self-supervised (ss) framework. [10] proposed sparsity-invariant upsampling, average, and joint concatenation and convolution operations for the sparse input and feature maps. [12] proposed the use of depth coefficients as sparse depth representation and cross-entropy loss to avoid depth mixing. [26] included depth-normal constraints while training a unified CNN framework. [6] adopted the confidence propagation technique of [4] for RGB guided LiDAR depth completion. [1] fuses 2D and 3D features obtained using 2D and continuous convolutions, respectively, to complete sparse depth map. [25] used two parallel networks to predict local and global depth maps along with their confidences which were further used to fuse the two maps. A convolutional spatial propagation network (CSPN) was proposed in [3] to learn the affinity among neighboring pixels for diffusion of depth from known to missing depth regions. [2] facilitates dynamic adjustment of context and computational resources at each pixel in the proposed CSPN++ depth refinement module. [17] proposed a DCNN where depth is interpreted as closest depth plane distance from a specified plane and a residual value. The refinement module proposed in $FCFR$-Net [18] carries out energy-based fusion by pooling pixels that correspond to maximum energy in a local patch centered at the pixel. However, it remains challenging to automate the entire pipeline from scratch with highly sparse LiDAR depth data as input.

3 Residual Dense Blocks

Residual dense blocks (RDBs) have been extensively used for image super-resolution tasks [28] due to their capacity to extract abundant local features via densely connected convolutional layers. Local feature fusion in RDB facilitates richer contextual information flow from the preceding convolutional layers to the current layer. Besides, such residual and dense connections improve gradient flow and stabilize the training of deep models. However, they are not well-explored for the task of sparse depth completion. We, thereby, use sequence of RDBs in the decoder stream of the coarse dense depth completion network to adaptively learn the effective features from not just the currently fused multi-modal features but also from the previously fused hierarchical features.

4 Iterative Propagation

An iterative network with shared weights per iterations is equivalent to a recurrent network. Hence, this kind of training strategy not only saves memory but also allows increased number of training iterations per epoch. Our decoder network can be treated as cascade of K parts, $g_1, g_2, ..., g_K$, such that $f_1 = g_1(x), f_2 = g_2(g_1(x)), ..., f_K = g_K(g_{K-1}(g_{K-2}(...)))$. Here, K is the number of iterations carried out by the iterative propagation scheme. The crux of our iterative propagation method is to update x in the direction that improves the depth completion results with semi-dense ground truth depth maps as targets.

5 Design Overview

Our sparse to dense deep CNN architecture follows a meticulously designed two-stage pipeline. The overall pipeline of the proposed two-stage depth completion network is demonstrated in Fig. 2.

Coarse Depth Completion Network (*CDCN*): The depth and the RGB encoders consist of a cascade of simple yet non-trivial residual blocks (RB) for feature extraction. The first convolutional layer and the residual convolutional layer inside the residual block (pink box shown in Fig. 2) have the same stride (s) as shown above the RB blocks in Fig. 2, while the second convolutional layer has a stride of 1. Stride 2 (s2) is used for downsampling. Fusion of features from the RGB guidance branch is a concatenation of the encodings whereas, feature fusion from the depth encoder branch is carried out using element-wise feature addition rather than concatenation as desired dense depth output has the similar features to that of the input sparse depth map. The combined features are passed through a stack of combination of residual dense blocks (RDBs) [13]. Each RDB consists of densely connected three 3×3 conv2D and a 1×1 conv2D layers. Three up-convolutional blocks (Upconv) are used that perform 8x upscaling of the encoded feature maps to produce the upsampled coarse depth map at the output which has the same resolution as that of the input.

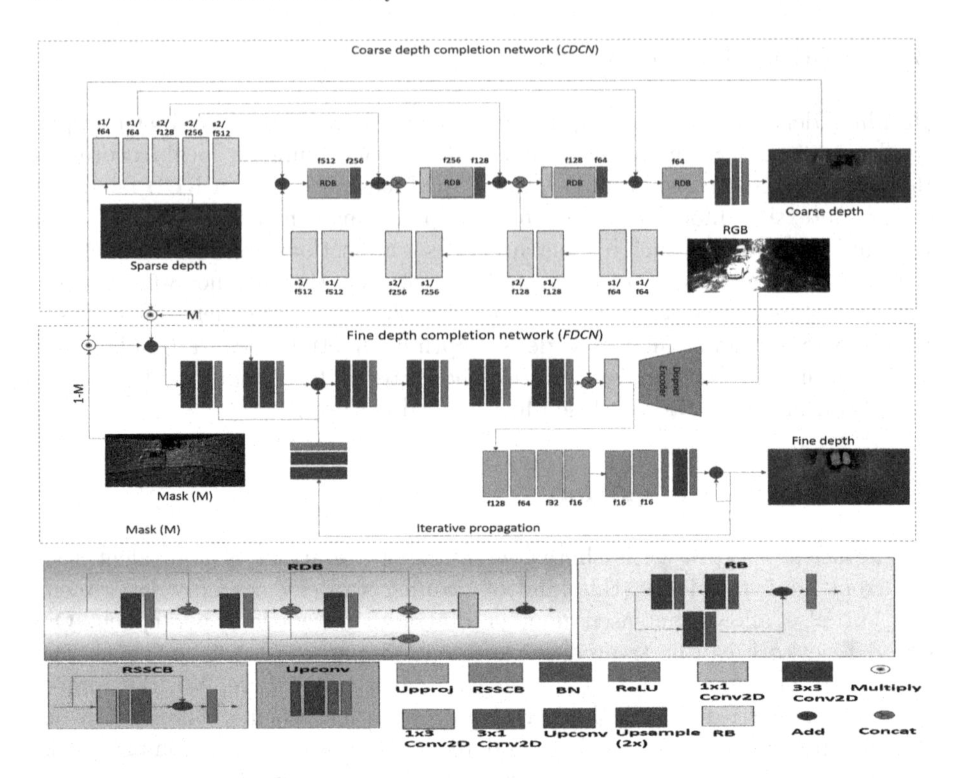

Fig. 2. Schematic of our proposed network architecture for LiDAR depth completion.

Fine Depth Completion Network (*FDCN*): The fine depth completion network is relatively easier to learn, given the coarse depth map as a prior for fine dense depth generation. Hence, we employ basic 2D convolutions in the depth encoder in the second stage for fine depth feature extraction. The encoder of the pretrained Dispnet model [29] is used as an RGB feature extractor. The depth and RGB features are fused using simple concatenation and a subsequent 1×1 convolution layer. Four "Upproj" layers are then used to gradually upsample the fused feature maps in the decoder stream of the fine depth completion network. "Upproj" denotes the up-projection layer introduced in [16] as a fast upsampling strategy. Subsequently, we use two residual spatially separable convolutional blocks (RSSCB) in our hardware-effective refinement stage that exploit the horizontal and vertical dimensions of the refined fused feature maps. RSSCB blocks are based on spatially separable convolutional layers with horizontal and vertical kernels of sizes 1×3 and 3×1, respectively. RSSCB modules are equivalent to RB modules, but they are more computationally efficient with fewer parameters. Lastly, a 3×3 2D convolutional layer is used to obtain the final dense depth map at the output of the fine depth completion network. The core part of the fine depth completion network is an iterative feedback mechanism enclosed in the decoder part of the network. We convert the dense depth map estimated by the previous iteration into a latent representation and feed it into the decoder network together with the color embeddings to refine the predicted depth iteratively. The last iteration of a single training step generates the final upsampled and refined dense depth map.

6 Loss Function

The training loss used for LiDAR depth map restoration consists of two prediction error terms $\mathcal{L}_p^c$, $\mathcal{L}_p^f$ at the coarse (c) and fine (f) levels, respectively.

$$\mathcal{L}_p^c = \sum_{j=1}^{N} \left\|(z_j^c - d_j) \odot M_d\right\|^{\gamma} \tag{1}$$

$$\mathcal{L}_p^f = \sum_{j=1}^{N}\sum_{k=1}^{K}(\rho_k \left\|z_{jk}^f - d_j) \odot M_d\right\|^{\gamma} \tag{2}$$

where z and d denote the output and semi-dense ground truth depth map, respectively. $M_d(M_d = 1, \text{if } d > 0 \text{ else } 0)$ denotes the ground-truth validity mask and k denotes the iterative feedback step. $\odot$ stands for element-wise multiplication. N and K denote the total number of depth pixels and iterations, respectively. The depth values obtained from the output of the last iteration is penalized more compared to the outputs of the previous iterations, i.e., $\rho_K > \rho_{K-1} > ... > \rho_1$. The overall objective function is a combination of $\mathcal{L}_p^c$ and $\mathcal{L}_p^f$ as follows:

$$\mathcal{L} = \lambda_1 \mathcal{L}_p^c + \lambda_2 \mathcal{L}_p^f \tag{3}$$

where $\lambda_{1,2}$ adjusts the weights between the terms of the loss function.

7 Experiments

7.1 Dataset and Training Details

We use KITTI depth completion benchmark [24] with 86.89K RGB-D pairs with semi-dense (30% dense) ground truth depth maps for training and 1K validation data for testing. Data augmentation was carried out during training phase through random cropping, scaling, and horizontal flipping. We have trained our network in three stages. In the first stage, we have trained $CDCN$ alone. In the second stage, we freeze the layers of $CDCN$ and update only $FDCN$ weights. At last, we fine-tune both the networks together in an end-to-end manner. We use an ADAM [14] optimizer ($\beta 1 = 0.9$, $\beta 2 = 0.999$) with a initial learning rate of 0.0001 and batch size of 4. We set the γ value to 1 (Eqs. (1) and (2)) in the supervised learning objective function. $\lambda_{1,2}$ in Eq. (3) are set to 1. In Eq. (2), $\rho_k = [0.6, 0.8, 1.0]$ for $k = [1, 2, 3]$, respectively.

Table 1. Comparisons with the state-of-the-art methods on KITTI depth completion validation set [24], ranked by RMSE (units mm). ↓ indicates that lower values are better.

Method	RMSE↓	MAE↓	iRMSE↓	iMAE↓
EncDec-Net[EF] [6]	1007.71	236.83	2.75	0.99
EncDec-Net[LF] [6]	1053.91	295.92	3.42	1.31
MS-Net[EF] [6]	932.01	209.75	2.64	0.92
MS-Net[LF] [6]	908.76	209.56	2.50	0.90
DC-3coeff [12]	1212.00	241.00	–	–
DC-all [12]	1061.00	252.00	–	–
STD (ss) [19]	1384.85	358.92	4.32	1.60
STD (s) [19]	858.02	311.47	3.06	1.67
Depth-Normal [26]	811.07	236.67	2.45	1.11
FuseNet [1]	785.00	217.00	2.36	1.08
DeepLiDAR [21]	**687.00**	215.38	2.51	1.10
CSPN++ [2]	725.43	207.88	–	–
PR [17]	867.12	204.68	**2.17**	0.85
TWISE [11]	879.40	**193.40**	2.19	**0.81**
FCFR [18]	802.62	224.53	2.39	1.00
Ours (s)	842.82	228.61	2.74	1.02

7.2 Evaluation Metrics

Following the existing LiDAR depth completion algorithms [5,20], we report combination of standard metrics for evaluation during comparison and ablation study: mean absolute relative error (REL): $REL = \frac{1}{n}\Sigma_{i=1}^{n}(|\hat{f}_i - f_i|)/f_i$, mean absolute error (MAE): $MAE = \frac{1}{n}\Sigma_{i=1}^{n}|(\hat{f}_i - f_i)|$, inverse RMSE (iRMSE): $iRMSE = \sqrt{\frac{1}{n}\Sigma_{i=1}^{n}(\frac{1}{\hat{f}_i} - \frac{1}{f_i})^2}$, Inverse MAE (iMAE): $iMAE = \frac{1}{n}\Sigma_{i=1}^{n}|(\frac{1}{\hat{f}_i} - \frac{1}{f_i})|$, root mean squared error (RMSE): $RMSE = \sqrt{\frac{1}{n}\Sigma_{i=1}^{n}(\hat{f}_i - f_i)^2}$, and δ_t. δ_t denotes the percentage of depth pixels in the output depth map which have relative error less than a predefined threshold t and, hence, must satisfy $max(\hat{f}_i/f_i, f_i/\hat{f}_i) < t$.

7.3 Comparison with the State-of-the-Art

Quantitative Comparison: In Table 1, we report quantitative results of our method as well as the published state-of-the-art (SoTA) approaches [1,2,6,11,12,17–19,21,26] on the KITTI validation dataset [24]. MS-Net [6] achieves their best results with late fusion (*LF*) compared to early fusion (*EF*) of RGB-D data. Sparse-to-dense (*STD*) method [19] trained their networks using both self-supervised (*ss*) and supervised (*s*) learning. *FCFR*-Net [18] achieves an

RMSE of 802.62 at the expense of slightly greater inference time and model parameters, as notified by the authors in [18], and is shown in Table 2.

Table 2. Comparison of inference time (s/frame) and model parameters with the SoTA two-stage LiDAR depth completion method of *FCFR*-Net [18]. The input size is of resolution 352 × 1216.

	FCFR [18]	*CDCN*+*FDCN*	*FDCN*
Runtime (s)	0.13	0.22	**0.07**
#Params (M)	50.59	41.88	**18.48**

Table 3. Effect of using *FDCN* as a plugin refinement module to baseline STD method [19]. ↓ indicates that lower values are better.

Method	RMSE↓	MAE↓	iRMSE↓	iMAE↓
STD (ss) [19]	1384.85	358.92	4.32	1.60
STD (ss) + *FDCN*	**1264.59**	**398.97**	**3.95**	**1.57**

Unlike [18], our *CDCN* is trained from scratch, facilitating the automatic generation of coarse depth map at the intermediate level. Whereas the coarse depth maps in [18] were obtained from the outputs of [19] for all training and testing purposes. Once trained, our *FDCN* can also be used as a plug-in to existing methods as shown in Table 3.

Visual Comparison: We present qualitative comparison results in Fig. 3 against SoTA methods [6,19,22,24] whose test codes and best pretrained weights on KITTI data are publicly available. Our model is able to produce better depth completion results (shown in (Fig. 3 (h))) than that of *STD* [19], MS-Net [6], SparsConvs [24] and DFuseNet [22] preserve the general structures more appropriately. Specifically, our method recovers the depth of small/thin objects more accurately and produces clearer and sharper boundaries of the cars and the poles than the aforementioned methods.

7.4 Ablation Experiments

Importance of Two-Stage Learning Framework: The first, second, and fourth rows in Table 4 show the significance of the two-stage learning paradigm. Significant improvements have been brought about by stacking *FDCN* to *CDCN*. The results in the second row in Table 4 were obtained by freezing *CDCN* and training *FDCN* alone. While the results in the fourth row were obtained by finetuning *CDCN* and *FDCN* together. Figure 4 shows some visual qualitative results on the KITTI depth completion validation set [24] to demonstrate the improvement in the quality of the completion results obtained with our two-stage dense depth map prediction algorithm.

Effect of Iterative Refinenment: The quantitative results in the third and fourth rows of Table 4 validate the use of iterative residual refinement in *FDCN*. Noticeable improvements in the error metrics could be observed by iteratively refining the dense depth estimate of *FDCN*.

Importance of RGB Guidance: We fuse the RGB and depth streams in *CDCN* and *FDCN* in order to combine the LiDAR data and RGB images in

Fig. 3. Visual comparison of the results for LiDAR depth map completion. (a) RGB image, (b) Sparse LiDAR depth map (dilated for visualization). Completed depth maps obtained by (c) STD (ss) [19], (d) STD (s) [19], (e) MS-Net[LF] [6], (f) SparsConvs [24], (g) DFuseNet [22], and (h) proposed method.

the feature space, leading to better results, as can be seen from the first and third rows in Table 5. The visual results for this ablation study are presented in Fig. 5. Ground truth depth being semi-dense, ground truth information for the occlusion boundaries are not available during training. Hence, at testing time, the missing information at the occlusion boundaries could not be recovered from the initial training epochs when using only sparse depth as input without any RGB guidance, as shown in Fig. 5. However, we can successfully estimate those missing regions by running the same number of epochs with the help of RGB guidance.

Importance of RDB: We present comparison results in the second and third rows of Table 5, between outputs obtained using simple convolution layers and that obtained using RDB at the decoder stage. Clearly, the use of RDB boosts the performance level of the proposed model. The visual results are shown in Fig. 6.

Table 4. Quantitative ablation study to analyze the effect of two-stage learning framework and iterative learning scheme. ↑ indicates higher the better, ↓ indicates lower the better.

	RMSE↓	MAE↓	REL↓	$\delta_{1.25}$ ↑	$\delta_{1.25^2}$ ↑	$\delta_{1.25^3}$ ↑
CDCN	2764.06	1167.83	0.074	0.9654	0.9891	0.9937
CDCN (freezed) + *FDCN* w Iter. Ref.	1411.17	494.30	0.039	0.9889	0.9913	0.9952
finetuned *CDCN* + *FDCN* w/o Iter. Ref.	997.34	371.95	0.024	0.9913	0.9958	0.9975
finetuned *CDCN* + *FDCN* w Iter. Ref.	**842.82**	**228.61**	**0.011**	**0.9980**	**0.9994**	**0.9997**

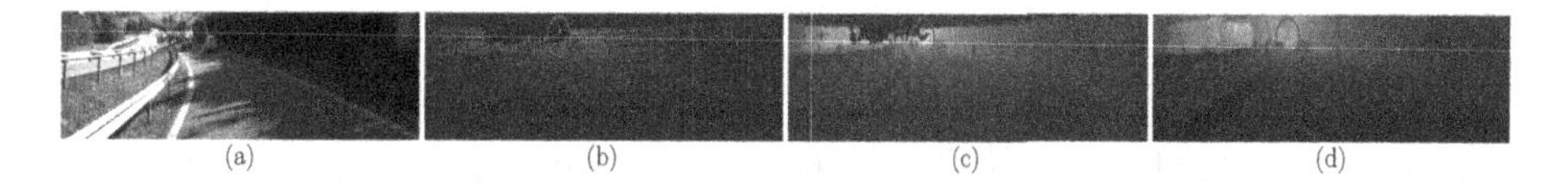

Fig. 4. Qualitative ablation study to analyze the impact of adopting two-stage depth completion framework. (a) RGB, (b) sparse depth map (dilated for visualization), (c) coarse depth map, (d) Fine depth map.

Table 5. Quantitative ablation study to analyze the impact of RGB guidance and RDBs on *CDCN*. ↑ indicates that the higher values are better, and ↓ indicates that the lower values are better.

	RMSE↓	MAE↓	REL↓	$\delta_{1.25}$ ↑	$\delta_{1.25^2}$ ↑	$\delta_{1.25^3}$ ↑
w/o RGB guidance	4030.56	2374.91	0.139	0.8229	0.9821	0.9929
Simple convolution (w/o RDBs)	3780.04	2005.22	0.116	0.8996	0.9858	0.9924
w/ RGB guidance + w/ RDBs	**2764.06**	**1167.83**	**0.074**	**0.9654**	**0.9891**	**0.9937**

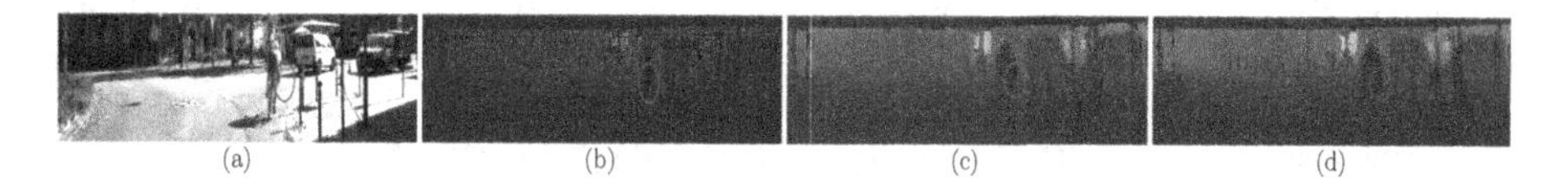

Fig. 5. Qualitative ablation study to analyze the impact of using RGB guidance in *CDCN*. (a) RGB, (b) sparse depth map (dilated for visualization). Coarse depth maps obtained (c) without RGB guidance and (d) with RGB guidance.

Fig. 6. Qualitative ablation study to analyze the impact of using RDB modules in the decoder stream of *CDCN*. (a) RGB, (b) sparse depth map (dilated for visualization). Coarse depth maps obtained by replacing RDBs with (c) simple 2D convolutional blocks and by using the (d) RDBs as proposed in the decoder of *CDCN*.

8 Conclusion

This work proposes a two-stage LiDAR depth completion method where dense depth map is generated using two consecutive subnetworks: coarse depth completion network ($CDCN$) and fine depth completion network ($FDCN$). Our fully trained $FDCN$ may be potentially useful to further refine the completion results of the existing methods. We conducted qualitative and quantitative comparisons with several SoTA LidAR depth completion algorithms which shows that our proposed method produces competitive results.

References

1. Chen,, Y., Yang, B., Liang, M., Urtasun, R.: Learning joint 2D–3D representations for depth completion. In: IEEE International Conference on Computer Vision, pp. 10023–10032 (2019)
2. Cheng, X., Wang, P., Guan, C., Yang, R.: CSPN++: learning context and resource aware convolutional spatial propagation networks for depth completion. In: AAAI Conference on Artificial Intelligence. vol. 34, pp. 10615–10622 (2020)
3. Cheng, X., Wang, P., Yang, R.: Learning depth with convolutional spatial propagation network. IEEE Trans. Pattern Anal. Mach. Intell. **42**(10), 2361–2379 (2019)
4. Chodosh, N., Wang, C., Lucey, S.: Deep convolutional compressed sensing for LiDAR depth completion. In: Jawahar, C.V., Li, H., Mori, G., Schindler, K. (eds.) ACCV 2018. LNCS, vol. 11361, pp. 499–513. Springer, Cham (2019). https://doi.org/10.1007/978-3-030-20887-5_31
5. Eldesokey, A., Felsberg, M., Khan, F.S.: Propagating confidences through CNNs for sparse data regression. In: British Machine Vision Conference, p. 14 (2018)
6. Eldesokey, A., Felsberg, M., Khan, F.S.: Confidence propagation through CNNs for guided sparse depth regression. IEEE Trans. Pattern Anal. Mach. Intell. **42**(10), 2423–2436 (2019)
7. Ferstl, D., Reinbacher, C., Ranftl, R., Rüther, M., Bischof, H.: Image guided depth upsampling using anisotropic total generalized variation. In: IEEE International Conference on Computer Vision, pp. 993–1000 (2013)
8. Hambarde, P., Dudhane, A., Patil, P.W., Murala, S., Dhall, A.: Depth estimation from single image and semantic prior. In: IEEE International Conference on Image Processing, pp. 1441–1445 (2020)
9. Hambarde, P., Murala, S.: S2DNET: depth estimation from single image and sparse samples. IEEE Trans. Comput. Imaging **6**, 806–817 (2020)
10. Huang, Z., Fan, J., Cheng, S., Yi, S., Wang, X., Li, H.: HMS-NET: hierarchical multi-scale sparsity-invariant network for sparse depth completion. IEEE Trans. Image Process. **29**, 3429–3441 (2019)
11. Imran, S., Liu, X., Morris, D.: Depth completion with twin surface extrapolation at occlusion boundaries. In: IEEE Conference on Computer Vision and Pattern Recognition, pp. 2583–2592 (2021)
12. Imran, S., Long, Y., Liu, X., Morris, D.: Depth coefficients for depth completion. In: IEEE Conference on Computer Vision and Pattern Recognition, pp. 12438–12447 (2019)
13. Khan, M.F.F., Troncoso Aldas, N.D., Kumar, A., Advani, S., Narayanan, V.: Sparse to dense depth completion using a generative adversarial network with intelligent sampling strategies. In: Proceedings of the 29th ACM International Conference on Multimedia, pp. 5528–5536 (2021)

14. Kingma, D.P., Ba, J.: Adam: a method for stochastic optimization. In: International Conference on Learning Representations (2015)
15. Knutsson, H., Westin, C.F.: Normalized and differential convolution. In: IEEE Conference on Computer Vision and Pattern Recognition, pp. 515–523 (1993)
16. Laina, I., Rupprecht, C., Belagiannis, V., Tombari, F., Navab, N.: Deeper depth prediction with fully convolutional residual networks. In: International Conference on 3D Vision, pp. 239–248 (2016)
17. Lee, B.U., Lee, K., Kweon, I.S.: Depth completion using plane-residual representation. In: IEEE Conference on Computer Vision and Pattern Recognition, pp. 13916–13925 (2021)
18. Liu, L., Song, X., Lyu, X., Diao, J., Wang, M., Liu, Y., Zhang, L.: FCFR-Net: feature fusion based coarse-to-fine residual learning for depth completion. In: AAAI Conference on Artificial Intelligence, vol. 35, pp. 2136–2144 (2021)
19. Ma, F., Cavalheiro, G.V., Karaman, S.: Self-supervised sparse-to-dense: self-supervised depth completion from lidar and monocular camera. In: International Conference on Robotics and Automation, pp. 3288–3295 (2019)
20. Ma, F., Karaman, S.: Sparse-to-dense: depth prediction from sparse depth samples and a single image. In: IEEE International Conference on Robotics and Automation, pp. 1–8 (2018)
21. Qiu, J., et al.: DeepLiDAR: deep surface normal guided depth prediction for outdoor scene from sparse LiDAR data and single color image. In: IEEE Conference on Computer Vision and Pattern Recognition, pp. 3313–3322 (2019)
22. Shivakumar, S.S., Nguyen, T., Miller, I.D., Chen, S.W., Kumar, V., Taylor, C.J.: DFuseNet: deep fusion of RGB and sparse depth information for image guided dense depth completion. In: IEEE Intelligent Transportation Systems Conference, pp. 13–20 (2019)
23. Tsuji, Y., Chishiro, H., Kato, S.: Non-guided depth completion with adversarial networks. In: International Conference on Intelligent Transportation Systems. pp. 1109–1114 (2018)
24. Uhrig, J., Schneider, N., Schneider, L., Franke, U., Brox, T., Geiger, A.: Sparsity invariant CNNs. In: International Conference on 3D Vision, pp. 11–20 (2017)
25. Van Gansbeke, W., Neven, D., De Brabandere, B., Van Gool, L.: Sparse and noisy LiDAR completion with RGB guidance and uncertainty. In: International Conference on Machine Vision Applications, pp. 1–6 (2019)
26. Xu, Y., Zhu, X., Shi, J., Zhang, G., Bao, H., Li, H.: Depth completion from sparse LiDAR data with depth-normal constraints. In: IEEE International Conference on Computer Vision, pp. 2811–2820 (2019)
27. Zhang, Y., Funkhouser, T.A.: Deep depth completion of a single RGB-D image. In: IEEE Conference on Computer Vision and Pattern Recognition, pp. 175–185 (2018)
28. Zhang, Y., Tian, Y., Kong, Y., Zhong, B., Fu, Y.: Residual dense network for image super-resolution. In: IEEE Conference on Computer Vision and Pattern Recognition, pp. 2472–2481 (2018)
29. Zhou, T., Brown, M., Snavely, N., Lowe, D.G.: Unsupervised learning of depth and ego-motion from video. In: Proceedings of the IEEE Conference on Computer Vision and Pattern Recognition, pp. 1851–1858 (2017)

3D Multi-voxel Pattern Based Machine Learning for Multi-center fMRI Data Normalization

Anoop Jacob Thomas[1(✉)] and Deepti R. Bathula[2]

[1] Department of Computer Science and Engineering, Indian Institute of Information Technology Tiruchirappalli, Tiruchirappalli 620012, Tamil Nadu, India
anoopjt@iiitt.ac.in

[2] Department of Computer Science and Engineering, Indian Institute of Technology Ropar, Rupnagar 140001, Punjab, India
bathula@iitrpr.ac.in

Abstract. Multi-center fMRI studies help accumulate significant number of subjects to increase the statistical power of data analyses. However, the seemingly ambitious gain is hindered by the fact that differences between centers have significant effects on the imaging results. We present a novel machine learning (ML) based technique, which uses non-linear regression with multi-voxel based anatomically informed contextual information, to help normalize multi-center fMRI data to a chosen reference center. Accuracy graphs were obtained by thresholding the estimated maps at high p-values of $p < 0.001$ after kernel density estimation. Results indicate significant reduction in spurious activations and more importantly, enhancement of the genuine activation clusters. Group level ROI based analysis reveals changes in activation pattern of clusters that are consistent with their role in cognitive function. Furthermore, as the mapping functions exhibit the tendency to induce sensitivity to the regions associated with the task they can help identify small but significant activations which could otherwise be lost due to population based inferences across centers.

Keywords: Multi-center · fMRI · Variability · Inter-scanner · Machine learning · Multi-voxel pattern · Contextual information · Correction functions

1 Introduction

Functional Magnetic Resonance Imaging (fMRI) is a relatively new form of neuroimaging [1]. Compared to other innovations in the domain, the blood oxygenation level dependent (BOLD) method of fMRI has several significant advantages: it is non-invasive, does not involve radiation or radio-active tracers and has excellent spatial and good temporal resolution. These attractions have made fMRI

B. Raman et al. (Eds.): CVIP 2021, CCIS 1568, pp. 516–528, 2022.
https://doi.org/10.1007/978-3-031-11349-9_45

a popular tool for imaging brain function both in clinical and research arenas. Current applications of fMRI include: brain function mapping, surgical planning, assessment of drug interventions and diagnosis/monitoring of neuro-degenerative disorders.

However, fMRI also suffers from problems that make the signal estimation a difficult task. Crucial aspects to consider include variation in hemodynamic response across voxels, low signal-to-noise ratio (SNR) and low contrast to noise ratio (CNR). These issues are compounded by the fact that BOLD signal change associated with neural activity is quite subtle, e.g., only about 1–2% at 1.5T [2]. Therefore, even moderate levels of noise in the data significantly impair the effectiveness of algorithms for detection and localization of brain activation regions. To compensate for low SNR, fMRI data is averaged over repeated measurements of physiologic activity and smoothed using spatial filtering techniques. However, the improvement in SNR provided by such post processing technique still does not render the data fit for studying human subjects individually. Consequently, many studies rely on pooling data across subjects to achieve more robust activation results. Such studies address questions regarding brain activation effects in populations of subjects [3].

Multi-center fMRI studies not only help accumulate significant number of subjects to increase the statistical power [4] of data analyses but also improve the generalizability of the results by incorporating subjects from potentially diverse demographic distributions. However, the seemingly ambitious gain is hindered by the fact that differences between centers have significant effects on the imaging results (Fig. 1). The variability in centers occurs due to heterogeneous scanner hardware, scanning protocols and environmental noise, with main factors being magnetic field strength, geometry of field in-homogeneity, transmit and receive coil characteristics, k-space trajectory, k-space filtering methods etc.

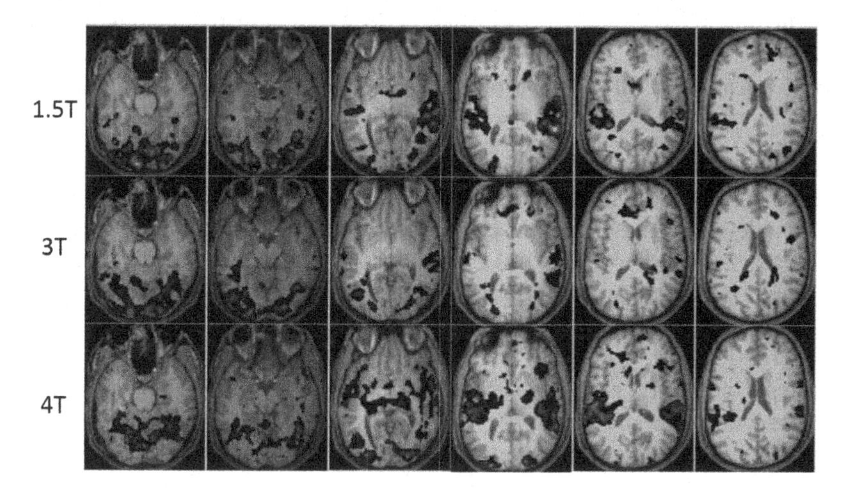

Fig. 1. Activation maps showing variation in the functionally activated regions ($p < 0.05$) for the same sensorimotor task of a single subject at three different scanners (three rows) [z = 40, 48, 54, 61, 70, 76]

Not only fMRI, but structural MR images are also affected by the scanner-to-scanner variations. Kelemen et al. [5] explored the use of correction functions to reduce the differences in structural MR images from multi-center MRI studies. Inspired by the idea of using correction functions, we explore the use of machine learning (ML) techniques to help normalize multi-center fMRI data. Specifically, we use learning algorithms to find mapping functions that can correct voxel based statistical activation parameters using multi-voxel contextual features obtained at a source site for a specific task in order to estimate what they would have been if the scans had been acquired at a chosen reference scanner. We use two complementary learning algorithms: artificial neural networks and random forests to learn the mapping functions. Although less reproducible across centers, activation maps/patterns are the most commonly used criterion to assess the results of fMRI analyses [6]. Hence, we chose to evaluate the performance of the correction functions in terms of number of activated voxels or activation pattern before and after correction with respect to a chosen reference activation map. Our results signify the potential of ML algorithms in reducing the inter-scanner variability in multi-center fMRI activations.

2 Materials and Methods

2.1 Dataset

We used the FBIRN Phase 1 Traveling Subjects dataset [7] where five healthy subjects were imaged on two occasions at 10 different scanners (ranging from 1.5 to 4.0T) located in geographically different locations. As this particular dataset was created specifically to help assess test-retest and between-site reliability of fMRI, it is conducive for the development of methods for calibration of between center differences in fMRI results. All five subjects were English speaking, right-handed males with mean age of 25.2. Though the scanners varied in field strength, vendor, coil type and functional sequence, a standard scanning protocol was used at all the centers to acquire a total of 20 scans per participant.

Sensorimotor Task. We chose the Sensorimotor (SM) task to study the inter-scanner differences in fMRI activations as it is considered as one of the most reproducible tasks [8]. Based on the simple block design paradigm, each block of this task involved 15 s (5 TRs) of rest (staring at fixation cross) followed by 15 s (5 TRs) of sensorimotor activity where subjects were instructed to tap their fingers bilaterally in synchrony with binaural tones, while watching 3 Hz flashing checkerboard. There were eight full cycles of rest-active blocks followed by a five TR rest period at the end; for a total of 85 TRs (4.25 min). There were at least eight sensorimotor scans per subject from the two visits to the scanner. Some of the subjects visited the scanners four times.

2.2 Data Analysis

Preprocessing. Data processing was accomplished using Nipype [9] that allows for pipelining of neuroimaging work-flow by leveraging different software packages. Firstly, large spikes in each run of sensorimotor data were removed using AFNI [10]. Despiking was followed by motion correction using SPM8 [11] where the functional scans were spatially registered to the mean functional image constructed for each subject, so as to spatially align all the scans of a subject. The data were then slice-time corrected using SPM8 followed by detrending of the time-series signals with a second order polynomial using AFNI. Temporal high-pass filtering with a lower cut-off frequency of 0.02 Hz using FSL [12] was performed to remove the physiological noise like cardiac and respiration. Finally the data was smoothed using a $7 \times 7 \times 7$ mm kernel using SPM8 to increase the SNR.

Additionally, as the hemodynamic response functions in white matter (WM) and gray matter (GM) are characteristically different, *a priori* knowledge regarding the tissue type of voxels can play an important role in reducing the differences in activations from multi-center studies. Consequently, the anatomical (T1) scan for each subject was co-registered to the mean-functional image and segmented using VBM8 [13] to produce tissue-probability maps (GM, WM and CSF).

Statistical Analysis. In this work, level 1 statistical image analysis of individual subject data was performed using Keith Worsley's FMRISTAT package [14] (http://www.math.mcgill.ca/keith/fmristat/). Each individual run of sensorimotor data was analyzed using General Linear Model (GLM) with correlated errors. The block design based experimental paradigm for the SM task was convolved with a hemodynamic response function modelled as a double gamma function [15] to produce the main regressor. Based on univariate analysis, regression coefficients at each voxel were estimated using least squares fitting of the individual time series with the design matrix. Finally, the estimated coefficients along with the corresponding variances for each scan were used to obtain a run-level T-statistic map for activation detection. Based on the assumption that variation between the individual runs of the SM task within a session is minimal, the results of individual run-level analysis are combined using fixed-effects model to generate the session-level statistical activation map (standard practice in all neuroimaging software). Furthermore, as the two visits are scheduled on consecutive days, session-level results are combined using random-effects model to obtain the site-level statistical activation map for each of the subjects.

2.3 Machine Learning

A number of supervised learning methods have been introduced in the last decade. In an effort to find correction functions to reduce inter-scanner variability in activations detected from fMRI data, we explore the potential of two classic and complementary machine learning techniques. An overview of the proposed methodology is depicted in Fig. 2.

Techniques like Artificial Neural Networks and Random Forests were explored, with two different features sets which were manually created.

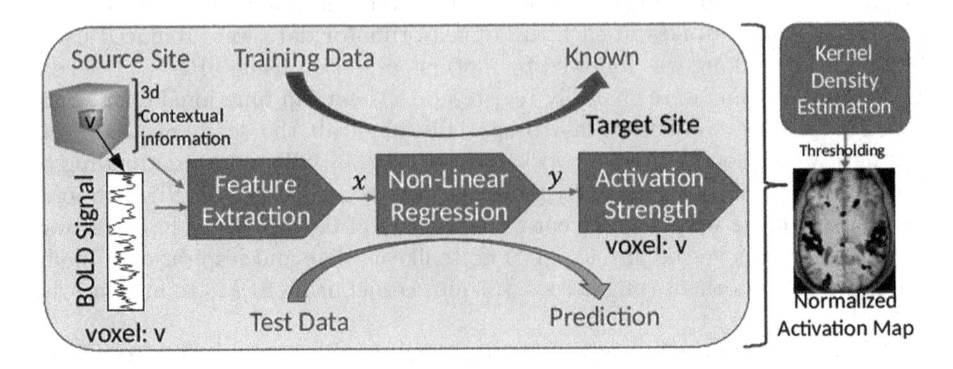

Fig. 2. Block diagram of the proposed correction function estimation technique

Feature Extraction. Current work investigates the performance of the learning algorithms with two different input feature vectors. While one feature set leverages only site-level statistical activation parameters, the other utilizes raw-time series signals as well. Both the feature sets incorporate multi-voxel anatomical information and exploit contextual information from the 3-D neighborhood of each voxel as described below, producing multi-voxel feature space:

- *Statistical Feature (SF) Set:* Leverages site-level statistical activation parameters. For every voxel, the feature vector includes the following 164 parameters extracted from its $3 \times 3 \times 3$ neighborhood: site-level t-statistics (27), site-level regression coefficients for the main regressor (27), temporal standard deviations (27), gray matter probabilities (27), white matter probabilities (27), CSF probabilities (27) and the mean and the standard deviation (2) of the site-level t-statistics in the neighborhood.
- *Extended Feature (EF) Set:* Incorporates raw time-series signals. For each voxel, in addition to the site-level statistical activation parameters, gray matter, white matter and CSF probabilities extracted from its $3 \times 3 \times 3$ neighborhood, the extended feature set includes raw time-series of all the 8-runs of the SM task along with run-level and session-level statistical activation parameters.

3 Experimental Results

Machine learning algorithms for estimating the correction functions were implemented using Theano [16] which is a CPU and GPU math expression compiler for Python. Experiments were run on NVIDIA Quadro K5200 GPU which has 2304 cores and 8GB of memory for training the neural networks. For optimization, Nesterov's Accelerated Gradient (NAG) was used because of its theoretical promise to avoid over-stepping and obtaining a better solution than the classical Stochastic gradient Algorithm (SGD) [17].

In ANNs, the hidden layers used the ReLU (Rectified Linear Unit) activation function and the output layer used a linear output function. ReLU, defined as $f(x) = \max(0, x)$, does not face the vanishing gradient problem like other activation functions like sigmoid, etc. The number of hidden nodes and learning rate were optimized through a grid-search by internal cross-validation on the training set under different modeling scenarios. From our experiments, it was observed that learning rate of 0.001, momentum of 0.9, 40 and 500 nodes in the hidden layer for statistical feature set and extended feature set respectively gave the best results.

Random forest regression was implemented using scikit-learn [18]. In our experiments with SF we use 70 trees and in the experiments with EF we use 500 trees, with square root of the number of features as the number of features learned by each tree. These parameters were chosen empirically and since it already is an average across many participating decision trees no hyper parameter optimization was performed.

Performance of the correction functions in terms of estimating the activation strength is measured using mean square error (MSE) between the estimated and reference statistical parametric maps (t-statistics) and compared with the uncorrected version. Accuracy in terms of activation voxels obtained by thresholding the estimated statistical parametric map after kernel density estimation is also taken into consideration. Furthermore, we use the receiver operating characteristic (ROC) curve and area under the curve (AUC) statistics to evaluate the performance of the mapping function by varying threshold values.

Although many factors influence the imaging results at a particular center, magnetic field strength is a significant component. Furthermore, inter-subject variance plays a major role in the context of any fMRI dataset being considered. Consequently, we designed the following experiments to understand the variance due to these factors and to assess the potential of ML algorithms to deal with increasingly challenging multi-center scenarios.

3.1 Multiple Subjects - Multiple Centers (Same Field Strength)

In this experiment, we simulate a multi-center study scenario where all the centers have the same magnetic field strength. The main aim of this experiment is to gauge the performance of the correction functions in the presence of inter-subject variability only as no variation in field strength is accounted for. In other words, this experiment was designed to evaluate the transferability of the models learnt between subjects.

3.2 Multiple Subjects - Multiple Centers (Different Field Strengths)

As most multi-center imaging studies involve centers with different magnetic field strengths, we emulate an even more practical multi-center study scenario in this experiment. Consequently, this particular experiment evaluates how the features learnt from a particular subject-center combination generalize to other subjects and centers.

For instance, we estimate the correction function to map subject 5's data collected from center 3 that hosts a 1.5T scanner to the same subject's data collected at center 3 with 4.0T scanner (reference site). The estimated function is then used to map other subjects' data collected at centers with 1.5T scanners to the reference map. MSE values are computed by comparing each subject's estimated new statistical parametric map with their corresponding t-statistic map generated at the reference site. Accuracy graphs were obtained by thresholding the estimated maps at $p < 0.001$ after kernel density estimation. As the graphs in Fig. 3 illustrate, most of the models provide significant reduction in MSE and improvement in accuracy, indicating that the model learned from a specific subject-site combination translates well to other subjects and centers with same magnetic field strength to overcome minor variations in hardware, software and environmental factors. Figure 4a depicts the AUC statistic calculated from the ROC analysis with varying activation thresholds. Figure 4b shows a sample ROC curve for Subject 3's data collected at center 9 that hosts a 1.5T scanner with significant threshold range ($p < 0.05$) highlighted.

It can be noticed from the above results that, two common ways to determine the accuracy of a prediction model, MSE and predictive accuracy do not always agree when it comes to identifying an optimal learning model. As they compute related but complementary metrics, it is quite possible for one model to have better MSE than a second model, but have a worse predictive accuracy than the second model. Although most of the models provide improvement compared to uncorrected maps, the performance of the models is not consistent across the subjects and the improvement in accuracy is quite moderate in some subjects. Furthermore, it can be noticed that RF based models with extended features seem to fare better than other models, while ANNs with statistical features seems to perform the worst.

For a more qualitative evaluation, Fig. 5 provides a visual comparison of activations detected in subject 6 at center 10 both before and after correction relative to that subject's reference map. It can be noticed that the correction function remains true to the observed data and judiciously reduces false positives (example highlighted in yellow) and converts false negatives (example highlighted in magenta). Overall, significant improvement can be observed in the reduction of spurious clusters and consolidation of the genuine clusters.

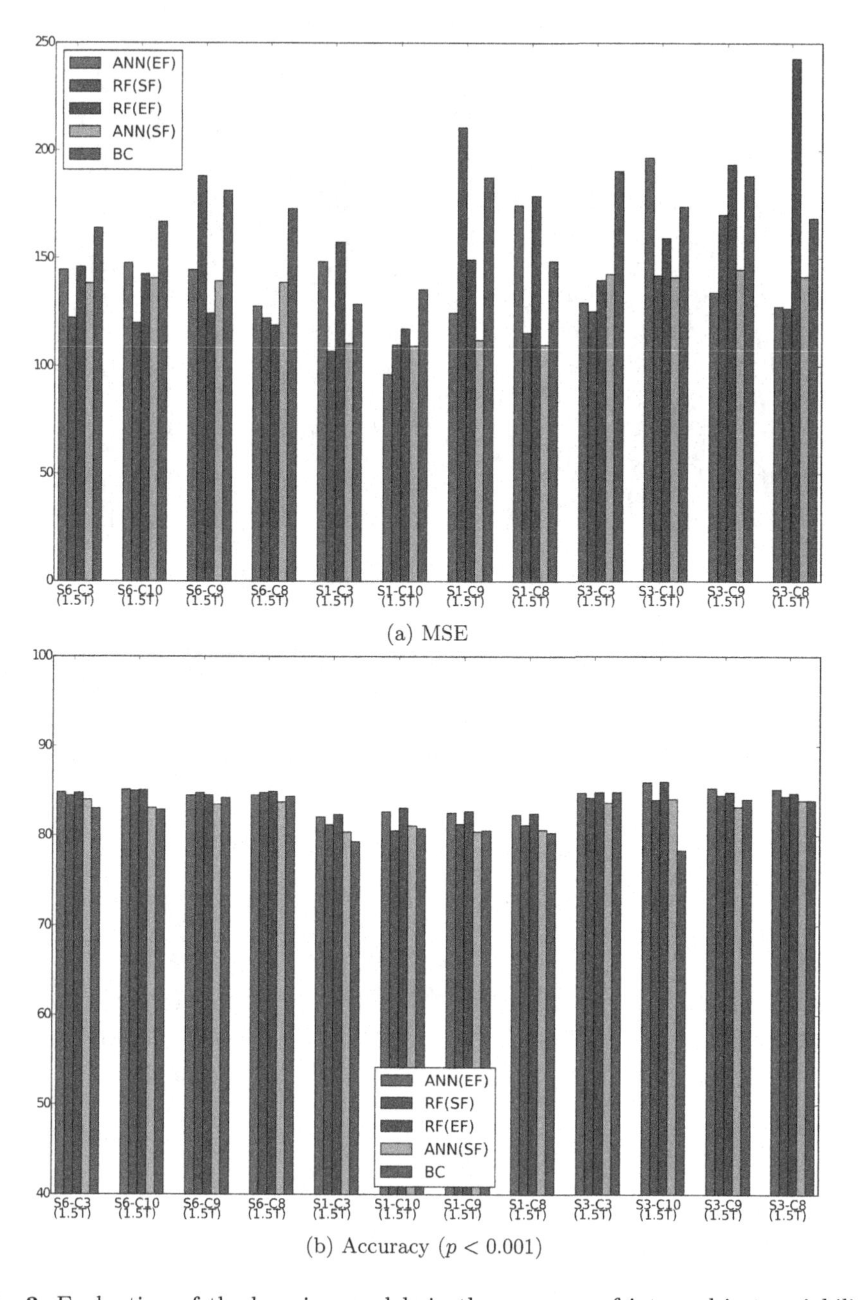

(a) MSE

(b) Accuracy ($p < 0.001$)

Fig. 3. Evaluation of the learning models in the presence of inter-subject variability using MSE and Accuracy: Results of applying the estimated correction function (Subject 5: C3-1.5T → C3-4.0T) (Notation: SX-CY-Z implies subject X's data from center Y with field strength Z) to other subjects at other centers with 1.5T scanners. Abbreviation: BC - Before Correction

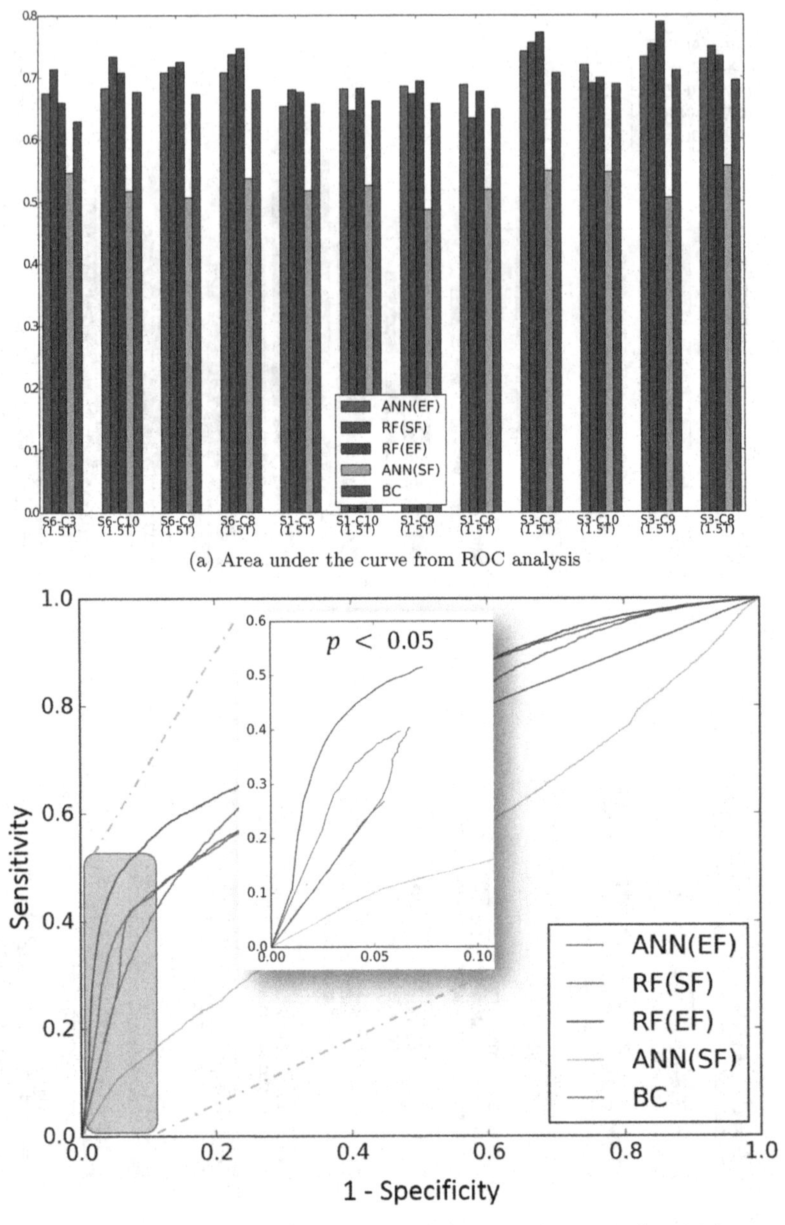

(a) Area under the curve from ROC analysis

(b) Sample ROC curve highlighting the standard significant range ($p < 0.05$)

Fig. 4. Evaluation of the learning models in the presence of inter-subject variability using AUC and ROC: Results of applying the estimated correction function (Subject 5: C3-1.5T → C3-4.0T) to other subjects at other centers with 1.5T scanners. Sample ROC curve was generated from applying this correction function to Subject3: C9-1.5T. Abbreviation: BC - Before Correction

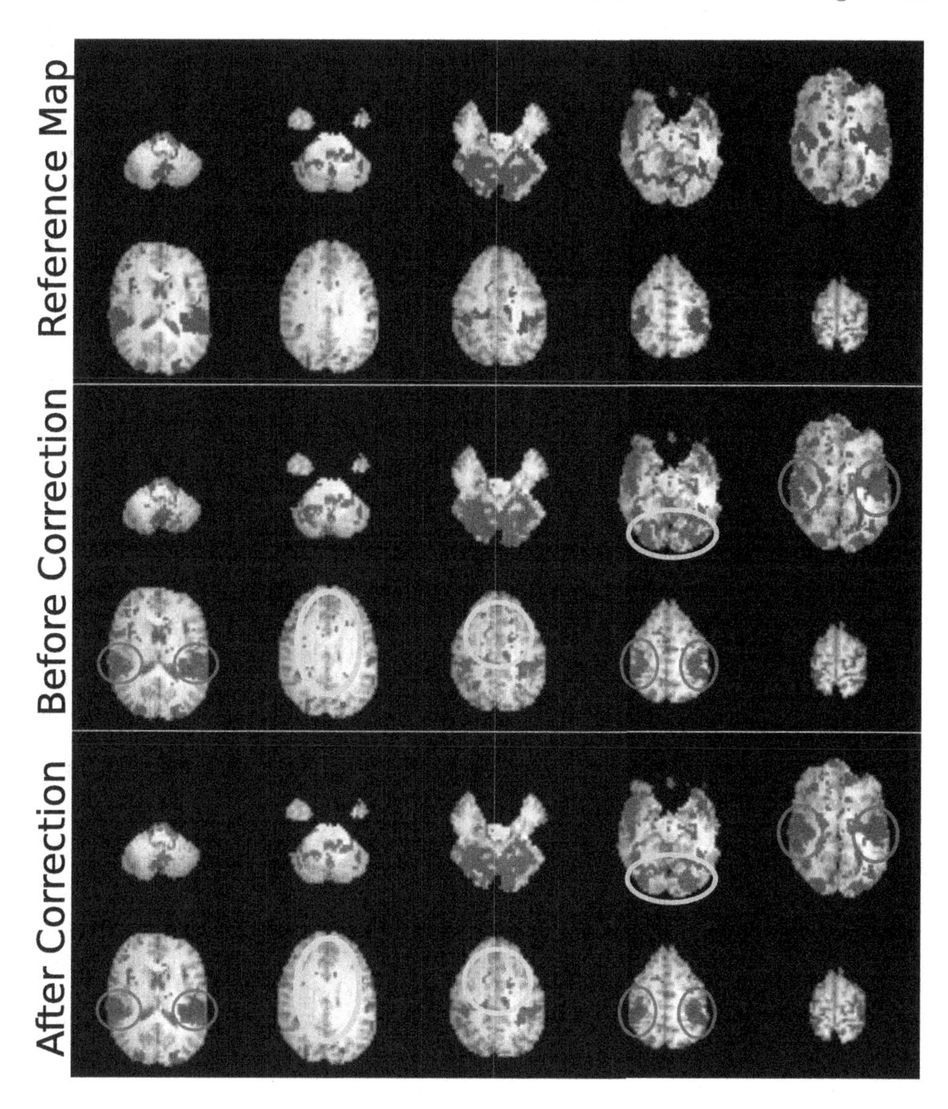

Fig. 5. Activation maps ($p < 0.001$) showing reference, before correction and after correction activations for S6-C10-1.5T (source) with respect to S6-C03-4.0T (reference). Model trained on (Subject 5: C3-1.5T → C3-4.0T) using ANN with extended features. Slices shown: (Top row) z = [3, 6, 9, 12, 15] (Bottom row) z = [18, 21, 24, 27, 30] Color codes: Blue - True Positives, Red - False Positives and Green - False Negatives. (Color figure online)

4 Discussion

Related Work. Only [19] and [20] have attempted fMRI data normalization. However, a direct comparison with these methods is not feasible as the approaches are different. While those studies focused on normalizing activation related parameters such as signal-to-fluctuation noise ratio (SFNR), smoothness, etc., the main contribution of this work is to enhance the comparability of activation patterns across sites. Our previous effort [21,22] involved using linear regression models to capture relationship between source and target site statistical activation parameters, treating each voxel as independent unit (without contextual information). Those preliminary findings instigated the use of non-linear regression models using multi-voxel features for substantial improvement and comprehensive analysis reported in this article.

5 Conclusion

In summary, this work demonstrates the potential of machine learning based algorithms to generate correction functions to map fMRI activations collected at any source site to a chosen reference site using the multi-voxel contextual information. Accuracy graphs were obtained by thresholding the estimated maps at high p-values of $p < 0.001$ after kernel density estimation. These improvements translate to significant reduction in spurious activations and more importantly, enhancement of the genuine activation clusters and ROIs in accordance with their role in cognitive function. With the potential to induce sensitivity to the regions associated with the task, ML based correction functions using multi-voxel contextual information can help identify small but significant activations which could otherwise be lost at the group level analysis stage. Due to the complex nature of the fMRI data, however, more work in this area is required to characterize the mapping functions by applying to larger fMRI datasets and to attain a standardized protocol for normalizing fMRI data across multiple centers.

Acknowledgements. Financial assistance from DST (grant no. SB/FTP/ETA-353/2013) New Delhi, India to DRB. The data used in this study was acquired through and provided by the Biomedical Informatics Research Network under the following support: U24-RR021992, Function BIRN and U24 GM104203, Bio-Informatics Research Network Coordinating Center (BIRN-CC).

References

1. Ogawa, S., Lee, T.M., Kay, A.R., Tank, D.W.: Brain magnetic resonance imaging with contrast dependent on blood oxygenation **87**(24), 9868–9872 (1990). http://www.pnas.org/content/87/24/9868
2. Parrish, T.B., Gitelman, D.R., LaBar, K.S., Mesulam, M., et al.: Impact of signal-to-noise on functional MRI. Magn. Reson. Med. **44**(6), 925–932 (2000)
3. Beckmann, C.F., Jenkinson, M., Smith, S.M.: General multilevel linear modeling for group analysis in fMRI. Neuroimage **20**(2), 1052–1063 (2003)
4. Horn, J.D.V., Grafton, S.T., Rockmore, D., Gazzaniga, M.S.: Sharing neuroimaging studies of human cognition **7**(5), 473–481 (2004). https://doi.org/10.1038/nn1231. http://www.nature.com/neuro/journal/v7/n5/abs/nn1231.html

5. Kelemen, A., Liang, Y.: Multi-center correction functions for magnetization transfer ratios of MRI scans. J. Health Med. Inform. **3**, 2 (2011)
6. Machielsen, W.C., Rombouts, S.A., Barkhof, F., Scheltens, P., Witter, M.P.: fMRI of visual encoding: Reproducibility of activation. Hum. Brain Mapp. **9**(3), 156–164 (2000). https://doi.org/10.1002/(SICI)1097-0193(200003)9:3⟨156::AID-HBM4⟩3.0.CO;2-Q
7. Keator, D.B., et al.: The function biomedical informatics research network data repository. Neuroimage **124**, 1074–1079 (2016)
8. Gountouna, V.-E., et al.: Functional magnetic resonance imaging (fMRI) reproducibility and variance components across visits and scanning sites with a finger tapping task, NeuroImage **49**(1), 552–560 (2010). https://doi.org/10.1016/j.neuroimage.2009.07.026. http://www.sciencedirect.com/science/article/pii/S1053811909007988
9. Gorgolewski, K., et al.: Nipype: a flexible, lightweight and extensible neuroimaging data processing framework in python. Front. Neuroinform. **5**, 13 (2011)
10. Cox, R.W.: AFNI: software for analysis and visualization of functional magnetic resonance neuroimages. Comput. Biomed. Res. **29**(3), 162–173 (1996)
11. Worsley, K.J., Friston, K.J.: Analysis of fMRI time-series revisited-again. Neuroimage **2**(3), 173–181 (1995)
12. Smith, S.M., et al.: Advances in functional and structural MR image analysis and implementation as FSL. Neuroimage **23**, S208–S219 (2004)
13. Pereira, J., Acosta-Cabronero, J., Pengas, G., Xiong, L., Nestor, P., Williams, G.: VBM with viscous fluid registration of gray matter segments in SPM. Front. Aging Neurosci. **5**, 30 (2012)
14. Worsley, K., et al.: A general statistical analysis for fMRI data **15**(1), 1–15 (2002) . https://doi.org/10.1006/nimg.2001.0933. http://www.sciencedirect.com/science/article/pii/S1053811901909334
15. Poldrack, R.A., Mumford, J.A., Nichols, T.E.: Handbook of Functional MRI Data Analysis. Cambridge University Press, Cambridge (2011)
16. Theano Development Team: Theano: a Python framework for fast computation of mathematical expressions. arXiv e-prints abs/1605.02688. http://arxiv.org/abs/1605.02688
17. Sutskever, I., Martens, J., Dahl, G., Hinton, G.: On the importance of initialization and momentum in deep learning. In: Proceedings of the 30th International Conference on Machine Learning (ICML 2013), pp. 1139–1147 (2013)
18. Pedregosa, F., et al.: Scikit-learn: machine learning in Python. J. Mach. Learn. Res. **12**, 2825–2830 (2011)
19. Friedman, L., Glover, G.H.: The FBIRN consortium, reducing interscanner variability of activation in a multicenter fMRI study: controlling for signal-to-fluctuation-noise-ratio (SFNR) differences. NeuroImage **33**(2), 471–481 (2006). https://doi.org/10.1016/j.neuroimage.2006.07.012. http://www.sciencedirect.com/science/article/pii/S1053811906007944
20. Friedman, L., Glover, G.H., Krenz, D., Magnotta, V.: Reducing inter-scanner variability of activation in a multicenter fMRI study: role of smoothness equalization. NeuroImage **32**(4), 1656–1668 (2006). https://doi.org/10.1016/j.neuroimage.2006.03.062. http://www.sciencedirect.com/science/article/pii/S1053811906004435

21. Thomas, A.J., Bathula, D.: Reducing inter-scanner variability in multi-site fMRI data: exploring choice of reference activation map and use of correction functions. In: 2015 International Conference on Computing, Communication & Automation (ICCCA), pp. 1187–1192. IEEE (2015)
22. Thomas, A.J., Bathula, D.: Reducing inter-scanner variability in multi-site fMRI activations using correction functions: a preliminary study. In: Singh, R., Vatsa, M., Majumdar, A., Kumar, A. (eds.) Machine Intelligence and Signal Processing. AISC, vol. 390, pp. 109–117. Springer, New Delhi (2016). https://doi.org/10.1007/978-81-322-2625-3_10

Efficient Approximation of Curve-Shaped Objects in $\mathbb{Z}^2$ Based on the Maximum Difference Between Discrete Curvature Values

Sutanay Bhattacharjee[1](✉) and Shyamosree Pal[2]

[1] Department of Computer Science and Engineering, Indian Institute of Technology Madras, Chennai, India
cs21d005@smail.iitm.ac.in

[2] Department of Computer Science and Engineering, National Institute of Technology Silchar, Silchar, Assam, India
spal@cse.nits.ac.in

Abstract. In this paper, we propose a novel algorithm for the cubic approximation of digital curves and curve-shaped objects in $\mathbb{Z}^2$. At first, the discrete curvature value is computed for each point of the given digital curve, C, using the improved k-curvature estimation technique. Based on the estimated k-curvature value, the points are selected from C to obtain the resultant set of reduced points, C'. We use a set of cubic B-splines for the approximation of the given digital curve C. For the selection of control points, our algorithm works on a new parameter, *threshold*, defined as the maximum difference between discrete curvature values, based on which the control points are selected from the given digital curve, C, such that the maximum discrete curvature difference from the last selected point and the next point to be selected do not exceed the *threshold*. Further adjustments are made in the selection of control points based on the principle that high curvature areas of a digital curve represent more information whereas low curvature areas represent less information. Experimental results and comparisons with the existing algorithm on various digital objects demonstrate our approach's effectiveness. It has been observed that our algorithm generates better output for approximating real-world curves in which there are large number of control points, and the rate of curvature change is fast. Our algorithm also takes less computational time since it selects the control points of a digital curve in a single iteration.

Keywords: Cubic approximation · k-curvature · Threshold

1 Introduction

A digital curve C is defined as a sequence of digital points (i.e., points with integer coordinates) in which two points $(x, y) \in C$ and $(x', y') \in C$ are neighbors of each other, if and only if $\max(|x - x'|, |y - y'|) = 1$ [6]. In general,

B. Raman et al. (Eds.): CVIP 2021, CCIS 1568, pp. 529–541, 2022.
https://doi.org/10.1007/978-3-031-11349-9_46

a digital curve is specified by a set of coordinate positions called control points. These points can be fitted with piecewise-continuous, parametric polynomial functions, or straight-line (polygonal approximation). When the generated polynomial curve is plotted so that some of the control points are not on the curve path, the resulting curve is said to *approximate* the set of control points [4].

Our work aims to efficiently approximate a given digital curve C by selecting a subset of data points from C using which the original curve can be adequately represented. We employ a heuristic approach to select the subset of data points C', based on the discrete curvature values of the successive points of the given digital curve, C. The linear approximation is the simplest and fastest, but it does not accurately approximate a given curve. The cubic polynomials are usually used since the lower-degree polynomials provide much less flexibility in maintaining the shape of the curve, whereas higher degree polynomials can introduce undesirable niggles and hence require greater computation. Thus in this work, we have used a set of cubic B-splines in order to obtain the desired approximation. B-spline curves possess both parametric (C^0, C^1, and C^2) and geometric (G^0, G^1, and G^2) continuities [1], which ensure the smoothness and optimal exactness of the fitted curve against the given set of control points.

1.1 Existing Approximation Methods

Most of the earlier works on curve approximation incorporate two broad approaches as follows:

- One approach is to detect dominant points directly through angle or corner detection [8,13].
- The other approach is to obtain a piecewise linear polygonal approximation of a digital curve subjected to certain constraints [11,12].

Teh and Chin [13] used the former approach for determining dominant points on a digital curve, which requires no input parameter and remains reliable even when curves and curve-shaped objects of multiple sizes are present on the digital curve. They first determined the precise region of support for each point based on its local properties, then computed each point's curvature, and finally detected dominant points by the process of non-maxima suppression. Medioni and Yasumoto [8] developed an efficient algorithm to locate corner points and detected additional points with significant curvature between those corner points. It enabled them to obtain an accurate position of the corners and encode curve segments between them using B-splines.

On the other hand, Plass and Stone [11] used curve fitting with piecewise parametric cubic polynomial taking a set of sample points, with optional endpoint and tangent vector specifications. They have iteratively derived a single parametric cubic polynomial that lies close to the data points, defined by an error metric based on least squares. Asif Masood and Muhammad Sarfraz [7] have used decomposition of the outline into smaller curves using cubic Bezier curve for approximation. Then, they have determined a set of control points for approximation using the cubic Bezier curve.

Most of the curve approximation approaches were computationally expensive and had high computational complexity for real-time applications. However, significant improvement in this approach came from S. Pal and P. Bhowmick [10]. This work is based on the *improved k-curvature estimation* method proposed in [9]. Using the discrete curvature values at each point of the given digital curve, the authors have iteratively obtained a reduced set of points based on the principle that more information is stored in high curvature areas than low curvature areas of the digital curve. The authors have used cubic B-splines to approximate the given digital curve.

1.2 Discrete Curvature Estimation

Since the digital curves are represented in the digital plane, computing the discrete curvature of each point greatly aids in finding the significant or dominant points for a digital curve. The discrete curvature of a digital curve acts as a *signature of the curve*, and can be used for efficient approximation of a digital curve.

To incorporate the curvature calculus on discrete data for two-dimensional or three-dimensional images, most of the earlier works on curvature estimation such as [7,13] have modified the discrete space into classical continuous space by applying interpolation and parameterization of mathematical objects (i.e., B-splines, quadratic surfaces) on which the continuous curvature can be computed.

In recent years, some interesting works have been proposed that utilize curvature information to address a variety of real-world problems specified on digital curves. C Jin, X Wang, Z Miao and S Ma in 2017 have proposed a new road curvature estimation method for self-driving vehicles [5]. The authors have used a new contour-based clustering and a parallel-based voting system for detecting the lanes of roads. The curve fitting using least square error method is used to fit the lane models. M Haghshenas and R Kumar in 2019 have trained a curvature estimation model using machine learning approach for Coupled Level Set Volume of Fluid (CLSVOF) method [3]. They have generated three datasets for the curvature: curvature as a function of volume fraction, distance function, and volume and distance function.

On the contrary, this work uses the *improved k-curvature estimation* method ($\kappa^{(3)}(p_i, k)$) proposed in [9] for computing discrete curvature of each point of a digital curve. The proposed method uses the angular orientation of successive points given in Freeman chain code representation [2] to compute the discrete curvature which offers a largely lower computational time and is comparatively easier to implement since it involves addition and subtraction operations only.

2 Cubic Curve Approximation

Using curvature value as the deciding criteria for the selection of an ordered subset of control points C', representing the digital curve C, we have used a set of cubic B-splines in order to achieve an efficient approximation of curve C using the selected subset of points, C'. The selection of control points should be such that these points can adequately represent all the curve portions with minimum redundancy. For example, regions such as the tip of the mouth and tips of the ears of rabbit (Fig. 1(e)) require more control points for approximation as these regions possess high curvature values. Here, we have used a *heuristic approach* that judiciously selects the control points and hence performs approximation of a digital curve C.

2.1 Proposed Algorithm

Algorithm IMPROVED_CURVE_APPROX (C, n, $threshold$)
1: $C' \leftarrow empty$
2: $sod \leftarrow 0$
3: $residue \leftarrow 0$
4: $m \leftarrow$ number of points in C'
5: $max\_curv \leftarrow 0$
6: $max\_cd \leftarrow 0$
7: $mod\_curv \leftarrow 0$
8: $modified\_curvature \leftarrow 0$
9: Add start point p_0 of curve C to C', $C' \leftarrow p_0$
10: **for** $i \leftarrow 1$ **to** n
11: Set sum of difference, $sod = sod + |k^{(3)}(p_{i+1}, k) - k^{(3)}(p_i, k)|$
12: **if** $sod \geq threshold$
13: Insert p_{i+1} as the next control point of C', $C' \leftarrow C' \cup \{p_{i+1}\}$
14: $residue \leftarrow sod - threshold$
15: Store curvature value of p_{i+1} , $crv \leftarrow k^{(3)}(p_{i+1}, k)$
16: **for** $p_s \leftarrow p_i$ **to** p_{i+1}
17: **if** $max\_curv < k^{(3)}(p_s, k)$
18: $max\_curv \leftarrow k^{(3)}(p_s, k)$
19: $max\_cd \leftarrow max\_curv >> 3$
20: $mod\_curv \leftarrow max\_cd + crv$
21: $modified\_curvature \leftarrow mod\_curv >> 3$
22: $sod \leftarrow residue + modified\_curvature$
23: **for** $j \leftarrow 1$ **to** m
24: CUBIC_B-SPLINE (C', j)

Algorithm 1: Algorithm on selection of control points for cubic curve approximation based on the maximum difference between discrete curvature values ($threshold$).

Here, we present a novel algorithm IMPROVED_CURVE_APPROX for selecting the set of control points, C', based on the discrete curvature values of the points of the given digital curve and parameter, *threshold*, the maximum difference between discrete curvature values. The points obtained in C' are used for the approximation using cubic B splines. In our algorithm, we have obtained the discrete curvature values at all the points of the digital curve using improved k-curvature estimation equation ($k^{(3)}(p_i, k)$) proposed in [9] for $k = 16$ and *threshold* values 8, 10 and 12. The steps of the proposed algorithm (Algorithm 1[1]) are demonstrated in Fig. 1.

[1] Note that the right shifting of the result by 3 bits is an empirical step, which is found to provide better approximations and is thus preferred since shift operations take less time than usual multiplication or division operations.

The input to the proposed algorithm is the ordered set of points of the digital curve with the k-curvature of the points estimated using *improved k-curvature estimation*. The first step of our algorithm is the initialization of the resultant curve, C' as *empty* (Step 1). Then, we initialize the variables *sod*, $residue$, max_curv, max_cd, mod_curv, $modified\_curvature$ that will be used in our algorithm. After that, we add the start point (p_0) of the given curve C in resultant curve C' (Step 9, Fig. 1(c)). Now, in each iteration of the outer for loop (Step 10), we compute the discrete curvature value of each point of the given curve C using the improved k-curvature estimation equation ($k^{(3)}(p_i, k)$) [9]. While we are obtaining the curvature at each point, we also compute the difference between the curvature values of the successive pair of control points ($(p_i, p_{i-1}); (p_{i+1}, p_i); \cdots\cdots$) of the given curve C, and accordingly add the difference and store them into a variable, *sod* (sum of difference) in Step 11. Now, at the point at which the computed *sod* becomes greater than or equal to $threshold$, we insert the first control point of the pair ($(p_i, p_{i-1}); (p_{i+1}, p_i); \cdots\cdots$) in the resultant set of control points, C' (Fig. 1(d)). Next, we compute the difference between the previously calculated *sod* in Step 11 and the input $threshold$ value, and this difference is then stored in another variable $residue$ (Step 14).

In order to obtain a good approximation of the given curve C, more control points should be picked in high curvature regions since these regions contain vital information about the curve compared to low curvature regions. To capture the sharpness or turning effects of the given curve C, particularly of the high-curvature regions, we incorporate a *heuristic approach* for improved curve approximation in Steps 15–21. To store more information in the high curvature areas, we first store the curvature values of those control points already selected in the resultant curve, C' in *crv* (Step 15). Next, in the inner for loop (Step 16), we find the maximum curvature value among the consecutive control points already selected in the resultant curve C', max_curv, and right shift it by 3 bits, max_cd (Step 19). After that, the previously obtained max_cd value is added to *crv*, and the result is again right-shifted by 3 bits and saved as $modified\_curvature$. Finally, this $modified\_curvature$ value is added to the $residue$ value previously obtained in Step 14 and it is updated in *sod* (Step 22).

The set of all added points in resultant curve, C' gives the final list of m control points (Figs. 1(e)). Thus, m segments of cubic B-spline are drawn piecewise, taking four control points (p_j, $p_{(j+1)\,mod\,8}$, $p_{(j+2)\,mod\,8}$, $p_{(j+3)\,mod\,8}$) at a time from C' in Steps 23–24 (Figs. 1(f)).

The Figs. 1, 2, 3 and 4 illustrates the selection of control points and fitting of b-spline curves applying the proposed algorithm for $k = 16$ and $threshold = 8, 10$ and 12 for cubic curve approximation on some real-world digital curves. The curve approximated after applying the proposed algorithm almost resembles the original curve, as can be seen from the input and output curves.

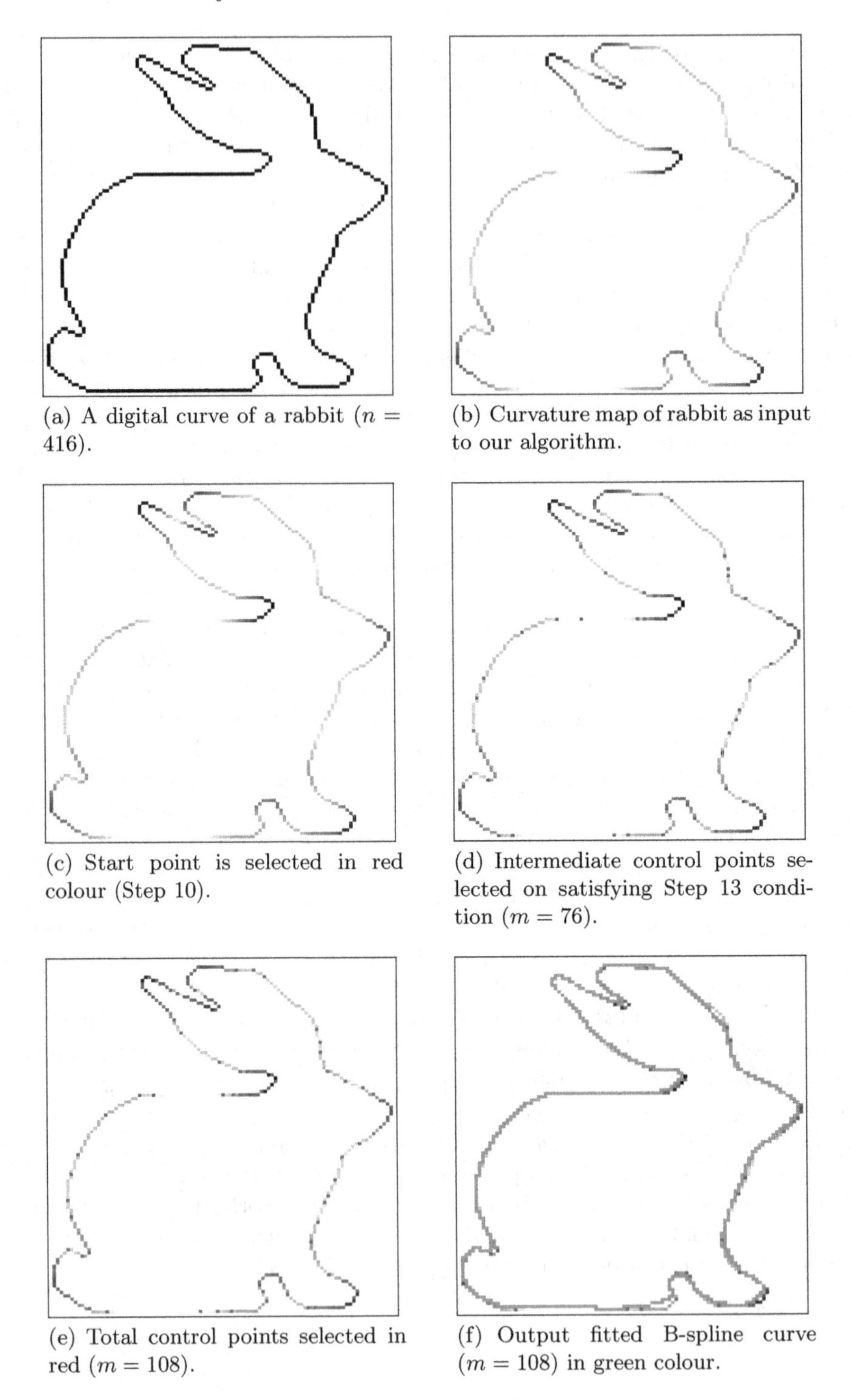

(a) A digital curve of a rabbit ($n =$ 416).

(b) Curvature map of rabbit as input to our algorithm.

(c) Start point is selected in red colour (Step 10).

(d) Intermediate control points selected on satisfying Step 13 condition ($m = 76$).

(e) Total control points selected in red ($m = 108$).

(f) Output fitted B-spline curve ($m = 108$) in green colour.

Fig. 1. Steps showing the cubic curve approximation on a real-world digital curve of "rabbit" applying proposed algorithm.

3 Observations and Results

We have implemented the proposed algorithm in C language on an Intel® Core^{TM} i7 – 4770 CPU @ 3.40 GHz × 8 processor using Linux 64–bit OS and version 16.04 LTS. Table 1 and 2 shows the results obtained after applying the proposed algorithm on various digital curves. Table 3 shows the comparisons between the proposed algorithm and the existing algorithm [10].

Table 1 shows the results obtained after applying the proposed approximation algorithm on some digital curves. The second column, n in the table, represents the total number of points in the given input curve C and m shows the reduced number of points in the resultant curve $C^{'}$, obtained after applying the algorithm. We have applied the proposed algorithm on these curves, considering threshold values 8, 10 and 12 and accordingly, the results are shown in the table below. It can be observed that increasing the *threshold* value leads to reduction in the number of control points, m in the resultant set of control points $C^{'}$ for each digital curve.

Table 1. Resultant no. of points and time taken by algorithm for some digital curves

Curve	n	Threshold = 8		Threshold = 10		Threshold = 12	
		m	CPU time (ms)	m	CPU time (ms)	m	CPU time (ms)
Rabbit	416	155	1.239	108	1.656	83	1.738
Elephant	457	242	2.113	168	2.017	126	2.010
Hand	503	209	1.851	151	2.314	113	1.944
Star	744	504	5.961	345	4.183	260	4.799
Clover	1116	178	3.335	136	3.082	111	2.738
T-Shirt	1227	115	2.782	88	2.402	72	2.434
Aeroplane	1428	179	3.196	134	3.083	108	3.122
Witch	1611	409	7.461	293	6.724	229	5.578
Dancer	1662	510	6.286	376	5.896	298	6.355
Graduate	1918	410	6.723	296	5.662	232	4.891
Guitarist	2167	508	7.156	392	8.183	321	6.191
Man	2506	705	8.95	519	8.496	402	7.832
Horse	3245	853	10.416	618	8.882	463	8.741
Stork	4396	861	10.694	623	9.883	478	9.647
Deer	4468	487	7.406	358	8.715	283	7.931

In order to quantitatively measure the quality of the output points, we have computed the point-wise error between the digital curve C and the approximated curve $C^{'}$ [13]. The *total error* column in Table 2 shows the total sum of deviation over all the points of a given digital curve. In order to calculate the *average error*, we have divided the total error by the total number of points of the given curve (n). In order to express the precision in error computation, we have computed another error norm called maximum error or Chebyshev error [13]. Table 2 below shows the total, average, and maximum errors for some digital curves.

Table 2. Error computation after applying the proposed algorithm considering *threshold* = 10 and $k = 16$

Curve	No. of input points (n)	Total error	Average error	Maximum error
Rabbit	416	7.483	0.01798	1.732
Hand	503	7.071	0.01406	1.732
Star	744	7.746	0.01041	1.414
Witch	1611	12.124	0.00752	1.923
Dancer	1662	12.884	0.00775	2.010
Guitarist	2167	13.711	0.00633	2.432

Table 3 shows the comparison of the proposed algorithm with the existing algorithm [10]. The second column, n in the table, represents the total number of points in the given input curve C and m shows the reduced number of points in the resultant curve $C^{'}$, obtained after applying the algorithm. We have compared the CPU time, reduced control points, m and the percentage of control points in both the algorithms, considering k as 16 and *threshold* to be 10.

Table 3. Comparison of the proposed algorithm with existing algorithm [10]

Curve	n	Existing algorithm (k = 2 to 32)			Proposed algorithm (Threshold = 10, k = 16)		
		m	CPU time (ms)	Control points	m	CPU time (ms)	Control points
Rabbit	416	171	3.074	41.11%	108	1.656	25.96%
Hand	503	186	3.596	36.98%	151	2.314	30.01%
Star	744	244	5.607	32.79%	345	4.183	46.37%
Witch	1611	443	13.642	27.49%	293	6.724	18.18%
Dancer	1662	510	15.914	30.68%	376	5.896	22.62%
Guitarist	2167	616	19.325	28.43%	392	8.183	18.08%

In each digital curve, it can be observed that the proposed algorithm uses lesser number of control points and lesser CPU time to approximate the given input curve than the existing algorithm [10]. The reason being that our point selection algorithm selects the necessary control points for approximation in one run of the entire digital curve. Whereas the algorithm in [10] makes a minimum of four runs of the entire curve and at each run estimates the k-curvature for a different k value to obtain the desired set of control points. Also, it can be observed that our algorithm outputs relatively lesser control points for approximating digital curves which contain large number of control points as compared to the existing algorithm [10]. This highlights the effectiveness of our approach for approximating real-world digital curves in which there are large number of control points, and having frequent curvature changes. A contrasting feature observed from the table above is that the digital object star requires more control points (345) for approximation in our algorithm as compared to the existing

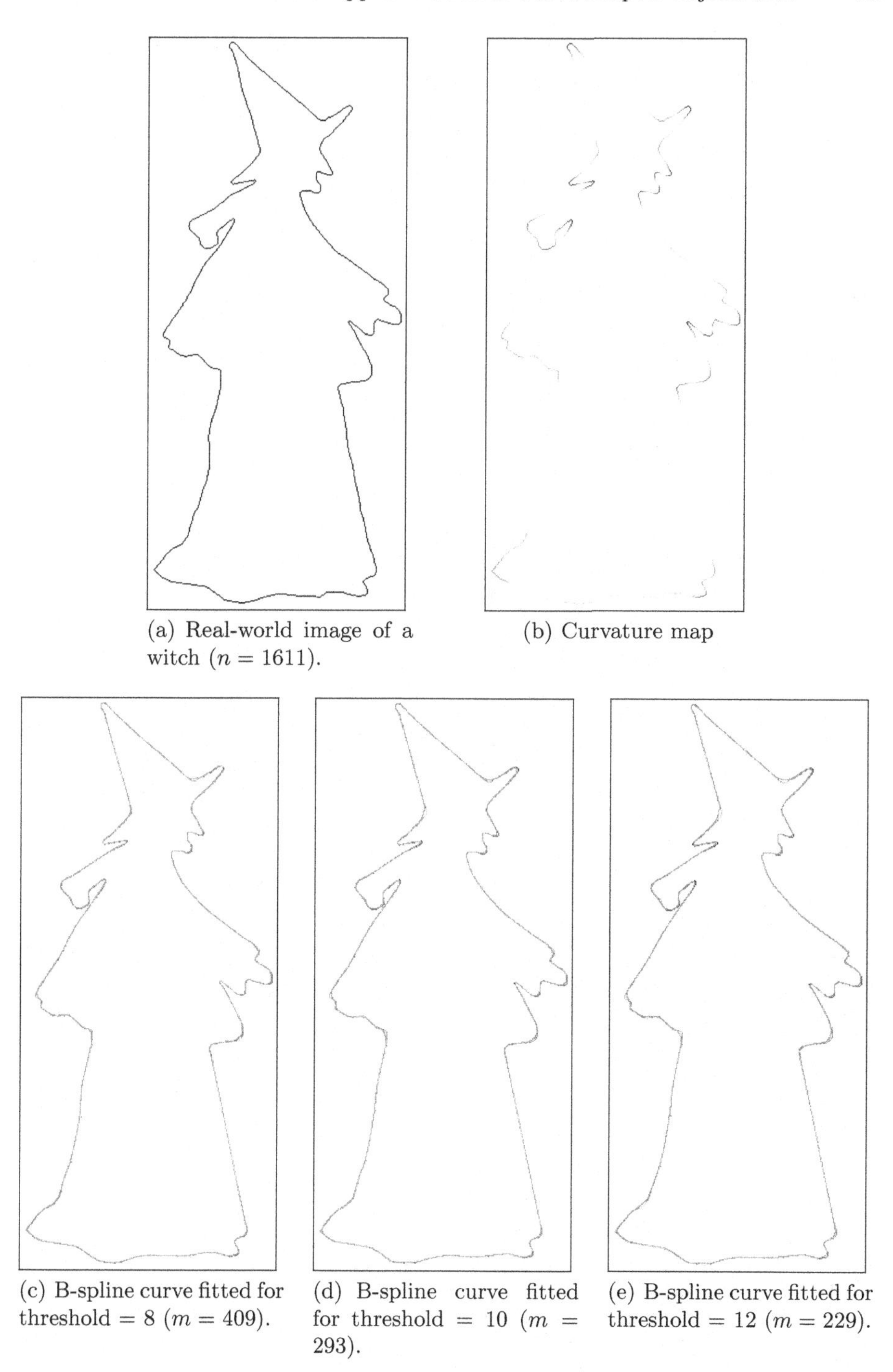

(a) Real-world image of a witch ($n = 1611$).

(b) Curvature map

(c) B-spline curve fitted for threshold = 8 ($m = 409$).

(d) B-spline curve fitted for threshold = 10 ($m = 293$).

(e) B-spline curve fitted for threshold = 12 ($m = 229$).

Fig. 2. Cubic approximation on digital curve representing a real-world image of "witch" applying the proposed algorithm for $k = 16$ and threshold values 8, 10 and 12.

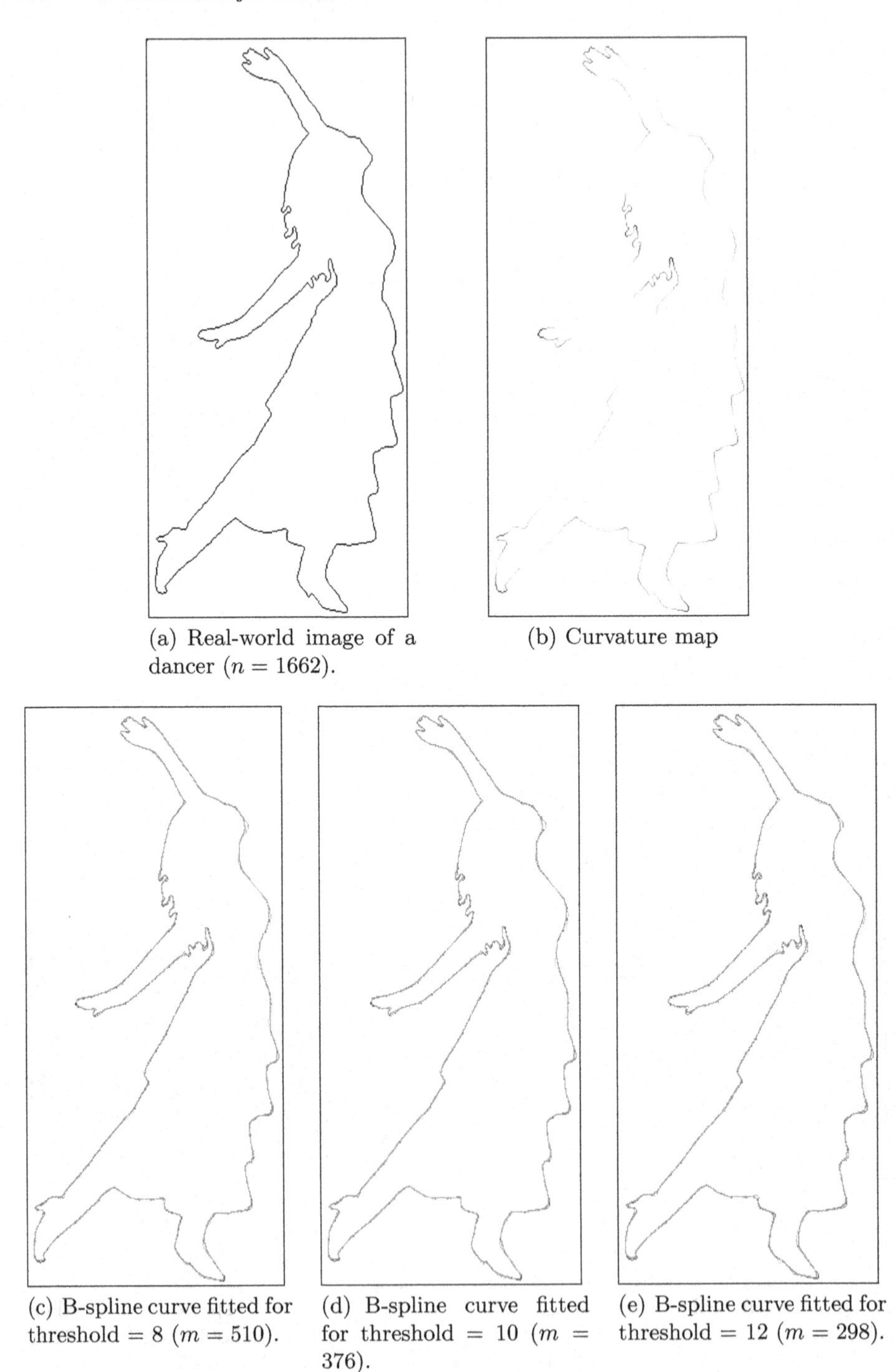

(a) Real-world image of a dancer ($n = 1662$).

(b) Curvature map

(c) B-spline curve fitted for threshold = 8 ($m = 510$).

(d) B-spline curve fitted for threshold = 10 ($m = 376$).

(e) B-spline curve fitted for threshold = 12 ($m = 298$).

Fig. 3. Cubic approximation on digital curve representing a real-world image of "dancer" applying the proposed algorithm for $k = 16$ and threshold values 8, 10 and 12.

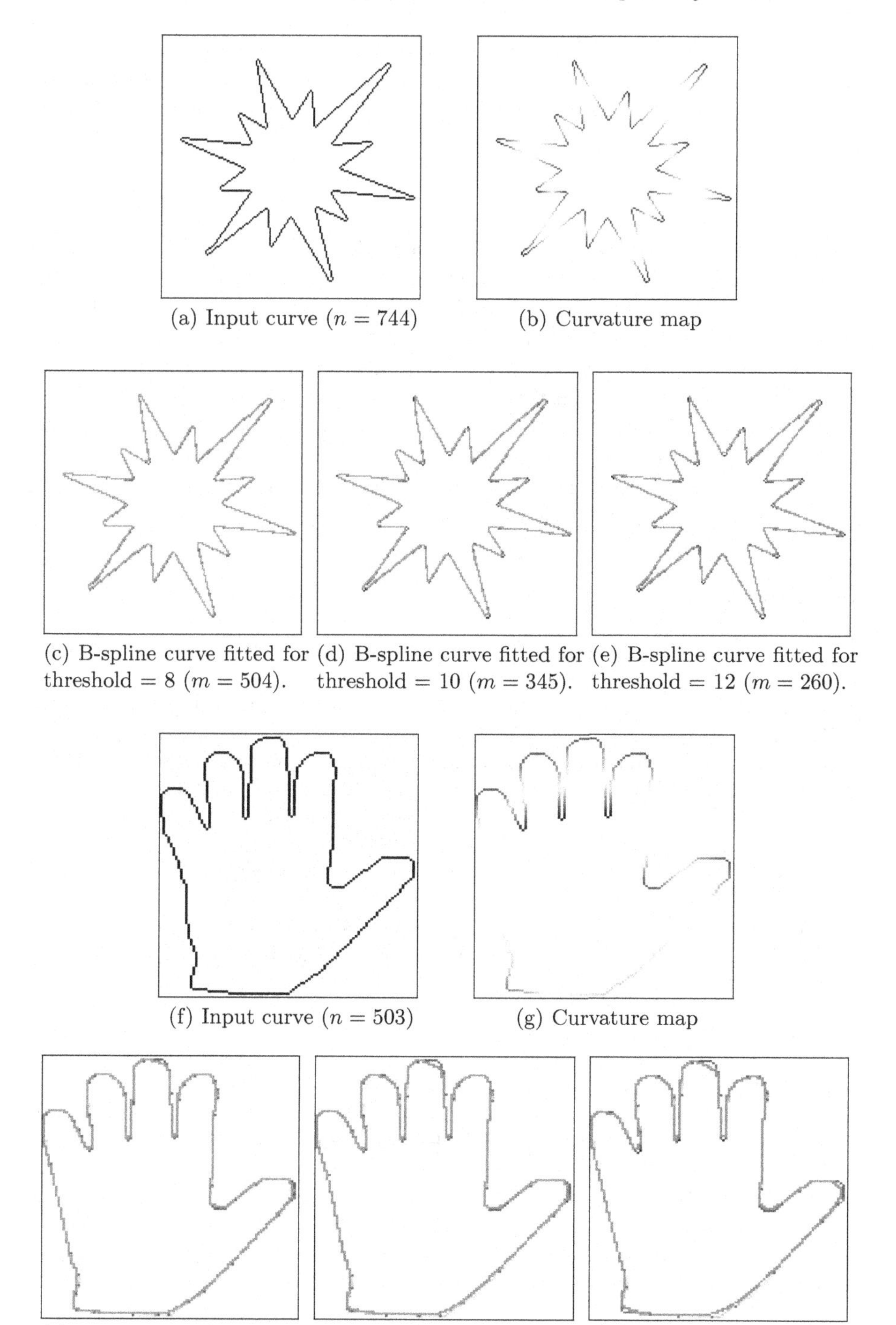

(a) Input curve ($n = 744$)

(b) Curvature map

(c) B-spline curve fitted for threshold = 8 ($m = 504$).

(d) B-spline curve fitted for threshold = 10 ($m = 345$).

(e) B-spline curve fitted for threshold = 12 ($m = 260$).

(f) Input curve ($n = 503$)

(g) Curvature map

(h) B-spline curve fitted for threshold = 8 ($m = 209$).

(i) B-spline curve fitted for threshold = 10 ($m = 151$).

(j) B-spline curve fitted for threshold = 12 ($m = 113$).

Fig. 4. Cubic approximation on real-world digital curves of "hand" and "star" applying the proposed algorithm for $k = 16$ and threshold values 8, 10 and 12.

algorithm (244). It may be due to the fact that star has more straight edges compared to curved ones and our algorithm is suited for performing approximation of digital objects having more curved edges such as in rabbit, dancer, guitarist.

4 Conclusion

Here we have proposed an easy and simplified approach for cubic approximation of real-world digital curve by first computing curvature value at each point of the input digital curve using the *improved k-curvature estimation* and then selecting control points on the resultant curve, on the basis of a new parameter, *threshold*, the maximum difference between discrete curvature values. Experimental results show that our algorithm almost reconstructs the original curve, giving a considerable reduction both in the number of control points and the computation time as compared to the existing algorithm [10]. The cubic approximation performed using the proposed algorithm help in the efficient reconstruction of the curve, which almost resembles the original digital curve.

References

1. Foley, J.D., et al.: Computer Graphics: Principles and Practice, vol. 12110. Addison-Wesley Professional, Reading (1996)
2. Freeman, H.: On the encoding of arbitrary geometric configurations. IRE Trans. Electron. Comput. **2**, 260–268 (1961)
3. Haghshenas, M., Kumar, R.: Curvature estimation modeling using machine learning for CLSVOF method: comparison with conventional methods. In: ASME-JSME-KSME 2019 8th Joint Fluids Engineering Conference. American Society of Mechanical Engineers Digital Collection (2019)
4. Hearn, D., Baker, M.P., et al.: Computer Graphics with OpenGL. Pearson Prentice Hall, Upper Saddle River (2004)
5. Jin, C., Wang, X., Miao, Z., Ma, S.: Road curvature estimation using a new lane detection method. In: 2017 Chinese Automation Congress (CAC), pp. 3597–3601. IEEE (2017)
6. Klette, R., Rosenfeld, A.: Digital Geometry: Geometric Methods for Digital Picture Analysis. Elsevier, Amsterdam (2004)
7. Masood, A., Sarfraz, M.: Capturing outlines of 2D objects with Bézier cubic approximation. Image Vis. Comput. **27**(6), 704–712 (2009)
8. Medioni, G., Yasumoto, Y.: Corner detection and curve representation using cubic B-splines. Comput. Vis. Graph. Image Process. **39**(3), 267–278 (1987)
9. Pal, S., Bhowmick, P.: Estimation of discrete curvature based on chain-code pairing and digital straightness. In: 2009 16th IEEE International Conference on Image Processing (ICIP), pp. 1097–1100. IEEE (2009)
10. Pal, S., Bhowmick, P.: Cubic approximation of curve-shaped objects in $\mathbb{Z}^2$: a generalized approach based on discrete curvature. J. Discrete Math. Sci. Cryptogr. **13**(5), 407–427 (2010)
11. Plass, M., Stone, M.: Curve-fitting with piecewise parametric cubics. In: Proceedings of the 10th Annual Conference on Computer Graphics and Interactive Techniques, pp. 229–239 (1983)

12. Sarfraz, M., Asim, M.R., Masood, A.: Piecewise polygonal approximation of digital curves. In: Proceedings of Eighth International Conference on Information Visualisation, IV 2004, pp. 991–996. IEEE (2004)
13. Teh, C.H., Chin, R.T.: On the detection of dominant points on digital curves. IEEE Trans. Pattern Anal. Mach. Intell. **11**(8), 859–872 (1989)

Exploring the Role of Adversarial Attacks in Image Anti-forensics

Krishan Gopal Sharma[1], Gurinder Singh[2](✉), and Puneet Goyal[2]

[1] Huazhong University of Science and Technology, Wuhan, China
i201921088@hust.edu.cn

[2] Indian Institute of Technology Ropar, Rupnagar, India
{gurinder.singh,puneet}@iitrpr.ac.in

Abstract. Deep learning (DL) has grown significantly in the field of image forensics. A lot of research has been going on to develop deep learning based image manipulation detection techniques. On the contrary, researchers are also challenging the robustness of these DL-based image forensic techniques by developing efficient anti-forensic schemes. This paper reveals the role of adversarial attacks in image anti-forensics with better human imperceptibility against recent general-purpose image forensic techniques. We propose an image anti-forensic framework by using recent adversarial attacks, i.e., Fast Gradient Sign Method (FGSM), Carlini and Wagner (C&W), and Projected Gradient Descent (PGD). Firstly, we have trained recent image forensic models on the BOSSBase dataset. Then, we generate adversarial noise by using the gradient of these image forensic models corresponding to each adversarial attack. Afterward, the obtained noise is added to the input image, resulting in the adversarial image corresponding to the particular attack. These adversarial images are generated by using the BOSSBase dataset and tested on the recent image forensic models. The experimental results show that the performance of the recent forensic models has decreased rapidly in the range from 50–75% against the different adversarial attacks, i.e., FGSM, C&W, and PGD. Furthermore, the high human imperceptibility of generated adversarial images is confirmed from the PSNR values.

Keywords: Anti-forensics · Adversarial attack · Image manipulation detection

1 Introduction

The advancement in multimedia services increases the frequency of manipulated images on social networking websites related to politics, promotion, and personal attacks. One can easily manipulate the image data due to the easy accessibility of the photo editing software and high-grade modern cameras. Thus, preserving the originality of digital images is one of the main concerns. Digital image forensic techniques are beneficial to generate information about the image and device used to capture that image. The inspection of this information concludes that

B. Raman et al. (Eds.): CVIP 2021, CCIS 1568, pp. 542–551, 2022.
https://doi.org/10.1007/978-3-031-11349-9_47

the considered image is doctored or not [19]. On the other hand, the goal of the anti-forensic techniques is to trick or mislead the forensic examiner by making challenges in the forgery detection process. The anti-forensic techniques can also assist the examiner by further providing flaws in conventional forensic techniques to improve digital forensics [20,21].

Numerous image anti-forensic techniques have been proposed in the literature based on contrast enhancement [3,4,18], median filtering [8,11,23], resampling [12,15], JPEG compression [5,9,10]. Nowadays, some works [14,16,22] have used Generative Adversarial Network (GAN) to design efficient image anti-forensic approaches. These GAN-based anti-forensic approaches outperform the traditional anti-forensic approaches in terms of forensic undetectability and image visual quality. But, we have to train the generator and discriminator simultaneously in GAN networks, which is a time-consuming process. Moreover, it is challenging for any anti-forensic technique to fool the recent deep learning based forensic methods [2,7,19,24]. These problems have been addressed in this paper by designing an anti-forensic framework based on efficient adversarial attacks, i.e., FGSM [13], C&W [6], and PGD [17]. To the best of our knowledge, this direction of using adversarial attacks in image anti-forensics has not been explored in the previous literature. We have evaluated our anti-forensic framework by considering recent DL-based image forensic models [2,7,19,24] for multiple image manipulation detection. In this paper, the suggested approach is evaluated by considering most of the image processing operations such as JPEG compression, Gaussian Blurring (GB), Adaptive White Gaussian Noise (AWGN), Resampling (RS), Median Filtering (MF), and Contrast Enhancement (CE). Moreover, we also considered recent anti-forensic schemes related to JPEG compression, MF, and CE as image processing operations for evaluation purposes. The experimental results show that the considered adversarial attacks can efficiently fool the recent forensic techniques when tested on multiple evaluated adversaries.

The main contributions of this paper are outlined as follows:

- We explore the role of adversarial attacks in image anti-forensics to fool the forensic detectors.
- We train recent general-purpose forensic models [2,7,19,24] by considering different image processing operations, including anti-forensic attacks on the BOSSBase dataset [1].
- We apply white-box adversarial attacks FGSM, C&W, and PGD to generate corresponding adversarial images datasets by considering different image processing operations. Note that these adversarial images are not used during the training of the forensic models.
- To show the effectiveness of our approach, we evaluated recent image forensic models on the generated adversarial images datasets. The results show that our proposed anti-forensic strategy efficiently fools the forensic models.
- We evaluate the peak signal-to-noise ratio (PSNR) of generated adversarial images, which shows that our generated images are unnoticeable and strongly misclassify the image forensic models.

2 Proposed Anti-forensic Framework Based on Adversarial Attacks

To achieve high image visual quality and forensic undetectability is the primary objective of image anti-forensics. The existing anti-forensic methods are easily circumvented by using higher-order statistical analysis and DL-based forensic techniques. Moreover, these anti-forensic methods are not able to maintain the images visual quality. Therefore, we explore the recent adversarial attacks in this paper to resolve these issues. We utilize three adversarial attacks i.e., FGSM, C&W, and PGD, for image anti-forensics against recent general-purpose image forensic techniques. The overall procedure of our method is shown in Fig. 1. The gradient of image forensic models is used by the different adversarial attacks to generate corresponding intelligent noise, as shown in Fig. 1. Afterward, this noise is added to the input image, which results in a final adversarial image.

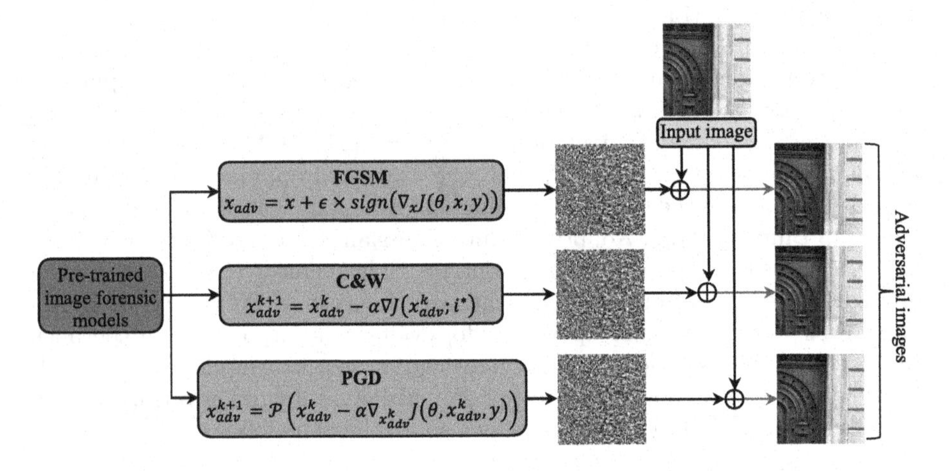

Fig. 1. Proposed anti-forensic framework based on adversarial attack.

Generally, in deep learning training, we minimize the loss function value for better classification results. But, the adversarial attacks work to maximize the loss function of a classifier by adding small perturbations. This forces the classifier to make false predictions. L-norm is used to restrict the level of added perturbation so that the generated images with noise are unnoticeable to the human eyes. The procedure of finding noise or perturbation (δ) can be considered as maximization problem expressed as:

$$\max_{\|\delta\|_p} J\left(\theta, x_{adv}, y\right) \tag{1}$$

where, $J\left(\theta, x_{adv}, y\right)$ denotes the loss function and $\|\delta\|_p$ is the constraint p-norm. Also, θ represents the model (f) parameters, x_{adv} is the adversarial image, and y denotes the corresponding labels. We aim to find the small perturbation for the input image. Therefore, we minimize $D(x, x_{adv})$ for the input image x, where D is the Euclidean distance metric between the input image and the adversarial image. This procedure can be described as:

$$minimize\left[\|\delta\|_p + \epsilon \times f\left(x_{adv}\right)\right], \quad x_{adv} = [0,1] \tag{2}$$

where, ϵ is a small constant value that limits the perturbation. Equation (2) is solved by using considered adversarial attacks as discussed in the following subsections.

2.1 Fast Gradient Sign Method

This method initially evaluates the gradient of the loss function corresponding to the input image, and then each pixel of the input image is modified based on the sign of the gradient. This attack used an L_∞ bounded constraint for the distance matrix, and for the gradient direction it uses a sign function. Gradient with respect to the input image is adjusted with a small step in the direction that will maximize the loss until it acquires the small perturbation. The resultant perturbed image can be obtained as [13]:

$$x_{adv} = x + \epsilon \times \text{sign}\left(\nabla_x J(\theta, x, y)\right) \tag{3}$$

2.2 Carlini & Wagner Attack

Carlini & Wagner's attack [6] is based on the optimization problem of wrong prediction objective and minimal distance. In other methods, softmax cross-entropy loss function is generally used, but this method is based on a logits-based loss function, and the optimization problem is solved using an Adam optimizer. Moreover, this method uses a binary search technique to find the optimal coefficient stabilizing the trade-off between wrong prediction and the distance between original and adversarial images. This method provides better adversarial examples with smaller perturbations defined as:

$$x_{adv}^{k+1} = x_{adv}^{k} - \alpha \nabla J\left(x_{adv}^{k}; i^*\right) \tag{4}$$

where, k is the number of iterations, α denotes the Gradient descent step size, which controls the rate of conversion, and i^* is the index of maximum response defined as:

$$i^* = \underset{y \neq t}{\text{argmax}}\left\{g_y\left(x_{adv}^{k}\right)\right\} \tag{5}$$

where, $g_y(.)$ denotes the discriminant function used to specify the decision boundaries of considered classes and t represents the target class.

2.3 Projected Gradient Descent

PGD [17] is the advanced iterative version of FGSM and is considered a universal attack among the first-order adversaries. It is important to note that PGD is better as compared to the FGSM. This happens because the clipping function enables the PGD to perform more iteration with a suitable step size. PGD attack uniforms the random perturbation, and then an adversarial image is generated by using the iterative step defined as:

$$x_{adv}^{k+1} = \mathcal{P}\left(x_{adv}^{k} - \epsilon \times \nabla_{x_{adv}^{k}} J\left(\theta, x_{adv}^{k}, y\right)\right) \tag{6}$$

where, P denotes the projection operator which clips the input according to the predefined perturbation range.

3 Experiment Results

The experimental tests are performed by considering the BOSSBase dataset [1]. The BOSSBase database contains 10,000 high-resolution grayscale PGM images. These images were produced from raw images by transforming them into PPM format with the help of UFRaw utility [11] and then converted into an 8-bit grayscale PGM format. Finally, a sub-image of size 512×512 is cropped from the center of all these grayscale images. For assessment purposes, these grayscale PGM images are converted into PNG format. We considered 4,167 and 250 images of size 512×512 from the BOSSBase dataset for training and testing, respectively, to create one original image dataset and 10 manipulated datasets based on image processing operations listed in Table 1. These images are processed to extract 4 patches of size 256×256 to generate 16,668 training and 1,000 testing images for each type of image manipulations. This results in a database of 1,94,348 grayscale images of size 256×256 for training and testing. Out of these images, 1,83,348 images (including 16,668 unaltered images) are used for training, and the remaining 11,000 images (including 1,000 unaltered images) are used for testing the considered neural networks. The image manipulation datasets are created by selecting the image operation parameter uniformly at random from the set of possible values listed in Table 1. The Keras deep learning framework is used to implement all of our neural networks. All the experiments are executed by using the Tesla V100 GPU with 16 GB RAM.

We also created a test dataset of adversarial images by using the FGSM, C&W, and PGD from the images of the above mentioned testing dataset to evaluate the robustness of trained forensic models against adversarial attacks. The value ϵ is set to 0.01 in the FGSM during the creation of adversarial images. In the case of C&W, we set $k = 5$ and $\alpha = 0.005$. On the contrary, the number of iterations k is set to 20 in PGD and $\epsilon = 0.01$. Table 2 provides the average detection accuracies of considered general-purpose image forensic techniques against different adversarial attacks for multiple image manipulation detection. The detection accuracy of forensic model Chen et al. [7] decreases by 51.59%, 51.38%, 53.12% against FGSM, C&W, and PGD, respectively. In the case of Bayar & Stamm [2], the detection

Table 1. Image processing or editing operations used to create the experimental datasets with arbitrary parameters for the training of general-purpose forensic approaches.

Image editing operations	Parameters
JPEG compression (JPEG)	$QF = 60, 61, 62, \ldots, 90$
Gaussian Blurring (GB)	$\sigma = 0.7, 0.9, 1.1, 1.3$
Adaptive White Gaussian Noise (AWGN)	$\sigma = 1.4, 1.6, 1.8, 2$
Resampling (RS) using bilinear interpolation	Scaling $= 1.2, 1.4, 1.6, 1.8, 2$
Median Filtering (MF)	Kernel $= 3, 5, 7, 9$
Contrast Enhancement (CE)	$\gamma = 0.6, 0.8, 1.2, 1.4$
JPEG anti-forensics (JPEGAF) [9]	$QF = 60, 61, 62, \ldots, 90$
JPEG anti-forensics (JPEGAF) [10]	$QF = 60, 61, 62, \ldots, 90$
Median filtering anti-forensics (MFAF) [11]	Kernel $= 3, 5, 7, 9$
Contrast enhancement anti-forensics (CEAF) [18]	$\gamma = 0.6, 0.8, 1.2, 1.4$

Table 2. Performance analysis of recent general-purpose image forensic approaches against normal testing dataset and adversarial images testing datasets based on FGSM, C&W, and PGD attacks.

Image forensic models	Accuracy (%) on normal test dataset	Accuracy (%) with adversarial images using FGSM	Accuracy (%) with adversarial images using C&W	Accuracy (%) with adversarial images using PGD
Chen et al. [7]	70.1	18.51	18.72	16.98
Bayar and Stamm [2]	85.52	21.73	17.54	17.01
Yang et al. [24]	93.19	18.99	23.63	17.99
Singh and Goyal [19]	96.1	27.09	23.81	22.29

accuracy decreases by 63.79%, 67.98%, and 68.51% against FGSM, C&W, and PGD, respectively. Similarly, the detection accuracy decreases by 74.20%, 69.56%, and 75.20% against FGSM, C&W, and PGD, respectively, for Yang et al. [24]. Moreover, the detection accuracy of the recent Singh and Goyal [19] forensic model decreases by 69.01%, 72.29%, and 73.81% against FGSM, C&W, and PGD, respectively. Therefore, these results confirm that the recent DL-based forensic models are perfectly fooled by considered adversarial attacks. Furthermore, we have evaluated the visual quality of generated adversarial images in terms of PSNR (dB) by considering normal images as reference. It is observed from Figs. 2 and 3 that the added noise in the resultant adversarial images is minimal. There are very few noise artifacts visible in the case of FGSM and PGD, as shown in Fig. 2. It can be seen from Fig. 2 that adversarial images corresponding to C&W have no visible noise. Similarly, it is clear from Fig. 3 that some noticeable noise patterns in the adversarial images correspond to FGSM and PGD. Moreover, it is observed that the C&W method achieves higher PSNR values as compared to FGSM and PGD attacks.

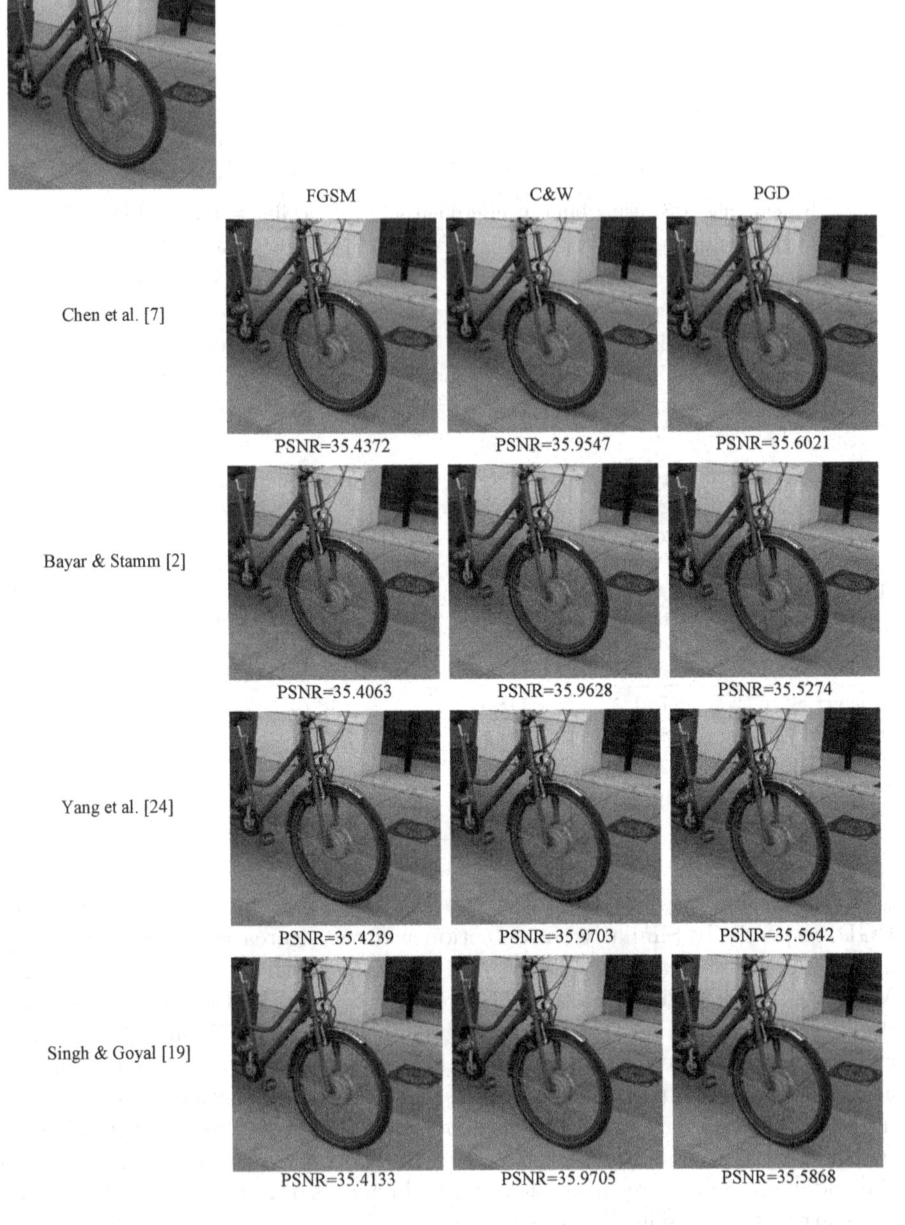

Fig. 2. Adversarial images with PSNR (dB) values correspond to Cycle image based on FGSM, C&W, and PGD attacks by considering recent image forensic techniques.

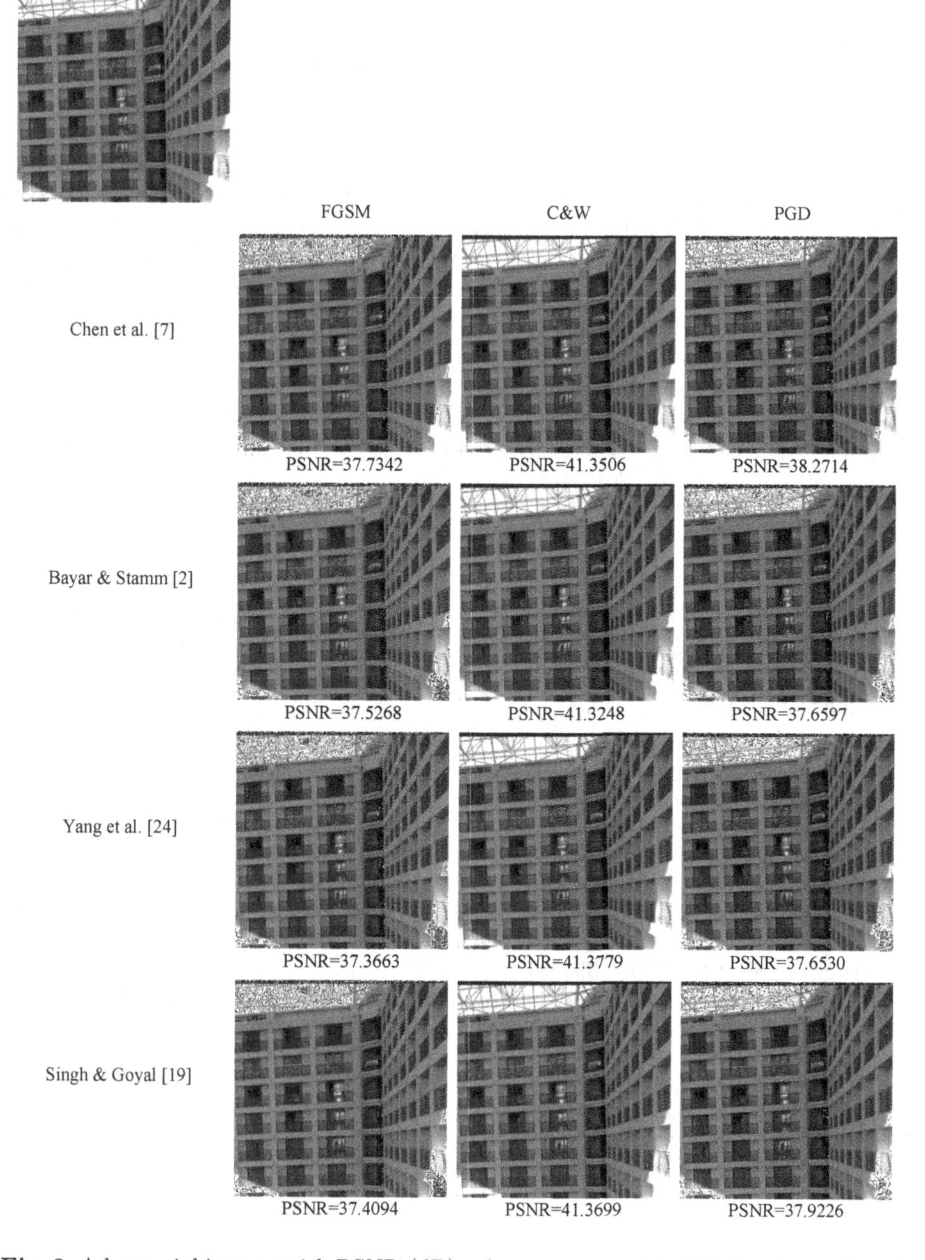

Fig. 3. Adversarial images with PSNR (dB) values correspond to Building image based on FGSM, C&W, and PGD attacks by considering recent image forensic techniques.

4 Conclusions

In this paper, we evaluate the robustness of various image forensic models against popular adversarial attacks by considering different image editing operations. The forensic undetectability against different forensic models is calculated for average detection accuracy for multiple image manipulation detection. The experiment results confirm that the adversarial attacks can fool the recent image forensic techniques, i.e., detection accuracy against these attacks decreases significantly. Higher PSNR values suggest that there are no significant visible changes in the generated adversarial images. Among these three adversarial attacks, C&W outperforms the other ones in terms of image visual quality. It is also observed that the PGD attack reduces the detection accuracy higher than the other attacks. In the future, we can design an adversarial attack capable of fooling forensic detectors with high human imperceptibility.

Acknowledgment. This work is supported by Indian Institute of Technology Ropar under ISIRD grant 9-231/2016/IIT-RPR/1395.

References

1. Bas, P., Filler, T., Pevný, T.: "Break Our Steganographic System": the ins and outs of organizing BOSS. In: Filler, T., Pevný, T., Craver, S., Ker, A. (eds.) IH 2011. LNCS, vol. 6958, pp. 59–70. Springer, Heidelberg (2011). https://doi.org/10.1007/978-3-642-24178-9_5
2. Bayar, B., Stamm, M.C.: Constrained convolutional neural networks: a new approach towards general purpose image manipulation detection. IEEE Trans. Inf. Forensics Secur. **13**(11), 2691–2706 (2018)
3. Cao, G., Zhao, Y., Ni, R., Tian, H.: Anti-forensics of contrast enhancement in digital images. In: Proceedings of the 12th ACM Workshop on Multimedia and Security. pp. 25–34 (2010)
4. Cao, G., Zhao, Y., Ni, R.R., Tian, H.W., Yu, L.F.: Attacking contrast enhancement forensics in digital images. Science China Inf. Sci. **57**(5), 1–13 (2014). https://doi.org/10.1007/s11432-013-4928-0
5. Cao, Y., Gao, T., Sheng, G., Fan, L., Gao, L.: A new anti-forensic scheme-hiding the single JPEG compression trace for digital image. J. Forensic Sci. **60**(1), 197–205 (2015)
6. Carlini, N., Wagner, D.: Towards evaluating the robustness of neural networks. In: 2017 IEEE Symposium on Security and Privacy (SP), pp. 39–57. IEEE (2017)
7. Chen, Y., Kang, X., Wang, Z.J., Zhang, Q.: Densely connected convolutional neural network for multi-purpose image forensics under anti-forensic attacks. In: Proceedings of the 6th ACM Workshop on Information Hiding and Multimedia Security, pp. 91–96 (2018)
8. Dang-Nguyen, D.T., Gebru, I.D., Conotter, V., Boato, G., De Natale, F.G.: Counter-forensics of median filtering. In: 2013 IEEE 15th International Workshop on Multimedia Signal Processing (MMSP), pp. 260–265. IEEE (2013)
9. Fan, W., Wang, K., Cayre, F., Xiong, Z.: A variational approach to JPEG anti-forensics. In: 2013 IEEE International Conference on Acoustics, Speech and Signal Processing, pp. 3058–3062. IEEE (2013)

10. Fan, W., Wang, K., Cayre, F., Xiong, Z.: JPEG anti-forensics with improved trade-off between forensic undetectability and image quality. IEEE Trans. Inf. Forensics Secur. **9**(8), 1211–1226 (2014)
11. Fan, W., Wang, K., Cayre, F., Xiong, Z.: Median filtered image quality enhancement and anti-forensics via variational deconvolution. IEEE Trans. Inf. Forensics Secur. **10**(5), 1076–1091 (2015)
12. Gloe, T., Kirchner, M., Winkler, A., Böhme, R.: Can we trust digital image forensics? In: Proceedings of the 15th ACM international conference on Multimedia, pp. 78–86 (2007)
13. Goodfellow, I.J., Shlens, J., Szegedy, C.: Explaining and harnessing adversarial examples. arXiv preprint arXiv:1412.6572 (2014)
14. Kim, D., Jang, H.U., Mun, S.M., Choi, S., Lee, H.K.: Median filtered image restoration and anti-forensics using adversarial networks. IEEE Signal Process. Lett. **25**(2), 278–282 (2017)
15. Kirchner, M., Bohme, R.: Hiding traces of resampling in digital images. IEEE Trans. Inf. Forensics Secur. **3**(4), 582–592 (2008)
16. Luo, Y., Zi, H., Zhang, Q., Kang, X.: Anti-forensics of JPEG compression using generative adversarial networks. In: 2018 26th European Signal Processing Conference (EUSIPCO), pp. 952–956. IEEE (2018)
17. Madry, A., Makelov, A., Schmidt, L., Tsipras, D., Vladu, A.: Towards deep learning models resistant to adversarial attacks. arXiv preprint arXiv:1706.06083 (2017)
18. Ravi, H., Subramanyam, A.V., Emmanuel, S.: ACE-an effective anti-forensic contrast enhancement technique. IEEE Signal Process. Lett. **23**(2), 212–216 (2015)
19. Singh, G., Goyal, P.: GIMD-Net: an effective general-purpose image manipulation detection network, even under anti-forensic attacks. In: 2021 International Joint Conference on Neural Networks (IJCNN), pp. 1–8. IEEE (2021)
20. Singh, G., Singh, K.: Improved JPEG anti-forensics with better image visual quality and forensic undetectability. Forensic Sci. Int. **277**, 133–147 (2017)
21. Singh, G., Singh, K.: Counter JPEG anti-forensic approach based on the second-order statistical analysis. IEEE Trans. Inf. Forensics Secur. **14**(5), 1194–1209 (2018)
22. Wu, J., Sun, W.: Towards multi-operation image anti-forensics with generative adversarial networks. Comput. Secur. **100**, 102083 (2021)
23. Wu, Z.H., Stamm, M.C., Liu, K.R.: Anti-forensics of median filtering. In: 2013 IEEE International Conference on Acoustics, Speech and Signal Processing, pp. 3043–3047. IEEE (2013)
24. Yang, L., Yang, P., Ni, R., Zhao, Y.: Xception-based general forensic method on small-size images. In: Pan, J.-S., Li, J., Tsai, P.-W., Jain, L.C. (eds.) Advances in Intelligent Information Hiding and Multimedia Signal Processing. SIST, vol. 157, pp. 361–369. Springer, Singapore (2020). https://doi.org/10.1007/978-981-13-9710-3_38

A Novel Artificial Intelligence-Based Lung Nodule Segmentation and Classification System on CT Scans

Shubham Dodia[1(✉)], B. Annappa[1], and Mahesh A. Padukudru[2]

[1] Department of Computer Science and Engineering, National Institute of Technology Karnataka, Surathkal, India
shubham.dodia8@gmail.com, annappa@ieee.org

[2] Department of Respiratory Medicine, JSS Medical College, JSS Academy of Higher Education and Research, Mysuru, Karnataka, India

Abstract. Major innovations in deep neural networks have helped optimize the functionality of tasks such as detection, classification, segmentation, etc., in medical imaging. Although Computer-Aided Diagnosis (CAD) systems created using classic deep architectures have significantly improved performance, the pipeline operation remains unclear. In this work, in comparison to the state-of-the-art deep learning architectures, we developed a novel pipeline for performing lung nodule detection and classification, resulting in fewer parameters, better analysis, and improved performance. Histogram equalization, an image enhancement technique, is used as an initial preprocessing step to improve the contrast of the lung CT scans. A novel Elagha initialization-based Fuzzy C-Means clustering (EFCM) is introduced in this work to perform nodule segmentation from the preprocessed CT scan. Following this, Convolutional Neural Network (CNN) is used for feature extraction to perform nodule classification instead of customary classification. Another set of features considered in this work is Bag-of-Visual-Words (BoVW). These features are encoded representations of the detected nodule images. This work also examines a blend of intermediate features extracted from CNN and BoVW characteristics, which resulted in higher performance than individual feature representation. A Support Vector Machine (SVM) is used to distinguish detected nodules into benign and malignant nodules. Achieved results clearly show improvement in the nodule detection and classification task performance compared to the state-of-the-art architectures. The model is evaluated on the popular publicly available LUNA16 dataset and verified by an expert pulmonologist.

Keywords: Lung cancer · Nodule classification · Nodule detection · Fuzzy C-Means clustering · Medical imaging · Bag-of-Visual-Words

B. Raman et al. (Eds.): CVIP 2021, CCIS 1568, pp. 552–564, 2022.
https://doi.org/10.1007/978-3-031-11349-9_48

1 Introduction

One of the most lethal cancers amongst all the cancer is lung cancer. Considering the statistics throughout the world, it nearly affects up to 12.9% of the overall population [1]. The investigation of lung cancer is predominantly conducted using medical imaging techniques like Chest radiographs (also popularly known as X-rays) or Computed Tomography (CT) scanning. Lung cancer screening is a time-consuming skill that requires expert radiologists. Not to mention, it also consists of intra-observer and inter-observer variability among the decisions of different radiologists in identifying nodules or tumors, which makes the task even more tedious [2,3]. Lung cancer is one of the highest mortality and morbidity rate, with no visible tumor presence symptoms until the patient has reached advanced stages. Therefore, early detection of cancer is one of the critical problems to be addressed. Computer-Aided Diagnosis (CAD) systems are developed to provide a second opinion to radiologists to overcome these issues. Automating this process will reduce the hassle for the doctors and radiologists and the quick and accurate diagnosis.

CAD systems work in two phases, and the initial phase involves the detection of pulmonary nodules in the CT scans. This is time-consuming with a high mistake rate since other anatomical structures are morphologically similar, leading to misunderstanding as a nodule. Therefore, this task requires careful examination and experienced radiologists to assign a candidate in the CT scan as a nodule. This task is named lung nodule detection. The final phase involves classifying the detected nodules into non-cancerous and cancerous. Not all nodules present in the thoracic region are cancerous nodules. The classification of the nodules is mainly based on the nodule size and malignancy type. Specific guidelines are provided by the Lung CT Reporting and Data Systems (Lung-RADS) that need to be followed in the follow-up strategy to categorize the lung nodules [3].

Related Works. In recent literature, Deep Convolutional Neural Networks (DCNN) are used to detect, segmentation, and classify lung nodules. Shen et al. [4] proposed a deep learning model based on crop operation, avoiding the typical segmentation of nodules. The multi-crop CNN model (MC-CNN) is used to identify the malignancy rate in the lung nodules. The model resulted in robust performance even after the exclusion of the segmentation step for lung malignancy analysis. The literature mentioned above used two-dimensional CT scans to perform any lung cancer diagnosis task. However, one crucial aspect overseen in two-dimensional CT scans is volumetric information extracted from three-dimensional CT scans. Few CAD systems have used three-dimensional CT scans to extract different attributes such as nodule shape, texture, type, etc., using volumetric information. This improved the CAD system's performance and resulted in a more accurate diagnosis [3,5,6].

Our Contribution. In this work, a novel segmentation method named Elagha initialization based Fuzzy C-Means clustering (EFCM) is proposed to perform segmentation of nodule regions from a given CT scan. Followed by nodule detection, a hybrid blend of features extracted from segmented images is used to classify lung nodules. The deep architectures are being used for various medical imaging tasks such as detecting nodules, segmenting the nodule regions, and classifying them into cancerous and non-cancerous. In our work, deep architecture is used as a feature extraction technique. The features from different intermediate layers from a deep CNN are validated for the lung nodule classification's best performance. Another set of features is also used to achieve better discriminative information in the nodule structure, Bag-of-Visual-Words (BoVW) features. These features build a visual dictionary for various input data, and using those dictionary values, the features of unseen test images are extracted. The combination of BoVW and CNN features outperformed the results of the individual feature representations and state-of-the-art nodule classification systems. The results achieved are verified by an expert pulmonologist.

2 Materials and Methods

2.1 Materials

The dataset utilized for the evaluation of the proposed method is Lung Nodule Analysis (LUNA-2016). This dataset is a curated version of the publicly available lung cancer CT dataset LIDC-IDRI. The number of CT scans in the dataset is 888, which includes 5,51,065 nodule candidates. There are a total of 1186 positive nodules in the dataset [7]. The LUNA16 dataset does not contain the malignancy rate of the nodules. However, the dataset includes the scans taken from the LIDC-IDRI dataset [8]. Each radiologist's malignancy score is provided in a Comma Separated Value (CSV) file for all the nodule scans. Based on all four radiologists' average malignancy score, the nodule is assigned as a benign nodule or a malignant nodule.

2.2 Methods

Architecture. Figure 1 demonstrates the architecture of the proposed method. The CT scans consist of candidates that need to be identified as nodules and non-nodules, which is performed using the EFCM method. Once the nodules are categorized, it needs to be further classified into benign and malignant nodules. The nodule can be assigned a malignant label based on the malignancy score provided by expert radiologists. The average score of all the radiologists is calculated, and the nodule is assigned with the corresponding label. Once we get the two classes, the classification is performed using two sets of feature representations, BoVW, deep features, and the combination of both features. The classifier used is the SVM for the final decision.

Image Enhancement. Enhancing the images draws more attention towards certain characteristics of an image, making the images more precise, sharp, and detailed. This, in turn, can be used for better analysis and information extraction from the images. In this work, the Histogram Equalization (HE) technique is used where the contrast is altered by adjusting the intensity of the image, which provides an enhanced CT scan image. The comparative frequency of occurrence of different gray levels in the image is represented in the histogram [9].

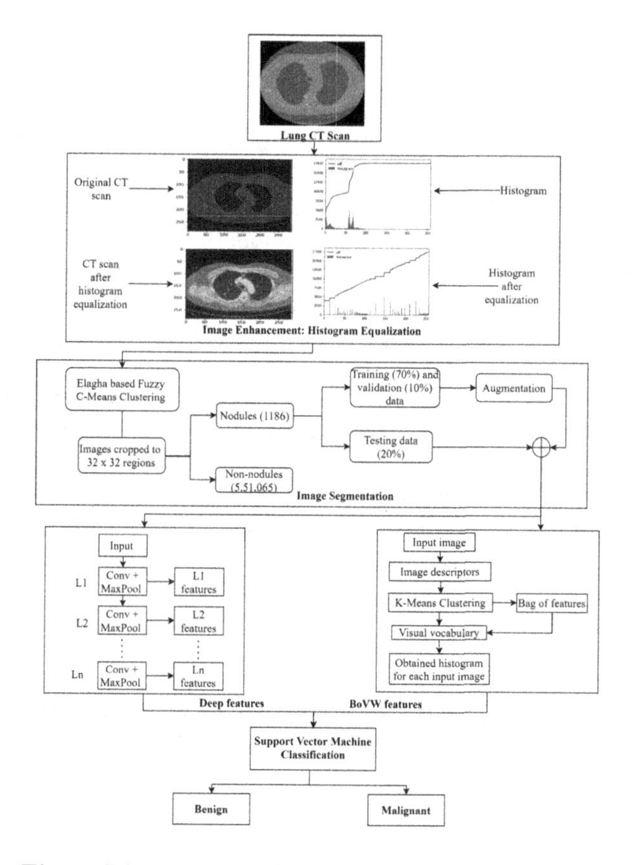

Fig. 1. Schematic architecture of the proposed method

The histogram $h(r_k)$ of an image consists of a L total intensity values r_k within a range of [0–255] (refer Eq. 1).

$$h(r_k) = n_k \tag{1}$$

where, n_k is the number of pixels with an intensity value of r_k in the image.

The histogram can be obtained by plotting the $p(r_k)$ which is shown in below Eq. 2:

$$p(r_k) = \frac{h(r_k)}{number\ of\ rows(M) * number\ of\ columns(N)} = \frac{n_k}{MN}; k = 0, 1, 2, ..., (L-1) \tag{2}$$

The HE of an image is a transformation function i.e., Cumulative Distribution Function (CDF) is given in below Eqs. 3 and 4:

$$cdf(k) = \sum_{i=0}^{k} P_r(r_i), i = 0, 1, ..., L-1 \tag{3}$$

$$s(k) = T(r_k) = \left\lfloor (L-1) \sum_{i=0}^{k} p_i \right\rfloor = \left\lfloor \frac{(L-1)}{MN} \sum_{i=0}^{k} n_i \right\rfloor ; k = 0, 1, ..., (L-1) \tag{4}$$

Image Segmentation. The proposed work uses a novel clustering approach named Elagha initialization based Fuzzy C-Means clustering (EFCM) to segment the nodule region from the given input CT scan. At first, FCM partitions the image into several clusters, and then the cluster centroids are selected randomly to compute the Euclidean distance. This random selection of initial centroids may lead to the local optimum solution. Thus, to overcome this drawback, Elagha initialization is used for the initialization of centroids. It generates the initial centroids based on the overall shape of the data. This modification in traditional FCM is termed EFCM.

EFCM method divides the input image (X) into M clusters such that $x_j = x_1, x_2, ..., x_m$. Then, Elagha initialization calculates the initial cluster centroids by identifying the boundaries of data points and divides them into F rows and F columns to calculate the initial centroids. The width w_j and height h_j of the grid cell is computed as shown in Eqs. 5 and 6:

$$w_j = \frac{w_{j,max} - w_{j,min}}{F} \tag{5}$$

$$h_j = \frac{h_{j,max} - h_{j,min}}{F} \tag{6}$$

where, $w_{j,max}$ and $w_{j,min}$ represents the maximum and minimum widths, $h_{j,max}$ and $h_{j,min}$ signifies the maximum and minimum heights respectively. The N number of initial cluster centroids (c_i) is given by Eq. 7,

$$c_i = \frac{w_j}{2} + \frac{h_j}{2}, i = 1, 2, ..., N \tag{7}$$

After initialization of centroids, the membership function calculation of each pixel is done using Eq. 8:

$$\mu_{ij} = \frac{1}{\sum_{i=1}^{N} \sum_{j=1}^{M} \left(\frac{1}{d_{ij}} \right)^{\frac{2}{q-1}}} \tag{8}$$

where, q indicates the power exponent, d_{ij} is the Euclidean distance between samples x_j and cluster centroid c_i and is given by, Eq. 9:

$$d_{ij} = \sqrt{\sum_{i=1}^{N} \sum_{j=1}^{M} (x_j - c_i)^2} \tag{9}$$

The objective function ξ used for the initialization of FCM algorithm is given by Eq. 10:

$$\xi = \sum_{i=1}^{N} \sum_{j=1}^{M} \mu_{ij} d_{ij}^2 \tag{10}$$

The clusters are formed for nodule and non-nodule regions into separate groups based on the Euclidean distance. The output of the EFCM algorithm is a segmented image consisting of lung nodule regions.

Bag-of-Visual Words (BoVW). Learning feature representations from images using the BoVW method is a two-tiered process. The information from the segmented images is extracted from a pre-generated codebook or dictionary consisting of low-level local features, also known as visual words. The image descriptors used in this work are SIFT features. A visual dictionary is represented using a histogram named "Bag of Visual Words (BoVW)," which is used as a mid-level feature representation [10]. The words in the image mean information in a patch of an image. The patch size must be larger than a few pixels to retrieve more and better information, as it should consider key parts like corners or edges.

SIFT operation is based on the local edge histogram technique. The SIFT technique is one of the popular methods that work very effectively for the BoVW method. Densely sampled SIFT features are extracted from the images. K-means algorithm is used to get cluster centers to generate a visual codebook or dictionary on these features. A histogram is built to the nearest code in the codebook based on the number of occurrences of a feature in each image. The image is then divided into sub-regions of size 2×2, and histograms are built for each sub-region. Once all the histograms are generated, all the sub-region histograms are concatenated to form a single feature vector.

Deep Features. In medical imaging, deep architectures are mostly used for final decision-making. However, in this work, deep architecture is used as a feature representation. The deep learning models are well-known for learning hierarchical information from the input images. The higher the layers, the more information the network learns. This novel set of features are used for classifying cancerous and non-cancerous nodules. Images of both categories are trained separately using a deep CNN architecture, and intermediate features of both the classes are extracted. The deep features learn better representations as the network gets deeper.

Nodule Classification. Classification of detected nodules into cancerous and non-cancerous is performed using an SVM classifier. The model is trained using BoVW, deep features, and the combination of these features. The kernel used for SVM is linear. The hyperparameters set for the SVM model are cross-validation parameter set to 5 and the cost parameter set to 0. The model is tested using probability estimates generated from the trained model for the classification.

3 Results and Discussion

3.1 Nodule Segmentation

A novel segmentation EFCM approach is proposed in this study to get the region of interest from the input CT image, which is the nodule region. The results obtained from the proposed method are illustrated in Fig. 2. The figure depicts five input CT scans, along with their ground truths and predicted masks. The visual representation of the results clearly shows that the proposed model significantly predicts the mask of the nodule. This yields an exact nodule region extraction from the input CT scan eliminating the additional background noise in the image. This segmentation of the nodule helps in getting better nodule information.

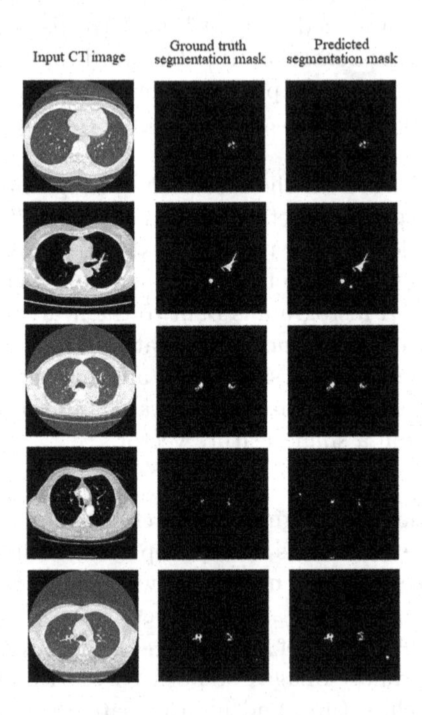

Fig. 2. Segmentation results achieved using EFCM model

Figure 3 shows the dominance of the proposed EFCM segmentation method. The proposed model obtains the Dice Score Coefficient (DSC) of 97.10%, whereas existing methods obtain lower values, such as U-Net of 80.36%, V-Net (92.86%), Fully Connected Network (FCN) U-Net (91.20%), and Mask Region-based CNN (Mask RCNN) (71.16%). Also, the Intersection-over-Union (IoU) of the proposed technique is 91.96%, but the existing methods show lesser values. Likewise, the Sensitivity (SEN) value of 95.35% makes the proposed model preferable to the current techniques. On the other hand, the proposed system attained the Positive

Predictive Value (PPV) of 96.30%, which is higher than the existing methods. This higher DSC, IoU, SEN, and PPV show the improved performance of the proposed EFCM segmentation model.

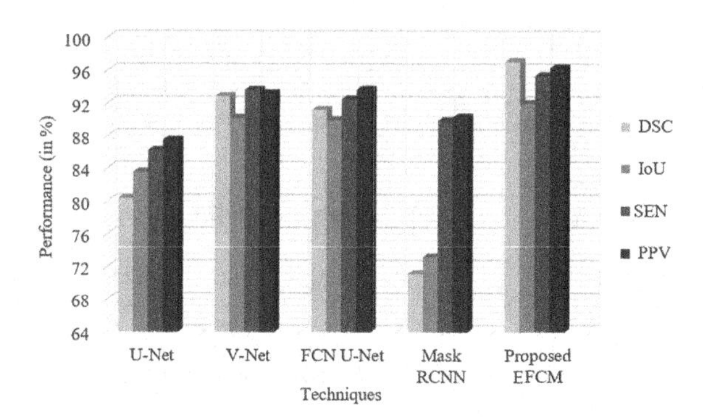

Fig. 3. Comparison of proposed EFCM model with existing segmentation techniques

The evaluation of the nodule detection system for the LUNA16 dataset is performed using a primary performance metric named False Positives per scan (FPs/scan). Figure 4 illustrates the Free-Response Receiver Operating Characteristic (FROC) curve achieved for the proposed EFCM method. The graph depicts that the proposed method resulted in low FPs/scans, proving it is a better performing system. The FPs/scan result for the proposed EFCM model is 2.7 FPs/Scan with a sensitivity of 95.35%.

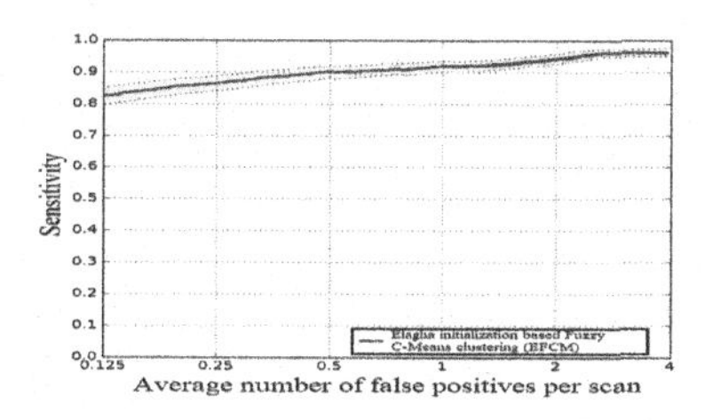

Fig. 4. Free-response receiver operating characteristic (FROC) curve for the performance of proposed EFCM at 2.7 FPs/Scan

The proposed EFCM segmentation method is compared with the existing lung nodule detection systems in Table 1. The methods considered for comparison are mostly deep learning architectures such as U-Net [15], dual branch residual network [17], convolution neural network [21], deep Fully Convolution Networks (FCN) [19], receptive field-regularized (RFR) V-Net [22], and so on. In recent trends, deep learning architectures have taken over image segmentation techniques. However, in the proposed method, a clustering approach for segmentation attained better results in both DSC and IoU.

Table 1. Comparison of the proposed system with the state-of-the-art lung nodule detection systems

Authors	Methods used	DSC	Authors	Methods used	IoU
Ronneberger et al. [15]	U-Net	94.97%	Wu et al. [16]	Pulmonary Nodule Segmentation Attributes and Malignancy Prediction	58.00%
Cao et al. [17]	Dual branch residual network	82.74%	Aresta et al. [18]	iW-Net	55.00%
Roy et al. [19]	Deep FCN	93.00%	Messay et al. [20]	Regression Neural Network	74.00%
Wang et al. [21]	Central focused CNN	82.15%	Wang et al. [21]	Central focused CNN	71.00%
Dodia et al. [22]	RFR V-Net	95.01%	Dodia et al. [22]	RFR V-Net	83.00%
Proposed method	EFCM	97.10%	Proposed method	EFCM	91.96%

3.2 Nodule Classification

The performance of the lung nodule classification system is evaluated on the publicly available LUNA16 dataset. The accuracy obtained for the lung nodule classification task is 96.87%. The performance metrics considered for the evaluation of the proposed method are accuracy, error rate, specificity, sensitivity, FPR, and F-score. The results are presented in Table 2.

Table 2. Performance of the lung nodule classification for the proposed system

Performance metric	BoVW	Deep features	BoVW + Deep features
Accuracy (in %)	93.48	95.32	**96.87**
Error rate (in %)	6.52	4.68	**3.13**
Specificity (in %)	93.04	95.01	**96.60**
Sensitivity (in %)	93.94	95.61	**97.15**
FPR	0.0696	0.0499	**0.0340**
F-score	0.9337	0.9522	**0.9681**

A layer-wise feature comparison is performed for deep and BoVW + deep features. Figure 5 illustrates the accuracy values obtained for different intermediate layers for deep features. The number of layers considered is from 1 to 10. It can be noticed that the performance of the system increases with the increase in layers. After 7th layer, there is a degradation of accuracy observed in the figure. This is due to overfitting the model for more layers as complexity increases, and less data is available to train the deep architecture. Therefore, the number of layers must be monitored to prevent the model from overfitting.

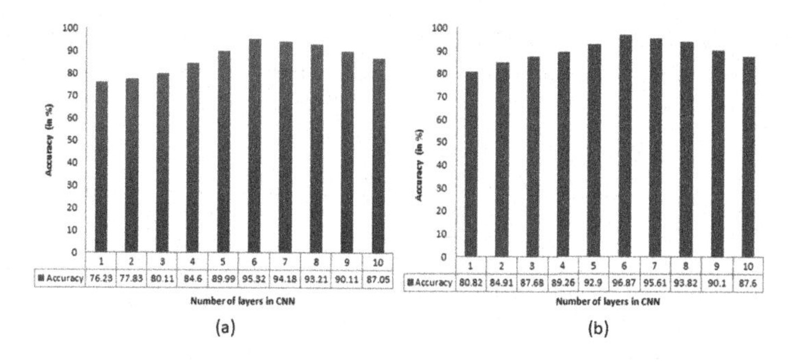

Fig. 5. Layer-wise accuracy values for (a) Deep features, (b) BoVW + Deep features

The proposed method is compared with state-of-the-art lung nodule classification systems. The systems previously proposed for performing lung nodule classification system utilizes deep learning architectures such as Artificial Neural Network (ANN), CNN, multi-scale CNN, Stacked Auto Encoder (SAE), etc. The results are presented in Table 3. It can be noted from the table that the proposed method achieved better performance as compared to the state-of-the-art lung nodule classification systems. Another major issue in training these networks is it is computationally expensive as it requires a lot of time to train a deep model. However, in our method, we used CNN as a feature extractor rather than a classifier. It does not require much time to extract intermediate features. It is also computationally less expensive as the classifier used does not require much time to learn BoVW and deep features. Because BoVW features are encoded, representations do not take up more time for calculation.

Table 3. Comparison of the proposed system with the state-of-the-art lung nodule classification systems.

Authors	Methods used	Accuracy	Sensitivity	Specificity
Silva et al. [11]	Taxonomic indexes and phylogenetic trees, SVM	88.44%	84.22%	90.06%
Song et al. [12]	CNN, DNN, SAE	84.15%	N/A*	N/A
Shen et al. [13]	Multi-scale CNN, Random forest	86.84%	N/A	N/A
Guptha et al. [14]	Super-Resolution CNN, SVM	85.70%	N/A	N/A
Shaukat et al. [23]	Intensity, shape, texture features and ANN	93.70%	95.50%	94.28%
Proposed method	BoVW + Deep features, SVM	96.87%	97.15%	96.60%

N/A*-Not Available

Figure 6 illustrates the quantitative analysis of the three feature representations using the Receiver-Operating Characteristics (ROC) curve. The ideal system provides an Area Under Curve (AUC) of 1. The AUC values attained for BoVW, deep, and BoVW + Deep features are 0.83, 0.88, and 0.92, respectively. The classification performance analysis is presented for the SVM classifier. It can be noted from the Figure that BoVW + Deep features resulted in the highest AUC. The feature combination worked effectively to improve the performance of the system.

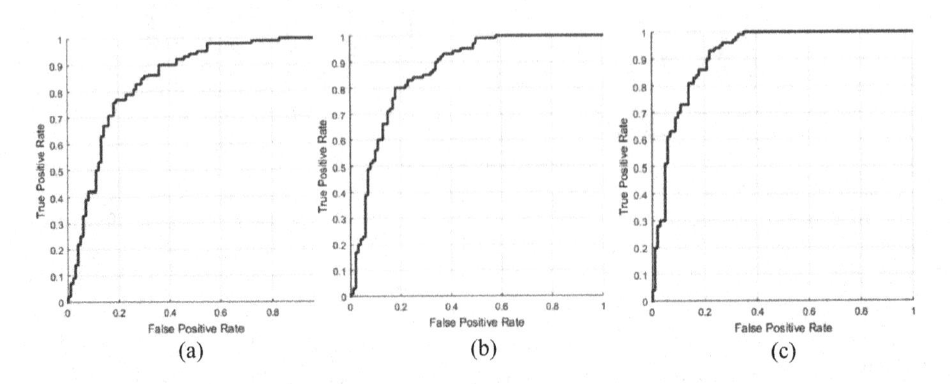

Fig. 6. Receiver-Operating Characteristics (ROC) curves for (a) BoVW, (b) Deep features, (c) BoVW + Deep features

4 Conclusion

Lung cancer is considered to be one of the deadliest diseases. In this work, a novel clustering-based segmentation method named EFCM is proposed to extract lung nodules from the given CT scan. A hybrid of two different types of feature representations for lung nodule classification is proposed in the work. The method glorifies that deep learning can be used as a classifier and as a suitable feature extractor. The segmentation method introduced in this work performs better than existing segmentation methods in terms of DSC, IoU, and PPV. The proposed method acknowledges that a combination of certain feature representations can enhance the system's performance in terms of various evaluation metrics such as accuracy, sensitivity, etc. It also reduces the computational cost of the system by reducing the system's learning parameters. The proposed system effectively combines the encoded feature representation method BoVW and deep features extracted from intermediate layers of a CNN. The performances obtained in the medical imaging tasks are rather critical and also require quicker output. The proposed method provided better and faster results than other CAD systems proposed for the lung nodule classification tasks. In future work, other sets of feature representations and also deeper networks with larger datasets can be explored for performing lung nodule classification.

References

1. Ferlay, J., et al.: Cancer incidence and mortality worldwide: sources, methods and major patterns in GLOBOCAN 2012. Int. J. Cancer **136**(5), E359–E386 (2015)

2. Shen, S., Han, S.X., Aberle, D.R., Bui, A.A.T., Hsu, W.: An interpretable deep hierarchical semantic convolutional neural network for lung nodule malignancy classification. arXiv:1806.00712 (2018)
3. Abid, M.M.N., Zia, T., Ghafoor, M., Windridge, D.: Multi-view convolutional recurrent neural networks for lung cancer nodule identification. Neurocomputing **453**, 299–311 (2021)
4. Shen, W., et al.: Multi-crop convolutional neural networks for lung nodule malignancy suspiciousness classification. Pattern Recogn. **61**, 663–673 (2017)
5. Kang, G., Liu, K., Hou, B., Zhang, N.: 3D multi-view convolutional neural networks for lung nodule classification. PLOS One **12**(11), 1–21 (2017)
6. Dou, Q., Chen, H., Yu, L., Qin, J., Heng, P.: Multilevel contextual 3-D CNNs for false positive reduction in pulmonary nodule detection. IEEE Trans. Biomed. Eng. **64**(7), 1558–1567 (2017)
7. Setio, A.A.A., et al.: Validation, comparison, and combination of algorithms for automatic detection of pulmonary nodules in computed tomography images: the LUNA16 challenge. Med. Image Anal. **42**, 1–13 (2017)
8. Armato, S., III., et al.: The Lung Image Database Consortium (LIDC) and Image Database Resource Initiative (IDRI): a completed reference database of lung nodules on CT scans. Med. Phys. **38**, 915–931 (2011)
9. Salem, N., Malik, H., Shams, A.: Medical image enhancement based on histogram algorithms. Procedia Comput. Sci. **163**, 300–311 (2019)
10. Sundarambal, B., Subramanian, S., Muthukumar, B.: A hybrid encoding strategy for classification of medical imaging modalities. J. Ambient Intell. Humaniz. Comput. **12**(6), 5853–5863 (2020). https://doi.org/10.1007/s12652-020-02129-1
11. da Silva, G.L.F., de Carvalho Filho, A.O., Silva, A.C., de Paiva, A.C., Gattass, M.: Taxonomic indexes for differentiating malignancy of lung nodules on CT images. Res. Biomed. Eng. **32**(3), 263–272 (2016)
12. Song, Q., Zhao, L., Luo, X., Dou, X.: Using deep learning for classification of lung nodules on computed tomography images. J. Healthc. Eng. **2017**, 1–7 (2017)
13. Shen, W., Zhou, M., Yang, F., Yang, C., Tian, J.: Multi-scale convolutional neural networks for lung nodule classification. In: Ourselin, S., Alexander, D.C., Westin, C.-F., Cardoso, M.J. (eds.) IPMI 2015. LNCS, vol. 9123, pp. 588–599. Springer, Cham (2015). https://doi.org/10.1007/978-3-319-19992-4_46
14. Gupta, A., Das, S., Khurana, T., Suri, K.: Prediction of lung cancer from low-resolution nodules in CT-scan images by using deep features. In: International Conference on Advances in Computing, Communications and Informatics (ICACCI), pp. 531–537 (2018)
15. Ronneberger, O., Fischer, P., Brox, T.: U-net: convolutional networks for biomedical image segmentation. arXiv:1505.04597, pp. 234–241 (2015)
16. Wu, B., Zhou, Z., Wang, J., Wang, Y.: Joint learning for pulmonary nodule segmentation, attributes and malignancy prediction. arXiv:1802.03584, pp. 1109–1113 (2018)
17. Cao, H., et al.: Dual-branch residual network for lung nodule segmentation. Appl. Soft Comput. **86**, 105934 (2020)
18. Aresta, G., et al.: iW-Net: an automatic and minimalistic interactive lung nodule segmentation deep network. Sci. Rep. **9**(1), 1–9 (2019)
19. Roy, R., Chakraborti, T., Chowdhury, A.S.: A deep learning-shape driven level set synergism for pulmonary nodule segmentation. Pattern Recogn. Lett. **123**, 31–38 (2019)

20. Messay, T., Hardie, R.C., Tuinstra, T.R.: Segmentation of pulmonary nodules in computed tomography using a regression neural network approach and its application to the lung image database consortium and image database resource initiative dataset. Med. Image Anal. **22**, 48–62 (2015)
21. Wang, S., et al.: Central focused convolutional neural networks: developing a data-driven model for lung nodule segmentation. Med. Image Anal. **40**, 172–183 (2017)
22. Dodia, S., Basava, A., Mahesh, P.A.: A novel receptive field-regularized V-net and nodule classification network for lung nodule detection. Int. J. Imaging Syst. Technol. **32**, 88–101 (2021)
23. Shaukat, F., Raja, G., Ashraf, R., Khalid, S., Ahmad, M., Ali, A.: Artificial neural network based classification of lung nodules in CT images using intensity, shape and texture features. J. Ambient Intell. Humaniz. Comput. **10**, 4135–4149 (2019)

Author Index

Abraham, Bejoy I-328
Abraham, Shilpa Elsa II-85
Aetesam, Hazique II-159
Agrawal, Ramesh Kumar II-350
Ahila Priyadharshini, R I-425
Ambati, Anirudh I-71
Annappa, B. II-552
Arivazhagan, S. I-95, I-425
Arun, M. I-425
Arvind, Pratul II-262
Ayala, Diego I-244

Bamoriya, Pankaj I-363
Bandar, Sahil Munaf I-269
Bandyopadhyay, Oishila I-221
Bansal, Palak II-273
Bartakke, Prashant II-97
Basu, Subhadip II-26
Bathula, Deepti R. II-516
Bhagat, Sandesh I-489
Bhalerao, Karuna I-187
Bhangale, Kishor B. II-192
Bharti, Aditya I-256
Bhattacharjee, Sutanay II-529
Bhaumik, Jaydcb I-151
Bhosale, Surendra I-187
Bhowmik, Mrinal Kanti I-118
Bhurchandi, Kishor M. II-219
Bhuyan, M. K. II-438
Biradar, Kuldeep II-426
Bole, Arti II-147

Chakraborty, Neelotpal II-26
Chandanwar, Aditya II-147
Chaudhari, Nilam I-281
Chaudhary, Sachin II-36, II-206, II-415, II-450, II-483
Chavez, Danilo I-244
Chigarapalle, Shoba Bindu II-319
Chintakunta, Pranav Kumar II-299
Chirakkal, Sanid I-281
Choudhuri, Rudrajit I-561
Chowdhury, Ananda S. I-151

Das, Nibaran I-10, I-458
Das, Shome S. I-209
Dasgupta, Anindita II-299
Dastidar, Shuvayan Ghosh I-10, I-458
Deshmukh, Preeti II-472
Deshmukh, Reena Abasaheb I-58
Devi, Tulasi Gayatri I-468
Dey, Somnath I-574
Dodia, Shubham II-552
Dube, Sachin II-426
Dubey, Shiv Ram I-33, I-71
Dutta, H Pallab Jyoti II-438
Dutta, Kalpita I-458

Gaikwad, Arun N. II-472
Gajjar, Manan I-375
Galshetwar, Vijay M. II-36, II-206, II-415
Ganesh, Kavya I-21
Gawhale, Kunal II-147
Ghosh, Debashis II-308
Goel, Nidhi I-439
Gongidi, Santhoshini II-1
Gour, Mahesh I-83
Goyal, Puneet I-45, II-542
Gula, Roman II-48
Gupta, Ajeet I-221
Guru, D. S. I-525, II-243

Halder, Amiya I-561
Hambarde, Praful I-489, II-329
Handa, Palak I-439
Haritha, K. M. I-232
Haswani, Vineet I-489
Hazarika, Abhinaba II-60
Hazra, Sumit II-403
Holambe, R. S. I-407
Hu, Min-Chun II-122

Indu, S. I-439
Iyshwarya Ratthi, K. I-316

Jagtap, Jayant I-139
Jain, Anamika I-198
Jain, Samir I-538

Jain, Sweta I-83
Jaiswal, Garima II-14
Jansi Rani, A. I-316
Jawahar, C. V. I-256, I-512, II-1, II-135
Jayaraman, Umarani I-175
Jayavarshini, R. I-550
Jeena, R. S. I-232
Jidesh, P. II-60
Jindal, M. K. I-107
Jobin, K. V. II-135
Joshey, Allen I-269
Joshi, Mahesh I-574

Kalidas, Yeturu I-58
Kalyan, Baddam I-33
Kamble, Ravi I-489
Kandiyal, Himanshu I-398
Karthiga, A. I-550
Kaur, Rupinder Pal I-107
Khandnor, Padmavati II-273
Khanna, Pritee I-363, I-501, II-338
Khare, Nishant II-338
Kiruthika, S. II-181
Kojjarapu, Satya Raj Vineel I-33
Kokare, Manesh I-489
Kolhar, Shrikrishna I-139
Kovoor, Binsu C. II-85
Kowndinya, Boyalakuntla I-58
Krishnan, Palani Thanaraj I-21
Kulkarni, Sahana II-147
Kumar, Anmol II-122
Kumar, B. Vinoth I-550
Kumar, Dhirendra II-350
Kumar, Jatin II-362
Kumar, Jayendra II-262
Kumar, Kalagara Chaitanya I-387
Kumar, Munish I-107
Kumar, Pawan I-221
Kumar, Puneet II-350, II-378
Kumar, Shashi II-159
Kumar, Uttam II-288, II-299
Kumar, Vidit II-494
Kumar, Vishal I-340
Kumari, Pratibha II-122
Kundu, Mahantapas I-458
Kushnure, Devidas T. II-110

Lala, Aryan I-387
Lalka, Labhesh II-147
Laskar, R. H. II-438
Laxman, Kumarapu I-33

Maheshwari, Chetan I-221
Majhi, Snehashis II-171
Maji, Suman Kumar II-159
Makandar, Sahil Salim I-269
Malik, Roopak II-329
Mallika II-308
Mandal, Srimanta I-281, I-375
Mapari, Rahul G. II-192
Marasco, Emanuela I-351
Masilamani, V. II-181
Mastan, Indra Deep II-362
Mazumdar, Bodhisatwa I-574
Medhi, Moushumi II-504
Meena, Tamanna II-273
Metkar, Suhit II-147
Misra, Arundhati I-281
Mitra, Suman K. I-281
Mohammad Aaftab, V. II-460
Moharana, Sukumar II-230
Moharir, Rutika II-230
Mollah, Ayatullah Faruk II-26
Mondal, Ajoy I-512, II-135
Mondal, Bisakh I-10
Mounir, Ramy II-48
Mulchandani, Himansh I-294
Mundu, Albert I-340
Munjal, Rachit S. II-230
Murala, Subrahmanyam II-73, II-329

Nair, Praveen II-288
Nanda, Ekjot Singh II-415
Nandan, Sowmith II-288
Nandy, Anup II-403
Napte, Kiran M. II-192
Narmadha, R. I-550
Narute, Bharati II-97
Nasipuri, Mita I-458
Nath, Malaya Kumar I-1
Nayak, Deepak Ranjan II-171

Ojha, Aparajita I-363, I-501, I-538, II-338
Otham, Asem I-351

Padukudru, Mahesh A. II-552
Pal, Shyamosree II-529
Palak II-483

Panikar, Abhijith II-147
Pant, Bhaskar II-494
Pareek, Jyoti I-306
Patel, Nilay I-294
Patil, Nagamma I-468
Patil, Omkar I-164
Patil, P. S. I-407
Patil, Prashant W. II-36, II-206
Patil, Shital I-187
Paunwala, Chirag I-294
Pavuluri, LillyMaheepa I-1
Philipose, Cheryl Sarah I-468
Phutke, Shruti S. II-73
Pinapatruni, Rohini II-319
Pisipati, Manasa II-403
Pokar, Hiren I-294
Prabhu, Arun D. II-230
Pradhan, Pyari Mohan II-308
Prajna, S. I-525
Prashanth, Komuravelli I-58
Puhan, Amrit II-403
Pundhir, Anshul II-378
Pundir, Arun Singh II-391
Putrevu, Deepak I-281

Raghav, Amartya II-450
Raghavendra, Anitha II-243
Rahane, Ameet I-512
Rai, Sharada I-468
Ramachandran, Sivakumar I-328
Raman, Balasubramanian II-378, II-391
Raman, Shanmuganathan II-362
Ramena, Gopi II-230
Randive, Santosh Nagnath II-192
Randive, Santosh II-147
Rao, Mahesh K. II-243
Rastogi, Vrinda II-14
Rathi, Vishwas I-45
Raveeshwara, S. II-254
Reddy, Dileep I-130
Robles, Leopoldo Altamirano I-244
Rodge, Vidya I-58

Saha, Anirban II-159
Saha, Priya I-118
Sahay, Rajiv Ranjan II-504
Sahayam, Subin I-175
Saini, Mukesh II-122
Sakthivel, Naveena I-550
Sarada, Yamini II-262
Saranya, M. I-550
Sarkar, Ram II-26
Sarkar, Sudeep II-48
Sastry, C. S. I-398
Scherer, Rafał II-403
Seal, Ayan I-538
Sethi, Jagannath I-151
Shah, Rinkal I-306
Shankar, Uma I-83
Sharma, Arun II-14
Sharma, Harshit II-273
Sharma, Krishan Gopal II-542
Sharma, Manoj I-387
Sharma, Mansi II-460
Sharma, Paras II-273
Sharma, Rahul II-273
Shebiah, R. Newlin I-95
Shekar, B. H. II-254
Sheorey, Tanuja I-501
Siddhad, Gourav I-363
Silambarasan, J. I-175
Singh, Ajitesh I-294
Singh, Deepak I-512
Singh, Gurinder II-542
Singh, Krishna Pratap I-198
Singh, Prashant II-262
Singh, Rinkal I-294
Singh, Satish Kumar I-198, I-340
Singh, Shweta I-447
Singha, Anu I-118
Smitha, A. II-60
Sreeni, K. G. I-232
Srivastava, Manish Kumar I-130
Srivastava, Sahima II-14
Subramanian, Anbumani I-512
Suguna, Kota Yamini II-438

Talbar, Sanjay N. II-110
Teja, Dande I-58
Thakur, Poornima Singh I-501, II-338
Theuerkauf, Jörn II-48
Thomas, Anoop Jacob II-516
Tiwari, Ashish I-269
Tripathi, Vikas II-494
Tyagi, Dinesh Kumar II-426

Umapathy, Snekhalatha I-21

Vasamsetti, Srikanth I-447
Velaga, Ramya I-58
Venkata Vamsi Krishna, Lingamallu S. N. II-299
Verma, Deepak II-378
Vijay, Chilaka I-58
Vinay Kumar, N. I-525
Vineeth, N. B. I-256
Vipparthi, Santosh Kumar II-426
Vishwakarma, Anish Kumar II-219
Vurity, Anudeep I-351
Vyas, Ritesh I-387
Vyshnavi, M. I-328

Waghmare, L. M. I-407
Wane, Kishor B. II-192

Yeturu, Kalidas I-130
Yogameena, B. I-316

Zacharias, Joseph I-232
Zalte, Rujul II-147

Printed in the United States
by Baker & Taylor Publisher Services